Wireless Ad Hoc and Sensor Networks

Wireless Ad Hoc and Sensor Networks describes the theory of ad hoc networks. It also demonstrates techniques for designing efficient algorithms and systematically analyzing their performance.

Li develops the fundamental understanding required to tackle problems in these networks by first reviewing relevant protocols, then formulating problems mathematically, and solving them algorithmically. Wireless MAC protocols, including various IEEE 802.11 protocols, 802.16, Bluetooth, and protocols for wireless sensor networks are treated in detail. Channel assignment for maximizing network capacity is covered; topology control methods are explored at length; and routing protocols for unicast, broadcast, and multicast are described and evaluated. Cross-layer optimization is also considered.

The result is a detailed account of the various algorithmic, graph-theoretical, computational-geometric, and probabilistic approaches to attack problems faced in these networks, delivering an understanding that will allow readers to develop practical solutions for themselves. This title is an invaluable resource for graduate students and researchers in electrical engineering and computer science departments, as well as for practitioners in the communications industry.

XiangYang Li is currently an associate professor of computer science at the Illinois Institute of Technology. He also holds a visiting professorship or adjunct-professorship at TianJing University, WuHan University, and NanJing University, in China. He was awarded his Ph.D. in 2001 from the Department of Computer Science at the University of Illinois at Urbana-Champaign. A leading researcher in the field of wireless networks, he has made important contributions in the areas of network topology and routing. His current research interests include cooperation, energy efficiency, and distributed algorithms for wireless ad hoc and sensor networks.

Wireless Ad Hoc and Sensor Networks

Theory and Applications

XIANGYANG LI

Illinois Institute of Technology

CAMBRIDGE
UNIVERSITY PRESS

CAMBRIDGE UNIVERSITY PRESS
Cambridge, New York, Melbourne, Madrid, Cape Town, Singapore, São Paulo, Delhi

Cambridge University Press
32 Avenue of the Americas, New York, NY 10013-2473, USA

www.cambridge.org
Information on this title: www.cambridge.org/9780521865234

First published 2008

Printed in the United States of America

A catalog record for this publication is available from the British Library

Library of Congress Cataloging in Publication Data

Li, XiangYang.
  Wireless ad hoc and sensor networks: theory and applications / XiangYang Li.
      p. cm.
  Includes bibliographical references and index.
  ISBN 978-0-521-86523-4 (hbk.)
  1. Wireless communication systems.   2. Sensor networks.   I. Title.

  TK5103.2.L546   2008
  621.384–dc22   2007037219

ISBN   978-0-521-865234 hardback

To my wife, Min
my daughter, Sophia
my son, Kevin
and my families

Contents

Preface

Introduction

In the next generation of wireless communication systems, there will be a need for the rapid deployment of independent mobile users. Significant examples include establishing survivable, efficient, dynamic communication for emergency/rescue operations, disaster relief efforts, and military networks. Such network scenarios cannot rely on centralized and organized connectivity and can be conceived as applications of mobile ad hoc networks (MANETs). A MANET is an autonomous collection of mobile users that communicate over relatively bandwidth-constrained wireless links. Because the nodes are mobile, the network topology may change rapidly and unpredictably over time. The network is decentralized; all network activity, including discovering the topology and delivering messages, must be executed by the nodes themselves; that is, routing functionality will be incorporated into mobile nodes.

In many commercial and industrial applications, we often need to monitor the environment and collect the information about the environment. In some of these applications, it would be difficult or expensive to monitor using wired sensors. If this is the case, wireless sensor networks in which sensors are connected by wireless networks are preferred. A wireless sensor network (WSN) consists of a number of sensors spread across a geographic area. Each sensor node has wireless communication capability and some level of intelligence for signal-processing and networking of data. A WSN could be deployed in wilderness areas for a sufficiently long time (e.g., years) without the need to recharge or replace the power supplies. Typical applications of WSNs include monitoring, tracking, and controlling.

The subject of wireless ad hoc networking and sensor networking is enormously complex, involving many concepts, protocols, technologies, algorithms, and products that work together in an intricate manner. The set of applications for MANETs is diverse, ranging from small, static networks that are constrained by power sources to large-scale, mobile, highly dynamic networks. Recently, wireless sensor networks have also been used in Supervisory Control and Data Acquisition (SCADA). SCADA systems are used to monitor or to control chemical or transport processes, in municipal water supply systems, control electric power generation, transmission, and distribution, gas and oil pipelines, and other distributed processes. The design of network protocols for these networks is a complex issue. Regardless of the application, MANETs and sensor networks need efficient distributed algorithms determining network organization, linking scheduling, and routing. However, determining feasible routing paths and delivering messages in a decentralized environment in which network topology fluctuates is not a well-defined problem. Although the shortest path (based on a given cost function) from a source to a destination in a static network is usually the optimal route, this idea is not easily extended to MANETs. Factors such as variable wireless link quality, propagation path loss, fading, multiuser interference, power expended, and topological changes become relevant issues. The network should be able to adaptively alter the routing paths to alleviate any of these effects. Moreover, in a military environment, preservation of security, latency, reliability, intentional jamming, and recovery from failure are significant concerns. Military networks are designed to maintain a low probability of intercept and/or a low probability of detection. Hence, nodes prefer to radiate as little power as necessary and transmit as infrequently as possible, thus decreasing the probability of detection or interception. A lapse in any of these requirements may degrade the performance and dependability of the network.

The basic goals of a wireless ad hoc sensor network generally depend on the application, but the following tasks are common to many networks:

1. *Determine the value of some parameter at a given location*: In an environmental network, one might want to know the temperature, atmospheric pressure, amount of sunlight, and relative humidity at a number of locations. This example shows that a given sensor node may be connected to different types of sensors, each with a different sampling rate and range of allowed values.
2. *Detect the occurrence of events of interest and estimate parameters of the detected event or events*: In the traffic sensor network, one would like to detect a vehicle moving through an intersection and estimate the speed and direction of the vehicle.
3. *Classify a detected object*: Is a vehicle in a traffic sensor network a car, a minivan, a light truck, a bus, and so on?
4. *Track an object*: In a military sensor network, one would like to track an enemy tank as it moves through the geographic area covered by the network.

In these four tasks, an important requirement of the sensor network is that the required data be disseminated to the proper end users. In some cases, there are fairly strict time requirements on this communication. For example, the detection of an intruder in a surveillance network should be immediately communicated to the police so that action can be taken. Because wireless sensors are often powered by batteries only, energy efficiency is critical for the lifetime of a wireless sensor network. Thus, a considerable amount of research has recently been devoted to developing energy-efficient protocols for wireless sensor networks. In addition to energy-efficient protocols, wireless ad hoc sensor network requirements include but are not limited to scalability (to support a large number of mostly stationary sensors for which networks of 10,000 or even 100,000 nodes are envisioned), network self-organization to support scalability and fault tolerance, collaborative signal-processing, and querying ability. Given the large number of nodes and their potential placement in hostile locations, it is essential that the network be able to self-organize; manual configuration is not feasible. Moreover, nodes may fail (either from lack of energy or from physical destruction), and new nodes may join the network. Therefore, the network must be able to periodically reconfigure itself so that it can continue to function. Individual nodes may become disconnected from the rest of the network, but a high degree of connectivity must be maintained. Another factor that distinguishes wireless sensor networks from MANETs is that the end goal is detection/estimation of some events of interest and not just communications. To improve the detection/estimation performance, it is often quite useful to fuse data from multiple sensors. This data fusion requires the transmission of data and control messages, and so it may put constraints on the network architecture. A user may want to query an individual node or a group of nodes for information collected in the region. Depending on the amount of data fusion performed, it may not be feasible to transmit a large amount of the data across the network. Instead, various local sink nodes will collect the data from a given area and create summary messages. A query may be directed to the sink node nearest the desired location.

Recent years have seen a great amount of research in wireless networks, especially wireless ad hoc networks. These works involve a number of theoretical aspects of computer science, including approximation algorithms, computational geometry, combinatorics, and distributed algorithms. Because of the limited capability of processing power, storage, and energy supply, many conventional algorithms are too complicated to be implemented in wireless ad hoc and sensor networks. Some other algorithms do not take advantage of the geometric nature of the wireless networks. Additionally, most of the currently developed location-based algorithms for wireless networks assume a precise position of each wireless node, which is impossible practically. The majority of the algorithms with theoretical performance guarantee developed in this area also assume that all nodes have a uniform transmission range. These algorithms will likely fail when nodes have disparate transmission ranges. In summary, the wireless ad hoc and sensor networks require efficient distributed algorithms with low computation complexity, low communication complexity, and low storage complexity. These algorithms are expected to take advantage of the geometry nature of the wireless ad hoc networks. Several fundamental questions should be answered: Can we improve the performance of traditional distributed algorithms, developed for wired networks, under wireless ad hoc networks? Does the position information of wireless nodes make a difference in algorithm performance? Much of the existing work in wireless ad hoc networking also assumes that each individual wireless node (possibly owned by selfish users) will follow prescribed protocols without deviation. However, each user may modify the behavior of an algorithm for self-interest reasons. How are desired global-system performances achieved when individual nodes are selfish?

This is a new book aimed at the teaching of wireless ad hoc and sensor networks from the algorithmic and theoretical perspective. The primary focus of the book is on the algorithms, especially efficient distributed algorithms, related wireless ad hoc protocols, and some fundamental theoretical studies of phenomena in wireless ad hoc and sensor networks. Many aspects of wireless networking are covered at the introductory level. I tried to cover as many interesting and algorithmic challenging topics related to wireless ad hoc and/or sensor networks as possible in this book. I know that several interesting topics and elegant algorithms are missing. Some are due to lack of space and some are due to the theme of the book. No judgment is implied for algorithms and protocols not covered in this book.

Audience

This book is intended for graduate students, researchers, and practitioners who are interested in obtaining a detailed overview of a number of various algorithmic, graph-theoretical, computational-geometric, and probabilistic approaches to attack certain challenging problems stemming from wireless networks, especially wireless ad hoc and sensor networks. Thus, when I wrote this book, I tried to cover many details for most of the algorithms studied. This book can, in general, serve as a reference resouce for researchers, engineers, and protocol developers working in the field of wireless ad hoc and/or sensor networks. Consequently, most of the chapters are written in such a way that they can be read and taught independently.

While I have tried to make the book (and most chapters) as self-contained as possible, some rudimentary knowledge of algorithm design and analysis, computational geometry, distributed systems, graph theory, linear algebra, networking protocols, and probability theory is required for reading this book.

Organization of the Book

This book essentially is organized based on the layers of wireless networking: the physical and medium-access-control (MAC) layers, the topology control functions that lie between the MAC and network routing layer, and the network routing layer.

The first part of the book presents introductory material that is necessary for the rest of the book.

Chapter 1 briefly reviews the history of wireless communications and discusses different wireless networks, such as infrastructure-based wireless networks (cellular networks) and infrastructureless wireless networks. Among infrastructureless networks, wireless ad hoc networks and wireless sensor networks are briefly discussed.

Chapter 2 covers some fundamentals of wireless transmissions. In this chapter, we study the interference constraints of wireless communications, the wireless propagation model, and the channel capacity of a wireless channel. We also define the communication graph and the interference graph (or conflict graph) induced by a wireless network. Because minimizing energy consumption is critical for the success of many wireless networks, we also review several energy-consumption models that are often used in the literature. Additionally, we discuss a number of mobility models to simulate mobile networks.

The second part of the book is mainly about the MAC protocols for wireless networks. We study CSMA, TDMA, and CDMA protocols.

Chapter 3 concentrates on the CSMA-based wireless MAC protocols. We study how hidden-terminal and exposed-terminal problems are addressed. We also briefly study several typical wireless MAC protocols such as IEEE 802.11 (or WiFi) protocols for wireless LANs, IEEE 802.16 (WiMAX) for mesh networks, and Bluetooth for wireless personal area networks. We briefly review some of the specific MAC protocols proposed for wireless sensor networks that integrate CSMA and TDMA.

Chapter 4 concentrates on the MAC protocols based on TDMA. These protocols assume that the time is slotted and that each link will be assigned some time slots, in which it can transmit data over this link. When a link is assigned a time slot, it is guaranteed that no wireless interference will occur when it uses this link at this time slot. This assignment is often 0/1: A slot either is assigned to a link or is not assigned. When a time slot is not assigned, a link cannot transmit at that specific time slot. We study some TDMA-based link-scheduling algorithms that can provide theoretical performance guarantees.

Chapter 5 concentrates on spectrum channel assignment for wireless networks (cellular networks and wireless ad hoc networks). We first study how to assign channels for a set of access networks such that the network capacity is maximized, or the number of assigned channels is minimized while certain capacity requirements are satisfied. We then study the results for spectrum channel assignment for ad hoc networks. The objective of a channel assignment could be to use the least number of channels to achieve a connected network while the channel availability and network interface constraints at all nodes are satisfied. We also study the transition phenomena of a number of network properties depending on the availability of a wireless spectrum.

Chapter 6 studies several algorithms for assigning a CDMA code to wireless networks when CDMA is supported.

The third part of the book is about topology control and power assignment for wireless networks.

Chapter 7 studies the construction of backbone for wireless networks. Backbone is especially useful for routing in mobile networks. We study several centralized and distributed algorithms that can construct a network backbone (i.e., a connected dominating set) whose size is within a constant factor of the optimum for wireless networks modeled by a unit disk graph. We also study some pure localized algorithms that have lower communication costs, although the theoretical constant-approximation ratio on the backbone size is not guaranteed.

Chapter 8 studies the construction of a backbone network when each wireless node has a weight denoting its cost of being at the backbone. The objective is to minimize the total weight of the backbone. We study several algorithms with good approximation ratios.

Chapter 9 studies topology-control algorithms that will construct flat network topologies with proved performance guarantees. Here, a network topology is said to be flat if every node in the network will assume the same role in network routing. Notice that for a backbone-based structure, the node on the backbone will forward the messages for nodes that are not on the backbone. We study efficient distributed algorithms that can construct energy-efficient network topologies.

Chapter 10 studies the power-assignment problems for wireless networks. Power assignment is selecting a transmission power for each node in the network such that the resulting communication network using the allocated transmission power has certain properties. The objective of a power assignment is often to minimize the total power used by all nodes or to minimize the maximum transmission power of all nodes. The latter is often easy to solve, based on a binary search on all choices of transmission power. We study algorithms that assign transmission powers such that the network is connected, k-connected, or consumes the least power for broadcast or multicast.

Chapters 11 and 12 are related to previous chapters but with different focuses. In these two chapters, we study the so-called transition phenomena of random wireless networks; in other words, the behavior of some certain parameters of the network when the number of nodes in the network goes to infinity. In Chapter 11, we study the critical transmission range r_n when a random network of n nodes distributed in a given region (typically, a unit square or a disk with unit area) is connected with high probability or k-connected with high probability. In Chapter 12, we study the critical node degree needed for producing a connected random network with high probability; the critical transmission range for connectivity in sparse networks or in mobile networks; the critical transmission range for a successful routing with high probability for certain localized routing algorithms; and the critical sensing range for covering a region with high probability.

The fourth part of the book is on routing protocols for wireless networks. We study routing protocols for unicast, multicast, and broadcast, and routing protocols with selfish agents.

Chapter 13 studies the energy-efficient unicast routing for wireless networks. We briefly review some typical proactive and reactive unicast routing protocols proposed in the literature, such as DSDV, OLSR, AODV, DSR, and opportunistic routing. We also study geographic routing protocols that utilize the geometry information of wireless nodes to improve the routing performance. Cluster-based hierarchical routing is also briefly discussed.

Chapter 14 studies energy-efficient routing protocols for broadcast and multicast. We first study some centralized algorithms for energy-efficient broadcast and multicast. These algorithms are often based on the node-weighted or link-weighted Steiner tree algorithms proposed in the literature. Later, we study several distributed or localized methods that are practically efficient.

Chapter 15 studies the routing from another point of view. In all previous protocols, it has been assumed that all wireless nodes will follow predescribed protocols without deviation. In practice, this may not be true, especially when wireless nodes are owned by individuals. In this chapter, we study how to design routing protocols when we know that individual nodes may not follow a routing protocol for their own benefit. We study this problem mainly using a game-theoretical approach, although a number of different approaches are also briefly discussed. In the game-theoretical-based approaches, wireless nodes will be compensated for their services to others. We study how each individual relay node is paid and how the payment to these nodes is implemented. For multicast, we also study algorithms that will fairly share the payments to relay nodes among potential receivers.

Chapter 16 studies how to improve the network through a cross-layer approach of jointly optimizing routing, link scheduling, and channel assignment. We formulate this problem as mixed-integer programming and then relax it to linear programming. By combining it with link scheduling, we show that the relaxed linear-programming formulation will find a solution that is at least a constant factor of the optimum for a number of network models.

The fifth and the last part of the book is devoted to studying a few other interesting topics in wireless networks; for example, location tracking, the performance of random networks, and security.

Chapter 17 studies finding the location of wireless sensor nodes and tracking the position of a moving object by using wireless sensor networks.

In previous chapters, especially Chapter 16, we study what maximum throughput is achievable by a given wireless network under a certain wireless interference model.

Chapter 18 concentrates on the asymptotic network capacity of a random network. We study how the capacity of wireless networks scale with the number of nodes in the networks (when given a fixed deployment region) or scale with the size of the deployment region (when given a fixed deployment density) for a number of operations, such as unicast and broadcast. We especially study a pioneering work by Gupta and Kumar on the network capacity of a random network for unicast. We also study the network capacity for broadcast under various channel models.

Chapter 19 concentrates on ensuring security in wireless networks. We mainly focus on some fundamentals of cryptography, some key-predistribution protocols, and some secure routing protocols proposed in the literature. Cryptography will provide us some fundamental tools such as symmetric-key and asymmetric-key encryption, digital signature, and hash functions to implement some security protocols. We then review some secure routing protocols proposed in the literature.

Acknowledgments

This work is supported in part by the National Basic Research Program of China (973 Program) under Grant No. 2006CB303000 and the National High Technology Research and Development Program of China (863 Program) under Grant No. 2007AA01Z180.

First, this book is in memory of Professor Chao-Ju (Jennifer) Hou from UIUC, a great researcher, an active leader, and a close colleague, who passed away due to breast cancer on December 2, 2007. I am indebted to her and many colleagues for background material and insightful input, which, in one way or another, have helped me to write certain parts of the book. I am also deeply grateful to many colleagues and peer researchers who shared with me in recent years the exciting tasks of studying algorithms and protocols with theoretical performance guarantees for wireless networks, whose research results have helped me in writing several chapters, and who provided feedback to early versions of this book. Especially, I would like to thank the following: Stefano Basagni, Gruia Călinescu, Guohong Cao, Marco Conti, Stephan Eidenbenz, Ophir Frieder, Jie Gao, Jennifer Hou, Jean-Pierre Hubaux, XiaoHua Jia, Sanjiv Kapoor, P. R. Kumar, Baochun Li, Erran Li, Qun Li, Xin Liu, YunHao Liu, Songwu Lu, Thyaga Nandagopal, Paolo Santi, Ivan Stojmenovic, Nitin Vaidya, Peng-Jun Wan, Roger Wattenhofer, Jie Wu, GuoLiang Xue, Frances Yao, and Chih-Wei Yi. I also would like to thank my (former and current) Ph.D. students at Illinois Institute of Technology: Professor Yu Wang, Dr. Weizhao Wang, Professor WenZhan Song, Kousha Moaveninejad, YanWei Wu, XuFei Mao, Ping Xu, Shajojie Tang, and XiaoHua Xu. Much of the material presented in this book is the fruit of our collaboration during the past few years.

The continued support and encouragement of my dad, KaiWen Li; my mom, LanHua Wu; my wife, Min Chen; my lovely daughter, Sophia Li; and my bright son, Kevin Li were essential ingredients in the completion of the book. I am grateful to them all. Without their unwearying love and patience, tireless encouragement, and full unconditional support, this book would never have been possible. I am also grateful for all my friends and colleagues who provided extraordinary insights, gave me helpful advice in research and life, and pointed me in the right direction.

Last but by no means least, I am deeply grateful to Anna Littlewood, Cambridge University Press, for her tireless effort in all matters relating to the preparation and production of many different versions of the manuscript for the book. I am also deeply grateful to Victoria Danahy and Barbara Walthall of Aptara Corp. for carefully copyediting all chapters of the book.

I am sure that I have missed many others, although not intentionally; I thank all of you.

XiangYang Li
Chicago, Illinois
February 2008

Abbreviations

1D	one-dimensional
2D	two-dimensional
2G	second-generation
3D	three-dimensional
3G	third-generation
ABR	associativity-based routing
ACK	acknowledgment (frame)
A/D	analog-to-digital (conversion)
AES	Advanced Encryption System
AFR	adaptive face routing
AHLoS	ad hoc localization system
AIFS	arbitration interframe space
Algorithm KV	algorithm of Khuller and Viskhin
AMPS	Advanced Mobile Telephone System
amp	amplifier
AoA	angle of arrival
AODV	ad hoc on-demand vector (routing)
AP	access point
APS	ad hoc positioning system
APX	approximable
APXH	APX-hard
AS	autonomous system
ATIM	ad hoc traffic indication map
ATM	asynchronous transfer mode
AWA	Accessos Web Alternativos
BAIP	broadcast average incremental power
BB	budget balance
BFS	breadth first search
BGP	border gateway protocol
BI	busy indication
BIP	broadcasting incremental power
BP	aBeacon period
BPS	bounded-degree planar spanner
BPSK	binary phase shift keying
BSC	base station controller

CA	collision avoidance (CSMA/CA often)
CBC	cipher-block chaining (mode)
CBT	core-based tree
CBTC	cone-based topology control
CCA	clear channel assessment
CCM	combined cipher machine
CCR	critical coverage range
CDMA	code-division multiple-access
CDS	connected dominating set
CEDAR	core-extraction distributed ad hoc routing
CFB	cipher-feedback (mode)
CF-End	contention-free end
CFP	contention-free period
CF-Poll	contention-free poll
CG	conflict graph
CGSR	cluster-head gateway switch routing
CM	cross-monotone
CNN	critical neighbor number
CP	contention period
CPA	closest point of approach
CPU	central processing unit
CRC	cyclic redundancy check
CSMA	carrier-sense multiple-access (protocol)
CSP	collaborative signal processing
CT2	cordless telephone
CTR	critical transmission range
CTS	clear-to-send (mechanism)
CW	contention window
D/A	digital-to-analog (conversion)
DAG	directed acrylic graph
D-AMPS	digital advanced mobile phone service
DARPA	Defense Advanced Research Projects Agency
DC	differential cryptanalysis
DCA	dynamic channel assignment
DCF	distributed coordination function
DECT4	digital European cordless telephone
Demod	demodulator
DES	data encryption standard
DG	disk graph
D-H	Diffie–Hellman
DIFS	distributed interframe space
DM	dense model
D-PRMA	distributed packet-reservation multiple-access (protocol)
DPT	distributed prediction tracking
DRAND	a protocol that technically is defined as distributed randomized TDMA scheduling

DREAM	distance routing effect algorithm for mobility
DS	dominating set
DSA	digital-signature algorithm
DSDV	destination-sequenced distance-vector [routing (protocol)]
DSL	digital subscriber line
DSN	Distributed Sensor Networks (program)
DSP	digital signal processing
DSR	dynamic source routing
DSSS	direct-sequence spread spectrum
DST	directed Steiner tree
DT	Delaunay triangulation
DV	distance vector
DVMRP	distance-vector multicast routing protocol
EAX	designation of a two-pass authenticated encryption scheme
EC	Euler circuit
ECB	electronic codebook (mode)
ECC	elliptic curve cryptography
EDCA	enhanced DCF channel access
EDGE	enhanced data rate for GSM evolution
EFF	Electronic Frontier Foundation
EIFS	extended interframe space
ELSD	equal link split downstream
EMST	Euclidean minimum spanning tree
ERNG	extended relative neighborhood graph
ETX	expected transmission count
ExOR	name given to an opportunistic multihop routing protocol
FDMA	frequency-division multiple-access
FFT	fast Fourier transform
FGSS	fault-tolerant global spanning subgraph
FHSS	frequency-hopping spread spectrum
FIPS	Federal Information Processing Standard
FLSS	fault-tolerant local spanning subgraph
FM	frequency modulation
FNR	farthest-neighbor routing
FP	final permutation
fPrIM	fixed-protocol-interference model
FPTAS	fully polynomial-time-approximation scheme
FSK	frequency-shift-keying
GC	graph coloring
GFR	greedy–face routing
GG	Gabriel graph
GOAFR	greedy other adaptive face routing
GPRS	General Packet Radio Service
GPS	global positioning system

GPSR	greedy perimeter stateless routing
GRG	geometric random graph
GSM	Global System for Mobile Communication
GTFT	method proposed in a paper
HC	hybrid coordinator
HCCA	HCF-controlled channel access
HCF	hybrid coordination function
HRMA	hop reservation multiple access
IARP	intrazone routing protocol
IBSS	independent basic service set
IBSSID	IBSS identifier
IC	incentive compatible
ICDS	induced connected dominating set (graph)
ID	identification
IEEE	Institute of Electrical and Electronics Engineers
IF	intermediate-frequency
iff	if and only if
IG	interference graph
IMBM	iterative maximum-branch minimization
IMRG	incident MST and RNG graph
IMS	IP (Internet Protocol) Multimedia Subsystem
IP	integer programming (formulation)
IP	Internet protocol
IP	initial permutation
IPTV	Internet protocol television
IR	individual rationality
IS	independent set
ISM	Industrial, Scientific, and Medical
ISP	Internet service provider
IT	information technology
IV	initial value
IV	initialization vector
kbps	kilobits per second
kbytes	kilobytes
kNN	k-nearest-neighbor (classifier)
LAN	local-area network
LAR	location-aided routing
LBM	location-based multicast
LC	linear cryptanalysis
LCP	least-cost path
LCPT	least-cost path tree
LDEL	local Delauney graph
LEARN	localized-energy-aware restricted neighborhood (routing)
LLACK	link-layer acknowledgment

LMST	localized minimum spanning tree
LNA	low-noise amplifier
LP	linear programming
LPL	low-power listening
LSS	local spanning subgraph
LST	least-cost Steiner tree
MAC	medium-access control
MAN	metropolitan-area network
MANET	mobile ad hoc network
MAP	maximum a posteriori probability
MATSF	name of a protocol proposed in a paper (from MANET time synchronization)
MBGP	multiprotocol extension for a border gateway protocol
MBS	Mobile Broadband System
MC-CDMA	multicode CDMA
MCDS	minimum connected dominating set
MCG	mutual-communication graph
MCMT	minimum-cost multicast tree
MCU	microcontroller unit
MDS	minimum dominating set
MEMS	Micro-Electro-Mechanical Systems
MFR	most-forwarding routing
MG	mutual-inclusion graph
MGC	minimum graph-coloring (problem)
MIB	management information base
MIMO	multiple-input multiple-output
MIP	multicast independent protocol
MIS	maximum independent set
ML	maximum-likelihood (classifier)
MMAC	multichannel MAC
MNP	monotone nonincreasing property
Mod	modulator
MOSPF	multicast open shortest path first
MPR	multipoint relay
MSC	mobile switching center
MST	minimum spanning tree
MUP	multiradio unification protocol
MVC	minimum vertex cover
MWCDS	minimum weighted connected dominating set
MWIS	maximum weighted independent set
MWVC	maximum weighted vertex cover
NAV	network allocation vector
NFR	no-free-rider
NIC	network interface card
NIST	National Institute of Standards and Technology
NNG	nearest-neighbor graph

NNR	nonnegative sharing
NP	nondeterministic polynomial
NPH	NP-Hard
NST	node-weighted Steiner tree
OAFR	other adaptive face routing
OCB	offset codebook (mode)
OFB	output feedback (mode)
OFDM	orthogonal frequency-division multiplexing
OFSF	orthogonal fixed-spreading-factor (code)
OLSR	optimized link-state routing (protocol)
OSPF	open shortest path first
OURS	optimal unicast routing system
OVSF	orthogonal variable-spreading-factor (code)
PA	power amplifier
PACS	personal-access communications systems
PAN	personal-area network
P-BIP	pruned broadcasting incremental power
PC	point coordinator
PCF	point coordination function
PCI	peripheral component interconnect
PDA	personal digital assistant
PhIM	physical-interference model
PHY	physical-layer (specification)
PI	planar and internal-node
PIFS	point-coordination-function interframe space
PIM-SM	protocol-independent multicast-sparse mode
PKCS	Public Key Cryptography Standard
P-MST	pruned minimum spanning tree
POMDP	partially observable Markov decision process
PP	primal linear programming
PrIM	protocol-interference model
PRMA	packet-reservation multiple-access (scheme)
PS	power-save (state or mode)
PSM	power-saving mode
P-SPT	pruned shortest-path tree
PSTN	public-switched telephone network
PTAS	polynomial-time-approximation scheme
PTC	polynomial-time computability
PTDMA	probabilistic time-division multiple access
QAM	quadrature amplitude modulation
QoS	quality of service
QPSK	quadrature phase-shift keying
RAD	random-assessment delay
RAM	random-access memory

RBOP	related neighborhood-graph-based broadcast-oriented protocol
RF	radio frequency
RIP	routing information protocol
RNG	relative neighborhood graph
RON	resilient overlay network
RP	rendevous point
RPB	reverse-path-broadcasting (scheme)
RPF	reverse-path forwarding (lookup)
RREP	route reply
RREQ	route request
RSA	Rivest–Shamir–Adleman
RSS	received signal strength
RTS	request-to-send (mechanism)
RWP	random-waypoint (model)
Rx	receive
SBT	share-based tree
SCADA	supervisory control and data acquisition
SCH	set-cover hard
SIFS	short interframe space
SINR	signal-to-interference-noise ratio
SIR	signal-to-interference ratio
SMS	short messaging service
SOP	spectrum opportunity
SPAN	a topology maintenance protocol proposed by Chen *et al.* (2002)
SPF	shortest path first
SPS	Standard Positioning Service
SPT	shortest-path tree
SSCH	slotted seeded channel hopping (protocol)
SSR	signal stability routing
SSR	security stochastic routing
STASF	a synchronization protocol proposed in a paper by Zhou and Lai (2005)
SURAN	Survivable Radio Network (project)
SVM	support vector machine
TA	trust authority
TACS	Total Access Communications System
TATSF	a synchronization protocol proposed in a paper
TBTT	target Beacon transmission time
TC	traffic class
TCP	transmission control protocol
TDM	time-division multiplexing
TDMA	time-division multiple-access
TDoA	time difference of arrival
ToA	time of arrival
TORA	temporarily ordered routing algorithm
TSF	timing synchronization function
Tx	transmit

TxIM	transmitter-interference model
TxoP	transmit opportunity
UDG	unit disk graph
UDP	user data-gram protocol
UMTS	Universal Mobile Telecommunication System
UPVCS	undirected minimum-power k-vertex-connected subgraph
US	ultrasound
UWB	ultrawideband
UWCDS	unicast weighted connected dominating set
VC	Vapnik and Chervonenkis
VCG	Vickrey–Clarke–Groves (mechanism)
VCO	voltage-controlled oscillator
VHF	very-high-frequency
VMST	virtual minimum spanning tree
VoIP	voice over IP
VOR	VHF omnidirectional ranging (aircraft navigation system)
VoWIP	voice over wireless IP
WAN	wireless ad hoc network
WCDMA	wideband code-division multiple-access
WCDS	weighted connected dominating set
WEP	wired equivalent privacy (encryption)
WiFi	common name used to refer to a wireless local-area network
WiMAX	Worldwide Interoperability for Microwave Access
WINS	a type of sensor node by Rockwell
WLAN	wireless local-area network
WMAN	wireless metropolitan-area network
WMN	wireless mesh network
WPA	WiFi protected access (mode)
WPAN	wireless personal-area network
WRP	wireless routing protocol
WSN	wireless sensor network
WWAN	wireless wide-area network
WWiSE	Worldwide Spectrum Efficiency (standard)
YG	Yao graph

Part I

Introduction

1 History of Wireless Networks

1.1 Introduction

The wireless arena has been experiencing exponential growth in the past decade. We have seen great advances in network infrastructures, rapid growth of cellular network users, the growing availability of wireless applications, and the emergence of omnipresent wireless devices such as portable or handheld computers, personal digital assistants (PDAs), and cellular phones, all becoming more powerful in their applications. The mobile devices are becoming smaller, cheaper, more convenient, and more powerful. They can also run more applications on the network services. For example, mobile users can rely on their cellular phones to check e-mail and browse the Internet. They can do so from airports, railway stations, cafes, and other public locations. Tourists can use the global positioning system (GPS) terminals installed in cars to view driving maps and locate attractions. All these factors are fueling the explosive growth of the cellular communication market. As of 2006, the number of cellular network users approached two billion worldwide. Market reports from independent sources show that worldwide cellular users have been doubling every 1.5 years.

In addition to that of the traditional cellular networks, an exponential growth of the wireless access point (AP), which is a device that connects wireless communication devices together to create a wireless network, is also being experienced. The AP is usually connected to a wired network and can relay data between devices on each side. Many APs can be connected together to create a larger network, which is a so-called *ad hoc network*. Low-cost, easily installed APs grew rapidly in popularity in the late 1990s and early 2000s. According to a new research study from Pyramid Research, WiFi users will outnumber cellular users by 2007. This trend will put increasing pressure on wireless operators to bundle both types of access. Currently, most of the connections among wireless devices occur over fixed-infrastructure-based service providers or private networks. Although the research and development efforts devoted to traditional wireless networks are still considerable, the interest of the scientific and industrial community of telecommunications has recently shifted to more challenging ad hoc wireless networks, in which a group of (potentially mobile) units equipped with radio transceivers can communicate without any fixed infrastructure. We will soon see a convergence of seamless networks that will keep everyone connected from their home to their office and all points in between. In addition, with the breakdown of traditional communications infrastructures during the recent Hurricane Katrina catastrophe, the need for reliable connectivity in order for emergency responders to talk to each other is even greater.

1.2   Different Wireless Networks

A number of different wireless networks exist and can be categorized in various ways depending on the criteria chosen for their classification, such as network architecture and communication coverage area.

Based on Network Architecture

Wireless networks can be divided into two broad categories based on how the network is constructed, i.e., the underlying network architecture.

1. **Infrastructure-based networks:** An infrastructure-based network is a network that has a preconstructed infrastructure that is made of a fixed network structure (typically, wired network nodes and gateways). Network services are delivered via these pre-constructed infrastructures. For example, cellular networks are infrastructure-based networks, which are built from public-switched telephone network (PSTN) backbone switches, mobile switching centers (MSCs), base stations, and mobile hosts. Each node of the network has its specific responsibility in routing the data, and the connection establishment follows a strict signaling sequence among the nodes. Another example of infrastructure-based networks are wireless local-area networks (WLANs).

2. **Infrastructureless networks:** An infrastructureless network is a network that is formed dynamically through the cooperation of an arbitrary set of independent wireless devices. There is no prearrangement of the specific roles for each node. Typically, each node is assumed to be able to forward the data packets for any other node if it is asked to do so. Each node can independently make its own decision based on the network situation. Examples of infrastructureless wireless networks include mobile ad hoc networks and wireless sensor networks.

Another classification criterion for wireless networks is based on the communication coverage area of the networks.

Based on Communication Coverage Area

As with wired networks, wireless networks can be categorized into different types of networks based on the distances over which the data are transmitted.

1. **Wireless wide-area networks (WWANs):** WWANs are infrastructure-based networks that rely on networking infrastructures to enable mobile users to establish wireless connections over remote networks. These connections often could be over a very large geographic areas (across cities or even countries) through the use of multiple antenna sites or satellite systems maintained by wireless service providers. Examples of WWANs include cellular networks and satellite networks.

2. **Wireless metropolitan-area networks (WMANs):** WMANs are also infrastructure-based networks that enable users to establish broadband wireless connections among multiple locations within a metropolitan area without the high cost of laying fiber or copper cabling lines. Both radio waves and infrared light can be used in WMANs to transmit data. The U.S. Institute of Electrical and Electronics Engineers (IEEE) set up a specific 802.16 Working Group on Broadband Wireless Access Standards that develops standards and recommended practices to support the development and the deployment of WMANs.

3. **Wireless local-area networks (WLANs):** WLANs enable users to establish wireless connections within a local area, typically within a corporate or campus building or in a public space such as an airport. The connections are typically within a 100-m range. WLANs can operate in an infrastructure-based mode or in an infrastructureless mode. In the infrastructure-based mode, wireless stations connect to wireless APs that serve as bridges between the stations and an existing network backbone. In the infrastructureless mode, several wireless stations within a limited area form a temporary network without using the wireless APs if they do not require access to outside network resources. Typical examples of WLAN implementations include 802.11 (also called WiFi) and Hiperlan2.

4. **Wireless personal-area networks (WPANs):** WPAN technologies enable users to establish ad hoc wireless communication among personal wireless devices such as PDAs, cellular phones, or laptops that are within a personal operating space. A WPAN operates in infrastructureless mode, and the connections are typically within a 10-m range. Two key WPAN technologies are Bluetooth and infrared light. Bluetooth is a cable-replacement technology that uses radio waves to transmit data to a distance of up to 10 m, whereas infrared can connect devices within a range of 1 m. WPAN implementations often have low complexity, lower power consumption, and are interoperable with 802.11 networks.

1.2.1 Wireless Cellular Networks

First-Generation Mobile Systems

The first generation of analog cellular systems included the Advanced Mobile Telephone System (AMPS), which was made available in 1983. It was first deployed in Chicago, with a service area of 2100 square miles. AMPS offered 832 channels, with a data rate of 10 kilobits per second (kbps). Although omnidirectional antennas were used in the earlier AMPS implementation, it was realized that using directional antennas would yield better cell reuse. In fact, the smallest reuse factor that would fulfill the 18-db signal-to-interference and noise ratio (SINR) by use of 120-deg directional antennas was found to be 7. Hence, a 7-cell reuse pattern was adopted for the AMPS. Transmissions from the base stations to mobiles occur over the forward channel by use of frequencies between 869 and 894 MHz. The reverse channel is used for transmissions from mobiles to the base station, with frequencies between 824 and 849 MHz.

In Europe, the Total Access Communications System (TACS) was introduced with 1000 channels and a data rate of 8 kbps. AMPS and TACS use the frequency-modulation (FM) technique for radio transmission. Traffic is multiplexed onto a frequency-division multiple-access (FDMA) system. In Scandinavian countries, the Nordic Mobile Telephone is used.

Second-Generation Mobile Systems

Compared with first-generation systems, second-generation (2G) systems use digital multiple-access technology, such as time-division multiple access (TDMA) and code-division multiple access (CDMA). The Global System for Mobile Communications, or GSM, uses TDMA technology to support multiple users. Examples of 2G systems are the GSM, cordless telephone (CT2), personal-access communications systems (PACSs), and digital European cordless telephone (DECT4).

A new design was introduced into the MSC of 2G systems. In particular, the use of base station controllers (BSCs) lightens the load placed on the MSC found in first-generation systems. This design allows the interface between the MSC and the BSC to be standardized. Hence, considerable attention was devoted to interoperability and standardization in 2G systems so that a carrier could employ different manufacturers for the MSCs and BSCs. In addition to enhancements in MSC design, the mobile-assisted handoff mechanism was introduced. By sensing signals received from adjacent base stations, a mobile unit can trigger a handoff by performing explicit signaling with the network.

2G protocols use digital encoding and include the GSM, digital AMPs (D-AMPS) (TDMA), and CDMA (IS-95). 2G networks are in current use around the world. The protocols behind 2G networks support voice and some limited data communications, such as faxing and short messaging services (SMSs), and most 2G protocols offer different levels of encryption and security. Although first-generation systems support primarily voice traffic, 2G systems support voice, paging, data, and fax services.

2.5G Mobile Systems

The move into the 2.5G world began with the General Packet Radio Service (GPRS). GPRS is a radio technology for GSM networks that adds packet-switching protocols, a shorter setup time for Internet service provider (ISP) connections, and the possibility of charging by the amount of data sent rather than by connection time. Packet switching is a technique whereby the information (voice or data) to be sent is broken up into packets of, at most, a few kilobytes each, which are then routed by the network between different destinations based on addressing data within each packet. Use of network resources is optimized as the resources are needed only during the handling of each packet.

The next generation of data heading toward third-generation (3G) and personal multimedia environments builds on the GPRS and is known as the enhanced data rate for GSM evolution (EDGE). EDGE is also a significant contributor in 2.5G. It allows GSM operators to use existing GSM radio bands to offer wireless multimedia Internet-protocol-(IP-) based services and applications at theoretical maximum speeds of 384 kbps with a bit rate of 48 kbps per time slot and up to 69.2 kbps per time slot in good radio conditions. EDGE will let operators function without a 3G license and compete with 3G networks offering similar data services. Implementing EDGE will be relatively painless and will require relatively small changes to network hardware and software because it uses the same TDMA frame structure, logic channel, and 200-kHz carrier bandwidth as today's GSM networks. As EDGE progresses to coexistence with 3G wideband CDMA (WCDMA), data rates of up to asynchronous-transfer-mode- (ATM-) like speeds of 2 Mbps could be available.

The GPRS will support flexible data transmission rates as well as a continuous connection to the network. The GPRS is the most significant step toward 3G.

Third-Generation Mobile Systems

3G mobile systems face several challenging technical issues, such as the provision of seamless services across both wired and wireless networks and universal mobility. In Europe, there are three evolving networks under investigation: Universal Mobile Telecommunications Systems (UMTSs), Mobile Broadband Systems (MBSs), and WLANs.

The use of hierarchical cell structures is proposed for IMT2000. The overlaying of cell structures allows different rates of mobility to be serviced and handled by different cells. Advanced multiple-access techniques are also being investigated, and two promising proposals have evolved, one based on WCDMA and another that uses a hybrid TDMA–CDMA–FDMA approach.

1.2.2   Wireless Access Points

A wireless AP is a device that connects wireless communication devices together to create a wireless network. The AP is usually connected to a wired network and can relay data between devices on each side. Many APs can be connected together to create a larger network that allows "roaming." In contrast, a network in which the client devices manage themselves is called an ad hoc network.

Low-cost, easily installed APs grew rapidly in popularity in the late 1990s and early 2000s. These devices offered a way to avoid tangled messes of cables associated with typical Ethernet networks of the day. Wireless networks also allowed users greater mobility, freeing individuals from the need to be stuck at a computer cabled to the wall. On the industrial and commercial side, wireless networking had a big impact on operations: Employees were often equipped with portable data terminals integrating bar-code scanners and wireless links, allowing them to update work-in-progress and inventory in real time.

One IEEE 802.11 AP can typically communicate with 30 client systems within a radius of 100 m. However, the communication range can vary a lot, depending on such variables as indoor or outdoor placement, height above ground, nearby obstructions, type of antenna, the current weather, operating radio frequency, and power output of the device. The range of APs can be extended through the use of repeaters and reflectors, which can bounce or amplify radio signals that ordinarily could not be received. Some experiments have been carried out to allow wireless networking over distances of several kilometers.

A typical corporate use of an AP is to attach it to a wired network and then provide wireless client adapters for users who need them. Within the range of the AP, the wireless end-user has a full network connection with the benefit of mobility. In this instance, the AP is a gateway for clients to access the wired network. Another use is to bridge two wired networks for which cable is not appropriate; for example, a manufacturer can wirelessly connect a remote warehouse's wired network with a separate (though within line of sight) office's wired network.

An AP may also act as the network's arbitrator, negotiating when each nearby client device can transmit. However, in the vast majority of currently installed IEEE 802.11 networks, this is not the case, as a distributed pseudo-random algorithm is used instead.

Limitations

There are only a limited number of frequencies legally available for use by wireless networks. Usually, adjacent APs will use different frequencies to communicate with their clients in order to avoid interference between the two nearby systems. Wireless devices are able to "listen" for data traffic on other frequencies and can rapidly switch from one frequency to another to achieve better reception on a different AP. However, the limited number of frequencies becomes problematic in crowded downtown areas with tall buildings housing multiple APs because there can be enough overlap between the wireless networks to cause interference.

Wireless networking is far behind wired networking in terms of bandwidth and throughput. Whereas (as of 2007) typical wireless devices for the consumer market can reach speeds of 11 (IEEE 802.11b) or 54 megabits per second (Mbit/s) (IEEE 802.11a, IEEE 802.11g), wired hardware of similar cost reaches 1000 Mbit/s (Gigabit Ethernet). One impediment to increasing the speed of wireless communications is that WiFi uses a shared communications medium, so the actual usable data throughput of an AP is somewhat less than half the over-the-air rate. Thus, a typical 54-Mbit/s wireless connection actually carries TCP/IP (TCP stands for transmission control protocol) data at 20 to 25 Mbit/s. Because users of legacy wired networks are used to the faster speeds, people using wireless connections are anxious to see the wireless networks catch up.

Security

Another issue with wireless access in general is the need for security. Many early APs were not able to discern whether a particular user was authorized to access the network. Although this problem reflects issues that have long troubled many types of wired networks (it has been possible in the past for individuals to plug computers into randomly available Ethernet jacks and get access to the network), this was usually not a significant problem because many businesses had reasonably good physical security. However, the fact that radio signals bleed outside of buildings and across property lines means that physical security is not as much of a deterrent to war drivers.

In response, several new security technologies have emerged. One of the simplest techniques involves only allowing access from certain medium-access control (MAC) addresses. However, MAC addresses can be easily spoofed, leading to the need for more advanced security measures. Many APs incorporate a wired equivalent privacy (WEP) encryption, but that also has been criticized by many security analysts as not good enough (the U.S. FBI demonstrated the ability to break WEP protection in 3 min). Newer (as of 2005) encryption standards available on wireless APs and client cards include WiFi protected access, WPA and WPA2 modes, both of which offer substantial improvements in security. The WiFi alliance has announced the inclusion of additional Extensible Authentication Protocol (EAP) types to its certification program for WPA- and WAP2-Enterprise. Also, a newer system for authentication, IEEE 802.1x, promises to enhance security on both wired and wireless networks. Wireless APs that incorporate technologies like these often also have routers built in, so they are somewhat more accurately described as wireless gateways.

1.2.3   Wireless Ad Hoc Networks

A wireless ad hoc network is a collection of autonomous nodes or terminals that communicate with each other by forming a multihop radio network and maintaining connectivity in a decentralized manner. The wireless nodes communicate over wireless links; thus, they have to contend with the effects of radio communication, such as noise, fading, and interference. In addition, the links typically have less bandwidth than in a wired network. Each node in a wireless ad hoc network functions as both a host and a router, and the control of the network is distributed among the nodes. The network topology is in general dynamic, as the connectivity among the nodes may vary with time because of node departures, new node arrivals, and the change of environments. Hence, there is a need for efficient routing protocols to allow the nodes to communicate over multihop paths. Some of these features are characteristic of the type of packet radio networks that were studied extensively in the 1970s and 1980s. Recently, the wireless ad hoc networking research has received much attention from academia, industry, and government. Because these networks pose many complex issues, there are many open problems for research and opportunities for making significant contributions.

There are two major types of wireless ad hoc networks: mobile ad hoc networks and smart sensor networks.

Mobile Ad Hoc Networks

A mobile ad hoc network (MANET) is a self-configuring wireless network composed of wireless devices. Figure 1.1 illustrates an example of an ad hoc network formed by eight laptop computers. In the figure, two computers are connected by a line if they can communicate directly with each other by using their wireless cards. In this case, we say they are within the transmission range of each other. The wireless devices are free to move randomly and organize themselves arbitrarily. Consequently, the network topology may change rapidly and unpredictably. A MANET network may operate in a

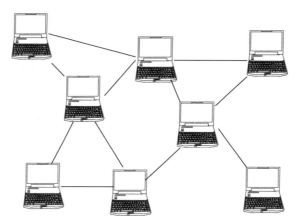

Figure 1.1 An ad Hoc network example.

stand-alone fashion or may be connected to the larger Internet. Because of their minimal configuration and quick deployment, ad hoc networks are often suitable for emergency situations like natural or human-induced disasters, military conflicts, emergency medical situations, and so on.

The earliest MANETs were called "packet-radio" networks, first sponsored by the U.S. Defense Advanced Research Projects Agency (DARPA) in the early 1970s. It is interesting to note that some early packet-radio systems predated the Internet and, indeed, were part of the motivation of the original Internet protocol (IP) suite. Later DARPA experiments included the Survivable Radio Network (SURAN) project, which took place in the 1980s. The third wave of academic activity on wireless ad hoc networks started in the 1990s, especially with the wide usage of inexpensive 802.11 radio cards for personal computers.

The popular IEEE 802.11 ("WiFi") wireless protocol incorporates an ad hoc networking system when no wireless APs are present. In an IEEE 802.11 system, each node transmits and receives data but does not route anything between the network's systems. Notice that it is possible to design higher-level protocols to aggregate various IEEE 802.11 ad hoc networks into MANETs.

Because of the growing interests in establishing survivable, efficient, dynamic communication for emergency/rescue operations, disaster relief efforts, and military networks, there is a strong need for the rapid deployment of independent mobile users. Obviously, we cannot rely on a centralized and organized network structure for these application scenarios. A MANET is an autonomous collection of mobile users that communicate over relatively bandwidth-constrained wireless links, for which all network activity including discovering the topology and delivering messages must be executed by the nodes themselves.

The design of network protocols for these networks is a complex issue. A unique characteristic of wireless networks is that the radio signal sent out by a wireless terminal will be received by all the terminals within its transmission range and also possibly causes signal interference to some terminals that are not intended receivers. In other words, the communication channels are shared by the wireless terminals. Thus, one of the major problems facing wireless networks is the reduction of capacity because of interference caused by simultaneous transmissions. Using multiple channels and multiple radios can alleviate but not eliminate the interference. This raises the scalability issue of all wireless networks (MANETs, WSNs).

Regardless of the application, MANETs need efficient distributed algorithms to determine network organization, link scheduling, and routing. However, determining feasible routing paths and delivering messages in a decentralized environment where network topology fluctuates over time is not an easy problem, and, to some extent, it is even not a well-defined problem. Although the shortest path (based on a given cost function) from a source to a destination in a static wired network is usually the optimal route, this idea is not easily extended to MANETs. A number of unique characteristics of wireless networks make the "simple" optimal unicast routing much harder. For example, various factors, such as variable wireless link quality, propagation path loss, fading, multiuser interference, power expended, and topological changes, become relevant issues. Notice that finding the path (or even multiple paths) with the largest throughput to connect a given pair of source and target nodes in a wireless network is already a nondeterministic-polynomial- (NP-) hard problem even if only the wireless interference (interpath interference and intrapath interference) is considered. Moreover, in many applications such as a military environment, preservation of security, achieving small latency, reliability, preventing intentional jamming, and recovery from failure are significant concerns. This will make the design of a good wireless protocol much harder. Additionally, in certain applications (especially military networks), we need to maintain a low probability of intercept and/or a low probability of detection. Hence, nodes prefer to radiate as little power as necessary and transmit as infrequently as possible. A lapse in any of these requirements may degrade the performance and dependability of the network. Although there are so many challenges in designing secure and efficient wireless ad hoc networks, this book is not intended to (and, clearly, it is impossible to) solve all important and interesting problems here. Some of the algorithmic and graph theoretical issues that can form a foundation for further study of some of the problems not addressed here are covered.

Wireless Sensor Networks

Most sensors are electrical or electronic, although other types exist. A sensor is a type of transducer. Sensors are either direct indicating (e.g., a mercury thermometer or electrical meter) or are paired with an indicator [perhaps indirectly through an analog-to-digital (A/D) converter, a computer, and a display]. Sensors are heavily used, in addition to other applications, in medicine, industry, and robotics. With the technical progress, more and more sensors are manufactured with Micro-Electro-Mechanic-Systems (MEMS) technology. This often offers the potential of reaching a much higher sensitivity. A good sensor obeys the following rules:

1. The sensor should be sensitive to the measured property.
2. The sensor should be insensitive to any other property.
3. The sensor should not influence the measured property.

In the ideal situation, the output signal of a sensor is exactly proportional to the value of the measured property.

Distributed, wireless, microsensor networks will enable myriad applications for sensing and controlling the physical world. A WSN is a network made of hundreds or thousands of devices using sensors (also called nanocomputers) to monitor different conditions, such as temperature, sound, vibration, pressure, motion, or pollutants, at different locations. Usually these devices are small and inexpensive, so that they can be produced and deployed in large numbers. For example, for the field of computer science, most sensors are made by two companies, **xbow** and **moteiv**. One of the main differences between MANETs and WSNs is that the wireless sensors often have severely constrained resources in terms of energy, memory, computational speed, and bandwidth. The sensor nodes are self-contained units equipped with a radio transceiver, a small microcontroller, and an energy source, usually a battery. Recently, acoustic sensors have also been built for underwater monitoring. In most WNSs, the sensors typically rely on each other to transport data to a monitoring computer. The nodes dynamically self-organize their network topology based on various network conditions, rather than having a preprogrammed network topology. Because of the limitations that are due to battery life, nodes are built with power conservation in mind and generally spend large amounts of time in a low-power "sleep" mode or processing the sensor data. Thus, each sensor has wireless communication capability and some level of intelligence for signal processing and networking of the data. The wireless ad hoc sensor networks offer certain capabilities and enhancements in operational efficiency in civilian applications as well as in assisting in the national effort to increase alertness to potential terrorist threats. For almost all WSNs, there are three essential functions: sensing, communications, and computation (hardware, software, algorithms).

Modern research on sensor networks started around 1980 at DARPA: the Distributed Sensor Networks (DSN) program. Smaller computing chips, more capable sensors, wireless networks, and other new information technologies (ITs) are pushing the development of sensor networks.

There are many ways to classify the WSNs. One way is whether the nodes are individually addressable, and another is whether the data in the network are aggregated. Whether addressability is needed depends on the applications. For example, the sensor nodes in a parking-lot network should be individually addressable, so that one can determine the locations of all the free spaces. On the other hand, if one wants to determine the temperature in a corner of a room, then addressability may not be so important. The ability of the sensor network to aggregate the data collected can greatly reduce the number of messages that need to be transmitted across the network. In the majority of tasks of a WSN, an important requirement is that the required data be disseminated to the proper end-users. In some cases, there are fairly strict time requirements on this communication. For example, the detection of an intruder in a surveillance network should be immediately communicated to the police so that action can be taken.

To design an efficient and secure WSN, we often need to address a number of challenging issues.

Aside from the deployment of sensors on the ocean surface or the use of mobile, unmanned, robotic sensors in military operations, most nodes in a smart sensor network are stationary. Networks of 10,000 or even 100,000 nodes are envisioned, so scalability is a major issue. The algorithms and protocols designed for WSNs should bound the communication cost with respect to the network size.

Because in many applications the sensor nodes are often powered by batteries and will be placed in a remote area, recharging the batteries of a node may not be possible. In this case, the lifetime of a node may be determined by the battery life. Consequently, the minimization of energy expenditure is crucial. There are a considerable number of research papers in the literature that propose reducing the energy consumption of WSNs by use of various approaches such as topology control, data aggregation, data compression, energy-efficient MAC protocols, and smart use of some of the properties of batteries (a battery will last longer if it is not used continuously for a long time).

Given the large number of nodes and their potential placement in hostile locations, it is essential that the network be able to self-organize. Moreover, nodes may fail (either from lack of energy or from physical destruction and so on), and new nodes may join the network. Therefore, the network must be able to periodically reconfigure itself so that it can continue to function. Individual nodes may become disconnected from the rest of the network, but a certain connectivity of the network (or large portion of the network) must be maintained.

Another unique factor that distinguishes WSNs from MANETs is that the end goal is the detection and estimation of some events of interest and not just communications. To improve the detection–estimation performance, it is often quite useful to fuse data from multiple sensors. This data fusion requires the transmission of data and control messages, and so it may put constraints on the network architecture. Another important phenomenon is that we need to be able to distinguish between false data collected and the data reflecting a real emergency (e.g., a fire) in certain area. For example, a high temperature in an area reported by a sensor may indicate that there is a fire or may be due to errors in sensing or processing.

All WSNs should provide querying ability. A user may want to query an individual node or a group of nodes for information collected in the region. Depending on the amount of data fusion performed, it may not be feasible to transmit a large amount of the data across the network. Instead, various local sink nodes will collect the data from a given area and create summary messages. A query may be directed to the sink node nearest to the desired location.

The last, but not least, important challenge is the interoperability issue. With the coming availability of low-cost, short-range radios, along with advances in wireless networking, it is expected that WSNs will become commonly deployed. In these networks, each node may be equipped with a variety of sensors, such as acoustic, seismic, infrared, and still/motion videocamera. These nodes may be organized in clusters and coordinate with each other such that a locally occurring event can be detected by most if not all of the nodes in a cluster. These nodes will collaborate to make certain local decisions based on the information and decisions collected from each individual node within the cluster. One node may act as the cluster master, and it will coordinate these efforts.

Mesh Networks

Wireless mesh networking is mesh networking implemented over a WLAN. This type of Internet infrastructure is decentralized, relatively inexpensive, and very reliable and resilient because each node need transmit only as far as the next node. Nodes act as repeaters to transmit data from nearby nodes to peers that are too far away to reach, resulting in a network that can span large distances, especially over rough or difficult terrain. Extra capacity can be installed by the addition of more nodes or the use of more channels. Mesh networks may involve either fixed or mobile devices. Wireless mesh networks (WMNs) are being used as the last mile for extending the Internet connectivity for mobile nodes. Many U.S. cities (e.g., Medford, Oregon; Chaska, Minnesota; and Gilbert, Arizona) have already deployed mesh networks. Accesos Web Alternativos (AWA), the Spanish operator of WLAN networks, will roll out commercial WLANs and WMNs for voice and data services. Several companies such as MeshDynamics have recently announced the availability of multihop multiradio mesh network technology. These networks behave almost like wired networks because they have infrequent topology changes, limited node failures, and so on. For WMNs, the aggregate traffic load of each routing node also changes infrequently.

The choice of radio technology for WMNs is crucial. In a traditional wireless network in which laptops connect to a single AP, the more laptops connected, the less bandwidth available for each user. This is because the devices share a fixed bandwidth amount. With mesh technology and adaptive radio, devices in a mesh network will connect only with other devices that are in a set range. The advantage is that like a natural load-balancing system, the more devices, the more bandwidth that becomes available, provided that the number of hops in the average communications path is kept low. To prevent an increased hop count from canceling out the advantages of multiple transceivers, one common type of architecture for a mobile mesh network includes multiple fixed-base stations that will provide gateways to services, wired parts of the Internet, and other fixed-base stations.

1.3   Conclusion

Many people were involved in the invention of radio transmission of information as we know it today. Despite this, during its early development and long after wide acceptance, disputes persisted as to who could claim sole credit for this invention. James Clerk Maxwell performed the theoretical physical research that correctly predicted the existance of radio (and all other electromagnetic) waves. David E. Hughes was the first to transmit Morse code by radio, but scientists of his time were not quick to recognize Maxwell's theories nor Hughes' experiments. Heinrich Rudolf Hertz was the experimental physicist who first created radio waves in a controlled manner. Later developments are greater or lesser engineering developments of their work. In late 1896 or early 1897, Nikola Tesla (10 July 1856–7 January 1943) received wireless signals transmitted from the Houston Street lab in New York City to West Point. Marconi began to conduct experiments, building much of his own equipment in the attic of his home at the Villa Griffone in Pontecchio, Italy. Marconi transmitted radio signals for about a mile at the end of 1895. In 1904, Marconi got his own patent, declaring principles that Tesla had developed. The issue of patent infringement by Marconi was addressed in a lawsuit brought by Tesla in 1915. In 1943, the Supreme Court of the United States credited Nikola Tesla as being the inventor of the radio. The first radio telephone network for commercial use was made available to consumers by the Bell Telephone Company in the early 1950s. In 1971, the world's first WLAN, named ALOHAnet, was developed by researchers at the University of Hawaii. These days, the use of GSM, PCS, and WiFi has spread to almost every corner in the world. In the past 10 years, wireless ad hoc networks and recently WSNs have drawn a considerable amount of research interests from various research fields. WSNs build a connection between the physical world that still has many unknowns left for exploring and the digital world dominated by the Internet. It is expected that WSNs will dramatically improve our understanding in many fields. In this chapter, only a brief review of the development history of wireless networks was given and some categorization of wireless networks was presented. A more detailed review on this topic can be found in many great books and on Internet websites.

Problems

1.1 What are the major differences between wired networks and wireless networks? Why do some problems that are easy to solve for wired networks becomes NP-hard for wireless networks? Can you list a few such questions?

1.2 In what band do most cellular and WLAN systems operate?

1.3 Find the relationship among bandwidth, information capacity, and the signal-to-interference-noise ratio.

1.4 What are the advantages and disadvantages of WiFi?

1.5 Compare and contrast the advantages and disadvantages of communicating via MANETs and via cellular networks.

1.6 What are the advantages and disadvantages of operating in unlicensed bands?

1.7 What are the advantages and disadvantages of operating in licensed bands?

1.8 What are the differences among MANETs, WLANs, WSNs, and WMNs?

1.9 Find and read good survey articles about MAC protocols, routing protocols, and mobility management protocols for MANETs.

1.10 Find and read good survey articles about MAC protocols, routing protocols, localization protocols, data processing, and aggregation protocols, and energy-efficient topology control protocols for WSNs.

1.11 Find and read good survey articles about MAC protocols, routing protocols, and cross-layer design protocols for WMNs.

1.12 What are the next-generation wireless networks? What is opportunistic spectrum usage? What is the current status? What are the major challenges in networks with opportunistic spectrum usage?

2   Wireless Transmission Fundamentals

In this chapter, some simple but also widely accepted models of wireless ad hoc networks are introduced. Notice that WSNs comprise a special subclass of wireless ad hoc networks; thus, when we use the term "wireless ad hoc networks," we also include WSNs if not specifically clarified. In this chapter, the main focus is on the wireless channel model, the interference model, the energy-consumption model, and the mobility model.

2.1   Wireless Channels

The main difference between wireless networks and traditional wired networks is that the wireless devices in a network communicate over wireless channels via wireless transceivers. Thus, to understand wireless ad hoc networks and design efficient protocols and algorithms for wireless networks, we need to understand the characteristics of wireless communications. An important building block of wireless ad hoc network studies is thus the wireless channel model. In the literature, there are a number of wireless channel models proposed and the model presented in this chapter is based on the material contained in Rappaport (1996) and Santi (2005b).

It is widely assumed that a radio channel from a transmitting wireless device u to a receiving wireless device v is established if and only if the received power of the radio signal at node v is above a certain threshold. Let $\mathbf{p}(u, v)$ denote the power assigned to node[1] u to transmit a signal from u to v. We always assume that this power can maintain a reasonably good communication link quality[2] from node u to node v. This power $\mathbf{p}(u, v)$ could be fixed throughout the network operations if no power-control techniques are employed, or it could be changed dynamically when it is needed by the power-control techniques or to ensure energy-efficient routing. It is well known that wireless propagation suffers from severe attenuation. Let $\|uv\|$ denote the Euclidean distance between two wireless nodes u and v. If node u transmits at a power $P_t(u)$, the power of the signal received at node v is assumed to be

$$P_r(v) = \frac{P_t(u)}{g(u, v)},$$

---

[1] In this book, the term *node* often represents a network device, *vertex* is a graph term, and *point* is a geometry term. We often interchange them if no confusion is caused.

[2] In practice, it often means that the link error probability is not larger than a certain threshold.

where $g(u, v)$ is the wireless gain between node u and v $[1/g(u, v)$ is often called *path loss* in the literature]. It is commonly assumed in the literature that we can always correctly decode the signal when the received power $P_r(v)$ satisfies $P_r(v) \geq \beta_0 N_0$, where β_0 is the required minimum *signal-to-interference-noise ratio* (SINR) and N_0 is the strength of the ambient noise. Here, the constant β_0 is wireless technology and device dependent. Thus, by assuming that the node u transmits at power $P_r(u) \geq \beta_0 N_0 g(u, v)$, it is assumed in the literature that we can guarantee that node v will receive the signal correctly.

Notice that, in practice, this is not the case. When a node u transmits at a power p to another node v, the link (u, v) has a packet-error probability dependent on the transmission power p. Notice that the packet-error probability also depends on other factors, such as the environment, the digital modulation techniques, and so on.

Modeling the path loss has historically been one of the most difficult tasks of the wireless-system designer because it depends on many parameters such as location, time, weather, and so on. A radio propagation model is an empirical mathematical formulation for the characterization of radio-wave propagation as a function of frequency, distance, and other conditions. A single model is usually developed to predict the behavior of propagation for all similar links under similar constraints. In the remainder of this section, we review some of the widely used radio propagation models used in the literature.

2.1.1   Free-Space Propagation Model

The free-space propagation model assumes the ideal propagation condition that there is only one clear line-of-sight path between the transmitter and receiver. In other words, this model can be used to predict radio-signal propagation when the path between the transmitter and the receiver is clear and without obstruction. Let P_t be the power used by the transmitter to transmit the radio signal. Let $P_r(d)$ be the power of the radio signal received by a node located at distance d from the transmitter. Under the free-space propagation model, we have

$$P_r(d) = \frac{P_t G_t G_r \lambda^2}{(4\pi)^2 d^2 L},\tag{2.1}$$

where G_t is the transmitter antenna gain, G_r is the receiver antenna gain, $L \geq 1$ is the system loss factor independent of the propagation, and λ is the wavelength in meters. By ignoring the specific characteristics of the transmitter and the receiver, we can simplify Eq. (2.1) as follows:

$$P_r(d) = \frac{C_f P_t}{d^2},\tag{2.2}$$

where C_f is a constant depending on the characteristics of the transmitter and the receiver. Here, f stands for *free space*. In certain scenarios (e.g., the devices are uniform), it is often assumed in the literature that $C_f = 1$. Notice that it is common to select $G_t = G_r = 1$ and $L = 1$ in NS$_2$ simulations.

When the sensitivity of the receiver node is given as β_0 and the background noise and interference are denoted as N_0, we can state that the transmitted message can be correctly received if and only if the distance d satisfies

$$d \leq \sqrt{\frac{C_f P_t}{\beta_0 N_0}}.$$

In other words, the free-space model basically represents the communication range as a disk (or sphere in three dimensions) around the transmitter with radius $\sqrt{\frac{C_f P_t}{\beta_0 N_0}}$. If a receiver is within the disk, it receives all packets. Otherwise, it loses all packets.

In practice, the free-space propagation model is valid only for values of d that are relatively far from the transmitting antenna. For values of d within the so-called close-in distance d_0, the path loss can be assumed to be constant.

2.1.2   Two-Ray Ground Model

For the free-space propagation model, it assumes that the single direct path between the transmitter and the receiver is the only physical means of propagation of the radio signal. In practice, it is rarely the case and thus the free-space model is often inaccurate, although it is widely adopted in the literature. The two-ray ground-reflection model considers both the direct path and a ground-reflection path. It is shown in Rappaport (1996) that this model gives a more accurate prediction at a long distance than the free-space model. The received power at distance d is predicted by

$$P_r(d) = \frac{P_t G_t G_r h_t^2 h_r^2}{d^4 L}, \tag{2.3}$$

where h_t and h_r are the heights of the transmitting and receiving antennas, respectively. Notice that in some literature [e.g., Rappaport (1996)], it is assumed that $L = 1$. To be consistent with the free-space propagation model, some works also add L here.

Equation (2.3) shows a faster power loss than the model for the free-space propagation model [Eq. (2.1)] as distance increases. However, the two-ray model does not give a good result for a short distance because of the oscillation caused by the constructive and destructive combination of the two rays. Instead, the free-space model is still used when d is small. When the distance d between the transmitter and the receiver is relatively large ($d \gg \sqrt{h_t h_r}$), we can abstract the features of the radio transmitters and receivers and get the following simplified formula:

$$P_r(d) = \frac{C_t P_t}{d^4}, \tag{2.4}$$

where C_t is a constant depending on the characteristics of the transmitter and the receiver. Here, t stands for *two-ray ground*.

When the sensitivity of the receiver node is given as β_0 and the background noise and interference are denoted as N_0, we can state that the transmitted message can be correctly received if and only if the distance d satisfies

$$d \leq \sqrt[4]{\frac{C_t P_t}{\beta_0 N_0}}.$$

In other words, the free-space model basically represents the communication range as a circle (or sphere in three dimensions) around the transmitter with radius $\sqrt[4]{\frac{C_t P_t}{\beta_0 N_0}}$. If a receiver is within the circle, it receives all packets. Otherwise, it loses all packets.

Obviously, the main difference between the free-space propagation model and the two-ray ground model is that the signal falloff is proportional to the distance raised to the fourth power in the two-ray ground model here, whereas it is the distance raised to the square. Therefore, a crossover distance d_c is computed in these two models. When $d < d_c$, the free-space-propagation model is used, and the two-ray ground model is used otherwise. At the crossover distance d_c, these two models should give the same result. Thus, we can compute d_c as

$$\frac{4\pi h_t h_r}{\lambda}.$$

2.1.3   The Log-Distance Path-Loss Model

The log-distance model is derived from the combination of analytical and empirical methods. The log-distance path-loss model is an indoor radio-propagation model that predicts the path loss a signal encounters inside a building over a distance. The log-distance path-loss model is formally expressed as

$$L = L_0 + N \log \frac{d}{d_0} + X_g,$$

where L is the total path loss inside a building [with units of decibels (dB)], L_0 is the path-loss at reference distance (usually, 1 km or 1 mile with units of dB), N is the path-loss distance exponent times 10, and X_g is a Gaussian random variable with zero mean and σ standard deviation. The coefficients N and σ depend on the environment and also on the frequency of the radio.

The log-distance model can be seen as a generalization of both the free-space and the two-ray-ground propagation model. In other words, the average long-distance path loss is proportional to the separation distance d raised to a certain exponent α, which is called the *path-loss exponent* or *distance-power gradient*. In most literature, it is assumed that

$$P_r(d) \propto \frac{P_t}{d^\alpha}.$$

In other words, the log-distance model also represents the communication range as a disk (or sphere in three dimensions) around the transmitter with radius $\sqrt[\alpha]{\frac{C_f P_t}{\beta_0 N_0}}$. If a receiver is within the disk, it receives all packets. Otherwise, it loses all packets. The exact value of α depends on the environmental conditions, and it has been evaluated in many scenarios. See Rappaport (1996) for more information about the empirical values of α.

2.1.4   Large-Scale and Small-Scale Variations

Notice that all the previously discussed propagation models predict the *average* received power at a certain distance from the transmitter. In practice, the intensity of the received signal is often denoted by a random variable, and its actual value can vary a lot from the predicted average value. Thus, probabilistic models have been used to account for this time- and location-dependent wireless channel. In a probabilistic propagation model, the coverage region of a radio (i.e., the region where a receiver can get the signal correctly) is no longer a disk. Two different classes of probabilistic propagation models have been discussed in Rappaport (1996). One of the models is the *large-scale model*, which predicts the variations of the signal intensity over large distances. The other one is the *small-scale model*, which predicts the variations of the signal intensity over very short distances. They are also called *multipath-fading models*. The shadowing model extends the ideal circle model to a richer statistic model: Nodes can probabilistically communicate only when near the edge of the communication range.

One of the most important large-scale models is the shadowing model, in which the path loss at distance d is modeled as a random variable with log-normal distribution centered about the mean value. The most important fading model is the Rayleigh model, which models small-scale variations of the signal intensity according to a random variable with Rayleigh distribution. See Rappaport (1996) for a more detailed discussion of the radio propagation models.

Observe that accounting for large-scale and small-scale variations of the radio signal is very complicated and renders the link model tightly coupled with the specific application environment. Thus, in this section and the remainder of the book, we model the wireless channel by using the log-distance path-loss model, which already extracts a considerable number of characteristics of the environment. This assumption is often a standard in conducting the theoretical and algorithmic study of wireless ad hoc networks (especially in the area of the topology control, power assignment, and so on).

2.2   The Wireless Communication Graph

The communication graph defines the network topology formed by a set of wireless devices; that is, the set of wireless links that these wireless devices can use to communicate with each other, possibly using multihop paths. Based on the discussion of the radio-signal propagation models in the previous section, obviously whether two devices (called nodes also hereafter) u and v can form a communication link (u, v) depends on (1) the relative Euclidean distance between these two nodes, (2) the transmitting power used by the transmitter to send the signal, and (3) the surrounding environment, which will determine which propagation model can be used.

Assume that there is a set $V = \{v_1, \ldots, v_n\}$ of n communication wireless terminals deployed in a region Ω (which could be some area in a two-dimensional plane or a region in a three-dimensional space). We also abuse the notation a little by using v_i to denote not only the identification of a wireless node but also the geometry position of this node.

If nodes are mobile, the physical node location is time-dependent. We always assume that the nodes will move within the original deployed region Ω. For simplicity, we use $v_i(t)$ to denote the location of node v_i at time t.

The complete communication graph is a *directed* graph $G = (V, E)$, where $V = \{v_1, \ldots, v_n\}$ is the set of terminals and E is the set of possible *directed* communication links between pairs of wireless terminals. Let $E^-(u)$ denote the set of directed links that end at node u [i.e., (w, u)] and let $E^+(u)$ denote the set of directed links that start at node u [i.e., (u, v)].

Every terminal v_i has a transmission range $R_T(v_i)$ such that the necessary condition for a terminal v_j to receive correctly the signal from v_i is $\|v_i - v_j\| \leq R_T(v_i)$, where $\|v_i - v_j\|$ is the Euclidean distance between v_i and v_j. In other words, the transmission range of a node v_i denotes the maximum distance within which the data transmitted by node v_i can be correctly received by the receiver. Given the transmission range r of a node v_i, the definition of the transmission region depends on the dimension of the network deploy region. In the case of one-dimensional networks, the transmission region is simply the segment of length $2r$ centered at node v_i; in the case of two-dimensional networks, the transmission region is simply the disk of radius r centered at node v_i; in the case of three-dimensional networks, the transmission region is simply the sphere of radius r centered at node v_i. Notice that here the transmission range of a node v_i is dependent on the transmission power of this node and the propagation model used. Throughout this book, we always assume that all nodes adopt the same signal-propagation model. Thus, the transmission range of any node is uniquely determined by its transmission power, and sometimes they are interchangeably used in the rest of this book. In most of the results presented in this book, we assume that $\|v_i - v_j\| \leq R_T(v_i)$ is *also* the sufficient condition for $(v_i, v_j) \in E$.

When all nodes have the same transmission range, then the resulting communication graph is often called a *unit disk graph* (UDG). In other words, we normalize the transmission range of each node to *one unit* and, consequently, two nodes can communicate with each other directly if and only if their distance is no more than one unit.

When different nodes may have different transmission powers (and thus different transmission ranges), several different models of communication graphs are used in the literature. One model requires that only undirected links be used to support communication. Thus, there is an *undirected link* $v_i v_j$ in the communication graph G if and only if node v_j is inside the transmission region of node v_i and v_i is inside the transmission region of node v_j. In other words, both directed links (v_i, v_j) and (v_j, v_i) exist. This model is called the *mutual-inclusion* graph model in X.-Y. Li *et al.* (2005c). Mathematically, a link $v_i v_j$ is included in the communication graph G if and only if the Euclidean distance $\|v_i - v_j\| \leq \min[R_T(v_i), R_T(v_j)]$. Another model will use all directed links for communication. Thus, when node v_j is inside the transmission region of node v_i, the communication graph will include a *directed* link (v_i, v_j).

Notice that, in practice, $\|v_i - v_j\| \leq R_T(v_i)$ is *not* the sufficient condition for $(v_i, v_j) \in E$. For some results presented in this book, we assume the latter case. Some links do not belong to G because of either the physical barriers or the selection of routing protocols. This model has been used to study various problems in the literature, e.g., Kumar *et al.* (2005) and W. Wang *et al.* (2006b). We always use (v_i, v_j) to denote the *directed* link (v_i, v_j) hereafter if the communication graph is assumed to have directed links; i.e., node v_j is inside the transmission region of node v_i. We simply use $v_i v_j$ to denote an undirected link between two nodes; i.e., node v_i can directly receive the signal correctly from v_j and vice versa. When this is the case, we call the network a *general geometry graph*.

2.3   Power Assignment and Topology Control

A wireless node can receive the signal from another node if it is within the transmission region of the sender. Otherwise, they communicate through multihop wireless links by using intermediate nodes to relay the message. A larger transmission range of a wireless node means that more neighbors can communicate directly, but it costs more energy. Energy conservation is a critical issue in a wireless ad hoc network for the individual node and the network because the nodes are powered by small batteries only. For example, in a battlefield scenario, soldiers may not have time to replace or recharge the batteries of their wireless devices, and running out of batteries means a loss of all of their communication capacity. Thus, a considerable amount of research efforts focus on designing minimum-power-assignment algorithms to save energy for typical network tasks such as broadcast transmission (Clementi *et al.*, 2001b; Huiban and Verhoeven, 2004; Wan *et al.*, 2002b; Wieselthier *et al.*, 2000), routing (Srinivas and Modiano, 2003), connectivity (Althaus *et al.*, 2003; Blough *et al.*, 2002; Chen and Huang, 1989; Clementi *et al.*, 2000c; Kirousis *et al.*, 2000), and fault tolerance (Călinescu and Wan, 2003; Cheriyan *et al.*, 2002; Hajiaghayi *et al.*, 2003).

Power Assignment

We assume that the power w_{uv} needed to support the communication between two nodes u and v is a monotone increasing function of the Euclidean distance $\|uv\|$. In other words, $w_{uv} > w_{xy}$ if $\|uv\| > \|xy\|$ and $w_{uv} = w_{xy}$ if $\|uv\| = \|xy\|$. For example, in the literature it is often assumed that $w_{uv} = c + \|uv\|^\alpha$, where c is a positive constant real number, and real number $\alpha \in [2, 5]$ depends on the transmission environment. We also assume that all nodes have omnidirectional antennas; i.e., if the signal transmitted by a node u can be received by a node v, then it will be received by all nodes x with $\|ux\| \leq \|uv\|$. In addition, all nodes can adjust the transmission power dynamically. Specifically, each node u has a maximum transmission power $\mathbf{P}_{\max}$ and it can adjust its power to be exactly w_{uv} to support the communication to another node v. Consequently, if all wireless nodes transmit in their maximum power, they define a network that has a link uv iff $w_{uv} \leq \mathbf{P}_{\max}$. When nodes adjust their power dynamically, we say that a node u can reach a node v in an *asymmetric* communication model if node u transmits at a power of at least w_{uv}. Notice that here, in asymmetric communications, node v may transmit at a power less than w_{vu} and thus cannot reach u. We say that a node u can reach a node v in a *symmetric* communication model if both nodes u and v transmit at a power of at least w_{uv}.

An observation of power adaption is that the network topology is entirely dependent on the transmission range of each individual node. Links can be added or removed when a node adjusts its transmission range. A *power assignment* $\mathcal{P}$ is an assignment of power setting $\mathcal{P}(v_i)$ to wireless node v_i. Given a power assignment $\mathcal{P}$, we can define an induced direct communication graph $\overrightarrow{G}_\mathcal{P}$ in which there is a directed edge $\overrightarrow{uv}$ if and only if $w_{uv} \le \mathcal{P}(u)$. We define the induced undirected communication graph $G_\mathcal{P}$ in which there is an edge uv if and only if $w_{uv} \le \mathcal{P}(u)$ and $w_{vu} \le \mathcal{P}(v)$. We hereby refer to $G_\mathcal{P}$ as the *induced communication graph* by power assignment $\mathcal{P}$. If all wireless nodes transmit at their maximum power $\mathbf{P}_{max}$, the induced communication graph is called the *original communication graph* (UDG), which provides information about all possible topologies in accordance with characteristics of the wireless environment and node power constraints. In other words, all possible achievable network topologies are subgraphs of the original communication graph.

On the other hand, given a subgraph $G = (V, E)$ of the original communication graph, we can also extract a minimum power assignment $\mathcal{P}_G$, where

$$\mathcal{P}_G(u) = \max_{\{v | uv \in E\}} w_{uv},$$

to support the subgraph. We call this $\mathcal{P}_G$ an *induced power assignment* from G.

There are a number of optimization criteria studied in the literature for power assignment. The *min–max power-range assignment* is to find a power assignment whose maximum power is minimized among all possible power assignments that can achieve a certain network property (e.g., the induced communication graph is connected). The *min–total power-range assignment* is to find a power assignment whose total assigned power to all nodes is minimized among all possible power assignments that can achieve a certain network property (e.g., the induced communication graph is connected). A number of network properties have been studied in the literature for power assignment such as connectivity, two-connectivity, and generally k-connectivity. Both centralized and distributed (or even localized) algorithms have been proposed.

Because of the importance of energy efficiency in wireless ad hoc networks, minimum power assignments for different network issues have been addressed recently. Research efforts have focused on finding the minimum power assignment so that the induced communication graph has some "good" properties in terms of network tasks such as disjoint paths, connectivity, or fault tolerance. The minimum-energy-connectivity problem was first studied by Chen and Huang (1989), in which the induced communication graph is strongly connected while the total power assignment is minimized. They showed that this problem is NP-hard. Recently, this problem was intensively studied, and many approximation algorithms were proposed when the network is modeled by use of symmetric links or asymmetric links (Althaus *et al.*, 2003; Blough *et al.*, 2002; Călinescu *et al.*, 2003; Clementi *et al.*, 2000c; Kirousis *et al.*, 2000; Ramanathan and Rosales-Hain, 2000). Along this line, several authors (Călinescu and Wan, 2003; Cheriyan *et al.*, 2002; Hajiaghayi *et al.*, 2003) considered the minimum total power assignment while the resulting network is k-strongly connected or k-connected. This problem has been shown to be NP-hard too. Solving this problem can improve the fault tolerance of the network. Clementi *et al.* (2000a, 2000b, 2000c) also considered the minimum-energy-connectivity problem while the induced communication graph has a diameter bounded by a constant h. Lloyd *et al.* (2002) proposed one general framework that leads to an approximation algorithm for minimizing total power assignment. Using the framework, they proposed a new two-connected approximation method for min–total power assignment. Krumke *et al.* (2003) also studied the minimum power assignment so that networks satisfy specific properties such as connectivity, bounded diameter, and minimum node degree. Other relevant work in the area of power assignment (also called energy efficiency) includes energy-efficient broadcasting and multicasting in wireless networks. The problem, given a source node s, is to find a minimum power assignment such that the induced communication graph contains a spanning tree rooted at s. This problem was proved to be NP-hard. In Clementi *et al.* (2001a), Huiban and Verhoeven (2004), Wan *et al.* (2002b), and Wieselthier *et al.* (2000), the authors presented some heuristic solutions and gave some theoretical analysis. Recently, Srinivas and Modiano (2003) also studied finding k-disjoint paths for a *given* pair of nodes while minimizing the total node power needed by nodes on these k-disjoint paths. An excellent survey of some recent theoretical advances and open problems on energy consumption in ad hoc networks can be found in Clementi *et al.* (2002).

Asymptotic Power Assignment

The previous discussion on power assignment assumed that there is *one fixed* network as the input. In the literature, there are a number of studies on power assignment that assume that the networking nodes could be mobile or there are a set of network instances as the input (whose number could be infinity). When the network nodes are mobile, a range assignment $\mathcal{P}$ is said to be *connected* at time t if the induced communication graph at time t is connected. Notice that the power assignment $\mathcal{P}$ could be a function of time t.

The universal minimum power used by all wireless nodes such that the induced network topology is connected is called the *critical power*. Determination of the critical power in which the wireless nodes are statically distributed was studied by several researchers recently (Gupta and Kumar, 1998; Ramanathan and Rosales-Hain, 2000; Sanchez *et al.*, 1999). Both Ramanathan and Rosales-Hain (2000) and Sanchez *et al.* (1999) use the power assignment induced by the longest incident edge of the Euclidean minimum spanning tree over wireless nodes V. Although determining the critical power for static wireless ad hoc networks is well studied, it remains to study the critical power for connectivity for mobile wireless networks. Because the wireless nodes move around, it is impossible to have a unanimous critical power to guarantee the connectivity for all instances of the network configuration. Thus, we need to find a critical power, if possible, at which each node has to transmit to guarantee the connectivity of the network almost surely (i.e., with a high probability sufficiently close to 1). For simplicity, we assume that the wireless devices are distributed in a unit square (or disk) according to some distribution function (e.g., uniform distribution or Poisson process). Additionally, we assume that the movement of wireless devices still keeps them the same distribution (uniform or Poisson process). Gupta and Kumar (1998) showed that there is a critical power almost surely when the wireless nodes are randomly and uniformly distributed in a unit area disk. The result by Penrose (1998) implies the same conclusion. Moreover, Penrose (1998) gave the probability of the network's being connected if the transmission radius is set as a positive real number r and the number of nodes n goes to infinity.

Let $G(V, r)$ be the graph defined on V with edges $uv \in E$ if and only if $\|uv\| \leq r$. Here, $\|uv\|$ is the Euclidean distance between nodes u and v. Let $\mathcal{G}_\Omega(\mathcal{X}_n, r_n)$ be the set of graphs $G(V, r_n)$ for n nodes V that are uniformly and independently distributed in a two-dimensional region Ω, which could be a unit-area disk $\mathcal{D}$ or a unit square $\mathcal{C}$ with its center at the origin. The problem considered by Gupta and Kumar (1998) is then to determine the value of r_n such that a random graph in $\mathcal{G}_\mathcal{D}(\mathcal{X}_n, r_n)$ is asymptotically connected with probability one as n goes to infinity. Let $P_{\Omega,k}(\mathcal{X}_n, r_n)$ be the probability that a graph in $\mathcal{G}_\Omega(\mathcal{X}_n, r_n)$ is k-connected. Then Gupta and Kumar (1998) showed that if $(n\pi)r_n^2 = \ln n + c(n)$, then $P_{\Omega,1}(n, r_n) \to 1$ iff $c(n) \to +\infty$ as n goes to infinity. The result by Penrose (1998) implies a stronger result: If $(n\pi)r_n^2 = \ln n + \alpha$, then $P_1(n, r_n) = e^{-e^{-\alpha}}$ as n goes to infinity.

Topology Control

It is common to separate the network design problem from the management and control of the network in the communication network literature. The separation is very convenient and helps to significantly simplify these two tasks, which are already very complex on their own. Nevertheless, there is a price to be paid for this modularity because the decisions made at the network design phase may strongly affect the network management and control phase. In particular, if the issue of designing efficient routing schemes is not taken into account by the network designers, then the constructed network might not be suited for supporting a good routing scheme. For example, a backbone-like network topology is more suitable for a hierarchical routing method than a flat network topology.

Topology control is to select either a subset of wireless devices or a subset of communication links that will be used for the network operations such as routing. Notice that power assignment by use of a smaller transmission power of some nodes also will remove some links from the original communication graph induced by use of the maximum transmission power. Topology control often has more choices by intentionally not using some physical links for routing.

A wireless ad hoc network needs some special treatment because it intrinsically has its own special characteristics and some unavoidable limitations compared with those of wired networks. For example, wireless nodes are often powered by batteries only, and they often have limited memories. A transmission by a wireless device is often received by many nodes within its vicinity, which possibly causes signal interferences at these neighboring nodes. On the other hand, we can also utilize this property to save the communications needed to send some information. Unlike most traditional static communication devices, the wireless devices of MANETs often are moving during the communication. Therefore, it is more challenging to design a network topology for wireless ad hoc networks that is suitable for designing an efficient routing scheme to save energy and storage memory consumption than it is to design one for the traditional wired networks. To simplify the question so we can derive some meaningful understanding of wireless ad hoc networks, we assume that the wireless nodes are quasi-static for a period of time. Then, in technical terms, the question we deal with is whether it is possible (and, if possible, then how) to design a network that is a subgraph of the original communication graph (mostly a UDG), such that it ensures both attractive network features such as bounded node degree, low-stretch factor, a linear number of links, and attractive routing schemes such as localized routing with guaranteed performances.

Unlike the wired networks that typically have fixed network topologies, each node in a wireless network can potentially change the network topology by adjusting its transmission range and/or selecting specific nodes to forward its messages, thus controlling its set of neighbors. The primary goal of topology control in wireless networks is to maintain network connectivity, optimize network lifetime and throughput, and make it possible to design power-efficient routing. Not every connected subgraph of the UDG plays the same important role in network designing. One of the perceptible requirements of topology control is to construct a subgraph such that the shortest path connecting any two nodes in the subgraph is not much longer than the shortest path connecting them in the original UDG. This aspect of path quality is captured by the *stretch factor* of the subgraph. A subgraph with a constant stretch factor is often called a *spanner*, and a spanner is called a *sparse spanner* if it has only a linear number of links. Here, we review and study how to construct a sparse network topology efficiently for a set of static wireless nodes.

Restricting the size of the network has been found to be extremely important in reducing the amount of routing information. The notion of establishing a subset of nodes that perform the routing has been proposed in many routing algorithms (Das and Bharghavan, 1997; Sinha et al., 1999; Stojmenovic et al., 2002; Wu and Li, 2001). These methods often construct a virtual backbone by using the connected dominating set (CDS) (Alzoubi et al., 2002a; 2002c; Wan et al., 2002a), which is often constructed from a dominating set or a maximal independent set. For a full review of the state of the art in constructing the backbone, see X.-Y. Li (2002).

The other imperative requirement for network topology control in wireless ad hoc networks is the fault tolerance. To guarantee a good fault tolerance, the underlying network structure must be k-connected for some $k > 1$; i.e., given any pair of wireless nodes, there need to be at least k disjoint paths to connect them. By setting the transmission range sufficiently large, the induced UDG will be k-connected without doubt. Because energy conservation is important to increase the life of the wireless device, then the question is how to find the minimum transmission range such that the induced UDG is multiply connected.

Limitations

For simplicity, it is traditionally assumed that the transmission region of each wireless node is a disk with unit radius. Here, a disk centered at a node u with radius r_u, denoted by $D(u, r_u)$, is the set of points whose distance to u is at most r_u. Thus, all nodes together define a UDG as a communication graph. However, graphs representing communication

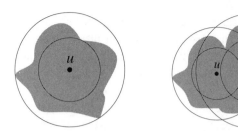

Figure 2.1   The transmission region of a node is modeled by a quasi-disk.

links are rarely specified as the UDG. Different nodes may have different transmission radii and, more important, the transmission region of a node is never a perfect disk. In other words, the main weakness of this point graph model (see Sen and Huson, 1997, for more details) is the assumption of perfectly regular radio coverage. Although this model is quite realistic in open-air and flat environments, ad hoc and sensor networks are likely to be used in various different situations, such as indoor or urban scenarios or under harsh conditions. In other words, in real-life situations, it is more likely that the radio coverage region is highly irregular because of the reflection influence of walls, buildings, and the interference with other infrastructures.

Considering this imperfect transmission region, previous routing algorithms, which guarantee packet delivery by use of some planar subgraph as network topology, are likely to fail because there might be no planar subgraph of the communication graph or some links might be missing. In the worst case, the communication graph could be very complicated. However, to have some meaningful study, including all these details in the network model, would make it extremely difficult and complex to study the performances of designed protocols and algorithms. For this reason, despite its limitations, the point graph model is still widely used in the study of wireless ad hoc network properties.

In addition to this point graph model, several enhanced models have been proposed in the literature to improve this. One model is the so-called quasi-disk-graph model (Barriere *et al.*, 2001; Moaveninejad *et al.*, 2005). Assume that each node u has a maximum transmission radius R_u and a minimum transmission radius r_u. These two thresholds depend on both the environment and the mobile hosts' technology. Thus, the transmission region of a node u is contained inside disk $D(u, R_u)$ and contains the disk $D(u, r_u)$. See Figure 2.1 for an illustration. If the Euclidean distance between two mobile hosts u and v exceeds the value $\min(R_u, R_v)$, they cannot communicate with each other directly (i.e., exchange messages). Conversely, two mobile hosts are always mutually reachable if their Euclidean distance is below the value $\min(r_u, r_v)$. Otherwise, they may or may not be mutually reachable.

2.4   The Wireless Interference Graph

Each terminal v_i also has an interference range $R_I(v_i)$ such that terminal v_j is interfered by the signal from v_i whenever $\|v_i - v_j\| \leq R_I(v_i)$ and v_j is not the intended receiver. Here, we assume that no node that transmits on a certain frequency can, at the same

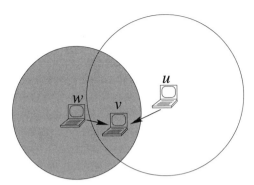

Figure 2.2 Interference happens at node v when the transmission region (denoted by the shaded disk) of node w intersects with the interference region of node u.

time, receive on the same frequency. Thus, we assume an interference occurs if the transmission region of one node (node w here) intersects with the interference region of another node (node u here). See Figure 2.2 for an illustration. Most researchers, for simplicity, treat the transmission region of a node as its interference region. However, this simplification is not accurate in practice. The interference range $R_I(v_i)$ is not necessarily the same as the transmission range $R_T(v_i)$. Typically, $R_T(v_i) < R_I(v_i) \leq cR_T(v_i)$ for some constant $c > 1$. We call the ratio between them the *interference–transmission ratio* for node v_i, denoted as $\gamma_i = [R_I(v_i)/R_T(v_i)]$. In practice, $2 \leq \gamma_i \leq 4$. For all wireless nodes, let $\gamma = \max_{v_i \in V}[R_I(v_i)/R_T(v_i)]$. Further, for a number of protocols, the actual simultaneous transmitting nodes must be separated by a distance called the *carrier-sensing range*. The carrier-sensing range for a node u is the largest range D such that a node that is of distance D away from u can still sense that the channel is busy when u is transmitting. Typically, the carrier-sensing range is larger than the interference range. For some theoretical studies, we need to use this carrier-sensing range to model the "interference" if two simultaneously transmitting nodes must be separated by at least their carrier-sensing range.

Two different types of interference have been studied in the literature: namely, *primary interference* and *secondary interference*. Primary interference occurs when a node transmits and receives packets at the same time; secondary interference occurs when a node receives two or more separate transmissions. Here, all transmissions could be intended for this node, or only one transmission is intended for this node (thus, all other transmissions are interference to this node). In addition to these interferences, there could have been some other constraints on the scheduling; e.g., the radio networks that deploy the IEEE 802.11 protocol with the request-to-send and clear-to-send (RTS/CTS) mechanism will pose some additional constraints. Several different interference models have been used to model the interferences in wireless networks. These are briefly reviewed in the following subsections.

Protocol-Interference Model (PrIM)

This model was first proposed in Gupta and Kumar (1999). In this model, a transmission by a node v_i is successfully received by a node v_j iff the intended destination

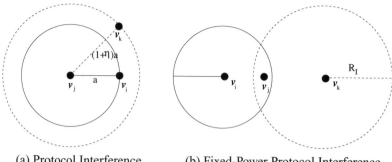

(a) Protocol Interference   (b) Fixed-Power Protocol Interference

Figure 2.3 (a) Protocol-interference model: Node v_j will be interfered by node v_k when $\|v_k - v_j\| \leq (1 + \eta)\|v_i - v_j\|$, where v_i is sending data to v_j and v_k is sending to other nodes. (b) Fixed-power-interference model: Node v_j will be interfered by node v_k when $\|v_k - v_j\| \leq R_I(v_k)$. Here, the dotted circle denotes the largest interference range of a node.

v_j is sufficiently apart from the source of any other simultaneous transmission; i.e., $\|v_k - v_j\| \geq (1 + \eta)\|v_i - v_j\|$ for any node $v_k \neq v_i$. Here, the constant $\eta > 0$ models situations in which a guard zone is specified by the protocol to prevent a neighboring node from transmitting on the same channel at the same time. See Figure 2.3(a) for an illustration. This model *implicitly* assumed that each node v_k will adopt the power-control mechanism when it transmits signals. Simulation analysis (Gronkvist and Hansson, 2001) as well as analytical results (Behzad and Rubin, 2003) indicate that the protocol-interference model does not necessarily provide a comprehensive view of reality because of the aggregate effect of interference in wireless networks. However, it does provide some good estimations of interference and, most important, it enables a theoretical performance analysis of a number of protocols designed in the literature. This model was used in Kumar *et al.* (2005) to study the throughput optimization for wireless networks.

Fixed Power-Protocol-Interference Model (fPrIM)

We assume that a node will *not* dynamically change its power based on the intended receiver in a packet level. However, we do assume that each node v_i has its own fixed transmission power and thus a fixed transmission range $R_T(v_i)$. We also assume that each node v_k has an *interference range* $R_I(v_k)$ such that any node v_j will be interfered by the signal from v_k if $\|v_k - v_j\| \leq R_I(v_k)$ and node v_k is sending a signal to some node other than v_j. In other words, the transmission from v_i to v_j is viewed successful if $\|v_k - v_j\| > R_I(v_k)$ for every node v_k transmitting in the same time slot using the same channel. See Figure 2.3(b) for an illustration.

RTS/CTS Model

This model was also studied previously [e.g., Alicherry *et al.* (2005)]. When using the RTS/CTS mechanism, a transmitter first sends a RTS frame before sending a data frame. The intended receiver then responds with a CTS frame indicating that the transmitter can send the data frame. Within the CTS frame, the receiver provides a value in the duration

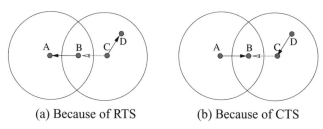

(a) Because of RTS   (b) Because of CTS

Figure 2.4 Communication restriction by RTS/CTS. © 2006, ACM.

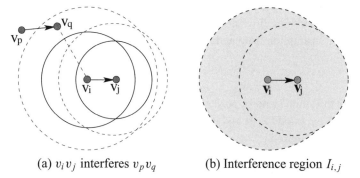

(a) $v_i v_j$ interferes $v_p v_q$ (b) Interference region $I_{i,j}$

Figure 2.5 RTS/CTS Interference Model. Here, the shaded disks denote the interference region defined by a link (v_i, v_j). © 2006, ACM.

field of the frame header that holds off other stations from transmitting for a certain time. For every pair of transmitter and receiver, all nodes that are within the interference range of either the transmitter or the receiver cannot transmit. Figure 2.4(a) shows the case in which communication from B to A and C to D cannot take place simultaneously because of RTS. Figure 2.4(b) shows the case in which communication from A to B and D to C cannot take place simultaneously because of CTS. Although RTS/CTS is not the interference itself, for convenience in our notation, we treat the communication restriction that is due to RTS/CTS as the *RTS/CTS-interference* model. Thus, for every pair of simultaneous communication links, say $v_i v_j$ and $v_p v_q$, it should satisfy the conditions that (1) they are distinct four nodes, i.e., $v_i \neq v_j \neq v_p \neq v_q$; and (2) v_i and v_j are not in the interference ranges of v_p and v_q, and vice versa. The *interference region*, denoted by $I_{i,j}$, of a link $v_i v_j$ is the union of the interference region of nodes v_i and v_j. When a directed link $v_i v_j$ (or $v_j v_i$) is active, all simultaneously transmitting links $v_p v_q$ cannot have an end point inside the area $I_{i,j}$. Notice it is possible that neither v_p nor v_q is in $I_{i,j}$, but $v_p v_q$ still interferes with $v_i v_j$ because v_i or v_j may be inside $I_{p,q}$. In other words, the conflict graph could be an asymmetrical directional graph. See Figure 2.5 for an illustration. Notice that when the carrier-sensing range is at least twice the transmission and all nodes have the same transmission range, then it can be shown that the RTS/CTS mechanism to address the hidden-terminal problem will be redundant because nodes will not transmit anyway when it detects that the channel is busy using carrier sensing.

Transmitter Interference Model (TxIM)

This model was introduced by Yi *et al.* (2003) to analyze the capacity of random ad hoc networks. In this model, a transmission from a node v_i is successful if and only if, for any other transmitter v_k, $d(v_i, v_k) \geq (1 + \eta)[R_T(v_i) + R_T(v_k)]$. Here, η is a system parameter.

Physical-Interference Model (PhIM)

In this model, the SINR is used to describe the aggregate interference in the network. Let $S_{i,j}$ be the received signal strength at node v_j (the signal is initiated by node v_i) and let $N_{-i,j}$ be the aggregate signal strength at v_j by all nodes other than v_i that are transmitting in the same time slot on the same channel, i.e., $N_{-i,j} = \sum_{v_k \neq v_i} S_{k,j}$. The SINR of the transmission from node v_i at node v_j is defined as

$$\text{SINR}_{i,j} = \frac{S_{i,j}}{N_{-i,j} + N_0},$$

where N_0 is the ambient-noise power level. The transmission from a node v_i is successfully received by a node v_j if and only if the SINR at node v_j is at least the minimum SINR threshold required by node v_j.

Notice that the fPrIM model and the RTS/CTS model are different. For example, in Figure 2.4(a), links BA and CD can be assigned the same time slot and channel in the PrIM but not in the RTS/CTS model. A similar statement holds for links AB and DC in Figure 2.4(b).

Given a communication graph $G = (V, E)$, we use the *conflict graph* (sometimes called the interference graph) (e.g., Jain *et al.*, 2003) F_G to represent the interference in G. Each vertex (denoted by $\mathbf{L}_{i,j}$) of F_G corresponds to a directed link (v_i, v_j) in the communication graph G. There is an *edge* between vertex $\mathbf{L}_{i,j}$ and vertex $\mathbf{L}_{p,q}$ in F_G if and only if the corresponding link (v_i, v_j) (also denoted as $\mathbf{L}_{i,j}$) conflicts with link (v_p, v_q) because of interference. Here, two links conflict with each other if the senders of both links transmit simultaneously; at least one of the receivers of these two links will not be able to get the signal correctly. Recall that whether two links conflict depends on the interference model used underneath (e.g., the PrIM, the fPrIM, the TxIM, or the RTS/CTS model). Thus, for a given communication graph G, the interference graph F_G may be different. Observe that we cannot characterize the physical interference model by using a conflict graph (or interference graph). To avoid the confusion, we use F_G^P to denote the interference graph under the PrIM and F_G^{D2} to denote the interference graph under the RTS/CTS model.

2.5 Related Graph Problems and Geometry Concepts

Given a communication graph or an interference graph derived from a wireless network description, almost all problems can be reduced to some classical questions on the communication graph or interference graph. For completeness, some concepts are briefly reviewed here.

A subset of vertices in a graph G is an *independent set* (IS) if for any pair of vertices, there is no edge between them. It is a *maximal IS* if no more vertices can be added to it to form a larger IS. It is a *maximum IS* (MIS) if no other IS has more vertices. The goal of the MIS problem is to compute, given a graph, a subset of pairwise unconnected vertices with the maximum cardinality. The independence number, denoted by $\alpha(G)$, of a graph G is the cardinality of the MIS. If every node has a weight, then the IS with the maximum total node weight is called a *maximum weighted independent set* (MWIS).

A subset S of V is a *dominating set* (DS) if each node u in V is either in S or is adjacent to some node v in S. Nodes from S are called dominators, whereas nodes not in S are called dominatees. Clearly, any maximal IS is a DS. A subset C of V is a *CDS* if C is a DS and C induces a connected subgraph. Consequently, the nodes in C can communicate with each other without using nodes in $V - C$. A CDS is also called a *backbone* in this chapter. A DS with minimum cardinality is called a *minimum* DS (MDS). A CDS with minimum cardinality is the *minimum CDS* (MCDS). In wireless ad hoc networks, assume that each node u has a cost $c(u)$. Then, a CDS C is called *weighted CDS* (WCDS). A subset C of V is a *minimum WCDS* (MWCDS) if C is a WCDS with minimum total cost.

Given a graph $G = (V, E)$, a subset $V' \subseteq V$ of vertices is a *vertex cover* if for every edge in E, V' contains at least one of its end vertices. In other words, every edge is dominated by some node from V'. The goal of the *minimum-vertex-cover* (MVC) problem is to find a vertex cover with the minimum cardinality.

The *graph-coloring* (GC) problem (often called *vertex-coloring* problem) is to assign each vertex a color. A vertex coloring is *valid* if two adjacent vertices are assigned different colors. The *minimum GC* (MGC) problem is to use minimum number of colors so a valid vertex coloring can be obtained. Given a graph G, we often use $\chi(G)$ to denote the minimum number of colors needed by any valid vertex coloring of its vertices. A list-coloring problem is, assuming that each node has a list of colors that it can choose from, to choose a color from its list such that the chosen colors for every adjacent node are different. A number of variations of list-coloring problems have been studied in the literature.

A graph G is called k-node-connected if there are k-node-disjoint paths connecting every pair of nodes in the network. Here, two paths connecting a given source node s and a given target node t is node-disjoint if they do not share any internal nodes, except the source and target nodes. The node connectivity of a graph G is the largest k such that G is k-node-connected. Similarly, a graph G is called k-edge-connected if there are k-edge-disjoint paths connecting every pair of nodes in the network. Here, two paths connecting a given source node s and a given target node t are edge-disjoint if they do not share any edge. The edge connectivity of a graph G is the largest k such that G is k-edge-connected.

A graph is a *d-inductive* graph if there is a rank of its vertices such that the ith-ranked vertex is connected to at most d vertices with larger rank. The *inductivity* of a graph G is the minimum d such that graph G is d-inductive. A number of coloring methods have performance bounds that depend on the d value found for a d-inductive graph.

Given a graph G of n vertices and a problem P, let $\text{OPT}_P(G)$ denote an optimum solution of problem P when the input graph is G. For a maximization problem P (e.g., MIS), an algorithm is a ρ-approximation algorithm for problem P if, given any input graph G, it runs in polynomial time and always computes a solution that is at least $1/\rho\text{OPT}_P(G)$. For a minimization problem P (e.g., MVC), an algorithm is a ρ-approximation algorithm for problem P if, given any input graph G, it runs in polynomial time and always computes a solution that is at most $\rho\text{OPT}_P(G)$. An algorithm is a *polynomial-time-approximation scheme* (PTAS) if, for any additional parameter $\varepsilon > 0$, it runs in a time polynomial of n and always computes a solution that is at least $(1/1 + \varepsilon)\text{OPT}_P(G)$ for a maximization problem P and at most $(1 + \varepsilon)\text{OPT}_P(G)$ for a minimization problem P. An algorithm is a *fully PTAS* (FPTAS) if, for any additional parameter $\varepsilon > 0$, it runs in a time polynomial of n and $(1/\varepsilon)$ and always computes a solution that is at least $1/(1 + \varepsilon)\text{OPT}_P(G)$ for a maximization problem P and at most $(1 + \varepsilon)\text{OPT}_P(G)$ for a minimization problem P. The class APX (an abbreviation of an "approximable") is the set of NP-optimization problems that allow polynomial-time-approximation algorithms with an approximation ratio bounded by a constant (or constant-factor-approximation algorithms for short). Unless $P = \text{NP}$, it can be shown that there are problems that are in the APX class but not in the PTAS; that is, problems that can be approximated within some constant factor, but not every constant factor. A problem is said to be *APX-hard* if there is a PTAS reduction from every problem in the APX class to that problem and to be *APX-complete* if the problem is APX-hard and also in the APX classs. Consequently, that a problem is APX-hard is generally bad news because it denies the existence of a PTAS, which is one of the most useful sorts of approximation algorithms. In other words, if a problem is APX-hard, unless $\text{NP} = P$, there is some $\varepsilon > 0$ such that no polynomial-time algorithm can find a solution that is within $1 - \varepsilon$ of the optimum for every possible input.

2.5.1   Geometry Concepts

Let $P = \{p_1, \ldots, p_k\}$ be a finite set of points in d-dimensional space R^d and their location vectors $\mathbf{x}_i \neq \mathbf{x}_j \; \forall \, i \neq j$. The region given by

$$V(p_i) = \left\{ \mathbf{x} \mid \|\mathbf{x} - \mathbf{x}_i\| \leq \|\mathbf{x} - \mathbf{x}_j\| \; \forall \, j \neq i \right\}$$

is called the Voronoi region (Voronoi box) associated with point p_i and

$$\mathcal{V}(P) = \bigcup_{i=1}^{k} V(p_i)$$

is said to be the Voronoi diagram (or Voronoi tessellation) of P. A Voronoi region is formed through the intersection of planes and is therefore a general irregular polyhedron. The facets of the Voronoi regions correspond in the dual graph to the Delaunay edges that connect the points of P.

Assume that the set P of points is *in a general position*; i.e., no $(d + 2)$ points are located on the same d-dimensional sphere's surface and no $(d + 1)$ points are located on the same d-dimensional hyperplane. A Delaunay triangulation, denoted by $\text{Del}(V)$, of a set of points P contains an edge pq if and only if the Voronoi regions centered at points $p \in P$ and $q \in P$ intersect. Another interpretation of Delaunay triangulation is as follows. For a set P of points in (d-dimensional) Euclidean space, a Delaunay triangulation is a triangulation $\text{Del}(P)$ such that no point in P is inside the circumhypersphere of any simplex in $\text{Del}(P)$. For example, a Delaunay triangulation for a set P of points in the plane is a triangulation $\text{Del}(P)$ such that no point in P is inside the circumcircle of any triangle in $\text{Del}(P)$. Delaunay triangulations maximize the minimum angle of all the angles of the triangles in the triangulation for two-dimensional points. Given a set P of n two-dimensional points, a Delaunay triangulation $\text{Del}(P)$ can be constructed in time $O(n \log n)$.

2.6   Energy-Consumption Models

Efficient usage of energy is one of the primary concerns in designing wireless ad hoc networks, especially the WSN. Thus, it is fundamental to model energy consumption accurately of a wireless device for all operations such as transmitting, receiving, sleeping, and so on. The energy-consumption models of wireless ad hoc networks and WSNs are discussed separately because the features of these two networking devices could be quite different.

2.6.1   Wireless Ad Hoc Networks

Wireless ad hoc networks, depending on the application scenarios, could be composed of various wireless devices such as laptops, cellular phones, PDAs, smart appliances, and so on. Additionally, for some applications, a wireless ad hoc network can be composed of heterogeneous devices. Given this diversity, typically we focus on only the power consumption of the wireless transceivers. In other words, we concentrate on reducing the energy consumed in communications (including transmitting and receiving), not on the energy consumed in computing and so on. The other reason is that the energy consumed in communications is a significant portion of the total energy consumed. For example, depending on the type of device, the energy consumed in communications varies from 15% (e.g., a laptop equipped with an IEEE 802.11 wireless card) to 35% (e.g., a PDA device) of the total energy consumption by a wireless device.

The IEEE 802.11 protocol is the prominent wireless protocol for WLANs and for forming wireless ad hoc networks. A good deal of research has been conducted to measure the energy consumption of commercial 802.11 wireless cards. In almost all measurements, an IEEE 802.11 wireless card has four different operation modes:

- **Idle:** The radio is turned on but it is not used for communications.
- **Transmit (Tx):** The radio is turned on and transmitting data packets.
- **Receive (Rx):** The radio is turned on and receiving data packets from other nodes or APs.
- **Sleep:** The radio is turned off.

Table 2.1 Power consumption of a CISCO Aironet IEEE 802.11 card

	Idle power (mA)	Tx power (mA)	Rx power (mA)
802.11a	203	554	318
802.11b	203	539	327
802.11g	203	530	282

As an example, Table 2.1 shows the power consumption of a CISCO Aironet IEEE 802.11 a/b/g card. Power consumption in the sleep mode is not reported here. Notice that for most of the studies, the absolute value of the power consumption is not important. The ratios among different operation modes are more important in studying the performances of the majority of energy-optimization protocols. Also notice that many protocols and algorithms (especially the power-assignment algorithms and topology-control algorithms) for wireless ad hoc networks are based on the ability of each wireless node to dynamically adjust its transmitting power. This feature is indeed available on some commercial IEEE 802.11 cards. On the other hand, the transmitting–receiving power is only a part of the total power consumed by a wireless card. The wireless card also consumes significant energy to just start up the other components. Furthermore, for the majority of theoretical studies in the literature for power assignment and topology control, it is assumed that each wireless node can adjust its power to infinitely small precision. In practice, most of the wireless cards can adjust their power to only a certain set of discrete values.

2.6.2 Wireless Sensor Networks

Sensor networks are typically composed of homogeneous devices, which are usually very simple. In addition, because the majority of sensors are designed in the research community, their energy-consumption features are very well known. See Raghunathan *et al.* (2002) for a number of energy-consumption measurements of wireless sensors.

Typically, a sensor node includes a sensor digital signal processing (DSP) unit for data processing; D/A and A/D for digital-to-analog and analog-to-digital conversions, respectively; and a wireless transceiver for data communication. The sensor/DSP, D/A, and A/D operate at low frequency and consume less than 1 mW. This is over an order of magnitude less than the power consumption of the transceiver. Therefore, the energy models studied in the literature usually ignore the contributions from these components.

When the transceiver is first turned on, it takes some time for the frequency synthesizer and the voltage-controlled oscillation (VCO) to lock to the carrier frequency. The start-up energy can be modeled as follows:

$$E_{\text{start}} = P_{\text{LO}}t_{\text{start}}, \qquad (2.5)$$

Table 2.2 Measured power consumption of Rockwell's WINS sensor node

MCU mode[a]	Sensor mode	Radio mode	Power (mW)
On	On	Tx (power 36.3 mW)	1080.5
On	On	Tx (power 0.12 mW)	771.1
On	On	Rx	751.6
On	On	Idle	727.5
On	On	Sleep	416.3
On	On	Removed	383.3
Sleep	On	Removed	64.0

[a] MCU: microcontroller unit.

where P_{LO} is the power consumption of the synthesizer and the VCO. The term t_{start} is the required settling time. The radio-frequency (RF) building blocks, including the power amplifier (PA), the low-noise amplifier (LNA), and the mixer have a negligible start-up time and therefore can remain in the off state during the start-up mode.

The active components of the receiver include the LNA, a mixer, a frequency synthesizer, the VCO, an intermediate-frequency (IF) amplifier (amp), and a demodulator (Demod). The receiver energy consumption can be modeled as follows:

$$E_{rx} = (P_{LO} + P_{RX})t_{rx}, \tag{2.6}$$

where P_{RX} includes the power consumption of the LNA, mixer, IF amp, and Demod. The receiver power consumption is dictated by the carrier frequency and the noise and linearity requirements.

The transmitter includes the modulator (Mod), frequency synthesizer, VCO (shared with the receiver), and PA. The data modulates the VCO and produces a frequency-shift-keying (FSK) signal at the desired data rate and carrier frequency. A simple transmitter energy model is shown in Eq. (2.7). The modulator consumes very little energy and therefore can be neglected:

$$E_{tx} = (P_{LO} + P_{PA})t_{tx}. \tag{2.7}$$

P_{LO} can be approximated as a constant. P_{PA} depends on additional factors and needs to be modeled more carefully.

Table 2.2 reports the power dissipation of Rockwell's WINS sensor node. See http://wins.rsc.rockwell.com/ for more details. A device is composed of three main components: the microcontroller unit MCU, the sensing apparatus (sensor), and the wireless radio for transmitting. For a wireless radio, there are four different operations: transmitting (Tx), receiving (Rx), idle, and sleep. For the energy consumed by only the wireless radio, we have the following ratios: sleep: idle: Rx: Tx $\simeq$ 0.09:1:1.07:2.02. The maximum transmitting power of a WINS sensor is 36.3 mW and the minimum transmitting power is 0.12 mW. Notice that the power used for receiving compared with the power used for transmitting is not negligible at all; they are actually almost the same.

2.7   Mobility Models

In an ad hoc wireless network, nodes may move freely within the field. For a pair of nodes to communicate, a route must be formed between intermediary nodes. For this type of network, it is very important to model node positions and movement, as the transmission range is generally fairly small in relation to the size of the field. Node mobility is a prominent feature of ad hoc networks and is also a fundamental feature in studying the performances of various protocols (especially routing protocols, MAC protocols, and so on) for MANETs. Furthermore, because real-world implementations of MANETs are rare, it is very difficult to obtain real-life moving patterns. Thus, a common approach in the literature (e.g., Jardosh *et al.*, 2003) is to use synthetic mobility models and measure the performances of protocols by simulations. Recently, there were also some theoretical studies of various performances of certain wireless protocols for which synthetic mobility models were used.

Because most of the mobility models used in the literature are synthetic, Santi (2005b) summarized two obvious properties that these models should satisfy:

- **Resemble real-life movement:** Because the mobility model is synthetic, it is natural to require that the model be as close to practice as possible. On the other hand, given the wide range of possible wireless networking applications and scenarios, it is virtually impossible to design a universal mobility model that can resemble all various movement patterns. However, when we perform a simulation study of a protocol, the chosen synthetic mobility model should be representative of the application scenario.

- **Simple enough for easy implementation:** Because synthetic mobility models are developed to capture and abstract certain movement patterns in real-life applications and to provide a convenient tool to implement the simulation software, the mobility models developed should be simple enough such that they can be easily integrated into the simulation software. Furthermore, the mobility model itself should also require as few computations as possible to keep the running time of the simulation software reasonable. Sometimes synthetic mobility models are also used to provide some theoretical studies of the performances of some protocols. In this case, the developed mobility model should allow some meaningful theoretical studies without losing some fundamental network characteristics and key properties of movement patterns. Then these theoretical studies could be used to provide some insight into optimizing the protocols' performances by revising the protocols developed for ad hoc networking.

Obviously, these two objectives could conflict with each other in most scenarios: The more realistic the mobility model is, the more details the model must include and thus the more complicated it is to implement and study. Then we often need to find a certain middle ground between them: Sometimes we need to compromise more representation and sometimes we need to compromise more simplicity.

The mobility models used in simulations can be roughly divided into two categories: *individual* and *group based*. In the individual models, the movement of each node is modeled independently of any other nodes in the simulation. In the group mobility models, there is some relationship among a certain subset of nodes and their movements throughout the cells or field.

Random-Walk and Random-Direction Models

The simplest mobility model used is known as the random walk. In the most common version of this model, a direction and a speed are chosen by a node from the intervals [minspeed, maxspeed] and [0, 2π], respectively. The node will move along this direction using this speed until either a predefined time elapses or a predefined distance is moved, and then a new direction and speed are chosen randomly. This model is well described in many prior papers, such as that of Camp *et al.* (2002). When the node hits the boundary of the deployment region, the node either reflects on the boundary or wraps around the boundary by treating the region as a torus.

Several other variations of so-called random models exist (e.g., random direction); as in the preceding two models, various parameters of motion are chosen from a uniform distribution and then nodes move at a constant speed until some condition is met.

Random-Waypoint Model

The random-waypoint (RWP) model is by far the most commonly used mobility model for ad hoc networks. One of the reasons for its popularity may be the fact that it is implemented in many wireless simulation tools such as NS2 and GloMoSim. The RWP model was first introduced by Johnson and Maltz (1996) to study the performances of dynamic source routing (DSR). The RWP model can be regarded as an extension of the random-walk model, in which a node waits for a predetermined pause time and then chooses a destination randomly. The node moves toward this destination at a random speed chosen from [minspeed, maxspeed]. On arriving at the destination, the node pauses again and repeats the process. Typically, the pause time is a predetermined value. This model is widely used in ad hoc wireless network routing protocols and is described by Broch *et al.* (1998).

Given its popularity, the RWP model has been widely studied in the literature. It was recently discovered (e.g., in Bettstetter *et al.*, 2003, and Resta and Santi, 2002) that the long-term node spatial distribution of mobile nodes when the RWP model is used is concentrated in the center of the deployment region, which is a so-called sojourn density distribution. In addition, the average nodal speed (defined as the average of node velocities at a given instant of time) decreases over time (Yoon *et al.*, 2003). Thus, RWP mobility models must be carefully simulated for wireless ad hoc networks. In particular, the network performance should be evaluated only after a certain period.

Random-Trip Mobility Model

Because the RWP mobility model cannot provide stationary states, it has been generalized to slightly more realistic models. For example, Bettstetter *et al.* (2003) extended the RWP model by allowing each node to choose its pause time from an arbitrary probability distribution instead of from a fixed, predetermined value. They also selected a random fraction of the network nodes to remain stationary for the entire simulation.

The random-trip model (Boudec and Vojnovic, 2005) is a generic mobility model that generalizes the random waypoint and random walk to realistic scenarios. As with a wide class of mobility models, the random-trip model is many existing mobility models in one, plus some new ones. For example, it includes the RWP in a general connected domain, a restricted RWP, and a random-walk model with either wraparound or reflection. If well defined, a random-trip model features a unique steady-state distribution.

A trip is the combination of a duration and a path. The position $X(t)$ of the mobile model at time t is defined iteratively as follows. There is a set $T_n \in R, n \in Z$ of transition instants, such that $T_0 \leq 0 < T_1 < T_2 < \ldots$. At time T_n, a path $P_n \in \mathcal{P}$ and a trip duration $S_n \in R^+$ are drawn according to some specified trip-selection rule specific to the model. The next transition instant is $T_{n+1} = T_n + S_n$, and the position of the mobile is $X(t) = P_n(\frac{t-T_n}{S_n})$ for $T_n \leq t \leq T_{n+1}$.

Markov Mobility Models

Markov (and other state-dependent) mobility models comprise a large class of mobility models used in both cellular and ad hoc network mobility modeling. The simplest model, which is two-dimensional, assigns a probability to moving left, moving right, and staying stationary. A simple version of this model, used in cellular network simulations, has nodes moving to adjacent cells with a set probability, often described in state tables.

Smooth Random-Mobility Model

The smooth random-mobility model (Bettstetter, 2001) is basically an extension of the simpler random-walk model. Two independent stochastic processes are used to trigger direction and speed changes. The new speeds (for example) are chosen from a weighted distribution of preferred speeds. Following such a trigger, the speed (or direction) changes as determined by a Poisson process.

Another similar model is the random Gauss–Markov mobility model, which is presented in Liang and Haas (1999) as an improvement over the smooth random-mobility model. A node's next location is predicted (or generated) by its past location and velocity. Depending on the parameters set, this allows modeling along a spectrum from random walk to fluid flow. The Gauss–Markov mobility model can eliminate the sudden stops and sharp turns encountered in the random-walk mobility model.

Group Mobility

The mobility models described so far concentrate on the movement of individual nodes. However, in many situations, nodes could move in groups (e.g., tourists). Group mobility models are, in general, less well studied than independent models. This is no doubt due in part to the fact that such detailed models are not necessary in cellular networks, where these details can be abstract. Perhaps the simplest and most common example of this is the fluid-flow mobility model (see Xie *et al.*, 1993, for details). In this model, the motion of nodes between cells is modeled as a set of constant-velocity fluid-flow equations. In group-based mobility models (see Hong *et al.*, 1999, e.g.), a small subset of network nodes could be defined as the set of *group leaders*. Other nodes are (randomly) assigned to one of the group leaders, thus forming several groups. The group leader will move according to one of the previously defined mobility models for individuals. The other group member will follow the group leader: The speed and direction of each member is a random small perturbation of the group leader's values. When two groups cross, a member is sometimes allowed to switch the group with certain probability.

Other Issues

Contrary to the case of the RWP model or the random-direction mobility model, there is another class of Brownian-like mobility models that resemble nonintentional movement. In these Brownian-like movements, the position of a node at a given time step depends on the node position in the previous time step with a certain probability. In other words, no explicit definition of moving direction and speed is required in these models.

Although not relevant to all mobility models, there are other factors that relate to the simulation of node mobility. Various parameters relating to the field itself may influence mobility. For example, by applying constraints to the paths that nodes may follow, the simulation may be radically affected. Thus, some mobility models limit nodes to virtual city streets in a field.

Another consideration is what effect (if any) the edges of a field may have. In the real world, there may not be any effective boundary on a field, or all nodes may stay within a certain field (e.g., a campus). For simulation, though, other possibilities are considered: A toroidal field maps a rectangular field onto a torus; some models specify that nodes stop, rebound, or leave the field on reaching the edge of the boundary.

2.8   Conclusion

This chapter presents a number of models for wireless ad hoc networks, including the wireless channel model, the communication graph, the interference model and the interference graph, the energy-consumption model, and the mobility model for mobile networks. Wireless communication is a complicated matter that cannot be simply described by a few formulas and models. This book is not intended as a complete study of wireless communication. There are a number of excellent books that are specifically for wireless channel models and other wireless characteristics, e.g., *Wireless Communications* by Rappaport (1996), *Introduction to Wireless and Mobile Systems* by Agrawal and Zeng, and *Wireless Networking* by Thurwachter.

Problems

2.1 Review the propagation models studied in this chapter and find more material about wireless propagation models from other textbooks.

2.2 It is widely assumed that the power needed to support a link uv is proportional to $\|uv\|^\alpha$ for some $\alpha \geq 2$. Here, $\|uv\|$ is the Euclidean distance between node u and v. Will this model have any problems when we study dense networks in which the number of nodes inside a unit area could be arbitrarily large?

2.3 In this chapter, we defined the communication graph and interference graph. What are the major differences between these communication graphs and general graphs? How many vertices could an interference graph have if the original wireless network has n wireless nodes and a single channel?

2.4 Assume that a set of wireless nodes is randomly deployed in a square region with side length of a meters. The interference range of each node is b meters, where $0 < b < a$. What is the maximum number of nodes that can transmit simultaneously without causing interference at receivers? Write your formula in terms of a and b.

2.5 What are the main differences between the PrIM used in the literature and the fPrIM discussed in this chapter?

2.6 Assume that we use the physical model to model the channel capacity and that the path loss is proportional to the separation distance d raised to a certain exponent α. In other words, when a node u sends with power P, the strength of the signal received by a node v is $P/\|uv\|^\alpha$. Also assume that the background noise is 0. Assume that v wants to receive data from node u. To compute the SINR at a node v, we need to consider all possible transmitting nodes. Let P_w be the transmitting power of a transmitting node w. Let $\mathcal{A}$ be the set of all actively transmitting nodes at that time. Here, the actively transmitting nodes will guarantee that the SINRs at all receiving nodes are at least the threshold for correctly decoding the data; i.e., they are "interference-free." What will be the largest ratio of the following two SINRs computed?

(a) $\text{SINR}_{\text{all}} = \frac{P_u \|uv\|^{-\alpha}}{\sum_{w \in \mathcal{A}} P_w \|wv\|^{-\alpha}}$,

(b) $\text{SINR}_{\text{cut}} = \frac{P_u \|uv\|^{-\alpha}}{\sum_{w \in \mathcal{A}, \|wv\| \leq R} P_w \|wv\|^{-\alpha}}$.

You can assume that there are numerous nodes transmitting. Will the result change if the noise is not 0?

2.7 What are the similarities and differences between topology control and power assignment?

2.8 For topology control, what kind of properties should the resulted network topology have for supporting the operations of other layers?

2.9 Designing energy-efficient structures for certain operations, such as unicast, broadcast, or multicast, has been widely studied in the literature. Most of the theoretical results do not consider the energy cost incurred by nodes for receiving and listening that are not the intended receivers. What will be possible approaches to address this issue? How do we address this when it is assumed that TDMA link scheduling is used? How do we address this issue when carrier-sense multiple access/collision-avoidance (CSMA/CA) MAC is used and the network density is known a priori?

2.10   Recall the interference models briefly reviewed in this chapter; how can you further enhance them to closely reflect the wireless communication practice while making it still manageable for algorithm design?

2.11   Consider the following scenario. Inside a museum, there is a collection of visitors, and each visitor has a wireless PDA that can form ad hoc networks. What will be an appropriate mobility model for such a networking example?

2.12   Consider another scenario. Assume that, in the near future, each car will have a wireless terminal that can form ad hoc networks with other cars within certain distances. What will be an appropriate mobility model for such a networking example? Note that cars can drive only on roadways.

2.13   Assume that two nodes x and y are separated by a distance d initially. Both nodes move at speed v, but the direction of the movement is always chosen randomly at any time instant. Estimate the probability that nodes x and y will be within distance r at some time before time t. You first estimate such a probability (by giving some nontrivial lower bound on the probability) by assuming that movement is on a two-dimensional plane.

2.14   Continue from the preceding question. Study this probability by assuming that the movement will be restricted to the surface of a three-dimensional sphere. Can you estimate such a probability for other mobility models?

Part II

Wireless MACs

3 Wireless Medium-Access Control Protocols

3.1 Introduction

A MAC protocol is used to address resolving potential contention and collision when the communication medium is used. Many MAC protocols have been proposed for wireless networks (e.g., Bharghavan *et al.*, 1994; Fullmer, 1998; Fullmer and Garcia-Luna-Aceves, 1995; Garcia-Luna-Aceves and Tzamaloukas, 1999; Garcias and Garcia-Luna-Aceves, 1996; Lin and Gerla, 1997b; Lu *et al.*, 1999; Vaidya *et al.*, 2000), which often assume a common channel shared by mobile hosts.

Contention-Based MAC

The MAC protocol is essential for stations that share a common broadcast channel. CSMA protocols (Kleinrock and Tobagi, 1975) have been used in a number of packet-radio networks in the past (Leiner *et al.*, 1987). These protocols attempt to prevent a station from transmitting simultaneously with other stations within its transmitting range by requiring each station to listen to the channel before transmitting. Unfortunately, the performance of the CSMA protocol suffers from *hidden-terminal* problems and *exposed-terminal* problems substantially. To remedy these problems, several approaches (Bambos and Kandukuri, 2000; Bharghavan *et al.*, 1994; Colvin, 1983; Fullmer and Garcia-Luna-Aceves, 1995; Karn, 1990; Monks *et al.*, 2001) were proposed in the literature. Karn (1990) proposed a protocol called MACA that attempts to detect collision at the receiver by establishing a request-response dialog between senders and intended receivers. When a sending station wants to transmit, it sends an RTS to the receiver, which responds with a CTS if it receives the RTS correctly. On receipt of the CTS, the sender sends its queued data packet(s). If the sender does not receive a CTS after a time out, it resends its RTS and waits a little longer for a reply. Any station that overhears an RTS or CTS packet directed elsewhere inhibits its transmitter for a specified time. Several other MAC protocols based on similar RTS/CTS exchanges, or RTSs followed by pauses, have been proposed (Bharghavan *et al.*, 1994; Colvin, 1983) for single-channel wireless networks. However, none of these protocols specifies any provision to prevent data packets from colliding with control packets (RTS and CTS packets). Fullmer and Garcia-Luna-Aceves (1995) introduced and analyzed a new type of channel-access discipline for single-channel networks that permits a station to acquire control of the channel dynamically before transmitting data packets. The acquisition strategy is also based primarily on a request-response (RTS/CTS) control dialog between a sender and an intended proxy receiver. The main difference is that carrier sensing is used to increase channel throughput substantially.

It is more challenging to design an efficient MAC in networks with spectrum opportunities (SOPs) because secondary users cannot exchange local information before agreeing on a communication channel and coordinating transmissions among secondary users becomes more challenging. Several specific MAC protocols have been proposed in the literature. Zhao *et al.* (2005) proposed a decentralized cognitive MAC protocol that allows secondary users to recognize a SOP and transmit based on a partial observation of the instantaneous spectrum availability. Their approach is based on the framework of the partially observable Markov decision process (POMDP). They derived decentralized strategies for the secondary users to decide which channel to sense and access for the maximization of the overall network throughput. Several MAC protocols (Tang and Garcia-Luna-Aceves, 1999; Tzamaloukas and Garcia-Luna-Aceves, 2001) were proposed based on the frequency-hopping spread spectrum (FHSS). This approach is probability based and depends on the network traffic. Tang and Garcia-Luna-Aceves (1999) proposed a multichannel protocol called hop reservation multiple access (HRMA). HRMA uses a slow FHSS, and the hosts hop from one channel to another according to a predefined hopping pattern. When two hosts agree to exchange data by an RTS/CTS handshake, they stay in a frequency hop for communication. Other hosts continue hopping, and more than one communication can take place on different frequency hops. Tzamaloukas and Garcia-Luna-Aceves (2001) proposed a similar approach in which the receiver initiated channel hopping with dual polling. Both the schemes can be implemented with only one transceiver used for each host, but they apply to only frequency-hopping networks and cannot be used in systems using other mechanisms, such as a direct-sequence spread spectrum (DSSS). Wu *et al.* (2000) proposed a protocol that assigns channels dynamically, in an on-demand style. In their protocol, they maintain one dedicated channel for control messages and other channels for data and thus assume that each host has two transceivers so that it can listen on the control channel and the data channel simultaneously. RTS/CTS packets are exchanged on the control channel, and data packets are transmitted on the data channel. In the RTS packet, the sender includes a list of preferred channels. On receiving the RTS, the receiver decides on a channel and includes the channel information in the CTS packet. Then, DATA and ACK packets are exchanged on the agreed data channel. This protocol does not need synchronization and can utilize multiple channels with little control message overhead. When the number of channels is large, which is typically the case for spectrum sharing, the control channel can become a bottleneck. So and Vaidya (2004b) presented another approach that requires only one transceiver for each host. However, they do require nodes to be synchronized so that every node starts each beacon interval at about the same time. At the start of each beacon interval, every node listens on a common channel to negotiate channels in the ad hoc traffic indication map (ATIM) window. After the ATIM window, nodes switch to their agreed channel and exchange messages on that channel for the rest of the beacon interval.

TDMA-Based Node and/or Link Scheduling

Even if we can perfectly solve the hidden-terminal problem and exposed-terminal problem, as pointed out by several authors (Bianchi, 2000; Medepalli and Tobagi, 2005) for a MAC layer protocol that can avoid interference, the throughput of the network decreases as the number of simultaneous communications increases. Another approach, *link scheduling*, has received great attention over the years. In link scheduling, it is guaranteed that a scheduled transmission on a link $x \rightarrow y$ will not result in a collision at either node x or node y. Many scheduling problems in wireless networks have been shown to be NP-complete, including TDMA broadcast scheduling (Ephremedis and Truong, 1990) and link scheduling (Arikan, 1984; Even *et al.*, 1984; Nelson and Kleinrock, 1985). Indeed, for some of these problems, even polynomial algorithms with constant-approximation ratios appear highly unlikely for general graphs. For graphs modeling wireless networks, such as disk graphs (DGs) (Hale, 1980) and planar point graphs (Sen and Huson, 1996), the link scheduling and coloring problems still remain NP-complete. Scheduling has been studied extensively in the past few years (Bao and Garcia-Luna-Aceves, 2001, 2002; Behzad and Rubin, 2003; Chlamtac and Farago, 1994; Gandham *et al.*, 2005; Gronkvist and Hansson, 2001; Liu and Lloyd, 2001; Lloyd and Ramanathan, 1992; Nelson and Kleinrock, 1985; Ramanathan, 1999; Ramanathan and Lloyd, 1992, 1993; Salonidis and Tassiulas, 2005; Sen and Huson, 1997; Sen and Malesinska, 1997) because of its application to assigning time slots in TDMA MAC protocols that eliminate collision and guarantee fairness. Scheduling can be reduced to different coloring problems: *edge coloring* and *vertex coloring*.

3.2   IEEE 802.11 Architecture and Protocols

IEEE 802.11, or WiFi, denotes a set of WLAN standards developed by Working Group 11 of the IEEE LAN/MAN Standards Committee (IEEE 802). The term is also used to refer to the original 802.11. The 802.11 family currently includes six over-the-air modulation techniques that all use the same protocol; the most popular (and prolific) techniques are those defined by the a, b, and g amendments to the original standard; security was originally included and was later enhanced by the 802.11i amendment. Other standards in the family are service enhancement and extensions or corrections to previous specifications. 802.11b was the first widely accepted wireless networking standard, followed (somewhat counterintuitively) by 802.11a and 802.11g.

Other than being a solution for pure wireless ad hoc networking, the IEEE 802.11 ad hoc technology also constitutes an important and promising building block for solving the last-mile problem. The transmission range is limited because the RF energy disperses as the distance from the transmitter increases. In addition, even though a WLAN using 802.11 operates in a unregulated spectrum, the transmitter power is limited by regulatory bodies. 802.11b and 802.11g standards use the unlicensed 2.4-GHz band. The 802.11a standard uses the 5-GHz band. Operating in an unregulated frequency band, 802.11b and 802.11g equipment can incur interference from microwave ovens, cordless phones, and other appliances using the same 2.4-GHz band. IEEE 802.11 protocols can operate at several bit rates but, because the transmitter power is limited, the transmission range decreases when the data rate is increased.

The original standard defines CSMA/CA as the media-access method. A significant percentage of the available raw channel capacity is sacrificed (via the CSMA/CA mechanisms) in order to improve the reliability of data transmissions under diverse and adverse environmental conditions. At least five different, somewhat interoperable, commercial products appeared using the original specification. A weakness of this original specification was that it offered so many choices that interoperability was sometimes challenging to realize. It is really more of a "metaspecification" than a rigid specification, allowing individual product vendors the flexibility to differentiate their products. Legacy 802.11 was rapidly supplemented (and popularized) by 802.11b.

IEEE 802.11 specifies both the MAC layer and the physical layer. The MAC layer offers two different types of service: a contention-based service provided by the *distributed coordination function* (DCF) and a contention-free service provided by the *point coordination function* (PCF). These services are made available on top of various physical layers. Three different technologies have been specified in the standards: infrared, FHSS, and DSSS. The DCF service provides the basic access method of the 802.11 MAC protocol and is based on the CSMA/CA scheme. The PCF is actually implemented on top of DCF service, and it uses a point coordinator that cyclically polls wireless stations in the network to give them the opportunity to transmit. Notice that the PCF cannot be adopted in the ad hoc mode.

802.11b and 802.11g divide the spectrum into 14 overlapping, staggered channels whose center frequencies are 5 MHz apart. It is a common misconception that channels 1, 6, and 11 (and, if available in the regulatory domain, channel 14) do not overlap and those channels (or other sets with similar gaps) can be used such that multiple networks can operate in close proximity without interfering with each other. However, this statement is somewhat oversimplified. The 802.11b and 802.11g standards do not specify the width of a channel; rather, they specify the center frequency of the channel and a spectral mask for that channel. In reality, if the transmitter is sufficiently powerful, the signal can be quite strong. For example, a powerful transmitter on channel 1 can easily overwhelm a weaker transmitter on channel 6. In one lab test, throughput on a file transfer on channel 11 decreased slightly when a similar transfer began on channel 1, indicating that even channels 1 and 11 can interfere with each other to some extent. Therefore, it is incorrect to say that channels 1, 6, and 11 do not overlap. It is more correct to say that given the separation among channels 1, 6, and 11, the signal on any channel should be sufficiently attenuated to minimally interfere with a transmitter on any other channel. Although the statement that channels 1, 6, and 11 are "nonoverlapping" is incomplete, the 1, 6, 11 guideline has merit. If transmitters are closer together than channels 1, 6, and 11 (e.g., 1, 4, 7, and 10), overlap between the channels will probably cause unacceptable degradation of signal quality and throughput. The channels that are available for use in a particular country differ according to the regulations of that country. In the United States, for example, FCC regulations allow only channels 1–11 to be used.

3.2.1   Various IEEE 802.11 Protocols

802.11a

The 802.11a amendment to the original standard was ratified in 1999. The 802.11a standard uses the same core protocol as the original standard, operates in 5-GHz band, and uses a 52-subcarrier orthogonal frequency-division multiplexing (OFDM) with a maximum raw data rate of 54 Mbit/s. 802.11a has 12 nonoverlapping channels, 8 dedicated to indoors and 4 to point-to-point. Because the 2.4-GHz band is heavily used, using the 5-GHz band gives 802.11a the advantage of less interference. However, this high carrier frequency also brings disadvantages. It restricts the use of 802.11a to almost line of sight; thus, it may require the use of more APs. In addition, 802.11a cannot penetrate as far as 802.11b because it is absorbed more readily. 802.11a products started shipping in 2001, lagging behind the 802.11b products because of the slow availability of the 5-GHz components needed to implement products. 802.11a was not widely adopted overall because of the already widely adopted 802.11b, some of its disadvantages, poor initial product implementations (making its range even shorter), and regulations. Of the 52 OFDM subcarriers, 48 are for data and 4 are pilot subcarriers with a carrier separation of 0.3125 MHz (20 MHz/64). Each of these subcarriers can be binary-phase-shift keying (BPSK), quadrature-phase-shift keying (QPSK), 16 QAM, or 64 QAM (QAM is quadrature amplitude modulation). The total bandwidth is 20 MHz with an occupied bandwidth of 16.6 MHz. The advantages of using OFDM include reduced multipath effects in reception and increased spectral efficiency.

802.11b

802.11b has a maximum raw data rate of 11 Mbit/s and uses the same CSMA/CA media-access method defined in the original standard. Because of the CSMA/CA protocol overhead, in practice the maximum 802.11b throughput that an application can achieve is about 5.9 Mbit/s over the TCP and 7.1 Mbit/s over the user data-gram protocol (UDP). 802.11b is a direct extension of the DSSS modulation technique defined in the original standard. It is usually used in a point-to-multipoint configuration, wherein an AP communicates via an omnidirectional antenna with one or more clients that are located in a coverage area around the AP. With high-gain external antennas, the protocol can also be used in fixed point-to-point arrangements, typically at ranges up to 8 km. 802.11b suffers interferences from other products operating in the 2.4 GHz band. The 802.11b card provides *adaptive rate selection*: It can operate at 11 Mbit/s but will scale back to 5.5, then 2, then 1 Mbit/s if signal quality decreases. The reason is as follows. The lower data rates use less complex and more redundant methods of encoding the data. Thus, they are less susceptible to corruption that is due to interference and signal attenuation.

802.11g

802.11g is the third modulation standard ratified. This specification under the name of WiFi has been implemented all over the world. It works in the 2.4-GHz band (like 802.11b) but operates at a maximum raw data rate of 54 Mbit/s, or about 24.7 Mbit/s net throughput, like 802.11a. It is fully backward compatible with 802.11b and uses the same frequencies. Although 802.11g held the promise of higher throughput, actual results were mitigated by a number of factors: conflict with 802.11b-only devices, exposure to the same interference sources as 802.11b, limited channelization (only three fully nonoverlapping channels, like 802.11b), and the fact that the higher data rates of 802.11g are often more susceptible to interference than those of 802.11b, causing the 802.11g device to reduce the data rate to effectively the same rates used by 802.11b.

802.11n

This new amendment to the 802.11 standard for WLANs was started around early 2004. The real data throughput is estimated to reach a theoretical 540 Mbit/s (which may require an even higher raw data rate at the physical layer) and should be up to 10 times faster than 802.11a or 802.11g and nearly 40 times faster than 802.11b. It is projected that 802.11n will also offer a better operating distance than current networks. There are two competing proposals of the 802.11n standard, expected to be ratified: WWiSE (worldwide spectrum efficiency), backed by companies including Broadcom and TGn Sync backed by Intel and Philips.

802.11n builds on previous 802.11 standards by adding multiple-input multiple-output (MIMO) and OFDM. MIMO uses multiple transmitter and receiver antennas to allow for increased data throughput through spatial multiplexing and increased range by exploiting the spatial diversity.

3.2.2   Distributed Coordination Function

The basic 802.11 MAC layer uses the DCF to share the medium between multiple stations. DCF relies on CSMA/CA and optional 802.11 RTS/CTS to share the medium between stations. Because of this approach, RF interference is probably the biggest problem. If a source of RF interference (e.g., a cordless phone or another WLAN) is present, the DCF can block stations from transmitting for as long as the interfering signal is present. The stations sense enough energy on the medium and wait patiently, in most cases for just a few seconds or minutes. Of course, this causes the throughput of the network to drop significantly. From this observation, we should perform an RF site survey in the facility before installing a WLAN.

According to the DCF, a station must sense the channel to determine whether there is any other station transmitting data, before it starts transmitting data. If the wireless medium is sensed to be idle for an interval longer than the *distributed interframe space* (DIFS), the station can continue with its transmission. If the medium is busy, the transmission will be deferred until the end of the ongoing transmission. A random interval, typically referred to as *backoff time*, is then selected. It is used to initialize the *backoff timer*. The backoff timer is decreased whenever the channel is sensed as idle, and the backoff timer will be stopped (without increasing or decreasing) when a transmission is sensed on the channel. The backoff timer will be reactivated when the channel is sensed as idle again for a time duration of more than a DIFS. A station is allowed to transmit its data frame when the backoff timer reaches the value zero. Notice that the backoff timer is *slotted*: It is an integer number of slots that are initially randomly and uniformly chosen in the interval $(0, CW - 1)$. Here, CW is the *contention window*, also referred to as the *backoff window*. At the first transmit attempt, $CW = CW_{min}$, and this value is doubled at each retransmission attempt, with a value up to CW_{max}. The exact values of CW_{min} and CW_{max} depend on the physical layer underneath. For example, when a FHSS physical layer is used, values of CW_{min} and CW_{max} are 16 and 1024, respectively.

Clearly, it is possible that two or more stations can start transmitting simultaneously, and then a collision will occur. In the CSMA/CA scheme, a station is not able to detect a collision by hearing its own transmission. Thus, an immediate positive acknowledgment scheme needs to be employed to ascertain the successful reception of the data frame. For 802.11, on reception of the data frame correctly, the receiving station initiates the transmission of an acknowledgment frame (ACK) after a time interval called the *short interframe space* (SIFS). By letting the SIFS be shorter than the DIFS, the receiving station will have a priority over other possible potential transmitting stations that want to send data frames. Notice that the receiving station will *not* send an ACK to the source if the data frame is corrupted. The receiving station can use the cyclic redundancy check (CRC) algorithm for error detection. If the ACK is *not* received by the source station, the data frame is assumed to have been lost or corrupted, and the source station will schedule a retransmission of this data frame.

After the source station detects an erroneous data transmission, the station must remain idle for at least an *extended interframe space* (EIFS) interval before it reactivates the backoff algorithm. In other words, the EIFS should be used by the DCF whenever the physical layer indicates that the past frame transmission did not result in a correct reception of a complete MAC frame. Reception of an error-free frame during the EIFS resynchronizes the station to the actual busy–idle state of the medium. Thus, the EIFS is terminated and normal medium access by the DIFS, and backoff if needed, will continue following the reception of that frame.

3.2.3   Problems and Solutions for the Ad Hoc Model

Now we discuss some problems that can arise in wireless ad hoc networks (WANs). The characteristics of the wireless medium make the wireless networks fundamentally different from the wired networks. There are a number of characteristics that make the designing of efficient and secure wireless networks more challenging. First, the wireless medium is significantly less reliable than the wired medium. Second, the wireless medium has neither absolute nor readily observable boundaries outside of which stations are known to be unable to receive network frames. The wireless channel is also unprotected from outside signals. Additionally, the wireless channels have time-varying and asymmetric propagation properties. In a WAN that relies on a carrier-sensing random-access protocol, like the IEEE 802.11 DCF protocol, the wireless medium characteristics cause complicated phenomena such as the hidden-terminal problem and the exposed-terminal problem. If many stations communicate at the same time, many collisions

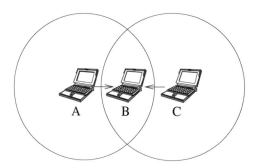

Figure 3.1 The hidden-terminal problem.

will occur, which will lower the available bandwidth (just as in Ethernet, which uses CSMA/CD). And there is no notion of high- or low-priority traffic. It is worse that once a station "wins" access to the medium, it may keep the medium for as long as it chooses. If a station has a low bit rate (1 Mbit/s, e.g.), then it will take a long time to send its packet, and all other stations will suffer from this. More generally, there are no quality-of-service (QoS) guarantees.

The hidden-terminal problem can be described as follows. Both terminal A and C want to communicate with terminal B at the same time. Terminal B is in the transmission range of terminal A and C, and neither is terminal A in terminal C's transmission range nor is terminal C in terminal A's transmission range. Thus, it is possible that both terminals A and C sense the link idle and transmit data to terminal B. Thus, the packets from terminals A and C will collide at terminal B. See Figure 3.1 for an illustration of the hidden-terminal problem.

The hidden-terminal problem can be alleviated by extending the DCF basic mechanism through a *virtual carrier-sensing* mechanism (also referred to as a floor-acquisition mechanism) that is based on two control frames: RTS and CTS. According to this mechanism, before transmitting data frames, the source node sends an RTS and the destination node then replies with a CTS to indicate that it is ready to receive the data frames. Both the RTS and CTS frames contain the total duration of the planned transmission of data frames; that is, the overall time interval needed to transmit the data frame and the related ACK. This information can be read by any listening station that will use this information to set up a timer called the *network allocation vector* (NAV). These nodes suspend transmission for the specified time indicated in the RTS/CTS frames. In other words, when the NAV timer is greater than 0, the listening station must refrain from accessing the wireless medium. These frames are atomic units of the MAC protocol. Stations that hear the RTS delay transmitting until the CTS frame. If they do not hear CTS, they transmit. The stations that hear CTS suspend transmission until they hear acknowledgment from the receiving station. Thus, by using the RTS/CTS mechanism, stations *may* become aware of transmissions from hidden terminals and learn how long the channel will be used for these ongoing transmissions.

The RTS/CTS mechanism can be disabled by an attribute in the management information base (MIB). The RTS/CTS mechanism can be disabled when there is a low

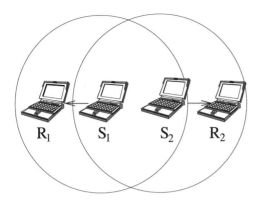

Figure 3.2 The exposed-terminal problem.

demand for bandwidth, if the stations are concentrated in an area where all are able to hear the transmissions of every station and where there is not much contention for the channel.

A similar problem is known as the *exposed-terminal* problem. See Figure 3.2 for an illustration. The exposed-terminal problem can be described as follows. Consider an example of four nodes labeled R_1, S_1, S_2, and R_2, where R_1 and R_2 are receivers and S_1 and S_2 are senders. The two receivers are out of range from one another, yet the two transmitters in the middle are in range of each other and one of the receivers. Here, if a transmission between S_1 and R_1 is taking place, node S_2 is prevented from transmitting to R_2 because it will conclude that it will interfere with the transmission by its neighbor S_1 beause it can sense the signal sent by S_1. In this case, node S_2 is called the *exposed node*. On the other hand, to increase the throughput, the exposed node S_2 should be allowed to transmit in a controlled fashion without interfering with the ongoing transmission between S_1 and R_1. One possible solution is the use of the IEEE 802.11 RTS/CTS mechanism. When a potential sending node S_2 hears an RTS from a neighboring node but not the corresponding CTS, node S_2 can conclude that it is an exposed node and is permitted to transmit to other neighboring nodes. Notice that it is possible that the transmission from S_2 to R_2 could be interfered by the transmission from S_1 if sending node S_1 did not provide certain information to help S_2 make decisions. In other words, node S_2 can transmit without interference if S_1 also did not hear the corresponding CTS from node R_2.

3.2.4   Ad Hoc Networking Support

We now discuss how two or more 802.11 stations can set up a wireless ad hoc network. In the IEEE 802.11 standard, an ad hoc network is called an *independent basic service set* (IBSS). An IBSS enables two or more wireless 802.11 stations to communicate with each other without the intervention of either a centralized AP or an infrastructure networking architecture.

The IEEE 802.11 standards support the peer-to-peer-mode IBSS, which is an ad hoc network with all its stations within one another's transmission range. An IEEE

802.11-based MANET has been proposed to extend the coverage of the network. Because of the flexibility of the CSMA/CA protocol, it is sufficient that all stations within the IBSS be synchronized to a common clock to ensure that stations can transmit and receive data correctly. Clock synchronization is also important for frequency hopping, power management, and many basic operations of 802.11 networks. The 802.11 standard specifies a *timing synchronization function* (TSF) to synchronize clocks among stations. In an infrastructural network, the clock synchronization is easily provided by the AP by letting all stations synchronize their clocks to the clock of the AP. While in an IBSS, clock synchronization is achieved through a distributed algorithm by transmitting special frames, called beacons, that contain the timing information. The TSF-based time synchronization requires two fundamental functionalities – namely, *synchronization maintenance* and *synchronization acquirement*.

Synchronization Maintenance

In the 802.11 TSF, each station maintains a TSF timer counting in increments of microseconds with modulus 2^{64}. Stations expect to receive beacons at a nominal rate defined by the aBeaconPeriod (BP) parameter. All stations in the IBSS compete for beacon transmission every BP second. The time BP defines the length of beacon intervals or periods. This value, established by the station that initiates the IBSS, is then used by all the other stations joining the IBSS and defines a series of *target beacon transmission times* (TBTTs) exactly BP time units apart. Stations use their TSF timers to determine the beginning of beacon intervals or periods. At the beginning of each BP, there is a *beacon generation window* consisting of $W + 1$ slots, where W is set as twice the minimum value of the CW. Each station calculates a random *beacon delay timer* uniformly distributed in $[0, W]$ and it is scheduled to transmit a beacon frame when no beacon has been received before the beacon delay timer expires. If a beacon arrives before the random beacon delay timer has expired, the station cancels the pending beacon transmission and the remaining random delay. On receiving a beacon, a station sets its TSF timer to the time stamp of the received beacon if the value of the time stamp is later than the station's TSF timer. It is important to note that clocks move only forward and never backward. The maximum clock offset between stations can be more than 4000 µs for a large IBSS running the 802.11 TSF.

Several protocols have been proposed during the past few years to resolve the scalability issue in a single-hop ad hoc network. The maximum clock offset is contained in less than 125 µs in the TATSF protocol (Zhou and Lai, 2004). Another solution, called SATSF, was proposed by Zhou and Lai (2005a). The maximum clock offset can be controlled in less than 25 µs. The SATSF allows the fastest station(s) to compete every beacon period and inhibit slower stations from beacon contention. It achieves very accurate clock synchronization through a bounded-frequency-adjustment scheme. However, these solutions cannot be adopted directly for MANETs. Zhou and Lai (2005b) proposed a new clock-synchronization protocol in a MANET, called MATSF, by implementing three strategies. Beacon transmission prioritization ensures good scalability. Bounded-frequency adjustment helps MATSF achieve good accuracy and long-term stability. The construction of a dominating set allows our protocol to converge faster than existing protocols. For MATSF, the maximum clock offset can be controlled within 50 μs.

Synchronization Acquirement

This functionality is used when a new station wants to join an already existing IBSS. A station discovers IBSSs by a scanning procedure of the wireless medium during which the receiver of the station is tuned to different radio frequencies, looking for particular control frames. When the scanning procedure does not find an IBSS, the station may start the creation of a new IBSS. The scanning could be active or passive.

In passive scanning, the station listens to the channels for a beacon frame. Remember that a beacon frame here contains not only the timing information for synchronization but also the complete set of IBSS parameters [e.g., the IBSS identifier (IBSSID), the BP parameter, the data rates that can be supported, and so on].

In active scanning, the station will generate *probe* frames and process the received *probe response* frames. The station that decides to initiate an active scanning has a list of radio frequencies (ChannelList) that will be scanned during the procedure. For each channel, a probe with a broadcast destination will be sent by the DCF access method, and a *ProbeTimer* will be started simultaneously. If no response is received before the ProbeTimer is increased to the minimum time, *MinChannelTime*, the next channel on the ChannelList will be scanned. If a response is received before ProbeTimer reaches MinChannelTime, the station will keep scanning the same channel until the timer reaches the *MaxChannelTime*. In an IBSS, only the station that generated the last beacon will respond to the probe request. It ensures that within an IBSS, at least one station (the station that sends out the last beacon will remain awake) is awake at any time step to respond to the probe request. Notice that it is possible that there is more than one station that responds to a probe request.

3.2.5   Power Management

For mobile wireless networks, the portable devices are often powered by batteries only and thus have limited energy resources. Power management is thus extremely important for extending the lifetime of the wireless network. The basic idea of power management in the MAC layer is to turn off the station whenever it does not have to be active. To prevent missing the data destined for it, each station wakes up from time to time. Thus, each station switches between the *power-save* (PS) state and the *awake* state. The PS state is also sometimes called the *doze* state, during which the station will not be able to transmit or receive. Obviously, in the ad hoc mode, PS strategies need to be completely distributed in order to maintain the self-organizing property of the IBSS. Multicast and frames destined to a power-conserving terminal are first announced during a period when all terminals are awake. In an IBSS, the ATIM window determines the time period during which all the stations, including those in PS mode, must be awake. In the ATIM window, only the beacon or ATIM frame will be transmitted. The length of the ATIM window is specified by the aATIMWindow parameter. It follows a TBTT and ends when the TSF timer modulo aBeaconPeriod is the same as the aATIMWindow. Directed ATIM frames are to be acknowledged by the destination terminal, whereas a multicast ATIM will not be acknowledged. Here, the station receiving the directed ATIM frame must acknowledge the reception and remain awake for the entire duration of the beacon interval, waiting for the announced data frame. If the acknowledgment frame does not arrive, the sending station will execute the backoff procedure for transmitting the ATIM frame. Data frames are transmitted at the end of the ATIM window according to the DCF access method. If a station does not receive any ATIM frame during the ATIM window, it can enter the doze state (i.e., power-saving state) at the end of the ATIM window.

3.2.6   Centrally Controlled Access Mechanism

Although the centralized controlled access mechanism contradicts the concept of ad hoc networking, the centralized controlled access mechanism by 802.11 is briefly reviewed here for the sake of completeness.

The IEEE 802.11 PCF protocol uses a poll-and-response protocol to eliminate the possibility of contention for the medium. A point coordinator (PC) controls the PCF. The PC is always located in an AP. Generally, the PCF operates by stations requesting that the PC register them on a polling list, and the PC then regularly polls the stations for traffic while also delivering traffic to the stations. The PCF is built over the DCF, and both operate simultaneously. The PCF uses point-coordination-function interframe space (PIFS) instead of DIFS. The PC begins a period of operation called the contention-free period (CFP), during which the PCF is operating. This period is called contention-free because access to the medium is completely controlled by the PC and the DCF is prevented from gaining access to the medium. The CFP occurs periodically to provide a near-isochronous service to the stations. The CFP also alternates with a contention period during which the normal DCF rules operate and all stations may compete for access to the medium. The standard requires that the contention period be long enough to contain at least one maximum length frame and its acknowledgment. The CFP begins when the PC gains access to the medium, using the normal DCF procedures, and transmits a beacon frame. Beacon frames are required to be transmitted periodically for the PC to compete for the medium. The traffic in the CFP consists of frames sent from the PC to one or more stations, followed by the acknowledgment from those stations. In addition, the PC sends a contention-free-poll (CF-Poll) frame to those stations that have requested contention-free service. If the station has data to send, then it responds to the CF-Poll. For medium-efficient utilization, it is possible to piggyback both the acknowledgment and the CF-Poll onto data frames. During the CFP, the PC ensures that the interval between frames is no longer than the PIFS to prevent a station operating under the DCF from gaining access to the medium. Until the CFP, the PC sends in a SIFS and waits for a response for the SIFS and tries again. The NAV prevents stations from accessing the medium during the CFP. The beacon contains the information about the maximum expected length of the CFP. The PIFS is used for those that did not receive the beacon. The PC announces the end of the CFP by transmitting a contention-free-end (CF-End) frame. It resets the NAV, and stations begin operation of the DCF independently.

3.2.7   Other Extensions and Enhancements

As we discussed previously, the legacy IEEE 802.11 protocol does not provide QoS guarantees. The 802.11e enhances the DCF and the PCF through a new coordination function: the hybrid coordination function (HCF). It provides QoS extensions to the 802.11 protocol and thus enables real-time audio and video streams to be given higher priority over regular data. Similar to those defined in the legacy of the 802.11 MAC, within the HCF, there are two methods of channel access: enhanced DCF channel access (EDCA) and HCF-controlled channel access (HCCA). Both EDCA and HCCA define traffic classes (TCs). For example, e-mails could be assigned to a low-priority class and a voiceover wireless IP (VoWIP) could be assigned to a high-priority class.

The enhancement to the DCF, the EDCA, introduces the concept of traffic categories. Each station has eight traffic categories, or priority levels. Using EDCA, stations try to send data after detecting that the medium is idle and after waiting a period of time defined by the corresponding traffic category, called the arbitration interframe space (AIFS). A higher-priority traffic category will have a shorter AIFS than a lower-priority traffic category. Thus, with the EDCA protocol, high-priority traffic has a higher chance of being sent than low-priority traffic because stations with lower-priority traffic must wait longer than those with high-priority traffic before trying to access the medium. In addition, each priority level is assigned a transmit opportunity (TXOP). A TXOP is a window of time during which a given station or AP has to send as many frames as possible. This helps minimize the problem of low-rate stations gaining an inordinate amount of channel time in the legacy of the 802.11 DCF MAC.

As a traditional DCF, to avoid collisions within a traffic category, the station counts down an additional random number of time slots, known as a CW, before attempting to transmit data. Remember that a decrease of the clock is performed only when the medium is sensed to be idle. It must be emphasized that no guarantees of service are provided by EDCA. However, EDCA does establish a probabilistic priority mechanism to allocate a bandwidth based on traffic categories.

For HCCA, the interval between two beacon frames is divided into two periods, the CFP and the contention period (CP). During the CFP, the hybrid coordinator (HC), which is also the AP, controls the access to the medium. During the CP, all stations function in the EDCA. The main differences with the PCF are as follows:

1. TCs are defined in HCCA.
2. The HC can coordinate the traffic in any fashion it chooses (not just round-robin in the PCF).
3. The stations give information about the lengths of their queues for each TC, which can be used by the HC to give priority to one station over another.
4. Stations are given a TXOP such that they may send multiple packets in a row, for a given time period selected by the HC. During the CP, the HC allows stations to send data by sending CF-Poll frames.

With the HCCA, the QoS can be configured with great precision. QoS-enabled stations have the ability to request specific transmission parameters, such as data rate and jitter. This allows advanced applications like voiceover IP (VoIP) and video streaming to work more effectively on a WiFi network. Typically, few (if any) APs currently available are enabled for HCCA because HCCA support is not mandatory for 802.11e APs.

3.3   WiMAX

WiMAX (Worldwide Interoperability for Microwave Access) is not a technology but rather a certification mark given to equipment that meets certain conformity and interoperability tests for the IEEE 802.16 family of standards. It is similar to WiFi in concept but has certain improvements that are aimed at improving performance and should permit usage over much greater distances.

Technical Advantages over WiFi

Because IEEE 802.16 networks use the same logical link controller (standardized by IEEE 802.2) as other local-area networks (LANs) and WANs, it can be both bridged and routed to them. An important aspect of the IEEE 802.16 is that it defines a MAC layer that supports multiple physical-layer (PHY) specifications. Enhancements to current and new technologies and potentially new basic technologies incorporated into the PHY can be used. A converging trend is the use of multimode and multiradio system-on-a-chip (SoCs) and system designs that are harmonized through the use of common MAC, system management, roaming, IP Multimedia Subsystem (IMS), and other levels of the system.

The MAC of WiMAX is significantly different from that of WiFi (and Ethernet, from which WiFi is derived). In WiFi, stations wishing to pass data through an AP are competing for the AP's attention on a random basis when a DCF is implemented. This can cause wireless nodes that are far from the AP to be repeatedly interrupted by closer nodes. This in turn will greatly reduce their throughput. This makes it difficult, if impossible, to maintain services such as VoIP or Internet protocol television (IPTV) that greatly depend on the QoSs. By contrast, the medium access among stations supporting WiMAX is more collaborative: The 802.16 MAC is a *scheduling* MAC in which the participating station has to compete only once (for initial entry into the network). After the competition, each wireless station is allocated a *time slot* by the base station. The scheduling algorithm allows the base station to control the QoS by balancing the assignments among the needs of the subscriber stations. The time slot remains assigned to the subscriber station, although it can enlarge and constrict. Other subscribers are not supposed to use the time slot, even if it is not used by the holder. This scheduling algorithm is stable under overload and oversubscription (unlike 802.11). It is also much more bandwidth-efficient.

The WiMAX specification improves on many of the limitations of the WiFi standard by providing increased bandwidth and range and stronger encryption. It provides a connectivity between network end points without the need for a direct line of sight in favorable circumstances. WiMAX makes clever use of multipath signals but does not defy the laws of physics. A recent addition to the WiMAX standard adds full mesh networking capability by enabling each WiMAX node to be able to simultaneously operate in "subscriber station" and "base station" mode. This will blur that initial distinction and allow for widespread adoption of WiMAX-based mesh networks and promises widespread WiMAX adoption. The 802.16 specification applies across a wide range of the RF spectrum. The original WiMAX standard, IEEE 802.16, specifies WiMAX in the 10–66-GHz range. Amendment 802.16a added support for the 2–11-GHz range, of which most parts are already unlicensed internationally. However, there is no uniform global licensed spectrum for WiMAX. The usage of unlicensed spectrum will clearly attract more business interests.

Applications for WiMAX

WiMAX is a WMAN technology that can connect IEEE 802.11 (WiFi) hotspots to each other and to other parts of the Internet and provide a wireless alternative to cable and digital subscriber lines (DSLs) for last-mile (last-kilometer) broadband access. IEEE 802.16 provides up to 50 km (31 miles) of linear service-area range and allows connectivity between users without a direct line of sight. This does not mean that users that are 50 km (31 miles) away without line of sight will have connectivity. The technology also provides shared data rates up to 70 Mbit/s.

It is also anticipated that WiMAX will allow interpenetration for broadband service provision of VoIP, video, and Internet access simultaneously. This should result in a better cost to both home and business customers. Even in areas without preexisting physical cable or telephone networks, WiMAX could allow access between anyone within range of each other. There is also an interesting potential for interoperability of WiMAX with legacy cellular networks. WiMAX antennas can "share" a cell tower without compromising the function of cellular arrays already in place. WiMAX antennae may be even connected to an Internet backbone by either a light fiber-optic cable or a directional microwave link. Cellular companies can use WiMAX as a means of increasing bandwidth for a variety of data-intensive applications. Notice that the cost effectiveness of WiMAX in a remote application could be higher.

Another application under consideration is gaming. For example, Sony and Microsoft are closely considering the addition of WiMAX as a feature in their next-generation game console. This will allow gamers to create ad hoc networks with other players. This may prove to be one of the "killer apps" driving WiMAX adoption: WiFi-like functionality with vastly improved range and greatly reduced network latency and the capability to create ad hoc mesh networks.

3.4   Bluetooth

Bluetooth is an industrial specification for WPANs. Bluetooth provides a way to connect and exchange information between devices like PDAs, mobile phones, laptops, personal computers, printers, and digital cameras via a secure, low-cost, globally available short-range radio frequency. Bluetooth lets these devices talk to each other when they come in range, even if they are not in the same room, as long as they are within up to 100 m (328 ft) of each other, depending on the power class of the product. The most common devices (class 2 devices) allow a quoted transmission distance of 10 m (32 ft). Bluetooth is also IEEE 802.15.1.

Bluetooth differs from WiFi in that the latter provides higher throughput and covers greater distances but requires more expensive hardware and higher power consumption. They use the same frequency range but employ different multiplexing schemes. Whereas Bluetooth is a cable replacement for a variety of applications, WiFi is a cable replacement only or LAN access.

For Bluetooth networks, there is a device playing the role of master and all other nodes are slave nodes. A Bluetooth device with a "master" role can communicate with up to seven slave devices. At any given time, data can be transferred between the master and one of the slaves. The master device switches rapidly to communicate from slave to slave in a round-robin fashion. Simultaneous transmission from the master to multiple slaves is possible but not used much in practice. These groups of up to eight devices (one master and seven slaves) are called *piconets*. Either device may switch the master/slave role at any time. The Bluetooth specification allows connecting two or more piconets together to form a *scatternet*, with some devices acting as a bridge by simultaneously playing the master role in one piconet and the slave role in another piconet.

Pairing

Pairs of devices may establish a trusted relationship by learning (by user input) a shared secret known as a "passkey." A device that wants to communicate only with a trusted device can cryptographically authenticate the identity of the other device. Trusted devices may also encrypt the data that they exchange over the air so that no one can listen in. The encryption, however, can be turned off, and passkeys are stored on the device's file system and not on the Bluetooth chip itself. Because the Bluetooth address is permanent, a pairing will be preserved even if the Bluetooth name is changed. Pairs can be deleted at any time by either device. Devices will generally require pairing or will prompt the owner before it allows a remote device to use any or most of its services.

Air Interface

The Bluetooth protocol operates in the license-free Industrial, Scientific, and Medical (IMS) band at 2.45 GHz. To avoid interfering with other protocols that use the 2.45-GHz band, the Bluetooth protocol divides the band into 79 channels (each 1 MHz wide) and changes channels up to 1600 times per second.

Possible Bluetooth Uses

One of the ways Bluetooth technology may become useful is in VoIP. Bluetooth may then end up being used for communication between a cordless phone and a computer listening for VoIP and with an infrared peripheral component interconnect (PCI) card acting as a base for the cordless phone. The cordless phone would then just require a cradle for charging. Bluetooth would naturally be used here to allow the cordless phone to remain operational for a reasonably long period. Developing a next-generation Bluetooth technology with ultrawideband (UWB) technology and delivering UWB speeds will enable Bluetooth technology to be used to deliver high-speed network data exchange rates required for VoWIP and music and video applications.

3.5   MAC Protocols for Wireless Sensor Networks

Standard MAC protocols developed for duty-cycled WSNs can be roughly categorized into synchronized and asynchronous approaches, along with hybrid combinations. These approaches are motivated by the desire to reduce idle listening, which is the time that the node is awake listening to the medium even though no packets are being transmitted to that node. Idle listening has been found to consume substantial energy both for MANETs and WSNs.

Synchronized protocols negotiate a schedule that specifies when nodes are awake and asleep within a frame. Specifying the time when nodes must be awake in order to communicate reduces the time and energy wasted in idle listening. Asynchronous protocols rely on *low-power listening* (LPL), also called *preamble sampling*, to link together a sender with data to a receiver who is duty cycling. Idle listening is reduced in asynchronous protocols by shifting the burden of synchronization to the sender. When a sender has data, the sender transmits a preamble that is at least as long as the sleep period of the receiver. The receiver will wake up, detect the preamble, and stay awake to receive the data. This allows low-power communication without the need of explicit synchronization between the nodes. The receiver wakes for only a short time to sample the medium, thereby limiting idle listening. Hybrid protocols also exist that combine a synchronized protocol with asynchronous LPL. A key advantage of asynchronous LPL protocols is that the sender and receiver can be completely decoupled in their duty cycles. The simplicity of this design removes the need for and the overhead introduced by synchronized wake/sleep schedules.

3.5.1   B-MAC

Polastre *et al.* (2004) proposed a MAC protocol for WSNs, called B-MAC, which is used as the default MAC for Mica2. Classical MAC protocols perform channel-access arbitration and are tuned for good performance over a set of workloads thought to be representative of the domain; for example, S-MAC (Ye *et al.*, 2004). S-MAC provides an RTS/CTS mechanism for channel arbitration and hidden-terminal avoidance, synchronization with its neighbors for low-power operation, and message fragmentation for efficiently transferring bulk data.

In contrast, the B-MAC protocol contains a small core of media-access functionality. B-MAC uses clear channel assessment (CCA) and packet backoffs for channel arbitration, link-layer acknowledgments for reliability, and LPL for low-power communication. B-MAC is only a link protocol, with network services like organization, synchronization, and routing built above its implementation. B-MAC has a set of interfaces that allows services to tune its operation in addition to the standard message interfaces. These interfaces allow network services to adjust B-MAC's mechanisms, including CCA, acknowledgments, backoffs, and LPL. By exposing a set of configurable mechanisms, protocols built on B-MAC make local policy decisions to optimize power consumption, latency, throughput, fairness, or reliability.

3.5.2   Z-MAC

Rhee *et al.* (2005) presented the design, implementation, and performance evaluation of a hybrid MAC protocol, called Z-MAC, for WSNs that combine the strengths of TDMA and CSMA while offsetting their weaknesses. Like CSMA, Z-MAC achieves high channel utilization and low latency under low contention and, like TDMA, achieves high channel utilization under high contention and reduces collision among two-hop neighbors at a low cost. A distinctive feature of Z-MAC is that its performance is robust to synchronization errors, slot-assignment failures, and time-varying channel conditions; in the worst case, its performance always falls back to that of CSMA. Z-MAC is implemented in TinyOS.

A common MAC paradigm in wireless networks is CSMA. It is popular because of its *simplicity, flexibility*, and *robustness*. It does not require much infrastructure support: no clock synchronization, no traffic load information, and global topology information are required, and dynamic node joining and leaving are handled gracefully without extra operations. These advantages, however, are not free, and they come at the cost of trial and error: A trial may cost access collision in which more than two *conflicting* nodes transmit at the same time, causing signal fidelity degradation at destinations. Although collision among one-hop neighbors can be greatly reduced by carrier-sensing before transmission, carrier-sensing does not work beyond one hop. Notice that collision can happen in any two-hop neighborhood of a node because of the hidden-terminal problem. The hidden-terminal problem causes a serious throughput degradation, especially in high-data-rate sensor applications. Although RTS/CTS can alleviate the hidden-terminal problem, it incurs high overhead [40% to 75% of the channel capacity in sensor networks (Woo and Culler, 2001)] because data packets are typically very small in sensor networks.

TDMA, on the other hand, can solve the hidden-terminal problem without extra message overhead because it can schedule transmission times of neighboring nodes to occur at different times. However, TDMA has many other disadvantages, as widely observed in the literature. First, finding an efficient time schedule in a scalable fashion is not trivial. It often requires a centralized node to find a collision-free schedule. Furthermore, developing an efficient schedule with a high degree of concurrency or channel reuse is very hard (the optimal solution is often NP-hard). In Chapter 4, we discuss in detail some algorithms that can produce efficient link schedulings under a number of assumptions.

Second, TDMA needs clock synchronization. Although clock synchronization is an essential feature of many sensor applications, tight synchronization incurs a high-energy overhead because it requires frequent message exchanges. Third, sensor networks may undergo frequent topology changes because of time-varying channel conditions, physical environmental changes, battery outage, and node failures. Handling dynamic topology changes is expensive, possibly requiring a global change. Fourth, it is difficult if not impossible to ascertain the interference relation among neighboring nodes because radio interference ranges are different from communication ranges, and some interfering nodes may not be in a direct communication range of each other. This phenomenon is called the *interference irregularity* in the literature. Therefore, any channel assignment that uses the communication ranges in place of interference ranges for building the conflict relations does not necessarily yield an interference-free schedule. Furthermore, as interference ranges and also channel conditions are highly time-varying, it is unlikely that one fixed schedule is sufficient to prevent collision all the time. Fifth, during low contention, TDMA gives much lower channel utilization and higher delays than CSMA because, in TDMA, a node can transmit only during its scheduled time slots, whereas in CSMA, nodes can transmit at any time as long as there is no contention. Notice that the scheduling methods studied in Chapter 4 are mainly for wireless mesh networks in which the clock synchronization is easier than that of sensor networks, and the channel often is almost saturated with a high data rate.

Z-MAC uses CSMA as the baseline MAC scheme but uses a TDMA schedule as a *hint* to enhance contention resolution. In Z-MAC, a time-slot assignment is performed at the time of deployment—higher overhead is incurred at the beginning. Its design philosophy is that the high initial overhead is amortized over a long period of network operation, eventually compensated for by improved throughput and energy efficiency. After the slot assignment, each node reuses its assigned slot periodically in every predetermined period, called a frame. We call a node assigned to a time slot an *owner* of that slot and the others the *nonowners* of that slot. There can be more than one owner per slot because DRAND allows any two nodes beyond their two-hop neighborhoods to own the same time slot.

Unlike TDMA, a node may transmit during *any* time slot in Z-MAC. Before a node transmits during a slot (not necessarily at the beginning of the slot), it always performs carrier-sensing and transmits a packet when the channel is clear. However, an owner of that slot always has higher priority over its nonowners in accessing the channel. The priority is implemented by adjusting the initial contention window size in such a way that the owners are always given earlier chances to transmit than nonowners. The goal is that during the slots when owners have data to transmit, Z-MAC reduces the chance of collision because owners are given earlier chances to transmit and their slots are scheduled a priori to avoid collision; but, when a slot is not in use by its owners, nonowners can steal the slot. This priority scheme has an effect of implicitly switching between CSMA and TDMA, depending on the level of contention. An important feature of this priority scheme is that the probability of owners accessing the channel can be adjusted independently from that of nonowners. They show that this feature contributes to increasing the robustness of the protocol to synchronization and slot-assignment failures while enhancing its scalability to contention.

By mixing CSMA and TDMA, Z-MAC becomes more robust to timing failures, time-varying channel conditions, slot-assignment failures, and topology changes than a stand-alone TDMA; in the worst case, it always falls back to CSMA. Because Z-MAC needs only local synchronization among senders in two-hop neighborhoods, Rhee *et al.* (2005) devised a simple local synchronization scheme in which each sending node adjusts its synchronization frequency based on its current data rate and resource budget.

There are a number of protocols proposed in the literature that combine CSMA and TDMA. For example, S-MAC (Ye *et al.*, 2004) and T-MAC (van Dam and Langendoen, 2003) are hybrids of CSMA and TDMA in that they also maintain the synchronized time slots. Nodes maintain periodic duty cycles to listen for channel activities and transmit data. Unlike TDMA, their slots can be much bigger than normal TDMA slots and synchronization failures do not necessarily lead to communication failure because they employ RTS/CTS. Because these protocols use RTS/CTS, the overhead of the protocols is quite high because most data packets in sensor networks are small. T-MAC (van Dam and Langendoen, 2003) improves the energy efficiency of S-MAC by forcing all the transmitting nodes to start transmission at the beginning of each active period. Polastre *et al.* (2004) proposed B-MAC, which is used as the default MAC for Mica2. B-MAC allows application to implement its own MAC through a well-defined interface. It also adopts LPL and engineers the CCA technique to improve channel utilization. B-MAC is shown to have higher throughput and better energy efficiency than S-MAC and T-MAC.

Seamlessly adapting the MAC behavior between TDMA and CSMA according to the level of contention was previously explored by Ephremides and Mowafi (1982) for a WLAN (or one-hop) environment using a scheme called probabilistic TDMA (PTDMA). As in TDMA, real time is slotted, and by adjusting the access probability of owners (p_a) and that of nonowners (p_b), PTDMA adapts the behavior of MAC between TDMA and CSMA, depending on contention. These probabilities are adjusted by a function $p_a + (M - 1)p_b = 1$, where M is the number of senders. PTDMA also assumes buffered senders in which all nodes have the same statistical arrival. In a network in which a subset of nodes is the only active data source (which is a common scenario in sensor networks), PTDMA may exhibit very low channel utilization and does not behave like CSMA.

The distributed packet-reservation multiple-access (D-PRMA) protocol (Alasti and Farvardin, 1999) extends the earlier centralized packet-reservation multiple-access (PRMA) scheme into a distributed scheme that can be used in WANs. D-PRMA is a TDMA-based scheme. The channel is divided into fixed- and equal-sized frames along the time axis. Each frame is composed of s slots, and each slot consists of m *minislots*. A minislot can be further divided into two control fields, RTS/BI and CTS/BI, where BI stands for busy indication. These control fields are used for slot reservation and for overcoming the hidden-terminal problem.

Any node that wants to transmit packets has to first reserve slots, if they have not been reserved already: The node will contend for the first minislot of each slot. If no node wins the first minislot, then the remaining minislots are continuously used for contention, until a contending node wins any minislot. The remaining minislots are granted to the node that wins the contention. Also, the same slot in each subsequent frame can be reserved for this winning terminal until it completes its packet-transmission session. Within a reserved slot, communication between the source and receiver nodes takes place by means of either TDMA or FDMA.

The competition for the first slot is similar to RTS/CTS protocol. A certain period at the beginning of each minislot is reserved for carrier-sensing. If a sender node detects the channel to be idle at the beginning of a slot (minislot 1), it transmits an RTS packet (slot reservation request) to the intended destination through the RTS/BI part of the current minislot. On successfully receiving this RTS packet, the receiver node responds by sending a CTS packet through the CTS/BI of the same minislot. If the sender node receives this CTS successfully, then it gets the reservation for the current slot and can use the remaining minislots. Otherwise, it continues the contention process through the subsequent minislots of the same slot.

3.5.3   X-MAC

Although the LPL approach is simple, asynchronous, and energy-efficient, the long preamble in LPL exhibits several disadvantages:

1. It is suboptimal in terms of energy consumption at both the sender and receiver.
2. It is subject to overhearing that causes excess energy consumption at nontarget receivers.
3. It introduces excess latency at each hop.

First, a receiver node typically has to wait the full preamble period until the preamble is finished before the data acknowledgment packet exchange can begin, even if the receiver has woken up at the start of the preamble. This wastes energy at both the receiver and the transmitter. Second, the LPL approach suffers from the overhearing problem, in which receivers that are not the target of the sender also wake up during the long preamble and have to stay awake until the end of the preamble to find out if the packet is destined for them. This wastes energy consumed by all nontarget receivers within transmission range of the sender. Third, because the target receiver has to wait for the full preamble before receiving the data packet, the per-hop latency is at least the length of the preamble. Over a multihop path, this latency can accumulate to become quite substantial.

Buettner *et al.* (2006) proposed a new approach to LPL called X-MAC, which employs a short preamble to further reduce energy consumption and to reduce latency. Their first idea is to embed address information of the target in the preamble so that nontarget receivers can quickly go back to sleep. This addresses the overhearing problem. The second idea is to use a strobed preamble to allow the target receiver to interrupt the long preamble as soon as it wakes up and determines whether it is the target receiver. This short, strobed-preamble approach reduces the time and energy wasted waiting for the entire preamble to complete. They demonstrated through implementation in a wireless sensor testbed that X-MAC results in significant savings in terms of both energy and per-hop latency. In addition, X-MAC includes an automated algorithm for adapting the duty cycle of the nodes to best accommodate the traffic load in the network.

3.5.4   Funneling-MAC

For WSNs, the combination of hop-by-hop communications and centralized data collection at a sink creates a choke point on the free flow of events out of the sensor network. For example, the funneling of events leads to increased transit traffic intensity and delay as events move closer toward the sink, resulting in significant packet collision, congestion, and loss. This will lead to limited application fidelity measured at the sink in the best scenario, whereas, in the worst case, it will lead to the congestion collapse of the sensor network. For sensors nearest to the sink, typically within a small number of hops, they will lose a disproportionate larger number of packets and consume significantly more energy than sensors farther away from the sink. Consequently, this will shorten the operational lifetime of the overall network if the MAC layer is not carefully designed to address the above asymmetric traffic load. The region of the funnel is called the *intensity region* in Ahn *et al.* (2006). Mitigating the funneling effect represents an important challenge to the sensor network community. Ahn *et al.* (2006) proposed a new MAC protocol called *funneling-MAC* to address this challenge.

The funneling-MAC represents a hybrid (schedule-based) TDMA and (contention-based) CSMA/CA MAC scheme that operates in the intensity region of the event funnel. Pure CSMA/CA operates network-wide in addition to acting as a component of the funneling-MAC that operates in the intensity region. The funneling-MAC mitigates the funneling effect by using local TDMA scheduling in the intensity region only, providing additional scheduling opportunities to nodes closer to the sink, which typically carry considerably more traffic than nodes farther away from the sink. Observe that the nodes near the sink often have a high data rate (the data produced by themselves and the data produced by their downstream nodes that they need to relay to the sink). TDMA-based CA often is best utilized in this high-data-rate scenario. Because the intensity region is close to the sink, the burden of managing TDMA scheduling of sensor events in the intensity region falls on the sink node and not on resource-limited sensor nodes. This will also significantly reduce the TDMA scheduling overhead. The funneling-MAC localized TDMA is triggered by a beacon broadcast by the sink. All sensor nodes perform CSMA by default unless they receive a beacon; they are then called *f-nodes*. The sink regulates the boundary of the intensity area by controlling the transmission power of the beacon. The nodes that received the beacon consider themselves to be in the intensity region and *f*-nodes. These *f*-nodes can perform TDMA while the nodes that do not receive the beacon (e.g., those nodes outside the intensity region) perform CSMA. In funneling-MAC, *f*-nodes rely on the beacon sent to activate TDMA and regulate the boundary of the intensity region for clock synchronization. As soon as a node receives a beacon, it becomes an *f*-node and synchronizes with other *f*-nodes by initializing its clock. The propagation delay of a beacon is on the scale of microseconds in WSNs, whereas the accuracy of synchronization required for the funneling-MAC is on the scale of milliseconds, so beacon-based synchronization can keep the synchronization tight enough to perform TDMA scheduling. The beacon is sent only when it is necessary and in an on-demand basis. The beacon is not sent when the network is idle or receiving very low traffic. Note that every *f*-node keeps a timer that expires if the *f*-node does not receive a beacon for a period longer than the beacon interval. When the timer expires, the node performs pure CSMA.

The scheduling of funneling-MAC is path-oriented. An *aggregated path* is defined as a path that results from the merging of two or more paths on or before entering the intensity region. The funneling-MAC treats an aggregated path as a single-path entry. The sink monitors the incoming data packet and keeps track of the incoming traffic rate for each path, along with the path head ID and number of hops. The sink keeps the traffic rate on a per-path basis in the path table. The sink computes the schedule by allocating time slots on a per-path rather than on a per-node basis. Notice that the TDMA schedule, studied later, is based on either nodes or links. In funneling-MAC, for example, assume that the traffic rate of a path is k and the number of hops of the path is h. In the funneling-MAC method, the sink should allocate every node in the path with $\lfloor k \rfloor$ slots so the sink allocates $\lfloor k \rfloor \times h$ slots to the path. This implicitly assumes that all links on that path will conflict with each other; i.e., no spatial channel reuse is possible. If the traffic rate of a path is less than 1, the sink allocates h slots to the path. To enhance the throughput inside the funnel area, the sink has to consider spatial reuse. It is very difficult to design an optimal spatial reuse scheme without having the complete physical topology information of the network with regard to the communication regions and interference regions. The funneling-MAC takes a simple suboptimal approach and reuses the same slot if two nodes are more than two hops away from each other. In other words, they assume that the interference is always two-hop interference. In this case, f-nodes are unlikely to interfere because one of the nodes may back off because, in the funneling-MAC, carrier-sensing is used even for the scheduled access. Once the f-nodes receive a schedule packet, they synchronize their communication to the funneling-MAC framing structure.

3.6   Conclusion

In this chapter, a brief review was given for a number of MAC protocols designed for wireless networks, including WANs, WMNs, and WSNs. Among the MAC protocols, we studied contention-based MAC protocols, mainly based on CSMA, TDMA-based MAC protocols, and some hybrid MAC protocols that use various combinations of the CSMA MAC and TDMA MAC. Specifically, we reviewed IEEE 802.11, WiMAX, Bluetooth, B-MAC, Z-MAC, X-MAC, and funneling-MAC. We also discussed the advantages of using some hybrid MAC protocols by using a combination either in the temporal domain or in the spatial domain.

Problems

3.1 Compare and contrast FDMA-, TDMA-, and CSMA/CA-based MAC protocols by discussing the advantages and disadvantages of each protocol.

3.2 What are the potential advantages of hybrid TDMA and CSMA/CA MAC protocols?

3.3 Assume all stations can hear all other stations. One station wants to transmit and sense the carrier idle. Why can a collision still occur after the start of transmission?

3.4 How do IEEE 802.11 WLANs address the problem of unique receivers?

3.5 List the unique characteristics of MANETs, WLANs, WSNs, and WMNs that require some special considerations when a MAC protocol is designed for them.

3.6 TDMA scheduling is to assign time slots to nodes such that the nodes assigned to the same time slot will not cause interference when they transmit. Let $X(e, t) \in \{0, 1\}$ be the indicator of whether link e will be active at time slot t. For different interference models, write the constraint on $X(e, t)$ for all links $e \in E$ such that X will result in valid scheduling. You could consider the physical-interference model, the PrIM, or the two-hop interference model.

3.7 List a set of questions a MAC protocol should address when the network has multiple channels.

3.8 Assume that TDMA scheduling is used for a multihop multichannel network. Let $X(e, f, t) \in \{0, 1\}$ be the indicator of whether link e will be active at time slot t using a channel f. For different interference models, write the constraint on $X(e, \mathbf{f}, t)$ for all links $e \in E$ such that X will result in valid scheduling. You could consider the physical-interference model, the PrIM, or the two-hop interference model.

3.9 Study influences of various phenomena, such as node mobility, network topology, network node density, TCP congestion window size, and the interaction between MAC and TCP, on the performance of IEEE 802.11 protocol.

3.10 Assume the following wireless channel model: For each node, there is a transmission range R_T, a physical carrier-sensing range R_C, and an interference range R_I, with $R_T < R_I < R_C$. Let d be the distance between a node and the sender S. When $d \leq R_T$, the node will correctly get the data from the sender S. When $R_T < d < R_C$, the node is not able to get the data correctly; however, it can observe that the channel is busy and thus can defer its own transmissions. When $d > R_C$, this node can start to transmit simultaneously with the sender; however, the channel they observe may be affected by the energy radiated by the sender S. Further, when $d < R_C + R_T$, some interference phenomena may occur. This depends on the interference range R_I (which itself is difficult to model). Several interesting observations can be derived by use of this model. Use this to analyze the hidden-station phenomenon.

4 TDMA Channel Assignment

4.1 Introduction

Wireless multihop radio networks such as ad hoc, mesh, or sensor networks are formed of autonomous nodes communicating via radio. Wireless networks have drawn a great deal of attention in recent years because of their potential applications in various areas. For example, WMNs are being used as the last mile for extending the Internet connectivity for mobile nodes. These wireless mesh or sensor networks behave almost like wired networks because they have infrequent topology changes, limited node failures, and so on. For WMNs or WSNs, the aggregate traffic load of each routing node changes infrequently also. A unique characteristic of wireless networks is that the radio sent out by a wireless terminal will be received by all the terminals within its transmission range and also possibly cause signal interference to some terminals that are not intended receivers. In other words, the communication channels are shared by the wireless terminals. Thus, one of the major problems facing wireless networks is the reduction of capacity because of interference caused by simultaneous transmissions. Using multiple channels and multiple radios can alleviate, but not eliminate, the interference. To achieve robust and collision-free communication, there are two alternatives. One is to utilize a random-access MAC layer scheme; this was discussed in detail in Chapter 3. The other is to carefully construct a transmission schedule. One variant, link scheduling in the context of time-division multiplexing (TDM), is the subject of this chapter.

Throughout this chapter, we assume that the time is slotted and synchronized. A link scheduling assigns each link a set of time slots $\subset [1, T]$ on which it will transmit (T is the scheduling period). A link scheduling is *interference aware* (or called *valid* in this book) if a scheduled transmission on a link $x \rightarrow y$ will not result in a collision at either node x or node y or at any other node of the network. In this context, two types of collisions must be avoided: namely, primary interference and secondary interference. Link scheduling has received great attention from both the networking and theory fields (Alicherry *et al.*, 2005; Kodialam and Nandagopal, 2003, 2004, 2005; Krumke *et al.*, 2001; Kumar *et al.*, 2004, 2005; Moscibroda and Wattenhofer, 2005a; Ramanathan, 1999) in the past few years because of its application to assigning time slots in TDMA MAC protocols that eliminate collision and guarantee fairness. Many scheduling problems in wireless networks have been shown to be NP-complete, including TDMA broadcast scheduling (Ephremedis and Truong, 1990) and link scheduling (Arikan, 1984; Even *et al.*, 1984). For some of these problems, even polynomial-time algorithms with constant-approximation ratios appear unlikely for general graphs.

The majority of the previous studies on link scheduling either assume a very general graph model or assume a very specific graph model such as a UDG. It is widely accepted in the wireless networking community that neither a general graph model nor a UDG model accurately captures unique properties of wireless networks. A general graph model could not capture a certain geometry property of wireless networks; e.g., two nodes must be within a certain distance to be able to communicate directly (or one node's transmission could interfere with the other node's reception). A UDG model is idealistic because, in practice, two nearby nodes may still be unable to communicate for various reasons such as barrier and path fading. In this chapter, we study some efficient centralized and distributed algorithms (Kumar *et al.*, 2005; W. Wang *et al.*, 2006b) that can obtain a valid link scheduling with theoretically proven performances for a more realistic wireless network model. The studied algorithms (see the aforementioned references for more details) address the link scheduling in the following networking model: (1) each node has its own transmission power and thus its own transmission range; (2) that the receiver must be within the transmission range of the sender is only a necessary (but not sufficient) condition for two nodes to communicate directly; i.e., two nearby nodes may still be unable to communicate directly; and (3) if a node v is within a certain distance of a sender u, then the transmission by u will interfere with the reception of node v. In summary, the communication graph could be an arbitrary geometry graph. Notice that in the literature, several similar realistic models (e.g., Jain *et al.*, 2001; Kodialam and Nandagopal, 2005; Kumar *et al.*, 2005), using weighted and unweighted flows, modeling the interference range to be different from the transmission range, and so on have been proposed and modeled, and heuristic algorithms have been given for each or all of these. The main characteristics of the algorithms studied in this chapter are that they can provide theoretical bounds for link-scheduling algorithms in these cases.

In several wireless networks (e.g., mesh, sensor networks), we can estimate the traffic demand by each wireless node. Thus, from a given routing algorithm, we can predict the average traffic load $\ell(e)$ on each link e of the network. We can then design link-scheduling algorithms to meet this traffic demand, if possible. In the algorithms studied in this chapter, we model this by assuming that each link e has an integral *weight* $w(e)$ specifying the number of slots it needed in a scheduling period to support its traffic load. Here, $w(e) = \lceil T \frac{\ell(e)}{c(e)} \rceil$, where $c(e)$ is the capacity of link e if there is no interference and T is a given period for a schedule. Notice that here, $w(e)$ is an estimated average load of the link e under certain routing algorithms. Later, in Chapter 16, it is shown how to find a flow assignment and the corresponding link scheduling such that the network throughput is approximately maximized. In certain networks, it is difficult if not impossible to estimate the load of every link. We then assume that each node needs at least one time slot for transmission, and the objective is to design a scheduling that minimizes T.

We mainly study link scheduling for two kinds of interference models: the RTS/CTS model and fPrIM. For both models, we investigate both centralized and distributed link-scheduling algorithms that use time slots at most a constant factor of the optimum. For the fPrIM, the algorithms require that the interference range of a node be larger than its transmission range, which is always true in practice (the interference range of a node is about twice its transmission range). One of the studied distributed algorithms has not only low communication complexity but also has good performance guarantee that is only logarithmic of the ratio between the maximum and minimum interference ranges. We also study in detail how to prove asymptotical optimal bounds for the performance. To build connections with the joint routing and link scheduling (studied in Chapter 16), we also study both necessary and sufficient conditions for schedulable flows under various interference models.

All the algorithms studied here will preserve the independence between layers. In other words, it is assumed that there is already an existing routing algorithm that will select a path for every pair of source and destination nodes. The performance guarantee of methods studied here is *independent* of the routing algorithm when the routing is given. The results presented here will later be extended in Chapter 16 to the scenario in which we want to maximize the throughput by optimizing the routing and TDMA link scheduling together.

4.2   System Model and Assumptions

Interference issues have been studied extensively recently because it is widely believed that reducing the interference can increase the overall performance of a wireless network. There are different approaches to reducing the interference, including scheduling on the MAC layer, route selection on the routing layer, and power control on the physical layer. In this section, we first discuss in detail the network and interference models we use and formally define the problem that we study in this chapter.

4.2.1   Network and Interference Models

Network Model

In this chapter, we assume that there is a set V of communication terminals deployed in a plane. Each wireless terminal is equipped with only a *single* radio interface. The complete communication graph is a *directed* graph $G = (V, E)$, where $V = \{v_1, \ldots, v_n\}$ is the set of terminals and E is the set of possible directed communication links. Every terminal v_i has a transmission range t_i such that the necessary condition for a terminal v_j to receive correctly the signal from v_i is $\|v_i - v_j\| \leq t_i$, where $\|v_i - v_j\|$ (sometimes we denote it as $d_{i,j}$ for simplicity) is the Euclidean distance between v_i and v_j. Notice that $\|v_i - v_j\| \leq t_i$ is not a sufficient condition for $(v_i, v_j) \in E$. Some links do not belong to G because of either the physical barriers or the selection of routing protocols. We always use $\mathbf{L}_{i,j}$ to denote (v_i, v_j) hereafter. Each terminal v_i also has an interference range r_i such that v_j is interfered by the signal from v_i if $\|v_i - v_j\| \leq r_i$ and v_j is not the intended receiver. The interference range r_i is not necessarily the same as the transmission range t_i. Typically, $r_i > t_i$. We call the ratio between them the *interference–transmission ratio* for node v_i, denoted as $\gamma_i = (r_i/t_i)$. In practice, $2 \leq \gamma_i \leq 4$. For all wireless nodes, let $\gamma = \max_{v_i \in V}(r_i/t_i)$.

Interference Models

To schedule two links at the same time slot, we must ensure that the schedule will avoid the interference. Two different types of interference have been studied in the literature: namely, *primary interference* and *secondary interference*. Primary interference occurs when a node transmits and receives packets at the same time. Secondary interference occurs when a node receives two or more separate transmissions. Here, all transmissions could be intended for this node or only one transmission could be intended for this node (thus, all other transmissions are interference to this node). In addition to these interferences, there could have been some other constraints on the scheduling; e.g., the radio networks that deploy the IEEE 802.11 protocol with the RTS/CTS mechanism will pose some additional constraints. Several different interference models have been used to model the interferences in wireless networks. See Chapter 2 for a discussion of some of the interference models. In this chapter, we mainly focus on link scheduling for the fPrIM model and the RTS/CTS model.

4.2.2   Problem Formulation

Assume that the communication links in the wireless network are predetermined, either by some existing routing protocol such as ad hoc on-demand vector (AODV) routing or DSR, or can be predicted from the existing routes. Given a communication graph $G = (V, E)$, we use the *conflict graph* (e.g., Jain *et al.*, 2003) F_G to represent the interference in G. Each vertex (denoted by $\mathbf{L}_{i,j}$) of F_G corresponds to a directed link (v_i, v_j) in the communication graph G. There is an *edge* between vertex $\mathbf{L}_{i,j}$ and vertex $\mathbf{L}_{p,q}$ in F_G if and only if $\mathbf{L}_{i,j}$ conflicts with $\mathbf{L}_{p,q}$ because of interference. Recall that whether two links conflict depends on the interference model used underneath (e.g., the fPrIM or the RTS/CTS model). Thus, for a given communication graph G, the interference graph F_G may be different. To avoid confusion, we use F_G^P to denote the interference graph under the fPrIM and F_G^{D2} to denote the interference graph under the RTS/CTS model.

The objective of link scheduling is to give each link $\mathbf{L} \in G$ a transmission schedule $S(\mathbf{L})$, which is the list of time slots it could send packets such that the schedule is interference-free and the overall throughput of the network is maximized. Let $X_{e,t} \in \{0, 1\}$ be the indicator variable that is 1 iff e will transmit at time t. We focus on periodic schedules in this chapter. A schedule is *periodic* with period T if, for every link e and time slot t, $X_{e,t} = X_{e,t+iT}$ for any integer i. For a link e, let $I(e)$ denote the set of links e' that will cause interference if e and e' are scheduled at the same time slot. A schedule S is *interference-free* if $X_{e,t} + X_{e',t} \leq 1$ for any $e' \in I(e)$. In graph-theory terminology, the interference-free link-scheduling problem is essentially the (list) *vertex coloring* of F_G.

When the traffic load of links is unknown, the objective of link scheduling is to find a scheduling with the minimum period χ such that every link is scheduled at least once. If we schedule all links within a period χ such that no two links in same time slot interfere with each other, then at least one packet can be delivered over each communication link in every χ time slots. Thus, $1/\chi$ can be used to estimate the *throughput* of the network based on this schedule. The second case is that the average traffic load $\ell(e)$ of each link is known in advance. We model this by assuming that each communication link e [vertex in the conflict graph or sometimes called interference graph (CG)] has a *weight* $w(e)$ specifying the minimum number of time slots it required in each period for sending packets. Here, $w(e) = \lceil T \frac{\ell(e)}{c(e)} \rceil$, where $c(e)$ is the capacity of link e if there is no interference and T is the given period for a schedule. Our main focus in this chapter is how to schedule the communication links in an interference-free manner such that the throughput of the network is maximized; i.e., with the smallest T.

Notice that for simplicity we assume that there is only a single channel in the network. All the results in this chapter can be extended to the case in which multiple channels are available, as in Alicherry *et al.* (2005). If nodes have a set of preassigned channels for each link, then the link scheduling with multiple channels is just the simple union of a set of schedulings, in which each scheduling is for all links using the same channel. Here, it is implicitly assumed that the number of channels assigned to each link uv is no more than the number of network interface cards (NICs) either node u or node v has. However, notice that the static assignment of correct channels to appropriate links is a bigger factor in determining the performance of the corresponding link scheduling. If links can dynamically switch channels, then the greedy algorithms will find the channel with the smallest available time slot for each link to be scheduled and similar performance guarantees hold.

4.3   Centralized Scheduling

In this section, we first study several centralized algorithms for link scheduling under different interference models. These algorithms were first proposed in W. Wang *et al.* (2006b). The performances of centralized algorithms are then used as a certain benchmark to evaluate the performances of the distributed algorithms to be studied later. Recall that the TDMA link-scheduling problem is essentially the vertex-coloring problem in the interference graph (IG) that describes the conflict relations among simultaneous transmiting links. It is a NP-hard problem to find minimum vertex-coloring for an arbitrary input graph. A first-fit heuristic has been a major vertex-coloring method in the literature. First-fit coloring will order the vertices in the input graph (in this case, the IG) and then process vertices in sorted order. The ith vertex will be assigned the smallest color that will not cause any conflict with its neighboring vertices; i.e., not used by all its processed neighboring vertices. It has been shown in the literature that there is one ordering of vertices such that first-fit will result in an optimum vertex-coloring. It was also proved that for any d-inductive graph, *any* first-fit coloring heuristic will result in a coloring using colors at most $O(d \log n)$ times the optimum. Furthermore, for any d-inductive graph, there is one first-fit coloring heuristic that will result in a coloring using colors at most $O(d)$. Here, a graph is, a d-inductive graph if there is a rank of its vertices such that the ith-ranked vertex is connected to at most d vertices with a larger rank. The algorithms presented in the following subsections are mainly to find an ordering of vertices in a graph G such that the graph G is d-inductive and d is of the order of $\chi(G)$.

4.3.1   Scheduling under the Protocol-Interference Model

A number of centralized algorithms for link scheduling have been proposed in the literature (e.g., Alicherry *et al.*, 2005; Kumar *et al.*, 2005). The scheduling algorithms in Kumar *et al.* (2005) study link scheduling for the traditional PrIM and TxIM and assume the same networking model as used throughout this chapter. A common approach is to assign each link the best possible channels (smallest time slots here) by a greedy heuristic. The difference between them is the processing order of links: In Kumar *et al.* (2005), links with smaller lengths are processed first, whereas in Alicherry *et al.* (2005), links are processed in an arbitrary order (because it uses UDG models for both communication and interference). In this chapter, we first review the methods developed in Kumar *et al.* (2005) for the PrIM. Assume that, for a schedule, we already know the schedule period T; in other words, all time slots must be assigned in $[1, T]$ and then repeated after that.

Algorithm 1 Centralized scheduling under the PrIM

Input: A communication graph $G = (V, E)$ of m links.
Output: An interference-free link scheduling.

1: Sort all links E in nondecreasing order of the Euclidean edge length. Let $\mathbf{L}_1$, $\mathbf{L}_2, \cdots, \mathbf{L}_m$ be the sorted list of m wireless links.
2: **for** $i = 1$ to m **do**
3:    Assign link $\mathbf{L}_i$ the smallest $w(\mathbf{L}_i)$ time slots not yet assigned to any link in $I(\mathbf{L}_i)$, which is the set of links e' that will cause interference if e' and $\mathbf{L}_i$ transmit at the same time slot. Notice that here the time slots assigned are not necessarily consecutive for link $\mathbf{L}_i$.
4: **end for**

The following theorem was proved by Kumar *et al.* (2005).

THEOREM 4.1 *Algorithm 1 and its distributed implementation produces a conflict-free schedule; i.e., for any two interfering links e_1 and e_2, the assigned time slots are disjoint.*

This theorem is straightforward from the algorithm if there are enough time slots left for each link when the schedule of previous links is given. The main concern now is to make sure that there are enough time slots left based on the flow requirement of all links. The following technical lemma was proved by Kumar *et al.* (2005) to address this.

LEMMA 4.1 *The link-flow-scheduling algorithm and its distributed implementation produce a valid schedule for traffic flow ℓ if the following holds:*

$$\forall e \in E, \quad \frac{\ell(e)}{c(e)} + \sum_{e' \in I_{\geq}(e)} \frac{\ell(e')}{c(e')} \leq 1.$$

Here, $I_{\geq}(e)$ denotes the subset of links from $I(e)$ that have Euclidean length at least that of link e. Notice that because the algorithm processes the links in nonincreasing order of the Euclidean length, when we process a link e_i, the processed links from $I(e_i)$ must be exactly $I_{\geq}(e_i)$. Then, the lemma implies that there are enough time slots for link e_i.

To ensure that the algorithm will achieve a flow that is within a constant factor of the optimum, the following technical lemma was proved again.

LEMMA 4.2 *A flow ℓ is schedulable (called stable) by a TDMA MAC schedule if the following condition holds:*

$$\forall e \in E, \quad \frac{\ell(e)}{c(e)} + \sum_{e' \in I_{\geq}(e)} \frac{\ell(e')}{c(e')} \leq c,$$

where c is a constant depending on the interference model. Specifically, $c = 5$ for the TxIM and $c \leq 4(1 + \frac{1}{\eta})^2$ for the PrIM.

Recall that under the PrIM, a transmission by a node v_i is successfully received by a node v_j iff the intended destination v_j is sufficiently apart from the source of any other simultaneous transmission; i.e., $\|v_k - v_j\| \geq (1 + \eta)\|v_i - v_j\|$ for any node $v_k \neq v_i$. Here, $\eta > 0$ is a constant.

The basic idea of distributed implementation of the preceding algorithm is to partition the transmission range of nodes into buckets $[R_{\max}, \frac{R_{\max}}{2}]$, $[\frac{R_{\max}}{2}, \frac{R_{\max}}{2^2}]$, $[\frac{R_{\max}}{2^2}, \frac{R_{\max}}{2^3}]$ $\cdots$ $[\frac{R_{\max}}{2^i}, \frac{R_{\max}}{2^{i+1}}]$ $\cdots$. Here, $R_{\max}$ is the maximum transmission range. Notice that the actual algorithm proposed by Kumar *et al.* (2005) used a slightly different partition. Then, nodes participate in assigning time slots only if the nodes in buckets with a larger transmission radius are finished. For nodes in the same bucket, with probability $1/d$, each node v will randomly elect itself as the node to assign time slots, where d is the number of neighboring nodes in the same bucket as this node. When only one node elects itself in its neighborhood, node v will assign time slots using a greedy approach: assign the earliest available time slots. Then, node v will inform neighboring nodes about its selected time slots.

4.3.2   Scheduling under the RTS/CTS Model

The centralized algorithm studied in this chapter, under the RTS/CTS model, will process links in a special order, as in Hochbaum (1983). The basic idea is to first sort links as follows: Every time we pick a link, say **L**, from the remaining graph that has the smallest number of interfered links in the remaining graph and then remove **L** from this graph; we repeat this until the graph becomes empty. We then assign time slots to links in the reverse order of picked links using the smallest time slot available (i.e., not used by interfering links). In summary, a link e with larger $I(e)$ will more likely be processed earlier by the algorithm. Algorithm g summarizes our method.

Algorithm 2 Centralized scheduling under the RTS/CTS model

Input: A communication graph $G = (V, E)$ of m links.

Output: An interference-free link scheduling.

1: Construct the conflict graph F_G^{D2} and let graph $G' = F_G^{D2}$.
2: **while** G' is not empty **do**
3:    Find the vertex with the *smallest* total degree in G' and remove this vertex from G' and all its incident edges. Let $\mathbf{L}_k$ denote the $(m - k + 1)$th vertex removed and the degree of $\mathbf{L}_k$ in graph G' just before it is removed be its δ-*degree*.
4: **end while**
5: Process links from $\mathbf{L}_1$ to $\mathbf{L}_m$ and assign to each $\mathbf{L}_k$ the smallest time slot not yet assigned to any of its neighbors in F_G^{D2}. Here, m is the total number of links in G (equivalently, the total number of vertices in F_G^{D2}).

Before the performance of the centralized algorithm is analyzed, some necessary definitions and properties needed to prove the performance of the algorithms are presented. Given a communication link $\mathbf{L}_{i,j}$, we define the *interference radius of link* $\mathbf{L}_{i,j}$ as $r_{i,j} = \max\{r_i, r_j\}$. If $r_i > r_j$ or $r_i = r_j$ and the ID of node v_i is larger than the ID of node v_j, then v_i is called the *head* (denoted as $h_{i,j}$) of link (v_i, v_j) and v_j is the *tail* (denoted as $t_{i,j}$) of this link. Notice that here, the *head* of a link is not necessarily the sender of the directed communication link. Given a node v_k, we use $R(v_k, x)$ to denote the disk centered at v_k and with radius xr_k. A node v_k interferes a node v_i if node v_i is inside the interference region [i.e., disk $R(v_k, 1)$] of node v_k. We say a link $\mathbf{L}_{p,q}$ interferes a node v_k if either v_p or v_q interferes v_k. For a given v_k, we use $N_{\geq}(v_k, \alpha)$ to denote the set of nodes satisfying that (1) each of their interference radii is at least r_k, and (2) each of them interferes some nodes in $R(v_k, \alpha)$. Notice that a node from $N_{\geq}(v_k, \alpha)$ could be arbitrarily far away from node v_k. Similarly, for a link $\mathbf{L}_{i,j}$, let $R(\mathbf{L}_{i,j}, x)$ denote the union of two disks centered at v_i and v_j, respectively, with radius $x \cdot r_i$ and $x \cdot r_j$, respectively. Let $N_{\geq}(\mathbf{L}_{i,j}, \alpha)$ denote the union of node sets $N_{\geq}(v_i, \alpha)$ and $N_{\geq}(v_j, \alpha)$. The following theorem estimates the local chromatic number based on node degree.

THEOREM 4.2 *(W. Wang et al., 2006b) For a given node v_k and any node set $V_k \subseteq N_{\geq}(v_k, \alpha)$ with constant α, there exists a subset V_k' of V_k with cardinality $|V_k|/C_\alpha$ such that each node interferes with each other, where $C_\alpha \leq (6\alpha + 1)^2 + 11$.*

Proof. We consider a partition of V_k: the nodes inside and outside region $R(v_k, 3\alpha)$, denoted by V_k^1 and V_k^2, respectively.

First, we consider the node set V_k^1. Using a simple area argument, we find that there are at most $\frac{\pi[(3\alpha + \frac{1}{2})r_k]^2}{\pi(\frac{1}{2}r_k)^2} = (6\alpha + 1)^2$ disks with radius $r_k/2$ that can be placed inside the disk $R(v_k, 3\alpha)$. See Figure 4.1 for illustration. Thus, there exists a node set in V_k^1 with a size of at least $|V_k^1|/(6\alpha + 1)^2$ such that each node in the set interferes with each other.

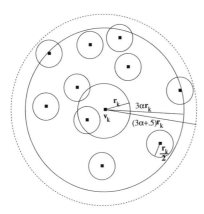

Figure 4.1 Number of small independent disks. © 2006, ACM.

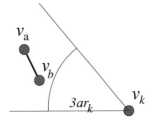

(a) Divide the space into 11 cones   (b) Two nodes interfere in the same cone

Figure 4.2 Illustration of the partition of the region. © 2006, ACM.

Second, we consider the node set V_k^2. We divide the whole space into 11 equal cones using 11 rays from v_k, as shown Figure 4.2(a). If v_a and v_b are in the same cone, then $\angle v_a v_k v_b < 33°$. Let $d_{a,b} = \|v_a - v_b\|$. Because $v_a \in N_{\geq}(v_k, \alpha)$, v_a interferes with some nodes in $R(v_k, \alpha)$, $d_{a,k} \leq r_a + \alpha r_k$. Similarly, $d_{b,k} \leq r_b + \alpha r_k$. Thus, $\max\{d_{a,k}, d_{b,k}\} \leq \max\{r_a, r_b\} + \alpha r_k$. On the other hand, because both v_a and v_b are outside $R(v_k, 3\alpha)$, $\min\{d_{a,k}, d_{b,k}\} \geq 3\alpha r_k$. As shown in Figure 4.2(b), for v_a and v_b,

$$d_{a,b}^2 < d_{a,k}^2 + d_{b,k}^2 - 2\cos(33°)d_{a,k}d_{b,k}$$

$$= \max\{d_{a,k}, d_{b,k}\}^2 + \min\{d_{a,k}, d_{b,k}\}^2 - \frac{5}{3}\max\{d_{a,k}, d_{b,k}\}\min\{d_{a,k}, d_{b,k}\}$$

$$\leq \max\{d_{a,k}, d_{b,k}\}\left(\max\{d_{a,k}, d_{b,k}\} - \frac{2}{3}\min\{d_{a,k}, d_{b,k}\}\right)$$

$$\leq (\max\{r_a, r_b\} + \alpha r_k)(\max\{r_a, r_b\} + \alpha r_k - 2\alpha r_k)$$

$$\leq \max\{r_a, r_b\}^2 - \alpha^2 r_k^2 < \max\{r_a, r_b\}^2.$$

The transition between the second and third inequalities is because $\max\{d_{a,k}, d_{b,k}\} \leq \max\{r_a, r_b\} + \alpha r_k$ and $\min\{d_{a,k}, d_{b,k}\} \geq 3\alpha r_k$. Thus, v_a interferes with v_b. Therefore, each pair of nodes in the same cone interfere with each other. This proves that there exists a node set in V_k^2 with a size of at least $|V_k^2|/11$ such that the nodes in the set interfere with each other.

Consequently, there exists a node set with a size of at least

$$\max \left\{ |V_k^1|/(6\alpha+1)^2, \ |V_k^2|/11 \right\} \geq \frac{|V_k^1| + |V_k^2|}{(6\alpha+1)^2 + 11} = \frac{|V_k|}{C_\alpha},$$

such that all nodes in the set interfere with each other. Here, $C_\alpha \leq (6\alpha+1)^2 + 11$, and we call it the α-hop interference number. Notice that $(6\alpha+1)^2 + 11$ is an upper bound on C_α, and it can be improved by use of a tighter analysis. ∎

From Theorem 4.2, we can conclude that if there are more than $C_\alpha + 1$ nodes in $N_\geq(v_k, \alpha)$, then there are at least two nodes in $N_\geq(v_k, \alpha)$ that interfere with each other. Notice that Theorem 4.2 works for interference on nodes only. For a link $e = \mathbf{L}_{i,j}$, let $I_\geq(e)$ be the links e' interfering with e under the RTS/CTS model and whose radius is not smaller than e. The following theorem shows a counterpart that works for links also. In other words, for a given link $\mathbf{L}_{i,j}$ and a node set $V_{ij} \subseteq N_\geq(\mathbf{L}_{i,j}, \alpha)$, there exists a subset V_{ij}' of V_{ij} with cardinality $|V_{ij}|/(2C_{\alpha+1})$ such that all links with end points from V_{ij}' interfere with each other.

THEOREM 4.3 *(W. Wang et al., 2006b) For a given link $e = \mathbf{L}_{i,j}$, at least $|I_\geq(e)|/(2C_1)$ time slots are needed to schedule all links in $I_\geq(e)$.*

Proof. For each link $\mathbf{L}_{p,q} \in I_\geq(e)$, without loss of generality, we assume that $r_p \geq r_q$. Recall that $e' = \mathbf{L}_{p,q}$ and e interfere by definition. We discuss by cases as follows:

Case 1: The interference region of v_p covers either v_i or v_j.

Case 2: The interference region of node v_p can cover neither v_i nor v_j, and v_q is *outside* the union $R(\mathbf{L}_{ij}, 1)$ of the interference region of v_i and v_j. Clearly, in this case, v_p must also be outside of $R(\mathbf{L}_{ij}, 1)$. Because e and e' interfere, it must be that the interference region of v_q covers either v_i or v_j.

Case 3: The interference region of node v_p can cover neither v_i nor v_j, and v_q is *inside* the union $R(\mathbf{L}_{ij}, 1)$ of the interference region of v_i and v_j. Then, v_p will interfere a dummy node v_q.

In summary, we conclude that at least one end node of $\mathbf{L}_{p,q}$ interferes with some nodes in region $R(\mathbf{L}_{i,j}, 1)$; i.e., the head of $\mathbf{L}_{p,q}$ is in $N_\geq(\mathbf{L}_{i,j}, 1)$. Recall that $N_\geq(\mathbf{L}_{i,j}, 1) = N_\geq(v_i, 1) \bigcup N_\geq(v_j, 1)$. The head of $\mathbf{L}_{p,q}$ is either in $N_\geq(v_i, 1)$ or $N_\geq(v_j, 1)$. Without loss of generality, we assume that at least $|I_\geq(e)|/2$ heads of the links in $I_\geq(e)$ are in $N_\geq(v_i, 1)$. From Theorem 4.2, there are at least $|I_\geq(e)|/(2C_1)$ heads that interfere with each other. Thus, there are at least $|I_\geq(e)|/(2C_1)$ links in $I_\geq(e)$ that interfere with each other. This finishes the proof. ∎

Consequently, we have the following necessary condition for any interference-free link scheduling under the RTS/CTS model:

LEMMA 4.3 *(W. Wang et al., 2006b) For any time slot τ, any valid RTS/CTS interference-free link scheduling S must satisfy*

$$X_{e,\tau} + \sum_{e' \in I_\geq(e)} X_{e',\tau} \leq 2C_1,$$

where $I_\geq(e)$ are the links interfering with e whose radius is not smaller than e.

Notice that the preceding theorems hold for any multihop wireless networks in which both the transmission range and interference range could be heterogeneous and some links could be missing for various reasons. If the interference range is homogeneous, then the constant C_α could be improved.

Let $\delta(F_G^{D2})$ be the *maximum δ-degree* of all links $\mathbf{L}_k$ in Steps 2 and 3 of Algorithm 2. Obviously, Algorithm 2 uses at most $\delta(F_G^{D2}) + 1$ colors. We now prove that Algorithm 2 has the following performance guarantee.

THEOREM 4.4 *(W. Wang et al., 2006b) Under the RTS/CTS model, Algorithm 2 needs at most $2C_1\delta_{\mathrm{opt}}$ time slots for all links without interference, where δ_{opt} is the minimum schedule period T.*

Proof. Let H be the vertex-induced subgraph of F_G^{D2} such that each vertex in H has a degree at least $\delta(F_G^{D2})$. The existence of H is straightforward from the definition of $\delta(G)$. Without loss of generality, let $\mathbf{L}_{i,j}$ be the vertex in H with the smallest interference range. From Theorem 4.3, there exists a clique of a size of at least $\frac{\delta(F_G^{D2})+1}{2C_1}$ in F_G^{D2}. The optimal solution thus needs at least $\frac{\delta(F_G^{D2})+1}{2C_1}$ colors. Algorithm 2 uses at most $\delta(F_G^{D2}) + 1$ colors. This finishes the proof. ∎

4.3.3   Scheduling under the fPrIM

Kumar *et al.* (2005) studied scheduling under a different PrIM (with parameter η): A transmission by a node v_i is successfully received by a node v_j iff $\|v_k - v_j\| \geq (1 + \eta)\|v_i - v_j\|$ for any node $v_k \neq v_i$. This needs every node to dynamically change its transmission power based on the receiving node. Recall that in this chapter, we assume that any node will have a fixed transmission power. It is not difficult to design network examples in which the methods (processing links in the order of decreasing length) developed by Kumar *et al.* (2005) will not work under the model used in this chapter.

Under the RTS/CTS model, we essentially showed that the optimal color assignment needs at least $\delta(F_G^{D2})$ colors. Note that when the graph is modeled by a UDG, $\delta(F_G^{D2})$ is essentially $\Delta(F_G^{D2})$, where $\Delta(F_G^{D2})$ is the maximum degree of the conflict graph F_G^{D2}. Thus, almost any greedy-based coloring method (using at most $\Delta(F_G^{D2}) + 1$ colors) has a constant-approximation ratio. Several previous works claimed the same result {that the optimal coloring needs $\Theta(\Delta(F_G^P))$ colors} under the fPrIM model and proposed some algorithms to color the communication graph G by use of $O(\Delta(F_G^P))$ colors, where $\Delta(F_G^P)$ is the maximum degree of the conflict graph F_G^P under the fPrIM model. We can also define $\delta(F_G^P)$ as the maximum δ-degree of the F_G^P that we can compute by applying Steps 2 and 3 of Algorithm 2 on F_G^P. However, as shown later, there are examples of communication graphs whose optimal coloring needs constant colors, whereas both $\Delta(F_G^P)$ and $\delta(F_G^P)$ are $O(n^{1-\epsilon})$ for any $0 \leq \epsilon < 1$ if all nodes have the same transmission range and $t_i = r_i = r$. This shows that any greedy algorithm that uses $\Theta(\Delta(F_G^P))$ or even $\Theta(\delta(F_G^P))$ colors could be very bad compared with the optimal solution.

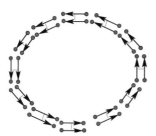

Figure 4.3 Bad example of a simple greedy algorithm. © 2006, ACM.

Now such an example as in Figure 4.3 is described. Here, all nodes have the same transmission range and interference range r. The links form several groups such that all links in each group are parallel and each link has length r. The groups are placed in a cyclic manner such that any sender of one group interferes with all receivers in the previous group and does not interfere with any other receivers in other groups. The number of links in each group is $n^{1-\epsilon}$, and there are n^{ϵ} groups. Here, $0 < \sum < 1$ is a constant. Obviously, in the graph F_G^P, the degree of each vertex (corresponding to a physical link) is $n^{1-\epsilon}$. Thus, $\Delta(F_G^P) = \delta(F_G^P) = n^{1-\epsilon}$. On the other hand, we can use at most three colors to color all the links without conflict: We color groups in clockwise order, and all links in the same group are assigned the same color that is the smallest available.

The preceding example shows that it is unclear whether Algorithm 2 can find a scheduling that approximates the optimal solution in which the interference range equals the transmission range (the proof of Theorem 4.4 does not extend to this scenario). Fortunately, the ratio of the interference range over the transmission range is usually around two in practice. Next, we utilize this property to design efficient link scheduling with a constant-approximation ratio.

Given any two nodes $\mathbf{L}_{i,j}$ and $\mathbf{L}_{p,q}$ in CG F_G^P such that v_j and v_q are receivers, if $\mathbf{L}_{i,j}$ and $\mathbf{L}_{p,q}$ interfere with each other, then it is possible that (1) v_i interferes v_q, or (2) v_p interferes v_j, or (3) both. If v_p interferes v_j, then we treat the link between $\mathbf{L}_{i,j}$ and $\mathbf{L}_{p,q}$ as an *incoming link* for $\mathbf{L}_{i,j}$. Similarly, if v_i interferes v_q, we treat the link as an *outgoing link* for $\mathbf{L}_{i,j}$. Let $d_{i,j}^{\text{in}}(F_G^P)$ and $d_{i,j}^{\text{out}}(F_G^P)$ be the incoming and outgoing degrees, respectively, of $\mathbf{L}_{i,j}$ in the conflict graph F_G^P. The number of incoming links of a vertex in F_G^P is its incoming degree, and the number of outgoing links is its outgoing degree. Similarly, we define $\Delta^{\text{in}}(F_G^P)$ and $\Delta^{\text{out}}(F_G^P)$ as the maximum incoming and outgoing degrees, respectively, in graph F_G^P. When $\gamma_i > 1$ for each node v_i, we will show that the optimal coloring needs at least $\Theta(\Delta^{\text{in}}(F_G^P))$ colors, where the hidden constant depends on $\min_i \gamma_i$ (which is typically two in practice).

LEMMA 4.4 *(W. Wang et al., 2006b) Consider any communication link $\mathbf{L}_{i,j}$, where v_j is the receiver. Consider two links $\mathbf{L}_{p,q}$ and $\mathbf{L}_{s,t}$ that are $\mathbf{L}_{i,j}$'s incoming links in CG F_G^P, where v_q and v_t are the receivers. If $\angle v_q v_j v_t \leq \arcsin \frac{\gamma-1}{2\gamma}$, then link $\mathbf{L}_{p,q}$ interferes with link $\mathbf{L}_{s,t}$.*

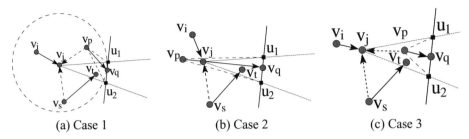

(a) Case 1   (b) Case 2   (c) Case 3

Figure 4.4 Links in a small neighborhood will interfere with each other in a PrIM. © 2006, ACM.

Proof. Draw two rays $v_j v_a$, $v_j v_b$ emanating from node v_j such that $\angle v_a v_j v_b = $ arcsin $\frac{\gamma-1}{2\gamma}$ and v_q, v_t are in the cone as shown in Figure 4.4(a). Without loss of generality, we assume that $\|v_j - v_q\| \geq \|v_j - v_t\|$. Draw a circle $\mathcal{C}$ centered at v_j with radius $\|v_j - v_q\|$. Let $u_1 u_2$ be the line passing v_q that is tangent to circle $\mathcal{C}$ and u_1, u_2 be the intersections of this line with lines $v_j v_a$, $v_j v_b$, respectively. Because $\angle u_1 v_j v_q \leq$ arcsin $\frac{\gamma-1}{2\gamma}$, we have

$$\|u_1 - v_q\| \leq \|v_j - v_q\| \frac{\gamma-1}{2\gamma} \leq 2r_p \frac{\gamma-1}{2\gamma} = r_p \frac{\gamma-1}{\gamma}.$$

Thus, $\|v_p - u_1\| \leq \|v_p - v_q\| + \|u_1 - v_q\| \leq r_p \frac{1}{\gamma} + r_p \frac{\gamma-1}{\gamma} = r_p$. Similarly, $\|v_p - u_2\| \leq r_p$. As follows, we prove that node v_p interferes with v_t by cases.

Case 1: $v_p u_1 u_2 v_j$ is a convex quadrangle, as shown in Figure 4.4(a). In this case, v_t is inside either triangle $v_p v_j u_2$ or triangle $v_p u_1 u_2$. Because both $\|v_p - u_1\|$, $\|v_p - u_2\|$ and $\|v_p - v_j\|$ are not greater than r_p, we have $\|v_p - v_t\| \leq r_p$.

Case 2: v_j is inside $\triangle u_1 u_2 v_p$, as shown in Figure 4.4(b). In this case, v_t is inside triangle $\triangle u_1 u_2 v_p$. Then, it is easy to show that

$$\|v_p - v_t\| \leq \max\{\|v_p - u_1\|, \|v_p - u_2\|\} \leq r_p.$$

Case 3: v_p is inside $\triangle u_1 u_2 v_j$, as shown in Figure 4.4(c). In this case, v_t is inside one of the three triangles: $\triangle u_1 u_2 v_p$, $\triangle u_1 v_j v_p$, or $\triangle u_2 v_j v_p$. Similarly, we have $\|v_p - v_t\| \leq r_p$.

Obviously, the preceding three cases cover all possible situations. This proves that link $\mathbf{L}_{p,q}$ interferes with $\mathbf{L}_{s,t}$. ∎

Similar to Lemma 4.3, we have the following necessary condition for interference-free link scheduling under the fPrIM model.

LEMMA 4.5 *(W. Wang et al., 2006b) For any time slot τ, any valid interference-free link scheduling $\mathcal{S}$ under fPrIM must satisfy*

$$X_{e,\tau} + \sum_{e' \in I^{in}(e)} X_{e',\tau} \leq \left\lceil \frac{2\pi}{\text{arcsin} \frac{\gamma-1}{2\gamma}} \right\rceil,$$

where $I^{in}(e)$ is the set of incoming links of e that interfere e.

This is because for all incoming neighboring links of link e, Lemma 4.4 implies that there are at most $\left(\frac{2\pi}{\arcsin\frac{\gamma-1}{2\gamma}}\right)$ links that can be scheduled at any same time slot. We now review the main theorem about optimum coloring for the fPrIM model with $\gamma_i > 1$.

THEOREM 4.5 *(W. Wang et al., 2006b) Optimal vertex-coloring for conflict graph F_G^P needs $\Theta(\Delta^{\text{in}}(F_G^P))$ colors if $\min_i \gamma_i$ is some constant > 1.*

Proof. For any link $\mathbf{L}_{i,j}$ such that v_j is the receiver, we partition the space by using b equal-sized cones apexed at node v_j, where $b = \lceil\frac{2\pi}{\arcsin\frac{\gamma-1}{2\gamma}}\rceil$. From the pigeonhole principle, $\mathbf{L}_{i,j}$ has at least $d_{i,j}^{\text{in}}(F_G^P)/b$ links whose receivers are in the same cone. From Lemma 4.4, all links in the same cone interfere with each other. Thus, $\mathbf{L}_{i,j}$ has at least $d_{i,j}^{\text{in}}(F_G^P)/b$ incoming links such that they interfere with each other. It implies that any valid coloring will use at least $d_{i,j}^{\text{in}}(F_G^P)/b$ among the incoming neighbors of link $\mathbf{L}_{i,j}$. Thus, the optimal coloring needs at least $\Delta^{\text{in}}(F_G^P)/b + 1$ colors. ∎

Note that $\Delta(F_G^P)$ could be arbitrarily larger than $\Delta^{\text{in}}(F_G^P)$. Thus, a simple greedy algorithm that uses $\Delta(F_G^P)$ colors does not have a constant approximation ratio; e.g., the algorithm proposed in Alicherry *et al.* (2005) for the UDG networking model. It is known that optimal coloring can be obtained by using the greedy approach on a certain ordering of vertices in F_G^P. Next, we study a centralized scheduling method (Algorithm 3) that needs at most $2\Delta^{\text{in}}(F_G^P) + 1$ colors, with a careful selection of link ordering.

Algorithm 3 Centralized scheduling under the fPrIM

Input: A communication graph $G = (V, E)$ of m links.

Output: An interference-free link scheduling.

1: Construct the conflict graph F_G^{D2} under the fPrIM and let graph $G' = F_G^{D2}$.
2: **while** G' is not empty **do**
3:     Find the link $\mathbf{L}_{i,j}$ with the *largest* $d_{i,j}^{\text{in}}(G') - d_{i,j}^{\text{out}}(G')$ in G' and remove this vertex from G' and all its incident edges. Let $\mathbf{L}_k$ denote the kth vertex removed.
4: **end while**
5: Process the sequences of links $\mathbf{L}_k$ from $\mathbf{L}_m$ to $\mathbf{L}_1$. Assign each link $\mathbf{L}_k$ the smallest time slot not yet assigned to any of its neighbors in F_G^P.

THEOREM 4.6 *(W. Wang et al., 2006b) Algorithm 3 uses at most $2\Delta^{\text{in}}(F_G^P) + 1$ colors.*

Proof. The key observation is that in any directed graph, the sum of all vertices' incoming degrees equals the sum of outgoing degrees. For the link $\mathbf{L}_{i,j}$ with the largest $d_{i,j}^{\text{in}}(G') - d_{i,j}^{\text{out}}(G')$ in G', we must have $d_{i,j}^{\text{in}}(G') \geq d_{i,j}^{\text{out}}(G')$. Thus, when we assign color (or time slot) for the link $\mathbf{L}_{i,j}$, the subgraph induced by all the links that have already been processed is exactly the subgraph G' right before vertex $\mathbf{L}_{i,j}$ was removed in the **while** loop of Algorithm 3. Therefore, there are at most $2d_{i,j}^{\text{in}}(G')$ adjacent neighbors of $\mathbf{L}_{i,j}$ in F_G^P that have already been processed. In other words, the smallest time slot assigned to $\mathbf{L}_{i,j}$ is at most $2d_{i,j}^{\mathrm{in}}(G') + 1$, which is at most $2d_{i,j}^{\mathrm{in}}(F_G^P) + 1$. This proves that we need at most $2\,\Delta^{\mathrm{in}}(F_G^P) + 1$ time slots for an interference-free schedule. ∎

4.4   Distributed Algorithms

In Section 4.3, we studied two centralized algorithms that achieve constant approximation ratios for different interference models. In a wireless network, a centralized algorithm may not be possible, and, even if possible, because of the dynamic features of wireless networks, it is inefficient to update the coloring with a centralized algorithm. Thus, in this section, we design efficient distributed algorithms to obtain a valid coloring with good performance guarantee.

4.4.1   Algorithm for the RTS/CTS Model

In the literature, several distributed algorithms have been proposed for vertex-coloring. The first solution is to simply apply a distributed vertex-coloring method on the CG F_G. Recall that all previous distributed algorithms work for the general graph. By taking advantage of special properties of the CG defined here, we are able to obtain a deterministic distributed coloring algorithm that colors the links with $O(\Delta(F_G^{D2}))$ colors in almost constant time when the interference ranges are homogeneous. On the other hand, as shown in the centralized algorithm, the optimal color is $\Theta(\delta(F_G^{D2}))$, which could be much smaller than $\Delta(F_G^{D2})$ when interference ranges are heterogeneous. Thus, simply applying a coloring algorithm with ratio $\Theta(\Delta(F_G^{D2}))$ may not achieve a good performance. Our first instinct is to design a distributed version of Algorithm 2. However, finding the node with the global maximum degree iteratively does not seem promising for a distributed algorithm. Thus, we need to find some lower bound for the optimal color other than $O(\delta(F_G^{D2}))$.

Given two nodes v_i and v_j, we say that v_i *precedes* v_j if and only if $r_i > r_j$ or $r_i = r_j$ and $i > j$. Given a pair of links $\mathbf{L}_{i,j}$ and $\mathbf{L}_{p,q}$ with different heads $h_{i,j} \neq h_{p,q}$, we say that $\mathbf{L}_{i,j}$ *precedes* $\mathbf{L}_{p,q}$ if $r_{i,j} > r_{p,q}$ or $r_{i,j} = r_{p,q}$ and $h_{i,j} > h_{p,q}$. Recall that $r_{i,j} = \max\{r_i, r_j\}$. We also say that the corresponding vertex $\mathbf{L}_{i,j}$ precedes $\mathbf{L}_{p,q}$ in the conflict graph in this case. For a vertex $\mathbf{L}_{i,j}$ in graph F_G^{D2}, let $d_{i,j}^{\geq}(F_G^{D2})$ be the number of adjacent vertices that precede $\mathbf{L}_{i,j}$, which is called the *efficient degree* of $\mathbf{L}_{i,j}$. From Theorem 4.3, there are at least $d_{i,j}^{\geq}(F_G^{D2})/(2C_1)$ vertices adjacent to and preceding $\mathbf{L}_{i,j}$ that form a clique in which each vertex (i.e., the corresponding link in the communication graph) interferes with each other. Let $\phi(F_G^{D2}) = \max_{\mathbf{L}_{i,j}} d_{i,j}^{\geq}(F_G^{D2})$, then Theorem 4.3 shows that the optimal coloring algorithm needs at least $\phi(F_G^{D2})/(2C_1)$ colors. Thus, a coloring algorithm using at most $\Theta(\phi(F_G^{D2}))$ colors has a constant-approximation ratio. Unlike the centralized Algorithm 2, in which the lower bound of $\delta(F_G^{D2})$ could not be found by using only local information, the lower bound of $\phi(F_G^{D2})$ could be easily obtained by any link $\mathbf{L}_{i,j}$ that simply counts the number of interfering links that precede itself; i.e., with a larger link interference radius. Algorithm 4 illustrates the distributed coloring method (developed in W. Wang *et al.*, 2006b) that uses at most $\phi(F_G^{D2})$ colors.

Algorithm 4 Distributed coloring algorithm for the RTS/CTS model

---

Input: A communication graph $G = (V, E)$.
Output: A valid coloring of all links.

1: Each node v_i collects all communication links, say H_i, that contain v_i as the head; i.e., all links $\mathbf{L}_{i,j}$ with $r_i \geq r_j$.
2: Each node v_i collects all communication links, denoted by M_i, that are not in H_i and interfere with some links H_i.
3: Node v_i finds M_i^+, which is the subset of links in M_i that precedes *every* link in H_i and let $M_i^- = M_i - M_i^+$.
4: Node v_i sets all links in M_i^+ as uncolored.
5: **while** some links in M_i^+ are uncolored **do**
6:     Node v_i listens messages from other nodes.
7:     **if** v_i receives a message Color(p, q, k) **then**
8:         Node v_i marks $\mathbf{L}_{p,q}$ with color ID k if $\mathbf{L}_{p,q}$ is in M_i^+.
9:     **end if**
10: **end while**
11: **for** each node v_j in H_i **do**
12:     Find the color with minimum color ID, say k, that is not used by any link that is conflicted with $\mathbf{L}_{i,j}$. Color link $\mathbf{L}_{i,j}$ with color ID k.
13:     Sends the message Color(i, j, k) to all heads of the links adjacent to $\mathbf{L}_{i,j}$ in M_i^-.
14: **end for**

THEOREM 4.7 *(W. Wang et al., 2006b) Algorithm 4 computes a valid coloring using at most $\phi(F_G^{D2})$ colors, which is asymptotically optimal.*

Proof. First, we show that the algorithm does terminate. Because it is straightforward that the number of nodes in H_i is bounded by $\phi(F_G^{D2})$, the **for** loop terminates in $O(n)$ iterations. Thus, the maximum time needed for all other processes other than the **while** loop is bounded by a finite time T, and then we need to show that the **while** loop does terminate for any node v_i. Let $(v_{\sigma_1}, v_{\sigma_2}, \ldots, v_{\sigma_n})$ be the sorted list of nodes in the decreasing order of their interference range. Thus, v_{σ_i} precedes v_{σ_j} if and only if $i < j$. Because v_{σ_1} precedes every other node, $M_{\sigma_1}^+$ is empty and v_{σ_1} colors all links that are adjacent to v_{σ_1} in time T. Now consider the node v_{σ_2} and $M_{\sigma_2}^+$. If $\mathbf{L}_{p,q} \in M_{\sigma_2}^+$, then either v_p or v_q is v_{σ_1}. Thus, all links in $M_{\sigma_2}^+$ are colored. Therefore, all links that are adjacent to v_{σ_2} are colored before time $2T$. Similarly, all links that are adjacent to v_{σ_j} are colored before time jT. Thus, all links are colored in time nT. It is straightforward to show that by assuming that coloring one link takes a unit time, the running time of this algorithm is at most m, where m is the number of directed communication links.

Second, we show that the computed coloring is valid; i.e., no two conflict links have the same color. Consider conflict links $\mathbf{L}_{i,j}$ and $\mathbf{L}_{p,q}$; we again discuss by cases:

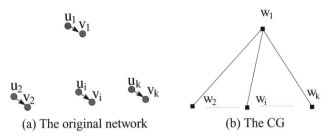

(a) The original network   (b) The CG

Figure 4.5 Δ could be $\Theta(n)$ times the number of colors used by Algorithm 4. © 2006, ACM.

Case 1: $\mathbf{L}_{i,j}$ and $\mathbf{L}_{p,q}$ have the same head. Without loss of generality, we assume that $v_i = v_p$ is the head of the links. Thus, both $\mathbf{L}_{i,j}$ and $\mathbf{L}_{p,q}$ are in H_i. Therefore, $\mathbf{L}_{i,j}$ and $\mathbf{L}_{p,q}$ have different colors.

Case 2: $\mathbf{L}_{i,j}$ and $\mathbf{L}_{p,q}$ have different heads. Then, without loss of generality, we can assume that $h_{i,j} = i$, $h_{p,q} = p$, and v_i precedes v_p. Because $\mathbf{L}_{i,j} \in M_p^+$, $\mathbf{L}_{i,j}$ is colored before M_p^+ becomes empty. Thus, $\mathbf{L}_{p,q}$ is colored after $\mathbf{L}_{i,j}$ is. Therefore, when v_p colors $\mathbf{L}_{p,q}$, it uses a color that is different from the color of $\mathbf{L}_{i,j}$ based on the algorithm.

Third, it is straightforward that Algorithm 4 uses at most $\phi(F_G^{D2})$ colors; i.e., it has a constant-approximation ratio. ∎

Notice that in Algorithm 4, we start to color a link after all interfering links preceding it are colored. Thus, in the worst case, it may take time $O(n)$ to color all the links, where n is the number of nodes in the network. Here, we assume that in one time unit, a node can color all its incident links. Compared with previous polylogarithmic-time-distributed coloring algorithms that color the graph using $\Delta(F_G^{D2})$ colors, Algorithm 4 may take a longer time. However, the following example shows that $\Delta(F_G^{D2})$ could be as large as $O(n)$ times of the color used by Algorithm 4, where n is the number of the nodes in the original network. In Figure 4.5(a), there are k pairs of transmission links $u_1 v_1, \ldots, u_n v_n$. Nodes u_1, v_1 have interference range 1 and all other nodes have interference range ϵ, where ϵ is a small positive constant such that node u_i does not interfere v_j for $i, j > 1$. The corresponding CG is shown in Figure 4.5(b). It is not difficult to see that we need only two colors while the degree of $\mathbf{L}_{1,1}$ is $n - 1$. In other words, compared with previous polylogarithmic-time methods with $\Omega(n)$ approximation ratios, the method presented previously has a constant-approximation ratio when a larger worst-case running time is used.

4.4.2 Faster Algorithm for the RTS/CTS Model

Although Algorithm 4 computes a coloring that is at most constant times of the optimal, it may need a linear number of rounds to compute the coloring. In certain circumstances, we would prefer the distributed algorithms that run fast over the distributed algorithms that have good performance as long as the fast distributed algorithm does not perform much worse. Next, we study another distributed algorithm presented in W. Wang *et al.* (2006b) that computes the coloring very fast with a good performance guarantee of

$O(\log(\psi) + 1)$, where ψ is the ratio between the maximum interference range over the minimum interference range among all nodes.

Algorithm 5 Fast distributed coloring algorithm for the RTS/CTS model

---

Input: A communication graph $G = (V, E)$.
Output: A valid coloring of the communication graph.

1: Node v_i computes a subset, say H_i, of all communication links containing v_i such that link $\mathbf{L}_{i,j} \in H_i$ if and only if $r_i > r_j$.
2: **while** node v_i failed to obtain the channel **do**
3:   Node v_i monitors the channel and competes for the channel.
4: **end while**
5: **for** each link $\mathbf{L}_{i,j} \in H_i$ **do**
6:   Color link $\mathbf{L}_{i,j}$ with the smallest color ID, say k, that is not used by any link that conflicts with $\mathbf{L}_{i,j}$.
7:   Broadcasts the message $\mathsf{Color}(i, j, k)$ to each head of links that conflict with $\mathbf{L}_{i,j}$.
8: **end for**

Algorithm 5 assumes that there is a certain competition-based MAC layer (e.g., 802.11 with RTS/CTS) available for a node to obtain the channel. We use this MAC mechanism to obtain a link scheduling that is efficient and interference-free. Algorithm 5 is very simple and can be implemented without much additional computation on each node. However, the proof of the performance guarantee is not straightforward. To prove the main theorem, we need some notation in order to extend Theorems 4.2 and 4.3. For a given node v_k, let $N_{\geq}(v_k, \alpha, \beta)$ be a node set composed of the nodes satisfying the following conditions:

1. Each interference radius is at least r_k/β.
2. Each interferes some nodes in $R(v_k, \alpha)$.

Let $N_{\geq}(\mathbf{L}_{i,j}, \alpha, \beta)$ be the union of $N_{\geq}(v_i, \alpha, \beta)$ and $N_{\geq}(v_j, \alpha, \beta)$. The proofs of the Lemmas 4.6 and 4.7 are similar to the proofs of Theorems 4.2 and 4.3, respectively, and thus are omitted here.

LEMMA 4.6 *(W. Wang et al., 2006b) For any node v_k and any set $V_k \subseteq N_{\geq}(v_k, \alpha, \beta)$, there exists a subset V_k' of V_k with a cardinality of at least $\lceil |V_k|/C_{\alpha,\beta} \rceil$ such that nodes in V_k' interfere with each other; $C_{\alpha,\beta} = (6\alpha\beta + 1)^2 + 11$.*

LEMMA 4.7 *(W. Wang et al., 2006b) For any link $\mathbf{L}_{i,j}$ and any set $V_{ij} \subseteq N_{\geq}(\mathbf{L}_{i,j}, \alpha, \beta)$, there exists a subset V_{ij}' of V_{ij} with a cardinality of at least $\lceil V_{ij}/(2C_{\alpha+1,\beta}) \rceil$ such that links in V_{ij}' interfere with each other.*

Let $\Delta(\alpha, \beta) = \max_{\mathbf{L}_{i,j}} |N_{\geq}(\mathbf{L}_{i,j}, \alpha, \beta)|$ and $\chi(F_G^{D2})$ be the optimal number of colors. Based on Lemma 4.7, the following theorem is straightforward for any fixed α, β:

THEOREM 4.8 *(W. Wang et al., 2006b) $\chi(F_G^{D2}) \geq \lceil \Delta(\alpha, \beta)/(2C_{\alpha+1,\beta}) \rceil$.*

THEOREM 4.9 *(W. Wang et al., 2006b) Algorithm 5 computes a coloring that is at most $O(\log(\psi) + 1)$ times of optimum $\chi(F_G^{D2})$.*

Proof. Without loss of generality, let link $\mathbf{L}_{i,j}$ be the link that has the maximum color ID, say g. To prove the theorem, we show that $g \le 2C_{1,2}(\log(\psi) + 1)\chi$. In what follows, we prove it by contradiction and, for the sake of contradiction, we assume that $g > 2C_{1,2}(\log(\psi) + 1)\chi$.

We first argue that for any $0 \le k \le \log(\psi)$, there exists a link $\mathbf{L}_{i^{(k)}, j^{(k)}}$ such that $r_{i^{(k)}, j^{(k)}} < r_{i,j}/2^k$ and its color ID is not smaller than $g - 2C_{1,2}k\chi$. We prove this argument by induction on k.

If $k = 0$, then the argument trivially holds. Assume that for $k \le p$ the argument holds. From Theorem 4.8, by letting $\alpha = 0$ and $\beta = 2$, $\chi \ge \Delta(0, 2)/(2C_{1,2})$. In other words, the number of links that interfere or are interfered by link $\mathbf{L}_{i^{(p)}, j^{(p)}}$ and whose radii are not smaller than $r_{i^{(p)}, j^{(p)}}/2$ is at most $2C_{1,2}\chi$. Thus, there must exist a link $\mathbf{L}_{i^{(p+1)}, j^{(p+1)}}$ such that

1. $\mathbf{L}_{i^{(p+1)}, j^{(p+1)}}$ interferes or is interfered with by $\mathbf{L}_{i^{(p)}, j^{(p)}}$.
2. $r_{i^{(p+1)}, j^{(p+1)}} < r_{i,j}/2^{p+1}$.
3. $\mathbf{L}_{i^{(p+1)}, j^{(p+1)}}$'s color ID is at least $g - 2C_{1,2}(p + 1)\chi$.

This finishes the induction.

Thus, let $k = \lfloor\log(\psi)\rfloor$; link $\mathbf{L}_{i^{\lfloor\log(\psi)\rfloor}, j^{\lfloor\log(\psi)\rfloor}}$ has a color ID no smaller than $g - 2C_{1,2}\lfloor\log(\psi)\rfloor\chi$. This implies that $\mathbf{L}_{i^{\lfloor\log(\psi)\rfloor}, j^{\lfloor\log(\psi)\rfloor}}$ has at least $2C_{1,2}\chi + 1$ adjacent links. Because $r_{i^{\lfloor\log(\psi)\rfloor}, j^{\lfloor\log(\psi)\rfloor}} < r_{i,j}/2^{\lfloor\log(\psi)\rfloor}$ and $r_{p,q} \ge r_{i,j}/2^{\log(\psi)}$, all links that interfere or are interfered with by link $\mathbf{L}_{i^{\lfloor\log(\psi)\rfloor}, j^{\lfloor\log(\psi)\rfloor}}$ have an interference radius of at least $r^{i^{\lfloor\log(\psi)\rfloor}, j^{\lfloor\log(\psi)\rfloor}}/2$. From Lemma 4.7, $\chi \ge \lceil\frac{2C_{1,2}\chi + 1}{2C_{1,2}}\rceil = \chi + 1$, which is a contradiction. Thus, $g \le 2C_{1,2}(\log(\psi) + 1)\chi$. This finishes the proof. ∎

4.4.3 Distributed Algorithm for the fPrIM

From Theorem 4.6, any coloring algorithm that uses $O(\Delta^{\text{in}}(F_G^P))$ colors under the fPrIM has a constant-approximation ratio. In Algorithm 3, we present a centralized method that computes the coloring that needs at most $O(\Delta^{\text{in}}(F_G^P))$ colors. Here, we review the distributed method (Algorithm 6) that bears the similar idea of the centralized method (Algorithm 3).

Regarding the distributed method (Algorithm 6), we have the following theorem:

THEOREM 4.10 *(W. Wang et al., 2006b) Algorithm 6 computes a valid coloring with at most $2\Delta^{\text{in}}(F_G^P) + 1$ colors with $O(m)$ messages, where m is the number of communication links.*

Proof. Notice that for each link $\mathbf{L}_{i,j}$, it uses the smallest color that is not used by any links in $S_{i,j}$. Because the number of incoming links is not smaller than the outgoing links in $S_{i,j}$, link $\mathbf{L}_{i,j}$ is colored with a color that is not greater than $2d_{i,j}^{\text{in}}(F_G^P) + 1$. Thus, Algorithm 6 computes a valid coloring with at most $2\Delta^{\text{in}}(F_G^P) + 1$ colors. Note

---

Algorithm 6 Distributed scheduling for the fPrIM

Input: A communication network $G = (V, E)$.

Output: A valid coloring of all links.

1: Assign each communication link a label WHITE.
2: The header of each communication link $\mathbf{L}_{i,j}$ collects all incoming links and outgoing links, denoted by $M_{i,j}^{in}$ and $M_{i,j}^{out}$.
3: **while** link $\mathbf{L}_{i,j}$ is WHITE **do**
4:     Link $\mathbf{L}_{i,j}$ monitors the channel.
5:     If some link e in $M_{i,j}^{in} \bigcup M_{i,j}^{out}$ announces that it becomes GRAY with time stamp k, link $\mathbf{L}_{i,j}$ locally stores the label of link e as GRAY and the time stamp k.
6:     **if** the number of WHITE links in $M_{i,j}^{in}$ is not smaller than the number of WHITE links in $M_{i,j}^{out}$, **then**
7:         Link $\mathbf{L}_{i,j}$ competes for the channel.
8:         **if** Link $\mathbf{L}_{i,j}$ obtains the channel, **then**
9:             Link $\mathbf{L}_{i,j}$ labels itself GRAY with a time stamp $t + 1$, where t is the maximum time stamp of all GRAY links stored locally. Here, $t = 0$ if no GRAY links are stored. Link $\mathbf{L}_{i,j}$ sends to all adjacent links in F_G^P the message that $\mathbf{L}_{i,j}$ becomes GRAY with the time stamp $t + 1$. Link $\mathbf{L}_{i,j}$ makes a list of links $S_{i,j}$ composed of the current WHITE links in $M_{i,j}^{in} \bigcup M_{i,j}^{out}$.
10:        **end if**
11:    **end if**
12: **end while**
13: **while** there exists some links in $S_{i,j}$ not colored **do**
14:    Link $\mathbf{L}_{i,j}$ listens to the announcement. If a link e' in $S_{i,j}$ announces its color, then link $\mathbf{L}_{i,j}$ locally updates the status of e' as colored together with the color of e'.
15: **end while**
16: Link $\mathbf{L}_{i,j}$ colors itself using the smallest color available that will not produce any conflict with links in $S_{i,j}$. It then sends to all adjacent links in F_G^P without a color the message about its current color assigned.

---

that each link $\mathbf{L}_{i,j}$ announces only twice in the distributed scheduling algorithm: when it becomes GRAY and when it is colored. Thus, the overall message complexity is $O(m)$. ∎

4.5   Weighted Coloring and Schedulable Flows

4.5.1   Scheduling with Traffic Load

In a TDMA system, the minimization of the number of colors is closely related to the maximization of the network throughput. One intrinsic assumption behind the idea of coloring is that each communication link has the same packet arrive rate; i.e., the number of traffic flows that need to go through each communication link is the same. However,

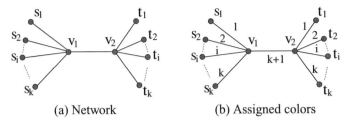

(a) Network   (b) Assigned colors

Figure 4.6 Simple example: Unweighted coloring is inefficient. © 2006, ACM.

this is not likely to be true, and it is possible that some communication links carry more traffic than others.

Consider a simple example of a multihop wireless network composed of k source and destination pairs (s_i, t_i), as shown in Figure 4.6(a). For simplicity of our presentation, we assume that every node in the network can transmit at a bps if it uses all time slots. Observe that we need at least $k + 1$ colors, which we can obtain by assigning color i to links $s_i v_1$ and $v_2 t_i$ and color $k + 1$ to link $v_1 v_2$, as in Figure 4.6(b). This implies that communication link $v_1 v_2$ can transmit once every $k + 1$ time slots. However, the path between each source to destination pair needs to go through link $v_1 v_2$. Thus, link $v_1 v_2$ becomes the bottleneck and the overall network throughput is only $a/k + 1$ bps. For each source–destination pair, its throughput is approximately $\frac{a}{k(k+1)}$ bps, which is inefficient. Thus, we need to generalize the coloring that can take the traffic rate on each communication link into account. In this chapter, we use the *weighted coloring* to capture this, which is defined as follows.

DEFINITION 4.1 *A graph $G = (V, E)$, where V is the set of vertices and E is the set of links. Every link $e_i \in E$ has an integral weight $c_i \geq 0$. A weighted link coloring is an assignment of at least w_i distinct colors to each link e_i such that no two links sharing the same color interfere with each other.*

By introducing the notation of weighted coloring, we can assign different weights to different communication links. For example, given a set of k flow requirements f_i from s_i to t_i, $1 \leq i \leq k$, a certain routing algorithm will determine the routing path for each flow. The weight of a link e is then the total flow passing through e divided by the bandwidth $c(e)$ of link e; i.e., $w_e = \frac{\sum_{f_i : f_i \text{ using } e} f_i}{c(e)}$. Let us see how the weighted coloring can help to improve the throughput by using the example shown in Figure 4.6. By assigning weight 1 to each link $s_i v_1$, $v_2 t_i$ for $1 \leq i \leq k$, and k to $v_1 v_2$, obviously a valid $2k$ coloring can be obtained. It is not difficult to observe that the total throughput is now $a/2$ bps and each communication pair has a throughput of $a/2k$. This increases the throughput obtained from the unweighted coloring by an order of k. In the following discussion, it is shown how to obtain a valid weighted coloring based on the unweighted coloring. It is then shown that Algorithm 7 has a performance guarantee that is not worse than that of the unweighted coloring algorithm $\mathcal{A}$.

Algorithm 7 Weighted coloring algorithm based on unweighted coloring algorithm $\mathcal{A}$

Input: A communication graph $G = (V, E)$ with weight on each link and an unweighted coloring algorithm $\mathcal{A}$.

Output: A valid coloring of the links.

1: Build the conflict graph F_G based on original graph G and the interference model. Assign weight $\mathbf{c}_{i,j}$ to vertex $\mathbf{L}_{i,j} \in F_G$.

2: Construct a new conflict graph F'_G from F_G as follows: For each vertex $\mathbf{L}_{i,j}$ with weight $\mathbf{c}_{i,j}$, we create $\mathbf{c}_{i,j}$ vertices, $\mathbf{L}^1_{i,j}, \mathbf{L}^2_{i,j}, \ldots, \mathbf{L}^{\mathbf{c}_{i,j}}_{i,j}$ and add them to F'_G. Add to graph F'_G the edges connecting $\mathbf{L}^a_{i,j}$, $\mathbf{L}^b_{i,j}$ for $1 \leq a < b \leq \mathbf{c}_{i,j}$. Add to graph F'_G an edge between $\mathbf{L}^a_{i,j}$ and $\mathbf{L}^b_{p,q}$ if and only if there is an edge between $\mathbf{L}_{i,j}$ and $\mathbf{L}_{p,q}$ in graph F_G.

3: Run the unweighted vertex-coloring algorithm $\mathcal{A}$ on F'_G.

4: Assign link $\mathbf{L}_{i,j}$ all the colors that are used by $\mathbf{L}^k_{i,j}$ for $1 \leq k \leq \mathbf{c}_{i,j}$ in F'_G.

THEOREM 4.11 *(W. Wang et al., 2006b) If $\mathcal{A}$ uses at most α times of the optimal colors for unweighted coloring, then Algorithm 7 also needs at most α times of the optimal colors for weighted coloring.*

Proof. Notice that for any valid weighted coloring for F_G, $\mathbf{L}_{i,j}$ is assigned at least $\mathbf{c}_{i,j}$ colors. By assigning each vertex $\mathbf{L}^k_{i,j}$ in F'_G a distinct color that is assigned to $\mathbf{L}_{i,j}$, we obtain a valid unweighted coloring for F'_G. Thus, $\chi(F'_G) \leq \chi(F_G)$. Here, $\chi(F'_G)$ is the minimum number of colors needed for unweighted coloring in F'_G, and $\chi(F_G)$ is the minimum number of colors needed for weighted coloring in F_G. Because $\mathcal{A}$ will return a coloring with at most $\alpha\chi(F'_G)$ colors, Algorithm 7 produces a coloring with at most $\alpha\chi(F'_G) \leq \alpha\chi(F_G)$ colors. This finishes the proof. ∎

The basic idea of Algorithm 7 is to create a clique of size $w_{i,j}$ for each link $\mathbf{L}_{i,j}$ and color the new graph by using unweighted coloring method $\mathcal{A}$. Although this gives a general framework to design weighted coloring, its time complexity could be large if the weight is large. Fortunately, Algorithm 7 could be simplified without much overhead compared with that of the unweighted algorithm: The main idea is to assign colors for one link at once; instead of assigning one time slot to a link $\mathbf{L}_k$, we assign w_k time slots to link $\mathbf{L}_k$ when processing link $\mathbf{L}_k$. As an example, we modify Algorithm 5 to obtain a fast weighted coloring (Algorithm 8). We show that Algorithm 8 has the same performance guarantee as Algorithm 5.

THEOREM 4.12 *(W. Wang et al., 2006b) Algorithm 8 finds a coloring that needs at most $O(\log(\psi) + 1)$ times of optimum.*

Proof. Let $\mathcal{A}_w$ be the coloring algorithm obtained by applying Algorithm 7 based on Algorithm 5. Observe that the coloring of $\mathcal{A}_w$ is nondeterministic; i.e., the output could be different because of the randomization introduced by the different processing time of different nodes. However, it is true that the output of Algorithm 8 is one of the possible outputs of $\mathcal{A}_w$. From Theorem 4.11, any coloring output by $\mathcal{A}_w$ is at most

---

Algorithm 8 Fast distributed weighted coloring algorithm

Input: A communication graph $G = (V, E)$.

Output: A valid coloring of links in the communication graph.

1: Node v_i computes a subset, say H_i, of all communication links containing v_i such that link $\mathbf{L}_{i,j} \in H_i$ if and only if $r_i > r_j$.

2: **while** node v_i failed to obtain the channel **do**

3:   Node v_i monitors the channel and competes for the channel.

4: **end while**

5: **for** each link $\mathbf{L}_{i,j} \in H_i$ **do**

6:   Color link $\mathbf{L}_{i,j}$ with the first-fit $\mathbf{c}_{i,j}$ colors that are not used by any link that interferes or is interfered by $\mathbf{L}_{i,j}$. Here, the assigned colors are not required to be continuous.

7:   Broadcasts the message $\mathsf{Color}(i, j, k)$ to each head of links that conflict with $\mathbf{L}_{i,j}$.

8: **end for**

$O(\log(\psi) + 1)$ times the optimal. Thus, Algorithm 8 computes a coloring that needs at most $O(\log(\psi) + 1)$ times optimal color. ∎

Similarly, we can modify Algorithm 2 and Algorithm 4 to obtain efficient weighted coloring methods with the same time complexities and approximation ratios. The details are omitted here.

4.5.2 Necessary and Sufficient Conditions for Schedulable Flows

Similar to Alicherry *et al.* (2005), Kodialam and Nandagopal (2003), and Kumar *et al.* (2005), we also make the connection with flows on the links of a wireless network G and the link scheduling. We give both a necessary and a sufficient condition on the link flows such that interference-free link scheduling is feasible. Recall that we use $\ell(e)$, $c(e)$ to denote the load and the capacity of link e, respectively. From Lemma 4.3 and Theorem 4.4, Theorem 4.13 follows:

THEOREM 4.13 *(W. Wang et al., 2006b) Under the RTS/CTS model, any link flow ℓ that permits interference-free link scheduling must satisfy the constraint $\frac{\ell(e)}{c(e)} + \sum_{e' \in I_{\geq}(e)} \frac{\ell(e')}{c(e')} \leq 2C_1$. On the other hand, if $\frac{\ell(e)}{c(e)} + \sum_{e' \in I_{\geq}(e)} \frac{\ell(e')}{c(e')} \leq 1$, then any link flow ℓ permits interference-free link scheduling.*

Similarly, under the fPrIM, we have the following theorem:

THEOREM 4.14 *(W. Wang et al., 2006b) Under the fPrIM, any link flow ℓ that permits interference-free link scheduling must satisfy the following constraint: $\frac{\ell(e)}{c(e)} + \sum_{e' \in I^{\text{in}}(e)} \frac{\ell(e')}{c(e')} \leq \lceil \frac{2\pi}{\arcsin \frac{\gamma - 1}{2\gamma}} \rceil$. On the other hand, if $\frac{\ell(e)}{c(e)} + \sum_{e' \in I^{\text{in}}(e)} \frac{\ell(e')}{c(e')} \leq 1$, then any link flow ℓ permits interference-free link scheduling.*

The proofs of the preceding theorems are similar to those of Alicherry *et al.* (2005), Kodialam and Nandagopal (2003), and Kumar *et al.* (2005) for other interference and networking models and are thus omitted here because of space limitation. Similar theorems can be obtained for networks with multiple channels and multiple radios.

4.6　Further Reading

Scheduling has been studied extensively in the past few years because of its application for assigning time slots in TDMA MAC protocols that eliminate collision and guarantee fairness. Scheduling can be reduced to different coloring problems: *edge-coloring* and *vertex-coloring*. Many assignment problems in wireless networks have been shown to be NP-complete, including TDMA broadcast scheduling (Ephremedis and Truong, 1990) and link scheduling (Arikan, 1984; Even *et al.*, 1984). Indeed, for some of these problems, even polynomial algorithms with constant-approximation ratios appear highly unlikely (i.e., unless $P =$ NP) for general graphs. For graphs that model wireless networks, such as DGs (Hale, 1980) and planar point graphs (Sen and Huson, 1996), link-scheduling and coloring problems still remain NP-complete.

Edge-coloring, in which every edge corresponds to a valid communication link, is a natural way to capture the link-scheduling problem. An edge-coloring is *valid* if no two incident edges share the same color. Vizing's theorem (Berge, 1973) states that a valid edge-coloring for an *indirected* graph can be obtained by using at most $\Delta + 1$ colors, where Δ is the maximum node degree in the graph. On the other hand, any edge-coloring needs at least Δ colors. Surprisingly, it is NP-hard to decide whether it is possible to color a general graph by using Δ colors. Any edge-coloring that uses $\Theta(\Delta)$ colors is close to the optimal. Panconesi and Srinivasan (1992) proposed a randomized distributed-edge-coloring method that uses at most $2\Delta + 1$ colors.

To some extent, this captures some transmission restrictions in WANs and WSNs in which no node can receive or send at the same time slot, but it did not address some other interferences, such as secondary interference. When one has a valid edge-coloring, it can be easily mapped to a TDMA scheduling. However, it is possible that two communication links sharing the same color still interfere with each other in a wireless network. To remedy this, Gandham *et al.* (2005) proposed using a two-phase scheduling method: In the first phase, a distributed valid edge-coloring is obtained; in the second phase, a valid scheduling that takes into account the secondary interference is obtained. The scheduling method in the second phase decomposes the conflicting communication links with the same color to several trees and uses two time slots for each tree. In essence, the work of Gandham *et al.* (2005) is based on the PrIM. The overall scheduling in Gandham *et al.* (2005) provides a performance guarantee only when the conflicting links form a tree. Jain *et al.* (2003) proposed a new concept, the *conflict graph*, that captures the interference in WANs and WSNs and is interference-model-independent. Given a communication graph, a CG is a graph in which each vertex corresponds to a (directed) communication link in the original graph. There is an edge between two vertices in the CG if and only if they interfere with each other in the original communication graph. By using the CG concept, interference-aware scheduling is related to the vertex-coloring problem.

Vertex-coloring is one of the most fundamental NP-hard problems in graph theory and has been thoroughly studied. A vertex-coloring is *valid* iff any two adjacent vertices receive different colors. The minimum number that is needed for valid vertex-coloring for a graph G is known as the *chromatic number* $\chi(G)$. It is known that for a general graph, the chromatic number cannot be approximated within $n^{1-\varepsilon}$ for any $\varepsilon > 0$, unless ZPP = NP (Feige and Kilian, 1998). Here, ZPP (zero-error probabilistic polynomial time) is the complexity class of problems for which a probabilistic Tuning Machine exists with the following properties:

1. It always returns the correct YES or NO answer.
2. The running time is unbounded but is polynomial on average for any input.

For vertex-coloring of a general graph G, it was proved by Szekeres and Wilf (1968) that every graph G can be colored using $\delta(G) + 1$ colors. Here, $\delta(G)$ denotes the largest $d > 0$ such that G contains a subgraph H in which each node has a total degree at least d. Then, Hochbaum (1983) presented a method to find the value of $\delta(G)$ and color G using $\delta(G) + 1$ colors in $O(|V| + |E|)$ time. Ramanathan (1999) proposed a unified framework for TDMA-, FDMA-, and CDMA-based multihop wireless networks. They also proposed a time-slot assignment to edges; the number of time slots required is at most $O(\theta)$ times the optimum, where θ is the thickness of a graph (i.e., the minimum number of planar graphs into which the network can be decomposed). Krumke *et al.* (2001) proposed efficient approximation algorithms for the distance-two vertex-coloring problem for various geometric graphs, including (r, s)-civilized graphs, planar graphs, and graphs with bounded genus. Kumar *et al.* (2004) studied packet scheduling under the RTS/CTS interference model and gave polylogarithmic/constant-factor approximation algorithms for various families of DGs and randomized near-optimal approximation algorithms for general graphs.

Several distributed algorithms that use $O(\Delta)$ colors have been proposed in the literature. A $(\Delta + 1)$ coloring can be computed in time $O(\log n + \Delta)$ (Panconesi and Rizzi, 2001) or $O(\Delta \log n)$ (Goldberg *et al.*, 1987). Marco and Pelc (2001) proposed a distributed algorithm that computed an $O(\Delta)$ coloring in time $O(\log n)$. All of the preceding distributed algorithms do not take interference into account and are based on the message-passing model (Peleg, 2000), which implies that the actual time used in a wireless environment could be much larger (Moscibroda and Wattenhofer, 2005a). Recently, Moscibroda and Wattenhofer (2005a) proposed an $O(\Delta)$ distributed-coloring method with time complexity $O(\Delta \log n)$. It is worth pointing out that the coloring in Moscibroda and Wattenhofer (2005a) considered a simple interference model and the time was close to the time needed in practice. However, the coloring method in Moscibroda and Wattenhofer (2005a) is based on the assumption that the wireless ad hoc network can be modeled as a UDG; i.e., their method will return a coloring that guarantees only that any nodes that are adjacent in the UDG will get different colors; nodes that are not adjacent in UDG may get the same color. In addition, they assumed that all nodes have the same transmission range and the interference range of each node is same as its transmission range. This is different from the interference-free scheduling studied in this chapter.

Kodialam and Nandagopal (2004) studied the effect of interference on the achievable rate region in multihop wireless networks. They treated the interference models as linear constraints and solved the flow problem by using a linear program. The same authors (2003) also considered the problem of jointly routing the flows and scheduling transmissions to achieve a given rate vector by using the PrIM. They developed necessary and sufficient conditions for the achievable rate vector. They formulated the problem as a linear programming (LP) problem and implemented primal–dual algorithms for solving the problem. They solved the scheduling problem as a graph edge-coloring problem by using existing greedy algorithms. Then, in 2005, they extended their work to multiradio multichannel WMNs. Again, they provided a relaxation of their model to a LP that gives the necessary conditions for a valid feasible solution. They used this LP solution to assign channels to the links and also to schedule the time slots in which each link and channel is active by using the greedy approach.

Kumar *et al.* (2005) considered the throughput capacity of wireless networks between given source–destination pairs for various interference models. They developed analytical performance-evaluation models and distributed algorithms for routing and scheduling that incorporate fairness, energy, and dilation (path-length) requirements and provide a unified framework for utilizing the network close to its maximum throughput capacity. Alicherry *et al.* (2005) mathematically formulated the joint channel assignment and routing problem in multiradio mesh networks by taking into account the interference constraints, the number of channels in the network, and the number of radios available at each mesh router. They established necessary and sufficient conditions under which an interference-free link communication schedule could be obtained, and they designed a simple greedy algorithm to compute such a schedule. They used a flow transformation technique to design an efficient channel-assignment algorithm, and they showed that their algorithm for the joint channel-assignment, routing, and scheduling problem is a constant-factor-approximation algorithm. Notice that the studied network in Alicherry *et al.* (2005) is restricted to be a UDG; i.e., the uniform interference range is assumed to be a fixed multiple of the uniform communication range. Recently, L. Chen *et al.* (2005, 2006) also studied the cross-layer optimization of congestion control and routing together with the scheduling problem under both primary and secondary interference.

4.7   Conclusion and Remarks

In this chapter, we considered the problem of obtaining good interference-aware link scheduling for a wireless network to maximize the throughput of the network. We used link coloring to resolve this and assumed a general model for wireless networks; i.e., nodes could have different transmission ranges and different interference ranges, and a link uv may not exist even if $\|uv\|$ is less than the transmission range of node u. We studied both centralized algorithms and efficient distributed algorithms that use time slots within a constant factor of the optimum. It was also pointed out that simple link coloring does not imply a good throughput when each link has a weight; we then studied a general weighted link-coloring problem by reviewing efficient algorithms to obtain TDMA link scheduling with proven performances.

There are still a number of challenging questions left for future research. The first question is how to efficiently collect the information about the interfering links of a given link in a wireless networking environment. This is not an issue in the previous studies because they assume a UDG model and assume the same interference range for all nodes. The second question is how to improve the overall time complexity of the distributed algorithms studied in this chapter. The results presented by Moscibroda and Wattenhofer (2005a) may give some insight on this, but it is not obvious because the model used in this chapter is more complicated than the model used in Moscibroda and Wattenhofer (2005a). We suspect the existence of polylogarithmic-time-distributed algorithms for problems studied in this chapter under the unstructured environment (Moscibroda and Wattenhofer, 2005a). The third issue to study is link scheduling in an asynchronized environment. We believe that these methods still apply with small modifications. The last but not least question to study is link scheduling in a dynamic environment where the traffic load on links could have some small changes.

Problems

4.1   Compare and contrast the advantages and disadvantages of TDMA link scheduling and CSMA/CA MAC. What are the main technique difficulties in implementing TDMA link scheduling for wireless ad hoc networks? How are such technique difficulties addressed in the literature?

4.2   It has been proved that for the general graph-coloring problem, we cannot design a polynomial-time algorithm to find a coloring that is within $|V|^{1/7-\varepsilon}$ for any $\varepsilon > 0$ if some complexity hierarchy holds. See "Free Bits, PCPs and Non-approximability—Towards Tight Results," SIAM *J. Comput* **27**, 804–915, by M. Bellare, O. Goldreich, and M. Sudan (1998), for details. In this chapter, we essentially study several algorithms that can find a good coloring within a constant factor of the optimum. Does this form a counterexample to the results by Bellare *et al.*? What special properties of wireless network models used here allow us to design constant-approximation algorithms for coloring?

4.3   Assume that there is a disk $\mathcal{D}$ centered at the origin with radius 2. How many nonoverlapping disks can you place such that each of them has radius 1 and is centered inside the disk $\mathcal{D}$? How many such disks can you place if the disk $\mathcal{D}$ has a radius $k > 1$?

4.4   Assume that there is a set of n disks with radius $1/2$ contained inside a two-dimensional square region Ω. Design a polynomial-time algorithm with approximation ratio 3 to find a maximum independent set of disks if you know the geometry information of all disk centers. Here, two disks are independent if they do not overlap. Design a polynomial-time algorithm to find a maximum independent set of this set of disks if the area of region Ω is bounded by a constant C. Assume that the geometry positions of all disk centers are given. Now we draw a dimensional grid, called the (s, t)-grid, formed by the horizontal lines $y = s + ik$, for $i \in Z$ and vertical lines: $y = t + jk$, for $j \in Z$. Here, $k > 2$ is a fixed integer. Given the (s, t)-grid, we remove all disks that intersect any line of the grid and then compute a MIS of the remaining disks. Prove that there is a choice of (s, t)-grids such that the MIS just computed is at least $1 - 1/k$ times the optimum MIS for the original problem.

4.5 A first-fit heuristic has been shown to be useful in designing algorithms for a number of coloring questions. Consider a set of disks $\mathcal{D} = \{D(v_i, r_i) \mid v_i \in V\}$ with centers at V, where $D(v_i, r_i)$ is the disk centered at node v_i with radius r_i. A disk graph $G = (V, E)$, defined over V, has an edge (v_i, v_j) if and only if two disks $D(v_i, r_i)$ and $D(v_j, r_j)$ overlap. The question is to find a minimum coloring of the graph G. Prove that the following first-fit approach gives a five-approximation: (1) sort the vertices in nonincreasing order of the degree; and (2) process the vertices in such an order and assign the current node the smallest color that is not used by its neighbors in G. This is often called the *smallest-degree-last* version of the first-fit approach.

4.6 Continue from the previous question; assume that the disks are coming one by one and we have to assign a color to a disk when it arrives. This is often called *on-line coloring of DGs*. Assume that we always use the first-fit color approach: Assign the disk the smallest color that is not used by its neighbors in G. Prove that there is a sequence of arrival of the disks such that the first-fit heuristic produces an optimum coloring. Prove that the first-fit heuristic will produce a coloring that is at most $O(\min(\log n, \log \varrho))$ times of the optimum, where ϱ is the ratio of the largest radius over the smallest radius. Construct a DG of n disks and a sequence of arrival of disks such that the approximation ratio of the first-fit approach is large. Can you construct an example such that the first-fit approach will have an approximation ratio $\Omega(\min(\log n, \log \varrho))$?

4.7 Assume that wireless nodes have different interference ranges. Assume that each node v has interference range $R_I(v)$. Consider a specific node v and a set of wireless nodes U such that for each $u \in U$, $\|uv\| \le R_I(v)$. Let $R = \min_{u \in U} R_I(u)$ and assume that $R_I(v) = 1$. Give an asymptotic upper bound on the number of simultaneous transmissions by nodes in U such that their interference ranges do not overlap.

4.8 Design network examples in which the methods processing links in the order of decreasing length, developed by Kumar *et al.* (2005), will not work under the model used in this chapter.

5 Spectrum Channel Assignment

5.1 Introduction

In Chapter 4, we basically studied how to assign time slots to communication links in quasi-static networks (e.g., WMNs and WSNs) such that the scheduled transmissions are interference-free and the scheduling period is minimized (thus maximizing the throughput). We assumed that there is only one spectrum used by all links in the network. In this chapter, we mainly focus on multichannel networks when there are multiple spectrums available for wireless terminals in the network. The wireless devices may have multiple wireless NICs or just a single wireless NIC. Because close-by mesh routers compete for certain wireless channels, the capacity of a WMN will be increased tremendously when the single channel is extended to multiradio, multichannel, and multihop. For example, if two nodes v_i and v_j could communicate with each other by channel f_1, and both nodes have at least one more available NIC that could operate on another channel f_2, if f_2 is also available for both nodes, v_i and v_j could use both f_1 and f_2 to communicate simultaneously. When such cases are applied to more wireless nodes, the throughput of the wireless network will be increased tremendously.

On the other hand, with recent fast-growing spectrum-based services and devices, the remaining spectrum available for future wireless services is being exhausted, known as the *spectrum-scarcity problem*. This is a serious problem in several aspects: Cell phone companies have to spend billions of dollars to bid for a 20–30-MHz 3G spectrum; it is difficult for new innovative technologies to enter the market competitively; it is difficult to provide a certain QoS for systems operated in unlicensed spectrums. The current fixed-spectrum-allocation scheme leads to significant spectrum *white spaces* (a spectrum that is not used at all for some period of time), in which many allocated spectrum blocks are used only in certain geographical areas and/or in brief periods of time. For example, the Shared Spectrum Company performed spectrum occupancy measurements, determining the spectrum occupancy in each band (30–3000 MHz) at multiple locations from January 2004 until August 2005, for the National Science Foundation. The company found that (1) the average occupancy over all of the locations is 5.2%; (2) the maximum total spectrum occupancy is 13.1% (New York City); and (3) the minimum total spectrum occupancy is 1% (National Radio Astronomy Observatory). This indicates that the spectrum usages vary dramatically in time, geographic locations, and frequency; and there is at least 87% of white space, which was measured in New York City. Thus, a huge amount of precious spectra (below 5 GHz), perfect for wireless communications that are worth billions of dollars, sit there silently.

It is widely agreed that opportunistic or dynamic spectrum usage will significantly mitigate the spectrum scarcity. Recently, more and more researchers have been concentrating on how to reuse the licensed spectrums for secondary users when the licensed spectrum is idle or when secondary users use licensed spectrums within some special interference temperature such that the secondary users interfere with the primary users in the range of interference temperature. However, series of technologies (topology of wireless network, routing, link scheduling, and so on) become more complex and difficult because of the properties of wireless networks (nodes), such as lower power, interferences, and some other physical barriers; using multiple channels and radios could mitigate such interference and improve the throughput to some extent, but it could not eliminate interference completely. All in all, if we do not handle them carefully, the performance improvement (capacity) brought about by multichannel and multiradio will be influenced greatly or even degrade.

In this chapter, we address the problem of channel assignment in wireless networks with multiradio and multichannels under some network models and interference models. We assume that each wireless node u has a given $\kappa(u)$ number of NICs and a subset of channels on which this node may operate. We need to choose a channel for each NIC of every node [assigning $\kappa(u)$ channels to node u] such that the resultant network has certain properties such as being connected or two interfering nodes are assigned different channels. Two different network models will be studied in this chapter. The first networking model assumes that the wireless routers will serve as APs only for some clients and will not form a network themselves. For this networking model, clearly it is sufficient to require that the spectra assigned to interfering wireless routers be different. When this is impossible, we may want to minimize the total interference produced from the assignment. The second networking model assumes that the wireless routers will form an ad hoc network (e.g., a mesh network) and then need to communicate with each other. One reasonable objective here could then be assigning spectrum channels to wireless routers such that the resultant network is connected. Here, two nodes u and v can form a communication link (u, v) if they share a common assigned spectrum channel.

We will prove that it is NP-hard to decide if we can assign spectrum channels to nodes such that the network is connected while the total assigned channels to every node do not exceed the number of NICs that node has. We then study several novel methods for assigning channels. One of the methods is based on node clustering: We first partition nodes into a number of clusters and connect these clusters via a backbone; we then find a channel assignment for nodes on the backbone such that the backbone is connected; and last, for each cluster, we perform a local channel assignment such that all nodes in a cluster will form a connected subnetwork. One of the advantages of using the cluster-based channel assignment is that the method can adapt to the dynamic channel availability. When a node's channel availability changes, we need to update the channel assignment for only one cluster if the node is not on the backbone. If it is on the backbone, we need additionally to update the channel assignment for only backbone nodes, who number far less than the total number of wireless nodes in the network. Extensive simulations showed that the backbone-based channel assignment method will induce a connected network without violating the NIC constraints at all nodes with high probability.

We also study how to assign channels to obtain a connected network such that the fewer used channels are as few as possible. By using fewer channels, we leave more freedom for future channel assignments when new nodes join the network. We also study how to assign channels such that the resultant network supports a large number of simultaneous transmissions. This, in turn, will intuitively increase the network throughput of the resulted network.

5.2   Network System Model

First, the network model used throughout this chapter is presented. Consider a set $V = \{v_1, v_2, \cdots, v_n\}$ of n wireless nodes distributed in a plane. We first assume that each node v_i has a transmission range $R_T(i)$ such that only those nodes within distance $R_T(i)$ from v_i could receive the signal from v_i correctly. Here, we emphasize that $\|v_j - v_i\| \leq R_T(i)$ is a necessary but not a sufficient condition for two nodes v_i and v_j to compose a physical communication link; e.g., such a link may not exist because of physical barriers. We also assume that each terminal v_i also has an interference range $R_I(i)$ such that every unintended receiver would be interfered by the signal from v_i when it is using the same channel as v_i simultaneously. Typically, $R_T(i) \leq R_I(i) \leq cR_T(i)$, where c is a constant factor and $2 \leq c \leq 4$ in practice. We use $G_0 = (V, E)$ to denote the complete communication graph, where E is the set of all communication links.

In addition, every wireless terminal (node) v_i has one or more NIC, and each could operate on one or more spectrum radios. Here, we use $I(v_i)$ to denote the number of NICs of a node v_i. Let $F = \{f_1, f_2, \cdots, f_k\}$ denote the set of K orthogonal (could be nonorthogonal in certain situations) channels that could be used by all nodes V. We generally assume that each NIC could operate on only a subset of F channels because of the an ith hardware constraints. Let $F(v, i)$ be the set of channels that can be operated by an ith NIC for node v, where $1 \leq i \leq I(v)$. Let $F(v)$ be the set of channels that can be used by node v via any of its NICs; i.e., $F(v) = \bigcup_{1 \leq i \leq I(v)} F(v, i)$. For each link $e = (u, v) \in E$, $F(e)$ denotes the set of channels that could be used by e; that is, $F(e) = F(u) \bigcap F(v)$. When $F(e)$ is empty, it means that the two end nodes of e cannot communicate directly over e under the radio and channel constraints. For simplicity, we abuse the notation a little bit; we use $G = (V, E)$ to denote the communication graph where E now denotes all the links $e = (u, v)$ such that u and v can communicate directly with each other under the radio and channel constraints of these two nodes.

5.3   List-Coloring for Access Networks

In this section, we focus on the opportunistic exploration of white space by users other than the primary licensed ones on a noninterfering or leasing basis. Such usage is being enabled by regulatory policy initiatives and radio technology advances. Opportunistic spectrum sharing is enabled by software-defined radio or cognitive radio technologies. Such technology advances provide the capability for a radio device to sense and operate on a wide range of frequencies using appropriate communication mechanisms and thus enable dynamic and more intense spectrum reuse in space, time, and frequency dimensions.

First, a model is introduced for channel availability observed by the secondary users. We abstract each secondary network topology into a graph, where vertices represent wireless users such as wireless routers, wireless lines, WLANs, or cells, and edges represent interferences between vertexes. In particular, if two vertices are connected by an edge in the graph, we assume that these two wireless components cannot use the same spectrum for communication simultaneously because of possible wireless interferences. In addition, we associate with each vertex a set that represents the available spectra at this location. Because of the differences in the geographical location of each vertex, the sets of spectra of different vertices may be different. Notice that in practice, a wireless component may observe time-varying channel availability because of the traffic load variation of the primary users. For simplicity, in this chapter we assume that the list of available spectra for a vertex remains the same for a reasonable duration of time. When the availability of spectra changes, we should either apply certain dynamic algorithms to update the spectra assignments or reassign the spectra to all nodes. Observe that the CG model used in this section is different from the CG graph model defined in Chapter 2. For CG modeling of the interferences among communication links, the vertices of the CG correspond to the set of wireless communication links; whereas, for the CG model here, the vertices correspond to the set of wireless terminals.

To characterize the interferences among wireless components (or terminals), we assume that each vertex in the CG is associated with an interference range, which specifies that all other vertices that are within the interference range to this vertex cannot be assigned the same spectrum as this vertex. For certain methods studied here, we may require knowledge of the geographical position of each wireless vertex. The objective of the spectra assignment for this model is to maximize the total number of spectra assigned to all vertices or the total bandwidth capacity of all assigned spectra.

5.3.1   Centralized Method for Maximizing Capacity

We first present a mathematical formulation of the optimization problem as follows. There is a graph $G = (V, E)$, where the vertices correspond to all wireless components and each edge uv denotes that u and v cannot be assigned the same channel. For each vertex v, there is a list $F(v) \subseteq \mathcal{F}$ of channels that can be assigned to v. Each vertex v is also associated with a number $\kappa(v)$ that typically denotes the maximum number of channels it can operate. In certain scenarios, this number $\kappa(v)$ could be infinity. For each channel f_i, let $c(f_i)$ denote the average capacity of this channel. The solution is to assign each vertex v a subset $\mathcal{A}(v)$ of at most κ channels from $F(v)$ such that $\mathcal{A}(u) \cap \mathcal{A}(v) = \emptyset$ for every edge uv in G. The objective is to maximize

$$\sum_v \sum_{f \in \mathcal{A}(v)} c(f).$$

When we want to maximize the total number of channels instead, we can simply set $c(f) = 1$ for all channels. This problem is a generalization of *list-coloring* in the literature, which requires that each vertex be properly colored with one color from its list. A graph is *k-choosable*, or *k-list-colorable*, if it has a list coloring for *every* collection of lists of k colors. The choosability, or list colorability, $\chi_l(G)$ of a graph G is the least number k such that G is k-choosable.

The problem of list-coloring a graph generalizes the vertex-coloring problem and hence is NP-hard. Given a graph of maximum degree Δ, one straightforward method of coloring it with $\Delta + 1$ colors is to consider the vertices one at a time and assign the considered vertex a color that is also different from the colors already assigned to their neighbors in the CG. Obviously, this simple coloring strategy will use at most $\Delta + 1$ colors to color all vertices. This simple strategy can be adapted to list-coloring as follows. For each vertex, we pick a color from its availability list that is different from the colors already assigned to its neighbors in the CG. Consequently, if the availability list $F(v)$ is of a size at least $d(v) + 1$ for every vertex v, we can always list-color the graph. Here, $d(v)$ is the degree of vertex v in the CG, G.

Observe that when all vertices have the same list of channel availability, the list-coloring of a graph is similar to what we studied in Chapter 4: We need to assign each vertex v $\kappa(v)$ in the CG (in which a vertex corresponds to a wireless terminal) a certain number of different spectra. Recall that for the problems studied in Chapter 4, we assign each vertex **L** in the CG (in which a vertex corresponds to a link in the original communication graph) a certain number of time slots. By replacing time slots with channels, we can apply those algorithms developed in Chapter 4 to solve the channel-assignment problems here. When the interference ranges of all wireless terminals are modeled by a perfect disk, it is not difficult to show that those algorithms will also give a theoretical performance guarantee: The number of used different channels [i.e., the size of $\bigcup_{v \in V} \mathcal{A}(v)$] is within a constant factor of the optimum.

For a general graph G, Marathe *et al.* (1995) use an algorithm by Hochbaum (1983) for coloring the graph using at most $\delta(G) + 1$ colors. Here, $\delta(G)$ is the maximum integer such that G contains a subgraph H in which every node has a degree of at least $\delta(G)$. Hochbaum provided an efficient algorithm to find such a $\delta(G)$ in polynomial-time. Marathe *et al.* (1995) proved that their method has a competitive ratio of 3 for the input of UDGs.

There are also a number of works devoted to on-line channel assignment that do not perform the reassignment of channels. In other words, once a channel is assigned to a node, the node will keep the channel. For the input of UDGs, Marathe *et al.* (1995) presented an on-line method with a competitive ratio of 6 for graph multicoloring (in which a node may require multiple colors). In other words, the total number of colors used, when the coloring requests of nodes come one by one, is no more than six times the minimum number of colors needed by *any* algorithm that knows all requests in advance. The basic idea of their method is to use the fact that for any node in the UGD, there are at most five independent neighbors. Then, the coloring method is just a simple first-fit greedy approach: assigning the current requesting node all the channels that are first-fits; i.e., the minimum numbered channels that are not used by its neighbors yet.

So far, the discussed algorithms will minimize the total number of channels. Observe that for wireless channel assignment, in certain applications, minimizing the number of used channels is not a concern; instead, maximizing the total channel capacity $\sum_v \sum_{f \in \mathcal{A}(v)} c(f)$ is preferred. We first show that maximizing the total channel capacity is an NP-complete problem. The proof is based on induction from the MWIS problem, which is an NP-complete problem. The MWIS problem is defined as this: Given a graph G and a weight for each vertex, we need to find an IS such that the total weight of vertices in this set is the maximum among all ISs. Here, a set of vertices is an IS if there is no edge in G connecting any two vertices in the set.

THEOREM 5.1 *Given an IG, G, and the list of available channels for each vertex, it is NP-complete to find a channel assignment that will maximize the total channel capacity of all assigned channels.*

Proof. We prove this by showing that if we solve this problem in polynomial time, then we can solve the MWIS problem in polynomial time; i.e., we reduce the maximum independent problem to this problem. We consider a special case of the channel-assignment problem in which $\kappa(v) = 1$ for every vertex v. Given a graph $G = (V, E)$ and the available channels $F(v)$ for each vertex $v \in V$, we construct another graph, H, as follows. For each vertex $v \in V$, we make $F(v)$ copies of v as $v_1, v_2, \ldots, v_{F(v)}$. The weight of copied node v_i is the capacity of the ith channel in $F(v)$. Here, we assume that the channels $F(v)$ comprise a sorted set. We add all such vertices to the graph H and all edges $v_i v_j$. In other words, each vertex v in G is mapped to a completed graph of $F(v)$ vertices. For two vertices u and v from G, we add an edge $u_j v_i$ to H only if $uv \in E$ and the jth channel in $F(u)$ is the same as the ith channel in $F(v)$. Here, our implicit assumption is that for wireless networks, different channels will *not* cause any interferences if they are used simultaneously.

We then show that graph G has an optimum channel assignment that has total channel capacity C if and only if the MWIS of H has weight C.

First, if G has a channel assignment with capacity C, we construct an IS whose total weight is C, as follows. For a vertex v from G, we choose vertex v_i from H to the IS if vertex v is assigned the ith channel in $F(v)$. Because $\kappa(v) = 1$, we can select only one vertex from among all vertices $v_1, v_2, \ldots, v_{F(v)}$. In addition, because the channel assignment is valid, for any possible interfering neighboring node u of the node v, the assigned channel to node u must be different from the channel assigned to v. This implies that if node v uses the ith channel in $F(v)$ and node u uses the jth channel in $F(u)$, these two channels must be different; i.e., there is no link $v_i u_j$ in H. Thus, the previously defined set is indeed an IS. Clearly, the total weight of such set is also C.

On the other hand, if H has an IS with total weight C, we can also easily construct a channel assignment whose total capacity is C also and the number of channels assigned to each vertex is 1.

Consequently, finding a channel assignment to maximize the total capacity is an NP-hard problem because all the preceding conversions can be done in polynomial time. The problem clearly is in NP because we can easily verify whether a given assignment has at least a certain capacity in polynomial time. This proves that the capacity-maximizing channel-assignment problem is NP-complete. ∎

Observe that given a general graph H, it is known in the literature that when the interference graph G is a general graph, it will be difficult to design an efficient algorithm for a MIS with a good approximation ratio [no better than $\Theta(n^{1-\epsilon})$ when some complexity hierarchy NP $\neq$ ZPP holds]. For some special graphs that can represent the conflict relations of wireless ad hoc networks, X.-Y. Li and Wang (2002, 2006) recently presented a PTAS for approximating the MWIS based on the shifting strategy proposed by Hochbaum (1983). Here, an algorithm is a PTAS if, given any sufficiently small $\epsilon > 0$, the algorithm can find a solution that is at least $1 - \epsilon$ times of the optimum solution in polynomial time of the size of the input but not necessarily in the polynomial time of $1/\epsilon$. It is called a FPTAS (a fully polynomial-time approximation scheme) if the time complexity is also a polynomial function of $1/\epsilon$. They considered a network in which different nodes may have different transmission radii and different interference radii. The graph model DG, studied in X.-Y. Li and Wang (2006) can be used to model the interference among different wireless terminals here. Two nodes u and v form an edge in the DG model if the interference region $I(u, v)$ {defined as $[D(u, r_u) \cap D(v, t_u)] \bigcup [D(u, r_u) \cap D(v, t_u)]$} is not empty. Here, $D(u, t_u)$ denotes the transmission disk centered at node u with radius t_u, and $D(u, r_u)$ denotes the interference disk of node u with radius $r_u \geq t_u$.

Thus, when every wireless terminal only needs one channel, we can use the approximated MWIS set of graph H constructed in the proof of Theorem 5.1. It will give a solution whose total achieved bandwidth capacity is at least $1 - \epsilon$ of that of the optimum channel assignment for any $\epsilon > 0$. When some terminals may require multiple channels, we can modify the construction of graph H as follows. For a node v, we assume that it needs $\kappa(v)$ channels assigned from its list $F(v)$ of available channels. Then, we first define a graph G' from G by replacing every node v in G with a completed graph of $\kappa(v)$ virtual vertices. After this, we follow the same procedure as in the proof of Theorem 5.1 to construct graph H from the graph G'. It is not difficult to prove that an IS in H indeed implies a valid channel assignment in G with the same capacity, and a valid channel assignment in G also implies an independent set in H with the same capacity. Thus, we have the following theorem:

THEOREM 5.2 *Given a network G and an arbitrary $\varepsilon > 0$, there is a polynomial-time algorithm that can find a channel assignment by using the available channels for every vertex such that the total capacity of the assignment is at least $1/1 + \epsilon > 1 - \epsilon$ times of the optimum. The algorithm runs in time $\frac{1}{\varepsilon^2} N^{O(\frac{1}{\varepsilon^4})}$, where $N = \sum_{v \in V} \kappa(v) \cdot |F(v)|$.*

Notice that the algorithm developed in X.-Y. Li and Wang (2006) essentially is dynamic programming and is centralized. It remains to future research to design a communication-efficient distributed method that can find a channel assignment with a PTAS or just a constant-approximation ratio. Here, we say a distributed method is communication efficient if the total communication cost is $\Theta(n)$ [if this is not achievable, we hope that the communication cost is at most $O(n \log n)$].

5.3.2   Distributed Coloring for Minimizing Channels

When nodes do not have any initial information about the network topology and also do not have an assigned channel for communication, these kinds of networks are called *unstructured* networks (Kuhn *et al.*, 2004b; Moscibroda and Wattenhofer, 2005a). This unstructured-network model captures the characteristics of newly deployed ad hoc and sensor networks; i.e., asynchronous wake-up, no collision detection, and scarce knowledge about the network topology. When a network is modeled as a graph with a *bounded independence*, Moscibroda and Wattenhofer (2005a) presented an algorithm that produces a correct coloring of the nodes with $O(\Delta)$ colors in time $O(\Delta \log n)$ with high probability, where n and Δ are the number of nodes in the network and the maximum nodal degree, respectively. In their algorithm, the number of locally used colors depends on only the local node density. Notice that a graph is called a *bounded-independence* network graph as long as the maximal number of mutually independent nodes in the two-hop neighborhood of any node is bounded by some arbitrary constant (denoted as κ_2). This bounded-independence model generalizes the frequently studied models for wireless networks, such as the UDG. Unlike the UDG or other explicit geometric graph models, however, this bounded-independence model can capture obstacles as well as physical signal-propagation aspects such as fading, reflection, or shielding.

In the rest of the section we review the distributed-coloring method presented by Moscibroda and Wattenhofer (2005a) for unstructured networks. Their algorithm partitions the nodes into different states. Before the algorithm is initiated at a node, the node is in the sleep state (denoted by $\mathcal{Z}$). On waking up, a node has no information as to whether it is the first to wake up or whether other nodes have been running the algorithm for a long time already. A node is called *sleeping* before its wake-up and *awake* thereafter. Only awake nodes can send or receive messages, and sleeping nodes are *not* woken up by incoming messages. The two extreme cases of this asynchronous wake-up model are as follows. First, all nodes start synchronously at the same time, or only one of the sleeping nodes wakes up while all others remain sleeping for a long time. For unstructured networks, nodes are unaware which (if any) of the two extreme cases holds. When a node is wake-up, it is further divided into three different states:

1. $\mathcal{A}_i$: A node is in state $\mathcal{A}_i$ if the node is verifying color i; i.e., it is trying to decide on color i.
2. $\mathcal{R}$: A node is in state $\mathcal{R}$ if it is requesting an *intracluster* color from their leader.
3. $\mathcal{C}_i$: A node is in state $\mathcal{C}_i$ if it has already made an irrevocable decision of using color i.

In their algorithm, they first partition the nodes into clusters and elect a node within a cluster as the leader of the cluster. The leader will be colored color 0. In other words, the node in state C_0 is called the *leader*. For each possible state, they provide an algorithm. On wake-up, a node starts at state $\mathcal{A}_0$, without having any knowledge about whether some of its neighbors have already started the algorithm beforehand.

The basic idea of their algorithm can be described briefly as follows. In the first stage, the nodes will elect a set of mutually independent *leaders* (nodes in state C_0) among themselves. These leaders are typically a maximal independent set for all nodes awake so far. Then, these leaders will assign themselves color 0, which clearly will not cause any color conflict. Each nonleader node will associate itself with a leader within its own neighborhood. A leader and all nodes associated with it will form a cluster. Then, the leader of a cluster will assign a unique *intracluster* color tc_v to every node v within the cluster. Notice that these colors assigned by leaders may not be a valid coloring because two neighboring nodes v_1 and v_2 in two different clusters may be assigned the same intracluster color by their corresponding leader. However, on the other hand, because the leaders are mutually independent and the graph is a bounded independence graph, for any nonleader node v_1, there is at most a constant κ_2 number of leaders that can assign intracluster color for *any* of its neighboring node v_2. In other words, for such nodes v_1, the number of neighboring nodes v_2 that are assigned the same intracluster color as v_1 is bounded by κ_2. Then, a verification procedure is proposed to resolve such possible conflicts.

The major difficulty of the preceding process stems from the fact that the nodes will wake up asynchronously and the clocks of different nodes are not synchronized. Therefore, the different phases (e.g., verification, requesting intracluster color) of different nodes may be arbitrarily intertwined or shifted in time. For example, while some nodes may still compete for becoming leader, their neighbors may already be much more advanced in their coloring process. Moreover, messages may be lost because of collisions at any time. To ensure both correctness and fast progress with high probability in this harsh environment, they use a technique of counters c_v for each node v (which are increased by 1 every time slot), critical ranges, and local competitor lists. Intuitively, this counter represents $v's$ progress toward deciding on color i and v selects i as soon as c_v reaches a certain threshold. To prevent two neighboring nodes from selecting the same color, their algorithm will make sure that as soon as a node v selects its color, all neighbors of v can be notified before their counter also reaches the threshold (thus making coloring decisions). It is not difficult to observe that there should be a large time interval between two neighboring nodes' counters reaching their thresholds. A simple idea is to let every node transmit its current counter with a certain probability, and a node receiving a higher counter will reset its counter to 0. Unfortunately, this simple method cannot avoid the chains of cascading resets, which may lead to a much longer time for nodes to get their colors. Notice that in the worst case, every "round," there may only have one node that can make decisions because of the possible cascading effects of resets. The idea by Moscibroda and Wattenhofer (2005a) is as follows. On receiving a message from a neighboring node, a node resets its counter only if its counter is within a critical range of the received counter; i.e., only nodes that may likely compete with the sending node for choosing colors. A careful selection of *critical range* will balance the requirement for more parallelism and the requirement to avoid cascading resets. They further refine this by using a local *competitor list* (containing the current counter values of neighboring nodes) to prevent nodes from resetting their counters to a value within the critical range of neighboring nodes and to ensure that all counters must remain relatively close to the verification threshold even after a reset.

DEFINITION 5.1 *The time complexity T_v of a node v running a distributed algorithm is defined as the number of time slots between the nodeòs waking up and the time it has made its irrevocable final decision (e.g., on its color for a distributed-coloring method). The algorithm's time complexity is the maximum number T_v over all nodes in the network.*

Notice that in the preceding definition and also the algorithm that we will study later, for the sake of simplicity, it is assumed that time is divided into discrete time slots that are synchronized between all nodes. This assumption is used merely for the purpose of facilitating the analysis; i.e., their algorithm does not rely on this assumption in any way, as long as the nodes' internal clock runs at the same speed.

Algorithm 9 Coloring algorithm: node v in state $\mathcal{A}_i$

On entering state $\mathcal{A}_i$ (when waking up, a node is initially in state $\mathcal{A}_0$)

1: $P_v \leftarrow \emptyset$;   {* v is passive *}
2: $\zeta_i \leftarrow 1$ if $i = 0$; otherwise $\zeta_i \leftarrow \Delta$.
3: $\mathcal{A}_{\text{succ}} \leftarrow \mathcal{R}$ if $i = 0$; otherwise $\mathcal{A}_{succ} \leftarrow \mathcal{A}_{i+1}$.
4: **for** $\lceil \alpha \Delta \log n \rceil$ time slots **do**
5:      **for each** $w \in P_v$ **do** $d_v(w) \leftarrow d_v(w) + 1$;
6:      **if** message $M_{\mathcal{A}}^i(w, c_w)$ received **then** $P_v \leftarrow P_v \cup \{w\}$; $d_w \leftarrow c_w$ **end if**
7:      **if** message $M_C^i(w)$ received **then** $state \leftarrow \mathcal{A}_{\text{succ}}$ $L(v) \leftarrow w$ **end if**
8: **end for**
9: $c_v \leftarrow \xi(P_v)$, where $\xi(P_v)$ is the maximum value such that $\xi(P_v) \leq 0$ and $\xi(P_v) \notin [c]_w - [\gamma \zeta_i \log n], c_w + [\gamma \zeta_i \log n]$ for **each** $w \in P_v$
10: **while** $state = \mathcal{A}_i$ **do**
11:     $c_v \leftarrow c_v + 1$;
12:     **for each** $w \in P_v$ **do** $d_v(w) \leftarrow d_v(w) + 1$;
13:     **if** $c_v \geq \lceil \sigma \Delta \log n \rceil$ **then** $state \leftarrow C_i$ **end if**
14:     **transmit** message $M_{\mathcal{A}}^i(v, c_v)$ with probability $\frac{1}{\kappa_2 \Delta}$.
15:     **if** message $M_C^i(w)$ received then $state \leftarrow \mathcal{A}_{\text{succ}}$ $L(v) \leftarrow w$ **end if**
16:     **if** message $M_{\mathcal{A}}^i(w, c_w)$ received **then**
17:         $P_v \leftarrow P_v \cup \{w\}$; $d_w \leftarrow c_w$;
18:         **if** $|c_v - c_w| \leq x \lceil \gamma \zeta_i \log n \rceil$ **then** $c_v \leftarrow \xi(P_v)$;
19:     **end if**
20: **end while**

On waking up, a node v enters state $\mathcal{A}_0$ and tries to become a leader. However, it will wait for $\lceil \alpha \Delta \log n \rceil$ times slots to let its already awake nodes have a chance to give

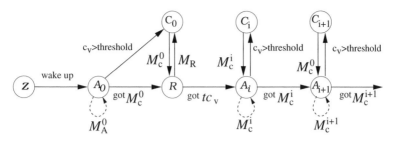

Figure 5.1 The sequence of states and the transition between states in the distributed-coloring algorithm. Each color i is represented by a state C_i.

their status. Thus, this node v will not wrongfully compete for the leader. As soon as it receives a message M_C^i from a neighboring node that has already joined C_i, v will join the succeeding state $\mathcal{A}_{\text{succ}}$. In other words, the colors assigned to nodes will be increased along the time. If no such message is received, node v will start to actively compete for using color i. To ensure that no two neighboring nodes are both competing for the same color i, a node will periodically send message M_A^i to neighboring nodes to state that they will try to enter state C_i (see Figure 5.1). Whenever the current node v receives a message M_C^i from a neighboring node w, v knows that it cannot verify color i anymore because it is already claimed by a neighboring node; thus, it will consequently move on to the next state $\mathcal{A}_{\text{succ}}$. For neighboring nodes that compete for using color i, node v will use P_v to store all such neighboring nodes as its *competitor list*. It also stores a local copy [denoted by $d_v(w)$ in the algorithm] of c_w for each such competitor node w. These local copies will be updated accordingly to keep track with the real current counter of c_w as much as possible. If its own counter c_v and the counter c_w are within the *critical range* of each other, v will reset its own counter to some special value $\xi(P_v)$. Here, the critical range is chosen as $[\gamma \zeta_i \log n]$ and $\xi(P_v)$ is defined such that the new counter is not within the critical range of *any locally* stored copy of neighboring counters. Observe that because the local copy of counters may be out-of-date, the selected value $\xi(P_v)$ may still fall in the critical range of some c_w for some neighboring node w.

If a node v finds that it has succeeded in getting its counter up to the *threshold value* $[\sigma \Delta \log n]$, it will then choose to color itself using color i and join state C_i. This method will guarantee that with high probability no two neighboring nodes will joint state C_i; i.e., the set of nodes induced by state C_i is independent.

Algorithm 10 Coloring algorithm: node v in state $\mathcal{R}$

On entering state $\mathcal{R}$:

1: **while** *state* $== \mathcal{R}$ **do**
2:    **transmit** message $M_\mathcal{R}(v, L(v))$ with a probability $\frac{1}{\kappa_2 \Delta}$.
3:    if message $M_C^0(L(w), w, tc_w)$ received **then** *state* $\leftarrow \mathcal{A}_{tc_v(\kappa_2+1)}$;
4: **end while**

The set of nodes in state C_0 plays a special role as the leaders of some clusters. Each nonleader node v in state $\mathcal{R}$ will send a request message $M_{\mathcal{R}}(v, w)$ for an intracluster color to its leader node w. The leader node w, on receiving such a request, will transmit for $[\beta \log n]$ time slots, with a probability of $1/\kappa_2$ a message $M_C^0(w, v, tc)$. Here, tc denotes the intracluster color assigned to node v by the leader node w, and this value will be increased for each subsequent requesting node. Requests are buffered in an internal queue Q. Because there could be as many as κ_2 neighboring nodes getting the same intracluster color as a node v, when a nonleader node v receives a message $M_C^0(w, v, tc_v)$ from its leader w, node v will try to compete for color $c = tc_v(\kappa_2 + 1)$ instead of competing for color tc_v directly. The advantage is to spread out the potential color assignment so that each of the neighboring nodes (including node v) with the same intracluster color will be able to resolve their conflict locally without affecting the coloring of other nodes. If competing for the color c is unsuccessful (a neighboring node with the same intracluster color already got c), this node v will enter state $\mathcal{A}_{c+1}$ and compete for color $c + 1$. It repeats this until it manages to get a color assigned. It is not difficult to prove that every node with the same intracluster color tc is capable of getting a color from $c, c + 1, c + 2, \ldots, c + \kappa_2$ with high probability. Remember that $c = tc(\kappa_2 + 1)$. Thus, with high probability, two neighboring nodes with different intracluster colors will *not* compete for the same color.

Algorithm 11 Coloring algorithm: node v in state C_i

On entering state C_i:

1:    $color_v \leftarrow i$;   {* v is active *}
2:    **if** $i > 0$ **then**
3:      **repeat transmitting** message M_C^i with probability $\frac{1}{\kappa_2 \Delta}$;
4:    **else if** $i == 0$ **then**
5:      $tc \leftarrow 0$; $Q \leftarrow \emptyset$;
6:      **repeat**
7:        **if** message $M_{\mathcal{R}}(w, v)$ received and $w \notin Q$ **then** add w to Q
8:        **if** Q is empty **then**
9:          **transmit** message $M_C^0(v)$ with probability $\frac{1}{\kappa_2}$;
10:       **else**
11:         $tc \leftarrow tc + 1$;
12:         Let w be the first element in Q;
13:         **for** $[\beta \log n]$ times slots **do**
14:           **transmit** message $M_C^0(v, w, tc)$ with probability $\frac{1}{\kappa_2}$;
15:         **end for**
16:         Remove w from Q;
17:       **end if**
18:      **until** true
19: **end if**

The algorithm's four parameters, α, β, γ, and σ, can be chosen so as to trade off the running time and the probability of correctness. The higher the parameters, the less likely the algorithm fails in producing a correct color. To obtain the high-probability result, the parameters are defined as

$$\alpha \geq 2\gamma\kappa_2 + \sigma + 1,$$
$$\beta \geq \gamma,$$
$$\gamma = \frac{5\kappa_2}{[\frac{1}{e}(1-\frac{1}{\kappa_2})]^{\kappa_1/\kappa_2}[\frac{1}{e}(1-\frac{1}{\kappa_2\Delta})]^{1/\kappa_2}},$$
$$\sigma = \frac{10e^2\kappa_2}{(1-\frac{1}{\kappa_2})(1-\frac{1}{\kappa_2\Delta})},$$

for $\Delta \geq 2$. The constants are chosen to guarantee a correct coloring and running time with a probability of at least $1 - O(\frac{1}{n})$. Simulation results by Moscibroda and Wattenhofer (2005a) show that in randomly distributed networks, significantly smaller values suffice. In fact, the constants are small enough to yield a practically efficient coloring algorithm for WANs and WSNs that can be employed for the purpose of initializing the network.

The following theorem proved in Moscibroda and Wattenhofer (2005a) shows that for all nodes in any given class, C_i forms an independent set at *all* time slots during the execution of the algorithm with high probability. This theorem establishes the correctness of their algorithm because if all color classes form an IS, the resulting coloring of nodes must be conflict-free (i.e., no two adjacent nodes getting the same color).

THEOREM 5.3 *For all i, the color class C_i forms an IS throughout the execution of the algorithm with a probability of at least $1 - 2n^{-3}$.*

The following lemmas proved in Moscibroda and Wattenhofer (2005a) are essential for bounding the time complexity of their algorithm.

LEMMA 5.1 *Let T_v^i denote the number of time slots a node v spends in state $\mathcal{A}_i$. With a probability of at least $1 - 3n^{-3}$, it holds that for* all *nodes v and i, $T_v^i \in O(\kappa_2^3 \Delta \log n)$.*

LEMMA 5.2 *Let T_v^R denote the number of time slots a node v spends in state $\mathcal{R}$. With a probability of at least $1 - 3n^{-3}$, it holds that for* each *node v, $T_v^R \leq (\gamma + \beta)\Delta \log n$.*

Then, Theorem 5.4 showed that their algorithm will terminate quickly:

THEOREM 5.4 *Every node decides on its color within time $O(\kappa_2^4 \Delta \log n)$ after its wake-up with a probability of at least $1 - 3n^{-1}$.*

The only thing remaining is a bound on the number of different colors assigned by the algorithm. If the vertex coloring in the graph is used for setting up a TDMA scheduling of node activities in a wireless network, for instance, the bandwidth assigned to a node v is related to its assigned time slot, which is often inversely proportional to the value of the highest color (time slot) in its neighborhood. Thus, generally, the colors assigned to each node should be as small as possible when a color C_i actually corresponds to the time slot i in a periodical time-division scheduling. The highest color assigned to a neighbor of a node v by the algorithm is dependent on only the local graph properties. This allows nodes located in low-density areas of the network to send more frequently than nodes in dense and congested parts. Let θ_v be the maximum node degree of all nodes within a two-hop distance of v, and let ϕ_v be the highest color assigned to all these nodes within two hops of v. Moscibroda and Wattenhofer (2005a) showed that with a probability of at least $1 - 2n^{-3}$, their algorithm produces a coloring such that, for all $v \in V$,

$$\phi_v \leq \kappa_2 \theta_v.$$

This implies that the algorithm will produce a coloring with at most $\kappa_2 \Delta$ colors with a probability of at least $1 - 2n^{-3}$.

The algorithm presented in Moscibroda and Wattenhofer (2005a) adopted a simple interference model. A node v receives a message in a time slot t only if exactly one node in its neighborhood transmits a message in this time slot. The message size of their algorithm is always limited to $O(\log n)$ bits per message. Notice that this neighborhood is also used as the communication graph model. In practice, the communication graph and the interference graph are often different. It is interesting to extend these results to the situations in which we are given a communication graph G_c for communication and an IG G_I modeling conflicting relations among links. Notice that for IG G_I, the vertices are the links in G_c and an edge xy in G_I denotes whether the corresponding two communication links (denoted by vertices x and y, respectively) will have any interference if they are actively transmitting simultaneously. The actual IG G_I depends on the underlying interference models used; see Chapter 2 for more discussion. In other words, a node u can send a message to a node v (i.e., a neighbor of u in graph G_c) only if no links conflicting with link uv are transmitting at the same time.

Recently, Kuhn and Wattenhofer (2006) studied the complexity of distributed coloring of a graph when originally all nodes were colored using at most m colors. They proved that for algorithms that run for a single communication round–i.e., every node of the network can send only its initial color to all its neighbors (which have direct communications)– the number of colors of the computed coloring has to be at least $\Omega(\Delta^2 / \log^2 \Delta + \log \log m)$. If one such round of algorithms is iteratively applied to reduce the number of colors step-by-step, they proved that the minimum rounds to obtain an $O(\Delta)$-coloring is $\Omega(\Delta / \log^2 \Delta + \log^{\star} m)$.

5.4   List-Coloring for Ad Hoc Networks

In the previous section, the channel assignment for wireless network does not need to form an ad hoc network. In certain scenarios, we may need to use the assigned channels to form communication links among the wireless terminals and then form certain network structures by using these links. In other words, the channel assignment for access-based networks requires that the channels assigned to adjacent vertices be *disjoint*. No communication links are needed to form between two nearby wireless terminals. In this section, we study the scenario in which we prefer that there be an *overlap* between the assigned channels of adjacent terminals. We call this kind of problem list-coloring for ad hoc networks.

We first formally define the channel-assignment problem when only network connectivity is required. Remember that throughout this chapter, we consider only *static* channel assignment; i.e., each NIC selects one and only one channel from its operable channels to operate. We notice that in practice, dynamic channel switching is possible for some networking cards. Dynamic channel switching clearly can potentially further improve the network's throughput performance. However, this is not without a price: The hardware needs to be specially designed in some circumstances, and the channel switching does have some overheads such as delay and synchronization between the transmitting node and the receiving node in which they need to switch to the same channel almost at the same time period. In light of these possible overheads, we mainly consider the static channel assignment in which each radio of every wireless node will operate on a predetermined channel. The channel-assignment problem is to determine which channel to operate exactly. Let $\mathcal{A}(v_i, k, f_j) \in \{0, 1\}$ be the channel-allocation function that denotes whether or not the kth NIC of node v_i selects to operate on channel f_j. Obviously, $\mathcal{A}(v_i, k, f_j) = 0$ if $f_j \notin F(v_i, k)$; $\mathcal{A}(v_i, k, f_{j_1}) + \mathcal{A}(v_i, k, f_{j_2}) \leq 1$ because one NIC of a node can operate on one channel under the static channel assignment. When a channel assignment is given, we use $\mathcal{A}(v_i) = \{f_j \mid \mathcal{A}(v_i, k, f_{j_1}) = 1$ for some NIC$\}$ to denote the channels selected by some NICs of node v_i.

Define the communication graph $G_A = (V, E')$ induced by the channel-assignment method $\mathcal{A}$ as follows: E' contains a link $e = (u, v)$ only if $e \in E$ and $\mathcal{A}(u) \cap \mathcal{A}(v) \neq \emptyset$. We say that a channel assignment $\mathcal{A}$ is *valid* if the communication graph G_A is a connected graph. The total channels used by all nodes under $\mathcal{A}$ are denoted as $F_A = \bigcup_{u \in V} \mathcal{A}(u)$. In this chapter, we first seek a channel assignment that produces a connected network while using as few channels as possible.

Complexity Result

First, given a wireless network and the radio and channel-availability constraints of all wireless nodes, we show that it is NP-complete to decide whether we have a channel assignment that results in a connected network without channel switching.

THEOREM 5.5 *It is NP-complete to decide whether we have a channel assignment that results in a connected network without channel switching (i.e., the assigned channels to each node is at most its number of NICs).*

Proof. Clearly, the problem is NP because we can check whether a given assignment is valid in polynomial time. We then show that it is NP-hard from the induction of a NP-complete problem given a graph G, find a spanning tree whose maximum degree is minimum among all spanning trees of G.

It is known that it is NP-complete to decide whether a given graph G has a spanning tree whose maximum node degree is at most a given integer D (Garey and Johnson, 1979). Given the graph $G = (V, E)$ and integer D, we construct a channel-assignment problem for a network G' as follows. The network $G' = (V', E')$ is the same as G, and each wireless node of V' will have D radios. There are m total channels, where each channel corresponds to an edge in the graph G. The set of channels that a wireless node can operate is the set of channels corresponding to all incident edges of this node in G. We then show that graph G has a spanning tree T with maximum degree D iff there is a channel assignment for G' that results in a connected network without channel switching.

First, we need to find if such a tree T does exist. Then, for the network G', our channel assignment will work as follows: A node u is assigned a channel represented by the edge e if e is an edge incident upon u in the tree T. Clearly, the total number of channels assigned to every node u is at most D, and the network induced by the channel assignment is obviously connected because T will be a subgraph of the induced network.

Second, we need to find if there is a channel assignment that induces a connected network. Because each node has at most D radios, the number of channels assigned to every node is at most D, and thus the induced network has the maximum degree D. Consequently, any spanning tree of the induced network is a spanning tree for G whose maximum degree is at most D. This finishes the proof. ∎

It is challenging to design efficient channel-assignment algorithms such that a certain network property is achieved. It remains an open problem whether we can design polynomial algorithms for channel assignment such that the resultant network is connected while the channel-switching times are minimized.

5.5   Transition Phenomena on Channel Availability

The first phenomenon we want to investigate is the effect of channel availability on the connectivity of network G. Assume that because of designing or environmental constraints, a wireless card can operate only on randomly selected c channels out of a total maximum of $|F| = k$. It is not difficult to see that when c is sufficiently smaller than k, the network is more likely to be disconnected, and when c is sufficiently close to k, the network is more likely to be connected (when the graph G is connected). We thus would like to study the asymptotical transition behavior of the network connectivity in terms of the relation of c and k and other parameters such as the number n of nodes and the normalized transmission range R_T.

Recent work (Gupta and Kumar, 1998; Sanchez et $al.$, 1999) has shown that when n nodes are randomly deployed in a certain region, there is a critical transmission power r_n required to ensure that two wireless nodes in the network could communicate with each other through one-hop or multihop paths with high probability if all nodes have fixed power r_n. In other words, such critical transmission power is also the threshold of keeping the whole network connected, if the area of the plane and the number of the nodes in the networks is fixed. Generally, it was known that, given n nodes V randomly deployed in a square region or disk region with unit area and every node with transmission range r_n, the UDG network $G(V, r_n)$ formed by V (which has a link uv iff $\|u - v\| \le r_n$) is connected with high probability if $\pi n r_n^2 > \log n$; it is disconnected with non-zero probability if $\pi n r_n^2 < \log n$.

For multihop, multiradio, multichannel wireless networks, several networking parameters will affect the network connectivity, such as the transmission range of wireless nodes, the number of radios available at every node, the number of channels that can be used by each radio, and so on. Traditional results on the critical transmission range for connectivity assumed that every wireless node has only one NIC and that there is only one channel from which to choose. In a wireless network with multiradios, we assume that the channels that can be operated by a radio of a specific wireless node is randomly chosen from a set of channels. Then, we would like to study whether there is a critical density of channels for the network connectivity. Clearly, we first need the transmission range r_n to satisfy $\pi n r_n^2 \geq \log n$.

In multiradio, multichannel networks, we generally assume that there are k channels $\mathcal{F}$ available globally and each radio will be able to operate on s channels randomly selected from $\mathcal{F}$. Notice that when two end nodes of a link uv in $G(V, r_n)$ do not have a common operable channel, then this link uv will not be operable by u and v, although nodes u and v are physically within the transmission range of each other. We define a network $G(V, r_n, s, k)$ as the set of links uv, where $\|u - v\| \leq r_n$ and u and v have a common channel. Clearly, if $s \geq (k/2)$, then it is guaranteed that two end nodes of a link uv in $G(V, r_n)$ will have a common channel to communicate; i.e., $G(V, r_n, s, k)$ is connected as long as $G(V, r_n)$ is connected. We want to know the critical value for s (with fixed r_n and k) such that the network $G(V, r_n, s, k)$ remains connected.

Obviously, a link uv remains in $G(V, r_n, s, k)$ if they have a common channel, whose probability is

$$P_{k,s} = 1 - \frac{\binom{k}{s}\binom{k-s}{s}}{\binom{k}{s}\binom{k}{s}}$$

if every node has only one radio. When every node has the same set of R distinctive radios, the probability that two nodes will not have a common channel among any of these R radios is

$$P_{k,s,R} = 1 - \left(\frac{\binom{k}{s}\binom{k-s}{s}}{\binom{k}{s}\binom{k}{s}}\right)^R.$$

Consequently, every link in $G(V, r_n)$ will survive in $G(V, r_n, s, k)$ with a probability of $P_{k,s,R}$; i.e., the network $G(V, r_n, s, k)$ can be roughly modeled as a Bernoulli model. From Yi (2005), we know that a Bernoulli network $G(V, r_n, p_e)$ is connected with high probability only if

$$\pi P_e n r_n^2 > \log n.$$

Here, a Bernoulli network $G(V, r_n, p_e)$ is defined over $G(V, r_n)$ where each edge from $G(V, r_n)$ is chosen independently with a probability p_e. Notice that the probability $P_{k,s,R}$ of whether two nodes share a common channel is not independent among edges. Thus, we cannot directly apply the result in Yi (2005). If we apply that result to estimate the transition phenomenon, we find that the network $G(V, r_n, s, k)$ is connected with high probability only if

$$\pi P_{k,s,R} n r_n^2 > \log n.$$

Figure 5.2 plots the transition phenomena of network connectivity on spectrum availability. Here, the x coordinate indicates the number of available channels that each node

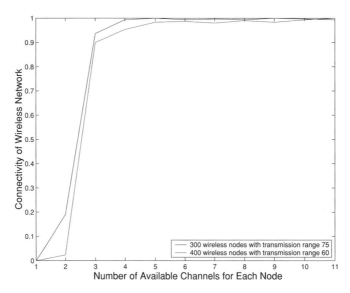

Figure 5.2 Transition phenomena of network connectivity on spectrum availability in wireless networks.

could operate on during a period of time, and the y coordinate denotes the connectivity of a whole wireless network. In this simulation, we randomly generate 300 and 400 wireless nodes, respectively, in a 500×500 unit region; the difference between these two cases is that the transmission range of the former is 75 units and that of the latter is 50 units. Typically, a unit represents about 1 m here. We assume that the maximum number of available channels on which wireless nodes could operate is 11, obeying the criteria of IEEE 802.11. We vary the number of available channels for wireless nodes from 1 to 11, and the results indicate that the connectivity of the wireless network increases sharply from 2 to 4 available channels. The whole wireless network is connected almost 100% after the available channels are equal to or greater than four.

5.6   Further Reading

List-coloring has been widely studied in the literature, especially for cellular networks. See Narayanan (2002) for an excellent survey of channel assignment by use of graph-coloring techniques. Garg *et al.* (1996a) presented distributed list-coloring for mobile base stations. List-coloring has also been used to study the channel assignment for wireless networks; e.g., Khanna and Kumaran (1998), Malesinska (1995), Mishra *et al.* (2005), and Ramachandran *et al.* (2006). McDiarmid and Reed studied weighted coloring for wireless networks when the wireless vertices are placed in a triangular lattice and each vertex v will require $w(v) \geq 0$ colors. They prove that it is NP-complete to determine if a given weighted graph is three-colorable. They further present a polynomial-time combinatorial algorithm that, on inputting a triangular lattice graph together with a corresponding weight vector w, finds a weighted coloring that uses at most $\frac{4\omega(G_w)+1}{3}$ colors, where $\omega(G_w)$ is the clique number of the replicated graph G_w. Here, graph G_w is obtained from G by replacing every vertex v in G with a complete graph on $w(v)$ vertices.

As we know, traditional multihop wireless networks comprise single radio components, and using multiradios and multichannels could efficiently enhance the throughput of the whole wireless network by exploiting the multiple channels and channel reuse opportunities. However, simply using multiradios and multichannels without an efficient channel-assignment algorithm cannot effectively exploit the increasing bandwidth available; it might even decrease the network performance because of channel-switching overhead and system-management overhead. As noted in Das *et al.* (2005), Leiner *et al.* (1987), and Tobagi (1987), there are a number of common issues concerning multiradio, multichannel, and multihop wireless networks, including network connectivity, sharing channels, switching channels dynamically, network topology, and so on.

Concerning the subject of multichannel, multihop wireless networks, some researchers have proposed several approaches. For example, Jain *et al.* (2000, 2001) proposed a CSMA protocol based on a modification of IEEE 802.11 for multihop wireless networks that uses multiple channels and a dynamic channel-selection method. Their main idea is to divide the available bandwidth into N channels, and the transmitting station selects the appropriate channel to transmit packets based on the interference power measurements on all available channels. So *et al.* (2004b, 2004c) proposed a MAC protocol requiring only one transceiver per host for WANs. This MAC protocol enables hosts to utilize multiple channels by switching channels dynamically, thus increasing the throughput of the wireless network.

Unlike those preceding approaches that need to modify the IEEE 802.11, Bahl *et al.* (2004) present a link-layer protocol called slotted seeded channel hopping (SSCH). SSCH increases the capacity of an IEEE 802.11 network by utilizing frequency diversity, and each node in the wireless network uses SSCH to decrease interference, thus improving the capacity of the wireless network. Das *et al.* (2005) present two mixed-integer LP models for solving the fixed-channel-assignment problem with multiple radios through finding fixed-channel-assignment strategies to maximize the number of simultaneous bidirectional links.

There was some other research concerning the use of multi-NICs for each node, such as those of Bahl *et al.* (2004) and Hsiao *et al.* (2001). The methods adopted by Hsiao *et al.* (2001) require each node to have as many NICs as its neighbors do and also require a sufficiently large number of available channels. Bahl *et al.* (2004) assume that there is an *a priori and identical* channel assignment to the NICs of each node, and the channel assignments for all wireless nodes are the same: binding #1 NIC with channel 1, #2 NIC with channel 2, and so on. However, as we know, such requirements are unrealistic because the number of NICs each wireless node has may be different and the same NIC could operate on only a fixed subset of channels because of the hardware constraints, such as 802.11X series NICs could operate on 11 channels. In this chapter, we assume that different wireless nodes may have different numbers of NICs and the types of NICs need not be uniform also. Furthermore, the actual available channels for different wireless nodes could be different, which is more general than previous studies and more close to practice.

Other research, such as that of Alicherry *et al.* (2005), committed a joint channel-assignment scheme for throughput optimization in multiradio WMNs. However, it assumes that all nodes have the same transmission range and considered only the static channel-assignment situation. In other words, they assume that for each node v, the number of available channels for it is no more than the number of NICs, $I(v)$. However, as we know, in practice, not only could the transmission ranges of wireless nodes be different but also nodes could adjust the channels they operate on dynamically.

5.7   Conclusion and Remarks

In this chapter, we studied the *static* channel-assignment problem by statically mapping a channel to each radio of the nodes. We first studied the channel assignment such that the channels assigned to adjacent nodes in a graph modeling the network should be disjoint. We assume that each node has a list of channels that it can operate and also that it has a fixed number of radios that it can use to operate statically assigned channels. For each channel, it has an average capacity associated with it. We show that it is NP-complete to find a channel assignment that can maximize the total capacity of all assigned channels while the channel assignment is valid. We also studied an efficient centralized method that can get an assignment that is arbitrarily close to the optimum.

We then studied another version of the spectrum assignment, in which we need to assign channels to nodes such that the resultant network is connected, where a link uv exists in the resultant network iff they have a common assigned channel and they are within the transmission ranges of each other. We showed that it is NP-complete to find a valid channel assignment such that the resultant network is connected. Here, a channel assignment is called valid if the channels assigned to each node number no more than its radios. We also studied a transition phenomenon: the network connectivity on the number of available channels at nodes. We theoretically analyzed the expected number of *randomly chosen* channels that a node needs to be able to operate such that it can result with high probability in a connected network. One important topic for future research is designing a centralized and a message-efficient distributed algorithm such that it has a theoretically proven worst-case performance guarantee for channel assignment, achieving network connectivity and maximizing network throughput.

Problems

5.1   Channel assignment has been widely studied for cellular networks. In this chapter, we studied the channel assignment for wireless access networks. What are the main differences between the channel assignments for cellular networks and for wireless access networks?

5.2 In this chapter, we show that there is a PTAS for the channel assignment that maximizes the total capacity. Design a simple greedy heuristic that can achieve a constant-approximation ratio when all interference disks are the same.

5.3 Prove that a 3×3 complete bipartite graph cannot be an interference graph of any six broadcast stations.

5.4 Finding a MWIS of a graph can be performed by using the following simple heuristic: We first sort all nodes in a nonincreasing order of the node weight; then we process nodes in such order and select the ith node only if it is not a neighbor of any previously selected node. How badly will such a heuristic perform for a general weighted graph? In other words, what is the worst-case approximation ratio of such a heuristic? Assume that the input graph is a UDG. Then, what is the worst-case approximation ratio of such a heuristic?

5.5 Consider a wireless ad hoc network. Two nodes u and v are connected in the interference graph if there is a node located inside the transmission disk of one node and located inside the interference disk of the other node. Assume that node u has the smallest transmission radius. Prove that there are at most a constant number of independent nodes that are neighbors of node u. What is the best upper bound you can prove here?

5.6 Assume that we designed a distributed-coloring algorithm that runs in only one communication round; i.e., every node of the network can send only its initial color to all its neighbors (which have direct communications). Prove that the number of colors of the coloring returned by the algorithm has to be at least $\Omega(\Delta^2 / \log^2 \Delta + \log \log m)$. Here, m is the number of colors used by nodes in a valid coloring before the algorithm and Δ is the maximum node degree.

5.7 Assume that we randomly put each of m balls into n bins, where $m \geq n$. What is the expected number of balls each bin will have? What is the probability that a specific bin will be empty? What is the expected number of bins that will be empty? What is the probability that at least $n/2$ of the bins will not be empty of balls?

5.8 Analyze the time complexity and message complexity of the distributed-coloring algorithms presented in Subsection 5.3.2.

6   CDMA Code Channel Assignment

6.1   Introduction

In Chapter 4, we basically studied how to assign time slots to links such that the simultaneous transmissions will be interference-free. In Chapter 5, we studied how to assign frequency channels to wireless terminals such that (1) they are interference-free; i.e., the spectrums assigned to nearby terminals will be disjoint; or (2) the links formed by the assigned channels will form a network with certain networking properties such as being connected. In this chapter, we study the channel assignment to wireless networks when the channels are defined by CDMA codes.

Code-division multiple access (CDMA) provides a higher capacity, flexibility, scalability, reliability, and security than conventional frequency-division multiple access (FDMA) and time-division multiple access (TDMA). It has already been widely deployed in 2G cellular communication systems and was proposed for the emerging and future wireless systems, including WLANs and wireless ad hoc networks. In a CDMA system, the communication channels are defined by pseudo-random codewords, which are carefully designed to cancel each other out as far as possible. Each communication utilizes the entire available spectrum, and every bit of data is multiplied by the codeword used by the communication channel. Thus, many duplicates of the same information are transmitted to ensure that at least one gets through. The number of duplicates, which is equal to the length of the codeword, is known as the *spreading factor*. The inverse of the length of the codeword is known as the *rate* of the codeword. There is a trade-off on the length of the codewords. On the one hand, longer codewords can increase the number of available channels and the robustness of the communications. On the other hand, longer codewords would result in lower data rates of the communication channels because the raw data rate seen by the user is the inverse of the codeword length.

Conventional CDMA used for voice communications in cellular systems is a constant rate in nature. Correspondingly, all codewords in the code have a fixed length. Such a code is referred to as the *orthogonal fixed-spreading-factor* (OFSF) code. In the past several years, data services have become increasingly important to cellular networks. Indeed, one major role of the 3G cellular systems is to support differentiated QoS guarantees for emerging multimedia applications, which are typically of variable data rates. The support of high-rate data service by the OFSF code can be achieved by assigning multiple codewords to a connection. This mode of operation is called multicode CDMA

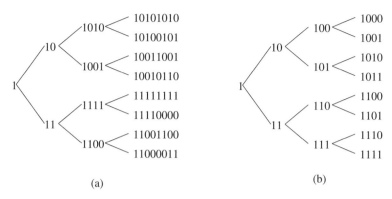

Figure 6.1 OVSF code: (a) code-tree structure; (b) binary color representation. © 2005, Springer.

(MC-CDMA). However, MC-CDMA requires multiple transceiver units at each node and thus introduces increased hardware complexity.

Motivated by the support of variable-rate data service at a low hardware cost, Adachi *et al.* developed a variable-length code, known as *orthogonal variable-spreading-factor* (OVSF) code, in 1997. The idea of the OVSF code is to allow the codewords in the code to have variable lengths, and a higher rate request is assigned a single shorter codeword. So when the OVSF code is used, only a single transceiver is required per node. The generation of the OVSF code can be depicted by the code-tree structure (Adachi *et al.*, 1997) shown in Figure 6.1(a). The code tree is a balanced binary tree whose vertices represent the codewords. The root, which is at level zero, is associated with the codeword 1. Recursively, if a vertex has the codeword c, then its two children have codewords cc and $c\bar{c}$, respectively, where $\bar{c}$ is the complement of c. Thus, at level l, there are 2^l codewords, each 2^l bits long. To simplify our notation, we represent the ith OVSF-CDMA codeword at layer l by an $l + 1$-bit binary representation, where the most significant bit is always 1 and the remaining l bits are the binary representations of integer $i - 1$. For example, the first codeword at layer 3 of the codeword tree is 10101010, which we denote as 1000 in our binary representation; the third codeword at layer 3 of the codeword tree is 10011001, which we denote as 1010 in our binary representations. See Figure 6.1(b) for the corresponding binary color representations of the CDMA codewords.

The OVSF code has two prominent features different from the OFSF code: (1) the number of codewords in an OVSF code is infinite, whereas the number of codewords in an OFSF code is finite; and (2) not every pair of codewords in an OVSF code is orthogonal to each other. Indeed, two OVSF codewords are orthogonal to each other if and only if neither is an ancestor or, equivalently, a prefix of the other. On the other hand, all codewords in an OFSF code are orthogonal to each other.

A wireless ad hoc network is a collection of radio nodes (transceivers) located in a geographic region. Each node is equipped with an omnidirectional antenna and has limited transmission power. A communication session is established either through a single-hop radio transmission, if the communication parties are close enough, or otherwise through relaying by intermediate nodes. A channel assignment to the nodes in a wireless ad hoc network should avoid two collisions. The *primary collision* occurs when a node simultaneously transmits and receives signals over the same channel, or two nonorthogonal channels in the case of OVSF-CDMA. The *secondary collision* occurs when a node simultaneously receives more than one signal over the same channel, or nonorthogonal channels in the case of OVSF-CDMA. Thus, to prevent the primary collision, two nodes can be assigned the same channel or two nonorthogonal channels if and only if neither of them is within the transmission range of the other. Similarly, to prevent the secondary collision, two nodes can be assigned the same channel or two nonorthogonal channels if and only if no other node is located in the intersection of their transmission ranges.

Given an OFSF-CDMA code assignment, its *throughput* is the sum of the rates of the assigned codewords, and its *bottleneck* is the minimum of the rates of the assigned codewords. The *throughput* of a wireless ad hoc network is then the maximum of the throughput over all possible conflict-free OFSF-CDMA code assignments to its nodes. Similarly, the *bottleneck* of a wireless ad hoc network is then the maximum of the bottleneck over all possible conflict-free OFSF-CDMA code assignments to its nodes. In this chapter, we study the OVSF-CDMA code assignments for wireless ad hoc networks with the objectives of maximizing the throughput or the bottleneck of the assignment or both.

In this chapter, we first establish the relation between the independence number and the throughput, and the relation between the bottleneck and the chromatic number. We show that the throughput of the optimum CDMA code assignment for a wireless ad hoc network is at least $5/8$ the independence number of the corresponding interference graph. The bottleneck of the optimum CDMA code assignment for a wireless ad hoc network is at least a small constant factor of the chromatic number of the corresponding interference graph. After that, we present several heuristics for conflict-free OVSF-CDMA codeword assignments. The obtained code assignments can achieve a throughput within a constant factor of the maximum throughput and/or a bottleneck within a constant factor of the maximum bottleneck. We also extend the results to a more general wireless ad hoc network model in which each wireless node has an interference disk inside which its signal will interfere, and it has a transmission disk inside which a node can receive its signal correctly.

It will be seen that the correctness of the methods presented in this chapter does not require that the transmission region of each wireless device be a disk centered at this node. The methods that we study apply to all wireless networks when the communication channels are varying with distance, time, and obstacles. The specific wireless network model used here enables us to prove only that these methods have theoretical performance guarantees. The correctness of these methods also does not depend on the node positions. The usage of the node positions enables us to design algorithms with better approximation ratios. The position error will not affect these methods as long as the position error will not change the topology of the network; i.e., the network topology derived from the perceived nodes' positions is the same as the actual physical network topology.

6.2   System Model and Assumptions

6.2.1   Problem Formulations

Let V be the set of radio nodes in a given wireless ad hoc network, and let r_v be the specified transmission radius of node v for each $v \in V$. For any pair of nodes u and v, we use $\|uv\|$ to denote their Euclidean distance. Then, we can obtain a geometric graph G over V by creating an edge between each pair of nodes (u, v) satisfying that either $\|uv\| \leq \max\{r_u, r_v\}$ or there is a node $w \in V \setminus \{u, v\}$ such that $\|uw\| \leq r_u$ and $\|vw\| \leq r_v$. The graph G is referred to as the *interference graph* (IG).

With the introduction of the interference graph, a conflict-free channel assignment in wireless ad hoc networks channelized by FDMA, TDMA, or OFSF-CDMA is equivalent to a proper vertex-coloring of the interference graph. However, such equivalency disappears if the wireless ad hoc network is channelized by OVSF-CDMA. Instead, a conflict-free channel assignment in a wireless ad hoc network channelized by OVSF-CDMA is equivalent to the following variant of vertex-coloring, referred to as *prefix-free vertex-coloring*, or simple *prefix-free coloring*, of the interference graph G: The colors are represented by positive binary numbers, as shown in Figure 6.1(b). Note that the first (i.e., leftmost) bit of every binary color is one, and a binary color at level l has $l + 1$ bits. Two binary colors are said to be *prefix-free* if neither is a prefix of the other. Then, two binary colors are prefix-free if and only if the corresponding codewords are orthogonal. A *prefix-free coloring* of G is a vertex-coloring such that any pair of adjacent vertices in G receives prefix-free colors.

We associate each binary color with a *rate* attribute, which is equal to the rate of the corresponding codeword. Thus, the rate of an i-bit binary color is equal to the 2^{-i+1}. Given a conflict-free CDMA/OVSF code assignment $\{c_v \mid v \in V, \forall \text{ link } uv, c_u \text{ and } c_v \text{ are orthogonal}\}$ of the IG, G, its *throughput* and *bottleneck* are defined as $\sum_{v \in V} 2^{-|c_v|+1}$ and $\min_{v \in V} 2^{-|c_v|+1}$, respectively, where $|c_v|$ denotes the number of bits of the color c_v. In other words, the throughput of a conflict-free CDMA/OVSF code assignment is the sum of the rates of the assigned codes, and its bottleneck is the minimum of the rates of the assigned codes. The *throughput of an IG*, G, denoted by $\tau(G)$, is then the maximum of the throughput over all possible conflict-free CDMA/OVSF code assignments of G. Similarly, the *bottleneck of an IG*, G, denoted by $\beta(G)$, is then the maximum of the bottleneck over all conflict-free CDMA/OVSF code assignments of G.

6.2.2   A Technical Lemma

In this subsection, we prove some important lemmas that will later be used to study some fundamental properties of an optimum OVSF-CDMA code assignment for wireless ad hoc networks.

Let T be a (rooted) binary tree. For each vertex v of T, the *level* of v in T, denoted by $\ell_T(v)$, is defined as the length of the path in T between the root and v. Thus, the level of the root is zero. A binary tree is *full* if every nonleaf vertex has exactly two children.

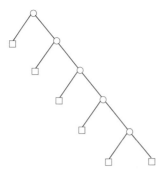

Figure 6.2 An extremely imbalanced full binary tree. © 2005, Springer.

A binary tree is *balanced* if the levels of all leaves differ by at most one. A binary tree is said to be *extremely unbalanced* if there are exactly two leaves at the maximum level and one leaf at any other level (see Figure 6.2).

Consider a finite set S of items in which each item s is associated with a positive weight $\omega(s)$. Let $\mathcal{T}_S$ denote the set of binary trees whose leaves are the items of S. For each tree T in $\mathcal{T}_S$, its throughput, denoted by $f(T)$, is defined by

$$f(T) = \sum_{s \in S} \omega(s) 2^{-\ell_T(s)}.$$

A tree in $\mathcal{T}_S$ is said to be optimal if its throughput achieves the maximum among all trees in $\mathcal{T}_S$. Obviously, any optimal tree must be full. Let T^* be an extremely unbalanced tree in $\mathcal{T}_S$ satisfying that the levels of the items sorted in the decreasing order of the weights monotonically increase. The next lemma states that T^* is optimal.

LEMMA 6.1 *T^* is an optimal tree in $\mathcal{T}_S$. If S is a finite set of items with weights $\omega_1 \geq \omega_2 \geq \cdots \geq \omega_k$, then its throughput is*

$$\sum_{i=1}^{k-1} \frac{\omega_i}{2^i} + \frac{\omega_k}{2^{k-1}}.$$

The proof of this lemma is similar to the proof of the correctness of Huffman code construction (see, e.g., Chapter 16 of Cormen *et al.*, 2001). It will use the following two lemmas.

LEMMA 6.2 *Let x and y be two items having the lowest weights. Then, there exists an optimal tree in which x and y appear as the sibling leaves of maximum level.*

Proof. The idea of the proof is to take an arbitrary optimal tree T and modify it to make a tree representing another optimal tree such that x and y appear as the sibling leaves of maximum level in the new tree. We use the swapping argument. Let a and b be two items that are sibling leaves of the maximum level in T (see Figure 6.3). Without loss of generality, we assume that $\omega(x) \leq \omega(y)$ and $\omega(a) \leq \omega(b)$. Then, $\omega(x) \leq \omega(a)$ and $\omega(y) \leq \omega(b)$. As shown in Figure 6.3, we exchange the positions in T of a and x to

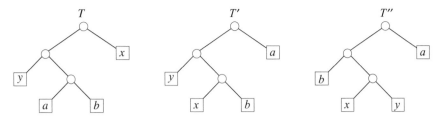

Figure 6.3 An illustration of the swap operations in the proof of Lemma 6.2. © 2005, Springer.

produce a tree T', and then we exchange the positions in T' of b and y to produce a tree T''. The difference in throughput between T and T' is

$$f(T) - f(T') = \sum_{s \in S} \omega(s)2^{-\ell_T(s)} - \sum_{s \in S} \omega(s)2^{-\ell_{T'}(s)}$$

$$= \omega(x)2^{-\ell_T(x)} + \omega(a)2^{-\ell_T(a)} - \omega(x)2^{-\ell_{T'}(x)} - \omega(a)2^{-\ell_{T'}(a)}$$

$$= \omega(x)2^{-\ell_T(x)} + \omega(a)2^{-\ell_T(a)} - \omega(x)2^{-\ell_T(a)} - \omega(a)2^{-\ell_T(x)}$$

$$= [\omega(a) - \omega(x)]\left(2^{-\ell_T(a)} - 2^{-\ell_T(x)}\right)$$

$$\leq 0,$$

because $\omega(a) \geq \omega(x)$ and $\ell_T(a) \geq \ell_T(x)$. Thus, $f(T) \leq f(T')$, which means that exchanging x and a does not decrease the throughput. Similarly, exchanging y and b does not decrease the throughput and, hence, $f(T') \leq f(T'')$. Therefore, $f(T) \leq f(T'')$. Because T is optimal, $f(T) = f(T'')$. Thus, T'' is an optimal tree in which x and y appear as the sibling leaves of maximum level, from which the lemma follows. ∎

The next lemma shows that the optimal tree has the optimal-substructure property.

LEMMA 6.3 *Let T be an optimal tree in $\mathcal{T}_S$. Consider any two items x and y that appear as the sibling leaves in T, and let z be their parent. Then, considering z as an item with weight $\omega(z) = \frac{\omega(x)+\omega(y)}{2}$, the tree T' obtained from T by putting z as the parent of a and y and them removing x and y is an optimal tree in $\mathcal{T}_{S'}$, where $S' = S - \{x, y\} \cup \{z\}$.*

Proof. We first show that the throughput $f(T)$ of T is equal to the throughput $f(T')$ of T'. For each $s \in S - \{x, y\}$, we have $\ell_T(s) = \ell_{T'}(s)$ and, hence, $\omega(s)2^{-\ell_T(s)} = \omega(s)2^{-\ell_{T'}(s)}$. Because

$$\ell_T(x) = \ell_T(y) = \ell_{T'}(z) + 1,$$

we have

$$\omega(x)2^{-\ell_T(x)} + \omega(y)2^{-\ell_T(y)} = [\omega(x) + \omega(y)]2^{-\ell_{T'}(z)-1}$$

$$= 2\omega(z)2^{-\ell_{T'}(z)-1}$$

$$= \omega(z)2^{-\ell_{T'}(z)}.$$

Thus, we conclude that $f(T) = f(T')$.

If T' is not an optimal one in $\mathcal{T}_{S'}$, then there exists a tree T'' in $\mathcal{T}_{S'}$ such that $f(T'') > f(T')$. Because z is treated as an item in S', it appears as a leaf in T''. If we add x and

y as the children of z in T'', then we obtain a tree in $\mathcal{T}_S$ with $f(T'') > f(T') = f(T)$, contradicting the optimality of T. Thus, T' must be optimal in $\mathcal{T}_{S'}$. ∎

Note that if x and y are the two items having the lowest weights, then the new item z has the lowest weight in the set S'. This fact, together with the preceding two lemmas, implies the correctness of Lemma 6.1.

6.3   Throughput and Bottleneck of General Graphs

The results in this section hold for general graphs. The concepts of prefix-free coloring, throughput, and bottleneck can be extended to general graphs. Let G be an arbitrary graph. Following the standard notation, we use $\chi(G)$ and $\alpha(G)$ to denote the chromatic number and the independence number, respectively, of G. However, two new rotations are also introduced. For any graph G, we use $\tau(G)$ and $\beta(G)$ to denote the throughput and bottleneck, respectively, of G. The main results of this section are the following relations among these four graph parameters.

THEOREM 6.1 *For any graph G,*

$$\alpha(G)/2 \leq \tau(G) \leq \alpha(G),$$
$$\beta(G) = 2^{-\lceil \log \chi(G) \rceil}.$$

The proof of the first part of Theorem 6.1 involves a new concept of *canonical* prefix-free coloring, which is subsequently defined herein. We observe that in any prefix-free coloring of G, all nodes receiving the same color form an IS of G. Thus, any prefix-free coloring of G can be regarded as a partition of $V(G)$ into ISs $V_1, V_2, \ldots, V_k$ followed by an assignment of colors to these independent sets as a whole. A prefix-free coloring of G is said to be *canonical* if it partitions $V(G)$ into ISs $V_1, V_2, \ldots, V_k$ with

$$|V_1| \geq |V_2| \geq \cdots \geq |V_k|$$

for some integer k, and assigns the color $1^i 0$ to all nodes in V_i for $1 \leq i \leq k-1$ and the color 1^k to all nodes in V_k. By definition, a canonical prefix-free coloring is fully determined by the partition of V into ISs. The next lemma states that there exists a canonical prefix-free coloring of G that achieves the maximum throughput.

LEMMA 6.4 *For any graph G, there is a canonical prefix-free coloring of G that achieves the maximum throughput.*

Proof. A prefix-free coloring that uses k different colors $c_1 < c_2 < \cdots < c_k$ is said to be *locally tight* if each node receiving a color c_i for some $i > 1$ has at least one neighbor receiving the color c_j for any $1 \leq j < i$. It is easy to see that every prefix-free coloring can be transformed to a locally tight one with the same or smaller throughput. Therefore, there is a prefix-free coloring that is locally tight and achieves the maximum throughput. Let OPT be such a prefix-free coloring. Assume that OPT uses k different colors $c_1 < c_2 < \cdots < c_k$. Because OPT is locally tight, these k colors are pairwise prefix-free. For each $1 \leq i \leq k$, let V_i denote the set of vertices that receive the color c_i. Then, the k subsets $V_1, V_2, \ldots, V_k$ form a partition of $V(G)$ into ISs. Now, we renumber them such that

$$|V_{1^*}| \geq |V_{2^*}| \geq \cdots \geq |V_{k^*}| .$$

Let OPT* be the prefix-free coloring that assigns the color $1^i 0$ to all nodes in $V_{i'}$ for $1 \leq i \leq k - 1$ and the color 1^k to all nodes in $V_{k'}$. Then, OPT* is a canonical prefix-free coloring. We will prove that the throughput of OPT* also achieves the maximum throughput by using Lemma 6.1.

To apply Lemma 6.1, we treat each subset V_i as an item with weight $\omega(V_i) = |V_i|$ and let $S = \{V_1, V_2, \ldots, V_k\}$. We define two trees T and T^* in $\mathcal{T}_S$ as follows. For each $1 \leq i \leq k$, we let P_i denote the path in the tree representation of binary colors shown in Figure 6.1 from the root to the tree vertex representing color c_i. Because the k colors $c_1, c_2, \ldots, c_k$ are pairwise prefix-free, the union of the k paths $c_1, c_2, \ldots, c_k$ is a binary tree with k leaves. For each $1 \leq i \leq k$, we place the item V_i to the leaf which comes from P_i. The resulting tree in $\mathcal{T}_S$ is then defined to be the tree T. The tree T^* is defined as the extremely unbalanced binary tree in $\mathcal{T}_S$ with the item V_{i^*} being the (unique) leaf at level i for each $1 \leq i \leq k - 2$ and the two items $V_{(k-1)^*}$ and V_{k^*} being the two leaves at level $k - 1$. Clearly, $f(T)$ equals the throughput of OPT and $f(T^*)$ equals the throughput of OPT*. By Lemma 6.1, $f(T) \leq f(T^*)$. Thus, the throughput of OPT is less than or equal to the throughput of OPT*. Because OPT achieves the maximum throughput, so too does OPT*. ∎

Proof of Theorem 6.1:   Now, we are ready to prove the first part of Theorem 6.1. First, we show that $\tau(G) \leq \alpha(G)$. Consider a canonical prefix-free coloring of G that achieves the maximum throughput $\tau(G)$. Assume that k colors are used. For each $1 \leq i \leq k$, let V_i be the set of nodes receiving the color $1^i 0$. Then,

$$\alpha(G) \geq |V_1| \geq |V_2| \geq \cdots \geq |V_k| .$$

Thus,

$$\tau(G) = \sum_{i=1}^{k-1} \frac{|V_i|}{2^i} + \frac{|V_k|}{2^{k-1}}$$

$$\leq \alpha(G) \left(\sum_{i=1}^{k-1} \frac{1}{2^i} + \frac{1}{2^{k-1}} \right)$$

$$= \alpha(G).$$

Second, we prove that $\alpha(G)/2 \leq \tau(G)$. Let V_1 be a MIS and let $\{V_2, \ldots, V_k\}$ be an arbitrary partition of $V \setminus V_1$ into ISs with

$$|V_2| \geq \cdots \geq |V_k| .$$

Then,

$$\alpha(G) = |V_1| \geq |V_2| \geq \cdots \geq |V_k| .$$

Consider the canonical prefix-free coloring of G determined by $V_1, V_2, \ldots, V_k$. Its throughput is

$$\sum_{i=1}^{k-1} \frac{|V_i|}{2^i} + \frac{|V_k|}{2^{k-1}} \geq \frac{|V_1|}{2} = \frac{\alpha(G)}{2}.$$

Therefore,

$$\tau(G) \geq \frac{\alpha(G)}{2}.$$

Next, we prove the second part of Theorem 6.1. First, we show that $\beta(G) \leq 2^{-\lceil \log \chi(G) \rceil}$. Consider any prefix-free coloring with maximum bottleneck $\beta(G) = 2^{-\ell+1}$ for some ℓ. Then, every color in this coloring is at most ℓ-bits long. We replace each ℓ'-bit color c with $\ell' < \ell$ by the ℓ-bit color $c0^{\ell-\ell'}$; i.e., the color we obtained from c by appending $\ell - \ell'$ zeros. This new coloring remains prefix-free and uses only ℓ-bit colors. Because the first bit of every ℓ-bit color is always one, the total number of ℓ-bit colors is at most $2^{\ell-1}$. Thus, $\chi(G) \leq 2^{\ell-1}$. This implies that $\lceil \log \chi(G) \rceil \leq \ell - 1$. Thus,

$$\beta(G) = 2^{-(\ell-1)} \leq 2^{-\lceil \log \chi(G) \rceil}.$$

First, we show that $\beta(G) \geq 2^{-\lceil \log \chi(G) \rceil}$. Consider any proper vertex-coloring of G by use of χ colors. These χ colors can all be represented by distinct $(1 + \lceil \log \chi(G) \rceil)$-bit binary colors. Thus,

$$\beta(G) \geq 2^{-\{1+\lceil \log \chi(G) \rceil\}+1} = 2^{-\lceil \log \chi(G) \rceil}.$$

This completes the proof of Theorem 6.1. ∎

Theorem 6.1 concentrates on constructing one maximal IS from the IG. Intuitively, if we also construct a good maximal IS for the remaining nodes, we could improve the performance bound on the throughput of the CDMA code assignment. The new approach will compute a maximal independent V_1' and then compute a maximal independent V_2' for the remaining nodes after v_1' is removed from G. The nodes in V_1' will receive a CDMA code 10 and the nodes in V_2' will receive a CDMA code 110. The following theorem shows that it indeed improves the approximation ratio:

THEOREM 6.2 *A ϱ-approximation algorithm for the MIS gives a $\frac{5}{8}\varrho$-approximation algorithm for the maximum throughput CDMA code assignment.*

Proof. Consider a canonical maximum independent decomposition $V_1, V_2, \ldots, V_k$ of all nodes V. Here, $|V_1'| \geq \varrho|V_1|$. Let $t_{i,j} = \frac{|V_i' \cap V_j|}{|V_j|}$; i.e., the portion of V_j is used in V_i'. After V_1' is generated, we know that the MIS in the remaining graph (removing V_1' and all its incident edges) has a size of at least

$$\max((1 - t_{1,1}) \cdot |V_1|, (1 - t_{1,2})|V_2|),$$

because $V_1 - V_1' \cap V_1$ and $V_2 - V_1' \cap V_2$ are still ISs. Notice that $t_{1,1}|V_1| + t_{1,2}|V_2| \leq V_1$. Then, $(1 - t_{1,1})|V_1| + (1 - t_{1,2})|V_2| \geq |V_2|$. It implies that V_2' has a size of at least $\varrho|V_2|/2$. Consequently, the throughput τ' generated by partition $V_1', V_2', \ldots, V_k', \ldots, V_{k_2}'$ is at least $\varrho \left(\frac{|V_1|}{2} + \frac{|V_2|}{2 \cdot 2^2} \right)$. Remember that the canonical coloring has throughput τ of at most $\frac{|V_1|}{2} + 2\frac{|V_2|}{2^2}$ from the fact that $|V_i| \le |V_2|$. From $|V_2| \le |V_1|$, it is easy to show that $\tau' \ge \frac{5}{8}\varrho\tau$. This finishes the proof. ∎

6.4   Approximation Algorithms for Interference Graphs

Throughout this section, we use V to denote the set of given radio nodes. All nodes in V are assumed to be located in a plane. For each node $v \in V$, its transmission radius is denoted by r_v. The nodes in V are said to have *quasi-uniform* transmission radii if the ratio of $\max_{v \in V} r_v$ to $\min_{v \in V} r_v$ is at most $1/\left(2 \sin \frac{360^o}{13}\right)$ and have *uniform* transmission radii if all r_v's are equal. We use G to denote the IG.

6.4.1   First-Fit Prefix-Free Coloring

First-fit coloring is a class of greedy algorithms for conventional (proper) vertex-coloring. Each first-fit coloring is associated with a vertex ordering and colors the vertices sequentially according to the associated vertex ordering by assigning each vertex the least possible color that will not cause conflict with its neighbors. A first-fit coloring of a graph G using k colors partitions V into k ISs $V_1, V_2, \ldots, V_k$, where V_i is the set of vertices receiving the ith color. Note that V_1—the set of vertices receiving the first (i.e., smallest) color—is always a maximal IS. In addition, for any $1 \le i < j \le k$, at least one vertex in V_j is adjacent to some vertex in V_i.

A first-fit coloring can be adapted for max-throughput prefix-free coloring in the following "unbalanced" manner. First, apply the first-fit coloring to obtain a proper vertex-coloring. Assume that k colors are used. Replace the ith color with the binary color 1^i0 for $1 \le i \le k - 1$, and replace the kth color with the binary color 1^k. Such prefix-free coloring is referred to as *unbalanced first-fit prefix-free coloring*.

A first-fit coloring can also be adapted for max-bottleneck prefix-free coloring in the following "balanced" manner. First, apply the first-fit coloring to obtain a proper vertex-coloring. Assume that k colors are used. Let T_k be a balanced full binary tree of k leaves. By mapping the root of T_k with the binary color 1, the k leaves of T_k correspond to k binary colors $c_1, c_2, \ldots, c_k$ in increasing order. For each $1 \le i \le k$, replace the ith color in the first-fit coloring with the binary color c_i. Such prefix-free coloring is referred to as *balanced first-fit prefix-free coloring*.

As with first-fit coloring, the performance of a first-fit prefix-free coloring depends on the associated vertex ordering. In this chapter, we consider the following three vertex orderings:

1. Radius-increasing ordering: In this ordering, the vertices are sorted in the increasing order of their transmission radii.
2. Radius-decreasing ordering: In this ordering, the vertices are sorted in the decreasing order of their transmission radii.
3. Lexicographic ordering: In this ordering, the vertices are sorted in the lexicographic order of their coordinates.

An *unbalanced first-fit prefix-free coloring in radius-increasing ordering* is proposed as a heuristic for max-throughput prefix-free coloring. Its performance is given in the following theorem:

THEOREM 6.3 *Unbalanced first-fit prefix-free coloring in radius-increasing ordering is a 26-approximation for max-throughput prefix-free coloring. If all nodes have quasi-uniform transmission radii, then it is a 24-approximation for max-throughput prefix-free coloring.*

Proof. Let V_1 be the set of vertices receiving the binary color 10. It was proved in Wan *et al.* (2006) that $|V_1| \geq \alpha(G)/13$. Thus, the throughput of the output prefix-free coloring is at least $|V_1|/2 \geq \alpha(G)/26$. By Theorem 6.1, $\alpha(G) \geq \tau(G)$. Thus, the throughput of the output prefix-free coloring is at least $\tau(G)/26$. This implies that unbalanced first-fit prefix-free coloring in radius-increasing ordering is a 26-approximation for max-throughput prefix-free coloring.

If all nodes have quasi-uniform transmission radii, then it was proved in Wan *et al.* (2006) that $|V_1| \geq \alpha(G)/12$. Using the same argument as in the previous paragraph, we can show that in this case, unbalanced first-fit prefix-free coloring in radius-increasing ordering is a 24-approximation for max-throughput prefix-free coloring. ∎

Balanced first-fit prefix-free coloring in radius-decreasing ordering is proposed as a heuristic for max-bottleneck prefix-free coloring. The following theorem gives an upper bound on its approximation ratio.

THEOREM 6.4 *Balanced first-fit prefix-free coloring in radius-decreasing ordering is a 16-approximation for max-bottleneck prefix-free coloring.*

Proof. Let k be the number of binary colors used by the output prefix-free coloring. Then, the number of bits in any of these k binary colors is at most $1 + \lceil \log k \rceil$. The bottleneck of the output prefix-free coloring is at least $2^{-\lceil \log k \rceil}$. It was proved in Wan *et al.* (2006) that $k \leq 13\chi(G)$. By Theorem 6.1, the bottleneck of the output prefix-free coloring is at least

$$2^{-\lceil \log(13\chi(G)) \rceil} \geq 2^{-\lceil \log 13 \rceil - \lceil \log \chi(G) \rceil}$$
$$= 2^{-\lceil \log \chi(G) \rceil}/16$$
$$= \beta(G)/16.$$

This implies that balanced first-fit prefix-free coloring in radius-decreasing ordering is a 16-approximation for max-throughput prefix-free coloring. ∎

When all nodes have uniform transmission radii, *unbalanced first-fit prefix-free coloring in lexicographic ordering* is proposed as a heuristic for max-throughput prefix-free coloring and *balanced first-fit prefix-free coloring in lexicographic ordering* as a heuristic for max-bottleneck prefix-free coloring. Their performances are given in the following theorem:

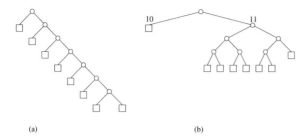

     (a)   (b)

Figure 6.4 Modification to the coloring by first-fit: (a) the original colors; (b) the new colors. © 2005, Springer.

THEOREM 6.5 *Assume that all nodes have uniform transmission radii. Then, unbalanced first-fit prefix-free coloring in lexicographic ordering is a 14-approximation for max-throughput prefix-free coloring, and balanced first-fit prefix-free coloring in lexicographic ordering is an 8-approximation for max-bottleneck prefix-free coloring.*

Proof. Let V_1 be the set of vertices receiving the binary color 10 in the output of unbalanced first-fit prefix-free coloring in lexicographic ordering. It was also proved in Wan *et al.* (2006) that $|V_1| \geq \alpha(G)/7$. Following the same argument as in the proof of Theorem 6.3, unbalanced first-fit prefix-free coloring in lexicographic ordering is a 14-approximation for max-throughput prefix-free coloring.

Let k be the number of binary colors used by the output of balanced first-fit prefix-free coloring in lexicographic ordering. It was also proved in Wan *et al.* (2006) that $k \leq 7\chi(G)$. Following the same argument as in the proof of Theorem 6.4, we can show that balanced first-fit prefix-free coloring in lexicographic ordering is an 8-approximation for max-throughput prefix-free coloring. ∎

We observe that an unbalanced first-fit prefix-free coloring achieves a good throughput but a very poor bottleneck. Indeed, every unbalanced first-fit prefix-free coloring always outputs extremely unbalanced coloring with colors corresponding to the leaves of the binary tree depicted in Figure 6.4(a). On the other hand, a balanced first-fit prefix-free coloring achieves a good bottleneck but may have a poor throughput. In what follows, we discuss how to modify them so as to achieve both good throughput and good bottleneck.

For disparate transmission radii, modified first-fit prefix-free coloring consists of two steps. In the first step, we apply the first-fit heuristic in the radius-*increasing* ordering to find a maximal IS. All nodes in the obtained maximal IS will receive the binary color 10. This first step ensures a good throughput. In the second step, we use first-fit coloring in the radius-*decreasing* ordering to find proper vertex-coloring of the remaining nodes. These colors will then be mapped to the binary colors that correspond to the leaves of a balanced full binary tree rooted at color 11 [see Figure 6.4(b)]. This second step ensures a good bottleneck. Such modified first-fit prefix-free coloring is referred to as *bicriteria first-fit prefix-free coloring in double radius ordering*. Its performance is given in the following theorem:

THEOREM 6.6 *Bicriteria first-fit prefix-free coloring in double radius ordering is a 26-approximation for max-throughput prefix-free coloring and a 32-approximation for max-bottleneck prefix-free coloring. If all nodes have quasi-uniform transmission radii, then it is a 24-approximation for max-throughput prefix-free coloring and a 16-approximation for max-bottleneck prefix-free coloring.*

The proof of Theorem 6.6 is similar to those of Theorems 6.4 and 6.3 and is omitted here.

For uniform transmission radii, we modify first-fit prefix-free vertex-coloring in lexicographic ordering as follows: We first apply first-fit in lexicographic ordering to find proper vertex-coloring. Then, the smallest color is mapped to binary color 10, and all other colors are mapped to the binary colors that correspond to the leaves of a balanced full binary tree rooted at color 11 [see Figure 6.4(b)]. Such modified first-fit prefix-free coloring is referred to as *bicriteria first-fit prefix-free coloring in lexicographic ordering*. Its performance is given in the following theorem:

THEOREM 6.7 *Assume that all nodes have uniform transmission radii. Then, bicriteria first-fit prefix-free coloring in lexicographic ordering is a 14-approximation for max-throughput prefix-free coloring and a 16-approximation for max-bottleneck prefix-free coloring.*

The proof of Theorem 6.7 is similar to that of Theorem 6.5 and is omitted here.

6.4.2   Tile Prefix-Free Coloring

In this subsection, we assume that all nodes have uniform transmission radii equal to one. A spatial divide-and-conquer heuristic, referred to as *tile prefix-free coloring*, is proposed. It is attractive because of its easy implementation, especially for dynamic and on-line prefix-free coloring and also distributed prefix-free vertex-coloring.

In this heuristic, we tile the plane into regular hexagons of sides equal to $1/2$ (see Figure 6.5). Each hexagon, or cell, is considered to be left-closed and right-open, with the top-most point included and the bottom-most point excluded (see Figure 6.6). Cells are further grouped into clusters of size 12 according to the pattern as shown in Figure 6.5. We then label the 12 hexagons in a cluster with the numbers 1–12 in an arbitrary pattern and repeat the same labeling for all clusters. Then, the distance between any two (half-closed and half-open) hexagons with the same label is greater than 2. Thus, colors can be spatially reused among the hexagons with the same label.

Now, for each $1 \leq i \leq 12$, let V_i denote the set of nodes within the hexagons labeled with i. We assign colors to the nodes such that for any $1 \leq i < j \leq 12$, the colors assigned to nodes in V_i are disjoint from the colors assigned to nodes in V_j. For this purpose, all nodes in a set V_i will receive colors that are descendants of some color c_i corresponding to a leaf in the balanced full binary tree with 12 leaves, as shown in Figure 6.7. For each V_i, we further partition into groups such that each group consists of nodes in V_i that are within a hexagon. Because the IG over all nodes in a group is a clique,

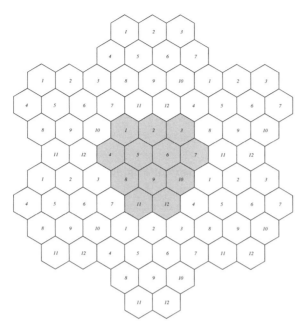

Figure 6.5 Tiling of the plane into hexagons with 12 hexagons per cluster. © 2005, Springer.

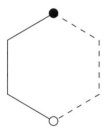

Figure 6.6 Half-closed, half-open hexagon. © 2005, Springer.

we apply a "shifted-down" version of the algorithm for prefix-free vertex-coloring of complete graphs to all nodes in a group. In other words, the coloring to nodes in each group of V_i corresponds to a *balanced* full binary tree rooted at c_i with one-to-one correspondence between the nodes and the leaves. With this coloring, the throughput of all nodes in a group of V_i is exactly the rate of c_i. Thus, to maximize the throughput, the mapping from V_i's to c_i's is chosen such that a set V_i with more groups will be mapped to a color c_i of shorter length.

The next theorem gives the performance of tile prefix-free coloring.

THEOREM 6.8 *Assume that all nodes have uniform transmission radii. Then, tile prefix-free coloring is a 12-approximation for max-throughput prefix-free coloring and a 16-approximation for max-bottleneck prefix-free coloring.*

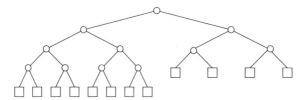

Figure 6.7 Each of the 12 colors corresponding to the 12 leaves is the prefix of the colors assigned to all nodes in some V_i. © 2005, Springer.

Proof. We first prove that tile prefix-free coloring is a 12-approximation for max-throughput prefix-free coloring. For each $1 \leq i \leq 12$, let g_i denote the number of hexagons labeled with i that contain at least one node. Note that in any prefix-free coloring, the total rates of the binary colors assigned to all nodes in a nonempty hexagon is at most one. Thus,

$$\tau(G) \leq \sum_{i=1}^{12} g_i.$$

Without loss of generality, assume that

$$g_1 \geq g_2 \geq \cdots \geq g_{12}.$$

Because in tile prefix-free coloring the total rates of binary colors assigned to all nodes in a nonempty hexagon labeled with i are exactly the rates of binary colors c_i, the throughput of tile prefix-free coloring is exactly

$$\frac{1}{8} \sum_{i=1}^{4} g_i + \frac{1}{16} \sum_{i=5}^{12} g_i.$$

Note that

$$\left(\frac{1}{8} \sum_{i=1}^{4} g_i + \frac{1}{16} \sum_{i=5}^{12} g_i \right) - \frac{1}{12} \sum_{i=1}^{12} g_i = \frac{1}{24} \sum_{i=1}^{4} g_i - \frac{1}{48} \sum_{i=5}^{12} g_i$$

$$\geq \frac{1}{24} 4 g_4 - \frac{1}{48} 8 g_5$$

$$= \frac{g_4 - g_5}{6}$$

$$\geq 0.$$

Therefore,

$$\frac{1}{8} \sum_{i=1}^{4} g_i + \frac{1}{16} \sum_{i=5}^{12} g_i \geq \frac{1}{12} \sum_{i=1}^{12} g_i \geq \frac{1}{12} \tau(G).$$

This implies that tile prefix-free coloring is a 12-approximation for max-throughput prefix-free coloring.

Next, we prove that tile prefix-free coloring is a 16-approximation for max-bottleneck prefix-free coloring. Let m be the largest number of nodes contained in a hexagon.

Then, each binary color used in tile prefix-free coloring has at most $5 + \lceil \log m \rceil$ bits. Thus, the bottleneck of tile prefix-free coloring is at least $2^{-4-\lceil \log m \rceil}$. On the other hand, $\chi(G) \geq m$. Thus, by Theorem 6.1,

$$\beta(G) = 2^{-\lceil \log \chi(G) \rceil} \leq 2^{-\lceil m \rceil}.$$

So, the bottleneck of tile prefix-free coloring is at least

$$2^{-4-\lceil \log m \rceil} \geq \frac{1}{16} \beta(G).$$

This implies that tile prefix-free coloring is a 16-approximation for max-bottleneck prefix-free coloring. ∎

6.4.3 Improved Approximation Ratio

In previous sections, several simple heuristics were presented that achieve constant-approximation ratios for the throughput and/or the bottleneck of optimal CDMA code assignment for a wireless ad hoc network. These heuristics have the advantage that they could be implemented efficiently and also in a distributed manner. In the remainder of the section, a new method with a better approximation ratio is presented. Theorem 6.2 implies that if we can find a maximal IS of the IG with approximation ratio ρ, then we could have a CDMA code assignment method whose throughput is at least $\frac{5}{8}\varrho$ of the optimum. The method works as follows:

1. Find a maximal IS, say V_1, from the IG, G, by using an ρ-approximation method and then assign CDMA code 10 to every node in V_1.
2. Remove node set V_1 from G and then find a maximal IS, denoted as V_2, from the remaining graph, again using a ρ-approximation method. Then, assign code 110 to every node in V_2.
3. For the remaining nodes, we could assign a CDMA code to them by using the first-fit heuristic.

The main task left is to find an algorithm with a small approximation ratio for the MIS problem for the IG. X.-Y. Li *et al.* (2002, 2006) showed that there is a PTAS for the MIS problem for the IG defined here. In other words, given any positive $0 < \epsilon < 1$, we can find a MIS whose size is at least $1 - \epsilon$ times of that of the optimum. Based on the results presented in X.-Y. Li *et al.* (2002, 2006), it is not difficult to show that the time complexity of the PTAS based on a shifting strategy is $O\left(\frac{1}{\epsilon^2} n^{O(\frac{1}{\epsilon^4})}\right)$, where n is the number of wireless nodes. Thus, we have the following theorem:

THEOREM 6.9 *The throughput of the CDMA code assignment for a WAN can be approximated within $\frac{5}{8}(1 - \epsilon)$ in time $O\left(\frac{1}{\epsilon^2} n^{O(\frac{1}{\epsilon^4})}\right)$.*

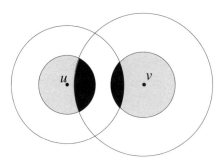

Figure 6.8 The black region denotes $I(u, v)$. Here, the lightly shaded smaller disks denote the transmission regions and the white larger disks denote the interference regions.

6.5   Maximum Weighted Independent Set for a General Wireless Network Model

From Chapter 4, Chapter 5, and the CDMA code assignment from this chapter, a good approximation of a MWIS is often needed to guarantee a certain performance of the algorithms studied. In the remainder of this chapter, we study an efficient distributed method that approximates the MIS and also an efficient centralized algorithm that computes a MWIS whose total weight is arbitrarily close to the optimum. The studied algorithms also use a more general wireless network model instead of the simple UDG model.

6.5.1   A General Wireless Network Model

So far, we have assumed that each wireless node v has a transmission range r_v such that all nodes inside the disk centered at v using r_v as radius will receive the signal from v and that this disk contains all possible nodes, with whom the transmission by node v will interfere. In practice, the region where the transmission of node v could interfere is often larger than the region where the transmission of node v could be received correctly. Thus, we study a more general wireless network model in which each wireless node v has two different ranges: the transmission range and the interference range. We define the *transmission radius* of node as the radius of the disk representing its transmission region. Similarly, the *interference radius* is the radius of the disk representing its interference region. Each node $v \in V$ has a transmission radius t_v and an interference radius r_v. We always assume that $t_v \leq r_v$.

Let $D(v, r)$ denote the disk centered at v with radius r. Then, each node defines two disks: the transmission disk $D(v, t_v)$ and the interference disk $D(v, r_v)$. The set of wireless node V defines two sets of disks $\mathcal{T} = \{D(v, t_v) \mid v \in V\}$ and $\mathcal{D} = \{D(v, r_v) \mid v \in V\}$ in a two-dimensional plane. Given two wireless nodes u and v, we define their intersection region $I(u, v)$ as $[D(u, t_u) \cap D(v, r_v)] \cup [D(u, r_u) \cap D(v, t_v)]$. See Figure 6.8 for an illustration of the intersection regions. Two nodes u and v can communicate with each other directly if they are inside the transmission disk of each other. The transmission that a node u interferes with is node v if v is inside the interference disk $D(u, r_u)$ of node u. Clearly, when $I(u, v)$ is null, nodes u and v can be assigned the same channel because the transmission of one of them cannot interfere with the transmission of the other. Most important, when $I(u, v)$ is not null, nodes u and v can still use the same channel if $I(u, v)$ does not contain any other wireless node inside. To capture this property, we define the IG, G, as follows: Two nodes u and v are connected in IG G iff there is a node from V inside $I(u, v)$. The chromatic number of the IG is exactly the minimum number of channels needed when we must assign a channel to each node so that all nodes can communicate simultaneously without causing interferences in our interference model.

In addition to this straightforward definition of an IG, several other graph models have been studied in X.-Y. Li *et al.* (2006). In the computational geometry community, the DG model has been widely studied.

Traditionally, a *disk graph* is the intersection graph of the set of disks each of which is centered at a unique node from a node set V; i.e., two nodes $u, v \in V$ are connected in the traditional DG if the two disks centered at u and v have a nonempty intersection. We first extend this conventional definition to wireless ad hoc networks in which each wireless node defines two disks: namely, the interference disk and the transmission disk. Here, the DG for wireless ad hoc networks has an edge uv if and only if the intersection area $I(u, v)$ is not empty. In other words, there is an edge uv if and only if $\|uv\| \leq \min(t_u + r_v, r_u + t_v)$. Here, $\|uv\|$ is the Euclidean distance between two wireless nodes u and v. Notice that the special case when $t_u = r_u$ for all wireless nodes u was studied in Erlebach *et al.* (2001) and X.-Y. Li and Wang (2002) recently, X.-Y. Li and Wang (2002) also studied some other geometry graphs derived from wireless networks when every node u has $t_u = r_u$.

In wireless ad hoc networks, if two nodes u and v are not connected in the DG, then they can transmit messages simultaneously without causing interference with each other under our interference model. Therefore, the chromatic number of the DG previously defined is an upper bound of the minimum number of frequencies (also called channels) needed when we must assign a channel to each node so that all nodes can communicate simultaneously without interference. Here, we assume that a wireless node can tune its receiving device to channels other than its transmission channel. Obviously, the DG model is an overestimation for interferences for ad hoc wireless networks. For example, when the intersection region $I(u, v)$ of two corresponding wireless nodes u and v is not empty, these two nodes can still use the same channel if the intersection region $I(u, v)$ does not contain any wireless node inside. Notice that the DG model can be used to model the scenario in which the channels assigned to two nodes should be disjoint if their interference regions overlap.

The IG model captures all links (u, v), where u and v cannot transmit simultaneously using the same channel. In actual wireless communications, two nodes u and v can communicate with each other directly if they are within the transmission range of each other. We call the graph formed by all such links (u, v) a *mutual-communication graph* (MCG). In other words, a link uv is kept in the MCG if and only if there is a physical symmetric link uv. For completeness, two more interesting new graph models were introduced in X.-Y. Li and Wang (2002) and X.-Y. Li *et al.* (2006): namely, a *mutual-inclusion graph* (MG) and a *conflict graph* (CG). We believe that these two new graph models may also find some applications later in wireless ad hoc networks. In the MG, two nodes u and v are connected if and only if the intersection region $I(u, v)$ contains both u and v inside, whereas in the CG model, two nodes u and v are connected if and only if the intersection region $I(u, v)$ contains at least one of u and v inside. Obviously, the chromatic number of the conflict graph is a lower bound of the minimum number of channels needed when we must assign a channel to each node so that all nodes can communicate without interferences.

By definition, MCG $\subseteq$ MG $\subseteq$ CG $\subseteq$ IG $\subseteq$ DG.

6.5.2   Simple Distributed Method for a MIS

We assume that we know the interference radius r_v of each wireless node v and the edge list of the underlying IG. If the graph structure is unknown, additionally we need the transmission radius t_v and the exact geometry location of each node v to construct it. An efficient method for approximating the MIS is presented that, in turn, will give an approximation of the throughput of the CDMA code assignment for this wireless ad hoc network. The method works as follows:

1. It finds the node with the smallest interference radius r_v, and adds it to the independent set.
2. It removes this node and all its adjacent nodes from the graph.
3. It repeats the preceding steps until the graph is empty.

Obviously, this algorithm computes a maximal IS with $O(n \log n)$ running time. Then, the following theorem that guarantees the quality of the computed IS is proved. It is actually shown that the IG model introduced here has a *hereditary* property: There is a node with a constant bounded number of independent neighbors and the subgraph, by removing this node and its neighbors, also does. Let IS be the computed IS of nodes.

THEOREM 6.10 *The computed IS has a size at least 1/40 of that of the MIS for the IG model previously defined.*

Proof. We prove this by using an area argument. Consider any node u selected by the algorithm. The nodes from an optimum solution, which are connected to u and removed by u, can be partitioned into two cases: outside D_u or inside D_u. Remember that here $D_u = D(u, r_u)$ is an interference disk in our notation.

First, consider the nodes outside D_u. Let $v_1, v_2, \ldots, v_k \notin D_u$ be the k-nodes from the optimum solution that are connected to u and are removed by the algorithm because of the removal of u. The selection of u implies that $r_{v_i} \geq r_u$ for all $i = 1, 2, \ldots, k$.

The fact that node $v_i, i = 1, 2, \ldots, k$, is connected to u implies that $D_i = D(v_i, r_{v_i})$ intersects with disk $D_u = D(u, r_u)$ because if they do not intersect, then obviously $I(u, v_i)$ is empty, which further implies that there is no edge uv_i in any of the graph models introduced here. In addition, the disk $D_i = D(v_i, r_{v_i})$ centered at $v_i, i = 1, 2, \ldots, k$, cannot contain any node $v_j, j \neq i$, inside because all disks centered at $i = 1, 2, \ldots, k$ are mutually independent in the corresponding interference graph. If v_j is inside

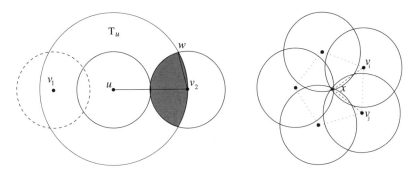

Figure 6.9 (Left) The intersection $T_u \cap D_i$ is bounded from below. (Right) The number of independent disks that cover any point x is bounded by 5.

$D_i = D(v_i, r_{v_i})$, then $I(v_i, v_j)$ contains v_j regardless of the transmission radii of v_i and v_j, implying that $v_i v_j$ is an edge in the interference graph.

Let B_u be the disk centered at u with radius $2r_u$ and $T_u = B_u - D_u$. Then, every disk $D_i, i = 1, 2, \ldots, k$, intersects T_u because it intersects D_u. It is not difficult to show that $T_u \cap D_i$ achieves the smallest area when v_i is on the boundary of B_u and $r_{v_i} = r_u$ (see Figure 6.9). We can show that $\angle w v_2 u > \frac{5}{12}\pi$. Thus, this smallest area is at least $\frac{5}{12}\pi$. Notice that the area of T_u is 3π.

Notice that the regions $T_u \cap D_i$ and $T_u \cap D_j$ for $1 \le i, j \le k$ may overlap. However, we will show that every point x is covered by at most five disks from $D_i, i = 1, 2, \ldots, k$ (see Figure 6.9). Assume that node x is covered by two disks, D_i and D_j; i.e., $\|x v_i\| \le r_{v_i}$ and $\|x v_j\| \le r_{v_j}$. Then, $\|v_i v_j\| > \max(r_{v_i}, r_{v_j})$ because D_i and D_j are independent in the corresponding IG, which implies that $\angle v_i x v_j > \frac{\pi}{3}$. Thus, x is covered by at most five independent disks. Therefore, by an area argument, we have $k \frac{5}{12}\pi < 5(3\pi)$. Thus, $k \le 35$.

Then, consider the nodes inside D_u. Let $v_1, v_2, \ldots, v_h \in D_u$ be the h nodes from the optimum solution that are connected to u and are removed by the algorithm because of the removal of u. Then, obviously, all disks centered at $v_i, i = 1, 2, \ldots, h$ contain node u. Because u is covered by at most five disks from the previous analysis, $h \le 5$.

Consequently, there are at most $35 + 5 = 40$ independent nodes removed when we remove all nodes adjacent to a node u selected by the algorithm. This finishes the proof of the theorem. ∎

By combining Theorems 6.10 and 6.2, we have the following theorem:

THEOREM 6.11 *The throughput of the OVSF-CDMA code assignment to a WAN in which each node has an interference disk and a transmission disk can be approximated within 1/64.*

6.5.3   Polynomial-Time-Approximation Scheme for a MIS

It is now shown how to design a PTAS for the IG of a general WAN model.

Assume that we are given a set $\mathcal{D} = \{D_1, D_2, \ldots, D_n\}$ of n interference disks in a two-dimensional plane, where disk D_i has interference radius r_i, center $v_i = (x_i, y_i)$, and

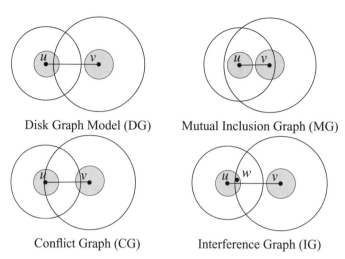

Disk Graph Model (DG)   Mutual Inclusion Graph (MG)

Conflict Graph (CG)   Interference Graph (IG)

Figure 6.10 Nonindependence in different graph models. The shaded smaller disks represent the transmission regions, and the nonshaded larger disks represent the interference regions. © 2005, ACM.

a weight $w(D_i) = w_i$. The weight is typically assigned to a node, but here we assume that it is assigned to the interference disk D_i for later convenience. The transmission radius of node v_i is t_i, which will be used only to determine if two nodes are connected in the corresponding graph model. For a subset of disks $U \subseteq \mathcal{D}$, let $w(U) = \sum_{D_i \in U} w(D_i)$; i.e., the summation of the weights of disks in U. Two disks D_i and D_j are said to be *independent* if the two nodes v_i and v_j are not connected in the corresponding graph model; otherwise, they are called *nonindependent*. Figure 6.10 shows the *nonindependence* in different graph models.

Similarly, we also adopt the shifting strategy (Hunt *et al.*, 1998) to develop a PTAS for the MWIS problem under various new graph models introduced in this chapter. Before presenting the method, we briefly review the shifting strategy (Hunt *et al.*, 1998) used to develop a PTAS for a MIS in the UDG. Here, we assume that the radius of each disk is $1/2$. The plane is assumed to be subdivided into a grid of size m by m for some integer m by a collection of vertical lines $x = im$ and horizontal lines $y = jm$. A subdivision is called (r, s)-shifting if it is formed by a collection of vertical lines $x = im + r$ and horizontal lines $y = jm + s$, where $0 \leq r, s < m - 1$. A *square* is formed by two consecutive vertical lines and two consecutive horizontal lines in (r, s)-shifting. For each square, an optimal solution of the MIS is obtained in polynomial time for all disks contained in the square but not intersecting the boundary of the square. The union of the MIS in all squares is returned as the final solution for this shifting, which clearly is an IS. Through the pigeonhole principle, it was shown that there is a shifting of the subdivision such that the size of the computed IS is at least $(1 - \frac{1}{m})^2$ of optimum.

We build the PTAS for approximating the MIS in various graph models introduced in this chapter based on the approach of Erlebach *et al.* (2001); i.e., to divide the interference disks into different levels according to their radii. At the same level, all nodes have *similar* interference radii; i.e., they are within a constant factor of each other.

As in Erlebach *et al.* (2001) and X.-Y. Li and Wang (2002), we scale all disks so that the largest disk has interference radius $1/2$. Let $r_{\min}$ be the smallest interference radius among all wireless nodes. Let $k > 1$ be a fixed integer (whose value will be specified later) and $\ell = \lfloor \log_{k+1} \frac{1}{2r_{\min}} \rfloor$. We partition the set of interference disks $\mathcal{D}$ into $\ell + 1$ levels such that level j, $0 \leq j \leq \ell$, consists of all disks D_i with interference radius satisfying $\frac{1}{(k+1)^{j+1}} < 2r_i \leq \frac{1}{(k+1)^j}$. Let $l(D_i)$ denote the level of disk D_i; i.e., $l(D_i) = \lfloor \log_{k+1} \frac{1}{2r_i} \rfloor$. Notice that we do not partition the transmission disks at all. Surprisingly, the sole partition of interference disks is enough to get a PTAS for us.

Similar to Erlebach *et al.* (2001), Hunt *et al.* (1998), and X.-Y. Li and Wang (2002), for each level j, we subdivide the plane into a grid by using a set of vertical lines $L_{j,v}$: $x = v \frac{1}{(k+1)^j}$, $v \in Z$, and a set of horizontal lines $H_{j,h}$: $y = h \frac{1}{(k+1)^j}$, $h \in Z$. Hereafter, j is called the *level* of the lines $L_{j,v}$ and $H_{j,h}$; v (and h) is called the *index* of the vertical (and horizontal) line $l_{j,v}$ (and $h_{j,h}$) at level j. An (r, s)-*shifting* of the subdivision is the grid defined by the set of vertical lines whose indices modulo k equal r and the set of horizontal lines whose indices modulo k equal s. It was proved in Erlebach *et al.* (2001) that a vertical line at level j of an (r, s)-shifting subdivision is also a vertical line at level $j + 1$ of the (r, s)-shifting subdivision.

Any two consecutive vertical lines at level j whose indices modulo k equal r, and any two consecutive horizontal lines at level j whose indices modulo k equal s, form a j-*square* in the (r, s)-shifting subdivision. See Figure 6.11 for an illustration of a 0-square for $r = s = 0$ and $k = 3$. In the figure, the solid lines are at level 0 and all dashed lines are at level 1. The j-squares are represented by thicker lines. The square represented by the thicker solid lines is a 0-square, and the squares represented by thicker dashed lines are 1-squares.

Clearly, any j-square S is subdivided into $(k + 1)^2$ $(j + 1)$-squares (by lines $L_{j+1,v}$ and $H_{j+1,h}$ at level $j + 1$). Notice that it contains only k^2 grids defined by lines at level j. These $(j + 1)$-squares S' are called the *children* of S, denoted by $S' \prec S$, and S is called the *parent* of S'. Obviously, any j-square S has length $\frac{k}{(k+1)^j}$. Notice that an interference disk at level j has radius r satisfying $\frac{1}{(k+1)^{j+1}} < 2r \leq \frac{1}{(k+1)^j}$. Thus, a j-square can contain some disks inside with a level at least $j - 1$, but not any disks inside with a level less than $j - 1$.

A disk D_i with center (x_i, y_i) and radius r_i is said to hit a vertical line at $x = a$ if $a - r_i < x_i \leq a + r_i$. Similarly, we say the disk D_i hits a horizontal line at $y = b$ if $b - r_i < y_i \leq b + r_i$. In other words, a disk hits a line if it intersects this line or it touches the line from the left or from the bottom.

For a MWIS problem, an interference disk D_i at level $l(D_i) = j$ is said to be *active* [respecting (r, s)-shifting] if it does not intersect the boundary of any j-square of the (r, s)-shifting subdivision. However, the definition of active disks for a minimum vertex cover (MVC) is different, which is discussed in detail later. Let $\mathcal{D}_S$ be the set of disks in $\mathcal{D}$ that are active for S. For a j-square S, let $\mathcal{D}_S^{<j}$ be the set of active disks with a level of less than j and intersecting with S. Similarly, we define $\mathcal{D}_S^{\leq j}$, $\mathcal{D}_S^{=j}$, $\mathcal{D}_S^{>j}$, and $\mathcal{D}_S^{\geq j}$ for active disks intersecting S and with a level of no more than j, equal to j, larger than j, and no less than j, respectively. For a j-square S, let $\text{OPT}_S^{<j}$ be the set of disks from $\text{OPT}(\mathcal{D}_S)$ with a level less than j and intersecting with S. Similarly, we define $\text{OPT}_S^{\leq j}$, $\text{OPT}_S^{=j}$, $\text{OPT}_S^{>j}$, and $\text{OPT}_S^{\geq j}$, respectively.

For each level j, let $\mathcal{D}_j(r, s)$ be the set of active interference disks at level j respecting (r, s)-shifting. We define $\mathcal{D}(r, s) = \cup_{j=0}^{\ell} \mathcal{D}_j(r, s)$; i.e., the union of active interference disks at all levels respecting (r, s)-shifting. Then, a j-square S is called *relevant* if $\mathcal{D}(r, s)$ contains at least one disk of level j that is inside S in the corresponding graph model. A more rigorous definition of *relevant*, which depends on the graph models, is given later. Let $\text{OPT}_{\text{IS}}(\mathcal{D}', G)$ denote the weight of the MWIS for a set of disks $\mathcal{D}'$ when the network is modeled by graph model G. We omit G and/or IS when it is clear from the context. The following proof is given by Erlebach *et al.* (2001) for the DG model and $r_u = t_u$ for each node u. It was found that the correctness of this lemma does not depend on the graph model. It is included here for completeness of presentation.

LEMMA 6.5 *Given a graph G, which is a DG, an IG, a MG, or a CG, there is at least one (r, s)-shifting, $0 \le r, s < k$, such that*

$$\text{OPT}(\mathcal{D}(r, s), G) \ge \left(1 - \frac{1}{k}\right)^2 \text{OPT}(\mathcal{D}, G).$$

Proof. Consider a MWIS $S^\star \subseteq \mathcal{D}$ for any graph model introduced here. Let $S_r^\star$ be the set of disks $D_i \in S^\star$ such that the disk D_i hits some vertical line $L_{j,v}$ at the level $j = l(D_i)$ whose index v modulo k equals r. Then, $\cup_{r=0}^{k-1} S_r^\star \subseteq S^\star$. In addition, $S_r^\star$, $0 \le r \le k - 1$ are pairwise disjoint. Thus, $\sum_{r=0}^{k-1} w(S_r^\star) \le w(S^\star)$. The pigeonhole principle implies that there is an index r_0 such that $w(S_{r_0}^\star) \le \frac{1}{k} w(S^\star)$. Let $S_{\overline{r_0}}^\star = S^\star - S_{r_0}^\star$. Therefore, $w(S_{\overline{r_0}}^\star) \ge (1 - \frac{1}{k}) w(S^\star)$.

Using the same technique, we can show that there is an index s_0 such that the set of disks from $S_{\overline{r_0}}^\star$, among which each D_i does not hit a horizontal line $H_{j,h}$ at its level $j = l(D_i)$ and the index h modulo k equals s_0, has a total weight of at least $(1 - \frac{1}{k}) w(S_{\overline{r_0}}^\star)$. Use $S_{\overline{r_0}, \overline{s_0}}^\star$ to denote such a set of disks; i.e., $S_{\overline{r_0}, \overline{s_0}}^\star = S_{\overline{r_0}}^\star - S_{\overline{r_0}, s_0}^\star$. Here, $S_{\overline{r_0}, s_0}^\star$ is defined similarly to $S_r^\star$ by replacing $S^\star$ with $S_{\overline{r_0}}^\star$.

Obviously, $S_{\overline{r_0}, \overline{s_0}}^\star$ is an IS for $\mathcal{D}(r_0, s_0)$. Thus, there is an (r_0, s_0)-shifting such that the weight of the MWIS in $\mathcal{D}(r_0, s_0)$ is at least $(1 - \frac{1}{k})^2$ of the optimum $\text{OPT}(\mathcal{D})$. ∎

The lemma implies the following corollary:

COROLLARY 6.1 *If we can solve the MWIS for disks inside each relevant j-square optimally, then we have a PTAS for the MWIS; i.e., setting $k = \frac{1 + \epsilon + \sqrt{1 + \epsilon}}{\epsilon}$ implies that $\text{OPT}(\mathcal{D}(r, s), G) \ge \frac{1}{1+\epsilon} \text{OPT}(\mathcal{D}, G)$.*

Before the PTASs for the MWIS for the graph models introduced in this chapter are shown, we first examine the structural properties of an optimum solution for all disks in $\mathcal{D}(r, s)$ for $0 \le r, s \le k - 1$. Hereafter, for convenience, we say that two interference disks are independent if the corresponding two nodes are not connected in the corresponding graph model.

Given a graph model, an optimum solution cannot contain any disk that hits a line at level 0 of the (r, s)-shifting subdivision. In other words, each disk of the optimum solution $\text{OPT}[\mathcal{D}(r, s)]$ is contained inside some 0-square. Moreover, the optimum

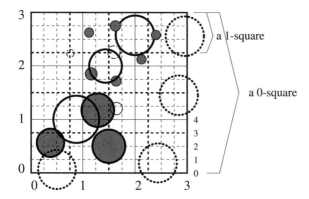

Figure 6.11 An optimum solution for a 0-square when the DG model is used and $t_v = r_v$ for each node v. © 2005, ACM.

solution can be divided into two subsets. One contains some independent interference disks at level 0, denoted by I_0. The other one contains independent interference disks at a lower level that are independent of any interference disk from I_0. By the definition of $\mathcal{D}(r, s)$, all interference disks in the second subset cannot intersect any lines, with level 1, of the (r, s)-shifting subdivision. In other words, each interference disk in the second subset is contained inside some 1-square. Figure 6.11 gives an example of optimum solution in a 0-square. Here, $k = 3$ and $r = s = 0$. The interference disks with the thickest boundary are at level 0. The interference disks that are not active are represented by a dashed boundary. The shaded interference disks are in the optimum solution.

The preceding partition of interference disks in the optimum solution in a 0-square can be performed recursively down to the squares at level ℓ as follows. Given a j-square S, let I be a set of independent interference disks of a level smaller than j, each of which intersects S. Let MWIS(S, I) be a MWIS of interference disks that are contained in S (i.e., must be of a level of at least j) and independent from the interference disks in I. Then, the union of MWIS($S, \emptyset$) for all relevant squares S without a parent must be the optimum solution for $\mathcal{D}(r, s)$.

We then discuss in detail how to compute MWIS(S, I) using dynamic programming. We assume that we have already computed the entry MWIS(S, I) for all squares S with a level of at least $j + 1$ and all appropriate ISs with I intersecting S. The interference disks in MWIS(S, I) can be divided into two subsets. One, denoted by X, contains some independent interference disks inside S with level j that are independent of interference disks from I. The other one contains independent interference disks with a level larger than j that are independent of any interference disk from I and X. By the definition of $\mathcal{D}(r, s)$, all interference disks in the second subset cannot intersect any lines, with level $j + 1$, of the (r, s)-shifting subdivision because we consider only active interference disks. In other words, each interference disk in the second subset is contained inside some $(j + 1)$-square S', which is contained in S. Thus, by properly choosing the set of interference disks X (i.e., interference disks inside S with level j and independent of I), we compute MWIS(S, I) from

$$\text{MWIS}(S, I) = \max_X \left\{ \left[\bigcup_{S' \prec S} \text{MWIS}(S', I_{S'} \cup X_{S'}) \right] \bigcup X \right\}.$$

Here, $I_{S'}$ is the subset of interference disks from I that intersect S'. $X_{S'}$ is defined similarly.

The algorithm processes all relevant squares in order of nonincreasing levels. For each j-square S and some appropriate IS I, MWIS(S, I) is computed by dynamic programming, as shown in Algorithm 12.

Algorithm 12 Approximate MWIS

Input: All n wireless nodes, their geometry locations, and the interference disks and the transmission disks defined by each node.

Output: A set of independent interference disks $\mathcal{I}$.

1: **for all** $j = \ell + 1$ down to 1 **do**
2:   **for all** square S with level j **do**
3:     Let R be all interference disks in $\mathcal{D}(r, s)$ of level $\leq j$ and intersecting S. For the mutual-inclusion graph model, we consider only disks of level $\leq j$ and with centers inside S.
4:     **for all** $J \subseteq R$ with at most C interference disks **do**
5:       **if** J is an independent set, **then**
6:         Let X be interference disks in J with level j.
7:         **for all** child square S' of S **do**
8:           Let I' be disks in J intersecting S'.
9:           Set $X = X \cup \text{MWIS}(S', I')$.
10:         **end for**
11:         Let I be disks in J with level less than j.
12:         **if** $w(X) > w(\text{MWIS}(S, I))$, **then**
13:           $\text{MWIS}(S, I) = X$.
14:         **end if**
15:       **end if**
16:     **end for**
17:   **end for**
18: **end for**
19: $\mathcal{I} \leftarrow \bigcup_{S:}$ S is relevant and does not have a parent MWIS$(S, \emptyset)$.

For the base situation, we can try all the possible ISs of interference disks to get the optimum solution in the largest level ℓ. The output of the algorithm is the union of the MWIS$(S, \emptyset)$, taken over all relevant squares S that do not have a parent. As Erlebach *et al.* (2001) did, we can easily show that the running time of this algorithm is $O(k^2 n^C)$. Here, C is the constant in Lemma 6.6. Notice that recently, Chan (2003) presented a PTAS for MWIS for DGs defined in Erlebach *et al.* (2001) with

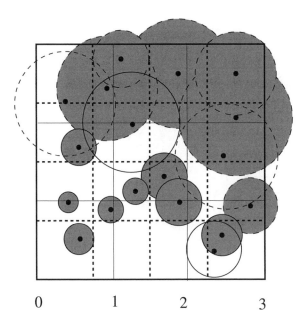

0   1   2   3

Figure 6.12 There is a constant number of independent interference disks at a level of at most j, intersecting a j-square S in the mutual-inclusion graph model. Only the interference disks of nodes are drawn. The transmission disks are omitted here. Here, $k = 3$ and the largest square is j-square. The smaller dashed square is $(j + 1)$-square. © 2005, ACM.

time complexity $n^{O(1/\epsilon)}$ for two-dimensional fat objects. It remains a future work if it also produces PTASs for the MWIS in the graph models introduced in this chapter.

It is not difficult to prove the correctness of the preceding dynamic programming approach. To guarantee that it runs in polynomial time of the number of interference disks n and k, we have to show that the size of $I_S \cup X_S$; i.e., the number of independent interference disks with a level of at most j and intersecting a j-square S is always bounded by a constant under the graph models introduced in this chapter.

LEMMA 6.6 *Let S be any j-square and let I be a set of independent interference disks with a level of at most j, each of which may connect to some interference disks contained in S. Then, there is a constant C depending on the graph model and k such that the cardinality of I is at most C.*

Proof. We prove this lemma individually for each graph model.

For the DG model and $t_v = r_v$ for each node v, Erlebach *et al.* (2001) gave a constant $C = \frac{4}{\pi}(k + 2)^2(k + 1)^2$ by using an area argument. For the DG model introduced here, we defer its analysis by first studying the other graph models.

For the MG model, the interference disk whose center is outside of S cannot connect with any interference disk contained inside S regardless of the size of their transmission radii. So, if we add them to MWIS(S, I), the resultant set is still guaranteed to be an IS of interference disks. Thus, to bound $|I|$, we have to consider only the interference

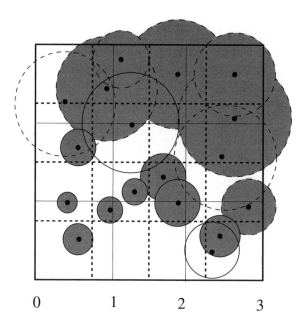

0   1   2   3

Figure 6.13 There are at a most a constant number of independent interference disks at a level of at most j, intersecting a j-square S in MG model. Only the interference disks are drawn. The transmission disks are omitted. Here $k = 3$ and the largest square is j-square. The smaller dashed squares are $(j + 1)$-squares. © 2005, ACM.

disks whose centers are inside S; i.e., the interference disks that can possibly "connect" with some interference disk contained inside S. Remember that in the MG model, we say two interference disks centered at nodes u and v are connected iff u and v are inside $I(u, v)$ (see Figure 6.12). Notice that all interference disks in I have a level of at most j, which implies that each interference disk in I has a diameter of at least $\frac{1}{(k+1)^{j+1}}$. Then, the distance between the centers of any two interference disks D_p and D_q from I is at least $\frac{1}{2(k+1)^{j+1}}$; otherwise, these two disks D_p and D_q will not be independent in the MG. The j-square S has a side length $\frac{k}{(k+1)^j}$. Therefore, there are at most a constant C_{MG} independent disks whose centers are inside S (see Figure 6.13). Here, by an area argument,

$$C_{\mathrm{MG}} \leq \left[\frac{k}{(k+1)^j}\right]^2 \bigg/ \left[\pi\left(\frac{1}{4(k+1)^{j+1}}\right)^2\right] = \frac{16k^2(k+1)^2}{\pi}.$$

For the CG model, again, we have to consider only all the independent interference disks that could connect to some interference disks contained in S. Remember that, here, two interference disks D_p and D_q are *independent* if v_p and v_q are not inside $I(v_p, v_q)$. Equivalently, v_p is not inside D_q and v_q is not inside D_p. Thus, the distance between the centers of any two disks from I is at least $\frac{1}{2(k+1)^{j+1}}$; i.e., the radius of the smallest possible interference disks of level j. Using the same area argument as for the MG, we can show that there are at most $\frac{16k^2(k+1)^2}{\pi}$ independent interference disks whose centers are inside S. Then, we concentrate on estimating how many independent

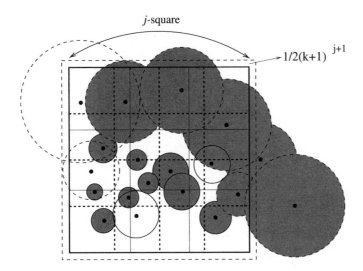

j-square

$1/2(k+1)^{j+1}$

Figure 6.14 There are at most a constant number of independent interference disks at a level of at most j, intersecting a j-square S in the CG model. Only the interference disks are drawn. The transmission disks are omitted. Here, $k = 3$ and the largest square is j-square. The smaller dashed squares are $(j + 1)$-squares. © 2005, ACM.

interference disks, denoted by I_O, there are such that (1) their centers are not inside S, (2) with a level of at most j, and (3) each one intersects S. We show that there are only a constant number of such interference disks by an area argument. Consider the four strips, denoted by $B(S)$, surrounding S with width $\frac{1}{2(k+1)^{j+1}}$. See Figure 6.14 for an illustration. For an interference disk $D_i \in I_O$, it is not difficult to show that $B(S) \cap D_i$ achieves the smallest area when v_i is on the boundary of $B(S)$ and $r_{v_i} = \frac{1}{2(k+1)^{j+1}}$. The smallest area of $B(S) \cap D_i$ is $\pi \frac{1}{8(k+1)^{2(j+1)}}$. Similar to Theorem 6.10, every point in $B(S)$ is covered by at most five interference disks from I_O. The area of $B(S)$ is $\frac{2k(k+1)+1}{(k+1)^{2(j+1)}}$. Thus, the size of I_O is at most

$$5 \frac{2k(k+1)+1}{(k+1)^{2(j+1)}} \Big/ \left[\pi \frac{1}{8(k+1)^{2(j+1)}} \right] = \frac{80k(k+1)+40}{\pi}.$$

Thus, the total number of independent interference disks I with a level of at most j and intersecting S is at most

$$C_{\mathrm{CG}} \leq \frac{16k^2(k+1)^2}{\pi} + \frac{80k(k+1)+40}{\pi} = \frac{16k^2(k+1)^2 + 80k(k+1)+40}{\pi}.$$

For IG models, because we already showed that CG $\subseteq$ IG, any IS in the IG is also an IS in the CG. For the CGs, we considered all the interference disks that intersect the j-square and for the IG model, we also have to consider all interference disks intersecting the j-square. Thus, C_{CG} is also an upper bound of the number of independent disks with a level of at most j intersecting S for the IG model.

For the DG model, there are two approaches to bound the number of independent interference disks with a level of at most j that intersect the j-square, denoted by C_{DG}.

One approach is to use the fact that $CG \subseteq DG$ and the set of interference disks from which to select an IS is the same for the DG model and the CG model. Thus, we have $C_{DG} \leq C_{CG}$. The other approach is to follow the analysis for the CG model. If two interference disks are independent, then the distance between their centers is at least the smaller radius of these two disks. Following the analysis for the CG, we will get the exact same bound C_{DG} for the DG model.

This finishes the proof of Lemma 6.6. ∎

Remark 1: Comparing "the preceding method" with the method presented in Erlebach *et al.* (2001), the main trick-parts of "preceding method" is the definition of an *active* disk respecting an (r, s)-shifting, and the *new* method (stated in Lemma 6.6) to include which disks should be considered when processing a square S. Notice that the simple extension of the method in Erlebach *et al.* (2001) will *not* work here because the number of independent interference disks that intersect a j-square in the MG model is not bounded by any constant. It is easy to show that the disks $D_i, i \geq 1$, with center $v_i = (2^{i-1}(1 + \epsilon)^i, 0)$ and radius $r_i = 2^{i-1}(1 + \epsilon)^i$, are independent; all such disks intersect any square containing the point $(0, 0)$. Here, ϵ is a small positive real number.

Remark 2: The other noticeable features of these algorithms are that although the underlying graph models require the transmission radius of each node (thus, the independence of nodes requires the nodes' transmission radii), these algorithms introduced a new concept of *independence* among the interference disks only. Consequently, in the analysis of the upper bound of the number of independent interference disks that *intersect* a j-square, the transmission radius t_v of every node v does not play any role here. Notice that we do not require any relations among the transmission radii and the interference radii of all nodes, except that the interference radius of each node must be at least its transmission radius.

Remark 3: The PTAS presented here can be extended further as long as Lemma 6.6 holds, even if the transmission regions and the interference regions are not disks.

6.6   Further Reading

All prior studies of prefix-free coloring have been restricted to *complete* graphs in the context of channel assignment to nodes in a single cell of an OVSF-CDMA cellular network (Amico *et al.*, 2002; Fantacci and Nannicini, 2000; Levcopoulos *et al.*, 1998; Minn and Siu, 2000). Prefix-free vertex-coloring of complete graphs is fairly easy. Indeed, because each node must receive a color different from those of the others, a prefix-free coloring can thus be represented by a binary tree with a one-to-one correspondence between the nodes (or their colors) and the leaves. Every binary tree with n leaves leads to a valid prefix-free coloring. If the binary tree is full, then the corresponding coloring achieves the maximum throughput, one. If the binary tree is full and balanced, the corresponding coloring achieves both maximum throughput and maximum bottleneck. Furthermore, if each node specifies a demand equal to a power of $1/2$, then, as an immediate application of Kraft's inequality, all demands can be satisfied if and only if the total demands are at most one. The dynamic reassignment of colors to meet a new demand is addressed in Minn and Siu (2000).

The minimum (proper) vertex-coloring of the interference graph has been studied in the context of channel assignment in WANs channelized by FDMA, TDMA, or OFSF-CDMA (Chlamtac and Farago, 1994; Chlamtac and Kutten, 1985; Ephremedis and Truong, 1990; Goldberg and Rao, 1997; Nelson and Kleinrock, 1985; Ramanathan and Lloyd, 1993; Ramaswami and Parhi, 1989; Sen and Huson, 1997; Sen and Malesinska, 1997; Stevens and Ammar, 1990; Wan *et al.*, 2006). The majority of these works simply presented networking protocols to obtain proper coloring without addressing the computational complexity or the theoretical performance. Sen and Huson (1997) proved the NP-hardness of the minimum vertex-coloring of the interference graph even when all nodes are located in a plane and have the same transmission radii. Sen and Malesinska (1997) made an attempt to analyze the approximation ratio of the classical first-fit coloring in smallest-degree-last ordering, which is due to Matula and Beck (1983) when applied to the IG. Unfortunately, their analysis turned out to be erroneous. Wan *et al.* (2006) recently provided a correct and tighter analysis of Matula and Beck's algorithm and several other approximation algorithms as well.

A problem related to the vertex-coloring of IGs is the *distance-2 vertex-coloring* of a graph (Krumke *et al.*, 2003). A *distance-2 vertex-coloring* of a graph G is a coloring of the vertices such that any two vertices separated by at most two hops receive different colors. In other words, it is a proper vertex-coloring of G^2, the *square graph* of G, the graph obtained by creating an edge between each pair of vertices of G whose graph distance in G is at most two. When all nodes have equal transmission radii, their IG happens to be the square of the UDG over these nodes; hence, in this case, the vertex coloring of the IG is the same as a distance-2 vertex coloring of a UDG (Clark *et al.*, 1990). However, when the nodes have disparate transmission radii, the IG may not be the square of any graph, as observed in Wan *et al.* (2006). Therefore, distance-2 vertex-coloring is, in general, different from the vertex-coloring of IGs.

To my best knowledge, there has been no attempt to maximize the throughput when coloring vertices. The only vertex-coloring problem that can be considered to be some-what related is the *minimum-chromatic-sum problem* (Bar-Noy *et al.*, 1998; Kubicka and Schwenk, 1989), which seeks a vertex-coloring of a given graph G, using natural numbers, such that the total sum of the colors of the vertices is minimized among all proper vertex-coloring of G. However, the maximum-throughput prefix-free vertex-coloring problem possesses several unique features that make it different from the minimum-chromatic-sum problem. First, the vertex-coloring must be prefix-free instead of being proper only. Second, the rate of the colors is different from the color number itself. Third, it is a maximization problem, whereas the minimum-chromatic-sum problem is a minimization problem.

Several results have been published for approximating the MIS for various geometric graphs. For UDGs, Marathe *et al.* (1995) gave simple centralized heuristics to approximate the MIS, the MVC, the minimum vertex-coloring, the minimum dominating set, and the minimum connected dominating set within the constants 3, $\frac{3}{2}$, 3, 5, and

10, respectively. Hunt *et al.* (1998) then presented the first PTASs to approximate the MIS, the MVC, and the minimum dominating set in UDG. Recently, Y. Wang *et al.* (2005a, 2005b) proposed a distributed algorithm to approximate a MWIS for wireless networks.

For traditional DGs, Marathe *et al.*. (1995) claimed that the MIS problem can be approximated within 5. Then, Erlebach *et al.* (2001) proposed an elegant PTAS for a MWIS and a maximum weighted vertex cover (MWVC) based on the shifting strategy proposed by Hochbaum and Maass (1985). The algorithm runs in time $\frac{1}{\varepsilon^2} n^{O(\frac{1}{\varepsilon^4})}$. Thus, when each node u has the same transmission range t_u and interference range r_u, we already have PTASs for both the MWIS and MWVC problems in the DG model.

Recently, X.-Y. Li and Wang (2002) also studied the MWIS and MWVC problems under the following graph models derived from wireless ad hoc networks: the IG, the MG, and the CG when the interference region of every node is the same as its transmission region. Both simple heuristics with constant-bounded-approximation ratios and PTASs are given for the MWIS and the MWVC under the graph models DG, IG, MG, and CG. Simple heuristics with constant-approximation ratios are presented for the minimum graph-coloring (MGC) problems over all the graph models. X.-Y. Li and Wang (2006) then presented PTASs for the MWIS and MWVC problems for all graph models when a node has a transmission range and a possibly larger interference range.

6.7   Conclusion and Remarks

In FDMA, TDMA, or OFSF-CDMA wireless ad hoc networks, a conflict-free channel assignment is equivalent to conventional (proper) vertex-coloring of the underlying inter-ference graphs. Because of the limited number of channels available in these networks, the cost metric of a conflict-free channel assignment in these networks is typically the same as the number of channels used. In OVSF-CDMA wireless ad hoc networks, a conflict-free channel assignment is no longer equivalent to conventional vertex-coloring of the underlying IGs. Indeed, because not every pair of OVSF codewords is orthogonal to each other, the channels assigned to any pair of nodes adjacent to each other in the IG must not only be different from each other but must also be orthogonal to each other. Because of this constraint, a new type of vertex-coloring, called prefix-free (vertex) coloring with positive binary numbers, was introduced. A conflict-free channel assign-ment in OVSF-CDMA wireless ad hoc networks is equivalent to a prefix-free coloring of the underlying IGs. Furthermore, because there are an infinite number of channels in OVSF-CDMA wireless ad hoc networks, the number of channels used is no longer a concern. Instead, the throughput and the bottleneck become appropriate cost metrics of a conflict-free channel assignment in OVSF-CDMA wireless ad hoc networks. Corre-spondingly, the concepts of the throughput and bottleneck of prefix-free coloring and the throughput and bottleneck of a graph were introduced. Two new maximization problems were also introduced: namely, max-throughput prefix-free coloring and max-bottleneck prefix-free coloring.

In this chapter, we first established two fundamental relations between the independence number and the throughput of a graph, and between the chromatic number and the bottleneck of a graph, respectively. After that, we proposed several algorithms for prefix-free coloring. Each of these algorithms is either a constant approximation for max-throughput prefix-free coloring, or a constant approximation for max-bottleneck prefix-free coloring, or constant approximations for both max-throughput prefix-free coloring and max-bottleneck prefix-free coloring at the same time. It is left as a future work whether we can design a PTAS for maximizing the throughput of the OVSF-CDMA code assignment for the IG of a wireless ad hoc network.

We then studied the approximation of the MWIS in a variety of graph models for wireless ad hoc networks. We showed that the MIS problem and the MGC problem can be approximated by simple approximation algorithms within 40, 40, 5, 5, and 5 for the DG, IG, CG, MG, and MCG, respectively. We also studied PTASs for the weighted versions of a MIS for all graph models. The time complexity of our PTASs is $O[\frac{1}{\epsilon^2} n^{O(\frac{1}{\epsilon^4})}]$.

One of the most challenging problems is to design a PTAS for DS and CDS problems for the introduced graph models, if it is possible. Notice that a PTAS for a MCDS has been proposed in Cheng *et al.* (2003) and Hunt *et al.* (1998) for a UDG when each node has only one transmission disk. However, little is known for the graph models presented in this chapter, even for the simpler case in which nodes have only transmission disks and different nodes may have different transmission radii.

Problems

6.1 What are the major advantages of CDMA channel-sharing compared with TDMA and FDMA? What are the major technique difficulties of using CDMA for wireless networks?

6.2 What are the relations between spread spectrum and CDMA?

6.3 Compare and constrast classic CDMA, IS-96 CDMA, WCDMA, 802.11 CDMA, and CDMA 2000.

6.4 What are the main differences in assigning the CDMA code to nodes and spectrum frequency to nodes?

6.5 Consider the following graph-coloring problem. Assume that we have a graph $G = (V, E)$ and a color-conflict graph $H = (C, F)$. Here, C is a set of all possible colors, and an edge $(c_1, c_2) \in F$ denotes that color c_1 and c_2 conflict with each other. The min-color with color-conflict problem is to assign each vertex in V a color such that two adjacent vertices in G will not be assigned a pair of conflict colors. Prove that the min-color with color-conflict problem is NP-hard. Design an efficient approximation algorithm for this problem when G is a DG and H is a tree.

6.6   Analyze the size of the table MWIS$(\cdot, \cdot)$ used for computing an approximation of the MWIS.

6.7   Given a graph $G = (V, E)$, a set $U \subset V$ of vertices is a vertex cover for G if, for every edge $e \in E$, one of its end nodes belongs to U. Design a PTAS to approximate a MWVC for a weighted UDG. For a general graph, design an efficient algorithm with approximation 2 for the vertex cover problem.

6.8   Assume that the input IG is not a bipartite graph. Thus, the graph cannot be colored using two colors. Assume that we first compute a MIS V_1' by using a ϱ-approximation algorithm, then compute a MIS V_2' of the remaining graph, and then repeatedly find the MIS until the remaining graph is empty. What is the relation between V_3 and V_3'? Can you improve the bound $\frac{5}{8}\varrho$ proved in Theorem 6.2 by studying the properties of IS V_3'?

Part III

Topology Control and Clustering

7   Clustering and Network Backbone

7.1   Introduction

The (localized) *topology-control* technique lets each wireless device (*locally*) adjust its transmission range and select certain neighbors for communication while maintaining a decent global structure to support energy-efficient routing and to improve the overall network performance. A distributed method is *localized* if it runs in a constant number of rounds (Naor and Stockmeyer, 1993). By enabling each wireless node to shrink its transmission power (which is usually much smaller than its maximum transmission power) enough to cover its farthest selected neighbor or selecting only a portion of nodes to forward data for others, topology-control schemes can not only save energy and prolong network life but can also improve the network throughput through mitigating the MAC-level medium contention. Unlike traditional wired and cellular networks, the movement of wireless devices during communication could change the network topology to some extent. Hence, it is more challenging to design a topology-control algorithm for ad hoc wireless networks: The topology should be locally and self-adaptively maintained at a low communication cost, without affecting the whole network's performance.

Wireless networks have drawn a good deal of attention in recent years because of the potential applications in various areas. Many routing protocols have recently been proposed for wireless ad hoc networks. The simplest routing method is to flood the message, which not only wastes the rare resources of wireless nodes but also diminishes the throughput of the network. One way to avoid flooding is to let each node communicate with only a selected subset of its neighbors or to use a hierarchical structure like Internet; e.g., a connected-dominating-set- (CDS-) based routing (Das and Bharghavan, 1997; Liang and Haas, 2000; Sivakumar *et al.*, 1998).

7.2   Network Models and Problem Formulation

In this section, some definitions and notation that are used later are given first. We assume that all wireless nodes are given as a set V of n points in two-dimensional space. Each wireless node has an omnidirectional antenna. This is attractive because a single transmission of a node can be received by all nodes within its vicinity. Each node has some computational power. We always assume that the nodes are almost static in a reasonable period of time. A communication graph $G = (V, E)$ over a set V of wireless nodes has an edge uv between nodes u and v iff u and v can communicate directly with each other; i.e., inside the transmission region of each other. Hereafter, we always assume that G is a connected graph. Let $d_G(u)$ be the degree of node u in a graph G and Δ be the maximum node degree of all wireless nodes [i.e., $\Delta = \max_{u \in V} d_G(u)$]. We assume that each wireless node u has a cost $c(u)$ of being in the backbone. Here, the cost $c(u)$ could be the value computed based on a combination of its remaining battery power, its mobility, its node degree in the communication graph, and so on. We discuss in detail several possible weight functions for different applications in Section 8.6. In general, smaller $c(u)$ means that the node is more suitable for being in the backbone. Let $\delta = \max_{i,j \in E} c(i)/c(j)$, where E is the set of communication links in the wireless network G. We call δ the *cost smoothness* of the wireless networks. When δ is bounded by some small constant, we say that the node costs are *smooth*. In this chapter, we study methods of constructing CDSs with only a minimum size.

We call all nodes within constant k-hops of a node u in the communication graph G the *k-local nodes* or *k-hop neighbors* of u, denoted by $N_k(u)$, which includes u itself. The k-local graph of a node u, denoted by $G_k(u)$, is the induced graph of G on $N_k(u)$; i.e., $G_k(u)$ is defined on vertex set $N_k(u)$ and contains all edges in G with both end points in $N_k(u)$.

The independence number, denoted as $\alpha(G)$, of a graph G is the size of the MIS of G. The *k-local independence number*, denoted by $\alpha^{[k]}(G)$, is defined as $\alpha^{[k]}(G) = \max_{u \in V} \alpha[G_k(u)]$. It is well known that for a UDG, $\alpha^{[1]}(\text{UDG}) \le 5$ and $\alpha^{[2]}(\text{UDG}) \le 18$.

A subset S of V is a *dominating set* (DS) if each node u in V is either in S or is adjacent to some node v in S. Nodes from S are called dominators, whereas nodes not in S are called dominatees. Clearly, any maximal IS is a dominating set. A subset C of V is a *connected dominating set* (CDS) if C is a dominating set and C induces a connected subgraph. Consequently, the nodes in C can communicate with each other without using nodes in $V - C$. A CDS is also called a *backbone* in this chapter. A DS with minimum cardinality is called a *minimum dominating set* (MDS). A CDS with minimum cardinality is the *minimum connected dominating set* (MCDS).

We assume that each node u has a cost $c(u)$. Then, a CDS C is called a *weighted connected dominating set* (WCDS). A subset C of V is a *minimum weighted connected dominating set* (MWCDS) if C is a WCDS with minimum total cost. Figure 7.1 illustrates an example of a CDS based on a DS. Here, the solid lines in the graph form the CDS graph, the square nodes are dominators or connectors, and the circular nodes are dominatees. The set of dominators and connectors forms a CDS.

In this chapter, we assume that the cost of every wireless node is 1 unit. In other words, we study the algorithms that minimize the size of the CDS. In Chapter 8, we study some algorithms that can approximately build a good weighted backbone when the cost of nodes is smooth.

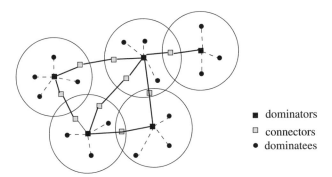

■ dominators
□ connectors
● dominatees

Figure 7.1 An example of a backbone. © 2005, ACM.

7.3   Centralized Algorithms for a Connected Dominating Set

7.3.1   Algorithms for General Graphs

In this section, we first study the centralized algorithms (e.g., the algorithm proposed by Guha and Khuller, 1996, 1998a) to construct a CDS whose size is comparable with the optimum solution. These algorithms also form a certain foundation for analyzing the performances of distributed algorithms that are studied later in this chapter and also in Chapter 8 for a weighted backbone. The problem of finding a CDS of minimum size is NP-complete even for UDGs (Clark *et al.*, 1990).

The MCDS set problem for general graphs is NP-complete, which can be proved by the induction from the set-cover problem. Notice that the set-cover problem was one of Karp's 21 NP-complete problems shown to be NP-complete in 1972. For the set-cover problem, Lund and Yannakakis (1994) showed that a set cover cannot be approximated in polynomial time to within a factor of $\frac{\log_2 n}{2} \simeq 0.72 \ln n$, unless the NP has quasi-polynomial-time algorithms. Feige (1998) improved this lower bound to $[1 - o(1)] \ln n$ under the same assumption, which essentially matches the approximation ratio achieved by the greedy algorithm. Alon *et al.* (2006) established a lower bound of $c \ln n$, where c is a constant, under a weaker assumption that $P \neq$ NP. Guha and Khuller (1996, 1998a) showed an approximation preserving reduction from the set-cover problem to the CDS problem, showing that it is hard to improve the approximation guarantee unless NP $\subseteq DTIME[n^{O(\log \log n)}]$.

Guha and Khuller (1996, 1998a) studied the approximation of the CDS problem for general graphs. They gave two different approaches; both of them guarantee an approximation ratio of $\Theta(H(\Delta))$. Because their approaches are for general graphs, they thus do not utilize the geometry structure if applied to wireless ad hoc networks. One approach is to grow a specially designed spanning tree that includes all nodes. The internal nodes of the spanning tree are selected as the final CDS. This approach has an approximation ratio of $2[H(\Delta) + 1]$. The other approach is first to approximate the DS and then connect the DS to a CDS. Guha and Khuller (1996) proved that this approach has an approximation ratio of $\ln \Delta + 3$. One can also use the Steiner tree algorithm to connect the dominators. This straightforward method gives an approximation ratio of $c[H(\Delta) + 1]$, where c is the approximation ratio for the unweighted Steiner tree problem. Currently, the best ratio is $1 + \frac{\ln 3}{2} \simeq 1.55$, which is due to Robins and Zelikovsky (2000).

The idea behind the spanning-tree-growing approach is as follows. It grows a tree T, starting from the vertex of the maximum node degree. At each step, we pick a vertex v in T and "scan it." When scanning a vertex, we add edges to T from v to *all* its neighbors that are currently not in T. In the end, we find a spanning tree T and pick the internal nodes of T as the CDS.

The scanning process works as follows. Initially, all vertices are unmarked (or marked white). When we scan a vertex (whose color is either gray or white), we color it black and mark all its neighbors that are not in T (by coloring them gray) and add them to T. Thus, the marked nodes that have not been scanned are always the leaves in T (these nodes have color gray). The algorithm continues scanning marked nodes until all the vertices are marked (gray or black). The set of black nodes (i.e., those already scanned) will be selected as the CDS when the tree T becomes a spanning tree (i.e., all nodes in G are marked black or gray). The main question is how to pick a vertex for scanning at every step. A natural choice that picks the vertex with the maximum number of unmarked (white) neighbors, unfortunately, does not work well in some instances, as shown in Guha and Khuller (1996). Then, Guha and Khuller defined a new operation of scanning a pair of *adjacent* vertices u and v. To scan a pair of such vertices, they need one of them to be *gray* (say, u) and the other to be *white* (say, v). To scan the pair of nodes, their algorithm first marks u black and marks all its white neighbors gray; then, it marks color node v black and all its white neighbors gray. The total number of nodes that are colored gray by scanning the pair of nodes is called the *yield* of a scan step. At each step, their algorithm will scan either a single vertex or a pair of vertices, whichever gives the higher yield. In other words, their algorithm is doing a one-step "look-ahead" by one extra vertex and is willing to scan a pair of vertices if this will yield a higher yield.

The preceding approach is summarized in Algorithm 13.

Then, Guha and Khuller proved the following theorem for this simple scanning method:

Algorithm 13 Construct a CDS by scanning (Guha and Khuller, 1996)

1: At the start of the first phase, all nodes are colored white.
2: **while** there is white nodes left **do**
3:     Find the vertex w that has the largest number of white neighbors among all vertices and also the pair of adjacent vertices u and v such that the union of their white neighbors has the largest size among all adjacent vertices u and v.
4:     Scan vertex w or the pair of vertices u and v, whichever will have the largest yield. When scanning a vertex, we mark its white neighbors to gray and mark itself to black.
5: **end while**

THEOREM 7.1 *(Guha and Khullar, 1996) Using the scanning rule just described yields a CDS of a size of at most* $2[1 + H(\Delta)]|\text{OPT}_{\text{DS}}|$, *where* OPT_{DS} *is the set of vertices in an optimal DS.*

An alternative approach to growing one connected tree is to grow separate components that form a DS and then connect them together. One straightforward approach is to find a DS by using a greedy heuristic and to use a Steiner tree approximation algorithm to connect the DS into a CDS. Because members of the optimum CDS along with the members of the DS found form a spanning tree, they can prove a performance guarantee of $c[1 + H(\Delta)]$, where c is the best approximation ratio for the unweighted Steiner tree problem, which currently is $c = 1 + \frac{\ln 3}{2} \simeq 1.55$ (from Robins and Zelikovsky, 2000).

The second algorithm by Guha and Khuller works as follows. The method (Algorithm 13) runs in two phases.

Algorithm 14 Construct a CDS by merging (Guha and Khuller, 1996)

1: At the start of the first phase, all nodes are colored white. Each time we include a vertex in the DS, we color it black. Nodes that are dominated by this node (i.e., all nodes that are adjacent to a black node) will be colored gray. In the first phase, the algorithm picks a node at each step and colors it black, coloring all adjacent white nodes gray. A *piece* is defined as a white node or a black connected component. Here, a *connected component* is an induced subgraph of the original graph; that is, connected. It is called a black connected component if all nodes are already colored *black*. At each step, we pick a node to color black that gives the maximum (nonzero) reduction in the number of pieces. It is easy to prove that at the end of the first phase, if no vertex gives a nonzero reduction to the number of pieces, then there are no white nodes left.

2: In the second phase, we need to connect the collection of black connected components. Then, the algorithm will simply pick a chain of two vertices to connect a pair of black components; we repeat this until one black component is formed. At last, the set of black vertices forms a connected component, which will serve as a CDS.

Notice that after the first phase, all nodes will be colored black or gray. Let C_1, $C_2, \ldots, C_k$ be the k-black-connected components formed after the first phase of the algorithm terminates. We define a virtual graph C by using C_i, $1 \le i \le k$, as its vertices. Two vertices C_i and C_j are connected by a link $C_i C_j$ if the graph distance between the black connected components C_i and C_j is at most 3. The weight of the link $C_i C_j$ is the graph distance between them in the original graph G. Here, the *graph distance* of two subsets C_i and C_j of vertices in a graph G is the length of the shortest path connecting some node from C_i and some node from C_j. It is well known that the graph C is connected. This can be proved by contradiction. If it is not connected, then the black connected components can be grouped into two subsets such that the graph distance between any component from one group to any component from another group is at least four. Pick a black connected component C_1 from the first group and C_2 from the other group such that their graph distance is the smallest. Let $v_1 v_2 \ldots v_h$ be the shortest path connecting them, where $v_1 \in C_1$ and $v_h \in C_2$ and $h \ge 5$. Let $a = \left[\frac{h}{2}\right]$. Node v_a is the middle node of the path. Notice that nodes $v_2, v_3, \ldots, v_{h-1}$ all have the color *gray*. For node v_a, it must have a neighboring node, say node u, whose color is *black* from the coloring procedure of the algorithm. Let C_p be the black connected component containing node u. It is easy to show that C_p cannot be C_1 or C_2 from the selection of C_1 and C_2 (otherwise, $v_1 v_2 \ldots v_h$ is not the shortest path connecting them). When C_p is different from C_1 and C_2, assume that C_p and C_1 are from the same group of black connected components. Then, the distance between C_p and C_2 is smaller than the distance between C_1 and C_2, which is a contradiction to the selection of C_1 and C_2 that has the smallest distance between components from one group and components from another group. Thus, the virtual group $\mathcal{C}$ is connected.

THEOREM 7.2 *(Guha and Khuller, 1996) The CDS found by Algorithm 14 is of a size of at most* $(\ln \Delta + 3)|\mathrm{OPT}|$.

Proof. This proof was originally presented in Guha and Khuller, 1996. Let a_i be the number of pieces left after the ith iteration. Clearly, $a_0 = n$. Because a node can connect up to Δ pieces at any stage, we have $|\mathrm{OPT}| \geq \frac{a_0}{\Delta}$. Consider the iteration $i + 1$. The optimum solution can connect a_i pieces. Hence, the greedy procedure is guaranteed to pick a node that connects at least $\lceil \frac{a_i}{|\mathrm{OPT}|} \rceil$ pieces. Then, we have the following recurrence relation about a_i:

$$a_{i+1} \leq a_i - \left\lceil \frac{a_i}{|\mathrm{OPT}|} \right\rceil + 1 \leq a_i \left(1 - \frac{1}{|\mathrm{OPT}|} \right) + 1.$$

This will give us the solution for a_{i+1} as

$$a_{i+1} \leq a_0 \left(1 - \frac{1}{|\mathrm{OPT}|} \right)^i + \sum_{j=1}^{i-1} \left(1 - \frac{1}{|\mathrm{OPT}|} \right)^j.$$

Recall that $a_0 \leq \Delta|\mathrm{OPT}|$. Notice that after $|\mathrm{OPT}| \ln \frac{a_0}{|\mathrm{OPT}|}$ iterations, the number of pieces left is less than $2|\mathrm{OPT}|$. For each node we choose, we reduce the number of pieces by at least one. Thus, at most $|\mathrm{OPT}|$ additional vertices will be picked (starting from the time that the number of black connected components is less than $2|\mathrm{OPT}|$) such that we will have at most $|\mathrm{OPT}|$ black connected components left. When there are at most $|\mathrm{OPT}|$ black connected components left, the number of additional vertices needed to connect them in the second phase is at most $2|\mathrm{OPT}|$. Thus, the total number of nodes in the CDS is at most

$$|\mathrm{OPT}| \ln \frac{a_0}{|\mathrm{OPT}|} + |\mathrm{OPT}| + 2|\mathrm{OPT}| \leq (\ln \Delta + 3)|\mathrm{OPT}|.$$

This finishes the proof of the theorem.   ∎

Notice that an approximation within $3 \ln \Delta$ is possible when each node in the graph has a nonnegative weight. In Chapter 8, we review such an algorithm in detail.

7.3.2   Algorithms for Graphs Modeling Wireless Ad Hoc Networks

For graphs that model wireless ad hoc networks, it is possible to design algorithms with much better approximation ratios; e.g., there exists a PTAS for approximating the MCDS problem for UDGs when the node positions are known in advance. Notice that given only a UDG $G = (V, E)$, it is NP-hard to find node positions for all nodes V such that the resultant UDG based on such position information is G.

By definition, any algorithm generating a maximal IS is a clustering method; i.e., any MIS is a DS. We first review the methods that approximate the MIS, the MDS, and the MCDS for UDGs. Hunt *et al.* (1998) and Marathe *et al.* (1995) studied the approximation of the MIS and the MDS for UDGs. They gave the first PTASs for a MDS in a UDG. The method is based on the following observations: A maximal IS is always a DS; given a square Ω with a fixed area, the size of any maximal IS is bounded by a constant C. Assume that there are n nodes in Ω. Then, we can enumerate all sets with a size of at most C in time $\Theta(n^C)$. Among these enumerated sets, the smallest dominating set is the MDS. Then, using the shifting strategy, they derived a PTAS for the MDS problem.

Because we have a PTAS for the MDS and the graph Virt$G$ connecting every pair of dominators within at most three hops is connected (Wang and Li, 2002b), we have an approximation algorithm [constructing a minimum spanning tree (MST) Virt$G$] for a MCDS with approximation ratio $3 + \epsilon$. Notice that Berman *et al.* (1998) gave a 4/3 approximation method to connect a DS, and Robins and Zelikovsky (2000) gave a 4/3 approximation method to connect an IS. Thus, we can easily have an $\frac{8}{3}(1 + \epsilon)$ approximation algorithm for a MCDS, which was reported in Alzoubi (2002). Recently, Cheng *et al.* (2003) designed a PTAS for a MCDS in a UDG. However, it is difficult to run their method efficiently in a distributed manner.

We now briefly review the PTAS of Cheng *et al.* (2003) to approximate the MCDS for a UDG. The general picture of this construction is as follows. First, they divide the space containing all vertices of the input UDGs into a grid of small cells. For each small cell, they take the nodes with distance h units away from the boundary of each cell (such an area is called the central area of a cell). Then, they optimally compute a minimum union of CDSs in each cell for connected components of the central area of the cell. The key observation is that the union of all such minimum unions is no more than the MCDS for the original whole UDG. Then, for vertices not in the central areas, they consider some additional vertices from the central area: They consider all nodes that are within a distance of no more than $h + 1$ units away from the boundary of the cell. They then use any constant-approximation algorithm of a MCDS for such a set of "cell boundary nodes." The final MCDS is just the union of the CDS for the central-area nodes and the CDS for the cell-boundary nodes. The tricky part again is to use the shifting argument proposed by Hochbaum to make sure that there is a total of at least half good partitions with good approximations overall.

7.4   Message Lower Bound for Distributed-Backbone Construction

In this chapter, we establish the $\Omega(n \log n)$ lower bound on the message complexity for distributed algorithms for leader election, spanning tree, and *nontrivial* CDS in WANs. Here, a CDS is nontrivial if it does not contain all nodes from the network. Notice that all nodes in the network clearly are a CDS. The reduction is first studied in Alzoubi *et al.*

(2002c) and Wan *et al.* (2002a) and is made from the following well-known bound on the message complexity of the distributed leader election in asynchronous ring networks with point-to-point transmission (Burns, 1980).

THEOREM 7.3 *(Burns, 1980). In asynchronous rings with point-to-point transmission, any distributed algorithm for leader election sends in at least $\Omega(n \log n)$ messages.*

The following two theorems were proved in Alzoubi *et al.* (2002c) and Wan *et al.* (2002a):

THEOREM 7.4 *In asynchronous wireless ad hoc networks whose UDG is a ring, any distributed algorithm for leader election sends at least $\Omega(n \log n)$ messages.*

Proof. Let $\mathcal{A}$ be any distributed algorithm for leader election in wireless ad hoc networks whose UDG is a ring. Let $\mathcal{A}'$ be the algorithm by replacing each wireless transmission with two point-to-point transmissions. Then, $\mathcal{A}'$ is a distributed algorithm for leader election in asynchronous rings with point-to-point transmission. Note that the algorithm $\mathcal{A}'$ sends twice the messages of those sent by $\mathcal{A}$. Thus, from Theorem 7.3, $\mathcal{A}$ must also send at least $\Omega(n \log n)$ messages.   ■

THEOREM 7.5 *In asynchronous wireless ad hoc networks whose UDG is a ring, any distributed algorithm for a spanning tree sends at least $\Omega(n \log n)$ messages.*

Proof. Let $\mathcal{A}$ be any distributed algorithm for a spanning tree in wireless ad hoc networks whose UDG is a ring. Note that any spanning tree of a ring consists of all edges in the ring except one. Thus, it has exactly two leaves, which are also neighbors. Thus, after a spanning tree is completed, the two leaves can exchange a message to select the leader between them according to some symmetry-breaking criterion; for example, by their IDs. After the leader is identified, it then notifies all other nodes in a linear number of messages. Thus, from algorithm $\mathcal{A}$, we can derive a distributed algorithm for leader election whose message complexity is $\Theta(n)$ more than the number of messages sent by $\mathcal{A}$. From Theorem 7.4, the message complexity of $\mathcal{A}$ is at least $\Omega(n \log n)$.   ■

A distributed algorithm for leader election in wireless ad hoc networks has been proposed in Cidon and Mokryn (1998). The algorithm proposed in Alzoubi *et al.* (2002c) and Wan *et al.* (2002a) has a message complexity $O(n \log n)$ and, therefore, is message efficient. Its actual implementation also constructs a spanning tree rooted at the leader.

THEOREM 7.6 *In asynchronous wireless ad hoc networks whose UDG is a ring, any distributed algorithm for a nontrivial CDS sends at least $\Omega(n \log n)$ messages.*

Proof. Let $\mathcal{A}$ be any distributed algorithm for a CDS in wireless ad hoc networks whose UDG is a ring. Note that for any nontrivial CDS, a ring consists of all nodes except either a unique node or two neighboring nodes. So, after a nontrivial CDS is completed, the leader can be elected as follows. A dominatee declares itself as the leader if both its neighbors are dominators, or one of its neighbor is a dominatee but has a larger ID. The leader then notifies all other nodes in a linear number of messages. Thus, from algorithm $\mathcal{A}$, we can derive a distributed algorithm for leader election whose message complexity is $\Theta(n)$ more than the number of messages sent by $\mathcal{A}$. From Theorem 7.4, the message complexity of $\mathcal{A}$ is at least $\Omega(n \log n)$.   ∎

Notice that this lower bound for the asymptotic message complexity for finding a CDS is true only if we always want a nontrivial CDS from the algorithm as the output. When we can tolerate the fact that the algorithm may output a trivial solution in some inputs, this lower bound will no longer hold, even if we consider all algorithms that have asymptotically optimum approximation ratios. For example, we will design algorithms that can find a CDS by using a linear number of messages, and the size of the output is always guaranteed to be within a small constant factor of the optimum.

7.5   Some Backbone-Formation Heuristics

7.5.1   Algorithm by Das *et al.*

The centralized version of the distributed algorithm proposed by Das *et al.* (1997) consists of the following three stages:

- The first stage finds an approximation to a MDS that is essentially the well-studied set-cover problem. Not surprisingly, the heuristic proposed by Das *et al.* (1997a, 1997b) is a translation of Chvátal's (1979) greedy algorithm for set cover and thus guarantees an approximation factor of $H(\Delta)$, where Δ is the maximum degree and $H()$ is the harmonic function.[1] Let U denote the dominating set output in this stage.
- The second stage constructs a spanning forest F. Each tree component in F is a union of stars centered at the nodes in U. The stars are generated when each dominatee node is allowed to pick up an arbitrary neighbor in U.
- The third stage expands the spanning forest F to a spanning tree T. All internal nodes in T form a CDS. It is easy to show that the CDS generated in this way contains at most $3|U|$ nodes and, therefore, the algorithm is a $3H(\Delta)$-approximation of a MCDS.

Das *et al.* showed by extensive simulations that this algorithm will find a CDS whose size is reasonably small. Alzoubi *et al.* (2002c) and Wan *et al.* (2002a) presented an example to show that in the worst case, the CDS found by the algorithm designed by Das *et al.* could be as large as $\Omega(\log n)$ times the minimum CDS for some UDGs modeling wireless networks. They further argued that the distributed implementation of the preceding greedy algorithm proposed in Das *et al.* (1997a, 1997b) has a very high time complexity and message complexity. Indeed, both time complexity and message complexity can be as high as $\Theta(n^2)$. They also noticed that such a distributed implementation is technically incomplete. For example, the distributed implementation consists of multiple stages, but

---

[1] $H(n) = 1 + \frac{1}{2} + \frac{1}{3} + \frac{1}{4} + \cdots + \frac{1}{n}$.

the implementation lacks mechanisms to bridge two consecutive stages; i.e., to let nodes know when to switch from one stage to the next stage. Thus, individual nodes have no way to tell when the next stage should begin. Although this technical incompleteness is possible to fix, it may take much effort. Furthermore, the approximation factor of the greedy algorithm is intrinsically poor. An interesting question is to study the average case performance of this method by assuming that the input wireless network nodes satisfy a certain distribution; e.g., uniform random distribution or Poisson distribution.

7.5.2   Algorithm by Wu and Li

The algorithm proposed by Das *et al.* is essentially a straightforward extension of the centralized method for constructing a CDS. Wu and Li recently proposed a sequence of methods that are purely localized algorithms. Remember that here, the localized algorithm for wireless ad hoc networks is defined as follows.

DEFINITION 7.1 *Assume that every node v has some information set Info(v) before a distributed algorithm $\mathcal{A}$ is performed. The distributed algorithm $\mathcal{A}$ for wireless ad hoc networks is a* localized algorithm *if every node u makes decisions purely on the information Info(v) collected from nodes v within a constant number of hops from u.*

Notice that for wireless networks, this definition is *different* from the traditional definition of a localized algorithm. For wired networks, a distributed algorithm is a localized algorithm if the algorithm will terminate after a constant number of rounds. With this traditional definition of a localized algorithm, it is impossible to design a localized algorithm for wireless ad hoc networks because of wireless interference. For example, assume that the information of each node will take one time slot to transmit. Then, if a node u has $d(u)$ neighbors, it will take at least $d(u)$ time slots for node u to collect the information Info(v) of all these neighbors because at any time slot, only one neighbor can communicate with u because of wireless interference. On the other hand, if node u uses only the information Info(v) of its one-hop neighbors, the algorithm is still considered to be a *localized algorithm* for wireless ad hoc networks throughout this book.

Whereas the algorithms proposed by Das *et al.* first find a DS and then grow this DS into a CDS, the algorithms proposed by Wu and Li (1999, 2000, 2001) and Wu *et al.* (2002) take an opposite approach. Their algorithms first find a CDS and then prune certain redundant nodes from the CDS. The initial CDS, say U, consists of all nodes that have at least two nonadjacent neighbors. A node u in U is considered *locally redundant* if it has either a neighbor in U with a larger ID that dominates all other neighbors of u or two adjacent neighbors with larger IDs that together dominate all other neighbors of u. The algorithm then removes all locally redundant nodes from U. This algorithm applies only to wireless ad hoc networks whose UDG is not a complete graph. Alzoubi *et al.* (2002c) and Wan *et al.* (2002a) presented a network example such that these algorithms, in the worst case, will find a solution whose size could be as large as $n/2$ times the optimum, where n is the total number of wireless nodes in the network.

The distributed implementation of the algorithms proposed in Wu and Li (1999, 2000, 2001) and Wu *et al.* (2002) typically runs in two phases. In the first phase, each node first broadcasts to its neighbors the entire set of its neighbors, and, after receiving this adjacency information from all neighbors, it declares itself as dominator if and only if it has two nonadjacent neighbors. These dominators form the initial CDS. In the second phase, a dominator declares itself as a dominatee if it is locally redundant. Note that a dominator can find whether it is locally redundant from the adjacency information of all its neighbors. It was shown in Alzoubi *et al.* (2002c) and Wan *et al.* (2002a) that the message complexity of this algorithm is $\Theta(m)$, where m is the number of edges in the graph modeling the wireless network, because each edge contributes two messages in the first phase.

7.5.3   Algorithm by Stojmenovic *et al.*

Stojmenovic *et al.* (2002) presented three synchronized distributed algorithms for constructing a CDS in the context of clustering and broadcast activity. In each of the three algorithms, the CDS consists of two types of nodes: the *cluster-heads* and the *border-nodes*. The cluster-heads form a maximal IS; i.e., a special dominating set in which any pair of nodes is nonadjacent in the graph modeling the wireless network. A node is a border-node if it is not a cluster-head, and there are at least two cluster-heads within its two-hop neighborhood. They find a set of cluster-heads based on a (total) ranking of nodes. Various different rankings have been used in the literature; e.g., the node's ID only (Gerla and Tsai, 1995; Lin and Gerla, 1997a), an ordered pair of degree and ID (Chen and Stojmenovic, 1999), and an ordered pair of degree and location (Stojmenovic *et al.*, 2002). The selection of the cluster-heads is given by a synchronized distributed algorithm, which can be generalized to the following framework:

- Initially all nodes are colored white.
- In each stage of the synchronized distributed algorithm, all white nodes that have the lowest rank among all white neighbors are colored black.
- Then, all white nodes adjacent to the black nodes are colored gray.
- The ranks of the remaining white nodes are updated.
- The algorithm ends when all nodes are colored either black or gray. All black nodes then form the cluster-heads.

Regardless of the choice of the ranking, the algorithms in Stojmenovic *et al.* (2002) have a $\Theta(n)$ worst-case approximation factor, which stems from the nonselective inclusion of all border-nodes: a node is selected as a border-node as long as it has at least two cluster-heads as neighbors. For example, consider a network formed by nodes $v_1, v_2, \cdots, v_n$ lying on a line segment and all possible links $v_i v_j$, except $v_1 v_n$, are communication links in the network. Further, assume that nodes v_1 and v_n have the two smallest ranks among all the nodes. It is easy to show that nodes v_1, v_n will be selected as cluster-heads and all other nodes will be selected as border-nodes. Obviously, if we let the non-cluster-head nodes collaboratively decide whether to serve as border-nodes, we could theoretically prove the performance of the algorithms. This is exactly what was done in Alzoubi *et al.* (2002a, 2002c), Wan *et al.* (2002a), and Y. Wang and Li (2002b). Notice that everything has a price: To make collaborative decisions, these non-cluster-heads will incur more communications. As will be shown later, these new algorithms do *not* significantly increase the message complexity: The number of additional messages is only $\Theta(n)$ when the network is modeled by UDGs.

Notice that in certain scenarios, these algorithms may fail to find a CDS in the worst case. Consider four nodes, v_1, v_2, v_3, and v_4, on a line segment in this order. Further, assume that only links v_1v_2, v_2v_3, and v_3v_4 exist in the communication graph and that the ranks of nodes are rank $(v_1) = 1$, rank $(v_2) = 3$, rank $(v_3) = 4$, and rank $(v_4) = 2$. It is easy to show that nodes v_1 and v_4 will be selected as the cluster-heads. However, neither node v_3 nor v_2 will be selected as border-nodes because neither of these two nodes has at least two neighbors in the cluster-heads.

7.6   Efficient Distributed-Nontrivial-Backbone-Formation Method

We first study some algorithms that will always construct a nontrivial CDS for a connected network with at least two nodes. The algorithms were first proposed by Alzoubi (2002), Alzoubi *et al.* (2002c), and Wan *et al.* (2002a). The distributed algorithm for a CDS consists of two phases. These two phases construct a maximal IS and a dominating tree, respectively. They are described and analyzed in the next three subsections.

7.6.1   MIS Construction

By definition of a maximal IS, any pair of nodes in a maximal IS is separated by at least two hops in the original graph. However, a subset of nodes in a maximal IS may be three hops away from the subset of the rest nodes in this maximal IS. Here, given a graph $G = (V, E)$ and two subsets $V_1 \subset V$ and $V_2 \subset V$, the *hop distance* between V_1 and V_2, denoted by $d_G(V_1, V_2)$, is defined as

$$d_G(V_1, V_2) = \min_{x \in V_1, y \in V_2} d_G(x, y),$$

where $d_G(x, y)$ is the minimum hop length of paths connecting nodes x and y in graph G. It has been a folklore result that for any maximal IS U and any node $v \in U$, the hop distance $d_G(\{v\}, U \setminus \{v\})$ is at most 3.

In this section, we study an algorithm in Alzoubi (2002) that guarantees that the hop distance between any node v in the maximal IS and the rest of the maximal IS nodes are *exactly* two hops. The basic idea of Alzoubi's method is to partition nodes into layers and then find the maximal IS layer by layer. Alzoubi's construction uses a carefully chosen rank definition. The ranking is induced by an arbitrary rooted spanning tree T, which can be constructed by the distributed leader-election algorithm in Burns (1980), with $\Theta(n)$ time complexity and $O(n \log n)$ message complexity. Notice that for a general wireless ad hoc network, all nodes are uniform; thus, we need a leader election to find the root, which takes $O(n \log n)$ messages. In some special networks, such as WSNs, we already have a base station that will serve as the sink node for all other wireless sensor nodes. Consequently, we do not need to run the leader-election algorithm. When a leader is known, we can construct a spanning tree by using only $O(n)$ messages by simple flooding; e.g., we can construct a breadth-first-search (BFS) tree by using $O(n)$ messages when a leader (root of BFS) is known. In such cases, the algorithm presented later actually has a message complexity of only $O(n)$ because the rest of the operations in constructing a CDS cost $O(n)$ messages.

Given a rooted spanning tree T, the (tree) level of a node is the number of hops in T between itself and the root of T. The level of the root is 0. The rank of a node is then given by the ordered pair of its level and its ID. Such ranking gives rise to a total ordering of the nodes in lexicographic order. The following distributed process enables each node to calculate its own rank and the number of lower-ranked neighbors. For consistency, we use the same notations as in Alzoubi (2002) to describe the method. Each node will maintain several local variables, as follows:

1. The variable x_1 counts the number of neighbors whose levels have not yet been identified; thus, its initial value is set as the number of its neighbors in the communication graph G.
2. The variable x_2 counts the number of children who have not yet reported the completion and is thus initialized to the number of children in the tree T.
3. The variable LevelList (which is a list) records the levels of its neighbors and is initially empty.
4. The local variable y stores the number of lower-ranked neighbors.

After the rooted spanning tree T is constructed, the root node announces its level 0 by broadcasting a Level message. On receiving a Level message, a node appends an entry consisting of the sender's ID and level to LevelList and then decreases variable x_1 by 1. When variable $x_1 = 0$, it sets variable y to the number of lower-ranked neighbors that can be calculated from LevelList. If the sender of the Level message is its parent in T, it sets its own level to one plus the sender's level and then announces this level by broadcasting a Level message. If the sender is a leaf in T (i.e., $x_2 = 0$ initially) and its own level has been determined, it transmits a LevelComplete message to its parent.

For any node, on receiving a LevelComplete message toward itself, the node reduces its local variable x_2 by 1. When $x_2 = 0$ after the update and the node is not the root, the node transmits a LevelComplete message to its parent and then resets x_2 to be the number of its children nodes in T. When the root node has its local variable $x_2 = 0$, the root node resets x_2 to the number of children. Observe that when $x_2 = 0$ happens for the root node, every node in the network now knows its own rank and the ranks of all its neighbors. Then, the root node will move to the next phase: constructing a MIS by a color-marking process.

All nodes are initially marked with white color and will be marked with either gray or black eventually. The nodes marked with black color will form the final MIS constructed by their algorithm. Each node also maintains a BlackList that records the IDs of its black neighbors. Note that the BlackList can contain at most five black nodes when the network is modeled by a UDG. This is based on the observation that for each node, any subset of its neighbors that is independent (i.e., no links among them) has a size of no more than 5.

The algorithm starts from the root node and will always choose the root node to the final MIS. The root first marks itself black and broadcasts a message BLACK. Here, the purpose of the message BLACK is to inform its neighbors that this node has already joined the MIS. Consequently, all its neighbors should exclude themselves from joining the MIS. On receiving a message BLACK, a node adds the sender's ID to BlackList, and if its color is still white, it marks itself gray and broadcasts a message GRAY that contains its level. On receiving a message GRAY, if the rank of the sender is lower than its own, a white node decreases the value y by 1; if $y = 0$ after the update, it marks itself black and broadcasts a BLACK message to all its neighbors. This means that a node v joins the MIS only if it has the largest rank among all its white neighboring nodes. When a leaf node is marked with either gray or black, it transmits a message Mark-Complete to its parent. On receiving a MARK-COMPLETE message toward itself, a node reduces the variable x_2 by 1; if $x_2 = 0$ after the update and the node itself is not the root, the node transmits a message Mark-Complete to its parent. By the time the local variable $x_2 = 0$ at the root, all nodes have been marked with either gray or black; thus, the root will move on to the construction of the CDS. Observe that the message Mark-Complete is not needed to ensure that the maximal IS is correctly constructed. The main purpose of this message is to let the root node know that the maximal IS has been constructed.

The construction of the CDS in the next phase relies on the following property of the black nodes.

THEOREM 7.7 *All black nodes form a maximal IS U and, for any node v in the maximal IS U, $d_G(\{v\}, U \setminus \{v\}) = 2$.*

Proof. Let $U = \{u_i \mid 1 \leq i \leq k\}$, where u_i is the ith node that is marked black. From the construction, any pair of black nodes not adjacent to each other and thus U is a MIS. For any $1 \leq j \leq k$, let H_j be the graph over $\{u_i \mid 1 \leq i \leq j\}$ in which a pair of nodes is connected by an edge if and only if their graph distance in G is two; i.e., they are connected by a shortest path of two hops. We then prove by induction on j that H_j is connected. Because H_1 consists of a single vertex, it is connected trivially. Assume that H_{j-1} is connected for some $j \geq 2$. When the node u_j is marked black, its parent in T must be already marked gray. Thus, there is some node u_i with $1 \leq i < j$ that is adjacent to the parent of node u_j in T. So, (u_i, u_j) is an edge in H_j. Because H_{j-1} is connected, so must be H_j. Therefore, by induction, H_j is connected for any $1 \leq j \leq k$. This implies that for any node v in the MIS U, $d_G(\{v\}, U \setminus \{v\}) = 2$. This finishes the proof. ∎

7.6.2   Connected-Dominating-Set Construction

After a maximal IS is found with the preceding approach, we connect it to a tree, and this tree will serve as a CDS of the input graph. To construct a tree, denoted by $T^\star$, each node will maintain several additional local variables, as follows:

1. A local boolean variable z that is initialized to 0 and set to 1 after the node joins the tree $T^\star$. In other words, z is an indicator of whether this node is in the final CDS.
2. A local variable Parent that stores the ID of its parent in $T^\star$ and is initially empty.
3. A variable childrenList that records the IDs of its children in $T^\star$ and also is initially empty.

The root of $T^\star$ is a (gray) neighbor of the root of T, which has the largest number of black neighbors. To select the root for $T^\star$, the root of T also maintains a variable root and a variable degree that is initialized to 0. The root of T first resets the local variable x_1 to the number of its neighbors and then broadcasts a QUERY message. On receiving a QUERY message, a (gray) node transmits to the sender of the QUERY message a REPORT message that contains the number of its black neighbors. On receiving a REPORT message toward itself, the root of T reduces its variable x_1 by one, and if the number of black neighbors of the sender is greater than the value of degree, it resets degree to the number of black neighbors of the sender and also resets the variable root to the ID of the sender. If $x_1 = 0$ after the update, the root of T transmits a ROOT message to the node whose ID is stored in the local variable root. On receiving the ROOT message toward itself, a node becomes the root of $T^\star$. It sets $z = 1$ and then broadcasts an INVITE2 message. All other nodes join the tree $T^\star$ according to the following rules:

- On receiving an INVITE2 message, a black node with $z = 0$ sets $z = 1$ and parent to the ID of the sender, transmits a JOIN message toward the sender, and then broadcasts an INVITE1 message.
- On receiving an INVITE1 message, a gray node with $z = 0$ sets $z = 1$ and parent to the ID of the sender, transmits a JOIN message toward the sender, and then broadcasts an INVITE2 message.
- On receiving a JOIN message toward itself, a node adds the ID of the sender to childrenList.

Here, "broadcast message" means that a node sends a message to all its neighbors within its transmission range.

Theorem 7.7 guarantees that whenever there is any black node outside the current $T^\star$, at least one black node would join $T^\star$. Thus, eventually all black nodes will join $T^\star$. Consequently, all gray nodes will join $T^\star$ eventually. The internal nodes of $T^\star$ form a CDS.

7.6.3   Properties of Constructed Backbone

We now analyze the message complexity and time complexity of this method as done by Alzoubi *et al.* (2002c) and Wan *et al.* (2002a). After the rooted spanning tree T is constructed, the MIS construction in the first phase additionally uses linear messages and takes at most linear time. The construction of the dominating tree $T^\star$ also uses linear messages and takes at most linear time. Thus, besides the construction of the tree T, the algorithm uses $O(n)$ messages and takes $O(n)$ time. Because the algorithm in Burns (1980) used for the construction of T has $O(n \log n)$ message complexity and $O(n)$ time complexity, the previously studied algorithm has $O(n \log n)$ message complexity and $O(n)$ time complexity overall. Note that the message complexity of this algorithm is dominated by the construction of a rooted spanning tree. Wan *et al.* (2002a) also proved that the algorithm presented in the preceding subsection will construct a CDS whose size is at most eight times the size of a MCDS when the original communication graph is a UDG. Their proof is based on the following key lemma.

LEMMA 7.1 *The size of any IS in a UDG $G = (V, E)$ is at most $4opt + 1$, where opt is the size of the MCDS.*

Proof. Let U be *any* IS of V, and let T' be *any* spanning tree of the MCDS (denoted by opt). Consider an arbitrary preorder traversal of T', say $v_1, v_2, \ldots, v_{opt}$. We then partition the nodes in U to opt disjoint subsets. Let U_1 be the set of nodes in U that are adjacent to v_1. For any $2 \le i \le$ opt, let U_i be the set of nodes in U that are adjacent to v_i but none of the previous nodes $v_1, v_2, \ldots, v_{i-1}$. Then, $U_1, U_2, \ldots, U_{opt}$ form a partition of U as OPT is a DS. Because v_1 can be adjacent to at most five independent nodes, we have $|U_1| \le 5$. For any i with $2 \le i \le$ opt, at least one node from the following nodes $v_1, v_2, \ldots, v_{i-1}$ is adjacent to v_i, say, node v_j is inside the disk centered at node v_i. Notice that nodes from U_i cannot lie inside the unit disk centered at node v_j. Thus, U_i lies in a sector of at most $240°$ within the coverage range of node v_i. This implies that $|U_i| \le 4$. Therefore,

$$|U| = \sum_{i=1}^{opt} |U_i| \le 5 + 4(opt - 1) = 4opt + 1.$$

This completes the proof. ∎

7.7   Efficient Distributed-Backbone-Formation Method

In the previous section, an algorithm was presented that will always find a nontrivial CDS whose message complexity is $O(n \log n)$, which matches the lower bound of any distributed algorithm constructing a nontrivial CDS. When we do not require the constructed CDS to be always nontrivial, more efficient algorithms can be designed, which are studied in this section. These algorithms for building a CDS typically have two phases: clustering and finding connectors (also called *gateways*). The clustering algorithm basically finds a subset of nodes such that the rest of the nodes are visible to at least one of the cluster-heads. By definition, any algorithm generating a maximal IS is a clustering method. Various methods can then be used to connect the cluster-heads to form a connected graph. For the completeness of presentation, we review some priori arts on building a CDS, a MCDS, and a localized Delaunay graph. We interchange the terms *cluster-head* and *dominator*. The node that is not a cluster-head is also called an *ordinary* node or a *dominatee*. A node is called a *white* node if its status is yet to be decided by the clustering algorithm. Initially, all nodes are white. The status of a node, after the clustering method finishes, could be a *dominator* or a *dominatee*.

7.7.1 Clustering

Many algorithms for clustering have been proposed in the literature (Alzoubi, 2002; Alzoubi *et al.*, 2002c; Amis and Prakash, 2000; Amis *et al.*, 2000; Baker *et al.*, 1984; Baker and Ephremides, 1981; Basagni, 1999; Chlamtac and Farago, 1999; Das and Bharghavan, 1997; Lin and Gerla, 1997a; Lin *et al.*, 1991; Wu and Li, 2001). All algorithms assume that the nodes have distinctive identities (denoted by ID hereafter). We typically review those by Baker and Ephremides (1981), Baker *et al.* (1984), Alzoubi (2002), and Alzoubi *et al.* (2002b). For the sake of a general description of these priori arts, we summarize them using our own words. The well-known methods for building a DS typically use two messages, lamDominator and lamDominatee and have the following procedures:

- A white node claims itself to be a dominator if it has the smallest ID among all of its white neighbors, if there are any, and broadcasts lamDominator to its one-hop neighbors.
- A white node receiving the lamDominator message marks itself as dominatee and broadcasts lamDominatee to its one-hop neighbors.

The set of dominators generated by the preceding method is actually a maximal IS because no two adjacent nodes will be marked as dominators. Here, we assume that each node knows the IDs of all its one-hop neighbors, which can be achieved by requiring each node to broadcast its ID to its one-hop neighbors initially. This protocol can be easily implemented with synchronous communications as was done in Baker and Ephremides (1981) and Baker *et al.* (1984). If the number of neighbors of each node is known a priori, then this protocol can also be implemented with asynchronous communications. Here, knowing the number of neighbors ensures that a node gets all the updated information on its neighbors so it knows whether it has the smallest ID among all the white neighbors.

After clustering, one dominator node can be connected to many dominatees. However, it is well known that a dominatee node can be connected to at most only five dominators in the UDG model. For the completeness of presentation, a short proof is included here.

LEMMA 7.2 *For every dominatee node v, it can be connected to at most five dominator nodes in a UDG model.*

Proof. For the sake of contradiction, assume that a node v has six dominator neighbors. We know that the unit disk centered at v must have two dominator neighbors w and u; the angle $\angle wvu$ is at most $\pi/3$. So, the distance between w and u must be no more than one unit, which means that there is an edge between w and u in a UDG. This is a contradiction to the definition of a maximal IS. ∎

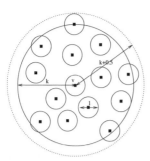

Figure 7.2 For every node v, the number of dominators within k-units is bounded by a constant ℓ_k. © 2005, ACM.

Generally, it is well known that for each node (dominator or dominatee), there are at most a constant number of dominators that are at most k-units away. *The following lemma bounds the number of dominators within k-units from a dominator node v.*

LEMMA 7.3 *For every* dominator *node v, the number of dominators inside the disk centered at v with a radius of k-units is bounded by a constant ℓ_k.*

Proof. Because any two dominators are *more than* one unit away, the half-unit disks centered at dominators are disjoint with each other. In addition, all such dominators should be in the *annulus* centered at v *of radii* 1 and k. Then, ℓ_k is bounded by how many disjoint half-unit disks we can park in the *annulus* centered at v *of radii* 0.5 and $k + 0.5$ (see Figure 7.2). We have $\ell_k < \frac{\pi(k+0.5)^2 - \pi(0.5)^2}{\pi(0.5)^2} = (2k+1)^2 - 1$ by using an area argument. When $k = 2, 3, 4$, we have $\ell_k < 24, 48, 80$. ∎

The bounds on l_k can be improved by a tighter analysis. The previous lemma implies that for every node v, the number of dominators within k-hops is bounded by a constant ℓ_k.

Almost all proposed clustering methods are similar to this synchronous protocol. The differences of the previous methods in approximating a MCDS lie in how to find gateways to connect these cluster-heads and whether they provide performance guarantees. For example, the basic algorithm for constructing a CDS proposed in Gerla and Tsai (1995) does not guarantee that the constructed clusters are connected. In some cases, it needs a *distributed gateway* to connect some clusters that are nonoverlapping. But, how to choose the distributed gateways was not specified. Additionally, no performance guarantee was proved. Baker and co-workers (1981, 1984) consider in detail how to select the gateway nodes to connect the clusters based on cases of overlapping clusters and nonoverlapping clusters. Here, two clusters (headed by two different cluster-heads) are said to be overlapping if there is at least one common dominatee node; they are said to be nonoverlapping if two dominatee nodes (one from each cluster) are connected. However, they did not prove the message complexity of their protocols or the approximation ratio of the generated CDS. Additionally, as they agreed, it may generate two or perhaps more gateway pairs for some nonoverlapping clusters pair. On the other hand, Alzoubi (2002) and Alzoubi *et al.* (2002b) proposed a localized method to find connectors by using a total of $O(n)$ messages and showed that the constructed CDS is within a constant factor of optimum. This property enables us to build a planar spanner in a total linear number of messages, which is crucial for wireless ad hoc networks because the communication is the highest energy-consuming operation. Actually, we will show that a modification of the method by Baker *et al.* also has a linear number of messages and the size of the constructed structure is also within a constant factor of optimum. We discuss in detail these two methods, which will be the first phase of our hybrid method for building a planar backbone.

7.7.2   Finding Connectors

The second step of CDS formation is to find some *connectors* (also called *gateways*) among all the dominatees to connect the dominators. Then, the connectors and the dominators form a *CDS* (also called a *backbone*).

Given a dominating set S, let Virt$G$ be the graph connecting all pairs of dominators u and v if there is a path in the UDG connecting them with at most three hops. It is well known that the graph Virt$G$ is connected.

THEOREM 7.8 *The graph VirtG is connected.*

Proof. We prove this by contradiction. Assume that Virt$G$ is not connected. Then, there are at least two components in Virt$G$. Let u and v be two different dominators from two different components, and the shortest path connecting them in G has the minimum number of hops among all pairs of dominators from different components. Then, the path connecting u and v has at least four hops (otherwise, u and v are connected in Virt$G$). Let node w be the node on the path that is at least two hops away from both u and v. Clearly, node w must be a dominatee node [otherwise, it is a contradiction to the selection of dominator pair (u, v)]. Then, let node x be the dominator of node w. Either x is not in the same connected component with u in Virt$G$ or is not in the same connected component with v in Virt$G$ (otherwise, u and v are connected in Virt$G$). In either case, we find a pair of dominators from different components that has a path with a smaller number of hops than that of the dominator pair (u, v).   ■

This observation is the basis for several algorithms (Baker and Ephremides, 1981; Baker *et al.* 1984; Gerla and Tsai, 1995; Lin and Gerla, 1997a) for a CDS, although no proofs were given in these previous results. It is natural to form a CDS by finding connectors to connect any pair of dominators u and v if they are connected in Virt$G$. This strategy was used in several previous methods (Alzoubi, 2002; Alzoubi *et al.*, 2002c; Baker and Ephremides, 1981; Baker *et al.*, 1984; Lin and Gerla, 1997a). Let $\Pi_{\text{UDG}}(u, v)$ be the path connecting two nodes u and v in a UDG with the smallest number of hops, where u has a smaller ID than v. Let's first consider how to connect two dominators within three hops. The method of Alzoubi (2002) and Alzoubi *et al.* (2002b) chose the connectors as follows: (1) if the path $\Pi_{\text{UDG}}(u, v)$ has two hops, then u selects the dominatee node that comes first to the notice to connect u and v; (2) if the path $\Pi_{\text{UDG}}(u, v)$ has three hops, then u selects the node, say w, that comes first to the notice such that w and v are two hops apart. Then, node w selects the node that comes first to the notice to connect w and v. Thus, basically, node u will decide the next node on the path to connect to another node v.

Notice that the preceding approach is different from the one adopted by Baker and co-workers (1981, 1984). In their protocols, they let the dominatee nodes decide whether or not they will serve as the connectors (gateways). For example, if a dominatee node finds that it is dominated by two nonadjacent dominators, say u and v, it claims itself as a candidate of the connectors for u and v. The node with the highest ID among the nodes in the intersection area covered by nodes u and v is chosen as the gateway node for the node pair u and v. In other words, they let the nodes in this intersection area elect the one with the highest ID, but no detailed protocol is given to do so. For the case of nonoverlapping clusters, a pair of adjacent dominatees (one from each cluster) needs to claim themselves as the candidates for the gateways of these two clusters. They always select the pair of dominatees with the largest sum of ID numbers. In the case of a tie, the pair involving the node with the highest ID number is chosen. However, here we may end up with two or perhaps more gateway pairs. The existence of one pair may not be known to both partners of the other pair (Baker and Ephremides, 1981). This cannot be avoided without increasing the communications (Baker and Ephremides, 1981).We modify the method of Baker *et al.* and show that it does approximate a CDS by using the linear number of communications. We then discuss in detail the approach to optimize the communication cost and the memory cost. It uses the following primitive messages (some messages are used in forming clusters):

- IamDominator(u): Node u tells its one-hop neighbors that u is a dominator.
- IamDominatee(u, v): Node u tells its one-hop neighbors that u is a dominatee of node v.
- TryConnector(u, w, v, i): Node w proposes to its one-hop neighbors that it could be one of the connectors to connect two dominators u and v. Integer i specifies whether it is the first or the second node on the path to connect u and v. If uv are two hops apart, then set $i = 0$.
- IamConnector(u, w, v, i): Node w tells its one-hop neighbors that it is the connector to connect two nodes u and v.

Notice that the message IamDominator(u) is broadcast at most only once by each node; the message IamDominatee(u, v) is broadcast at most only five times by each node u for all possible dominators v from Lemma 7.2. From Lemma 7.3, we know that TryConnector(u, w, v, i) is also broadcast at most a constant time by each node for all possible dominators u and v.

LEMMA 7.4 *Each node has to send out at most a constant number of messages in forming a CDS.*

Each node uses the following link lists:

- **Dominators:** It stores all dominators of u if there are any. Notice that if the node itself is a dominator, no value is assigned for **Dominators**.
- **2HopDominators:** It stores all dominators v that are two hops apart from u.

From Lemma 7.3, for each node u, there are at most ℓ_k number of dominators v that are k-hops apart from u. The size of each list is bounded by ℓ_1 and ℓ_2, respectively. Now we are in the position to discuss the distributed algorithm for finding connectors, which are built on the framework of Baker and co-workers (1981, 1984), Alzoubi (2002), and Alzoubi *et al.* (2002b). Assume that a maximal IS is already constructed by a cluster algorithm.

Algorithm 15 Finding connectors

1: Every dominatee w broadcasts message lamDominatee(w, v) indicating that w is a dominatee of v.

2: Every node x stores its two-hop-away dominators from the messages lamDominatee(w, v) broadcast by its neighbor w. Additionally, for each two-hop-away dominator node v, it stores a unique neighbor node w that connects it and v. Node w could be the one with the smallest ID or the largest remaining battery power.

3: Every dominatee node w broadcasts to its one-hop neighbors a message TryConnector(u, w, v, 0) for every dominator pair u and v (stored at Dominators).

4: If node w has the smallest ID among all its neighbors that sent TryConnector((u, $*$, v, 0)), then node w broadcasts lamConnector(u, w, v, 0).

5: Every dominatee node w broadcasts to its one-hop neighbors TryConnector(u, w, v, 1) for its dominator u and a two-hop-away dominator v. To save communications and decrease the number of connectors produced, we can let node w broadcast such a message if u has a smaller ID than v.

6: Similarly, if w has the smallest ID among all its neighbors sending TryConnector(u, $*$, v, 1), then node w broadcasts a message lamConnector(u, w, v, 1) to its one-hop neighbor.

7: Node w also sends message to the dominatee node x selected by w to connect v, asking it to be the connector. After receiving such a message, dominatee node x broadcasts to its one-hop neighbors a message lamConnector(u, x, v, 2) for the two-hop-away dominator u and its dominator v.

It is possible that there may have been multiple paths selected to connect two dominators u and v. See Figure 7.3 for an illustration. However, this increases the robustness of the backbone.

For each two-hop-away dominator pair u and v, there are at most two nodes claiming it to be connectors for them. This is because we can put at most two nodes inside the line defined by u and v such that they cannot hear each other directly. Notice that if two nodes can hear each other (i.e., neighbors), then they cannot both have the smallest ID among all their neighbors that sent TryConnector(u, $*$, v, 0). Thus, there are at most two connectors introduced for two dominators that are two hops apart. See the lefthand side of Figure 7.3 for an illustration.

For two dominators that are three hops away, it is obvious that there are at most five nodes sending out lamConnector(u, $*$, v, 1). Moreover, each such sent message will trigger at most another node to send out the message lamConnector(u, $*$, v, 2). Thus, there are at most five connectors introduced for two dominators. (This number can be improved by tighter analysis, but here we are interested in showing that it is bounded by a constant.)

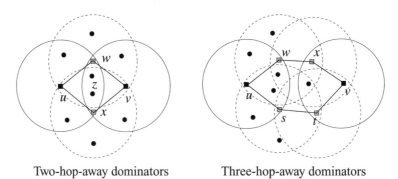

Two-hop-away dominators   Three-hop-away dominators

Figure 7.3 Multiple connectors found for one pair of dominators. © 2003, IEEE.

Consequently, the total number of connectors introduced is at most a constant factor of the number of dominators in the graph. It is well known that a MDS in a UDG can be approximated within a constant of 5. So, the preceding method will generate a CDS whose size is within a constant factor of the minimum. Additionally, it is obvious that the number of communications by each node is bounded by a constant: There are a constant number of dominator pairs (u, v), one within two hops and one within one hop of a dominatee node, and for each pair the communication is at most two: one message for claiming itself as connector candidate and one message for claiming itself (if necessary) as connector.

Additionally, instead of using the smallest ID to determine which node will serve as the connector for dominators pair u and v, the following criteria may perform better practically:

1. The node w that broadcasts the message TryConnector first for the same node pair serves as the connector, and the other node (of neighbors w) will not try to broadcast the message TryConnector for the same node pair. This strategy is feasible because only one node (within the transmission range) can send a message at any specific time for wireless ad hoc networks.
2. Or, the node with the largest weight will serve as the connector. Here, the weight could be its remaining battery power, the reciprocal of its moving speed (so the connector will not be recomputed frequently), or some other characteristics important for WANs. For more such criteria, see Basagni (1999) and Basagni et al. (1997).

The method discussed here only uses one-hop-neighbor information and a constant number of two-hop-neighbor information by the dominatee nodes to decide whether they could serve as connectors. Notice that the method of Alzoubi (2002) and Alzoubi et al. (2002b) uses one-hop-neighbor information and a constant number of two-hop- and three-hop-neighbor information to select connectors by the dominator nodes. It is interesting to see the practical performance differences of these two methods in a mobile environment. The method proposed here should perform better in terms of updating structures in a mobile environment.

The graph constructed by the preceding algorithm is called a CDS graph (or, the *backbone* of the network). If we also add all edges that connect all dominatees to their dominators, the graph is called an extended CDS, denoted by CDS′. The CDS induces a graph: Two nodes are connected if and only if their distance is no more than one unit. The induced graph is called induced connected dominating set (ICDS) graph. Obviously, the CDS is a subgraph of an ICDS. If we also add all edges that connect all dominatees to their dominators, the graph is called an extended ICDS, denoted by ICDS′. Later, we prove that both CDS′ and ICDS′ are the hop and distance spanners; both CDS and ICDS have a bounded node degree. Graphs ICDS and ICDS′ can be constructed with only one message for each node (to tell its neighbors whether it is a dominator, dominatee, or connector node) if a CDS is constructed.

7.7.3   The Properties of Backbone

This subsection shows that the CDS′ graph is a sparse spanner in terms of both hops and length; meanwhile, the CDS has a bounded node degree.

LEMMA 7.5 *The node degree of the CDS (the number of neighbors in the CDS) is bounded by a constant for any node in the CDS.*

Proof. Consider any node u. There are two cases: u is either a dominator node or a connector node.

For a dominator u, it can be connected only to some connectors w, which must have some dominators v that are one hop or two hops away from w. From Lemma 7.3, we know that the number of this kind of dominator v is bounded by ℓ_3. When v and u are two hops apart, there are at most two connectors introduced for them. When v and u are three hops apart, there are at most 10 connectors introduced for them and at most 5 of them connected to u. So, the degree of u is also bounded by $2\ell_2 + 5\ell_3$.

For a connector w, it can be connected to at most ℓ_1 dominator nodes and to some connectors. Each of these connectors p (within the transmission range of w) should be directly connected to some dominator q; then, the number of this kind of dominator q is bounded by ℓ_2. And, for each such dominator, it introduces at most $2\ell_2 + 5\ell_3$ connectors. So, the degree of w is bounded by a constant $\ell_1 + \ell_2(2\ell_2 + 5\ell_3)$.   ∎

The preceding lemma immediately implies that CDS is a sparse graph; i.e., the total number of edges is $O(k)$, where k is the number of dominators. Moreover, the graph CDS′ is also a sparse graph because the total number of the links from dominatees to dominators is at most $5(n − k)$. Notice that we have at most $n − k$ dominatees, each of which is connected to at most five dominators. However, the degree of some dominator node in CDS′ may be arbitrarily large because some dominator node could have many dominatees.

We then show that for every dominatee node v, there are at most a constant number of neighbors in the CDS. Remember that we select a maximal IS as the DS of the network and then select the connectors to connect these dominators to form a CDS. For node v, we select at most five nodes from its neighbors as dominators. The other

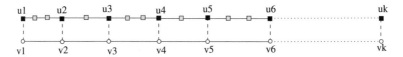

Figure 7.4 CDS′ is a hop spanner. © 2003, IEEE.

neighbors of v in the constructed CDS must be the connectors selected to connect some dominators, which can be at most two hops away from the node v. Notice that the number of dominators within two hops of v is at most ℓ_2. The number of connectors introduced by a dominator node is at most a constant ℓ_3. Thus, the total number of neighbors of v that are in the CDS is at most $\ell_1 + \ell_2\ell_3$. Thus, we have the following lemma:

LEMMA 7.6 *For any node not in the CDS, the number of neighbors in the CDS is bounded by a constant.*

After we construct the backbone CDS and the ICDS′ graph, if a node u wants to send a message to another node v, it follows this procedure: If v is within the transmission range of u, node u directly sends a message to v. Otherwise, node u asks its dominator to send this message to v (or one of its dominators) through the backbone. Then, we show that the CDS′ (plus all implicit edges connecting dominatees that are no more than one unit apart) is a good spanner in terms of both hops and length. In the following proofs, we use $\Pi_{G_h}(s, t)$ and $\Pi_{G_l}(s, t)$ to denote the shortest hop path and the shortest length path in a graph G from node s to node t. Let $l(\Pi)$ and $h(\Pi)$ be the length and the number of hops of path Π, respectively. The following proof was similar to that presented by Gao *et al.* (2001b). Alzoubi (2002) also gave a similar proof for his method. However, our proof shows that given any two nodes s and t, there is a *unique* path such that its length is no more than a constant factor of $l(\Pi_{UDG_l}(s, t))$, and its hops are no more than a constant factor of $h[\Pi_{UDG_h}(s, t)]$.

LEMMA 7.7 *The hops stretch factor of CDS′ is bounded by a constant* 3.

Proof. Assume that the shortest hop path from s to t in the UDG is $\Pi_{UDG_h}(s, t) = v_1v_2\ldots v_k$, where $v_1 = s$ and $v_k = t$, as illustrated by Figure 7.4. We construct another path in the CDS′ from s to t, and the number of hops of this path is at most $3k + 2$.

For each node v_i in the path $\Pi_{UDG_h}(s, t)$, let u_i be its dominator if v_i is not a dominator; otherwise, let u_i be v_i itself. Notice that there is a three-hop path $u_iv_iv_{i+1}u_{i+1}$ in the original UDG. Then, from Algorithm 15 in Subsection 7.7.2, we know there must exist one or two connectors connecting u_i and u_{i+1}. Obviously, nodes u_1 and u_k are connected by a path $\Pi_{CDS'}(u_1, u_k)$ in CDS′ using at most $3k$-hops. It implies that nodes s and t are connected by a path $\Pi_{CDS'}(s, t)$ [link su_1 followed by $\Pi_{CDS'}(u_1, u_k)$], followed by link u_kt with at most $3k + 2$ hops in CDS′. Thus, the hop-stretch factor of the CDS′ is bounded by 3 (with an additional constant of 2). ∎

LEMMA 7.8 *The length-stretch factor of CDS′ is bounded by a constant* 6.

Proof. Given any two nodes s and t such that $\|st\| > 1$, we will show that the path $\Pi_{\text{CDS}'}(s, t)$ constructed in the proof of Lemma 7.7 has a length at most six times the length of $\Pi_{\text{UDG}_l}(s, t)$.

First, for any path Π, $l(\Pi) \leq h(\Pi)$ because the length of every link is no more than one unit. From Lemma 7.7, we also know that $h[\Pi_{\text{CDS}'}(s, t)] \leq 3k + 2$, where k is the minimum number of hops needed to connect s and t; i.e., $k = h[\Pi_{\text{UDG}_h}(s, t)]$. Then,

$$l(\Pi_{\text{CDS}'}(s, t)) \leq h(\Pi_{\text{CDS}'}(s, t)) \leq 3k + 2.$$

Notice that in the shortest path $\Pi_{\text{UDG}_l}(s, t) = w_1 w_2 \ldots w_m$, the sum of each two adjacent links $w_{i-1} w_i$ and $w_i w_{i+1}$ must be larger than one; otherwise, we can use link $w_{i-1} w_{i+1}$ instead of $w_{i-1} w_i w_{i+1}$ to find a shorter path from the triangle inequality $\|w_{i-1} w_{i+1}\| \leq \|w_{i-1} w_i\| + \|w_i w_{i+1}\|$. Therefore,

$$l[\Pi_{\text{UDG}_l}(s, t)] > \lfloor h[\Pi_{\text{UDG}_l}(s, t)]/2 \rfloor.$$

Notice that $h(\Pi_{\text{UDG}_l}(s, t)) \geq h(\Pi_{\text{UDG}_h}(s, t)) = k$. So, $k < 2l(\Pi_{\text{UDG}_l}(s, t)) + 2$. Then,

$$l(\Pi_{\text{CDS}'}(s, t)) \leq 3k + 2 \leq 6l(\Pi_{\text{UDG}_l}(s, t)) + 6.$$

Consequently, the length-stretch factor of CDS is bounded by 6 (with an additional constant 6). Here, we are interested only in nodes s and t with $\|st\| > 1$. ∎

Similarly, we can show that the ICDS has a bounded node degree. Because CDS$'$ is a subgraph of ICDS$'$, the hop and length-stretch factors of ICDS$'$ are also at most 3 and 6, respectively.

LEMMA 7.9 *The node degree of the ICDS is bounded by a constant.*

Proof. For any dominator node u, it can connect only to connectors that are introduced by some dominator nodes within three hops of u. Notice that some connectors (within the transmission range of u) may be introduced by some dominator pairs (x, y) with $x \neq u$, $y \neq u$. However, x and y are still within three hops of u. Each dominator can introduce at most $5\ell_3$ connectors. Thus, the degree of a dominator node is at most $5\ell_3\ell_3$.

For a connector node w, it can connect to both connectors and at most five dominators. The connectors are introduced by some dominator nodes within two hops of w. There are at most ℓ_2 such dominators; each of them can introduce at most $5\ell_3$ connectors. Thus, the degree of a connector node is at most $5\ell_2\ell_3 + 5$.

Thus, the node degree in ICDS is bounded by $5\ell_3\ell_3$ because $5\ell_3\ell_3 > 5\ell_2\ell_3 + 5$. ∎

7.8   Linear-Programming-Based Approaches

Kuhn *et al.* (2006) studied the k-DS problem: Find a set of nodes such that each of the (other) nodes are dominated by at least k-nodes from this set. The set of such nodes is called a k-DS. They give upper bounds on the approximation ratio of distributed methods for the k-MDS problem in two different models: general graphs and UDGs. UDGs have been a popular model for modeling the characteristic wireless nature of communication in multihop radio networks. Because in reality signal propagation does often not form clear-cut disks, they also study the problem in general graphs, which can be regarded as pessimistic counterparts to the optimistic UDG model.

They first present a distributed approximation algorithm. For arbitrary values of t, the algorithm achieves an $O(t\Delta^{2/t}\log\Delta)$ approximation in time $O(t^2)$. The algorithm first approximates the fractional version of the k-MDS problem in a distributed way. The final solution is then obtained with a distributed form of randomized rounding. This approximation upper bound is particularly interesting in view of the approximation bound on distributed algorithms given in Kuhn et al. (2004a). The lower bound established in Kuhn et al. (2004a) implies that in $O(t)$ communication rounds, no (possibly randomized) algorithm can achieve an approximation ratio better than $(\Delta^{1/t}/t)$, even if the message size is unbounded and the nodes have unique identifiers. Specifically, this lower bound indicates that the time-approximation trade-off achieved by the algorithm presented in Kuhn et al. (2006) is not too far from the optimum.

The second algorithm presented in Kuhn et al. (2006) is a randomized algorithm that computes a k-fold DS in time $O(\log\log n)$ in UDGs, if nodes can sense distances to neighboring nodes. The algorithm achieves a constant-approximation ratio in expectation. Furthermore, the algorithm uses only small messages of size $O(\log n)$ bits. Observe that in most practical scenarios, the size of messages cannot be arbitrarily large. This algorithm for the UDG is based on an algorithm given in Gao et al. (2001a) for approximating the standard MDS problem.

For the k-DS problem, they basically form it as an integer programming problem and then relax it to linear programming (LP). Let $x_i \in \{0, 1\}$ denote whether node v_i is selected to a k-DS. For a node v_i, let $N(v_i)$ be the set of neighboring nodes of v_i in the communication graph G. Then, the primal linear programming (PP) for the k-DS problem can be easily formed as follows:

$$\min \sum_{n=1}^{n} x_i$$
$$s.t. \sum_{v_j \in N(v_i)} x_j \geq k_i, \quad \forall v_i,$$
$$x_i \in \{0, 1\}, \quad \forall v_i.$$

In the preceding LP formulation, we use k_i to denote the coverage requirement of the node v_i; i.e., the number of nodes in the k-DS that are neighbors of v_i. In this formulation, it is implicitly assumed that we need to satisfy the coverage requirement of *all* nodes in the network, not only the nodes that are not in the DS. This formulation can also be easily modified to describe the scenario when the objective function is to minimize the total weight $\sum_{i=1}^{n} w_i x_i$ of all selected nodes, where w_i is the weight of node v_i. Because the DS problem is well known to be NP-complete, it is impossible to solve the preceding integer programming (IP) formulation in polynomial time. One of the approaches typically used in the literature is to relax the constraints $x_i \in \{0, 1\}$ to

$0 \leq x_i \leq 1$. This relaxed version is often called the *fractional* DS problem formulated as follows:

$$(PP) \text{ for } k\text{-DS}$$

$$\min \sum_{i=1}^{n} x_i$$

$$s.t. \sum_{v_j \in N(v_i)} x_j \geq k_i, \quad \forall v_i,$$

$$0 \leq x_i \leq 1, \quad \forall v_i.$$

The distributed algorithm to compute a solution for the primal linear program (PP) is given by Algorithm 16. The basic structure of the algorithm and its analysis can be seen as a distributed version of the greedy k-MDS algorithm as described in Vazirani (2001),which is based on dual primal linear programming.

Dual LP for k-DS

$$\max \sum_{i=1}^{n} k_i y_i - z_i,$$

$$s.t. \sum_{v_j \in N(v_i)} y_j - z_i \leq 1,$$

$$0 \leq y_i, 0 \leq z_i, \quad \forall v_i.$$

Here, in the dual LP, the variable y_i corresponds to the inequality $\sum_{v_j \in N(v_i)} x_j \geq k_i$ and the variable z_i corresponds to the inequality $x_i \leq 1$. In the greedy algorithm for a DS, we start with an empty set S. In each step, a node with a maximal number of not yet completely covered neighbors is added to S.

As noticed in Kuhn *et al.* (2006), the main problem of applying the greedy algorithm in a distributed environment is how to efficiently deal with the scenarios when different nodes with equal or similar numbers of uncovered neighbors are capable of joining the DS. In other words, we have to solve a classical symmetry-breaking problem. Solving a fractional problem instead of its integral counterpart often is a good way to avoid most problems arising in the context of symmetry breaking. Intuitively, whenever there are q neighbors of a node u that could all join the DS according to the greedy condition, instead of selecting one of the q nodes, we can increment the x value of each of them by $1/q$. Recall that we first solve the fractional DS problem. Then, Kuhn borrowed an idea from the distributed algorithm for the standard DS problem (Kuhn and Wattenhofer, 2003) to solve the preceding problem. During the execution of their algorithm, all nodes are not covered (we say with color *white*); i.e., each node starts with value $x_i = 0$ and increases its x_i over time. A node v_i is said to be colored *gray* whenever the sum $\sum_{v_j \in N(v_i)} x_j$ exceeds 1; i.e., when the node is completely covered. The number of white neighboring nodes of a node v_i (including node v_i) at any given time is called the *dynamic degree* of node v_i and is denoted by $\tilde{\delta}_i$. Initially, all nodes are white and thus $\tilde{\delta}_i = \delta(v_i) + 1$, where $\delta(v_i)$ is the degree of node v_i.

Algorithm 16 will find a primal solution x and a dual solution (y, z) for the PP and dual LP. The following theorem was proved in Kuhn *et al.* (2006).

Algorithm 16 Distributed LP approximation of k-DS

1: $x_i \leftarrow 0$; $\tilde{\delta}_i \leftarrow \delta(v_i) + 1$; $col_i \leftarrow white$; $c_i \leftarrow 0$; $\alpha_{j,i} \leftarrow 0, \forall j$; $\beta_{j,i} \leftarrow 0, \forall j$;

2: **for** $p = t - 1$ to 0 by -1 **do**

3:      **for** $q = t - 1$ to 0 by -1 **do**

4:        **if** $x_i < 1$ and $\tilde{\delta}_i \geq (\Delta + 1)^{p/t}$ **then**

5:          $x_i^+ \leftarrow \min\{\frac{1}{(\Delta+1)^{q/t}}, 1 - x_i\}$;

6:          $x_i \leftarrow x_i + x_i^+$;

7:        **end if**

8:        **Send** $x_i, x_i^+, \tilde{\delta}_i$ to all neighbors;

9:        **if** $col_i = white$ **then**

10:         $c_i^+ \leftarrow \sum_{v_j \in N(v_i)} x_j^+$;

11:         $\lambda \leftarrow \min(1, \frac{k_i - c_i}{c_i^+})$;

12:         $c_i \leftarrow c_i + c_i^+$;

13:         **for all** $v_j \in N(v_i)$ **do**

14:           $\beta_{j,i} \leftarrow \beta_{j,i} + \frac{\lambda x_j^+}{(\Delta+1)^{p/t}}$;

15:           $\alpha_{j,i} \leftarrow \alpha_{j,i} + \lambda x_j^+$;

16:         **end for**

17:         **if** $c_i \geq k_i$ **then**

18:           $col_i \leftarrow gray$;

19:           $y_i \leftarrow \frac{1}{(\Delta+1)^{p/t}}$;

20:         **end if**

21:       **end if**

22:       Send col_i to all neighbors;

23:       $\tilde{\delta}_i \leftarrow |\{v_j \in N(v_i) \mid col_j = white\}|$, i.e., the number of remaining white neighboring nodes;

24:     **end for**

25: **end for**

26: $z_i \leftarrow \sum_{v_j \in N(v_i)} (\alpha_{j,i} y_j - \beta_{j,i})$.

THEOREM 7.9 *(Kuhn et al., 2006) For arbitrary t, Algorithm 16 computes a feasible solution for the PP in time $O(t^2)$. The approximation ratio of the algorithm is at most $t[(\Delta + 1)^{\frac{2}{t}} + (\Delta + 1)^{1/t}]$.*

The fractional solution computed in Algorithm 16 is then converted into an integral one (a valid DS) by applying a distributed variant of a standard randomized rounding scheme, summarized in Algorithm 17.

The following theorem proved by Kuhn *et al.* (2006) shows that as expected, the size of the obtained solution is only a factor of roughly $\log \Delta$ larger than the objective value of the solution for the LP.

THEOREM 7.10 *Kuhn et al. (2006) Starting with a ρ-approximate solution for the LP, Algorithm 17 computes an integral solution with approximation ratio $\rho \log \Delta + O(1)$ in constant time.*

Algorithm 17 Distributed randomized rounding

1: $p_i \leftarrow \min\{1, x_i \cdot \ln(\Delta + 1)\}$;

2: Set $x_i' \leftarrow 1$ with probability p_i and 0 with probability $1 - p_i$.

3: **Send** x_i to all its neighbors

4: **if** $\sum_{v_j \in N(v_i)} x_j' \le k_i$ **then**

5:     **Send** REQ to $k_i - \sum_{v_j \in N(v_i)} x_j'$ neighbors $v_\ell \in N(v_i)$ with $x_\ell' = 0$.

6: **end if**

7: **if** REQ is received **then**

8:     Set $x_i' \leftarrow 1$.

9: **end if**

Proof. Notice that a node v_i can be chosen as a dominator (i.e., x_i' is set as 1) only in Lines 2 and 5 of the algorithm. Let's introduce two variables X and Y, denoting the number of nodes joining the DS in Lines 2 and 5, respectively. Because x_i' is chosen to be 1 with probability $p_i \le x_i \ln(\Delta + 1)$, we have

$$E(X) = \sum_{i=1}^{n} E(x_i') = \sum_{i=1}^{n} p_i \le \sum_{i=1}^{n} x_i \ln(\Delta + 1) \le \ln(\Delta + 1)\rho\,\text{OPT},$$

where OPT denotes the size of the minimum k-DS. We then bound the expected value of Y. To do this, let us study the probability q_i that v_i is not covered k_i times after Line 2. If $k_i = 1$, it is easy to show that

$$q_i = \prod_{v_j \in N(v_i)} (1 - p_j)$$

$$\le \prod_{v_j \in N(v_i)} \left[1 - \frac{1 - \sum_{v_j \in N(v_i)} p_j}{|N(v_j)|} \right]^{|N(v_j)|}$$

$$\le \left[1 - \frac{\ln(\Delta + 1)}{|N(v_j)|} \right]^{|N(v_j)|}$$

$$\le e^{-\ln(\Delta+1)}$$

$$= \frac{1}{\Delta + 1}.$$

When $k_i > 1$, the probability that v_i is not covered k_i times is that it is covered 0, 1, 2, or up to $k_i - 1$ times. Then, based on the Chernoff bound (reviewed in Theorem 7.11), Kuhn *et al.* (2006) proved that

$$q_i \le \left(\frac{e^{-\frac{\ln(\Delta+1)-1}{\ln(\Delta+1)}}}{\left(\frac{1}{\ln(\Delta+1)}\right)^{\frac{1}{\ln(\Delta+1)}}} \right)^{k_i \ln(\Delta+1)} = O\left(\frac{1}{k_i(\Delta + 1)} \right)$$

Here, in the Chernoff bound, we select β such that $\beta p N(v_i) = k_i$, where $p = \sum_{v_j \in N(v_i)} p_j$. Note that $k_i \ge 2$. The expected value of Y can therefore be bound as follows:

$$E[Y] \leq \sum_{i=1}^{n} k_i q_i = \sum_{i=1}^{n} k_i O\left(\frac{1}{k_i(\Delta+1)}\right) = O\left(\frac{n}{\Delta+1}\right) = O(\text{OPT}).$$

Combining the bounds for $E[X]$ and $E[Y]$ completes the proof. ∎

There are many variations on the Chernoff bound. Let $p_1, p_2, \ldots, p_n$ be nonnegative real numbers with a value of at most 1. Let $p = \sum_{i=1}^{n} p_i/n$. Let $X_1, X_2, \ldots, X_n$ be n mutually independent variables with

$$\Pr[X_i = 1] = p_i, \quad \Pr[X_i = 0] = 1 - p_i,$$

and $X = X_1 + X_2 + \cdots + X_n$. Then, we have the following theorem:

THEOREM 7.11 *For any* $\beta \geq 1$,

$$\Pr[X \geq \beta p n] < (e^{\beta-1}\beta^{-\beta})^{pn}.$$

Here, X may be interpreted as the number of successes in n independent trials when the probability of successes in the ith trial is p_i.

Notice that these two algorithms implicitly assumed that every node knows the maximum degree Δ. This assumption can be removed by using techniques presented in Kuhn and Wattenhofer (2006).

7.9   Geometry-Position-Based Approaches

In Subsection 7.3.2, we briefly reviewed a number of centralized methods for constructing a CDS by using geometry information of all wireless nodes. Here, we briefly study some distributed methods for constructing a CDS by using geometry information.

Greedy Method Based on Node Positions

The first class of method follows the framework of constructing a maximal IS and then finding connectors for the final CDS. A typical method, using two messages IamDominator and IamDominatee, has the following procedures:

- Initially, all nodes will be marked white. Each node v will broadcast a message to its neighbors to inform its x coordinate.
- A white node marks itself as dominator if it has the smallest x value among all of its white neighbors, if there are any, and broadcasts IamDominator to its one-hop neighbors. If two neighboring nodes having the same x value, then the nodes with the smallest y value will mark itself as dominator.
- A white node receiving the IamDominator message marks itself as dominatee and broadcasts IamDominatee to its one-hop neighbors.
- When a white node v receives the IamDominatee message from a node u, it updates the status of node u as dominatee.

It is easy to show that the set of nodes marked as dominator will form a DS. Obviously, each node needs to send only two messages in the preceding procedure: one for stating its location and one for stating its status (i.e., dominator or dominatee).

LEMMA 7.10 *The size of the constructed DS is at most three times the MDS.*

Proof. Let OPT be a MDS of V, and let U be the DS constructed with the preceding approach. Consider an order of the nodes in OPT in the nondecreasing order of the x value, say $v_1, v_2, \ldots, v_{\text{opt}}$. We then partition the nodes in U to opt disjoint subsets. Let U_1 be the set of nodes in U that are adjacent to v_1. For any $2 \leq i \leq$ opt, let U_i be the set of nodes in U that are adjacent to v_i but none of the previous nodes $v_1, v_2, \ldots, v_{i-1}$. Then, $U_1, U_2, \ldots, U_{\text{opt}}$ form a partition of U because OPT is a DS. For any i with $1 \leq i \leq$ opt, we show that $|U_i| \leq 3$. There are two cases here. If node $v_i \in U_i$, obviously, $|U_i| = 1$ because U is an IS also. If node $v_i \notin U_i$, we claim that no node $w \in U_i$ has a larger x value than node v_i. Assume, for contradiction, that there is such a node w. Consider the moment when w is marked as dominator. Because v_i is not selected, at that moment, node v_i will have a smaller x value than node w and is a white neighbor of node w. This contradicts that node w is marked as dominator. Thus, U_i lies in a sector of at most $180°$ within the coverage range of node v_i. This implies that $|U_i| \leq 3$. Therefore,

$$|U| = \sum_{i=1}^{\text{opt}} |U_i| \leq 3\text{opt}.$$

This completes the proof.   ∎

When a DS is constructed, we can apply the previous method to find connectors to connect these dominators. It is obvious that the constructed CDS is still a constant approximation of the MCDS. A tighter bound on the approximation ratio of the constructed CDS obtained with the preceding approach is left as an exercise for readers.

Grid-Partition-Based Approach

Another approach to construct a CDS by using geometry information works as follows. Assume that the transmission range of all nodes is normalized as one unit. Imagine that the region is partitioned into grids of cells with cell size a. The essential ingredient of this approach is to select a properly such that the set of selected nodes will form a connected dominating set with small size. We choose $a = \sqrt{2}/2$. Then, some node from each cell will be chosen as the dominator in the CDS, if there is any node in the cell. A number of different ways could be used to choose such a node: the node with the largest residue power, the node with the least mobility, or randomly selected. Notice that when $a = \sqrt{2}/2$, each cell needs at most one node to dominate all nodes inside this cell. Two cells are said to be *nearby* if the distance between them is at most one. Each cell will have at most 20 *nearby* cells. Notice that for every edge uv in the UDG, u and v are either in the same cell or in two *nearby* cells. Edge uv is called the *crossing edge* if u and v are from different cells. Two nearby cells are called *connected* by a set S if there is a crossing edge uv, where u is a node from S in one cell and v is a node from S in another cell.

After dominators are selected, for each pair of *nearby* cells that are not connected by selected dominators and connectors yet, we select a crossing edge uv for these two cells if there are any, and mark u and v as connectors. Consequently, each cell will have at most 21 nodes selected to the CDS: one dominator node selected in the first step and at most 20 connectors in the second step, one for each nearby cell.

We now prove Lemma 7.11.

LEMMA 7.11 *The set of selected nodes (dominators and connectors) will form a CDS.*

Proof. Obviously, the selected nodes forming a DS. We need to prove that the selected nodes form a connected graph. To prove this, we need to show that for each edge $xy \in G$, we can connect them by using nodes selected previously. If both x and y are from a same cell, then obviously they can be connected via the dominator node in that cell. If nodes x and y are from nearby cells C_1 and C_2, respectively, we can connect them as follows. Let v_1 and v_2 be the connectors used to connect cells C_1 and C_2. Let u_1 and u_2 be the dominator nodes in cells C_1 and C_2, respectively. Then, we connect x to y via path $x u_1 v_1 v_2 u_2 y$. This proves that the selected nodes form a CDS. ∎

It is easy to prove that the CDS constructed with the preceding approach also is within a constant factor ($\leq 12 \times 21 = 252$) of the minimum CDS: Each disk centered at a node from the MCDS will intersect at most 12 cells. A careful analysis and more careful selection of connectors will significantly reduce the approximation ratio. More details are left as an exercise for the readers.

7.10   Further Reading

Many researchers proposed using the CDS as a virtual backbone for hierarchical routing in wireless ad hoc networks (Das and Bharghavan, 1997; Liang and Haas, 2000; Wu and Li, 2000, 2001). Efficient distributed algorithms for constructing CDSs in ad hoc wireless networks were well studied (Alzoubi *et al.*, 2002c; Baker and Ephremides, 1981; Baker *et al.*, 1984; Basagni, 1999; Chlamtac and Farago, 1999; Das and Bharghavan, 1997; Stojmenovic *et al.*, 2002; Wu and Li, 2001). The notion of cluster organization has been used for wireless ad hoc networks since their early appearance. Baker and co-workers (1981, 1984) introduced a "fully distributed linked cluster architecture" mainly for hierarchical routing and demonstrated its adaptivity to the network connectivity changes. The notion of the cluster has been revisited by Gerla and Tsai (1995) and Lin and Gerla (1997b) for multimedia communications with emphasis on the allocation of resources (namely, bandwidth and channel) to support the multimedia traffic in an ad hoc environment. Gao *et al.* (2001a) proposed a randomized algorithm for maintaining the discrete mobile centers; i.e., DSs. They showed that it is an $O(1)$ approximation to the optimal solution with very high probability, but the constant-approximation ratio is quite large. Recently, Alzoubi *et al.* (2002c) proposed a method to approximate a MCDS within eight whose time complexity is $O(n)$ and message complexity is $O(n \log n)$. Alzoubi (2002) continued to propose a localized method to approximate the MCDS by using linear number of messages.

Existing clustering methods first choose some nodes to act as coordinators (i.e., cluster-heads) of the clustering process. Then, a cluster is formed by associating the cluster-head with some (or all) of its neighbors. Previous methods differ on the criterion for the cluster-head selection, which is based on either the lowest (or highest) ID among all unassigned nodes (Baker *et al.*, 1984; Lin and Gerla, 1997b) or based on the maximum node degree (Gerla and Tsai, 1995) or based on some generic weight (Basagni, 1999) (i.e., the node with the largest weight will be chosen as the cluster-head). Notice that any maximal IS is always a DS. Several clustering methods essentially compute a maximal IS as the final cluster-heads.

7.11   Conclusion and Remarks

In this chapter, we studied algorithms to construct a network backbone by using a CDS graph. A communication-efficient localized algorithm was presented for approximating the MCDS within a constant factor of the minimum. The constructed CDS is efficient for both length and hops and has at most $O(n)$ edges and each node has a bounded degree. The algorithms constructing a CDS have the message complexity $O(n)$. Moreover, we showed that the number of messages sent by *each* node is bounded by a constant.

There are many interesting open problems left for further study. Remember that we use the following assumptions of wireless network models: omnidirectional antenna, single transmission received by all nodes within the vicinity of the transmitter, all nodes have the same transmission range, and nodes are static for a reasonable period of time. The problem will become much more complicated if we relax some of these assumptions. Another interesting open problem is to study the dynamic updating of the backbone efficiently when nodes are moving in a reasonable speed. It is interesting to see the practical performance differences of various methods approximating MCDS in a mobile environment. Further work should be done to lower the constant bounds given in this chapter by using a tighter analysis.

Notice that although the methods studied in this chapter and also the algorithms to be studied in Chapters 8 and 9 provide certain constant approximations of a number of properties for a number of structures, a strong assumption is that the network is static. It remains a fundamental challenge to design efficient dynamic algorithms to maintain these structures when the network nodes could be mobile. In other words, interest is not only in the efficient distributed construction of these approximations but also in efficient distributed schemes for the maintenance of these approximations when the network changes because of motion. To maintain a structure, we need a trade-off between the cost of updating the structure to reflect the new network structure that is due to mobility and the size of the structure compared with the optimum structure at any time instance.

Problems

7.1 Prove that for UDG G, $\alpha^{[2]}(G) \leq 18$. Here, $\alpha^{[k]}(G)(u)$ is the size of the MIS of nodes that are within k-hops of a node u in graph G and $\alpha^{[k]}(G) = \max_{u \in G} \alpha^{[k]}(G)(u)$.

7.2 Assume that $a_0 = n$ and $|\text{OPT}| \geq \frac{a_0}{\Delta}$. Further assume that

$$a_{i+1} \leq a_0 \left(1 - \frac{1}{|\text{OPT}|} \right)^i + \sum_{j=1}^{i-1} \left(1 - \frac{1}{|\text{OPT}|} \right)^j .$$

Prove that when $k \geq |\text{OPT}| \ln \frac{a_0}{|OPT|}$,

$$a_k \leq 2|\text{OPT}|.$$

7.3 Theorem 7.6 proved that any distributed algorithm that constructs a nontrivial CDS will have a time complexity of at least $\Omega(n \log n)$. Later, several distributed algorithms were also presented that can construct a good approximation of a CDS by using only a linear number of messages. Is there any conflict here among these two statements? Can you construct network examples such that Algorithm 15, together with a good distributed MIS algorithm described in Subsection 7.7.1, may construct a trivial CDS?

7.4 Assume that Algorithm 15 is used to find connectors after a MIS is constructed. Give an upper bound on the number of connectors a dominator node will have within its transmission range. Your bound should be as small as possible. Also construct a network example such that the number of connectors within a certain dominator node is as large as possible.

7.5 In Lemma 7.5, we proved that each node on the constructed CDS has a constant bounded number of neighbors in the CDS. Can you give a tighter bound on the node degree? Give a network example such that the maximum node degree of the ICDS graph is as large as possible.

7.6 For constructing a good DS, a distributed LP-based approach could be used. On the other hand, no distributed LP-based approach has been proposed for the CDS problem. What are the main technique difficulties of using a distributed LP approach for approximating a CDS here?

7.7 In Section 7.8, a distributed LP-based approach for approximating a DS was described. Assume that each node v_i has a weighted w_i and we want to approximate a minimum weighted DS. Can you convert the LP method of Kuan *et al.* (2006) to solve a minimum weighted DS problem?

7.8 Prove Theorem 7.11: For any $\beta \geq 1$,

$$\Pr[X \geq \beta pn] < (e^{\beta-1}\beta^{-\beta})^{pn},$$

where X_1, X_2, ..., X_n are n mutually independent variables with $\Pr[X_i = 1] = p_i$, $\Pr[X_i = 0] = 1 - p_i$, and $X = X_1 + X_2 + \cdots + X_n$.

7.9 Assume that a graph $G = (V, E)$ is connected and there is a subgraph $H \subseteq G$. Prove that H is a CDS if and only if the following two conditions are satisfied:

(a) For every node $v \in G$, it is connected to some node in H.

(b) For every edge $e = (u, v) \in G$, we can find a path in H to connect u and v.

7.10 For the grid-partition-based approach to find an approximated MCDS, we showed that the found CDS has a size within a constant factor of the optimum. What is the best approximation ratio you can prove for that method? Give a network example such that the ratio of the size of the CDS found using this approach over the size of the optimum solution is as large as possible. Can you use a different cell size to partition the region to improve the approximation ratio?

8 Weighted Network Backbone

8.1 Introduction

Most of the methods developed in the literature for backbone construction try to minimize the number of cluster-heads; i.e., the number of nodes in the backbone. However, in many applications of wireless ad hoc networks, minimizing the size of the backbone is not sufficient. For example, different wireless nodes may have different costs for serving as a cluster-head because of device differences, power capacities, and information loads to be processed. Therefore, in the rest of this chapter, for succinctness of presentation, we assume that each wireless node has a *generic cost* (or *weight*). The cost may also represent the *fitness* or *priority* of each node to be a cluster-head. Lower cost means higher priority. In practice, cost could represent the power-consumption rate of this node if a backbone with small power consumption is needed; the robustness of this node if a fault-tolerant backbone is needed; or a function of its security level if a secure backbone is needed. Y. Wang *et al.* (2005a) studied how to construct a sparse backbone efficiently for a set of weighted wireless nodes such that the total cost of the backbone is approximately minimized and there is a cost (or hops) *efficient* route connecting every pair of wireless nodes via the constructed network backbone. Here, a route is cost (or hops, respectively) *efficient* if its cost (or hops) is no more than a constant factor of the minimum cost (or hops) needed to connect the source and the destination in the original communication graph when all possible physical communication links are considered.

In this chapter, we study a novel distributed method in Y. Wang *et al.* (2005a, 2005b) to generate a weighted backbone with a good approximation ratio while having a small communication cost. The methods work not only for homogeneous networks but also for heterogeneous networks. The total cost of the constructed backbone is within $\min[4\delta + 1, 18\log(\Delta + 1)] + 10$ times of the optimum for homogeneous networks when all nodes have the same transmission range. Here, δ is the maximum ratio of costs of two adjacent wireless nodes and Δ is the maximum node degree in the communication graph. Notice that the advantage of our backbone is that the total cost is small compared with the optimum when either the costs of wireless nodes are smooth (i.e., two neighboring nodes' costs differ by a small constant factor) or the maximum node degree is low. The total number of messages of our method is $O(m)$ for any network composed of n wireless devices and m wireless links. The number of messages could be reduced if each node knows its own geometry location. With a small modification, the constructed

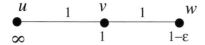

Figure 8.1 An example of why the first method fails to produce a low-cost WCDS. © 2005, ACM.

backbone is efficient for unicast: The total cost (or hop) of the least cost (or hop) path connecting any two nodes using a backbone is no more than three (or four) times the least cost (or hop) path in the original communication graph. This is significant because our backbone structure is much sparser than the original communication graph, which significantly reduces the cost of routing without losing much ground for the performance of the unicast.

8.2　Study of Typical Methods

Several methods have been proposed in the literature to find a small DS for homogeneous networks; most of them are based on greedy algorithms. Because in this chapter we are interested in distributed methods, we thus discuss mainly the priori distributed greedy methods here. If we insist on applying these distributed methods to approximate the minimum weighted DS, these methods may produce a backbone that is arbitrarily worse than the optimum. It is shown by examples that three classical methods do not generate a DS whose cost is always comparable with optimum in the worst case.

The first method to generate a DS is to generate a maximal IS as follows (Alzoubi *et al.*, 2002a; Chatterjee *et al.*, 2002). First, assume that all nodes are marked WHITE originally, which represents the fact that the node is not assigned any role yet. A node u sends a message lamDominator to all its one-hop neighbors if it has the smallest cost (ID is often used if every node has a unit cost) among all its WHITE neighbors. Node u also marks itself Dominator. When a node v receives a message lamDominator from its one-hop neighbors, node v then marks itself Dominatee. Node v then sends a message lamDominatee to all its one-hop neighbors. Clearly, the nodes marked with Dominator indeed form a dominating set. Notice that as assumed by all protocols, when a node u sends a message to all of its neighbors, its neighbors will get the message. In practice, because the links are not reliable, node u may have to transmit several times to ensure that all of its neighbors indeed receive the message correctly. For simplicity, we treat *all messages*, sent by node u to ensure that a correct message received by all of its neighbors, as *one metamessage*. The message complexity measures the number of metamessages used by a protocol.

Let's see by example why the produced DS may be arbitrarily larger than the optimum solution. Although the instance illustrated here uses a UDG as a communication graph, it is not hard to extend this to a general communication graph (see Figure 8.1 for an illustration). Assume that three wireless nodes, u, v, and w, are distributed along a line with one unit interval. The nodes' costs of u, v, and w are ∞, 1, and $1 - \epsilon$, respectively. The dominators selected by the first method are nodes w and u, and the total cost of the solution is ∞. However, the optimal solution is formed by v with a total cost of 1. The

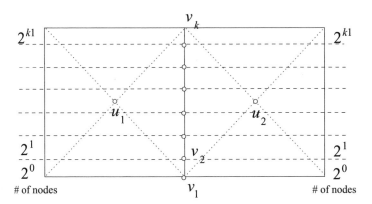

Figure 8.2 An example of why the second method fails to produce a low-cost WCDS. © 2005, ACM.

method presented in Wang *et al.* (2005a), which we study in detail later, produces a DS with a total cost of $2 - \epsilon$ for this example.

The second method of constructing a dominating set (Das and Bharghavan, 1997; Das *et al.*, 1997) is based on a minimum weighted set cover (Chvátal, 1979). The method can be described in a centralized way as follows: In each round, we select an unselected node i with the minimum ratio $c(i)/d_i$, where d_i is the number of nodes not covered by previously selected dominators. It is well known that this centralized method produces a DS set whose total cost is no more than $\log(\Delta + 1)$ times the optimum, where Δ is the maximum original degree of all nodes. Alzoubi *et al.* (2002c) gave an example (as in Figure 8.2) of a family of instances for which the size of the solution computed by the second method is larger than the optimum solution by a logarithm factor when all nodes have the same weight. Although the instance illustrated by Alzoubi *et al.* (2002b) uses a UDG as communication graph, we can obviously extend this to a general communication graph. In this example, all nodes have a unit weight. For details of this example, see Alzoubi *et al.* (2002c). Moreover, this method is expensive to implement in a distributed way. First, it is expensive to find node i with the minimum ratio $c(i)/d_i$ from among all unchosen nodes. Second, it is also expensive to update d_i, which is the number of neighbors not covered by previously selected dominators. The method described later will produce a DS whose size is no more than five times the optimum for a unit-weighted UDG. More important, that method is a fully distributed method.

The third method to select the DS is proposed by Bao and Garcia-Luna-Aceves (2003). Unlike the previous two methods, this is a fully localized method, and it can be executed in two rounds by use of a synchronous communication model. A node decides to become a dominator if either one of the following two criteria are satisfied: (1) the node has the smallest cost in its one-hop neighborhood; or (2) the node has the smallest cost in the one-hop neighborhood of one of its one-hop neighbors. It is shown by an example that the produced DS may be arbitrarily larger than the optimum solution. See Figure 8.3 for an illustration of an instance in a UDG. Assume that $2n + 1$ wireless nodes are distributed as shown in Figure 8.3. The nodes' costs of u_i, v_i, and w are 1, $1 - \epsilon$, and $1 - 2\epsilon$, respectively. The dominators selected by the third method are nodes w and v_i

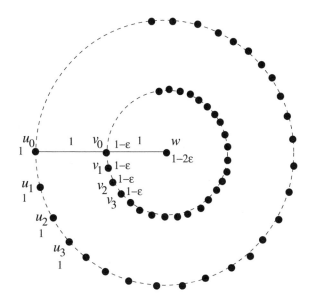

Figure 8.3 An example of why the third method fails to produce low-cost WCDS. © 2005, ACM.

$(0 \leq i < n)$, and the total cost of the solution is $n(1 - \epsilon) + 1 - 2\epsilon$. However, the optimal solution formed by node w and seven nodes from u_i has a total cost of $8 - 2\epsilon$. It is easy to show that seven unit disks centered at seven nodes among some u_i can cover all u_i. The method in Wang *et al.* (2005a) will produce an optimal DS in this special case.

8.3   Centralized Low-Cost Backbone-Formation Algorithms

First, an efficient central algorithm, given an arbitrary node-weighted graph G, is presented for constructing a WCDS whose cost is at most $3 \ln n$ times the optimum. The algorithm was proposed by Guha and Khuller (1998a). It first finds a DS and then connects the nodes in the DS by using an efficient Steiner tree algorithm as follows:

Step 1: Apply a weighted set-cover approximation algorithm to find a weighted DS. Here, a weighted set-cover instance is created by making each vertex an element, and each vertex corresponds to a weighted set that contains all its one-hop neighbors, in addition to itself. The weight of the corresponding set is the weight of this vertex. The greedy weighted set cover works as follows: In each iteration, it picks the set with the smallest ratio of its weight over the number of *new* elements it covers. Here, an element is called a *new* element covered by this set if this element is not covered by any previously selected set.

Step 2: Then, connect the vertices in the DS selected in the first step by using a node-weighted Steiner tree (NST) approximation algorithm (e.g., the one by Klein and Ravi, 1995) to connect all the vertices in the DS. In applying the NST algorithm, we make the weights of all vertices in the DS equal to *zero*. The DS constructed from Step 1 and the Steiner nodes found in this step form a CDS.

THEOREM 8.1 *(Guha and Khuller, 1998a) The weight of vertices in the CDS is at most* $3 \ln n w(OPT)$, *where OPT is a CDS, with the minimum weight in G, and w(U) denotes the total weight of a set of nodes U.*

Proof. The proof is due to Guha and Khuller (1998a). First, the weight of the vertices in the DS is at most $\ln \Delta w(OPT_{DS}) \leq \ln \Delta w(OPT)$, where OPT_{DS} is the *dominating set* of the graph G with the minimum weight. Notice that the approximation ratio of the NST algorithm by Klein and Ravi (1995) is $2 \ln k$, where k is the number of Steiner vertices, which is the size of the DS (denoted by $|DS|$) in this case. Consider the vertices in OPT. Clearly, these vertices in OPT and the vertices in the DS form a connected graph. Thus, OPT is a solution for the NST problem. Thus, the algorithm by Klein and Ravi will return a solution whose total weight is at most $2 \ln |DS| w(OPT)$. Thus, the total weight of the solution CDS is at most

$$w(DS) + 2 \ln |DS| \times w(OPT)$$
$$\leq \ln \Delta w(OPT) + 2 \ln n \times w(OPT)$$
$$\leq 3 \ln n \times w(OPT).$$

This finishes the proof. ∎

Cheng *et al.* (2003) proposed a PTAS for finding a CDS of a UDG when each node has a unit weight. In other words, given any $\epsilon > 0$, their method will return a CDS whose size is at most $1 + \epsilon$ times the minimum CDS in the time polynomial of n.

8.4   Efficient Distributed Low-Cost Backbone-Formation Algorithms

In this section, we study in detail a distributed algorithm (Y. Wang *et al.*, 2005a) that can construct a low-cost backbone (a WCDS) for a wireless ad hoc network G by assuming that each wireless node u has a cost $c(u)$ for being on the backbone. We prove that the total cost of the constructed backbone is no more than

$$\min \left\{ \alpha^{[2]}(G) \log(\Delta + 1), \left[\alpha^{[1]}(G) - 1 \right] \delta + 1 \right\} + 2\alpha^{[1]}(G)$$

times the optimum solution. Here, Δ is the maximum degree of all wireless nodes and $\delta = \max_{(i,j) \in E} c(i)/c(j)$, where E is the set of communication links in the wireless network. Notice that for homogeneous wireless networks modeled by a UDG, it implies that the backbone produced by this method has a cost no more than $\min[18 \log(\Delta + 1), 4\delta + 1] + 10$ times the optimum solution.

Here, we assume that each node knows the IDs and costs of all its one-hop neighbors, which can be achieved by requiring each node to broadcast its ID and cost to its one-hop neighbors initially. This protocol can be easily implemented by use of synchronous communications as did Baker and co-workers (1981, 1984). If the number of neighbors of each node is known a priori, then this protocol can also be implemented by use of asynchronous communications. This method has the following two phases:

1. The first phase (clustering phase) is to find a set of wireless nodes as the dominators.[1]
2. The second phase is to find a set of nodes, called *connectors*, to connect these dominators to form the final backbone of the network.

Notice that these two phases could interleave in the actual construction method. They are separated here just for the sake of easy presentations.

8.4.1   Finding Dominators

Let us first study how to construct a dominating set whose total cost is comparable with the optimum solution. Remember that we want to find a set of dominating nodes with the least cost. The simplest method to find a DS is to use the maximal IS because a maximal IS is always a DS. To find a DS with low cost, intuitively, we would have to use the node weight as selection criterion: A node is selected to a maximal IS if it has the lowest cost among all its neighbors who are *not* selected to the maximal IS yet.

As shown in the previous section, directly using a maximal IS as the DS may result in a CDS whose cost is arbitrarily larger than the optimum. To remedy this, we can conduct some local optimization as follows. For each node v in a maximal IS, we run a local greedy set-cover method on *local neighborhood* $N_2(v)$ to find some nodes GRDY$_v$ to cover all one-hop neighbors of v. If GRDY$_v$ has a total cost smaller than v, then we use GRDY$_v$ to replace v, which will further reduce the cost of the maximal IS. Otherwise, we just use node v to cover its one-hop neighbors.

Algorithm 18 illustrates the method of finding a DS that will approximate the minimum-cost DS.

For the example illustrated by Figure 8.1, the maximal IS will be two nodes w and u, whose cost is large. Node u is PossibleDominator and thus performs the local set cover. Clearly, $N_2(u) = \{u, v, w\}$ and $N_1(u) = \{u, v\}$. The local set cover will select v to cover all nodes in $N_1(u)$ because v covers both nodes in $N_1(u)$. Note that $c(v) < c(u)$, so node u will let v be a dominator. The other PossibleDominator w will keep itself as a dominator because the local set cover obtains a worse solution than itself. The final dominating set is then $\{v, w\}$, which is close to optimum $\{v\}$.

8.4.2   Finding Connectors

The second step of WCDS formation is to find some *connectors* (also called *gateways*) among all the dominatees to connect the dominators. Then, the connectors and the dominators form a CDS (also called backbone). Several methods (Alzoubi *et al.*, 2002a, 2002b; Baker and Ephremides, 1981; Baker *et al.*, 1984; Gerla and Tsai, 1995) have been proposed in the literature to find the connectors. However, all of these methods

---

[1] The terms *cluster-head* and *dominator* are interchanged. The node that is not a cluster-head is also called an *ordinary* node or *dominatee*. A node is called a *white* node if its status is yet to be decided by the clustering algorithm. Initially, all nodes are white. The status of a node, after the clustering method finishes, could be *dominator* or *dominatee*.

Algorithm 18 Construct a low-cost DS

1: First assume that all nodes are originally marked WHITE.
2: A node u sends a message ItryDominator to all its one-hop neighbors if it has the lowest cost among all its WHITE neighbors. Node u also marks itself PossibleDominator.
3: When a node v receives a message ItryDominator from its one-hop neighbors, node v then marks itself Dominatee. Node v then sends a message IamDominatee to all its one-hop neighbors.
4: When a node w receives a message IamDominatee from its neighbor v, node w removes node v from its list of WHITE neighbors.
5: Each node u marked with PossibleDominator collects the cost and ID of all of its two-hop neighbors $N_2(u)$.
6: Using the greedy method for minimum weighted set cover (like the second method), node u selects a subset of its two-hop neighbors to cover *all* the one-hop neighbors (including u) of node u. If the cost of the selected subset, denoted by $GRDY_u$, is smaller than the cost of node u, then node u sends a message YouAreDominator(w) to each node w in the selected subset. Otherwise, node u just marks itself Dominator.
7: When a node w receives a message YouAreDominator(w), node w marks itself Dominator.

---

consider only the unweighted scenario. We can show by examples that these methods generally do not produce a WCDS with a good approximation ratio.

Notice that because "dominating" is inherently a local property, finding a DS locally is thus possible; e.g., see Bao and Garcia-Luna-Aceves (2003). On the other hand, the "connectivity" is inherently a global property; it is not obvious that we can find a nontrivial CDS in a localized manner. A complete original graph is clearly a CDS, and we call it a trivial solution.

Given a DS S, let VirtG be the graph connecting all pairs of dominators u and v if there is a path in the original graph G connecting them with at most three hops. The following observation plays a key role in designing a localized method to find a CDS: graph VirtG is connected. It is natural to form a CDS by finding connectors to connect any pair of dominators u and v if they are connected in VirtG. This strategy was used in several previous methods (e.g., Alzoubi *et al.*, 2002a, 2002c; Baker and Ephremides, 1981; Baker *et al.*, 1984; Lin and Gerla, 1997a). The connector selection method for a WCDS is also based on this observation. First, we define two dominators u and v as *neighboring dominators* if they are at most three hops away; i.e., they are neighbors in the graph VirtG. Let LCP(u, v, G) denote the least-cost path $uv_1v_2 \cdots v_k v$ between vertices u and v on a weighted graph G, and let $\mathcal{L}(u, v, G)$ denote the total cost of nodes on path LCP(u, v, G) excluding u and v; i.e., $\mathcal{L}(u, v, G) = \sum_{1 \le i \le k} c(v_k)$. For every pair of neighboring dominators u and v, we will find the shortest path with at most three hops to connect them. The nodes on this shortest path will be assigned the role of connector.

The method to find connectors uses the following data structures and messages:

1. $D_k(v)$ is the list of dominators that are k-hops away from a node v.
2. $P_k(v, u)$ is the LCP from v to u using at most k-hops. Notice that u and v may be fewer than k-hops away.
3. OneHopDominatorList $[v, D_1(v)]$: Nodes $D_1(v)$ are the dominators of node v that are one-hop from v.
4. TwoHopDominator$[v, u, w, c(w)]$: Node u is a two-hop dominator of node v and the path uwv has the least cost.

Algorithm 19 reviews the method presented by Y. Wang *et al.* (2005a) to find connectors.

Algorithm 19 Low-cost connector selection

1: Every dominatee node v broadcasts to its one-hop neighbors the list of its one-hop dominators $D_1(v)$ using message OneHopDominatorList$[v, D_1(v)]$. When a node w receives OneHopDominatorList$[v, D_1(v)]$ from its one-hop neighbor v, it puts the dominator $u \in D_1(v)$ to $D_2(w)$ if $u \notin D_1(w)$. Update the path $P_3(z, u)$ as uvw if it has a lower cost.

2: When a dominatee node w received messages OneHopDominatorList from *all* its one-hop nodes, for each dominator node $u \in D_2(w)$, node w sends out message TwoHopDominator$[w, u, x, c(x)]$, where wxu is the LCP $P_2(w, u)$.

3: When a dominator z receives a message TwoHopDominator$[w, u, x, c(x)]$ from its neighbor w, it puts u to $D_3(z)$ if $u \notin D_2(z)$ and updates the path $P_3(w, u)$ as $uwxz$ if $c(w) + c(x)$ has a lower cost.

4: Each dominator u builds a virtual edge $\widetilde{uv}$ to connect each neighboring dominator v. The length of $\widetilde{uv}$ is the cost of path $P_3(u, v)$. Notice that here the cost of end nodes u and v is not included. All virtual edges form an *edge-weighted* virtual graph VirtG in which all dominators are its vertices.

5: Run a distributed algorithm to build a MST on graph VirtG. Let VMST denote MST(VirtG).

6: For any virtual edge $e \in$ VMST, select each of the dominatees on the path corresponding to e as a connector.

The graph constructed by combining all the dominators and the connectors selected by Algorithm 19 is called a WCDS graph (or *backbone*). Notice that because we run the MST on graph VirtG, the constructed backbone is a sparse graph; i.e., it has only a linear number of links.

8.5 Performance Guarantee

In this section, we first study the performances of the proposed weighted backbone structure in terms of the total node cost in the backbone. Then, by a small modification of the backbone-formation algorithm, we can make the weighted backbone more efficient for the unicast routing.

8.5.1 Total Cost of the Backbone

First, we would like to build a weighted backbone whose total node cost is as low as possible. We show that the backbone constructed by the previous method (Algorithms 18 and 19) is comparable with the optimum when the network is not dense or the costs of the nodes do not have a dramatic change; i.e., they are smooth. The following analysis is on homogeneous networks, but it can be extended to general heterogeneous networks without difficulty. Before the analytical result is described, we first review an important observation of a *dominating set* on a UDG, which will play an important role in the proofs later. After clustering, one dominator node can be connected to many dominatees. However, it is well known that a dominatee node can be connected to only at most five independent nodes in the UDG model. In other words, the 1-*local independence number* of a UDG, $\alpha^{[1]}$(UDG), is 5. Generally, it is well known that for each node, there are at most a constant number [$\alpha^{[k]}$(UDG)] of independent nodes that are at most k units away. The following lemma that bounds the number of independent nodes within k units from a node v is proved in Alzoubi et al. (2002a) by a simple area argument.

LEMMA 8.1 *For every node v, the number of independent nodes inside the disk centered at v with radius k-units, $\alpha^{[k]}(UDG)$, is bounded by a constant $\ell_k = (2k+1)^2$.*

The bounds on ℓ_k can be improved by a tighter analysis. Wan *et al.* (2004) gave a detailed proof to show that for a UDG, the number of independent nodes in a two-hop neighborhood (not including the one-hop neighbors) is at most 13, whereas the number of independent nodes in a one-hop neighborhood is at most 5. Therefore, there are at most 18 independent nodes inside the disk centered at a node v with radius 2; i.e., $\alpha^{[2]}$(UDG) = 18.

THEOREM 8.2 *Algorithm 18 constructs a DS whose total cost is no more than* $\min[18\log(\Delta+1), 4\delta+1]$ *times the optimum DS for networks modeled by a UDG.*

Proof. First, we prove that the total cost of the maximal IS (mIS) formed by all PossibleDominator nodes is no more than $4\delta + 1$ times the optimum. Assume node u is a node from the optimum DS OPT. If u is not a PossibleDominator node, then there are at most five PossibleDominator nodes around u. Let $v_1^u, v_2^u, \ldots, v_5^u$ denote them. The cost of one of these five nodes is lower than the cost of u; otherwise, node u will be selected as a PossibleDominator node. Without loss of generality (W.l.o.g.), let $c(v_1^u) \le c(u)$. We also know that $c(v_i^u) \le \delta c(u)$ for $2 \le i \le 5$. Thus,

$$\sum_{1 \le i \le 5} c\left(v_i^u\right) \le (4\delta + 1)c(u).$$

If we summarize the inequalities for all nodes in the optimum DS OPT, we get

$$\sum_{u \in \text{OPT}} \sum_{1 \le i \le 5} c\left(v_i^u\right) \le (4\delta + 1) \sum_{u \in \text{OPT}} c(u) = (4\delta + 1)c(\text{OPT}).$$

Notice that every node in the maximal IS will appear as v_i^u for at least one node $u \in$ OPT because OPT is a DS. Thus,

$$c(\text{mIS}) = \sum_{v \in \text{mIS}} c(v) \leq \sum_{u \in \text{OPT}} \sum_{1 \leq i \leq 5} c\left(v_i^u\right).$$

It follows that

$$c(\text{mIS}) \leq (4\delta + 1)c(\text{OPT}).$$

Then, we prove that the total cost of the nodes selected by the greedy method in Step 6 of Algorithm 18 is no more than $18\log(\Delta + 1)$ times the optimum. Assume that node u runs the greedy algorithm and gets the subset as GRDY_u, and the cost of the selected subset $c(\text{GRDY}_u)$ is at most $c(u)$. It is well known that the DS generated by the greedy algorithm for a set cover is no more than $\log f$ times the optimum if every set has at most f items. Here, we know that every dominator can cover at most Δ dominatees; thus, $c(\text{GRDY}_u) \leq \log(\Delta + 1)c(\text{LOPT}_u)$. Here, LOPT_u is an optimum DS [using nodes from $N_2(u)$] when the set of nodes to be covered is the one-hop neighborhood of u (including u). Assume that OPT_u is the subset of the global optimum solution, denoted as OPT, for the MWCDS that falls in the two-hop neighborhood of u; i.e., $\text{OPT}_u = \text{OPT} \cap N_2(u)$. Obviously, OPT_u is a DS for $N_1(u)$. Thus, we have $c(\text{LOPT}_u) \leq c(\text{OPT}_u)$ because LOPT_u is the local optimum. Therefore,

$$c(\text{GRDY}_u) \leq \log(\Delta + 1)c(\text{LOPT}_u) \leq \log(\Delta + 1)c(\text{OPT}_u).$$

Consider all nodes in the maximal IS; we get

$$c(\text{GRDY}) \leq \sum_{u \in \text{mIS}} c(\text{GRDY}_u) \leq \log(\Delta + 1) \sum_{u \in \text{mIS}} c(\text{OPT}_u).$$

Remember that for each node v, the number of independent nodes in the two-hop neighborhood of v is bounded by 18. Therefore, each dominator is counted at most 18 times (once for each node $u \in \text{mIS}$ that selects v to GRDY_u). Thus, $\sum_{u \in \text{MIS}} c(\text{OPT}_u) \leq 18c(\text{OPT})$.

For each node u in the mIS, we use either u as a dominator or use GRDY_u as a dominator, whichever has a lower cost. Then, the total weight of the final DS is at most

$$\sum_{u \in \text{mIS}} \min[c(u), c(\text{GRDY}_u)]$$

$$\leq \min\left[\sum_{u \in \text{mIS}} c(u), \sum_{u \in \text{mIS}} c(\text{GRDY}_u)\right]$$

$$\leq \min[4\delta + 1, 18\log(\Delta + 1)]c(\text{OPT}).$$

This finishes the proof. ∎

Notice that here the approximation ratio is $\min[18\log(\Delta + 1), 4\delta + 1]$. So, if one of $\log(\Delta + 1)$ and δ is a constant, the approximation ratio is a constant. Our analysis is also pessimistic. As our simulation shows, the practical performance is much better than this theoretical bound. It is easy to generalize the preceding result to heterogeneous networks.

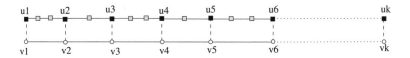

Figure 8.4 $\mathcal{L}(u, v, G) \geq 2\mathcal{L}(u, v, \text{Virt}G)$. © 2005, ACM.

THEOREM 8.3 *For a network modeled by a graph G, Algorithm 18 constructs a DS whose total cost is no more than* $\min[\alpha^{[2]}(G)\log(\Delta + 1), (\alpha^{[1]}(G) - 1)\delta + 1]$ *times the optimum.*

Now, we need to prove that the total cost of connectors selected by Algorithm 19 is also bounded. The following lemma about the relationship between $\mathcal{L}(u, v, G)$ and $\mathcal{L}(u, v, \text{Virt}G)$ is used in the proof.

LEMMA 8.2 *For any pair of dominators u and v,*

$$\mathcal{L}(u, v, \text{Virt}G) \leq 2\mathcal{L}(u, v, G).$$

Proof. Notice that the original graph is node-weighted whereas the virtual graph Virt$G$ is edge-weighted. Here, let $c(e)$ be the weight of edge $e = \widetilde{u_i u_j}$ and $c(e) = \mathcal{L}(u_i, u_j, G)$. We assume that path $uv_1v_2\cdots v_k v$ is the LCP connecting u and v in the original graph G, as shown in Figure 8.4.

For any dominatee node p in the original communication graph, it must be dominated by at least one dominator. Thus, we can assume that node u_i is node v_i's dominator, as shown in Figure 8.4. For dominators u_i and u_{i+1}, we argue that the length of $\widetilde{u_i u_{i+1}}$ is at most the summation of the cost of v_i and v_{i+1}. Notice that $u_i v_i v_{i+1} u_{i+1}$ is a three-hop path between u_i and u_{i+1} whose length is $c(v_i) + c(v_{i+1})$. Thus, the length of $\widetilde{u_i u_{i+1}}$ is at most $c(v_i) + c(v_{i+1})$. Thus, we have $c(\widetilde{u_i u_{i+1}}) \leq c(v_i) + c(v_{i+1})$ for $1 \leq i \leq k - 1$. Similarly, we also have $c(\widetilde{uu_1}) \leq c(v_1)$ and $c(\widetilde{u_k v}) \leq c(v_k)$. Summing all these inequalities, we get

$$\mathcal{L}(u, v, \text{Virt}G) \leq c(\widetilde{uu_1}) + c(\widetilde{v_k v}) + \sum_{i=1}^{k-1} c(\widetilde{u_i u_{i+1}}) \leq 2\sum_{i=1}^{k} c(v_i).$$

This finishes the proof.   ∎

In graph G, we set all dominators' costs to 0 to obtain a new graph G'. Assume that T_{opt} is the tree with the minimum cost that spans all dominators selected by Algorithm 18. The following lemma shows that there exists a tree T'_{opt} whose cost equals the cost of T_{opt} and every dominatee node u in T'_{opt} has a node degree of at most $\alpha^{[1]}(G)$.

LEMMA 8.3 *There exists a tree T'_{opt} in G' spanning all dominators selected in Algorithm 18, and the connectors in this tree have degrees of at most $\alpha^{[1]}(G)$.*

Proof. We prove this by construction. Consider any optimum cost tree T_{opt} spanning all dominators. In tree T_{opt}, assume there exist some connectors whose degrees are greater than $\alpha^{[1]}(G)$. We choose any one of them as the root. The depth of a connector is defined as the hops from this connector to the root in T_{opt}. We process all connectors u in T_{opt}

whose degree is greater than $\alpha^{[1]}(G)$ in an increasing order of their depths. Notice that, as we will see later, the depth of a node changes in our construction, but it will only increase. Assume that currently we are processing a node u with more than $\alpha^{[1]}(G)$ neighbors. Clearly, there are at least two neighbors of u in tree T_{opt} that are connected, say p, q. Notice either p's or q's depth is greater than u because u has only one parent. Without loss of generality, we assume that p's depth is bigger than u's depth. We then remove edge uq and add edge pq. Then, u's degree decreases by one while all other connectors whose depths are less than or equal to u's remain unchanged and p's degree increases by one. Notice that this will result in a new tree spanning all dominators while keeping the cost of the tree unchanged. Update the depth of node q and all nodes of the subtree rooted at q (the depths will increase by one). Repeat the preceding iteration until all nodes are processed. It is obvious that the preceding process will terminate. The resulting tree is T'_{opt}.   ∎

For tree T'_{opt}, we define its weight $\mathbf{c}(T'_{opt})$ as the sum of the cost of all connectors. We also define $\mathbf{c}(T) = \sum_{e \in T} c(e)$ for an edge-weighted tree T. The previous lemma implies that there is an optimum tree connecting all dominators with node degrees of at most five for networks modeled by a UDG.

THEOREM 8.4 *The connectors selected by Algorithm 19 have a total cost of no more than $2\alpha^{[1]}(G)$ times the optimum for networks modeled by G.*

Proof. Let K_G be another virtual complete graph whose vertices are all dominators selected in Algorithm 18 and whose edge lengths equal the cost of the LCP between two dominators on original graph G. In the subsequent discussion, we argue that the weight of the MST on graph K_G is at most $\alpha^{[1]}(G)$ times the weight of tree T'_{opt}.

For spanning tree T'_{opt}, we root it at an arbitrary node and duplicate every link in T'_{opt} (the resulting structure is called DT'_{opt}). Clearly, every node in DT'_{opt} has an even degree now. Thus, we can find an Euler circuit, denoted by $EC(DT'_{opt})$, that uses every edge of DT'_{opt} exactly once, which is equivalent to saying that every edge in $T'_{opt}(G)$ is used exactly twice. Consequently, every node v_k in $V(T'_{opt})$ is used exactly $d_{T'_{opt}}(v_k)$ times. Here, $d_G(v)$ denotes the degree of a node v in a graph G. Thus, the total weight of the Euler circuit is at most $\alpha^{[1]}(G)$ times $\mathbf{c}(T'_{opt})$; i.e.,

$$\mathbf{c}[EC(DT'_{opt})] \leq \alpha^{[1]}(G)\mathbf{c}(T'_{opt}).$$

Notice that here if a node v_k appears multiple times in $EC(DT'_{opt})$, its weight is also counted multiple times in $\mathbf{c}[EC(DT'_{opt})]$.

If we walk along $EC(DT'_{opt})$, we visit all dominators, and the length of any subpath between dominators u_i and u_j is no smaller than $\mathcal{L}(u_i, u_j, G)$. Therefore, the cost of $EC(DT'_{opt})$ is at least $\mathbf{c}[MST(K_G)]$ because $MST(K_G)$ is the minimum-cost tree spanning all dominators, and the edge $u_i u_j$ in $MST(K_G)$ corresponds to the path with the least cost between u_i and u_j. In other words,

$$\mathbf{c}[EC(DT'_{opt})] \geq \mathbf{c}[MST(K_{UDG})].$$

Consequently, we have

$$\mathbf{c}[\text{MST}(K_G)] \le \mathbf{c}[\text{EC}(DT'_{\text{opt}})] \le \alpha^{[1]}(G)\mathbf{c}(T'_{\text{opt}}). \tag{8.1}$$

Now, we prove that the weight of MST(VirtG) is at most two times the weight of MST(K_G). For any edge $e = u_i u_j \in \text{MST}(K_G)$, from Lemma 8.2, we have

$$c(e) \ge \mathcal{L}(u_i, u_j, G) \ge \frac{\mathcal{L}(u_i, u_j, \text{Virt}G)}{2}.$$

For each edge $e = u_i u_j \in \text{MST}(K_G)$, we connect them in graph VirtG by using path LCP(u_i, u_j, VirtG). This constructs a connected subgraph MST$'$ on graph VirtG whose cost is no greater than twice the weight of MST(K_G). Thus, we have

$$\mathbf{c}[\text{MST}(\text{Virt}G)] \le \mathbf{c}(\text{MST}') \le 2\mathbf{c}[\text{MST}(K_G)]. \tag{8.2}$$

The theorem follows from combining inequalities (8.1) and (8.2):

$$\mathbf{c}[\text{MST}(\text{Virt}G)] \le 2\mathbf{c}[\text{MST}(K_G)] \le 2\alpha^{[1]}(G)\mathbf{c}(T'_{\text{opt}}).$$

This finishes the proof of the theorem.   ■

Notice that Theorem 8.4 also implies the following side-product result: Given a group of receivers in a node-weighted network, the connectors found through the virtual MST (VMST) have a total cost of no more than $2\alpha^{[1]}(G)$ times the minimum-cost multicast tree. For the special case of a UDG, the total cost of the connectors is no more than 10 times that of the optimum multicast tree. Here, we assume that the receivers have cost 0.

Combining Theorems 8.3 and 8.4, we get the following theorem:

THEOREM 8.5 *For any communication graph G, our algorithm constructs a WCDS whose total cost is no more than*

$$\min[\alpha^{[2]}(G)\log(\Delta + 1), (\alpha^{[1]}(G) - 1]\delta + 1) + 2\alpha^{[1]}(G)$$

times the optimum.

Specifically, when the networks are modeled by a UDG, we have the following corollary.

COROLLARY 8.1 *For homogeneous wireless networks, our algorithm constructs a WCDS whose total cost is no more than* $\min[18\log(\Delta + 1), 4\delta + 1] + 10$ *times the optimum.*

8.5.2   Unicast Performance

After we construct the backbone WCDS, if a node u wants to broadcast a message, it follows the following procedure. If node u is not a dominator, then it sends the message to one of its dominators. When the message reaches the backbone, it will be broadcast along the VMST. In the previous section, we proved that the total cost of the WCDS is no more than a constant times the optimum, which implies that our structure is energy efficient for broadcast, when the cost of a node is its energy cost for sending a unit amount of data.

When considering unicast routing, we can modify our backbone formation algorithms by the following:

1. Removing steps 5, 6, and 7 (i.e., collecting two-hop information and running the greedy algorithm for the set cover) from Algorithm 18.
2. Modifying PossibleDominator to Dominator in step 2 of Algorithm 18.
3. Removing steps 5 and 6 (building the VMST) from Algorithm 19.

Notice that the changes to Algorithm 18 are not necessary, as we will see later. Let unicast WCDS (UWCDS) be the constructed backbone. If a node u wants to unicast a message, it follows the following procedure. If node u is not a dominator and node v is not a neighbor of u, node u sends the message to one of its dominators. Then, the dominator will transfer the message to the target or a dominator of the target through the backbone. Now, we prove that the backbone is a spanner for unicast application; i.e., every route in the constructed network topology is efficient. Remember, a route is *efficient* if its total cost (or total hop number) is no more than a constant factor of the minimum total cost (or total hop number) needed to connect the source and the destination in the original communication graph. The constant is called the cost (or hops) stretch factor.

We first prove that the backbone has a bounded cost-stretch factor.

THEOREM 8.6 *For any communication graph, the cost-stretch factor of the UWCDS is at most three.*

Proof. Consider any source node s and target node t that are not connected directly in the original communication graph G. Assume that the least-cost path LCP(s, t, G) from s to t in G is $\Pi_{G_h}(s, t) = v_1 v_2 \ldots v_k$, where $v_1 = s$ and $v_k = t$, as illustrated in Figure 8.4. We construct another path in the UWCDS from s to t, and the total cost of this path is at most three times the cost of the least-cost path LCP(s, t, G).

For any dominatee node p in the original communication graph G, we show that there must exist one dominator q whose cost is no greater than p's cost. First, from our selection procedure of the maximal IS, the fact that node p is not selected to the maximal IS implies that, at some stage, there is a neighbor, say u, with a smaller cost selected to the MIS, which will be PossibleDominator. Notice that this PossibleDominator node u may not appear in our final structure. However, this node is not selected only if $c(\text{GRDY}_u)$ is smaller than $c(u)$. Notice that, clearly, there is at least one node, say v, in GRDY$_u$ that dominates node p because p is a one-hop neighbor of node u and GRDY$_u$ covers all one-hop neighbors of u (including u). Clearly, all dominators in GRDY$_u$ have a cost of no more than $c(u)$ from $c(\text{GRDY}_u) \leq c(u)$. If node u is in the final structure, we set q as u; otherwise, we set q as node v. We call node q node p's *small dominator*. Notice that q and p can be the same node.

For each node v_i in the path LCP(s, t, G), let u_i be its small dominator if v_i is not a dominator; otherwise, let u_i be v_i itself. Notice that there is a three-hop path $u_i v_i v_{i+1} u_{i+1}$ in the original communication graph G. Then, from Algorithm 19, we know there must exist one or two connectors connecting u_i and u_{i+1}, and also the cost summation of these connectors is at most the cost summation of v_i and v_{i+1}. We define a path, denoted by LCP(s, t, UWCDS), to connect s and t in UWCDS as the concatenation of all paths LCP(u_i, u_{i+1}, VirtG), for $1 \leq i \leq k-2$, and a LCP (with $\leq$ two hops) connecting u_{k-1} and t. Remember that the path LCP(u_i, u_{i+1}, VirtG) is only the LCP among all paths connecting u_i and u_{i+1} using at most three hops.

We then show that the path LCP(s, t, UWCDS) has a cost of no more than three times of the path LCP(s, t, G), where LCP(s, t, G) is the LCP connecting s and t in the original communication graph G. Clearly,

$$\sum_{i=1}^{k-2} \mathcal{L}(u_i, u_{i+1}, \text{VirtG}) \leq c(v_1) + 2\sum_{i=2}^{k-2} c(v_i) + c(v_{k-1}).$$

Notice that in our unicast routing algorithm, when the target node t is within two hops of the dominator node u_{k-1}, node u_{k-1} will not send the data to dominator node u_k. Instead, if target t is a one-hop neighbor of node u_{k-1}, it will directly send data to node t; otherwise, node u_{k-1} will find a least-cost node, say w, to connect to the target node t directly. Obviously, $c(w) \leq c(v_{k-1})$ because node v_{k-1} connects u_{k-1} and target t. Thus, the total cost of the path in the constructed backbone is

$$\sum_{i=1}^{k-2} \mathcal{L}(u_i, u_{i+1}, \text{VirtG}) + \mathcal{L}(u_{k-1}, t, \text{VirtG}) + \sum_{i=1}^{k-1} c(u_i)$$

$$\leq c(v_1) + 2\sum_{i=2}^{k-2} c(v_i) + c(v_{k-1}) + c(v_{k-1}) + \sum_{i=1}^{k-1} c(v_i)$$

$$< 3\sum_{i=1}^{k-1} c(v_i).$$

This finishes the proof.   ∎

We can prove, similar to the proof in Alzoubi *et al.* (2002a), the following theorem:

THEOREM 8.7 *For any communication graph (not necessarily a UDG model), the hops stretch factor of the UWCDS is at most* 4^2.

8.5.3   Message Complexity

Compared with data processing, in data communication, the wireless node spends more energy. Here, we show that our algorithms are efficient in terms of communication complexity.

THEOREM 8.8 *Algorithm 18 uses $O(n)$ messages if the networks are modeled by UDG and the geometry information of all nodes is known.*

Proof. First, for messages ItryDominator and IamDominatee, every node at most sends out one of this kind of message. Thus, the total number of these two messages is $O(n)$.

---

[2] Actually, the bound is $3 + \frac{2}{k}$, where k is the number of hops of the shortest hop path in the original communication graph. The basic idea of the proof is similar to the idea used in the proof of Lemma 8.2 and illustrated by the example in Figure 8.4. Because one-hop neighbors can directly communicate with each other, for any nodes that are at least two hops away, the bound is 4.

Second, for each PossibleDominator node, it needs to collect the costs and IDs of all of its two-hop neighbors. This step may cost lots of communication [at most $O(m)$ messages when no geometry information is known, where m is the number of links in the original UDG]. Recently, Călinescu (2003) proposed a communication-efficient method [using $O(n)$ messages] to collect $N_2(u)$ for every node u when the geometry information is known for networks modeled by a UDG.

Third, after the greedy method is applied, node u may send a message YouAreDominator to node v, but because the number of independent nodes u in two hops of v is bounded by a constant, the total number of this kind of message is also $O(n)$.

Consequently, Algorithm 18 uses $O(n)$ messages. ∎

It is easy to show that Algorithm 18 uses $O(m)$ messages for a general network or if the geometry information of all nodes is unknown. For Algorithm 19, first, the number of messages in the first three steps is at most $O(m)$. Obviously, we can construct the MST on VirtG by using $O(m + n \log n)$ number of messages. In practice, we may not need to construct the MST exactly: A localized approximation of the MST (X.-Y. Li *et al.*, 2004c), which has a message complexity of only $O(n)$, may perform well enough. In addition, if the unicast is running only on the backbone, we can ignore the MST construction, and the message complexity is only $O(m)$.

8.5.4 Time Complexity

Considering the data processing at each wireless node, we also study the time complexity of our algorithms.

For Algorithm 18, the first four steps take at most $O(n)$ in time. To collect the information of two-hop neighbors, we apply the method proposed by Călinescu (2003), which also takes at most $O(n)$ time. Notice that the time complexity of the greedy method in Das and Bharghavan (1997) and Das *et al.* (1997) (based on the set-covering method in Chvátal, 1979) is at most $O(m\Delta)$, where m is the number of nodes participating in the algorithm and Δ is the maximum node degree. So, the sixth step of Algorithm 18 takes at most $O(\Delta_2\Delta)$, where Δ_2 is the maximum number of two-hop neighbors. Because $\Delta_2 \leq n$ and $\Delta_2 \leq \Delta^2$, the sixth step takes at most $O(\Delta^3)$ [or $O(n\Delta)$]. Therefore, the time complexity of Algorithm 18 is $O(n\Delta)$ in the worst case.

For Algorithm 19, the most time-consuming step is building a MST on VirtG. Obviously, we can construct the MST by using at most $O(m + n \log n)$ time.

8.6 Discussion

8.6.1 Practical Application

As mentioned in the introduction (Section 8.1), the proposed distributed algorithms can be used in wireless ad hoc networks to form a low-cost network backbone for a unicast routing or broadcasting application. The cost that we used as the input of our algorithms could be a *generic* cost, which is defined by various practical applications. Here, some possible weights are listed that may be used in wireless ad hoc networks.

Energy-Consumption Rate

Most backbone-based unicast routing or broadcasting protocols (Das and Bharghavan, 1997; Liang and Hass, 2000; Sivakumar *et al.*, 1998; Wu and Li, 2001) deliver packets only through the backbone or restrict the flooding packets in the backbone; thus, the nodes serving as cluster-heads or connectors in the backbone consume more energy than ordinary nodes. If we use the energy-consumption rate at each node as its weight, by using the proposed low-cost backbone-formation algorithm, we can achieve an energy-efficient backbone in which the total energy consumption of this backbone is at most constant times the energy consumption of the optimum. The unicast carried on the backbone is also power-efficient, compared with the lowest-energy-consumption path in the original communication graph.

Another way to build an energy-efficient backbone is to select nodes with the maximum amount of remaining energy (equivalently, the minimum amount of consumed energy if the initial energy of each node is the same).

Fault-Tolerant Rate

Fault tolerance is also an important issue in wireless ad hoc networks because nodes are mobile and in a dynamic environment. If each node estimates its probability of being faulty and we treat it as the weight, we can use our algorithm to build a fault-tolerant backbone for routing. We can evaluate the fault-tolerant rate by considering the mobility (stability, speed) of the node, the quality of links (link failures) around the node, the interference level at the node, or other metrics. Some research along this line has been done in Basagni (1999), Bettstetter and Krausser (2001), Kozat *et al.* (2001), and Min *et al.* (2004). Assume that p_i is the probability that the wireless node $v_i \in V$ will have a fault in computing or communicating with its neighbors. Two possible criteria could be used to measure the fault-tolerant quality of a backbone (i.e., a CDS $S \subset V$): $\sum_{v_i \in S} p_i$ or $\Pi_{v_i \in S} p_i$. In the first case, the cost (or weight) of node v_i is assigned as $c(v_i) = p_i$, whereas, in the latter case, the cost of v_i is assigned as $c(v_i) = \log p_i$. Then, building the most fault-tolerant backbone is equivalent to finding a CDS with the minimum total cost.

Security Level

Our algorithm can also be applied in designing secure routing protocols. Because ad hoc networks lack a central authority for authentication and key distribution, security is hard to achieve. Liu *et al.* (2004) proposed a dynamic trust model for an ad hoc network. Each node finds a security level by observing its neighbor. By using the security level information obtained from the method of Liu *et al.* (2004), we can apply our low-cost method to building a backbone for routing with high security. We could assign the cost to a node by using a method analogous to the case of fault-tolerance previously discussed.

Different metrics can be considered as the weight in our method, such as traffic load, signal overhead, battery level, and coverage. As done in Chatterjee *et al.* (2002), Chen *et al.* (2002), and Turgut *et al.* (2002), we can also use a combined weight function to integrate various metrics in consideration of forming a more robust and efficient backbone for wireless ad hoc networks in general applications.

Besides forming the backbone for routing or broadcasting, our cluster algorithm (Algorithm 18) can also be used in other applications. For example, Zheng *et al.* (2004) studied the *time-indexing* problem in sensor networks. To enable the time-indexed in-network storage of sensor data, they selected a subset of sensors; i.e., rendezvous points to collect, compress, and store sensor data from its neighborhood for predefined periods of time. To consider energy and storage balancing, we can apply our weighted cluster algorithm to select the rendezvous points. In another example (i.e., Kachirski and Guha, 2002), a simple cluster algorithm is used for selecting the mobile agents to perform intrusion detection in wireless ad hoc networks. We can also apply our method to their intrusion-detection system to achieve more robust and power-efficient agent selection.

8.6.2   Dynamic Update

After the generation of the weighted backbone, dynamic maintenance of the backbone is also an important issue because an ad hoc network could be highly dynamic. Two major events may cause the backbone to become obsolete: (1) *topology changes* that are due to node moving, node joining or leaving, or node failure; and (2) *weight changes* when weights are assigned based on some observed status of nodes. Notice that some of the practical weights we previously discussed change frequently, such as battery level and quality of links. Therefore, a dynamic update method for our backbone is needed. Usually, there are two kinds of update methods: an on-demand update or a periodical update. Most of the existing clustering algorithms are invoked periodically, whereas some algorithms (e.g., in Chatterjee *et al.*, 2002) perform the updating only when it is required (i.e., on demand). Our algorithm can adapt and combine both of these two update methods. If there is no major topology change or no remarkable weight change, no update will be performed until some preset timer expires. In other words, we perform our algorithm periodically with a preset time. The time could be set to be quite long depending on the types of the weight and applications. This kind of global update also ensures the load balance throughout the network. But, for some major topology changes (e.g., a cluster-head dying) or tremendous change of weights (e.g., a big drop in security level), an on-demand update is performed. Notice that because our algorithm is a localized algorithm,[3] the update process can be performed only in a local area where the change occurs. However, how to update the topology efficiently while preserving the approximation quality remains an open problem.

Given a structure or backbone, one may limit the number of structural changes. This naturally affects the approximation quality. There is, of course, a trade-off in the total number of structural changes and the approximations produced. This has been observed in Gao *et al.* (2001a) in the context of a bounded-degree algebraic motion model and the

---

[3] By using a *localized minimum spanning tree* instead of a MST, our distributed algorithm becomes a localized algorithm. We discuss it in detail in Subsection 8.6.3.

DS. The number of structural changes required for maintaining a certain approximation quality is another parameter of interest. As an example, we could study the trade-offs among the various parameters of the geometric structures by using the MIS in the UDG as a geometric structure. The MIS has been shown to be a good structure (Marathe *et al.*, 1995; Wan *et al.*, 2002a) for approximating several other structures and has also been used to derive good coloring for channel assignment. Another possible area of research is to explore the construction of other structures (e.g., backbones) that would be good candidates because of their low cost. For all dynamic algorithms, we need to balance the trade-offs between the quality of the structure maintained for a mobile network and the communication cost needed to maintain such a structure. Another criterion for a dynamic algorithm is the *stability* of the structure: the total number of structural changes the structure has to perform at any time instant when nodes move. In addition to quality and stability, *smoothness* is another performance metric of a suboptimal structure. It is measured by the communication cost and computation cost required by an individual update. Two types of smoothness are of interest. The worst-case smoothness has the worst communication cost and computation cost of an update; the average smoothness is the amortized communication cost and computation cost over all updates. Similar trade-offs exist among quality, cost, stability, and smoothness.

8.6.3   Practical Implementation

Because the distributed construction of the MST in Algorithm 19 is expensive, we implement the localized approximation of MST, the *localized minimum spanning tree* (LMST) (X.-Y. Li *et al.*, 2004c). For completeness, we define the LMST for a general edge-weighted graph G here.

DEFINITION 8.1 *The k-local minimum spanning tree (LMST$_k$) contains a* directed *edge* $\vec{uv}$ *if edge uv belongs to* $MST[N_k(u)]$.

For the edge-weighted graph Virt$G$, each dominator node u will first collect all dominator nodes that are at most k-hops away in Virt$G$. Typically, k is 1 or 2 in our methods. Node u then constructs the minimum spanning tree $MST[N_k(u)]$ and keeps all edges $uv \in MST[N_k(u)]$. The union of all such selected links is called the LMST, denoted by $LMST_k(G)$. Notice that here the weight of a link uv is the cost of the LCP (≤ 3 hops) connecting u and v in G. From the property of the MST, the following lemma is obvious:

LEMMA 8.4 *The global minimum spanning tree $MST(G)$ is a subgraph of the local minimum spanning tree $LMST_k(G)$.*

Unfortunately, in the worst case, the total cost of $LMST_k(G)$ could be arbitrarily larger than the cost of $MST(G)$. However, our simulations show that it is within a small constant factor on average. The advantage of using the LMST instead of the global MST is the significant reduction in the communication cost.

8.7   Further Reading

Efficient distributed algorithms for constructing CDSs in ad hoc wireless networks were well studied (Alzoubi *et al.*, 2002c; Baker and Ephremides, 1981; Baker *et al.*, 1984; Basagni, 1999; Das and Bharghavan, 1997; Stojmenovic, 2002; Wu and Li, 2001). The notion of cluster organization has been used for wireless ad hoc networks since their early appearance. Baker and co-workers (1981, 1984) introduced a fully distributed linked cluster architecture mainly for hierarchical routing and demonstrated its adaptivity to network connectivity changes. The notion of the cluster has been revisited by Gerla and Tsai (1995) and Lin and Gerla (1997a) for multimedia communications with the emphasis on the allocation of resources (namely, bandwidth and channel) to support the multimedia traffic in an ad hoc environment. Gao *et al.* (2001a) proposed a randomized algorithm for maintaining discrete mobile centers (i.e., DS). They showed that it approximates a *minimum dominating set* (MDS) within $O(1)$ with very high probability, but the constant-approximation ratio is quite large. Recently, Alzoubi *et al.* (2002c) and Wan *et al.* (2002a) proposed a method to approximate a *minimum connected dominating set* (MCDS) within eight whose message complexity is $O(n \log n)$ and time complexity is $O(n)$ for wireless networks modeled by UDGs. Alzoubi *et al.* (2002b) continued to propose a localized method for approximating the MCDS within a constant time by using a linear number of messages. Existing clustering methods first choose some nodes to act as coordinators of the clustering process; i.e., cluster-head. Then, a cluster is formed by associating the cluster-head with some (or all) of its neighbors. Previous methods differ on the criterion for the selection of the cluster-head, which is either based on the lowest (or highest) ID among all unassigned nodes (Baker *et al.*, 1984; Lin and Gerla, 1997a), or based on the maximum node degree (Gerla and Tsai, 1995), or based on some generic weight (Basagni, 1999) (the node with the largest weight will be chosen as cluster-head). Because any *maximal independent set* (mIS) is always a DS, several clustering methods essentially compute a mIS as the final cluster-heads. Alzoubi *et al.* (2002a) proposed building the local Delaunay graph on top of an approximated MCDS for efficient routing. Recently, Kuhn and Wattenhofer (2003) proposed a new distributed MDS approximation algorithm based on LP relaxation techniques. For an arbitrary parameter k and maximum degree Δ, their algorithm computes a DS of expected size $O(k \Delta^{2/k} \log \Delta |\text{MDS}|)$ in $O(k^2)$ rounds in which each node has to send $O(k^2 \Delta)$ messages of size $O(\log \Delta)$. Moreover, Kuhn *et al.* (2004a) further gave the time lower bounds for the distributed approximation of the MDS. Then, Kuhn *et al.* (2004a) showed how to compute a good DS in a harsh model in which there is no underlying MAC layer or asynchronous wake-up, and there is scarce knowledge about the network topology.

Recently, many proposed clustering algorithms (Bao and Garcia-Luma-Aceves, 2003; Basagni, 1999, 2001; Bettstetter and Krausser, 2001; Chatterjee *et al.*, 2002; Chen *et al.*, 2002; Heinzelman *et al.*, 2000; Kozat *et al.*, 2001; Liu and Gupta, 2004; Min *et al.*, 2004; Smaragdakis *et al.*, 2004; Turgut *et al.*, 2002) also considered different weights as a *priority criterion* to decide whether a node will be a cluster-head. Notice that the ultimate goal of the majority of protocols is still to minimize the size of the cluster (or backbone), not the total weight of the cluster (or backbone). For example, methods in Basagni (1999), Bettstetter and Krausser (2001), and Min *et al.* (2004) considered the stability or mobility of each node as the weight. They preferred the node with high stability and low mobility to be the cluster-head. However, the definitions of stability or the evaluation methods used are different. In Kozat *et al.* (2001), the authors also combined stability with the degree of each node as the weight. Higher priority is given to relatively stable and high degree nodes. Methods in Heinzelman *et al.* (2000) and Smaragdakis *et al.* (2004) considered clustering in heterogeneous sensor networks, in which each node has a different energy level. Most of them used the remaining energy or energy-consumption rate as the weight. Both Bao and Garcia-Luna-Aceves (2003) and Liu and Gupta (2004) considered two factors as priority: available energy and the speed, though they used different equations to combine them. Chatterjee *et al.* (2002) considered a combined weight metric for their clustering algorithm that takes into account several system parameters like the node degree, transmission power, mobility, and the battery power of the nodes. Similarly, Chen *et al.* (2002) also combined these four facts to be the weights for their clustering method; a nice literature review of cluster methods can be found in their work. Basagni *et al.* (2004) also showed the performance comparison of some proposed protocols for clustering and backbone formation. Most of these proposed weighted clustering algorithms applied the simple greedy algorithms in which the nodes with the highest priority (lowest cost) become cluster-heads. For example, the cluster method in Chatterjee *et al.* (2002) selects a node with the lowest cost from among its unchosen neighbors to serve as a cluster-head. These greedy heuristics work well in practice but, as we showed in Section 8.2, they may generate a backbone with a high cost compared with the optimum. Some of these methods (Heinzelman *et al.*, 2000; Smaragdakis *et al.*, 2004) are randomized algorithms: Nodes become cluster-heads randomly with a weighted election probability. Turgut *et al.* (2002) proposed a genetic algorithm to optimize cluster processing. None of these cluster methods guarantee any approximation ratio of the weighted cluster (or backbone) compared with the optimum. Notice that Basagni (2001) gave an algorithm to solve a MWIS in wireless networks, but here our solution for a cluster is a distributed approximation algorithm for the MWIS and the MCDS, which are well-known NP-hard problems. X.-Y. Li and Wang (2002) presented a centralized approximation algorithm for a MWIS for some special graphs. Guha and Khuller (1998a) studied centralized algorithms for weighted MCDSs in general graphs by combining a weighted set-cover approximation algorithm and a NST approximation algorithm; they achieved an approximation ratio of $3 \ln n$. Guha and Khuller (1999) further improved the approximation ratio to $1.35 \ln n$, which is the best-known ratio. In addition, any approximation algorithm with a ratio of α for the unweighted (connected) DS problem automatically gives a ratio of $\alpha\delta$ for the weighted version. In particular, the known PTAS for a DS in a UDG (Hunt *et al.*, 1998) implies that a weighted DS in a UDG can be approximated with a ratio of $(1 + \epsilon)\delta$ for arbitrary $\epsilon > 0$. Here, δ is the maximum ratio of costs of two adjacent wireless nodes.

8.8   Conclusion and Remarks

In this chapter, we studied how to construct a sparse structure for a network backbone in wireless ad hoc networks. A communication-efficient distributed algorithm was presented for the construction of a WCDS, whose total cost is guaranteed to be within a small constant factor of the minimum (when either δ or Δ is a constant). We also showed that with a small modification, the constructed backbone is efficient for both cost and hops (though losing the low-cost property). This topology can be constructed locally and is easy to maintain when the nodes move around.

There are many interesting open problems left for further study. Remember that we use the following assumptions on the wireless network model: omnidirectional antenna, a single transmission is received by all nodes within the vicinity of the transmitter, and nodes are static for a reasonable period of time. To prove that the backbone has a low cost, we also assume that all nodes have the same transmission range. Notice that the efficiency property for unicast does not require the communication graph to be a UDG. The problem will become much more complicated if we relax some of these assumptions. Another interesting open problem is to study the dynamic updating of the backbone efficiently when nodes are moving at a reasonable speed, although our cost function may integrate the mobility of the nodes. It is interesting to see the practical performance differences of all the proposed methods, such as the methods by Baker *et al.*, Alzoubi *et al.*, and our methods proposed here, in a mobile environment.

Problems

8.1   Consider a wireless network G in which the transmission ranges of all nodes are within a factor of 2 of each other. Give your best approximation on the value $\alpha^{[1]}(G)$ and $\alpha^{[2]}(G)$.

8.2   In the begining of Section 8.3, a centralized algorithm for constructing a low-cost CDS was described. In step 2 of the algorithm, when we connect the nodes in the DS by using Steiner nodes, we set the weight of the nodes in the DS to 0. Why do we have to do this? What will happen if we do not set the weights of these nodes to 0? Can we guarantee that the total cost of the final CDS is within $O(\log n)$ of the weight of the MWCDS?

8.3   In Algorithm 18, the set of nodes that marked themselves as PossibleDominator already form a DS. Show, by a network example, that the total weight of such a set of nodes could be arbitrarily larger than the minimum weighted DS: The ratio could be much larger than $\log n$. Also show, by a network example, that the ratio of the total weight of nodes marked with PossibleDominator over the total weight of the nodes marked with Dominator could be arbitrarily large.

8.4 In Algorithm 19, we defined the graph $\text{Virt}G$ as the graph over all dominators selected in Algorithm 18 and the edge weight of an edge uv is the cost of the minimum cost path connecting two dominators u and v. What is the largest ratio $\frac{w(\text{Virt}G)}{w(\text{MWCDS})}$ you can get by a network example? Here, $w(\text{Virt}G)$ is the total edge weight of graph $\text{Virt}G$ and $W(\text{MWCDS})$ is the weight of the MWCDS. Further, give a theoretical upper bound on $\frac{w(\text{Virt}G)}{w(\text{MWCDS})}$ by using δ, Δ, and/or n.

8.5 Compared with other distributed CDS algorithms, what are the advantages and disadvantages of the algorithms studied in this chapter? You can compare them in terms of message complexity, time complexity, and approximation ratio guarantees.

8.6 Consider a node-weighted wireless network $G = (V, e)$, where each node has a uniform transmission range. There is a source node s and a set of receivers $R \subset V$. Assume that each relay node $v \notin R$ will charge a cost $c(v)$ for relaying data. Nodes in R will not charge any cost for relaying data for nodes in R. Design an efficient algorithm to approximate the minimum-cost multicast tree connecting s and R.

8.7 Consider disk graphs defined as follows: There is a set of n disks D_i with center v_i and radius r_i, for $1 \leq i \leq n$. A disk graph G is defined over $V = \{v_1, v_2, \ldots, v_n\}$ and two nodes v_i and v_j are connected in G if and only if two disks D_i and D_j intersect. Assume that we want to find a MIS of G. What is the best approximation ratio you can get in polynomial time? Can you design a PTAS for such a problem? Assume that we want to find a MDS of G. What is the best approximation ratio you can get in polynomial time? Can you design a polynomial time algorithm with approximation ratio $\Theta(\log n)$? Can you design a polynomial time algorithm with approximation ratio $o(\log n)$?

9   Topology Control with Flat Structures

9.1   Introduction

Unlike in traditional fixed infrastructure networks, there is no centralized control over ad hoc networks, which consist of an arbitrary distribution of radios in a certain geographical area. An important requirement of wireless ad hoc networks is that they should be self-organizing; i.e., transmission ranges and data paths are dynamically restructured with changing topology. Energy conservation and network performance are probably the most critical issues in wireless ad hoc networks because wireless devices are usually powered by batteries only and have limited computing capability and memory. Recently, significant research (Grünewald *et al.*, 2002; L. Li *et al.*, 2001; X.-Y. Li *et al.*, 2001, 2002b; Rajaraman, 2002; Ramanathan and Rosales-Hain, 2000; Wang *et al.*, 2003; Wattenhofer *et al.*, 2001) has been conducted on designing a power-efficient network topology for wireless networks. Many research results applied a computational geometry technique (specifically, geometrical spanner) to achieve power efficiency. In this chapter, we review these approximation algorithms of a power spanner for ad hoc networks.

9.1.1   Ad Hoc Networks: Graph Model

A WAN consists of a set V of n wireless nodes distributed in a two-dimensional plane. Each node has the same *maximum* transmission range of R meters; e.g., a typical IEEE 802.11g WLAN adapter has a transmission range of around 100–500 m. By a proper scaling, we assume that all nodes have the maximum transmission range equal to one unit. These wireless nodes then define a *unit disk graph* UDG(V) in which there is an edge between two nodes if and only if their Euclidean distance is at most one. Notice that in practice, the transmission region of a node is not necessarily a perfect *disk*. As done in most results in the literature, for simplicity, we model it by *disk* in order to first explore the underlying nature of ad hoc networks. Hereafter, UDG(V) is always assumed to be connected. We also assume that all wireless nodes have distinctive identities (IDs). Additionally, we assume that each node knows the relative position of its one-hop neighbors. The relative position of neighbors can be estimated by the *direction of signal arrival* and the *strength of signal*. The geometry location of a wireless node can also be obtained by a localization method, as in Bulusu *et al.* (2000), Chen *et al.* (2003), Langendoen and Reijers (2003), Niculescu and Nath (2001), Savvides *et al.* (2001), and Zou and Chakrabarty (2003). We assume here that the localization is low cost or it is already required by some other protocols. Notice that the higher these localization costs, the less desirable the advocated approach to design of any protocol based on nodes' geometry location.

By one-hop broadcasting, each node u can gather the location information of all nodes within its transmission range. In the most common power-attenuation model, the power to support a link uv is assumed to be

$$\|uv\|^{\beta},$$

where $\|uv\|$ is the Euclidean distance between u and v and β is a real constant between 2 and 5, depending on the wireless transmission environment.

Throughout this chapter, we focus on only the transmission power of all nodes. This energy model accounts for only the emission power. This can be a good approximation of what happens in the case of long-range techniques, although the actual energy consumption is given by a fixed part (receiving power and the power needed to keep the electric circuits on) plus the emission-power component. In other words, we assume that the transmission range is large enough such that the emission power is the major component and the receiving power is negligible. Notice that even if the energy cost of receiving a packet is high, there are a number of ways of reducing this cost by reducing the number of packets received by but not intended for a node. It includes but is not limited to the following approaches: (1) signals are sent with special small preambles that identify the intended recipient; (2) the radios are frequency agile and can choose different frequency channels to communicate with different neighbors; (3) the radios have directional antennas that limit the volume over which their signals are received; (4) favoring routes that traverse sparser portions of the network; and (5) TDMA transmission scheduling.

9.1.2   Geometrical Spanner

Geometrical spanners have been studied intensively in computational geometry literature for years (Arya and Smid, 1994; Arya *et al.*, 1995; Bose *et al.*, 2002a; Chandra *et al.*, 1992; Karavelas and Guibas, 2001; Levcopoulos *et al.*, 1998; Yao, 1982). Let $G = (V, E)$ be an n-vertex-connected geometry graph. The distance in G between two vertices $u, v \in V$ is the total length (weight) of the shortest path between u and v and is denoted by $d_G(u, v)$.

DEFINITION 9.1 *A subgraph $H = (V, E')$, where $E' \subseteq E$ is a t-spanner of G if for every $u, v \in V$, $d_H(u, v) \leq t d_G(u, v)$, where t is a constant and is called the* length-stretch factor.

In other words, if H is a t-spanner of G, for any path $\mathsf{P}(u, v)$ in G from a node $u \in V$ to another node $v \in V$, there is always a path $\mathsf{P}'(u, v)$ in H that has a length of at most t times of $\mathsf{P}(u, v)$. The spanner H can be treated as a constant (t) approximation of the original graph G in the sense of the shortest path length. If the subgraph H is a sparse graph (i.e., the number of edges in H is linear with the number of vertices), it is called a *sparse spanner*.

9.1.3 Power Spanner

The *topology-control* techniques let each wireless device adjust its transmission range and select certain neighbors for communication, while maintaining a structure (network topology) that can support energy-efficient routing and improve the overall network performance. Remember that we use a UDG to model the original communication graph for an ad hoc network, and the UDG provides information about all possible topologies. Most topology-control methods construct a sparse subgraph of the UDG and then restrict the routings on the constructed subgraph to save energy. Not every connected subgraph of the UDG is suitable to be the network topology. A good network topology should be energy efficient; that is, the total power consumption of the shortest path (most power-efficient path) between any two nodes in the final topology should not exceed a constant factor of the power consumption of the shortest path in the original network. Given a path $v_1 v_2 \cdots v_h$ connecting two nodes v_1 and v_h, the energy cost of this path is $\sum_{j=1}^{h-1} \|v_j v_{j+1}\|^\beta$. The path with the lowest energy cost is called the shortest path in a graph. Borrowing the concept of spanner from computational geometry, we define the power spanner as follows:

DEFINITION 9.2 *A subgraph H is called a* power spanner *of a graph G if there is a positive real constant* $\rho_H(G)$ *such that for any two nodes, the power consumption of the shortest path in H is at most* $\rho_H(G)$ *times the power consumption of the shortest path in G. The constant* $\rho_H(G)$ *is called the* power-stretch factor.

Similarly, we can define the length-stretch factors $\ell_H(G)$ by setting $\beta = 1$, and we call a graph H a *length spanner* if the $\ell_H(G)$ is bounded by a constant. It is not difficult to show that for any $H \subseteq G$ with a length-stretch factor δ, its power-stretch factor is at most δ^β for any graph G. In particular, a graph with a constant-bounded length-stretch factor must also have a constant-bounded power-stretch factor, but the reverse is not true. Finally, the power-stretch factor has the following monotonic property: If $H_1 \subset H_2 \subset G$, then the power-stretch factors of H_1 and H_2 satisfy $\rho_{H_1}(G) \geq \rho_{H_2}(G)$.

Besides the power efficiency (power spanners), the network topology should also have some (or all) of the following desirable features: connected, sparse, planar, degree bounded, fault tolerant, and so on.

9.1.4 Efficient Localized Construction

Unlike traditional wired networks and cellular wireless networks, wireless devices in MANET are often moving during the communication, which could change the network topology to some extent. Hence, it is more challenging to design a topology-control algorithm for ad hoc networks: The topology should be locally and self-adaptively maintained without affecting the whole network, and the communication cost during maintaining should not be too high. In other words, the construction algorithm is preferred to be *localized*. Here, a distributed algorithm constructing a graph G is a *localized algorithm* if every node u can exactly decide all edges incident upon u based on the information only of all nodes within a constant hop of u. More important, we expect that the total

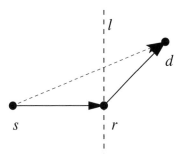

Figure 9.1 Relaying through another node r consumes less power than directly transmitting from s to d. © 2000, ACM.

communication cost of the algorithm is $O(n)$ messages, where each message is $O(\log n)$ bits.

9.1.5   Relay Regions

In the most common power-attenuation model, the signal power falls as $1/r^\beta$, where r is the distance from the transmitter antenna and β is a constant between 2 and 5, depending on the wireless transmission environment. This is typically called the *path loss*. To make this model meaningful, we always assume that the distance between any two nodes is at least one unit, so the preceding model does not violate the energy-consumption law. We also assume that the unit of power and the unit of distance between nodes satisfy the path-loss formula. The path loss normally depends on the heights of the transmit antennas as well as the transmitter–receiver separation. By simple geometry computing, it is easy to see that relaying a signal between nodes may result in lower power consumption than communication over large distances because of the nonlinear power attenuation.

As a simple illustration, consider three nodes s, r, and d on the plane as in Figure 9.1. Assume that all three nodes use identical transmitters and receivers and $\beta = 2$. The power to transmit a signal from s to d is therefore $\|sd\|^2$. If we use the node r to relay the signal, the total power used is $\|sr\|^2 + \|rd\|^2$, which is less than $\|sd\|^2$. In other words, if s wants to send a message to any node d lying in the right side of the line l, relaying through node r always consumes less power than directly transmitting to d.

There is also another source of power consumption we must consider in addition to the path loss. When a node receives a signal from another node, it must consume some power to receive, store, and then process that signal (Rodoplu and Meng, 1998). This additional power consumed at the receiver node is referred to as the *receiver power* at the relay node. Typically, every relay node consumes the same receiver power because of the nature of its operations. Hereafter, we denote such power by a constant c. Notice that additional power will also be consumed when the routing algorithm is running. In the design of modern processors, however, the power consumption required for such processing and computation can be made negligible compared with the transmission power and receiving power.

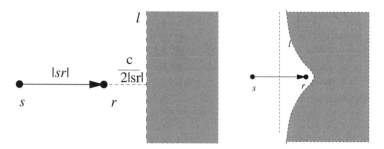

Figure 9.2 The relay regions $R(s, r)$ denoted by the shaded areas. Left $\beta = 2$; right $\beta = 4$.
© 2000, ACM.

Assume that node s wants to send a message to node d. Accordingly, node s is called the source (transmitter) and node d is called the destination (receiver). A node r could be used as a relay node if and only if

$$\|sd\|^{\beta} + c > \|sr\|^{\beta} + c + \|rd\|^{\beta} + c.$$

Notice that $\|sd\|^{\beta} + c$ is the power incurred if node s directly transmits a signal to node d, and $\|sr\|^{\beta} + c + \|rd\|^{\beta} + c$ is the power incurred if node s uses node r as the relay node for transmission from s to node d. Thus, given node s and node r, the locus node d, such that relaying through node r consumes less power than directly transmitting from s to d, is called the *relay region* of r for s (Rodoplu and Meng, 1998). Hereafter, we denote such a relay region as $R_{\beta,c}(s, r)$. When it is clear from the context, we drop the β and/or c from $R_{\beta,c}(s, r)$.

DEFINITION 9.3 (Relay region) *The relay region of a node* r *for a node* s *is defined as*

$$R_{\beta,c}(s, r) = \{x \mid \|sx\|^{\beta} > \|sr\|^{\beta} + \|rx\|^{\beta} + c\}.$$

Figure 9.2 illustrates typical relay regions in a propagation environment with $\beta = 2$ and $\beta = 4$. We now study in detail the mathematical formula to represent the relay region $R_{\beta,c}(s, r)$. Let (x_p, y_p) denote the position of a two-dimensional node p. Assume that node r has coordinates $(0, 0)$ and node s has coordinates $(-\|sr\|, 0)$. When $\beta = 2$, $d = (x_d, y_d) \in R(s, r)$ implies that

$$\begin{aligned} \|sd\|^2 &= (\|sr\| + x_d)^2 + y_d^2 \\ &> \|sr\|^2 + \|rd\|^2 + c \\ &= \|sr\|^2 + x_d^2 + y_d^2 + c. \end{aligned}$$

It implies that $x_d > \frac{c}{2\|sr\|}$. In other words, if $s = (-\|sr\|, 0)$ and $r = (0, 0)$, then

$$R_{2,c}(s, r) = \left\{ (x, y) \mid x > \frac{c}{2\|sr\|} \right\}.$$

Therefore, the boundary of the relay region $R_{2,c}(s, r)$ for any two nodes s and r is a line perpendicular to sr and node r has distance $\frac{c}{2\|sr\|}$ to the relay region. See Figure 9.2 for an illustration. When $\beta = 4$, we have

$$
\begin{aligned}
\|sd\|^4 &= \left((\|sr\| + x_d)^2 + y_d^2\right)^2 \\
&= \left(\|sr\|^2 + x_d^2 + y_d^2 + 2x_d\|sr\|\right)^2 \\
&> \|sr\|^4 + \|rd\|^4 + c \\
&= \|sr\|^4 + x_d^4 + y_d^4 + 2x_d^2 y_d^2 + c.
\end{aligned}
$$

It implies that

$$
(2x_d + \|sr\|)y_d^2 + 2x_d^3 + 3\|sr\|x_d^2 + 2\|sr\|^2 x_d > \frac{c}{2\|sr\|}.
$$

We now study the properties of the structure of the minimum-energy topology of a set of stationary nodes. For simplicity, let $E_{\beta,c}(s, r)$ be the complement of $R_{\beta,c}(s, r)$. The region $E_{\beta,c}(s, r)$ is called the enclosure region of node s by node r.

DEFINITION 9.4 (Enclosure region) *The enclosure region $E_{\beta,c}(s)$ of a node s is defined as*

$$
E_{\beta,c}(s) = \bigcap_{r \in T(s)} E_{\beta,c}(s, r).
$$

Notice that the preceding definition is analogous to the *Voronoi region* of a node s, which is defined as

$$
V(s) = \{x \mid \forall q \in V, \ \|xs\| < \|xq\|\}.
$$

Remember that here, $T(s)$ is the set of nodes lying within the transmission range of node s. A node u is said to be a *neighbor* of a node s if it is inside the enclosure region $E_\beta(s)$ of node s.

DEFINITION 9.5 (Neighbors) *The neighbors $N_{\beta,c}(s)$ of a node s are defined as*

$$
N_{\beta,c}(s) = \{u \mid u \in T(s), \ u \in E_{\beta,c}(s)\}.
$$

When it is clear from the context, we also drop the constant β and/or c from $E_{\beta,c}(s, r)$ and $N_{\beta,c}(s)$. As in Rodoplu and Meng (1998), here we define the enclosure graph as follows.

DEFINITION 9.6 (Enclosure graph) *The enclosure graph $G_e^{(\beta,c)} = (V, E)$ of a set of mobile nodes $\mathcal{V}$ is the directed graph whose vertices are $\mathcal{V}$ and whose edges are all (u, v), where $v \in N_{\beta,c}(u)$.*

When it is clear from the context, we also drop the β and/or c from $G_e^{(\beta,c)}$. It is easy to prove that the minimum-energy topology is always contained in the enclosure graph.

THEOREM 9.1 *The enclosure graph G_e contains the minimum-power topology.*

We then construct an example such that the enclosure graph is not equal to the minimum-energy topology. Figure 9.3 shows such an example when $\beta = 2$ and $c = 0$. It is not difficult to show that edge uv does not belong to the minimum-power topology because

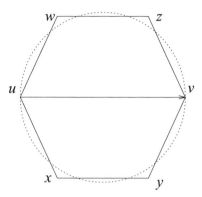

Figure 9.3 The enclosure graph and the minimum-power topology are not the same: The drawn graph is an enclosure graph, whereas the minimum-power topology does not have uv.

we have a path $uxyv$ consuming less power. However, as we show later, usually the number of edges in G_e is $O(n)$.

Let us first consider how to compute the enclosure graph G_β by using a centralized algorithm. One simple approach is as follows. For each pair of nodes u and v, compute the relay region $R(u, v)$ and $R(v, u)$. Then, by using an approach similar to computing the Voronoi diagram, we can compute the enclosure region for each node u. And all nodes covered by the enclosure region of u are connected to node u. The time complexity of the preceding approach could be as large as $O(n^3)$, where n is the number of all mobile nodes. This is apparently not practical for ad hoc networking.

When the receiver's cost c is negligible, we know that the minimum-power topology G_m is a subgraph of the Delaunay triangulation of all mobile nodes. Therefore, we can apply any $O(n \log n)$ time-complexity Delaunay triangulation algorithm to compute the Delaunay triangulation of all mobile nodes. Then, for each node u, we eliminate the nodes of Delaunay neighbors $D(u)$ that are in the relay region of other nodes from $D(u)$. The remaining nodes are $N_\beta(u)$. Notice that the fact that the average number of Delaunay neighbors $D(u)$ is at most six implies that the average time complexity to compute $N(u)$ from $D(u)$ is constant. Therefore, the average time complexity of the preceding algorithm with the Delaunay triangulation used to compute the enclosure graph is $O(n \log n)$ if the receiver's cost $c = 0$.

9.2   Current State of Knowledge

9.2.1   Unicast Topology

Several structures have been proposed for topology control in wireless networks. See Figure 9.4 for an illustration of some structures. The *relative neighborhood graph*, denoted by RNG(V) (Toussaint, 1980), consists of all edges uv such that the intersection of two circles centered at u and v and with radius $\|uv\|$ does not contain any vertex w from the set V. The *Gabriel graph* (Gabriel and Sokal, 1969) GG(V) contains an

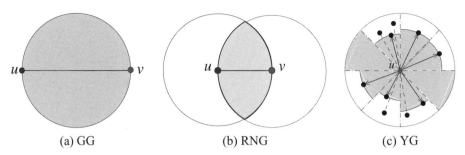

Figure 9.4 The definitions of RNG, GG, and YG. The shaded area is empty of nodes inside. © 2006, ACM.

edge uv if and only if disk(u, v) contains no other points of V, where disk(u, v) is the disk with edge uv as a diameter. For convenience, we also denote GG and RNG as the intersection of GG(V) and RNG(V) with UDG(V), respectively. Both GG and RNG are planar, are connected, and contain the *Euclidean minimum spanning tree* (EMST) of V if UDG is connected. RNG is not power-efficient for unicast because the power-stretch factor of RNG is $n - 1$ in the worst case (W. Wang *et al.*, 2003) and not bounded by a constant even for n nodes randomly distributed (our proof is similar to the proof in Bose *et al.*, 2002a, and is omitted here because of space limitations). Neither RNG nor GG is degree bounded. The *Yao graph* (Yao, 1982) with an integer parameter $k > 6$, denoted by YG$_k$, is defined as follows. At each node u, any k equally separated rays originated at u define k cones. In each cone, choose the shortest edge $uv \in$ UDG(V) from among all edges emanated from u, if there is any, and add a directed link $\vec{uv}$. Ties are broken arbitrarily or by ID. The resulting directed graph is called the *Yao graph*. It is well known that the Yao structure is power-efficient for unicast. Several variations (X.-Y. Li *et al.*, 2002c) of the Yao structure could also have a bounded logical node degree. However, not all Yao-related structures are planar.

L. Li *et al.* (2001) proposed the cone-based topology-control (CBTC) algorithm to first focus on several desirable properties, in particular an energy spanner with a bounded degree. It is basically similar to the Yao structure for topology control. Each node u finds a power $p_{u,\alpha}$ such that in every cone of degree α surrounding u, there is some node that u can reach with power $p_{u,\alpha}$. Here, nevertheless, we assume that there is a node reachable from u by the maximum power in that cone. Notice that the number of cones to be considered in the traditional Yao structure is a constant k. However, unlike the Yao structure, for each node u, the number of cones needed to be considered in the method proposed in L. Li *et al.* (2001) is 2Δ, where each neighboring node v could contribute two cones on both side of segment uv. Then, the graph G_α contains all edges uv such that u can communicate with v using power $p_{u,\alpha}$. They proved that if $\alpha \leq (5\pi/6)$ and the UDG is connected, then graph G_α is a connected graph. On the other hand, if $\alpha > (5\pi/6)$, they showed that the connectivity of G_α is not guaranteed by giving some counterexample (L. Li *et al.*, 2001). Unlike the Yao structure, the final topology G_α is not necessarily a bounded-degree graph.

Bose *et al.* (2002c) proposed a centralized method with running time $O(n \log n)$ to build a degree-bounded planar spanner for a two-dimensional point set. It constructs a planar t-spanner with low weight for a given node set V, for $t = (1 + \pi)C_{del} \simeq 10.02$, such that the node degree is bounded from above by 27. Hereafter, we use $C_{del} < 2.6$ to denote the spanning ratio of the Delaunay triangulation (Dobkin *et al.*, 1990; Keil and Gutwin, 1989, 1992). However, a straightforward distributed implementation of this centralized method takes $O(n^2)$ communications in the worst case for a set V of n nodes.

Y. Wang and Li (2003) proposed the first efficient distributed algorithm to build a bounded-degree planar spanner (BPS) for wireless ad hoc networks. Although their method can achieve three desirable features – planar, degree-bounded, and power-efficient – the theoretical bound on the node degree of their structure is a large constant. For example, when $\alpha = \pi/6$, the theoretical bound on the node degree is 25. In addition, the communication cost of their method can be very high, although it is $O(n)$ theoretically, which is achieved by applying the method in Călinescu (2003) to collect two-hop-neighbor information. The hidden constant is large, on the order of several hundred.

Recently, Song *et al.* (2004, 2005) proposed two methods to construct a BPS by applying the ordered Yao structures on a GG. They achieved better performance with a much lower communication cost, compared with the method in Y. Wang and Li (2003). One method in Song *et al.* (2004) costs only $3n$ messages for the construction and guarantees that there is at most one neighbor node in each of the $k = 9$ equal-sized cones. The trade-off here is that the structure presented in Song *et al.* (2004) is not a length spanner, whereas the structure presented in Y. Wang and Li (2003) is.

Notice that the structures constructed by the methods proposed in Song *et al.* (2004, 2005) are not guaranteed to be low-weighted. Both structures are planar and degree-bounded. The structure constructed in Song *et al.* (2004) is only a power spanner, whereas the structure constructed in Y. Wang and Li (2003) is also a length spanner. Notice that it is known that a length spanner is always a power spanner (X.-Y. Li *et al.*, 2002c). Later, we study the *first* method to construct a single topology that is planar, length spanner, bounded-degree, and low-weighted.

In summary, for energy-efficient unicast routing, it is preferred that the topology have the following features:

1. *Power spanner*: Formally speaking, a subgraph H is called a *power spanner* of a graph G if there is a positive real constant ρ such that for any two nodes, the power consumption of the shortest path in H is at most ρ times of the power consumption of the shortest path in G. Here, ρ is called the *power-stretch factor* or *spanning ratio*.

2. *Degree-bounded*: It is also desirable that the logical node degree in the constructed topology be bounded from above by a small constant. A small node degree could reduce the MAC-level contention and interference; it also may help to mitigate the well-known hidden- and exposed-terminal problems. In addition, a structure with a small degree will improve the overall network throughout (Kleinrock and Silvester, 1978). Bounded-degree structures also find applications in Bluetooth wireless networks because a *master* node can have only seven active slaves.

3. *Planar*: It is also preferred that network topology be planar (i.e., no two edges crossing each other in the graph) to enable some localized routing algorithms to work correctly

and efficiently, such as *greedy face greedy routing* (GFG) (Bose *et al.*, 2001), *greedy perimeter stateless routing* (GPSR) (Karp and Kung, 2000), *adaptive face routing* (AFR) (Kuhn *et al.*, 2002), and *greedy other adaptive face routing* (GOAFR) (Kuhn *et al.*, 2003a). Notice that with planar network topology as the underlying routing structure, these localized routing protocols guarantee message delivery without using a routing table: Each intermediate node can decide which logical neighboring node to forward the packet using only local information and the position of the source and the destination.

9.2.2   Energy-Efficient Broadcast Topology

Broadcasting is also a very important operation in wireless ad hoc/sensor networks because it provides an efficient way of communication that does not require global information and functions well with topology changes. For example, many unicast routing protocols (Jetcheva *et al.*, 2001; Park and Corson, 1997; Perkins and Bhagwat, 1994; Shah and Rabaey, 2002) for wireless multihop networks use broadcasting in the stage of route discovery. Similarly, several information-dissemination protocols in WNSs use some form of broadcast/multicast for solicitation or collection of sensor information (Heinzelman *et al.*, 1999; Intanagonwiwat *et al.*, 2003; Ye *et al.*, 2002). Because sensor networks mainly (Akyildiz *et al.*, 2002) use broadcast for communication, how to deliver messages to all the wireless devices in a scalable and power-efficient manner has drawn more and more attention. Not until recently have research efforts been made to devise power-efficient broadcast structures for WANs.

Notice that a broadcast routing protocol can be interpreted as *flood-based* broadcasting on a *certain* subgraph of the original communication networks because *any* broadcast routing can be viewed as an arborescence (i.e., a directed tree) T, rooted at the source node of the broadcasting, that spans all nodes. The tree T contains a directed edge $\vec{uv}$ if node v received the first copy of the broadcast data from node u. Once a broadcast structure H is constructed, the broadcast is a simple flooding on top of H: Once a node v gets the broadcast message from any of its *logical* neighbors, say u, for the first time, it will simply forward it to all of its *logical* neighbors (except maybe the node u) either through one-to-one or one-to-all communications. Let $f_H(p)$ denote the transmission power of the node p required by the broadcasting message on top of a broadcast structure H. We assume that the tree H is a directed graph rooted at the source of the broadcasting session: link $\vec{pq} \in H$ denotes that node p forwarded message to node q. For any leaf node p of H, clearly, we have $f_H(p) = 0$ because it does not have to forward the data to any other node. For any internal node p of H, $f_H(p) = \max_{\vec{pq} \in H} \|pq\|^\beta$ under our energy model if a one-to-all communication model is used, and $f_H(p) = \sum_{\vec{pq} \in H} \|pq\|^\beta$ under our energy model if a one-to-one communication model is used. The total energy required by H is $\sum_{p \in V} f_H(p)$. In the literature, the one-to-all communication model is typically assumed.

Minimum-energy broadcast routing in a simple ad hoc networking environment has been addressed in Clementi *et al.* (2001a), Kirousis *et al.* (2000), and Wieselthier *et al.*

(2000). It is known (Clementi *et al.*, 2001a) that the minimum-energy broadcast routing problem is NP-hard; i.e., it cannot be solved in polynomial time unless $P = \text{NP}$. Three greedy heuristics were proposed in Wieselthier *et al.* (2000) for the minimum-energy broadcast routing problem: the EMST, the shortest path tree (SPT), and the broadcasting incremental power (BIP). Wan *et al.* (2001, 2002b) showed that the approximation ratios of the EMST and BIP are between 6 and 12 and between 13/3 and 12, respectively; on the other hand, the approximation ratio of the SPT is at least $n/2$, where n is the number of nodes. The approximation ratio of the EMST was improved to 8 recently by Flammini *et al.* (2004). Unfortunately, none of the preceding structures can be formed and updated by use of only a linear number of messages or locally.

The RNG, which can be constructed locally, has been used for broadcasting in wireless ad hoc networks (Seddigh *et al.*, 2002). However, an example was given in X.-Y. Li (2003b) to show that the total energy used by broadcasting on the RNG could be about $O(n^{\beta})$ times the minimum energy used by an optimum method. Several practical broadcasting protocols (Cartigny *et al.*, 2003; Orecchia *et al.*, 2004; Wu and Dai, 2003) were proposed recently; however, none of them provided their theoretical performance bound on the energy consumption. In fact, X.-Y. Li (2003b) showed that *no* deterministic localized algorithm can find a structure that approximates the total energy consumption of broadcasting within a constant factor of the optimum. Furthermore, in the worst case, for *any* broadcast based on a locally constructed and connected structure, there is a network configuration of n nodes such that its energy consumption is $\Theta(n^{\beta-1})$ times optimum. On the other hand, given any low-weighted structure H – i.e., $\mathbf{c}(H) \leq O(1)\mathbf{c}(\text{EMST})$ – they proved the following lemma:

LEMMA 9.1 (X.-Y. Li *et al.*, 2003b) $\mathbf{c}_\beta(H) \leq O(n^{\beta-1})\mathbf{c}_\beta(\text{EMST})$, *where H is any low-weighted structure.*

Here, $\mathbf{c}(G)$ is the total length of the links in G (i.e., $\mathbf{c}(G) = \sum_{uv \in G} \|uv\|$) and $\mathbf{c}_\beta(G)$ is the total power consumption of links in G (i.e., $\mathbf{c}_\beta(G) = \sum_{uv \in G} \|uv\|^\beta$). Consequently, a low-weighted structure is asymptotically optimal for broadcasting among any connected structures built in a localized manner. Notice that the preceding analysis is based on the assumption that every link is used during the broadcast (i.e., one-to-one communication), such as using the TDMA scheme. Even considering that the broadcast signal sent by a node can be received by all nodes in its transmission region simultaneously (i.e., one-to-all communication), the preceding claim is also correct. The reason is basically as follows. Let $B_s(H)$ be the total energy consumed by broadcasting on a structure H with sender s using the one-to-all communication model. Clearly, *any* flood-based broadcast based on a structure H consumes energy at most $\sum_{e_i \in H} e_i^\beta$ if the message received by an intermediate node v is not forwarded to its parent; i.e., the node that just forwarded this message to v; and the total energy is at most $2\sum_{e_i \in H} e_i^\beta$ if an intermediate node v may also forward the message to its parent. On the other hand, the total energy $B_s(H)$ used by *any* structure H is at least $\sum_{e_i \in \text{EMST}} e_i^\beta/12$ (Wan *et al.*, 2002b). Thus,

$$B_s(\text{EMST}) \geq \sum_{e_i \in \text{EMST}} e_i^\beta/12 = \mathbf{c}_\beta(\text{EMST})/12.$$

Then, if H is a low-weighted structure, we have

$$B_s(H) \leq 2 \sum_{e_i \in H} e_i^{\beta} = O(n^{\beta-1}) \mathbf{c}_{\beta}(\text{EMST})$$

$$\leq 12O(n^{\beta-1})B_s(\text{EMST}).$$

Recall that $B_s(\text{EMST})$ is no more than a constant (≤ 8) times the optimum in a one-to-all communication model (Flammini *et al.*, 2004; Wan *et al.*, 2002b). Consequently, we have the following lemma:

LEMMA 9.2 *The broadcast based on any low-weighted structure H consumes energy at most $O(n^{\beta-1})$ times the optimum in both one-to-one and one-to-all communication models. The bound $O(n^{\beta-1})$ is tight.*

In summary, to enable energy-efficient broadcasting, it is preferred that the constructed topology be *low-weighted*, in addition to having the three properties given in Subsection 9.2.1 for unicast:

4. *Low-weighted*: The total link length of final topology is within a constant factor of that of the EMST.

Recently, several localized algorithms (X.-Y. Li, 2003b; X.-Y. Li *et al.*, 2004c) were proposed to construct low-weighted structures, which indeed approximate the energy efficiency of an EMST as the network density increases. However, none of them is power-efficient for unicast routing. Here, the first efficient distributed method for constructing a planar, bounded-degree spanner that is also low-weighted is presented.

9.3   Planar Structures

Many geometric routing algorithms require the planar topology to guarantee the message delivery, such as right-hand routing, GFG (Bose *et al.*, 2001), GPSR (Karp and Kung, 2000), and adaptive face routing (AFR) (Kuhn *et al.*, 2002). Therefore, in this section, we review the methods for building planar structures.

9.3.1   Relative Neighborhood Graph

Let $G = (V, E)$ be a geometric graph defined on vertex set V with edge set E. The *relative neighborhood graph*, denoted by RNG(G), is a geometric concept proposed by Jaromczyk and Toussaint (1992) and Toussaint (1980). It consists of all edges $uv \in E$ such that there is no point $w \in V$ with edges uw and wv in E satisfying $\|uw\| < \|uv\|$ and $\|wv\| < \|uv\|$. Thus, an edge uv is included if the intersection of two circles centered at u and v and with radius $\|uv\|$ do not contain any vertex w from the set V such that edges uw and wv exist in E. See Figure 9.4 (a) for an illustration. When G is a UDG, we use RNG(V) to denote the graph instead of RNG(G). RNG(V) is a planar graph (i.e., no two edges cross each other), which also implies its sparseness: $|\text{RNG}(V)| \leq 3n$, where n is the number of vertices. For an undirected and connected graph G, RNG(G) is connected and contains the MST of G. In other words, if the UDG(V) is connected, the RNG(V) is connected too. This ensures the connectivity of the ad hoc networks. Another important property is that RNG(V) can be constructed easily by a localized method. From the definition, each node needs information from only its one-hop neighbors to construct the RNG(V). The RNG(V) was used for efficient broadcasting (minimizing the number of retransmissions) in a one-to-one broadcasting model in Seddigh *et al.* (2002) and was used by the GPSR protocol (Karp and Kung, 2000) as the routing topology that guarantees the delivery of the packet. However, the analysis by Bose *et al.* (2002a) implied that the length-stretch factor of RNG(V) is at most $n - 1$. X.-Y. Li *et al.* (2001) and Y. Wang (2004) first studied and analyzed the power-stretch factors of RNG(V); they showed that the power-stretch factor of the RNG(V) is actually $n - 1$ by constructing an example. Thus, in summary, RNG(V) is not a power–length spanner; i.e., RNG(V) is not power-efficient for unicast routing in ad hoc networks.

The RNG structure can also be defined on any weighted graph G.

DEFINITION 9.7 *A link* uv *is not in RNG(G) if there are two links* uw $\in$ G *and* wv $\in$ G *such that both links have a weight smaller than the weight of* uv.

Obviously, when G is connected, RNG(G) is connected and contains the MST of G as a subgraph. See Kapoor and Li (2003) for other extensions of RNG, the GG, and the Delaunay graph to any general graph.

The idea of defining RNG on a weighted graph has also been used by Wattenhofer and Zollinger (2004) to design the XTC topology-control protocol. The weight of a network link is its "quality." A node u keeps a link to another node v iff there is no other node w such that the weights of links uw and wv are both no more than the weight of link uv. As proved in Wattenhofer and Zollinger (2004), the XTC protocol preserves network connectivity. This was also proved in Kapoor and Li (2003) when the original general graph G satisfies a certain property.

9.3.2   Gabriel Graph

Let disk(u, v) be a disk with diameter uv. Then, the *Gabriel graph* (Gabriel and Sokal, 1969) GG(G) contains an edge uv from G if and only if disk(u, v) contains no other vertex $w \in V$ such that there exist edges uw and wv from G. See Figure 9.4(b) for an illustration. When G is a UDG, we use GG(V) to denote the graph. GG(V) is also a popular planar graph. It is easy to show that RNG(V) is a subgraph of GG(V). Thus, for a connected graph G, GG(G) is also connected and contains the MST of G. GG(V) can be constructed easily with a localized method with one-hop-neighbor information. The GG was used as a planar subgraph in the face-routing protocol (Bose *et al.*, 2001; Datta *et al.*, 2002; Stojmenovic and Datta, 2002) and the GPSR routing protocol (Karp and Kung, 2000) that guarantee the delivery of the packet. The same analysis by Bose *et al.* (2002a) implied that the length-stretch factor of GG(V) is at most $[(4\pi\sqrt{2n - 4})/3]$. Recently, W. Wang *et al.* (2003) showed that the length-stretch factor of GG(V) is precisely $\sqrt{n-1}$. X.-Y. Li *et al.* (2001) and Y. Wang (2004) then proved that the power-stretch factor of any GG is one; i.e., all power-efficient paths are kept in GGs. Therefore, GG(V) is a power spanner but not a length spanner. For unicast routing, the GG is power-efficient.

9.3.3   Delaunay Triangulation

Although neither the GG nor the RNG is a length spanner, *Delaunay triangulation* is a well-known length spanner. Assume that there are no four vertices of V that are cocircular. A triangulation of V is a *Delaunay triangulation*, denoted by Del(V), if the circumcircle of each of its triangles does not contain any other vertices of V in its interior. A triangle is called the *Delaunay triangle* if its circumcircle is empty of vertices of V. See Figure 9.4(c) for an illustration. Dobkin *et al.* (1990) first proved that the Delaunay triangulation is a length spanner with a length-stretch factor bounded by $\frac{1+\sqrt{5}}{2}\pi$. Then, Keil and Gutwin (1992) improved the constant to be $\frac{4\sqrt{3}}{9}\pi$. Notice that the GG is a subgraph of the Delaunay triangulation and the power-stretch factor of GG is one. Thus, the power-stretch factor of Delaunay triangulation is also one because of the monotonic property of the power spanner.

Given a set of points V, let $U\text{Del}(V)$ be the graph of removing all edges of Del(V) that are longer than one unit; i.e., $U\text{Del}(V) = \text{Del}(V) \cap \text{UDG}(V)$. X.-Y. Li *et al.* (2002a) considered the *unit Delaunay triangulation* $U\text{Del}(V)$ for the planar spanner of a UDG, which is a subset of the Delaunay triangulation. It was proved in X.-Y. Li *et al.* (2002a) that $U\text{Del}(V)$ is a $\frac{4\sqrt{3}}{9}\pi$-length-spanner of UDG(V).

Although Delaunay triangulation is a well-known planar spanner, it is not appropriate to require a construction of the Delaunay triangulation in a wireless-communication environment because of the possible massive communications it requires. Notice that the circumcircle of a triangle can be very large (much larger than the transmission range of a wireless node); therefore, it may need global information to build the Delaunay triangulation. Recently, several published results (Gao *et al.*, 2001b; X.-Y. Li *et al.*, 2002a, 2002b) were proposed to build Delaunay triangulation or its relatives in a localized way. Here, we review one of these results.

9.3.4   Local Delaunay Graph

X.-Y. Li *et al.* (2002a, 2002b) gave a localized algorithm that constructs sequence graphs, called *localized Delaunay* $L\text{Del}^{(k)}(V)$, which are supergraphs of $U\text{Del}(V)$. Triangle $\triangle uvw$ is called a *k-localized Delaunay triangle* if the interior of the circumcircle of $\triangle uvw$ does not contain any vertex of V that is a k-neighbor of u, v, or w and all edges of the triangle $\triangle uvw$ have a length of no more than one unit. The *k-localized Delaunay graph* over a vertex set V, denoted by $L\text{Del}^{(k)}(V)$, has exactly all Gabriel edges in a UDG [edges in GG(V)] and edges of all k-localized Delaunay triangles. A local Delaunay graph always has a constant-bounded length-spanning ratio and power-spanning ratio. The localized algorithm for the construction of $L\text{Del}^{(k)}(V)$ is as follows.

Algorithm 20 Construct $L\text{Del}^{(k)}(V)$

1: Each node u first gathers the location information of its k-hop neighbors $N_k(u)$. It computes the Delaunay triangulation $\text{Del}[N_k(u)]$ of its k-neighbors $N_k(u)$, including u itself.

2: For each edge uv of $\text{Del}[N_k(u)]$, let $\triangle uvw$ and $\triangle uvz$ be two triangles incident upon uv. Edge uv is a Gabriel edge if both angles $\angle uwv$ and $\angle uzv$ are less than $\pi/2$. Node u marks all *Gabriel edges* uv, which will never be deleted.

3: Each node u finds all triangles $\triangle uvw$ from $\text{Del}[N_k(u)]$ such that all three edges of $\triangle uvw$ have a length of at most one unit. If angle $\pi/3$, node u broadcasts a message proposal(u, v, w) to form a k-localized Delaunay triangle $\triangle uvw$ in $L\text{Del}^{(k)}(V)$, and listens to the messages from other nodes.

4: When a node u receives a message proposal(u, v, w), u accepts the proposal of constructing $\triangle uvw$ if $\triangle uvw$ belongs to the Delaunay triangulation $\text{Del}[N_k(u)]$ by broadcasting message accept(u, v, w); otherwise, it rejects the proposal by broadcasting message reject(u, v, w).

5: A node u adds the edges uv and uw to its set of incident edges if the triangle $\triangle uvw$ is in the Delaunay triangulation $\text{Del}[N_k(u)]$ and both v and w have sent either accept(u, v, w) or proposal(u, v, w).

It was proved that the graph constructed by the preceding algorithm is $L\text{Del}^{(k)}(V)$. Indeed, for each triangle $\triangle uvw$ of $L\text{Del}^{(k)}(V)$, one of its interior angles is at least $\pi/3$ and $\triangle uvw$ is in $\text{Del}[N_k(u)]$, $\text{Del}[N_k(v)]$, and $\text{Del}[N_k(w)]$. So one of the nodes among $\{u, v, w\}$ will broadcast the message proposal(u, v, w) to form a k-localized Delaunay triangle $\triangle uvw$. Because $\text{Del}[N_k(u)]$ is a planar graph, and a proposal is made only if $\pi/3$, node u broadcasts at most six proposals, and each proposal is replied to by at most two nodes. Therefore, the total communication cost of steps 3–5 in Algorithm 20 is $O(n)$ messages (each message with $\log n$ bits).

As shown in X.-Y. Li *et al.* (2002a, 2002b), the graph $L\text{Del}^{(1)}(V)$ may contain some edges that intersect. On the other hand, $L\text{Del}^{(k)}(V)$ is a planar graph for any $k \geq 2$. Although $L\text{Del}^{(1)}(V)$ is not a planar graph, X.-Y. Li *et al.* (2002a) proved $L\text{Del}^{(1)}(V)$ has a thickness of 2, which implies that it is sparse. Because $U\text{Del}(V) \subseteq L\text{Del}^{(k)}(V)$, which is proved in X.-Y. Li *et al.* (2002a, 2002b), $L\text{Del}^{(k)}(V)$ is also a length spanner.

For ad hoc networks, we can construct $L\text{Del}^{(2)}(V)$, which is guaranteed to be a planar spanner of $U\text{Del}(V)$, but it is difficult to collect the two-hop neighbors for every node in $O(n)$ messages. A total communication cost of a simple broadcast approach to collect two-hop information is $O(m)$ messages, where m is the number of edges in $UDG(V)$ and could be as large as $O(n^2)$. Recently, Călinescu (2003) proposed an approach [using $O(n)$ messages total] that is based on the specific CDS introduced by Alzoubi *et al.* (2002b). Using this approach, Y. Wang and Li (2003) proposed an algorithm that can build $L\text{Del}^{(2)}$ in $O(n)$ messages; however, the hidden constant is still large. To reduce the total communication cost, X.-Y. Li *et al.* (2002a, 2002b) do not construct $L\text{Del}^{(2)}(V)$, and they extract a planar graph $PL\text{Del}(V)$ out of $L\text{Del}^{(1)}(V)$ instead. They provided a novel algorithm to make $L\text{Del}^{(1)}(V)$ planar by using linear communications after building

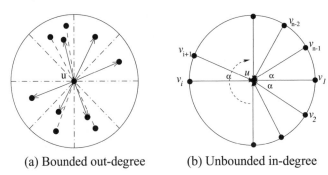

(a) Bounded out-degree   (b) Unbounded in-degree

Figure 9.5 Illustration of YG where $k = 8$: (a) bounded out-degree by Yao structure at node u; (b) unbounded in-degree at node u. © 2003, IEEE.

it. The final graph still contains $U\mathrm{Del}(V)$ as a subgraph. Thus, it is a spanner of the UDG.

9.4   Bounded-Degree Spanner and Yao's Family

For a topology of an ad hoc network, it is also desirable that the node degree in the constructed topology be small and bounded from above by a constant. A small node degree reduces the MAC-level contention and interference and also may help to mitigate the well-known hidden- and exposed-terminal problems. Especially in Bluetooth-based wireless networks, the *master* node degree is preferred to be at most 7, according to Bluetooth specifications. In addition, a structure with a small degree will improve the overall network throughout (Kleinrock and Silvester, 1978). However, not all of the previous planar topologies are degree-bounded. Thus, in this section, we review several degree-bounded spanners.

9.4.1   Yao Graph

The YG was proposed by Yao (1982) to construct a MST of a set of points in high dimensions efficiently. At a given node u, any $k \geq 6$ equally separated rays originating at u define k cones. In each cone, choose the closest node v within the transmission range of u, if there is any, and add a directed link $\overrightarrow{uv}$ [as shown in Figure 9.5(a)]. Ties are broken arbitrarily. The remaining edges are deleted from the graph. The resulting directed graph is called the Yao graph, denoted by $\overrightarrow{\mathrm{YG}}_k(G)$. If we add the link $\overrightarrow{vu}$ instead of the link $\overrightarrow{uv}$, the graph is denoted by $\overleftarrow{\mathrm{YG}}_k(G)$, which is called the *reverse* of the YG. Some researchers used a similar construction, called a θ-graph (Keil and Gutwin, 1992; Lukovski, 1999b); the difference is that in each cone, it chooses the edge that has the shortest projection on the axis of the cone instead of the shortest edge. The axis of a cone is the angular bisector of the cone.

The idea of applying Yao structure on $\mathrm{UDG}(V)$ to a bound node degree is very natural. Hereafter, we use $\overrightarrow{\mathrm{YG}}_k(V)$ to denote $\overrightarrow{\mathrm{YG}}_k[\mathrm{UDG}(V)]$. It is easy to prove that $\overrightarrow{\mathrm{YG}}_k(V)$

is connected and sparse. The $\overrightarrow{YG}_k(V)$ has a length-stretch factor of $\frac{1}{1-2\sin\frac{\pi}{k}}$. Thus, its power-stretch factor is no more than $(\frac{1}{1-2\sin\frac{\pi}{k}})^\beta$. X.-Y. Li *et al.* (2001) proved a stronger result: Its power-stretch-factor is at most $\frac{1}{1-(2\sin\frac{\pi}{k})^\beta}$.

Note that although the directed graph $\overrightarrow{YG}_k(V)$ has a bounded-power-stretch factor and a bounded out-degree k for each node, some nodes may have very large in-degrees. The node configuration given in Figure 9.5(b) will result in a very large in-degree for node v. The bounded out-degree gives us the advantage when applying several routing algorithms. However, an unbounded in-degree at node v will often cause large overhead at v. Therefore, it is often imperative to construct a sparse network topology such that both the in-degree and the out-degree are bounded by a constant while it is still power-efficient.

9.4.2   Symmetric Yao

A *symmetric Yao graph*, denoted by $YS_k(V)$, was proposed by X.-Y. Li *et al.* (2004a) and Y. Wang *et al.* (2003) to bound the node degree at most k. Each node u first applies the Yao structure. An edge uv is selected to graph $YS_k(V)$ if and only if both directed edges $\overrightarrow{uv}$ and $\overrightarrow{vu}$ are in the $\overrightarrow{YG}_k(V)$. Then, it is obvious that the maximum node degree is k. In X.-Y. Li *et al.* (2004a), the authors proved that $YS_k(V)$ is strongly connected if the UDG(V) is connected and $k \geq 6$. However, it was shown in Grünewald *et al.* (2002) that $YS_k(V)$ is not a spanner theoretically. They constructed a counterexample to show that $YS_k(V)$ may have large power- and length-stretch factors.

9.4.3   Sparse Yao

Another Yao-based algorithm was proposed by X.-Y. Li *et al.* (2002c) that constructs a sparser and bounded-degree topology. The basic idea is to apply a reverse Yao structure on $\overrightarrow{YG}_k$ to bound the in-degree. Node v chooses a node u from each cone, if there is any, so the directed link $\overrightarrow{uv}$ has the smallest length among all directed links $\overrightarrow{wv}$ in $\overrightarrow{YG}_k$ in that cone. The union of all chosen directed links is the final network topology, denoted by $\overrightarrow{Y\!Y}_k(V)$. Notice that in Grünewald *et al.* (2002) and Rührup *et al.* (2003), they reinvestigate $\overrightarrow{Y\!Y}_k(V)$ structure and call it the *sparsified Yao graph* or the *Yao–Yao graph*.

In X.-Y. Li *et al.* (2002c), the authors proved that $\overrightarrow{Y\!Y}_k(V)$ is strongly connected if the UDG(V) is connected and $k > 6$. It was proved in Y. Wang *et al.* (2003) that $\overrightarrow{Y\!Y}_k(V)$ is a spanner in civilized UDGs [also called λ-precision UDGs (Hunt *et al.*, 1998)]. Here, a UDG is a civilized graph if the distance between any two nodes in this graph is larger than a positive constant λ. X.-Y. Li *et al.* (2002c) and Y. Wang *et al.* (2003) conjectured that $\overrightarrow{Y\!Y}_k(V)$ also has constant-bounded length- and power-stretch factors theoretically in any UDG. Recently, Jia *et al.* (2003) and Schindelhauer *et al.* (2005) proved that $\overrightarrow{Y\!Y}_k(V)$ theoretically has a constant-bounded-power-stretch factor. However, it is still an open problem whether it is a length spanner.

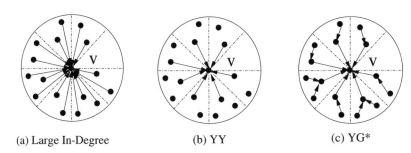

(a) Large In-Degree   (b) YY   (c) YG*

Figure 9.6 Illustration of sparse Yao graph and Yao and sink graphs: (a) star formed by links toward to v; (b) bound in-degree by reverse Yao structure; (c) directed tree $T(v)$ sink at v. © 2006, IEEE.

9.4.4   Yao and Sink Structure

Arya *et al.* (1995) gave an ingenious technique to generate a bounded-degree graph with a constant-length-stretch factor. X.-Y. Li *et al.* (2001) applied the same technique to construct a sparse network topology with a bounded-degree and a bounded-power-stretch factor from $\overrightarrow{YG}_k(V)$. The technique is to replace the directed star consisting of all links toward a node v in $\overrightarrow{YG}_k(V)$ [as shown in Figure 9.6(a)] with a directed tree $T(v)$ of a bounded degree with v as the sink [as shown in Figure 9.6(c)]. Tree $T(v)$ is constructed recursively. The algorithm is as follows. First, construct the graph $\overrightarrow{YG}_k(V)$. Each node v will have a set of incoming nodes $I(v) = \{u \mid \overrightarrow{uv} \in \overrightarrow{YG}_k(V)\}$. For each node v, use the following algorithm Tree($v,I(v)$) to build tree $T(v)$.

Algorithm 21 Construct $T(v)$ Tree($v,I(v)$)

1: To partition the unit disk centered at v, choose k-equal-sized cones centered at v: $C_1(v), C_2(v), \ldots, C_k(v)$.
2: Node v finds the nearest node $y_i \in I(v)$ in $C_i(v)$, for $1 \le i \le k$, if there is any. Link $\overrightarrow{y_i v}$ is added to $T(v)$ and y_i is removed from $I(v)$. For each cone $C_i(v)$, if $I(v) \cap C_i(v)$ is not empty, call Tree($y_i, I(v) \cap C_i(v)$) and add the created edges to $T(v)$.

The union of all trees $T(v)$ is called the *sink structure* $\overrightarrow{YG}_k^*(V)$. Notice that node v constructs the tree $T(v)$ and then broadcasts the structure of $T(v)$ to all nodes in $T(v)$. Because the total number of edges in the Yao structure is at most kn, where k is the number of cones divided, the total number of edges of $T(v)$ of all nodes v is also at most kn. Thus, the total communication cost of broadcasting the $T(v)$ to all its neighbors is still at most kn. The algorithm uses a directed tree $T(v)$ to replace the directed star for each node v. Therefore, if nodes u and v are connected by a path in $\overrightarrow{YG}_k$, they are also connected by a path in $\overrightarrow{YG}_k^*$. It is already known that $\overrightarrow{YG}_k$ is strongly connected if the UDG(V) is connected; so is $\overrightarrow{YG}_k^*$. In X.-Y. Li *et al.* (2001), the authors also proved that the power-stretch factor of $\overrightarrow{YG}_k^*(V)$ is at most $[\frac{1}{1-(2\sin\frac{\pi}{k})^\beta}]^2$ and the maximum degree of $\overrightarrow{YG}_k^*(V)$ is at most $(k+1)^2 - 1$.

9.4.5   Ordered Yao

Bose *et al.* (2002b) study a variant of θ-graphs called *ordered θ-graphs*. An ordered θ-graph of V is obtained by inserting the points of V in some order. When a point p is inserted, we draw the same cones around p and connect p to its closest previously inserted neighbor in each cone. An ordered θ-graph of V is dependent on the order imposed on V; different orderings of V can produce different graphs. Nevertheless, Bose *et al.* (2002b) show that ordered θ-graphs are also spanners, regardless of the ordering used.

In the same way, we can generally define an *ordered Yao graph* $YO_k(V)$ by some order π imposed on V. Using the same arguments as those in Bose *et al.* (2002b), we can prove that the power-stretch factor of the π-ordered Yao graph $YO_k(V)$ is at most $\frac{1}{1-(2\sin\frac{\pi}{k})^\beta}$ for any ordering π. However, the node degree of $YO_k(V)$ is *not* bounded by k because for node u, after a Yao structure is applied, other nodes can still add more edges to node u. The communications cost of this algorithm is $O(n)$ if an ordering is given. Notice that we can use local order π' instead of global order π. Because we can use $O(n)$ messages to build a local order, as we do in the localized algorithm in the next section, $YO_k(V)$ can be built with $O(n)$ messages.

Notice that in an ordered Yao graph, when we process a node, it considers only all previously processed nodes. If we change it to consider only all unprocessed nodes, the spanner proof still holds.

9.5   Bounded-Degree Planar Spanner

The structures discussed so far either have bounded-degree, are planar, or are spanners, but none of the structures has all three properties together. We now review some recent results that can locally construct a bounded-degree planar spanner for ad hoc networks.

9.5.1   Delaunay Triangulation plus Yao Structure

Bose *et al.* (2002c) proposed a centralized $O(n \log n)$ time algorithm that constructs a plan t-spanner for a given node set V, for $t \simeq 10.02$, such that the node degree is bounded from above by 27. As we know, this algorithm is the first method to compute a plane spanner of bounded degree. However, in their method, it is impossible to have a localized, evenly distributed version because they use BFS and many operations on polygons (e.g., degree-3 partitions). Notice that a breadth-first search may take $O(n^2)$ communications.

Inspired by the method of Bose *et al.*, X.-Y. Li and Wang (2004) recently proposed a centralized algorithm for building a planar spanner with a bounded-node degree for UDG(V). The basic idea of their method is to combine Delaunay triangulation and the ordered Yao structure (Bose *et al.*, 2002b). The algorithm is as follows. Here, we assume that each node u has a unique ID denoted by ID(u).

Notice that the algorithm uses *open* sectors, which means that in the algorithm, we do not consider adding the edges on the boundaries (i.e., any edge involving previously processed neighbors). In other words, the open cones do not include any edges uv_i.

This guarantees that the algorithm does not add any edges to node v_i after v_i has been processed. This approach bounds the node degree. In X.-Y. Li and Wang (2004), the authors proved that the maximum node degree of the graph $\mathrm{BPS}_c(V)$ is at most $[2\pi/\alpha]$. For example, when $\alpha = \pi/3$, the maximum node degree is at most 25. X.-Y. Li and Wang (2004) also proved that graph $\mathrm{BPS}_c(V)$ is a planar t-length-spanner, where $t = \max\{\frac{\pi}{2}, \pi \sin \frac{\alpha}{2} + 1\}C_{\mathrm{del}}$. Hereafter, we use C_{del} to denote the length-stretch factor of the Delaunay triangulation.

Algorithm 22 Centralized construction of bounded-degree planar length spanner

---

1: Compute Delaunay triangulation $\mathrm{Del}(V)$ and remove the edges whose lengths are longer than 1 in $\mathrm{Del}(V)$. Call the remaining graph the unit Delaunay triangulation $U\mathrm{Del}(V)$. For every node u, we know its unit Delaunay neighbors $N_U\mathrm{Del}(u)$ and its node degree d_u in $U\mathrm{Del}(V)$.

2: Find an order π of V as follows: Let $G_1 = U\mathrm{Del}(V)$ and let $d_{G,u}$ be the node degree of u in graph G. Remove the node u with the smallest value of $[d_{G_i,u}, \mathrm{ID}(u)]$ from G_i, let $\pi_u = n - i + 1$, and call the remaining graph G_{i+1}. Repeat this procedure for $1 \leq i \leq n$. Obviously, in ordering π, node u at most has five edges to its predecessors P_u in $U\mathrm{Del}(V)$. Here, x is a predecessor of y if $\pi_x < \pi_y$.

3: Let E and E' be the edge sets of $U\mathrm{Del}(V)$ and the desired spanner. Initialize $E' = \emptyset$, and all nodes in V are unprocessed. Then, do the same with the algorithm for the point set, for each node u in V, following the increasing order π, and run the following steps to add some edges to E':

    1. Node u uses its predecessors (i.e., processed unit Delaunay neighbors) in E to define at most five *open* sectors at node u [assume $v_1, \ldots, v_5$ are the processed neighbors of node u in $U\mathrm{Del}(V)$]. For each sector, we divide it into a minimum number of *open* cones of degree α, where $\alpha \leq \pi/3$.

    2. For each cone, let $s_1, s_2, \ldots, s_m$ be the ordered neighbors $N_U\mathrm{Del}(u)$ of u in this cone. That is, $s_1, s_2, \ldots, s_m$ are all unprocessed nodes that are connected by an edge of the unit Delaunay triangulation to u. For each cone, first add the shortest edge in E that is adjacent to u to the edge set E', then add to E' all the edges $s_j s_{j+1}$ between its geometrically ordered unprocessed neighbors in this cone, $1 \leq j < m$. Notice that here, such edges $s_j s_{j+1}$ are not necessarily in $U\mathrm{Del}(V)$; for example, when node u has a Delaunay neighbor x such that ux intersects edge $s_i s_{i+1}$ and $\|ux\| > 1$.

    3. Mark node u as processed.

    4. Repeat this procedure (steps 1–3) in order of π, until all nodes are processed.

4: Let $\mathrm{BPS}_c(V)$ denote the final graph formed by edge set E'.

9.5.2   Local Delaunay Graph plus Yao Structure

By using a local Delaunay graph and local ordering, Y. Wang and Li (2003) converted their centralized method to an efficient localized algorithm for building a BPS.

Note that the ordering computed by this method is not a global ordering. Some nodes may have the same order. However, no two neighboring nodes in $L\mathrm{Del}^{(2)}(V)$ receive the same order. Thus, after all nodes are ordered, the algorithm will process all nodes. Observe that the algorithm does not process two neighboring nodes at the same time. Assume that there are two nodes, say u and v, that are processed at the same time. Remember that the algorithm processes a node only if it has the highest ordering among its unprocessed neighbors. Thus, nodes u and v must receive the same order (i.e., $\pi_u = \pi_v$), which is impossible in the ordering method. Y. Wang and Li (2003), also proved that graph $\mathrm{BPS}(V)$ is a planar t-spanner, where $t = \max\{\frac{\pi}{2}, \pi \sin \frac{\alpha}{2} + 1\}.C_{\mathrm{del}}$. The proofs of the planar and spanner properties are much more complex than those of the centralized ones; refer to Y. Wang and Li (2003) for details. In addition, Algorithm 23 uses at most $O(n)$ messages, in which each message has $O(\log n)$ bits. However, the hidden constant could be as high as several hundred because the method needs to collect the two-hop information for every node.

9.5.3   Gabriel Graph plus Yao Structure

Remember that a GG is a planar-power spanner. To reduce the total communication cost, Song *et al.* (2004) proposed a new method by applying the ordered Yao structures on a GG to bound the node degree. Notice that the GG is much simpler and easier to build than a localized Delaunay graph. The algorithm is as follows.

It is proved in Song *et al.* (2004) that OrdYaoGG is a bounded-degree planar-power spanner. The power-stretch factor is at most $\frac{1}{1-(2\sin\frac{\pi}{k})^\beta}$ and the node degree is bounded from above by a positive constant $k + 5$, where $k > 6$ is an adjustable parameter. Moreover, they showed that the structure can be constructed using at most $24n$ messages, in which each message is $O(\log n)$ bits.

Furthermore, in the same chapter, the authors proposed another method to build a degree-bounded planar-power spanner, which can be constructed more easily and demands less communication cost during construction.

Algorithm 25 further reduces the communication cost during construction of a degree-bounded planar-power spanner to $3n$ messages because it does not demand local ordering before construction. It also reduces the degree bound to k and keeps all other nice properties, except that the theoretical power-stretch factor is relaxed to $\frac{\sqrt{2}^\beta}{1-(2\sqrt{2}\sin\frac{\pi}{k})^\beta}$, where $k > 8$ is an adjustable parameter.

Notice that both OrdYaoGG and SYaoGG are degree-bounded planar-power spanners, but they are not length spanners. However, $\mathrm{BPS}(V)$ is a degree-bounded planar-length spanner.

9.6   Low-Weighted Structures

Minimum-energy broadcast/multicast routing in ad hoc networking environments has been addressed in Clementi *et al.* (2001b) and Wieselthier *et al.* (2000). Three centralized

---

Algorithm 23 Localized construction of bounded-degree planar-length spanner

---

1: First, compute the planar localized Delaunay triangulation $L\text{Del}^{(2)}(V)$ [using the method in Călinescu (2003) to collect the location information of $N_2(u)$], so that every node u knows all of its neighbors $N_{L\text{Del}^{(2)}}(u)$ and its node degree $d(u)$ in $L\text{Del}^{(2)}(V)$. Assume a synchronized method is used to collect $N_{L\text{Del}^{(2)}}(u)$ for every node u.

2: Build a local order π of V as follows (every node u initializes $\pi_u = 0$; i.e., unordered):

    (a) If node u has $\pi_u = 0$ and $d(u) \leq 5$, then u queries[1] each node v, from its unordered neighbors, the current degree $d(v)$. If node u has the smallest ID among all unordered neighbors v with $d(v) \leq 5$, node u sets

$$\pi_u = \max\{\pi_v \mid v \in N_{L\text{Del}^{(2)}}(u)\} + 1,$$

    and broadcasts π_u to its neighbors $N_{L\text{Del}^{(2)}}(u)$.

    (b) If node u receives a message from its neighbor v saying that $\pi_v = k$, it updates its $d(u) = d(u) - 1$ and also updates the order π_v stored locally. So $d(u)$ represents how many neighbors are not ordered so far.
    If node u finds that $d(u) \leq 5$ and $\pi_u = 0$, it goes to step 2(a).
    When node u finds that $d(u) = 0$ and $\pi_u > 0$, it can go to step 3.

3: Build structures based on local order π as follows (i.e., initialize all nodes as unprocessed):

    (a) If an unprocessed node u has the highest local order in its unprocessed neighbors N_u in $L\text{Del}^{(2)}(V)$, let k be the number of processed neighbors[2] of u in $L\text{Del}^{(2)}(V)$. Assume that $v_1, v_2, \ldots, v_k$ are the processed neighbors of u in $L\text{Del}^{(2)}(V)$. Node u divides its transmission range into k *open* sectors cut by the rays from u to these processed neighbors. Then, divide each sector into a minimum number of *open* cones of degree at most α with $\alpha \leq \pi/3$. For each cone, let $s_1, s_2, \ldots, s_m$ be the ordered unprocessed neighbors of u in $N_{L\text{Del}^{(2)}}(u)$. For this cone, node u first adds an edge us_i, where s_i is the nearest neighbor among $s_1, s_2, \ldots, s_m$. Node u then tells $s_1, s_2, \ldots, s_m$ to add all the edges $s_j s_{j+1}$, $1 \leq j < m$. Node u marks itself processed and tells all nodes in $N_{L\text{Del}^{(2)}}(u)$ that it is processed.

    (b) If an unprocessed node v receives a message for adding edge vv' from its neighbor u, it adds edge vv'.

4: When all nodes are processed, the final network topology is denoted by $\text{BPS}(V)$.

---

greedy heuristic algorithms were presented in Wieselthier *et al.* (2000): a MST, a SPT, and a BIP. Wan *et al.* (2002b) showed that the approximation ratio of the MST-based

---

[1] If some unordered neighbor with $d(v) \leq 5$ has a smaller ID, we call such a query round a *failed round*. Node u performs a new round of queries only if it finds that the number of its unordered neighbors has been reduced in step 2(b). So, there are at most five rounds of queries.

[2] There are at most five processed neighbors because graph $L\text{Del}^{(2)}(V)$ is planar.

Algorithm 24 Localized construction of bounded-degree planar-power spanner

---

1: Each node self-constructs the Gabriel graph GG locally. Let $N_{GG}(u)$ be the neighbor set of nodes u in GG.

2: Each node u decides its order π locally using the same method in Algorithm 23 from Y. Wang and Li (2003).

3: All nodes self-form the final topology based on local order π as follows. Initially, all nodes are marked with WHITE color (i.e., unprocessed). Let $N_{OYGG}(u)$ be the set of neighbors of u in the final topology, which is initialized as $N_{GG}(u)$.

    (a) If node u is unprocessed (marked WHITE), and it has the largest order $\pi[u]$ among all its WHITE neighbors in $N_{GG}(u)$, it divides its transmission range (which is a unit disk centered at the node u) into k-equal-sized cones, and keeps one nearest WHITE neighbor $v \in N_{OYGG}(u)$ (if available) in each cone and deletes others. Node u marks itself BLACK (i.e., processed) and notifies all nodes in $N_{GG}(u)$ of the deleted edges through a broadcasting message UPDATEN. The message UPDATEN includes all unselected neighbors.

    (b) Once the node u receives the message UPDATEN for deleting edge vu from its neighbor v, it deletes the node v from its local list $N_{OYGG}(u)$.

4: When all nodes are processed, all the remaining edges $\{uv | v \in N_{OYGG}(u)\}$ form the final network topology OrdYaoGG.

---

approach is between 6 and 12 by assuming that the power needed to support a link uv is $\|uv\|^{\beta}$, where $\|uv\|$ is the Euclidean distance between u and v and β is a real constant between 2 and 5, depending on the wireless transmission environment. The best distributed algorithm (Faloutsos and Molle, 1995) can compute the MST in $O(n)$ rounds using $O(m + n \log n)$ communications for a general graph with m edges and n nodes. Obviously, a MST cannot be constructed in a localized manner; i.e., each node cannot determine which edge is in the defined structure by purely using the information of the nodes within some constant hops. Thus, several localized structures, such as RNG (Seddigh *et al.*, 2002), have been used for broadcasting. As shown in X.-Y. Li (2003b), the total energy used by the RNG-based approach could be about $O(n^{\beta})$ times optimum.

Recently, N. Li *et al.* (2003) proposed a MST-based method for topology control. Each node u first collects its one-hop neighbors $N_1(u)$. Node u then computes its minimum spanning tree $MST[N_1(u)]$ of the induced UDG on its one-hop neighbors $N_1(u)$. Node u keeps a directed edge uv if and only if uv is an edge in $MST[N_1(u)]$. They called the union of all directed edges the *local minimum spanning tree*, denoted by G_0. If only symmetric edges are kept, then the graph is called G_0^-; i.e., it has an edge uv iff both directed edge uv and directed edge vu exist. If the directions of the edges in G_0 are ignored, the graph is called G_0^+; i.e., it has an edge uv iff either directed edge uv or directed edge vu exists. They proved that the graph is connected and has a bounded degree of 6.

We actually can prove that graph G_0^- is also planar. For the sake of contradiction, assume that G_0^- is not planar and two edges uv and xy intersect each other. Assume that

---

Algorithm 25 New localized construction of a bounded-degree planar-power spanner

1: First, each node self-constructs the Gabriel graph GG locally.

2: All nodes together self-form the final topology as follows. Initially, each node u is marked with WHITE color (i.e., unprocessed) and initializes $N_{SYGG}(u)$ as the set of all the neighbor nodes in GG.

    (a) If a WHITE node u has the smallest ID among its WHITE neighbors in GG, it divides its transmission range into k-equal-sized cones, where $k > 8$ is an adjustable parameter. In each cone, node u checks whether there are some BLACK nodes in $N_{SYGG}(u)$ within the same cone.

        (1) Yes: Node u keeps the closest BLACK neighbor $v \in N_{SYGG}(u)$ among them and deletes all the other links in the cone.

        (2) No: Node u keeps the closest WHITE neighbor $v \in N_{SYGG}(u)$ (if available) among them and deletes all the other links in the cone.

    After processing all k cones, node u marks itself BLACK (i.e., processed), then notifies each deleted neighboring node v in GG by a broadcasting message UPDATEN.

    (b) Once a WHITE node v receives the message UPDATEN from a neighbor u in GG, it checks whether itself is in the nodes that are set for deleting. If so, it deletes the sending node u from $N_{SYGG}(v)$; otherwise, it marks u as BLACK in its local list $N_{SYGG}(v)$.

    (c) Once a BLACK node v receives the message UPDATEN from a neighbor belonging to $N_{SYGG}(v)$, it checks whether itself is in the nodes that are set for deleting. If so, it deletes the sending node u from $N_{SYGG}(v)$; otherwise, it marks u as BLACK in its local list $N_{SYGG}(v)$.

3: When all nodes are processed, all selected edges $\{uv | v \in N_{SYGG}(u)\}$ form the final network topology, denoted by SYaoGG.

---

the clockwise order of these four nodes is u, y, v, x. Obviously, one of the four angles, $\angle uxv$, $\angle xvy$, $\angle vyu$, and $\angle yux$, is at least $\pi/2$. Without loss of generality, assume that $\angle uxv \geq \pi/2$. Then, edge uv is the longest edge among triangle $\triangle uvx$. Thus, in the local minimum spanning tree $\text{MST}[N_1(u)]$, edge uv cannot appear because there is already a path uxv whose edges are all shorter than uv. Similarly, graph G_0^+ is a planar graph (by replacing the undirected edges with directed edges in the preceding proof).

We then construct an example such that the structures G_0^- and G_0^+ are not low weighted. Figure 9.7 illustrates such an example. Because it uses only one-hop information, at every node, the algorithm knows only that there is a sequence of nodes evenly distributed with small separation and another node that is one-unit away from the current node. It is easy to show that the final structure G_0^+ is exactly as illustrated in Figure 9.7. The MST will only use one horizontal link, whereas the LMST has $n/2$ horizontal links. It is easy to show that the total edge length of G_0 is $O(n)$ times that of the MST for this example.

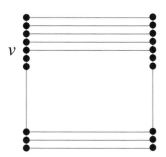

Figure 9.7 G_0 could consume arbitrarily large power for broadcasting compared with the optimum. © 2004, IEEE.

X.-Y. Li *et al.* (2004b) proposed several structures that can be constructed locally and are low-weighted. Here, a structure is called low-weighted if its total edge length is within a constant factor of the total edge length of the EMST. First, they proposed a family of structures [namely, a k-localized minimum spanning tree ($LMST_k$)] for topology control and broadcasting in wireless ad hoc networks. Let $N_k(u)$ be the set of nodes that are within k hops of node u in UDG. Here, $N_k(u)$ includes node u itself for the simplicity of notation later.

DEFINITION 9.8 *The* k-*local minimum spanning tree ($LMST_k$) contains a directed edge $\overrightarrow{uv}$ if edge* uv *belongs to* $MST[N_k(u)]$. *We further define two undirected variations,* $LMST_k^-$ *and* $LMST_k^+$. *Structure* $LMST_k^-$ *contains an edge* uv *if both directed edge* $\overrightarrow{uv}$ *and directed edge* $\overrightarrow{vu}$ *belong to* $LMST_k$. *Structure* $LMST_k^+$ *contains an edge* uv *if either* $\overrightarrow{uv}$ *or* $\overrightarrow{vu}$ *belongs to* $LMST_k$.

X.-Y. Li *et al.* (2004b) analytically proved that the node degree of the structure $LMST_k$ is at most 6; the $LMST_k$ is connected and planar; and, more important, the total edge length of the $LMST_k$ is within a constant factor of that of the MST when $k \geq 2$.

THEOREM 9.2 *All structures* $LMST_k^+$ *are low-weighted when* k ≥ 2.

X.-Y. Li *et al.* (2004b) further gave an efficient localized method to construct the $LMST_k$ using only $O(n)$ messages under a local broadcast communication model; i.e., the message sent by a node is received by all nodes within its transmission range. Second, they proposed another low-weighted structure, called the *Incident MST and RNG Graph* (IMRG), that can be constructed using at most $13n$ messages under the local broadcast communication model. Every node uses only its partial two-hop information to construct the structure IMRG. Notice that it was shown in X.-Y. Li (2003b) that some two-hop information is necessary to construct any low-weighted structure for a UDG. Third, X.-Y. Li *et al.* (2004b) studied the application of these structures for efficient broadcasting in WANs. Notice that Wan *et al.* (2002b) proved that the broadcasting based on the MST consumes energy within a constant factor of the optimum when *only* the energy consumed by the senders is considered. However, in practice, the receiver node also consumes energy to receive the signal. By assuming that the energy consumed by the receiver node is *no more than* the energy consumed by the sender, they then proved that the approximation ratio of the MST-based approach is still a constant when this more practical energy model is used. Because it is expensive to construct a MST in a distributed way, they use the newly proposed structures $LMST_k$ and IMRG to approximate it. Although a low-weighted structure cannot guarantee that the broadcasting based on it consumes energy within a constant factor of the optimum in the worst case, the energy consumptions when the structures $LMST_k$ ($k \geq 2$) and IMRG are used are theoretically within $O(n^{\beta-1})$ of the optimum in the worst case. This improves the previously known "lightest" structures, RNG and LMST, by a factor of $O(n)$. They further proved that these structures are asymptotically optimum for broadcasting among all *locally* constructed structures.

9.7   A Unified Structure: Energy Efficiency for Unicast and Broadcast

We now study the *first* communication-efficient algorithm that can construct a *unified* energy-efficient topology for unicast and broadcast in wireless ad hoc and sensor networks. In one single structure, the following network properties are guaranteed:

1. **Power-efficient unicast:** Given any two nodes, there is a path connecting them in the structure with a total power cost of no more than $2\rho + 1$ times of the power cost of any path connecting them in the original network. Here, $\rho > 1$ is some constant that will be specified later.
2. **Power-efficient broadcast:** The power consumption for broadcast is within a constant factor of optimum among all *locally* constructed structures. Notice that this new structure cannot guarantee that the energy consumption is within a constant factor of the *global* optimum. Essentially, we prove that the structure is *low-weighted*: Its total edge length is within a constant factor of that of the EMST.
3. **Bounded logical node degree:** Each node has to communicate with at most $k - 1$ logical neighbors, where $k \geq 9$ is an adjustable parameter.
4. **Bounded average physical node degree:** The average physical node degree is bounded from above by a small constant. Here, the physical degree of a node u in a structure H is defined as the number of nodes inside the disk centered at u with radius $\max_{uv \in H} \|uv\|$.
5. **Planar:** There are no edges crossing each other. This enables several localized routing algorithms, such as those in Bose *et al.* (2001), Karp and Kung (2000), and Kuhn *et al.* (2002, 2003c), to be performed on top of this structure to guarantee packet delivery without a routing table.
6. **Neighbors θ-separated:** The directions between any two logical neighbors of any node are separated by at least an angle θ, which, as we will see, reduces the signal interference. It also can be used to reduce the receiving power cost when a directional antenna is used.

In graph-theoretical terminology, given a UDG, we propose a communication-efficient distributed method to build a low-weighted planar-power spanner with a bounded logical node degree. Here, a geometric structure is called *low-weighted* if its total edge length is no more than a small constant factor of that of the EMST. To the best of our knowledge, it is the *first* known *communication-efficient* algorithm to construct such a *single* structure with all these desired properties. Previously, only a centralized algorithm was reported in Bose *et al.* (2002c). Moreover, by assuming that the ID and position of every node can be represented in $O(\log n)$ bits for a wireless network of n nodes, we show that the structure can be initially constructed using at most $5n$–$13n$ messages.

In addition, it was proved that the expected average node interference (which is defined as the number of nodes within its adjusted transmission range) in the structure is bounded by a small constant. This is significant on its own for the following reasons: It has been taken for granted that "*a network topology with a small logical node degree will guarantee a small interference*," and recently Burkhart *et al.* (2004) showed that this is not true generally. These results show that although generally a small logical node degree cannot guarantee a small interference, the expected average interference is indeed small if the logical communication neighbors are chosen carefully.

This new structure can be easily updated in a dynamic environment when a node moves or dies after the battery power is drained. When a node moves, the topology can be dynamically self-maintained without affecting the whole network because each node adjusts its transmission range and selects neighbors only according to its neighbor information.

To facilitate the efficient construction of such a unified energy-efficient topology, they first gave an improved method to construct a BPS by using relative positions only. The new structure has the same power-spanning ratio $\rho = \frac{\sqrt{2}^\beta}{1-(2\sqrt{2}\sin\frac{\pi}{k})^\beta}$ as the structure proposed in Song *et al.* (2004). Here, $k \geq 9$ is a customizable parameter. In addition, the directions between any two neighbors of each node are separated by at least a certain angle θ depending on the parameter k. Simulations show that the node interference in the new structure is indeed smaller than in the structure proposed in Song *et al.* (2004).

9.7.1 Power-Efficient Unicast: Spanner, Planar, and Bounded Degree

The ultimate goal of this chapter is to construct a unified topology that is power-efficient for both unicast and broadcast, in addition to being planar and having a constant-bounded logical node degree. To achieve this ultimate goal, in this section a new method is presented that can construct a power-efficient topology for unicast. We prove that the constructed structure is a power spanner and planar and has a bounded node degree. Furthermore, it has an extra property: Any two neighbors of each node are separated by at least a certain angle θ. Hereafter, we call it the Θ-*separation* property. As we will see later, this property further reduces the interference, especially when adopting directional antennas for transmission. This property also makes the proof much easier in that the structure constructed in the next section is also power-efficient for broadcast.

One possible way to construct a degree-bounded planar power spanner is to apply the Yao structure on a GG because the GG is already planar and has a power-stretch factor of exactly 1. X.-Y. Li *et al.* (2002b) showed that the final structure obtained by directly

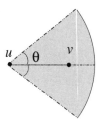

Figure 9.8 Node v's θ-dominating region with respect to node u. © 2005, ACM.

applying the Yao structure on the GG is a planar power spanner, called YaoGG, but its in-degree can be as large as $O(n)$. Song *et al.* (2004) proposed two new methods to bound the node degree by applying the ordered Yao structures on a GG. The structure SYaoGG in Song *et al.* (2004) guarantees that there is at most one neighbor node in each of the k-equal-sized cones. In this section, an improved algorithm is proposed to further reduce the medium contention by selecting fewer communication neighbors and making the separations between neighbors wider.

Before we give the algorithm, we first define a concept called the θ-dominating region.

DEFINITION 9.9 θ-Dominating region: *For each neighbor node* v *of a node* u, *the* θ-dominating region *of* v *is the* 2θ *cone emanating from* u, *with the edge* uv *as its axis.*

Figure 9.8 illustrates the θ-dominating region of a node v in the transmission disk of node u. By using the concept of a θ-dominating region instead of an absolute cone partition in SYaoGG (Song *et al.*, 2004), we are able to prove that any two neighbors of each node are guaranteed to be separated by at least an angle θ. The final topology is called SΘGG. Intuitively, the communication interference in SΘGG will be smaller than the interference in SYaoGG, which is also verified by simulations.

The basic idea of the method is as follows. Because the GG is planar and a power spanner, we remove some links of the GG to bound the nodal degree while not destroying the power-spanner property. The basic approach of bounding the nodal degree is only to keep some shortest link in the θ-dominating region for every node. We process the nodes in a certain order. A node is marked WHITE if it is unprocessed and is marked BLACK if it is processed. Originally, all nodes are marked WHITE. Initially, a node elects itself to start processing its neighbors if its ID[3] is smaller than all its unprocessed logical neighbors in the GG. Assume that a node u is to be processed. We further assume that there are already some processed logical neighboring nodes, say $v_1, \ldots, v_t$, among its neighbors in the GG. It keeps the link to the closest processed neighbor, say v_1, in the GG and removes all links to all neighbors in the θ-*dominating region* of v_1. In other

---

[3] It is not necessary to use ID here. We can also use some other mechanism to elect a certain node to perform the remaining procedures first. For example, we can use the RTS/CTS mechanism provided in the MAC layer to achieve this: The node that first successfully sent a RTS signal within its one-hop neighborhood will be elected. In this chapter, ID is used just for the sake of an easy presentation.

words, the neighbor v_1 dominates all other neighbors in its θ-dominating region. It then repeats the preceding procedure until no processed logical neighbors in the GG are left. Assume that node u also has some unprocessed logical neighbors; i.e., marked WHITE. The node u then keeps the link to the closest unprocessed neighbor, say w, in the GG if there is any, and then removes the links to all neighbors in the θ-*dominating region* of w. It then repeats the preceding procedure until no unprocessed neighbors in the GG are left. Node u then marks itself BLACK and then informs its logical neighbors in the GG about its change of status. The algorithm terminates when all nodes are marked processed. The remaining links form the final structure, called SΘGG.

In this new algorithm, a data structure is used: $N(u)$ is the set of neighbors of each node u in the final topology, which is initialized as the set of neighbor nodes in the GG. We are now ready to present Algorithm 26, which constructs a bounded-degree planar power spanner.

Algorithm 26 SΘGG: Power-efficient unicast topology

1: First, each node self-constructs the GG locally. Initially, all nodes mark themselves WHITE; i.e., *unprocessed*.

2: Once a WHITE node u has the smallest ID among all of its WHITE neighbors in $N(u)$, it uses the following strategy to select neighbors:

   (a) Node u first sorts all of its BLACK neighbors (if available) in $N(u)$ in distance-increasing order, then sorts all of its WHITE neighbors (if available) in $N(u)$ similarly. The sorted results are then restored to $N(u)$ by first writing the sorted list of BLACK neighbors and then appending the sorted list of WHITE neighbors.

   (b) Node u scans the sorted list $N(u)$ from left to right. In each step, it keeps the current pointed neighbor w in the list, while it deletes every *conflicted* node v in the remainder of the list. Here, if a node v is conflicted with w, it means that node v is in the θ-dominating region of node w. Here, $\theta = 2\pi/k$ ($k \geq 9$) is an adjustable parameter.

   Node u then marks itself BLACK (i.e.. *processed*) and notifies each deleted neighboring node v in $N(u)$ by a broadcasting message UPDATEN.

3: Once a node v receives the message UPDATEN from a neighbor u in $N(v)$, it checks whether itself is in the nodes set for deleting. If so, it deletes the sending node u from list $N(v)$; otherwise, it marks u as BLACK in $N(v)$.

4: When all nodes are processed, all selected links $\{uv | v \in N(u), \forall v \in GG\}$ form the final network topology, denoted by SΘGG. Each node can shrink its transmission range as long as it sufficiently reaches its farthest neighbor in the final topology.

Notice that the final topology based on a YG, such as SYaoGG (Song *et al.*, 2004), may vary as the choice of the direction of cones varies. Here, SΘGG does not rely on the absolute cone partition by adopting the new concept of the θ-dominating region. Hence, given the point set V, SΘGG is unique. In addition, the average node degree and interference and transmission ranges of SΘGG are expected to be smaller than those of

SYaoGG too. Furthermore, it is interesting to notice that the theoretical bound on the spanning ratio for SΘGG, which we can prove, is the same as that of SYaoGG, as proved later in Theorem 9.3.

LEMMA 9.3 (Song *et al.*, 2004) *Graph SΘGG is connected if the underlying graph GG is connected. Furthermore, given any two nodes* u *and* v, *there exists a path* {u, t₁, . . . , tᵣ, v} *connecting them such that all edges have a length of less than* $\sqrt{2}\|uv\|$.

The property that for any link uv there is a path connecting them such that the links on the path have a length of at most $\sqrt{2}\|uv\|$ is crucial for the later proof that Algorithm 27 builds a low-weighted BPS.

THEOREM 9.3 *The structure* $SΘGG$ *has a node degree of at most* $k-1$ *and is a planar power-spanner with neighbors* Θ-*separated. Its power-stretch factor is at most* $\rho = \frac{\sqrt{2}^{\beta}}{1-(2\sqrt{2}\sin\frac{\pi}{k})^{\beta}}$, *where* $k \geq 9$ *is an adjustable parameter.*

Proof. The proof will be similar to the proof of SYaoGG in Song *et al.* (2004). The only difference is that here, we used the concept of dominating cones instead of a YG. Although the power-stretch factor remains the same theoretically, the degree bound is reduced from k to $k-1$. Obviously, the links in SΘGG are Θ-separated; in other words, the direction of any two neighbors of a node is Θ-separated. ∎

9.7.2 Unified Power-Efficient Topology: Degree-Bounded Planar Spanner with Low Weight

So far, no communication-efficient topology control algorithm discussed has achieved all the desirable properties summarized in Section 9.2: *degree-bounded, planar, power spanner*, and *low-weighted*. These properties are not only interesting in terms of computational geometry but also have important applications in wireless networks, as discussed in Section 9.2: They enable energy-efficient unicast and broadcast routings in the same structure. Recall that the spanner property ensures that an energy-efficient path is always kept for any pair of nodes; hence, it is a necessary condition to support energy-efficient unicast. A low-weighted structure is optimal for broadcast among any connected structures built locally. Unfortunately, all the known spanners, including those of Yao (1982), the GG (Gabriel and Sokal, 1969), and the recently developed BPSs (Y. Wang and Li, 2003), SYaoGG, OrdYaoGG of Song *et al.* (2004), and SΘGG, are not low-weighted. As illustrated by an example in Song *et al.* (2004), the weight of any of them is at least $O(n)w(\text{EMST})$.

It is worth clarifying that in this subsection, we are interested in finding a subgraph to enable efficient broadcast routings, *even* those based on the simple-flooding method. We do *not* aim to substitute the known broadcasting protocols. In fact, the methods used in those broadcasting protocols (Cartigny *et al.*, 2003; Wu and Dai, 2003) can be applied to the low-weighted structures to conserve more energy. The main contribution of a low-weighted structure is that it bounds the worst-case performance for broadcasting.

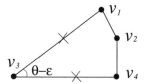

Figure 9.9 The graph could be disconnected if the previous method is applied to build a low-weighted structure on the SΘGG. © 2005, ACM.

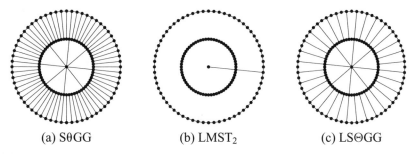

(a) SθGG   (b) LMST$_2$ (c) LSΘGG

Figure 9.10 Three different structures. © 2005, ACM.

Several known localized algorithms are given in X.-Y. Li (2003b) and X.-Y. Li *et al.* (2004c) to generate low-weighted graphs. In their algorithms, given a certain structure G, for any two links uv and xy of a graph G, they remove xy if xy is the longest link among quadrilateral $uvxy$. They proved that the final structures are low-weighted if G is a RNG$'$ (X.-Y. Li, 2003b) or a LMST$_2$ (X.-Y. Li *et al.*, 2004c). Obviously, they are not spanners. In fact, their techniques *cannot* be applied on a spanner graph to bound the weight without losing its spanner property. Figure 9.9 illustrates an example by applying their algorithms to SΘGG. The node ID of v_i is i, $\angle v_1 v_3 v_4 < \theta$, and $\|v_1 v_3\| > \|v_3 v_4\| > \max(\|v_1 v_2\|, \|v_2 v_4\|)$. While constructing the SΘGG, the first node v_1 selects $v_1 v_2$ and $v_1 v_3$ as its incident logical links, and node v_2 selects $v_2 v_1$ and $v_2 v_4$; then, node v_3 selects $v_3 v_1$ and deletes $v_3 v_4$. Hence, $v_3 v_4 \notin$ SΘGG. If the rule described in X.-Y. Li (2003b) and X.-Y. Li *et al.* (2004c) is applied, the link $v_1 v_3$ will also be deleted because $\|v_1 v_3\| > \max(\|v_1 v_2\|, \|v_2 v_4\|, \|v_3 v_4\|)$. Then, the graph will be disconnected. Then, we can conclude that the simple extension of methods in X.-Y. Li *et al.* (2004c) on top of the SΘGG does not even guarantee connectivity, not to mention the *power-spanner* property.

Indeed, the spanner property and low-weight property are *not* easily achieved at the same time. Intuitively, the spanner property requires keeping more links, whereas the low-weight property requires keeping fewer links from the original graph. In the subsequent text, a novel algorithm is described for building a low-weighted structure from SΘGG, while keeping enough links to guarantee the power efficiency. Figure 9.10 illustrates the differences among the LSΘGG, SΘGG, and LMST$_2$.

Algorithm 27 presents a method (Song *et al.*, 2004) that constructs a bounded-degree planar power spanner that is also low-weighted. Although this algorithm produces only a power spanner here, it can be extended to also produce the length spanner if it is needed. To get a length spanner, we construct the structure $L\text{Del}^2$ (defined in Li *et al.*, 2002b) instead of the GG used in this algorithm. It was proved in Li *et al.* (2002b) that $L\text{Del}^2$ is a planar length spanner and can be constructed with only $O(n)$ messages. The basic idea of this new method is as follows. Because the graph S⊖GG is already planar, a power spanner, and has a bounded degree, we remove some of its edges to guarantee that the resulting topology is low-weighted while not destroying the power-spanner property. Notice that removing edges will not break the planar property and the bounded-degree property. In all previous methods presented in the literature, a node x decides to remove or keep links that are incident upon x; i.e., it cares about only the incident edges. In the method presented here, however, a node x will decide whether to keep or remove links not only for incident links but also for the links that are incident upon one of its neighbors. To guarantee a low-weight property, the methods presented in X.-Y. Li (2003b) and X.-Y. Li *et al.* (2004c) remove some links from a certain structure such that the remaining links satisfy the *isolation property*: For each remaining link xy, the disk centered at the midpoint of xy by using a radius proportional to $\|xy\|$ does not intersect any other remaining links. They achieved this property by removing a link xy if there is another link uv such that xy is the longest link in the quadrilateral $uvyx$. However, this simple heuristic cannot guarantee the spanner property. Consider a link xy in some shortest path from s to t (see Figure 9.12 in the proof of Theorem 9.4 for an illustration). Link xy will be removed because of the existence of link uv. Link uv could also later be removed because of the existence of another link u_1v_1, which could also be removed because of the existence of another link u_2v_2, and so on. See Figure 9.11(b) for an illustration of the situation in which a sequence of links will be removed: All links u_iv_i for $i \geq 2$ will be removed. Consequently, the shortest path connecting nodes u_n and v_n could be arbitrarily long in the resulting graph.

Thus, instead of blindly removing all such links xy whenever they are the longest link in a quadrilateral $uvyx$, we keep such a link when some links in its certain neighborhood have been removed. To do so, among all links from a graph, such as S⊖GG – that is, a planar, bounded-degree, power-spanner – we *implicitly* define an IS of links. A link is in this IS, which will be kept, if it has the smallest ID among the unselected links from its neighborhood. Specifically, we implicitly define a virtual graph G' over all links in S⊖GG: the vertex set of G' is the set of all links in S⊖GG and two links xy and uv of S⊖GG are connected in G' if one end point of uv is in the transmission range of one end point of link xy (they will interfere with each other if transmitting simultaneously). For example, the links u_1v_1 and u_3v_3 are not independent in the network topology illustrated by Figure 9.11(a), whereas the links u_1v_1 and u_nv_n are independent. Notice that links u_1v_1 and u_1u_2 are independent because they do not form a four-vertex convex hull. Notice that in the method presented by Algorithm 27, we did not explicitly define such a graph G', nor did we explicitly compute the maximal IS of such a graph G'. We will prove that the selected IS of links in S⊖GG is indeed low-weighted and still preserves the power-spanner property, although with a larger power-spanning ratio. The new method will keep link u_1v_1 because it has the smallest ID among all links that are not independent. When link u_1v_1 is kept, all links that are not independent (here are

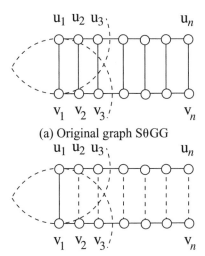

(a) Original graph SθGG

(b) Graph resulted using method in X.-Y. Li *et al.* (2004c)

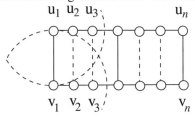

(c) Graph based on new method

Figure 9.11 A sequence of links is recursively removed. Here, the dashed links represent the links that are removed by a topology-control algorithm, the solid links represent the final structure constructed by a certain method. Here, we assume that $\|u_i v_i\| = R$ and the ID of link $u_i v_i$ is less than the ID of link $u_{i+1} v_{i+1}$. © 2005, ACM.

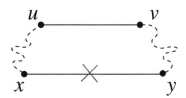

Figure 9.12 The shortest path between x and y is at most $(2\rho + 1)\|xy\|$ in the LSθGG if $xy \in$ SθGG. © 2005, ACM.

$u_2 v_2$ and $u_3 v_3$) will be removed. Then, link $u_4 v_4$ will be kept. The preceding procedure will be repeated until all links are processed. The final structure resulting from the new method is illustrated by Figure 9.11(c).

Obviously, the construction is consistent for two end points of each edge: If an edge uv is kept by node u, then it is also kept by node v. The ID of a link uv is defined as follows: $\text{ID}(uv) = \{\|uv\|, \min[\text{ID}(u), \text{ID}(v)], \max[\text{ID}(u), \text{ID}(v)]\}$. As we will see later, in the number 3 criterion of Algorithm 27,

$$\|xy\| > \max(\|uv\|, 3\|ux\|, 3\|vy\|)$$

is carefully selected.

Algorithm 27 Construct LSΘGG: Planar spanner with bounded degree and low weight

---

1: All nodes together construct the SΘGG in a localized manner, as described in Algorithm 26. Then, each node marks its incident edges in the SΘGG as *unprocessed*.

2: Each node u locally broadcasts its incident edges in the SΘGG to its one-hop neighbors and listens to its neighbors. Then, each node x can learn the existence of the set of two-hop links $E_2(x)$, which is defined as follows: $E_2(x) = \{uv \in SΘGG \mid u \text{ or } v \in N_{\mathrm{UDG}}(x)\}$. In other words, $E_2(x)$ represents the set of edges in SΘGG with at least one end point in the transmission range of node x.

3: Once a node x learns that its *unprocessed* incident edge xy has the smallest ID among all *unprocessed* links in $E_2(x)$, it will delete edge xy if there exists an edge $uv \in E_2(x)$ (here, both u and v are different from x and y), such that $\|xy\| > \max(\|uv\|, 3\|ux\|, 3\|vy\|)$; otherwise, it simply marks edge xy as *processed*. Here, assume that $uvyx$ is the convex hull of u, v, x, and y. Then, the link status is broadcasted to all neighbors through a message UPDATESTATUS(XY).

4: Once a node u receives a message UPDATESTATUS(XY), it records the status of link xy at $E_2(u)$.

5: Each node repeats the previous two steps until all edges have been *processed*. Let LSΘGG be the final structure formed by all remaining edges in SΘGG.

THEOREM 9.4 *The structure $LSΘGG$ is a degree-bounded planar spanner. It has a power-spanning ratio of $2\rho + 1$, where ρ is the power-spanning ratio of SΘGG. The node degree is bounded by $k - 1$, where $k \geq 9$ is a customizable parameter in SΘGG.*

Proof. The degree-bounded and planar properties are obviously derived from the SΘGG graph because we do not add any links in Algorithm 27. To prove the spanner property, we need only to show that the two end points of any deleted link $xy \in SΘGG$ are still connected in LSΘGG with a constant-spanning-ratio path. We will prove it by induction on the length of deleted links from SΘGG.

Assume that xy is the shortest link of SΘGG that is deleted by Algorithm 27 because of the existence of link uv with a smaller length. Obviously, path $x \rightsquigarrow y$ can be constructed through the concatenation of path $x \rightsquigarrow u$, link uv, and path $v \rightsquigarrow y$, as shown in Figure 9.12. Because $\|xy\| > \max(\|ux\|, \|vy\|)$ and link xy is the shortest among deleted links in Algorithm 27, we have $p(x \rightsquigarrow u) < \rho\|ux\|^\beta$ and $p(v \rightsquigarrow y) < \rho\|vy\|^\beta$. Hence, $p(x \rightsquigarrow y) < \|uv\|^\beta + \rho\|ux\|^\beta + \rho\|vy\|^\beta < (2\rho + 1)\|xy\|^\beta$.

Suppose that all the ith $(i \leq t - 1)$ deleted shortest links of SΘGG have a path connecting their end points with spanning ratio $2\rho + 1$. For the tth deleted shortest link $xy \in SΘGG$, according to Algorithm 27, it must have been deleted because of the existence of a link uv, such that $\|xy\| > \max(\|uv\|, 3\|ux\|, 3\|vy\|)$ in a convex hull

$uvyx$. Now, we have $p(x \rightsquigarrow u) < (2\rho + 1)\|ux\|^\beta$ and $p(v \rightsquigarrow y) < (2\rho + 1)\|vy\|^\beta$. Thus,

$$
\begin{aligned}
p(x \rightsquigarrow y) &= \|uv\|^\beta + p(u \rightsquigarrow x) + p(v \rightsquigarrow y) \\
&< \|uv\|^\beta + (2\rho + 1)\|ux\|^\beta + (2\rho + 1)\|vy\|^\beta \\
&< \|xy\|^\beta + (2\rho + 1)(\|xy\|/3)^\beta + (2\rho + 1)(\|xy\|/3)^\beta \\
&\le (2\rho + 1)\|xy\|^\beta.
\end{aligned}
$$

Thus, the LSⓄGG has a power-spanning ratio of $\le 2\rho + 1$.   ∎

We now show that graph LSⓄGG is low-weighted. To study the total weight of this structure, inspired by the method proposed in X.-Y. Li (2003b), we show that the edges in LSⓄGG satisfy the *isolation property* (Das *et al.*, 1995).

THEOREM 9.5 *The structure LSⓄGG is low-weighted.*

We continue to analyze the communication cost of Algorithms 26 and 27. First, clearly, building the GG in Algorithm 26 can be done using only n messages: Each message contains the ID and geometry position of a node. Second, to build SⓄGG, initially, the number of edges, say p, in a GG is $p \in [n, 3n - 6]$ because it is a planar graph. Remember that we will remove some edges from the GG to bound the logical node degree. Clearly, there are at most $2n$ such removed edges because we keep at least $n - 1$ edges from the connectivity of the final structure. Thus, the total number of messages, say q, used to inform the deleted edges from the GG is at most $q \in [0, 2n]$. Notice that $p - q$ are the edges left in the final structure, which are at least $n - 1$ and at most $3n - 6$. Third, in the marking process described in Algorithm 27, the communication cost of broadcasting its incident edges (or its neighbors) and updating link status are both $2(p - q)$. Therefore, the total communication cost is $n + 4p - 3q \in [5n, 13n]$. Then, the following theorem directly follows:

THEOREM 9.6 *Assuming that both the ID and the geometry position can be represented by* $\log n$ *bits each, the total number of messages used by constructing the structure LSⓄGG is in the range of* $[5n, 13n]$*, where each message has at most* $O(\log n)$ *bits.*

Compared with previous known low-weighted structures (X.-Y. Li, 2003b; X.-Y. Li *et al.* 2004c), LSⓄGG not only achieves more desirable properties but also costs much less in messages during construction. To construct a LSⓄGG, we need to collect only the information $E_2(x)$, which costs at most $6n$ messages. This algorithm can be generally applied to any known degree-bounded planar spanner to make it low-weighted while keeping all its previous properties, except increasing the spanning ratio from ρ to $2\rho + 1$ theoretically.

9.7.3   Expected Interference in Random Networks

This subsection is devoted to studying the average physical node degree of the new structure when the wireless nodes are distributed according to a certain distribution. For average performance analysis, we consider a set of wireless nodes distributed in a two-dimensional unit square region. The nodes are distributed according to either a uniform random-point process or a homogeneous Poisson process. A point set process is said to be a *uniform random-point process*, denoted by $\mathcal{X}_n$, in a region Ω if it consists of n-independent points, each of which is uniformly and randomly distributed over Ω. The standard probabilistic model of a *homogeneous Poisson process* is characterized by the property that the number of nodes in a region is a random variable depending on only the area (or volume in higher dimensions) of the region. In other words,

- The probability that there are exactly k nodes appearing in any region Ψ of area A is $\frac{(\lambda A)^k}{k!} e^{-\lambda A}$.
- For any region Ψ, the conditional distribution of nodes in Ψ given exactly k nodes in the region is *joint uniform*.

DEFINITION 9.10 *Given a structure* H, *the induced transmission range* $r_H(u)$ *is defined as* $\max_{uv \in H} \|uv\|$; *i.e., the longest edge of* H *incident upon* u. *The physical node degree of* u *in* H *is defined as the number of nodes inside the disk* $\mathcal{D}[u, r_H(u)]$. *The node interference, denoted by* $I_H(u)$, *of a node* u *in a structure* H *is simply the physical node degree of* u. *The maximum node interference of a structure* H *is defined as* $\max_u I_H(u)$. *The average node interference of a structure* H *is defined as* $\sum_u I_H(u)/n$.

THEOREM 9.7 *For a set of nodes produced by a Poisson point process with density n, the expected maximum node interference of* any *connected structure (e.g., EMST, GG, RNG, Yao, and LS$\ominus$GG) is at least* $\Theta(\log n)$.

Proof. Let $d_n(H)$ be the longest edge of a structure H of n points placed independently in two dimensions according to a standard Poisson distribution with density n. Obviously, $d_n(\text{EMST})$ is the smallest among *all* connected structures H. For simplicity, let $d_n = d_n(\text{EMST})$. Penrose (1997) showed that

$$\lim_{n \to \infty} \Pr\left(n\pi d_n^2 - \log n \le \alpha\right) = e^{-e^{-\alpha}}.$$

Notice that the probability $\Pr(n\pi d_n^2 - \log n \le \log n)$ will be sufficiently close to 1 when n goes to infinity, whereas the probability $\Pr(n\pi d_n^2 - \log n \le -\log\log n)$ will be sufficiently close to 0 when n goes to infinity. That is, with high probability, $n\pi d_n^2$ is in the range $[\log n - \log\log n, 2\log n]$.

Given a region with area A, let $m(A)$ denote the number of nodes inside this region by a Poisson point process with density δ. According to the definition of Poisson distribution, $\Pr[m(A) = k] = \frac{e^{-\delta A}(\delta A)^k}{k!}$. Thus, the expected number of nodes lying inside a region with area A is $E[m(A)] = \delta A$. For a Poisson process with density n, let uv be the longest edge of the EMST, and let $d_n = \|uv\|$. Then, the expectation of the number of nodes that fall inside $\mathcal{D}(u, d_n)$ is $E[m(\pi d_n^2)] = n\pi d_n^2$, which is larger than $\log n$ almost surely when n goes to infinity. That is, the expected maximum interference of the EMST is $\Theta(\log n)$ for a set of nodes produced according to a Poisson point process.

Because $d_n \leq d_n(H)$ for any connected structure H, the expected maximum node interference of any connected structure H (e.g., GG, RNG, Yao, or LSΘGG) is at least $\Omega(\log n)$. Thus, all commonly used structures for topology control in WANs have a large maximum node interference even for *randomly* deployed nodes. ∎

It is not difficult to show that the preceding theorem is also true when the nodes are distributed according to a uniform random distribution. Our following analysis will show that the average interference of all nodes of these structures is small for a randomly deployed network.

THEOREM 9.8 *For a set of nodes produced by a Poisson point process with density* n, *the expected average node interferences of a EMST and a RNG are bounded from above by some constants.*

Proof. Consider a set V of wireless nodes produced by a Poisson point process. Given a structure G, the interference $I_G(u_i)$ is the number of nodes inside the transmission region of node u_i. Here, the transmission region of node u_i is a disk centered at u_i with radius $r_i = \max_{u_i v \in G} \|u_i v\|$. Hence, the expected average node interference is

$$E\left[\frac{\sum_{i=1}^n I_G(u_i)}{n}\right] = \frac{1}{n}E\left[\sum_{i=1}^n I_G(u_i)\right] = \frac{1}{n}\sum_{i=1}^n E[I_G(u_i)]$$

$$= \frac{1}{n}\sum_{i=1}^n E[m(\pi r_i^2)] = \frac{1}{n}\sum_{i=1}^n (n\pi r_i^2) \leq 2\sum_{e_i \in G}(\pi e_i^2).$$

The last inequality follows from the fact that r_i is the length of some edge in G and each edge in G can be used by at most two nodes to define its radius r_i.

Let e_i, $1 \leq i \leq n-1$, be the length of all edges of the EMST of n points inside a unit disk. It was shown in Wan *et al.* (2002b) that $\sum_{e_i \in \text{EMST}} e_i^2 \leq 12$. Thus, the expected average node interference of the EMST structure is

$$E\left[\frac{\sum_{i=1}^n I_{\text{EMST}}(u_i)}{n}\right] \leq 2\sum_{e_i \in \text{EMST}}(\pi e_i^2) \leq 24\pi.$$

For a RNG, we define a diamond for each segment. The *open diamond* subtended by a line segment uv, denoted by $D(uv, \gamma)$, is a rhombus with sides of length $\|uv\|/(2\cos\gamma)$, where $0 \leq \gamma \leq \pi/3$ is a parameter. Similar to the proof of Wan *et al.* (2002b), here we can show that the diamonds $D(uv, \pi/6)$ do not overlap and $\sum_{e_i \in \text{RNG}} e_i^2 \leq 8\pi/\sqrt{3}$. This implies that

$$E\left[\frac{\sum_{i=1}^n I_{\text{RNG}}(u_i)}{n}\right] \leq 2\sum_{e_i \in \text{RNG}}(\pi e_i^2) \leq 16\pi^2/\sqrt{3}.$$

This finishes the proof. ∎

The following theorem was also proved in Song *et al.* (2004):

THEOREM 9.9 *Song et al. (2004) The expected average node interference of a* LS⊖GG *is bounded from above by a constant.*

9.7.4   Discussion

There are still lots of challenging questions we did not address in this chapter. First, throughout this chapter, we assumed that the emission power is the major component of power consumption. In some devices, the emission power is at the same level as that of the power needed for being idle or to receive packets. It is then necessary to design a structure with a theoretically proven worst-case performance under this new energy model when the receiving power is not negligible. Second, an implicit assumption of the power model is that each node can adjust its power to any specific value. In practice, the transmitter has to choose among a set of given discrete power values. It is already known that under this discrete power model, the minimum-energy broadcast problem is still NP-hard (Liang, 2002) and a method with approximation ratio $O(n^\epsilon)$ was proposed in Liang (2002) [the ratio is $O(\log^3 n)$ if identical nodes are used]. Then, the problem remaining is to close the gap between this huge ratio and the constant-approximation ratio (Wan *et al.*, 2002b) when the power is continuously adjustable. Third, although the algorithms proposed in this chapter use $O(n)$ messages, each with $O(\log n)$ bits, they are not *localized* algorithms. Recall that a distributed algorithm is localized for WANs if it uses the information only of nodes that are within certain constant hops from the running node (it does not need to run in constant time) independent of the size of the network (Kuhn *et al.*, 2004a; Naor and Stockmeyer, 1993). It would be interesting to study what kind of properties can be achieved locally (as in Naor and Stockmeyer, 1993) and what kind of properties cannot be achieved locally (as in Kuhn *et al.*, 2004a). If a certain combined property cannot be achieved locally, what will be the best achievable trade-off between time and approximation ratio? Currently, the locally achievable geometric properties include the planar spanner (X.-Y. Li *et al.*, 2002b), the bounded-degree spanner (X.-Y. Li *et al.*, 2002c), bounded degree and low weight (X.-Y. Li *et al.*, 2002c), and k-fault tolerant (N. Li and Hou, 2004; X.-Y. Li *et al.*, 2003b, 2003c).

9.8   Spanners for Heterogeneous Networks

Most prior work (Hu, 1993; L. Li *et al.*, 2001; X.-Y. Li *et al.*, 2001, 2002c; Ramanathan and Rosales-Hain, 2000; Wattenhofer *et al.*, 2001) on network topology control assumed that wireless ad hoc or sensor networks are modeled by *unit disk graphs* (UDGs); i.e., two mobile hosts can communicate as long as their Euclidean distance is no more than a threshold. However, practically, wireless ad hoc networks cannot be perfectly modeled as UDGs: The maximum transmission ranges of wireless devices may vary for various reasons, such as device differences and small mechanic/electronic errors during the process of transmitting even if the transmission powers of all devices are set the same initially. In Barriere *et al.* (2001) and Kuhn *et al.* (2003b), the authors extended the UDG into a new model, called *quasi unit disk graphs*, which are closer to reality than UDGs. In this chapter, we study a more generalized model. Each wireless node u may have its own transmission radius r_u. Then, heterogeneous wireless networks are modeled by mutual-inclusion graphs (MGs): Two nodes can communicate directly only if they are within the transmission range of each other; i.e., it has a physical link uv if and only if $\|uv\| \leq \min(r_u, r_v)$. Clearly, a UDG is a special case of a MG. Obviously, in the MG model for heterogeneous wireless ad hoc networks, a link uv is always symmetric; i.e., u and v can communicate directly with each other. We adopt this symmetric communication model because unidirected links in wireless networks have been shown to be costly (Prakash, 1999). A considerable amount of research has gone into topology control for wireless networks modeled by UDGs. Topology control for heterogeneous networks is even harder because many properties in homogeneous networks disappear in heterogeneous networks. Thus, we cannot simply extend the ideas from the well-studied topologies, such as GG, RNG, and YG, used in homogeneous networks to heterogeneous wireless networks.

In this section, we study several localized topology-control strategies in which every wireless sensor maintains communication with only a selected small subset of its physical neighbors in a heterogeneous network environment. Here, an algorithm is said to construct a topology H *locally* if every node u can decide which edges uv belong to H by using the information of only nodes within a constant number of hops of u. The global logical network topologies formed by these locally selected links are sparse and/or power efficient. By utilizing the wireless broadcast channel capability and assuming that a message sent by a sensor node will be received by all sensors within its transmission region with at most a constant number of transmissions, we prove that all the new methods use at most $O(n)$ total messages, in which each message has $O(\log n)$ bits. By further assuming that the transmission ranges of two neighboring nodes are within a constant-γ factor of each other (such a wireless network is called *smoothed*), we prove that some of the proposed structures also have a constant-bounded logical node degree.

9.8.1   Heterogeneous Wireless Network Model

A heterogeneous wireless network (e.g., a WAN) is composed of a set V of n wireless devices (called *nodes* hereafter) $v_1, v_2, \ldots, v_n$, in which each node v_i has its own maximum transmission power p_i. Thus, for simplicity, we assume that each mobile host v_i has its own transmission range r_i. The heterogeneous wireless ad hoc network is then modeled by a MG, in which two nodes v_i, v_j are connected if and only if they are within the transmission range of each other; i.e., $\|v_i v_j\| \leq \min(r_i, r_j)$. Previously, only a few methods were known for topology control when the networks were modeled as MGs. The correctness of the new methods does not require that each wireless node know the exact geometry position information. On the other hand, however, the spanner properties proved for our methods do depend on localization precision.

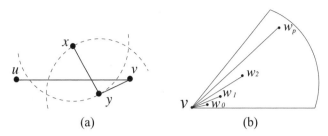

Figure 9.13 Limitations on heterogeneous networks: (a) planar topology does not exist; (b) degree of node v cannot be bounded by a constant. © 2006, ACM.

9.8.2 Limitations

In heterogeneous wireless ad hoc networks, a connected planar topology[4] does not necessarily exist. Figure 9.13(a) shows an example. There are four nodes, x, y, u, and v, in the network, where their transmission ranges are $r_x = r_y = \|xy\|$ and $r_u = r_v = \|uv\|$, node u is out of the transmission range of node x and y, and node v is in the transmission range of node y and out of the range of x. The transmission ranges of x and y are illustrated by the dashed ovals. According to the definition of a MG, there are only three edges, xy, vy, and uv, in the symmetric communication graph. Hence, no topology-control method can make the topology planar while keeping the communication graph connected. On the other hand, it is worth asking whether we can design a new routing protocol based on some pseudo-planar topologies. As we will see later, the pseudo-planar topology, GG(MG) and RNG(MG), proposed in this section has some special properties that are different from those of other general nonplanar topologies. For instance, two intersecting triangles cannot share a common edge. This is left as future work to further investigate them.

Another limitation for topology control in heterogeneous networks is that the node degree cannot be bounded by a constant if the ratio of the transmission radii of two neighboring nodes is unbounded. Figure 9.13(b) shows such an example. In the example, a node v has $p + 1$ incoming neighbors w_i, $0 \le i \le p$. Assume that each node w_i has a transmission radius $r_{w_i} = r_v/3^{p-i}$ and $\|vw_i\| = r_{w_i}$. Here, r_v is the transmission range of node v. Obviously, $\|w_i w_j\| > \min(r_{w_i}, r_{w_j})$ (i.e., any two nodes w_i, w_j) are not directly connected in a MG. Then, none of those edges incident upon v can be deleted; hence, there is no topology-control method that can bound the node degree by a constant without violating the connectivity. Consider the example illustrated by Figure 9.13(b); edges vw_i, $0 \le i \le p$, are all possible communication links. Thus, node v in any connected spanning graph has degree $p + 1$. On the other hand, as will be shown

---

[4] Here, we call a geometry structure *planar topology* if there are no physical edge intersections in the geometric structure. It is different with the standard definition of a *planar graph*, which allows redrawing the graph on a plane without crossing edges. One of the reasons that we use the special definition here is that in wireless networks, some geometric routing algorithms use the real geometric positions for routing, and they ask for the underlying geometric structure to be a planar topology without moving the node or using curved edges. Notice that when a standard definition of planar is used, every connected graph has a planar subgraph (e.g., a spanning tree).

in Subsection 9.8.5, in the worst case, any connected MG graph has degree $O(\log_2 \gamma)$, where $\gamma = \max_{v \in V} \max_{w \in I(v)} \frac{r_v}{r_w}$. Here, $I(v) = \{w \mid wv \in MG\}$. In the example, recall that $3^p r_{w_0} = r_v$ and, hence, γ equals to 3^p for this example. Thus, v has degree $\log_3 \gamma + 1 = O(\log_2 \gamma)$. In this chapter, we always assume that γ is a constant. It is practical because two wireless devices in a nearby region often have similar transmission ranges. Generally, we call a wireless network *smoothed* if γ is a constant for this network.

9.8.3   Heterogeneous Sparse Structure

In this subsection, a strategy is proposed for all nodes to self-form a sparse structure, called RNG(MG), based on the RNG. We will prove that the total number of links of this structure is $O(n)$.

DEFINITION 9.11 *[Structure RNG(MG)] A link $uv \in MG$ is kept in $RNG(MG)$ if and only if there is no other node w inside lune(u, v) and both links uw and wv are in MG. Here, lune(u, v) is the intersection of $\mathcal{D}(u, \|uv\|)$ and $\mathcal{D}(v, \|uv\|)$.*

Notice that the total communication cost of constructing RNG(MG) is $O(n \log n)$ bits, assuming that the radius and ID information of a node can be represented in $O(\log n)$ bits. In addition, the structure RNG(MG) is symmetric: If a node u keeps a link uv, node v will also keep the link uv. Thus, a node u does not have to tell its neighbor v whether it keeps a link uv.

It is not difficult to prove that structure RNG(MG) is connected by induction. On the other hand, as is the case in homogeneous networks (i.e., UDG mode), RNG(MG) does not have a bounded length-stretch factor or a constant-bounded power-stretch factor and does not have a bounded node degree. In this chapter, we show that RNG(MG) is a *sparse* graph: It has at most $6n$ links.

In the following, we define a new structure, called ERNG(MG), and present a localized algorithm to construct it.

DEFINITION 9.12 *[Structure ERNG(MG)] Each node u keeps the link to neighbor $v \in B(u)$ if and only if there is no other node w $\in B(u)$ inside lune(u, v) and both links uw and wv are in the MG. Here, $B(u) = \{v \mid r_v \geq r_u$ and uv $\in MG\}$ and lune(u, v) is the intersection of $\mathcal{D}(u, \|uv\|)$ and $\mathcal{D}(v, \|uv\|)$. All the links kept by all nodes form the final structure ERNG(MG).*

We now prove the following lemma:

LEMMA 9.4 *Structure ERNG(MG) has at most 6n links.*

Proof. Consider any node u. We show that u keeps at most six directed links uv, with $r_v \geq r_u$, emanated from u. Assume that u keeps more than six directed links. Obviously, there are two links uw and uv such that $\angle wuv < \pi/3$. Thus, vw is not the longest link in triangle $\triangle uvw$. Without loss of generality, we assume that $\|uw\|$ is the longest in triangle $\triangle uvw$. Notice that the existence of link uw implies that $\|uw\| \leq \min(r_u, r_w) = r_u$. Consequently, $\|vw\| \leq \|uw\| \leq \min(r_u, r_w)$. Thus, from the fact that $r_u \leq r_v$, we know $\|vw\| \leq \min(r_v, r_w)$. Hence, link vw does exist in the original communication MG. It implies that link uw cannot be selected to the ERNG. In other words, structure ERNG(MG) has at most 6n links.   ∎

Similar to Lemma 9.4, we can prove the following lemma:

LEMMA 9.5 *Structure RNG(MG) has at most* 6n *links.*

Proof. Imagine that each link uv $\in$ RNG(MG) has a direction as follows: $\overrightarrow{uv}$ if $r_u \leq r_v$. Then, similar to Lemma 9.4, we can prove that each node u only keeps at most six such imagined direct links. Thus, there are at most $6n$ links in RNG(MG).   ∎

9.8.4   Heterogeneous Power Spanner

In the previous subsection, we defined two structures based on the RNG. These structures are sparse; however, theoretically they could have an arbitrarily large power-spanning ratio. In this subsection, a strategy is given for all nodes to self-form a power-spanner structure, called GG(MG), based on the GG.

DEFINITION 9.13 [*Structure GG(MG)*] *A link* uv $\in$ *MG is kept in* $GG(MG)$ *if and only if there is no other node* w *inside* $\mathcal{D}(u, v)$ *and both links* uw *and* wv *are in the MG.*

It is not difficult to prove by induction that structure GG(MG) is connected if the original network is connected. In addition, because we remove a link uv only if there are two links uw and wv with w inside $\mathcal{D}(u, v)$, it is easy to show that the power-stretch factor of GG(MG) is 1. In other words, the minimum-power consumption path for any two nodes v_i and v_j in the MG is still kept in the GG(MG). Remember that here, we assume the power needed to support a link uv is $\|uv\|^\beta$, for $\beta \in [2, 5]$.

Similarly, we can define a structure called EGG(MG) as follows:

DEFINITION 9.14 [*Structure EGG(MG)*] *A link* uv $\in$ *MG is kept in* $EGG(MG)$ *if and only if there is* no *other node* w *inside* $\mathcal{D}(u, v)$ *such that* $r_u \leq r_w$.

On the other hand, as is the case in homogeneous networks (i.e., UDG mode), GG(MG) and EGG(MG) are not length spanners and do not have a bounded-node degree. Furthermore, it is unknown whether they are *sparse* graphs. Recently, it was proven in Kapoor and Li (2003) that GG(MG) has at most $O(n^{8/5} \log \gamma)$ edges, where $\gamma = \max r_u / r_v$.

Notice that the extension from the GG is nontrivial. In Kapoor and Li (2003), two structures defined as follows cannot guarantee the connectivity. In the first structure, called LGG$_0$(MG), they remove a link $uv \in$ MG if there is another node w inside $\mathcal{D}(u, v)$. In the second structure, called LGG$_1$(MG), they remove a link $uv \in$ MG if there is another node w inside $\mathcal{D}(u, v)$, and either link uw or link wv is in the MG.

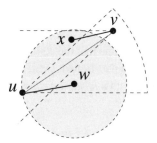

Figure 9.14 Simple extension of a Yao structure does not guarantee connectivity. © 2006, ACM.

9.8.5   Heterogeneous Degree-Bounded Spanner

Undoubtedly, as described before, we always prefer a structure that has more nice properties, such as degree bounded (stronger than sparse), power spanner, and so on. Naturally, we could extend the previous known degree-bounded spanner, such as the Yao-related structures, from homogeneous networks to heterogeneous networks. Unfortunately, even a simple extension of the Yao structure from a UDG to a MG does not guarantee the connectivity. Figure 9.14 illustrates such an example. Here, $r_u = r_v = \|uv\|$, $r_w = \|uw\|$, $r_x = \|vx\|$, and $\|uw\| < \|uv\|$, $\|uw\| < \|vw\|$, $\|vx\| < \|uv\|$, and $\|vx\| < \|ux\|$. In addition, v and w are in the same cone of node u, and nodes x and u are in the same cone of node v. Thus, the original MG contains links uv, uw, and vx only and is connected. However, when the Yao structure is applied on all nodes, node u will have information of only nodes v and w and it will keep links uw. Similarly, node w keeps link uw, node v keeps link vx, and node x keeps link xv; in other words, only links xv and uw are kept by the Yao method. Thus, applying a Yao structure disconnects nodes v and x from the other two nodes u and w. Consequently, we need more sophisticated extensions of the Yao structure to a MG to guarantee the connectivity of the structure.

In our first sparse spanner structure $\mathrm{EYG}_k(MG)$, unlike a traditional Yao structure, each node u keeps a node v as its communication neighbor if and only if uv is the shortest link among links between u and all nodes v_i in the same cone of u, and $r_{v_i} \geq r_u$. Formally speaking, it is defined as follows:

DEFINITION 9.15 [*Structure $EYG_k(MG)$*] *Each node* u *partitions its transmission region into k-equal-sized cones. In each cone, it keeps a communication neighbor* v *if* v *is the closest neighbor among all nodes* v_i *such that* $r_{v_i} \geq r_u$ *in the cone. Let* $\overrightarrow{EYG}_k(MG)$ *be the union of all chosen links. The undirected graph, by ignoring the direction of each link in* $\overrightarrow{EYG}_k(MG)$, *is called* $EYG_k(MG)$.

Notice that because node u chooses a node $v \in \mathcal{D}(u, r_u)$ with $r_v \geq r_u$, link uv is indeed a bidirectional link; i.e., u and v are within the transmission ranges of each other. Additionally, as we will see later, this strategy avoids possible disconnection by the simple Yao extension mentioned before. The localized construction algorithm is as follows:

Algorithm 28 Constructing EYG

1: Initially, each node u divides the disk $\mathcal{D}(u, r_u)$ centered at u with radius r_u by k-equal-sized cones centered at u. We generally assume that the cone is half-open and half-closed. Let $\mathbb{C}_i(u)$, $1 \leq i \leq k$, be the k-cones partitioned. Let $C_i(u)$, $1 \leq i \leq k$, be the set of nodes v inside the ith cone $\mathbb{C}_i(u)$ with a *larger or equal* radius than u. In other words,

$$C_i(u) = \{v \mid v \in \mathbb{C}_i(u), \ and \ r_v \geq r_u\}.$$

Initially, $C_i(u)$ is empty.

2: Each node u broadcasts a HELLO message with ID_u, r_u and its position (x_u, y_u) to all nodes in its transmission range.

3: **while** node u receives a HELLO message from some node v **do**

4:     Node u sets $C_i(u) = C_i(u) \bigcup \{v\}$, if node v is inside the ith cone $\mathbb{C}_i(u)$ of node u and $r_v \geq r_u$.

5:     Node u chooses a node v from each cone $C_i(u)$ such that the link uv has the smallest $ID(uv)$ among all links uv_j with v_j in $C_i(u)$, if there is any.

6: **end while**

7: Finally, each node u informs all one-hop neighbors of its chosen links through a broadcast message. Let $\overrightarrow{EYG}_k(MG)$ be the union of all chosen links. The final topology is the undirected graph by ignoring the direction of each link in $\overrightarrow{EYG}_k(MG)$ and is called $EYG_k(MG)$.

This is the main difference between this algorithm and the simple extension of the Yao structure discussed before, in which it considers all nodes v from which u can get a signal. In the algorithm, each node broadcasts only twice: one for broadcasting its ID, radius, and position; and the other for broadcasting the selected neighbors. Remember that it selects at most k-neighbors. Thus, each node sends messages at most $O((k + 1) \log n)$ bits. Here, we assume that the node ID and its position can be represented using $O(\log n)$ bits for a network with n wireless nodes. Obviously, we also have the following lemma:

LEMMA 9.6 *Structure* $EYG_k(MG)$ *has at most* kn *links, where* k > 6 *is a constant.*

The following theorems were proved in Li *et al.* (2005c, 2005d):

THEOREM 9.10 *The length-stretch factor of* $EYG_k(MG)$, $k > 6$, *is at most* $\ell = \frac{1}{1 - 2\sin(\frac{\pi}{k})}$.

THEOREM 9.11 *The power-stretch factor of structure* $EYG_k(MG)$, $k > 6$, *is at most* $\rho = \frac{1}{1 - (2 \sin \frac{\pi}{k})^\beta}$.

Notice that the node degree of the structure $EYG_k(MG)$ may not be bounded by a constant. Partitioning the space surrounding a node into k-equal-sized cones enables us to bound the node out-degree by using the Yao structure. Using the same space partition, a Yao–Yao structure (Li *et al.*, 2001, 2002b), produces a topology with a bounded in-degree when the networks are modeled by a UDG. A Yao–Yao structure (for a UDG) is generated as follows: A node u collects all its incoming neighbors v [i.e., $\overrightarrow{vu} \in \overrightarrow{YG}_k(V)$) and then selects the closest neighbor v in each cone $\mathbb{C}_i(u)$. Clearly, a Yao–Yao structure has a bounded degree of at most k. They also showed that another structure, YaoSink (Li et al., 2001, 2002b), has not only the bounded node degree but also a constant-bounded stretch factor. The network topology with a bounded degree can increase communication efficiency. However, these methods (Li et al., 2001, 2002b) may fail when the networks are modeled by a MG: They cannot even guarantee the connectivity, which is verified by the following discussion.

Assume that we have already constructed a connected directed structure $\overrightarrow{EYG}_k(MG)$. Let $I(v) = \{w \mid \overrightarrow{wv} \in \overrightarrow{EYG}_k(MG)\}$. In other words, $I(v)$ is the set of nodes that have directed links to v in $\overrightarrow{EYG}_k(MG)$. Let $I_i(v) = I(v) \cap \mathbb{C}_i(u)$; i.e., the nodes in $I(v)$ located inside the ith cone $\mathbb{C}_i(v)$. A Yao–Yao structure will pick the closest node w in $I_i(v)$ and add an undirected link wv to the Yao–Yao structure. The previous example in Figure 9.13(b) also illustrates the situation in which the Yao–Yao structure is not connected. In the example, a node v has $p + 1$ incoming neighbors w_i, $0 \le i \le p$. Assume that each node w_i has a transmission radius $r_{w_i} = r_v/3^{p-i}$ and $\|vw_i\| = r_{w_i}$. Obviously, $\|w_i w_j\| > \min(r_{w_i}, r_{w_j})$; i.e., any two nodes w_i, w_j are not directly connected in the MG. It is easy to show that the Yao structure $\overrightarrow{EYG}_k(MG)$ has only directed links $\overrightarrow{w_i v}$. Obviously, node v will only select the closest neighbor w_0 to the Yao–Yao structure, which disconnects the network. This same example can also show that the structure based on YaoSink (Li et al., 2001, 2002b) is also not connected for heterogeneous wireless ad hoc networks.

Thus, selecting the closest incoming neighbor in each cone $\mathbb{C}_i$ is too aggressive to guarantee the connectivity. Observe that in Figure 9.13(b), to guarantee the connectivity, when we delete a directed link $\overrightarrow{w_i v}$, we need to keep *some* link, say $w_j v$, such that $w_i w_j$ is a link in the MG. Thus, we further partition the cone into a limited number of smaller *regions* and we will keep *only* one node in each region (e.g., the closest node). Clearly, to guarantee that other nodes in the same region are still connected to v, we need to make sure that any two nodes w_i, $w_j \in I(v)$ that coexist in the same small region are directly connected in the MG. Consequently, if the number of regions is bounded by a constant, a degree-bounded structure could be generated. In the remainder of this subsection, a novel space-partition strategy satisfying this requirement is introduced.

For each node v, let $\gamma_v = \max_{w \in I(v)}, (r_v/r_w)$. Remember that all nodes in $I(v)$ have a transmission radius of at most r_v, so $\gamma_v \ge 1$. Let h be the positive integer satisfying $2^{h-2} < \gamma_v \le 2^{h-1}$. The proposed partition method works as follows:

Method 1 Partition Transmission Disks

1. Each node v divides each cone centered at v into a limited number of triangles and caps, as illustrated in Figure 9.15, where $\|va_i\| = \|vb_i\| = \frac{1}{2^{h-i}} r_v$ and c_i is the midpoint of the segment $a_i b_i$, for $1 \le i \le h$.
2. The triangles $\triangle va_1 b_1$, $\triangle a_i b_i c_{i+1}$, $\triangle a_i a_{i+1} c_{i+1}$, $\triangle b_i b_{i+1} c_{i+1}$, for $1 \le i \le h - 1$, and the cap $\widehat{a_n b_n}$ form the final space partition of each cone. For simplicity, we call such a triangle or the cap a *region*.

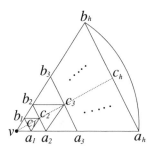

Figure 9.15 Extended Yao structure on heterogeneous networks: Further space partitions in each cone to bound in-degree. © 2006, ACM.

It was proved in X.-Y. Li *et al.* (2005c) that this partition indeed guarantees that any two nodes in any same region are connected in a MG.

LEMMA 9.7 *Assume that $k \geq 6$. Any two nodes* u, w $\in$ I(v) *that coexist in any one of the generated regions are directly connected in a MG; i.e.,* $\|uw\| < \min(r_u, r_w)$.

Using the space partition just discussed, a method is presented to locally build a sparse network topology with a bounded degree for heterogeneous WSNs. Here, we assume that $\gamma = \max_{v \in V} \gamma_v$ is bounded by some constant, where $\gamma_v = \max_{w \in I(v)}(r_v/r_w)$ and $I(v) = \{w \mid \overrightarrow{wv} \in \overrightarrow{EYG}_k(MG)\}$.

DEFINITION 9.16 [*Structure EYY$_k$(MG)*] *A link $\overrightarrow{uv}$ is kept by a node* v *if* u *is the closest neighbor in the corresponding region of* $\overrightarrow{EYG}_k(MG)$ *where* u *is located. The union of all chosen links is the final network topology, denoted by* $\overrightarrow{EYY}_k(MG)$. *We call it the* Extended Yao–Yao *graph. The structure EYY$_k$(MG) is the undirected graph by ignoring the direction of each link in* $\overrightarrow{EYY}_k(MG)$.

The following theorem was proved in X.-Y. Li *et al.* (2005c):

THEOREM 9.12 *(X.-Y. Li et al., 2005c) The out-degree of each node* v *in* $\overrightarrow{EYY}_k(MG)$, $k \geq 6$, *is bounded by* k *and the in-degree is bounded by* $(3[\log_2 \gamma_v] + 2)k$, *where* $\gamma_v = \max_{w \in I(v)}(r_v/r_w)$.

Let $\gamma = \max_v \gamma_v$. Obviously, the maximum node degree in graph EYY$_k$(MG) is bounded by $(3[\log_2 \gamma] + 3)k$. Notice that the structure EYY$_k$(MG) is a subgraph of the structure EYG$_k$(MG); thus, there are at most kn edges in EYY$_k$(MG). Consequently, the total communications to construct EYY$_k$(MG) in a distributed manner is at most $O(kn)$, where each message has $O(\log n)$ bits. It is interesting to see that the communication complexity does not depend on γ at all.

THEOREM 9.13 *(Li et al., 2005c) The graph EYY$_k$(MG), $k \geq 6$, is connected if the MG is connected.*

Although EYY$_k$(MG) is a connected structure, it is unknown whether it is a power or length spanner; it is left as a future work.

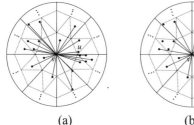
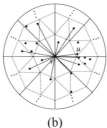
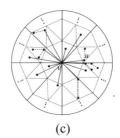

(a)   (b)   (c)

Figure 9.16 (a) In $EYG_k(MG)$, the star is formed by links toward v; (b) node v chooses the shortest link in $EYG_k(MG)$ toward itself from each region to produce $EYY_k(MG)$; (c) the sink structure at v in $EYY_k^*(MG)$. © 2006, ACM.

Bounded degree sparse spanner
In X.-Y. Li *et al.* (2001, 2002b), the authors applied the technique in Arya *et al.* (1995) to construct a sparse network topology in a UDG, a *Yao and sink graph*, which has a bounded degree and a bounded stretch factor. The technique is to replace the directed star in the YG consisting of all links toward a node v by a directed tree $T(v)$ with v as the sink. Tree $T(v)$ is constructed recursively. To apply this technique on a MG, we need a more sophisticated way to guarantee the connectivity. In X.-Y. Li *et al.* (2005c), the authors present a structure EYG* that can be constructed efficiently and is a sparse spanner whose degree is within a constant factor of optimum. The maximum node degree of the directed graph $\overrightarrow{EYG}_k^*(MG)$ is at most $k^2 + 3k + 3k[\log_2 \gamma]$. The total number of links in $EYG_k^*(MG)$ is at most kn. The length-stretch factor of $EYG_k^*(MG)$, $k > 6$, is at most $[\frac{1}{1-2\sin(\frac{\pi}{k})}]^2$. The power-stretch factor of the graph $EYG_k^*(MG)$, $k > 6$, is at most $[\frac{1}{1-(2\sin\frac{\pi}{k})^\beta}]^2$.

9.9   Fault-Tolerant Structures

In this section, we study how to control the network topology given an n-node network that is already k-fault tolerant. Because of the nodes' limited resource, the scalability is crucial for network operations. One effective approach is to maintain only a linear number of links by use of a localized construction method. However, this sparseness of the constructed network topology should not compromise on the fault tolerance or compromise too much on the power consumptions for both unicast and broadcast/multicast communications. We are interested in efficiently constructing a sparse network topology for a set of static wireless nodes such that every unicast route in the constructed network topology is power-efficient, in addition to being k-fault tolerant. Here, a route is *power efficient* for unicast if its power consumption is no more than a constant factor of the minimum power needed to connect the source and the destination. A network topology is said to be power-efficient if there is a power-efficient route to connect any two nodes in this topology.

Given a subgraph H, the power assigned to a node u is the minimum power needed by u to maintain the connectivity with all its neighbors in H; i.e.,

$$P_H(u) = \max_{(u,v)\in H} p(u, v).$$

Here, $p(u, v)$ is the power needed by node u to maintain the connectivity to node v. It is also called the weight of link (u, v). The *total power* needed to maintain the connectivity of all links in H is then defined as $p(H) = \sum_{u \in V} P_H(u)$. The *maximum power* needed to maintain the connectivity of all links in H is then defined as $P_{\max}(H) = \max_{u \in V} P_H(u)$.

In the literature, some researchers have tried to find a fault-tolerant structure (i.e., a k-connected subgraph of the graph modeling the wireless networks) that minimizes the total power of the structure, whereas some researchers have tried to find a fault-tolerant structure that minimizes the maximum power of the structure.

Lukovszki (1999b) gave a method to construct a spanner that can sustain k-nodes or link failures for a complete graph. One of the topology-control methods that we study is based on this method and the Yao structure (Yao, 1982). Recall that the *Yao graph* over a (directed) graph G with an integer parameter $p \geq 6$, denoted by $\overrightarrow{YG}_p(G)$, is defined as follows. At each node u, any p equal-separated rays originating at u define p equal cones. In each cone, choose the shortest (directed) edge $uv \in G$, if there is any, and add a directed link $\overrightarrow{uv}$. Ties are broken arbitrarily. Let $YG_p(G)$ be the undirected graph by ignoring the direction of each link in $\overrightarrow{YG}_p(G)$.

L. Li *et al.* (2001) proposed a structure that is similar to the Yao structure for topology control. Each node u finds a power $p_{u,\alpha}$ such that in every cone of degree α surrounding u, there is some node that u can reach with power $p_{u,\alpha}$. Then, the graph G_α contains all edges uv such that u can communicate with v by using power $p_{u,\alpha}$. It was proved in L. Li *et al.* (2001) that if $\alpha \leq (5\pi/6)$ and the UDG is connected, then graph G_α is a connected graph. On the other hand, if $\alpha > (5\pi/6)$, they showed that the connectivity of G_α is not guaranteed by giving some counterexamples (L. Li *et al.*, 2001). It is proved in L. Li *et al.* (2001) that G_α for $\alpha \leq (2\pi/3k)$ preserves the k-connectivity of G. As the work is extended from the CBTC algorithm, it shares the same assumption of a homogeneous network in which the maximal transmission power of each node is the same. Notice that they did not prove any bound on the total power or the maximum power assigned to all nodes.

Hajiaghayi *et al.* (2003) presented three approximation algorithms to find the minimum-power k-connected subgraph. The objective is to minimize the total power assigned to all nodes to ensure the k-connectivity of the resulting subgraph. Two global algorithms are based on existing approaches. The first gives an $O(k\alpha)$ approximation, where α is the best approximation factor [the best α so far is $O(\log k)$] for the k-UPVCS (undirected minimum-power k-vertex-contected subgraph) problem, which is defined as follows:

DEFINITION 9.17 *A k-UPVCS of a graph $G = (V, E)$ is a* k-*vertex-connected subgraph* $H = (V, F)$, $F \subseteq E$, *such that* $P(H) \leq P(H')$ *for any* k-*vertex-connected subgraph* $H' = (V, F')$, $F' \subseteq E$.

The other algorithm improves the approximation factor to $O(k)$ for general graphs. The third method is a distributed algorithm that gives an $O(k^{2\beta+2})$ approximation, where β is the exponent in the propagation model. This algorithm assumes that the edge weights of the network graph form a metric. For two-connectivity or three-connectivity, it first computes the MST of the input graph by using a distributed algorithm, which has the worst-case message complete $O(m + n \log n)$ and time complexity of $O(n)$ for a network of n nodes and m links. Then, it adds a path among the neighbors of each node in the returned MST tree. Because this method is based on the distributed MST algorithm, it has more maintenance overhead and delay when the topology has to be changed in response to node mobility or failure. Moreover, as pointed out in N. Li and Hou (2004), a closer investigation of the distributed algorithm reveals that the neighbors of a node on the MST may not be able to communicate with each other because of the limited transmission power. As a result, the arbitrary path connecting neighbors in the algorithm (Hajiaghayi *et al.*, 2003) may not exist in a network of low density. A counterexample was presented in N. Li and Hou (2004) to show that the two-UPVCS algorithm does not always preserve two-connectivity.

9.9.1   Fault Tolerance Based on a Geometric Approach

Li *et al.* (2001, 2002b) and Y. Wang and Li (2002a) proposed using the Yao structure on the UDG for topology control without sacrificing too much energy conservation. Some researchers used a similar construction, called a θ-graph (Lukovski, 1999b). The difference is that in each cone, the θ-graph chooses the edge that has the shortest projection on the axis of the cone instead of the shortest edge used by the Yao graph. Here, the axis of a cone is the angular bisector of the cone. For more details, refer to Lukovski (1999b). It is obvious that the Yao structure does not sustain k-faults in a neighborhood of any node because each node has only at most p neighbors and one neighbor selected in each cone at most. However, we can modify the Yao structure as follows such that the structure is k-fault tolerant.

Assume that the network is modeled by a UDG. Each node u defines any p equal-separated rays originating at u and thus defines p equal cones, where $p > 6$. In each cone, node u chooses the $k + 1$ closest node in that cone, if there is any, and adds directed links from u to these nodes. Ties are broken arbitrarily. Let $Y_{p,k+1}$ be the final topology formed by all nodes.

THEOREM 9.14 *The structure $Y_{p,k+1}$ can sustain* k-*node faults if the original UDG disk graph is* k-*node fault-tolerant.*

Proof. For simplicity, assume that all k-fault nodes $v_1, v_2, \ldots, v_k$ are neighbors of a node u. We show that the remaining graph of $Y_{p,k+1}$ (removed of nodes $v_1, v_2, \ldots, v_k$ and all links incident upon them) is still connected.

Notice that the original UDG is k-node fault-tolerant. Thus, the degree of each node is at least $k + 1$. Additionally, with the k-fault nodes $v_1, v_2, \ldots, v_k$ removed, there is still a path in the UDG to connect any pairs of remaining nodes. Assume that the path uses node u and has a link uw. We prove by induction that there is a path in the remaining graph to connect u and w.

If uw has the smallest distance among all pair of nodes, then uw must be in $Y_{p,k+1}$ and thus in the remaining graph.

Assume that the statement is true for a node pair whose distance is the rth shortest. Consider uw with the $(r + 1)$th shortest length.

If w is one of the $k + 1$ closest nodes to u in some cone, then link uw remains in the remaining graph. Otherwise, for the cone in which node w resides, there must be other $k + 1$ nodes that are closer to u than w and they are connected by u in $Y_{p,k+1}$. Because we have only k-failure nodes, at least one of the links of $Y_{p,k+1}$ in that cone will survive, say, link ux. It is easy to show that $\|xw\| < \|uw\| < 1$. Then, link uw can be replaced with link ux and a path from x to w with induction. This finishes the proof.

Notice that for the case in which the nodes removed are not all neighbors of the same node, the induction proof also holds. Induction is based on all pairs of nodes.   ■

Our techniques of constructing a k-connected subgraph of the UDG (assuming the UDG is already k-connected here) can be applied to a more general graph G if there is an embedding, denoted by $E(G)$, of G in the plane such that there is an edge in $E(G)$ iff their distance is not more than one unit. Notice that here, an embedding of G in the plane is to assign each vertex a two-dimensional position.

We now show that the preceding structure approximates the original UDG well. Let $\varrho_G(u, v)$ be the path found by a unicast routing method ϱ from node u to v in a weighted graph G, and let $\|\varrho_G(u, v)\|$ be the length of the path. The spanning ratio achieved by routing method ϱ is defined as $\max_{u,v} \|\varrho_G(u, v)\|/\|uv\|$. Notice that the spanning ratio achieved by a specific routing method could be much larger than the spanning ratio of the underlying structure. Nonetheless, a structure with a small spanning ratio is necessary for some routing method to possibly perform well.

THEOREM 9.15 *The structure $Y_{p,k+1}$ is a length spanner even with k-node faults.*

Proof. To prove the length-spanner property, it is easy to show that we only have to prove that each pair of nodes u and w with $\|uw\| \leq 1$ is approximated by a path with a length of no more than a constant factor, say β, of $\|uw\|$. The proof is similar to that of Theorem 9.14: We prove it by induction on the length of $\|uw\|$. Following the proof of Theorem 9.14, we have to show only that

$$\|ux\| + \beta\|xw\| \leq \beta\|uw\|$$

for any node x with $\|ux\| < \|uw\|$ and that x lies in the same cone as w does. Obviously, we need to set

$$\beta = \max_{\forall x, \|ux\| < \|uw\|} \frac{\|ux\|}{\|uw\| - \|xw\|}.$$

Notice that $\alpha = \angle wux < (2\pi/p)$. Then, a simple geometry reveals that $\beta = \max \frac{\cos\theta}{\cos(\theta+\alpha)}$, where $\theta = \frac{1}{2}\angle uwx \leq \frac{\pi-\alpha}{2}$. The minimum value for β is $\frac{1}{1-2\sin(\pi/p)}$. In other words, the spanning ratio of the remaining structure is at most β.   ■

Because of the limited power and resources of wireless nodes, wireless topologies always prefer to have a bounded node degree, such that every wireless node keeps only constant neighbors. The node degree of the structure $Y_{p,k+1}$ is at most $p(k + 1)$, where $p \geq 6$. Recently, Bahramgiri *et al.* (2002) showed how to decide the minimum transmission range of each node such that the resultant directed communication graph is k-connected. We can prove that their resultant graph is also a length spanner even

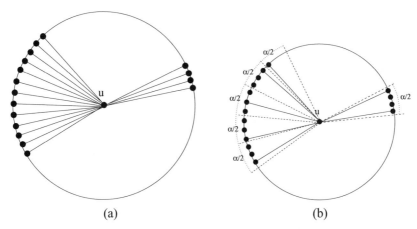

Figure 9.17 (a) Node u does not have a bounded degree in the graph generated by the protocol of Bahramgiri *et al.*; (b) new method to bound the node degree for the protocol of Bahramgiri *et al.* (2002) © 2004, ACM.

with k-node faults (the proof is omitted here because it is similar to the previous proof). However, their method does not bound the node degree. Figure 9.17(a) shows an example in which node u can have as many neighbors even after their method is applied. Now, we give a careful enhancement of their protocol to bound the node degree. In the method of Bahramgiri *et al.* (2002), they increase the power step by step until there is no gap greater than α between the successive neighbors or the power reaches the maximum power. They proved that if $\alpha \leq (2\pi/3k)$, then the resultant graph is k-connected. After applying their method, we can remove some links by the following method. For a node u, we divide its transmission range into $(4\pi/\alpha)$ equal cones (each cone having an angle $\alpha/2$). We select only one neighbor in each cone c, if there is any, and we delete all other links. However, if for a cone c, one of its adjacent cones, say b, does not have any neighbors of u, we select the boundary neighbor v such that vu forms the smallest angle with cone b; if both adjacent cones of c are empty, we select *two* neighbors in c (close to the two boundaries of cone c); if c does not have empty adjacent cones, we can select any one of the neighbors. See Figure 9.17(b) for an illustration. Because the gap between any two successive remaining neighbors is still not greater than α (except the empty cones), it is easy to show that the constructed graph is still k-connected if $\alpha \leq (2\pi/3k)$. The node degree is bounded by $(2\pi/\frac{\alpha}{2}) = (4\pi/\alpha)$. When $\alpha = (2\pi/3k)$, the node degree is bounded by $6k$, which is almost the same as that of the previous structure.

9.9.2 Fault Tolerance Based on a Greedy Approach

Assume that the network is modeled by a k-vertex-connected weighted graph G, in which the weight of each link denotes the energy cost to support the transmission of a unit amount of data over this link. N. Li and Hou (2004) first presented a centralized greedy algorithm, called a fault-tolerant global spanning subgraph (FGSS$_k$), that preserves k-vertex connectivity and is min–max optimal; i.e., its maximum power is no more than

$P_{max}(H)$ for any k-connected spanning subgraph H. From this algorithm, they then proposed a fully localized algorithm, called a fault-tolerant local spanning subgraph (FLSS$_k$) for topology control in wireless networks. "Fully localized" means that each node operates on the information locally collected. This feature enables the FLSS$_k$ to adapt to topology changes more easily. It can be proved that the FLSS$_k$ preserves k-vertex connectivity and maintains bidirectionality for all the links in the topology, while reducing the power consumption and improving the network capacity. They also prove that the FLSS$_k$ is min–max optimal among all strictly localized algorithms. The FGSS$_k$ is a generalized version of Kruskal's algorithm for $k \geq 2$. The algorithm is given in Algorithm 29.

Algorithm 29 FGSS$_k$

INPUT: $G(V, E)$, a k-connected simple graph;
OUTPUT: A k-connected spanning subgraph $G_k(V, E_k)$ of G;

1: $E_k \leftarrow \emptyset$;
2: Sort all edges in E in an ascending order of weight.
3: **for** each edge $e_i = (u, v)$ in the order **do**
4:     **if** node u is not k-connected to v in G_k **then**
5:        Add link e_i to graph G_k, i.e., $E_k \leftarrow E_k \cup \{(u, v)\}$;
6:     **else if** all nodes are in the same k-connected component **then**
7:        exit;
8:     **end if**
9: **end for**

For simplicity, let FGSS(G) denote the subgraph G_k computed by Algorithm 29 when the input is a graph G. Notice that by use of network flow techniques, a query on whether two vertices are k-connected in a graph can be answered in $O(k(n + m))$ time for any k, where n is the number of vertices and m is the number of edges in the graph. This is from the fact that an augmenting path can be found in time $O(n + m)$ if $\ell < k$ paths have been found in a k-connected graph. It can be proved by simple induction that the graph G_k computed by Algorithm 29 is k-connected if the given input graph G is k-connected.

The radius (or range) $P_H(u)$ of a node u in a graph H is defined as the maximum weight of all links in H that are incident upon u. Let SS$_k(G)$ be the set of all k-connected spanning subgraphs of G. N. Li and Hou (2004) proved that the FGSS$_k$ achieves the min–max optimality; i.e., $P_{max}(G_k) = \min_{H \in SS_k(G)} P_{max}(H)$. The proof is similar to the proof of correctness of Prim's algorithm for constructing a MST (Cormen *et al.*, 2001) and thus is omitted here.

We now review in detail the distributed method proposed in N. Li and Hou (2004) for constructing a k-connected structure. They proposed a localized, fault-tolerant topology control algorithm, called a fault-tolerant local spanning subgraph (FLSS$_k$). The topology is derived by having each node build its neighbor set and adjust its transmission power based on locally collected information. The algorithm consists of the following three phases:

1. **Information Collection:** Each node u collects local information of neighbors, such as their positions and IDs, and identifies the visible neighborhood N_u^V; i.e., the set of nodes that u can communicate with in one hop. For simplicity, we assume that $u \in N_u^V$. The visibility graph $G_u = (N_u^V, E_u)$ defined at a node u contains an edge (x, y) in E_u if and only if $x \in N_u^V$, $y \in N_u^V$, and (x, y) is a link in the original graph G.

2. **Topology Construction:** Each node defines, based on the information in N_u^V, the proper list of neighbors for the final topology.

3. **Construction of Topology with Only Bi-Directional Links (Optional):** Each node adjusts its list of neighbors to make sure that all the edges are bidirectional.

Given the visible neighborhood N_u^V, each node u builds its local spanning subgraph $S_u = [V(S_u), E_{u,k}]$ over N_u^V by using the algorithm of FGSS$_k$ described before, with one modification: The algorithm stops if u is k-connected to every other node in N_u^V, instead of requiring that all nodes be in the same k-connected component.

Algorithm 30 FLSS$_k$ at a node u

INPUT: $G_u = (N_u^V, E_u)$, the local visibility graph at node u.
OUTPUT: The local spanning subgraph $S_u = (N_u^V, E_{u,k})$ defined on N_u^V.

1: $E_{u,k} \leftarrow \emptyset$;
2: Sort all edges in E_u in an ascending order of weight.
3: **for** each link $e_i = (x, y)$ in the order **do**
4:     **if** node x is not k-connected to y in S_u **then**
5:         Add link e_i to graph S_u, i.e., $E_{u,k} \leftarrow E_{u,k} \cup \{(x, y)\}$;
6:     **else if** u is k-connected to every node $x \in N_u^V$ **then**
7:         exit;
8:     **end if**
9: **end for**

In the FLSS$_k$, node v is a neighbor of node u, denoted as $u \xrightarrow{\text{FLSS}} v$, if and only if $(u, v) \in E_{u,k}$. That is, v is a neighbor of u if and only if v is on u's local spanning subgraph S_u and is one hop away from u.

The network topology under the FLSS$_k$ is all the nodes in V and their individually perceived neighbor relations. Note that the topology is not a simple superposition of all local spanning subgraphs. In addition, the neighbor relation previously defined is not symmetric; i.e., $u \xrightarrow{\text{FLSS}} v$ does not necessarily imply that $u \xrightarrow{\text{FLSS}} v$.

DEFINITION 9.18 *The topology, G_{FLSS}, derived under the FLSS$_k$, is a directed graph* $G_{FLSS} = (V, E_{G_{FLSS}})$, *where* $E_{G_{FLSS}} = \{(u, v) \mid u \xrightarrow{\text{FLSS}v} v, u, v \in V\}$.

If only bidirectional links are allowed for communication, we can define a structure G_{FLSS}^- by removing the directed links as follows:

DEFINITION 9.19 *The topology, G_{FLSS}^-, derived under the FLSS$_k$, is an undirected graph* $G_{FLSS} = (V, E_{G_{FLSS}})$, *where* $E_{G_{FLSS}} = \{(u, v) \mid u \xrightarrow{\text{FLSS}} v, \text{ and } v \xrightarrow{\text{FLSS}} u, \forall u, v \in V\}$.

N. Li and Hou (2004) proved that if the original network G is a k-connected graph, then G_{FLSS} and G_{FLSS}^- are both k-connected. Notice that because G_{FLSS}^- is a subgraph of G_{FLSS}, the k-connectivity of G_{FLSS}^- implies the k-connectivity of G_{FLSS}. We can prove it as follows. For any undirected edge $uv \notin G_{\text{FLSS}}^-$, at least one of the directed edges (u, v) and (v, u) is not selected in G_{FLSS}; i.e., either node u did not select edge (u, v) in its local structure S_u, or node v did not select edge (v, u) in its local structure S_v. Without loss of generality, assume that node u did not select edge (u, v) in its local structure S_u. Then, when node u is processing the link (u, v), node v is already k-connected to node u in the partially built spanning graph S_u by node u. Notice that the weight of every edge on the k-disjoint paths that connect u and v in S_u is no more than the weight of uv. Based on Lemma 2 of N. Li and Hou (2004), u and v are still k-connected in G_{FLSS}^-.

Let $\text{LSS}_k(G)$ be all k-connected structures that can be constructed by any strictly localized algorithms. Here, a distributed algorithm is called strictly localized if, for every node u, it can use only the information derived from the local visibility graph G_u. Further, they proved the folllowing theorem:

THEOREM 9.16 *Among all strictly localized algorithms, $FLSS_k$ minimizes the maximum transmission radius (or power) of nodes in the network; i.e.,*

$$P_{\max}(G_{\text{FLSS}}) = \min\{P_{\max}(H) \mid H \in \text{LSS}_k(G)\}.$$

One important property of their method is that the correctness of all the preceding claims does not depend on the underlying graph model of the network: The graph modeling the network could be any arbitrary graph. When the weight of all links can be derived in other manners, it also does not require the nodes' position information.

9.10   Other Spanners

Besides the spanners (i.e., planar spanner, bounded-degree spanner, and bounded-degree planar spanner) we previously reviewed, there are many other spanners that have been studied for ad hoc network applications.

Fault-tolerant geometric spanners have been studied extensively (Czumaj and Zhao, 2003; Levcopoulos *et al.*, 1998, 2000), but most of the solutions are centralized methods that are too complex to have localized versions. Lukovszki (1999a) gave a method based on a θ-graph to construct a spanner that can sustain k-node or link failures for a complete graph. Similarly, X.-L. Li *et al.* (2003b) also proposed a method based on a Yao structure to build a fault-tolerant spanner. Bahramgiri *et al.* (2002) generalized the cone-based local heuristic of L. Li *et al.* (2001) and Wattenhofer *et al.* (2001) to achieve fault tolerance. We can prove that the resultant graph by Wattenhofer *et al.* (2001) is also a length spanner (the proof is similar to that of X.-Y. Li *et al.*, 2003b).

X.-Y. Li *et al.* (2005b) proposed a new bounded degree planar spanner that is also low-weighted. Here, a topology is low-weighted if the total link length of the topology is within a constant factor of that of the MST. For ad hoc networks, low-weighted structures enable energy-efficient broadcasting.

Table 9.1 Summary of the spanners

	Power	Length	Degree	Planar	Local
RNG	no	no	no	yes	yes
GG	yes	no	no	yes	yes
Del	yes	yes	no	yes	no
LDel	yes	yes	no	yes	yes
YG	yes	yes	no	no	yes
YG*	yes	yes	yes	no	yes
YY	yes	open	yes	no	yes
YS	no	no	yes	no	yes
YO	yes	yes	no	no	yes
BPS	yes	yes	yes	yes	yes
Ord Yao GG	yes	no	yes	yes	yes
SYAOGG	yes	no	yes	yes	yes

Burkhart *et al.* (2004) studied the spanners that can effectively constrain interference. Following this direction, X.-L. Li *et al.* (2005c) and Moscibroda and Wattenhofer (2005b) also studied more general low-interference topologies for ad hoc networks.

Schindelhauer *et al.* (2004, 2005) defined a new concept: a *weak spanner* in which the path can be arbitrarily longer than the original one but must remain within a disk or sphere of radius constant times the Euclidean distance between two vertices. They studied the relationship among length spanner, weak spanner, and power spanner. In addition, they proved that a sparse Yao graph is a weak spanner and a power spanner.

The spanners reviewed in this chapter are spanners for a UDG in which each node has the same transmission range. However, practically, the networks are never so perfect as UDGs. Kuhn *et al.* (2003b) modeled ad hoc networks as *quasi* UDGs and gave a spanner for quasi UDG. X.-Y. Li *et al.* (2005c) also studied spanners for networks with nonuniform transmission ranges in which different nodes could have different transmission ranges. They extended the Yao graph, sparsified Yao graph, and YaoSink graph to the nonuniform case.

9.11   Conclusion and Remarks

In this chapter, we studied several methods to construct geometric spanners as network topologies for ad hoc networks that can approximate the underlying communication graph well. A summary of properties of these spanners is given in Table 9.1. In this chapter, detailed algorithms, proofs, and simulation results were not given for all methods; please refer to the corresponding reference for more details.

Notice that a fundamental assumption of the majority topology-control protocols proposed in the literature is the assumption that every node u knows the relative positions of all its neighboring nodes. It is obvious that in the utilization of geometric knowledge, we need the availability of a positioning system that can provide precise or approximate location information. Practical position systems might not provide a precise position. Even a state-of-the-art global positioning system (GPS) will have a location error. In the literature, recently, there have been a number of studies that try to provide localizations without using a GPS. A challenging problem for topology control is the relaxations in precise position information and how to design effective structures even in these situations. The relaxations that can be used include relative distance ordering and relative direction ordering; i.e., topological information.

Problems

9.1 Given a connected UDG of n nodes, what is the maximum number of edges in the following graphs: RNG, GG, and Delaunay triangulation?

9.2 In the distributed construction of a localized Delaunay graph (Algorithm 20), a node u proposes a triangle $\triangle uvw$ only if $\angle vuw \geq \pi/3$ and all three edges of $\triangle uvw$ are at most 1. What will happen if we remove the condition $\angle vuw \geq \pi/3$? What is the worst-case communication complexity of the algorithm if that condition is removed?

9.3 In the construction of a YG, we require that the number of cones be larger than six; i.e., the angle of each cone is less than $\pi/3$. Can you construct an example of a network in which the YG will have an arbitrarily large spanning ratio when the angle of each cone is larger than $\pi/3$?

9.4 Construct an example of a network such that the local minimum spanning tree LMST, using only one-hop information, is not low-weighted. What is the largest ratio of the total edge length in the LMST over the total edge length in an EMST? Can you achieve a ratio $\Theta(n)$? Can you prove that the ratio $\Theta(n)$ is the largest possible ratio?

9.5 We already showed that Algorithm 27 will use at mosts $O(n)$ messages. What is the worst-case time complexity of this algorithm? In other words, assume that the clock of nodes are synchronized and in each time slot the communicating nodes must be interference-free.

9.6 Conduct simulations to study the performance of algorithms constructing RNG, GG, LDel, BPS, and LS$\ominus$GG when the input network is formed by nodes randomly placed.

9.7 Conduct simulations to study the worst length-spanning ratio of a RNG and a GG when the network is formed by n nodes randomly placed. You should plot the performance with respect to the increase of n. What is the trend you will observe when n is sufficiently large?

9.8 For all the algorithms presented in this chapter, we assumed that each wireless node knows its position. This may not be true in some application scenarios. Can you design efficient algorithms with similar performance guarantees for these questions? For example, design efficient algorithms that can construct a spanner when you know only the length of the links or even an approximation of the length of the links.

10   Power Assignment

Power is one of the most critical resources in wireless ad hoc networks when the wireless nodes are powered by batteries only. In this chapter, we study how to assign each wireless node a transmission power (level) such that the resulting communication graph has certain desired properties. Recently, much progress has been made on algorithmic and probabilistic studies of various power-assignment problems. These problems come in many flavors, depending on the power-requirement function and the connectivity constraint, and minimizing the total power consumption by all wireless nodes in a network is NP-hard for most versions. We study some of the best-known approximation algorithms for minimizing the total power consumption in the network and sketch useful heuristics with practical value. Observe that a majority of the power-assignment problems use the same network setting as some problems we studied in other chapters, especially about topology control; some questions are different (although they look similar). For example, in this chapter, we study the power-assignment problem by minimizing the total power while the resulting network is connected. A similar problem is to find a broadcast tree that has the minimum total power consumption. The difference here is that the leaf nodes are not required to transmit for broadcast applications, whereas all nodes are required to transmit for a tree spanning all nodes to result in a connected network.

10.1   Introduction

We consider a set $V = \{v_1, v_2, \ldots, v_n\}$ of n wireless nodes distributed in a two-dimensional plane. Let $\mathbf{w}(uv)$ be the power needed to support the communication from node u to node v. Notice that the power requirement $\mathbf{w}$ is said to be *asymmetric* if $\mathbf{w}(uv) \neq \mathbf{w}(vu)$ for some pair of nodes u and v. Otherwise, it is said to be *symmetric*. The power requirement is *arbitrary symmetric* if for an undirected graph, $H = (V, E)$, the power requirement is defined by a function $p : E \to R^+$. The power requirement is called Euclidean if it depends on the Euclidean distance $\|uv\|$. For example, in the literature, it is often assumed that $\mathbf{w}(uv) = c + \|uv\|^\beta$, where c is a positive constant real number, and real number $\beta \in [2, 5]$ depends on the transmission environment. Notice that the transmission power $\mathbf{w}(uv)$ could be a discrete value when there is only a finite number of power levels available because of hardware design. It is often assumed that the power $\mathbf{w}(uv)$ needed to support the communication from node u to node v is a monotone nondecreasing function of the Euclidean distance $\|uv\|$. In other words, $\mathbf{w}(uv) \geq \mathbf{w}(xy)$ if $\|uv\| > \|xy\|$ and $\mathbf{w}(uv) = \mathbf{w}(xy)$ if $\|uv\| = \|xy\|$. Notice that this assumption is not needed for most of the algorithms presented in this chapter.

We also assume that all nodes have omnidirectional antennas; i.e., if the signal transmitted by node u can be received by node v, then it will be received by all nodes x with $\|ux\| \leq \|uv\|$. Assume that each node u has a maximum transmission power $P_{\max}$. Consequently, if all wireless nodes transmit in their maximum power, they define a network that has a link uv iff $\mathbf{w}(uv) \leq P_{\max}$. This communication graph is called a unit disk graph (UDG).

In addition, all nodes can adjust the transmission power dynamically. Specifically, node u can adjust its power to be *exactly* $\mathbf{w}(uv)$ to support the communication to another node v. When nodes adjust their power dynamically, we say that node u can reach node v in an *asymmetric* communication model if node u transmits at a power of at least $\mathbf{w}(uv)$. Notice that here, in asymmetric communications, node v may transmit at a power less than $\mathbf{w}(vu)$ and thus cannot reach u. We say that node u can reach node v in a *symmetric* communication model if both nodes u and v transmit at a power of at least $\mathbf{w}(uv)$. Observe that in practice, a node may not adjust its power to any arbitrary value. A node can adjust its power to only a certain list of discrete power levels. The power assignment under this model has also been studied in several papers; e.g., see Liang and Haas (2000) and Lloyd *et al.* (2002) for a number of interesting problems.

An observation is that the network topology is entirely dependent on the transmission range of each individual node. Links can be added or removed when a node adjusts its transmission range. A *power assignment P* is an assignment of power setting $P(v_i)$ to every wireless node v_i. See Figure 10.1 for an example of power assignment to wireless nodes. Here, the weight of each link is the power needed to support a good communication on this link. In this specific example, the power requirement for links is symmetric. Given a power assignment P, we can define an induced direct communication graph $\overrightarrow{G}_P$ in which there is a directed edge $\overrightarrow{uv}$ if and only if $\mathbf{w}(uv) \leq P(u)$. We define the induced undirected communication graph G_P in which there is an edge uv if and only if $\mathbf{w}(uv) \leq P(u)$ and $\mathbf{w}(uv) \leq P(v)$; i.e., both directed links $\overrightarrow{uv}$ and directed links $\overrightarrow{vu}$ exist under this power assignment. We hereafter refer to G_P as the *induced communication graph*. See Figure 10.2 for the directed communication graph and undirected communication graph induced by the power assignment illustrated in Figure 10.1. If all wireless nodes transmit in their maximum power $P_{\max}$, the induced communication graph is called the *original communication graph*, which provides information about all possible topologies, in accordance with characteristics of the wireless environment and node power constraints. In other words, all possible achievable network topologies are subgraphs of the original communication graph.

On the other hand, given a subgraph $H = (V, E)$ of the original communication graph, we can also extract a minimum power assignment P_H, where

$$P_H(u) = \max_{\{v | uv \in E\}} \mathbf{w}(uv),$$

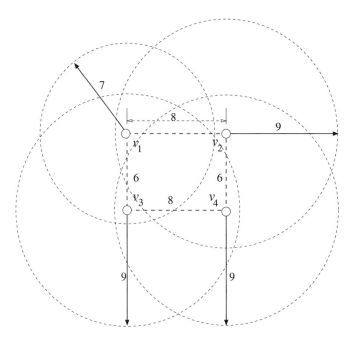

Figure 10.1 A power-assignment example.

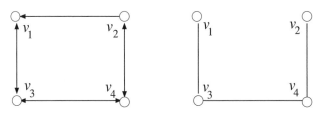

(a) Induced directed network.   (b) Induced undirected network.

Figure 10.2 Networks resulting from the power assignment.

to support the subgraph H. We call this power assignment P_H an *induced power assignment* from H. We call $P(H) = \sum_{u \in V} P_H(u)$ as the total power assignment based on structure H.

Notice that it is possible that a power assignment P will induce a strongly connected directed communication graph, but the induced undirected communication graph is not connected. Such a power assignment could be constructed as follows: For the node placement example illustrated in Figure 10.1, the power assigned to node v_4 is 7 instead of 9 and the power assigned to other nodes is fixed. In this case, the directed link $\overrightarrow{v_4 v_3}$ does not exist. The induced undirected communication graph has only links $v_1 v_3$ and $v_2 v_4$.

Typically, we require that the graph induced by the power assignment satisfy certain connectivity constraints. Assigning the power level to ensure that G_P, or $\overrightarrow{G}_P$, has certain graph properties is called the *power-assignment problem*. Many types of connectivity constraints have been studied in the literature and several are addressed later; e.g., the network $\overrightarrow{G}_P$ being strongly connected, the network G_P being connected, or the network G_P being two-connected, and so on. Given a connectivity constraint, the objective of power assignment is minimizing the total power assigned to all wireless nodes.

10.2 Power Assignment for Connectivity

10.2.1 Strong Connectivity

The study of the min-power power assignment for the strong connectivity problem was started by Chen and Huang (1989). Assuming symmetric power requirements, they prove that a MST of the original communication graph G has power of at most twice the optimum and, therefore, the MST algorithm has an approximation ratio of at most 2. For a symmetric power requirement $\mathbf{w}$ and any undirected structure $H = (V, E)$, we define $\mathbf{w}(H) = \sum_{uv \in E} \mathbf{w}(uv)$. Here, each edge $uv \in E$ is counted only once. This is often called the total link weight of the structure. For a directed structure $H = (V, E)$, $\mathbf{w}(H)$ is defined similarly and each directed link is counted once.

THEOREM 10.1 *(Chen and Huang, 1989) Assuming a symmetric link power requirement, i.e., $\mathbf{w}(uv) = \mathbf{w}(vu)$ for every pair of links* uv *and* vu, *the power assignment induced by the MST is a 2-approximation for the min-power power assignment for the strong connectivity problem.*

Proof. Let OPT be an optimum power assignment. Consider $\mathcal{G}$, the directed communication graph induced by OPT. Let s be an arbitrary node and B be a tree contained in $\mathcal{G}$ with edges oriented toward the root s. For example, such a tree B can be obtained from $\mathcal{G}$ by use of a breadth-first search (with all links oriented toward the root s). Notice that tree B is not strongly connected here. Observe that to get this tree B, every node $v \neq s$ with parent v' needs a power $p_B(v) = \mathbf{w}(vv')$. The total power assignment to induce the structure B is $P(B) = \sum_{v \neq s} p_B(v) = \mathbf{w}(B)$. On the other side, let us see the total power assignment induced by the MST. Then, we have:

$$P(\text{MST}) = \sum_{u \in V} P_{\text{MST}}(u) = \sum_{u \in V} \max_{uv \in \text{MST}} \mathbf{w}(uv) \leq \sum_{u \in V} \sum_{uv \in \text{MST}} \mathbf{w}(uv) = 2\mathbf{w}(\text{MST}),$$

because every edge in the MST is counted twice in $\sum_{u \in V} \sum_{uv \in \text{MST}} \mathbf{w}(uv)$. Consequently,

$$P(\text{MST}) \leq 2\mathbf{w}(\text{MST}) \leq 2\mathbf{w}(B) = 2P(B) \leq 2P(\mathcal{G}).$$

Here, $\mathbf{w}(\text{MST}) \leq \mathbf{w}(B)$ follows from the fact that the MST is the minimum spanning tree with edge weight $\mathbf{w}$. ∎

The example illustrated in Figure 10.3 shows that the ratio of 2 for the MST algorithm is actually tight. Consider $n = 2k$ points, $u_1, u_2, \ldots, u_{2k-1}, u_{2k}$, located on a single line such that the distance between consecutive points alternates between 1 and $\epsilon < 1$ and

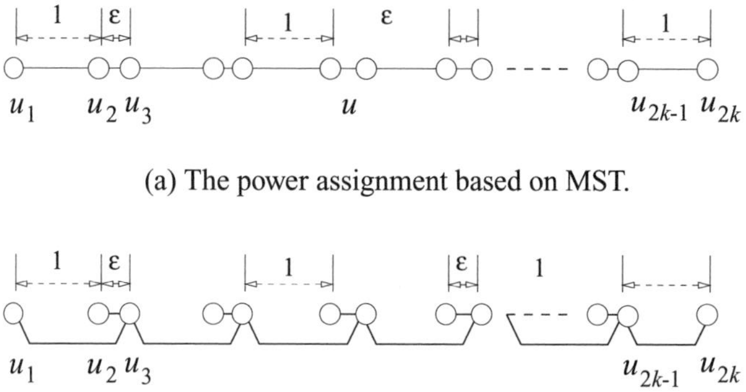

(a) The power assignment based on MST.

(b) A better power assignment.

Figure 10.3 Example in which the MST has a tight approximation ratio of 2.

assume that the power needed to support a link uv is $\|uv\|^2$. Then, the MST connects consecutive neighbors and has power $p(\mathrm{MST}) = n = 2k$ because each node will have a power assignment of 1. On the other hand, the tree T, with edges $u_{2i-1}u_{2i+1}$, for $1 \le i \le k-2$, edge $u_{2k-1}u_{2k}$, and edges $u_{2i}u_{2i+1}$, for $1 \le i \le k-1$, has power equal to $p(T) = (k-2)(1+\epsilon)^2 + (k-1)\epsilon^2 + 1$. When $n \to \infty$ and $\epsilon \to 0$, we find that $p(\mathrm{MST})/p(T) \to 2$.

We may wonder if we can design another approximation algorithm with a better approximation ratio. This seems to be difficult because min-power strong connectivity with symmetric power requirements is known to be APX-hard (approximation-hard). This means that there is an $\epsilon > 0$ such that the existence of an approximation algorithm with ratio $1 + \epsilon$ implies that $P = \mathrm{NP}$. Clementi $et\ al.$ (2001b) showed that min-power strong connectivity in the Euclidean case is NP-hard. Improving the approximation ratio to under 2 in the symmetric power-requirement case appears to be a very difficult problem. For the Euclidean power-requirement case, there is hope that one of the following two algorithms outputs a good solution: the $(5/3 + \epsilon)$-approximation algorithm presented in Althaus $et\ al.$ (2006), and Christofides' algorithm for the Traveling Salesman problem, followed by orienting all the edges of the Hamiltonian cycle to obtain a directed circuit.

So far, we always assumed that the power requirements are symmetric; i.e., the power needed to support a link uv is always the same as the power needed to support the link vu. The other scenario is that the power needed to support link uv may be different from the power needed to support link vu. In this case, the standard reduction of Călinescu $et\ al.$ (2003) shows that strong connectivity is as hard as the set-cover problem, which implies that there is no polynomial-time algorithm with approximation ratio $(1 - \epsilon) \ln n$ for any $\epsilon > 0$ unless $\{P = \mathrm{NP}\}$. Călinescu $et\ al.$ (2003) then presented a greedy approximation algorithm with ratio $2 \ln n + 3$ for strong connectivity with asymmetric power requirements. This algorithm picks an arbitrary vertex s, uses an approximation algorithm for broadcast from s (which we discuss later), and Edmunds' algorithm for a minimum-cost incoming branch rooted at s.

10.2.2   Symmetric Connectivity

In previous discussions, we tried to find power assignment P such that the induced graph $\overrightarrow{G}_P$ is strongly connected. Notice that strong connectivity guarantees only that for every node, there is a directed path to every other node in the graph. In many applications, we would like to have a path with only symmetric links because asymmetric links often create problems when acknowledgement is required. Given a directed graph $\overrightarrow{G}$, let G be the symmetric restriction of $\overrightarrow{G}$, where a link uv belongs to G if and only if both directed links $\overrightarrow{uv}$ and $\overrightarrow{vu}$ belong to the directed graph $\overrightarrow{G}$. Then, obviously, if G is connected, then $\overrightarrow{G}$ is strongly connected but not necessarily vice versa. It was observed in Althaus *et al.* (2006) that it is generally harder to ensure symmetric connectivity. In fact, the power for min-power strong connectivity can be half the power for min-power symmetric connectivity, as illustrated by an example in Călinescu *et al.* (2004) in a Euclidean case with $\beta = 2$. The example is constructed as follows. The terminal sets are formed by groups of nodes. The terminal set consists of n groups of $n + 1$ nodes each, located on the sides of a regular $2n$-gon with side length 1, where no two groups of nodes will be located on the adjacent sides of the regular $2n$-gon. Each group of nodes has two terminals that are within distance one of each other and $n - 1$ equally spaced nodes on the line segment between these two terminals. It is easy to see that the minimum power assignment that ensures strong connectivity assigns a power of 1 to one terminal in each group (this will ensure the connectivity of all terminals) and a power of $(1/n)^2$ to all other nodes in the group (this will ensure the connectivity of all nodes in this group to the terminal in this group). The total power assigned to all nodes and terminals is then $n + n^2(1/n)^2 = n + 1$. For symmetric connectivity, it is necessary to assign a power of 1 to all but two terminal nodes and of $(\frac{1}{n})^2$ to the remaining nodes, which results in a total power of $2n - 2 + (n^2 - n + 2)(\frac{1}{n})^2$. Observe that the total power needed for symmetric connectivity for this specific example is almost two times the power needed for strong connectivity when n is sufficiently large.

We first consider the scenario in which the power requirement is symmetric; i.e., the power needed to support a link uv is the same as the power needed to support the link vu. Note that the MST algorithm of the previous section produces a symmetric output and, therefore, the MST has an approximation ratio of 2 for symmetric connectivity in the symmetric power requirement case. Better approximation algorithms were presented in Althaus *et al.* (2006), with the best achieving a ratio of $5/3 + \epsilon$.

With an asymmetric power requirement [or even with symmetric power requirements modified by nonuniform efficiency (Călinescu *et al.*, 2003)], min-power symmetric connectivity is again as hard as the set-cover problem. Wireless nodes can have nonuniform transmission efficiency, and we use $e(v)$ to denote the efficiency of node v. Under the efficiency assumption, Călinescu *et al.* (2003) say that node u reaches node v with a power level of $p(u)$ if and only if

$$p(u) \geq \frac{\mathbf{w}(uv)}{e(u)s(v)}.$$

Here, wireless nodes can have a nonuniform power threshold for signal detection, and they call sensitivity $s(v)$ the threshold of node v. The algorithm subsequently described has an approximation ratio of at most $2 \ln n + 2$ for an symmetric connectivity with an asymmetric power requirement for a network of n nodes. It was proved by Călinescu *et al.* (2003) that Algorithm 31 has an approximation ratio of at most $2 \ln n + 2$ for a network of n nodes. For authors and readers familiar with the approximation algorithm for the weighted connected dominating set (WCDS) problem, the preceding algorithm is essentially some extension of the classic approximation algorithm for the minimum WCDS problem (Guha and Khuller, 1996).

Algorithm 31 Efficient power assignment for symmetric connectivity with asymmetric power requirement

---

1: Let G_0 be the collection of all vertices V, without any links. Let $i = 0$.
2: The algorithm starts iteration i with a graph G_i. Unless G_i is connected, a star S is computed such that it achieves the biggest reduction in the number of components divided by the power of the star; i.e., minimizing $p(S)/[d(S) - 1]$. The algorithm then adds the star (i.e., its set of edges) to G_i to obtain the graph G_{i+1}.

Here, a *star* S is a tree consisting of one center and several leaves adjacent to the center. Note that the power of the star S, denoted as $p(S)$, is the maximum power requirement of the links from the center to the leaves plus the sum of power requirements of the links from the leaves to the center. With respect to G_i, let $d(S)$ be the number of different components of G_i to which the vertices of the star belong. We need to find a star S minimizing $p(S)/(d(S) - 1)$.

10.2.3 Biconnectivity

Besides the strong connectivity and symmetric connectivity for a network, biconnectivity is also often studied in the literature to ensure a certain fault tolerance. Again, as with the connectivity requirement, there is also strong biconnectivity (which is simply called biconnectivity if it is clear from the context) and symmetric biconnectivity.

- **Strong biconnectivity** requires that for any two vertices u and v, the resulting directed communication graph $\overrightarrow{G}$ must contain at least *two* node-disjoint directed paths from u to v that do not share nodes except for end nodes u and v.
- **Symmetric biconnectivity** requires that the undirected communication graph G must be two-connected (i.e., removing any vertex from G still results in a connected graph).

Under the assumption of symmetric power requirement for links, Ramanathan and Rosales-Hain (2000) proposed the first heuristic for achieving strong biconnectivity. Their algorithm is based on the idea of Kruskal's algorithm for MSTs. We summarize their approach here in a slightly modified description.

Algorithm 32 Power assignment for strong biconnectivity

1: Sort the edges of the complete communication graph H when all nodes are using the maximum transmission power in nondecreasing cost order.

2: Use a simple binary search on indices to find the minimum index i such that the spanning subgraph, say Q, of H with the set of edges

$$\{e_1, e_2, e_3, \ldots, e_{i-1}, e_i\}$$

is biconnected.

3: Starting with e_i and going downward, remove from the spanning graph Q any edge e for which the remaining edges in Q still induce a biconnected spanning subgraph.

4: Assign power to vertices according to the final structure Q; that is, assign node v a power

$$p(v) = \max_{vu \in E(Q)} \mathbf{w}(vu).$$

5: For every vertex, reduce its power as long as the induced graph G_P based on current power assignment P is still biconnected.

Here, we observe that the last step of Algorithm 32 may be unnecessary. The reason is that if a node v can further reduce its power assignment $p(v)$, it implies that some of its incident edges (say, an edge e_j) can be removed without violating the biconnectivity property. If this is the case, then in step 3 of the preceding algorithm, we will remove link e_j when we process link e_j to check if we can remove it without violating the biconnectivity property of the remaining graph. Extensive simulations in Ramanathan and Rosales-Hain (2000) shows that this approach has good performance for randomly deployed networks. However, in the case of symmetric power requirements, Călinescu et al. (2004) claimed to have examples showing that the Ramanathan and Rosales-Hain heuristic has an approximation ratio of at least $n/2$ in the worst-case scenario. Its average case performance is unknown for symmetric power requirements. In addition, its approximation ratio in the Euclidean power-requirement case is also not known, and some extensive experiments show it is quite good in uniform random instances in the unit square.

Under the assumption of symmetric power requirement for directed links $\overrightarrow{uv}$ and $\overrightarrow{vu}$, Lloyd et al. (2002) proposed using the approximation algorithm of Khuller and Vishkin (1994) (which is hereafter referred to as Algorithm KV). Algorithm KV was designed for minimum-weight biconnected spanning subgraph. Lloyd et al. (2002) proved that their method achieves an approximation ratio of $2(2 - 2/n)(2 + 1/n) \simeq 8$ for min-power symmetric biconnectivity. Călinescu and Wan (2003) then showed that Algorithm KV actually has an approximation ratio of at most 4; i.e., the power of the resulted power assignment is within 4 of the power of the (possibly nonsymmetric) best solution of the min-power biconnectivity problem.

Algorithm KV is complicated and may be unsuitable for wireless networks. In the Euclidean case, Călinescu and Wan (2003) proposed a MST-augmentation-based approach (see Algorithm 33). This new algorithm also has a constant approximation ratio on the total power needed by all nodes to ensure biconnectivity; furthermore, it is much faster, simpler, and better suited for distributed implementation. This $O(n \log n)$-time algorithm first constructs a EMST T over all wireless nodes V. Then, at any nonleaf node v of T, a local EMST T_v over all the neighbors of v in T is constructed. The final structure, denoted by Q, is the union of EMST T and T_v for all nonleaf nodes v of the T. Each node is then assigned a power according to Q; i.e., the power assignment is induced by structure Q. Another advantage of this method is its independence of the path-loss exponent β.

Algorithm 33 EMST-based algorithm for biconnectivity with power $\mathbf{w}(uv) = \|uv\|^{\beta}$

1: Construct a EMST T of all wireless nodes V.
2: For each internal node v in T, let $N_T(v)$ be the set of neighbors of v in the tree T. Notice that the size of $N_T(v)$ is at most 6. Let T_v be the EMST of all nodes $N_T(v)$.
3: Let Q be the structure that has all edges of T and T_v for all nodes v.
4: The power assigned to a node v is then defined as $P(v) = \max_{vu \in Q} \mathbf{w}(vu)$.

10.2.4 k-Connectivity

In the literature, there are also a number of studies on a more general k-connectivity requirement for the network structure resulting from a power assignment. Most of the results in the literature concentrated on the edge connectivity.

1. **k-Edge-Connectivity:** A directed graph H is k-edge-connected if for any pair of nodes u and v, there are k-edge-disjoint paths from node u to node v and there are k-edge-disjoint paths from node v to node u.
2. **Symmetric k-Edge-Connectivity:** An undirected graph H is k-edge-connected if for any pair of nodes u and v, there are k-edge-disjoint paths between node u and node v.

The objective function of a power assignment for k-edge-connectivity is often to minimize the total power assigned to all nodes. The Ramanathan and Rosales-Hain heuristic can be applied to the k-edge-connectivity problem: The modification is for checking for k-edge-connectivity of the intermediate network structure instead of biconnectivity. Lloyd *et al.* (2002) also designed algorithms for minimizing the total power for the k-edge-connectivity problem by using the approximation algorithm in Khuller and Vishkin (1994) and proved an approximation ratio of $8(1 - 1/n)$ for min-power symmetric k-edge-connectivity. Călinescu and Wan (2003) further showed that the total power of the structure produced by the Algorithm KV is within $2k$ of the total power needed by the (possibly nonsymmetric) best solution for the k-edge-connectivity problem.

Notice that the previous discussions mainly focused on minimizing the *total* power assigned to all nodes while ensuring that the induced network structure has a certain property such as being (strongly) connected, being (strongly) two-connected, and so on. In certain situations, it is desired to minimize the maximum power assigned to all nodes instead, which could balance the power consumption of nodes to some extent. This set of problems (i.e., minimizing the total power while ensuring that the resulted network has a certain property $\mathbb{P}$) can be solved in polynomial time by use of a simple binary search method when the graph property $\mathbb{P}$ is monotone and can be tested in polynomial time.

DEFINITION 10.1 *A graph property $\mathbb{P}$ of a (directed or undirected) graph is* monotone *if the property continues to hold even when more links are added to the graph.*

For graphs modeling wireless ad hoc networks, adding links means that we need to increase the transmission power of some nodes. Then, for wireless ad hoc networks, we say that a property $\mathbb{P}$ of a (directed or undirected) graph is *monotone* if the property continues to hold when the powers assigned to some nodes are increased, whereas the powers assigned to the other nodes remain unchanged.

The majority of properties studied in the literature are monotone. For example, the k-edge-connected (or k-node-connected) property for undirected graphs is monotone because increasing the powers of some nodes while keeping the powers of other nodes unchanged may add only edges to the graph. However, properties such as *being a tree* or *having a bounded-node degree* are not monotone.

DEFINITION 10.2 *A graph property $\mathbb{P}$ of a (directed or undirected) graph is* polynomial-time testable *if there is a polynomial-time algorithm to test whether any given graph has this property $\mathbb{P}$.*

The majority of properties studied in the literature are polynomial-time testable. For example, the k-edge-connected (or k-node-connected) property for undirected graphs is polynomial-time testable by use of a simple max-network-flow technique. However, properties such as *having at least k-nodes that can transmit simultaneously without interference* or *a k-coloring for all vertices such that no two adjacent nodes get the same color* are not polynomial-time testable.

The following theorem is reported in Lloyd *et al.*, (2002):

THEOREM 10.2 *For any graph property $\mathbb{P}$ that is monotone and can be tested in polynomial time, the problems of minimizing the maximum power of a power assignment P whose induced graph G_P has property $\mathbb{P}$ can be solved in polynomial time.*

Proof. For a network of n nodes, there are at most $n(n-1)$ different values of the power needed to support the communication of all (directed or undirected) links. Let $\mathcal{P}$ be the set of all possible such values. Obviously, the optimum min–max power assignment is from $\mathcal{P}$. For each candidate power value $p \in \mathcal{P}$, the corresponding directed graph, in which each node is assigned a power p, can be constructed in $O(n^2)$ time. Let $T_{\mathbb{P}}(n)$ denote the time needed to test whether property $\mathbb{P}$ holds for a (directed or undirected) graph with n nodes. Thus, the time needed to test whether property $\mathbb{P}$ holds for each candidate solution value is $O(n^2 + T_{\mathbb{P}}(n))$. We can obtain optimal solution by sorting the $O(n^2)$ candidate solution values and using a binary search to determine the smallest value for which property $\mathbb{P}$ holds. It is easy to show that the total running time of this binary-search-based approach is $O((n^2 + T_{\mathbb{P}}(n)) \log n)$. Because $T_{\mathbb{P}}(n)$ is a polynomial, the algorithm runs in polynomial time. ∎

Unfortunately, this simple binary-search-based approach cannot be extended to the case in which the graph property $\mathbb{P}$ is *not* monotone. Lloyd *et al.* (2002) show that when the property $\mathbb{P}$ is "the resulting graph G is a tree," it is NP-complete to decide whether there is a power assignment such that the resulting graph G is a tree.

Notice that assigning *all* nodes the same power achieves a certain property $\mathbb{P}$. It is not difficult to construct examples such that we can reduce the power of some nodes without affecting the property $\mathbb{P}$ in the new network structure (induced by this new power assignment). Then, we may wonder if we can minimize the number of nodes with power equal to the maximum power. Unfortunately, it was shown in Lloyd *et al.* (2002) that for some graph properties $\mathbb{P}$, which are monotone and polynomial-time testable, it is NP-complete to find such an assignment. As an example, they show that it is NP-complete to find a power assignment that will minimize the number of nodes with maximum power while ensuring the graph property "*the diameter of G is at most 6.*"

10.3   Power Assignment for Routing

In previous sections, we reviewed some of the results in the literature concerning the power assignment such that the resulted network is (strongly) k-connected for $k \geq 1$. Connectivity ensures only that for every pair of nodes, there is at least one path connecting them. For wireless ad hoc networks, such a path between this pair of nodes may not be power-efficient compared with the power of the best path in the original communication graph when all nodes transmit at maximum power. In the remainder of the chapter, we concentrate on a power assignment such that the resultant network is power-efficient for routing (i.e., unicast, multicast, or broadcast) with the objective function of minimizing the total power or the maximum power of the power assignment.

10.3.1   Directed Unicast

We first study unicast routing. In this subsection, we assume that the input graph is a directed graph. We need to find a directed path P from a node u to another node v in the graph. Let $v_{i_1} v_{i_2} \cdots v_{i_{h-1}} v_{i_h}$ be such a path where $u = v_{i_1}$ and $v = v_{i_1}$. The total power to support a directed communication from u to v by such a path P is defined as $\sum_{j=1}^{h-1} p(v_{i_{j-1}}, v_{i_j})$, where $p(v_{i_{j-1}}, v_{i_j})$ is the power needed by node $v_{i_{j-1}}$ to support the communication to node v_{i_j}. Notice that the communication is directed: The power by a node $v_{i_{j-1}}$ is not required to be large enough to support its backward communication to node $v_{i_{j-2}}$. Given the power cost to support every possible directed link in the communication graph, the *min-power directed unicast* problem is to find a directed path P from node u to another node v with minimum power. This clearly can be solved by any shortest-path algorithm, such as Dijkstra's shortest-path algorithm, in the original communication graph.

10.3.2   Symmetric Unicast

We now study another variation of the unicast problem in which the power of a node needs to be large enough to support the backward communication; i.e., the power by a node v_{i_j} in a path $P = v_{i_1} v_{i_2} \cdots v_{i_{h-1}} v_{i_h}$ is

$$\max\{p(v_{i_j}, v_{i_{j-1}}), p(v_{i_j}, v_{i_{j+1}})\}$$

for $2 \leq j \leq h - 1$; the power by node v_{i_1} is $p(v_{i_1}, v_{i_2})$, and the power by node v_{i_h} is $p(v_{i_h}, v_{i_{h-1}})$. Given a path P, let $\|P\|_s$ denote the total power of all nodes needed to support both forward and backward communications. In other words,

$$\|P\|_s = \sum_{j=1}^{h} \max\{p(v_{i_j}, v_{i_{j-1}}), p(v_{i_j}, v_{i_{j+1}})\}.$$

The *min-power symmetric unicast* problem is then to find a path P from a node u to another node v with the minimum power $\|P\|_s$. It is not difficult to construct an example to show that the shortest path computed by Dijkstra's algorithm on the directed graph is not always the best path for the *min-power symmetric unicast* problem.

Althaus *et al.* (2006) presented a solution of the min-power symmetric unicast problem that first modifies the given graph $G = (V, E, c)$ and then applies Dijkstra's algorithm to the resultant directed graph H. We now describe the construction of the directed graph $H = (V', E'; c')$ and note that it does not assume that the cost function c is symmetric.

For any node $u \in V$, we sort all its adjacent vertices, say $\{v_1, v_2, \ldots, v_k\}$, in an ascending order of costs of links connecting them to u; i.e., we assume that $c(u, v_i) \leq c(u, v_{i+1})$ after sorting. The vertex u is replaced with a gadget as follows:

1. Each edge (u, v_i) is replaced with two vertices: $[u, v_i]$ and $[v_i, u]$.
2. For each node u, we connect all vertices $[u, v_i]$ by two directed paths:

$$P_1 = (u[u, v_1][u, v_2][u, v_3] \cdots [u, v_{k-1}][u, v_k]),$$
$$P_2 = ([u, v_k][u, v_{k-1}] \cdots [u, v_3][u, v_2][u, v_1]u).$$

3. The cost of all arcs on the path P_2 is zero. The costs of the arcs on path P_1 are $c(u, v_1), c(u, v_2) - c(u, v_1), \ldots, c(u, v_k) - c(u, v_{k-1})$, respectively; i.e., the cost of edge $[u, v_{j-1}][u, v_j]$ is $c(u, v_j) - c(u, v_{j-1})$ for $j \geq 1$.
4. Each edge (u, v) of G is replaced in H with one arc $([u, v], [v, u])$ of cost $c(v, u)$.

Althaus *et al.* (2006) proved that the shortest path from u to v in graph H computed by Dijkstra's algorithm indeed is the optimum path for *min-power symmetric unicast*.

10.3.3   Broadcast and Multicast

We now study other variations of routing, broadcast, and multicast.

1. **Directed broadcast:** We need to find a structure H that contains a directed path from a given node called the root to every other node in the network.
2. **Symmetric broadcast:** We need to find an undirected structure H that contains a path between a given node called the root and every other node in the network.
3. **Directed multicast:** Given a set $\mathbb{R}$ of terminals from V, we need to find a (directed) structure H that contains a directed path from a given node called the root to every other node in $\mathbb{R}$.
4. **Symmetric multicast:** Given a set $\mathbb{R}$ of terminals from V, we need to find an undirected structure H that contains a path between a given node called the root and every other node $\mathbb{R}$.

Min-power broadcast was first studied by Wieselthier *et al.* (2000). They proposed three heuristics but do not prove their approximation ratios. Three greedy heuristics were proposed in Wieselthier *et al.* (2000) for the minimum-energy broadcast routing problem: MST (minimum spanning tree), SPT (shortest-path tree), and BIP (broadcasting incremental power).

The MST heuristic first applies Prim's algorithm to obtain a MST and then orients it as an arborescence rooted at the source node. The power of a node is assigned accordingly: The power needed by a node u is the $\max_{(u,v)} \in \mathrm{MST} p(u, v)$. The SPT heuristic applies the Dijkstra's algorithm to obtain a SPT rooted at the source node. The power needed by a node u is $\max_{(u,v)} \in \mathrm{SPT} p(u, v)$. Notice that here, when assigning power to a node u, we need to consider only links connecting to this children v in a tree.

The BIP heuristic is the node version of Dijkstra's algorithm for the SPT. It maintains, throughout its execution, a single arborescence rooted at the source node. The arborescence starts from the source node, and new nodes are added to the arborescence one at a time on a minimum incremental cost basis until all nodes are included in the arborescence. The incremental cost of adding a new node to the arborescence is the minimum additional power increased by some node in the current arborescence to reach this new node. The implementation of BIP is based on the standard Dijkstra's algorithm, with one fundamental difference in the operation whenever a new node q is added. Whereas Dijkstra's algorithm updates the node weights (representing the currently known distances to the source node), BIP updates the cost of each link (representing the incremental power to reach the head node of the directed link). This update is performed by subtracting the cost of the added link pq from the cost of every link qr that starts from q to node r not in the new arborescence.

These three heuristics have been evaluated through simulations in Wieselthier *et al.* (2000), but little is known about their analytical performances in terms of the approximation ratio. Here, the approximation ratio of a heuristic is the maximum ratio of the energy needed to broadcast a message based on the arborescence generated by this heuristic to the least necessary energy by any arborescence for any set of points. For the Euclidean case in which the cost $p(u, v)$ is defined as $\|uv\|^{\beta}$ for some $\beta \geq 2$, Wan *et al.* (2000) studied the approximation ratios of the preceding three heuristics. An instance was constructed in Wan *et al.* (2000) to show that the approximation ratio of the SPT is as large as $n/2 - o(1)$. On the other hand, both MST and BIP have constant approximation ratios. Let

$$\sigma = \sup_{V \subset \mathcal{D}} \sum_{e \in MST(V)} \|e\|^{\beta},$$

where V is a set of (not infinite) nodes located in a disk $\mathcal{D}$ with radius 1, and MST(V) is the EMST of V. It was proved in Wan *et al.* (2000) that $6 \leq \sigma \leq 12$ for $\beta \geq 2$. In addition, any unidirectional broadcast routing has power of at least $c(\text{MST})/\sigma$, where $c(\text{MST})$ denotes the total link cost of links in the MST. This immediately implies that the approximation ratio of the MST is at most σ. Wan *et al.* (2000) also proved that $p(\text{BIP}) \leq c(\text{MST})$ and, therefore, BIP also has an approximation ratio of at most σ. Wan *et al.* (2000) also gave an example of networks such that the MST heuristic could return a tree whose cost for directional broadcast is six times the optimum, and also an example such that the BIP heuristic will give a tree whose cost for directional broadcast is 1-3/4 times the optimum. It was conjectured that $\sigma = 6$ and, therefore, MST has an approximation ratio of 6.

This conjecture was recently proved by Ambuhl (2005). Recently, Caragiannis *et al.* (2007) designed a new approximation algorithm for the minimum energy broadcast routing (MEBR) problem. Specifically, for any instance where a MST of the set of stations is guaranteed to cost at most ρ times the cost of an optimal solution for MEBR, their algorithm achieves an approximation ratio bounded by $2 \ln \rho - 2 \ln 2 + 2$. This implies a method for MEBR with approximation $2 \ln 3 + 2$ because the approximation ratio of EMST-based broadcast has approximation ratio $\rho = 6$.

When symmetric broadcast is needed (i.e., for a tree T), the power needed by a node v is the maximum power of all links from T incident upon v. We show that the MST heuristic is a 2σ approximation when the cost of a link uv is given by $\|uv\|^{\beta}$ for $\beta \geq 2$. Given a tree T,

- Let $c_B^D(T)$ be the cost of the tree if it was used for directional broadcast.
- Let $c_M^D(T)$ be the cost of the tree if it was used for directional multicast.
- Let $c_B^S(T)$ be the cost of the tree if it was used for symmetric broadcast.
- Let $c_M^S(T)$ be the cost of the tree if it was used for symmetric multicast.

First, given any tree T, we have $c_B^D(T) \leq c_B^S(T)$ and $c_M^D(T) \leq c_M^S(T)$ by definition of cost. Thus, the optimum cost, denoted as OPT_B^S for symmetric broadcast (or multicast), is at least the optimum cost OPT_B^D for directional broadcast (or multicast, respectively). Recall that $\text{OPT}_B^D \geq c(\text{MST})/\sigma$ (Wan *et al.*, 2000). Furthermore, we have $c_B^S(T) \leq 2c(T)$ and $c_M^S(T) \leq 2c(T)$ because each link is counted at most twice in computing $C_B^S(T)$

[and also $C_M^S(T)$]. Thus,

$$c_B^S(\text{MST}) \leq 2c(\text{MST}) \leq 2\sigma\text{OPT}_B^D \leq 2\sigma\text{OPT}_B^S.$$

A similar argument holds for symmetric multicast.

We still assume that the power requirement for a link uv is given by $\|uv\|^\beta$ for some $\beta \geq 2$. For multicast, an α-approximation Steiner tree algorithm gives an approximation ratio of $\alpha\sigma$ for the directional multicast. With the currently best known $\alpha = 1 + (\ln 3/2) + \epsilon$, we obtain an 9.3-approximation algorithm for the *directional multicast* problem and, thus, an 18.59-approximation algorithm for *symmetric multicast* problem.

We now study the case in which the power requirement for a link uv is *not* given by $\|uv\|^\beta$ for some $\beta \geq 2$. We assume that the power requirement is still symmetric. A standard reduction from the set cover in Călinescu *et al.* (2003) shows that no approximation ratio of better than $O(\log n)$ is possible. Algorithms (see, e.g., Călinescu *et al.*, 2003; Cheng *et al.*, 2005) based on some modification of the greedy algorithm for a CDS set have been proposed in the literature for achieving a $\Theta(\log n)$-approximation ratio for directional and/or symmetric multicast problems.

10.4   Further Reading

Research efforts have focused on designing minimum-power-assignment algorithms to save energy for typical network tasks such as broadcast transmission (Clementi *et al.*, 2001a; Huiban and Verhoeven, 2004; Wan *et al.*, 2000; Wieselthier *et al.*, 2000), routing (Srinivas and Modiano, 2003), connectivity (Althaus *et al.*, 2003; Blough *et al.*, 2002; Călinescu *et al.*, 2003; Chen and Huang, 1989; Clementi *et al.*, 2002; Kirousis *et al.*, 2000), and fault tolerance (Călinescu and Wan, 2003; Cheriyan *et al.*, 2002; Hajiaghayi *et al.*, 2003).

Because of the importance of energy efficiency in wireless ad hoc networks, minimum power assignments for different network issues have been addressed recently. Research efforts have focused on finding the minimum power assignment so that the induced communication graph has some "good" properties in terms of network tasks such as disjoint paths, connectivity, or fault tolerance. The minimum-energy-connectivity problem was first studied by Chen and Huang (1989), in which the induced communication graph is strongly connected while the total power assignment is minimized. This problem has been shown by them to be NP-hard. Recently, this problem has been heavily studied, and many approximation algorithms have been proposed when the network is modeled by symmetric links or asymmetric links (Althaus *et al.*, 2003; Blough *et al.*, 2002; Călinescu *et al.*, 2003; Clementi *et al.*, 2002; Kirousis *et al.*, 2000; Ramanathan and Rosales-Hain, 2000). Along this line, several authors (Călinescu and Wan, 2003; Cheriyan *et al.*, 2002; Hajiaghayi *et al.*, 2003) considered the minimum total power assignment while the resulting network is k-strongly-connected or k-connected. This problem has been shown to be NP-hard too. Solving this problem can improve the fault tolerance of the network. Clementi *et al.* (2000a, 2000b, 2000c) also considered the minimum-energy-connectivity problem in which the induced communication graph has a diameter bounded by a constant h. Lloyd *et al.* (2002) proposed one general framework that leads to an approximation algorithm for minimizing total power assignment. Using the framework, they proposed a new two-connected approximation method for power assignment. Krumke *et al.* (2003) also studied the minimum power assignment so that networks satisfy specific properties such as connectivity, bounded diameter, and minimum node degree. Other relevant work in the area of power assignment (also called energy efficiency) includes energy-efficient broadcasting and multicasting in wireless networks. The problem, given a source node s, is to find a minimum power assignment such that the induced communication graph contains a spanning tree rooted at s. This problem was proved to be NP-hard. Clementi *et al.* (2001a), Huiban and Verhoeven (2004), Wan *et al.* (2002b), and Wieselthier *et al.* (2000) presented some heuristic solutions and gave some theoretical analysis. Recently, Srinivas and Modiano (2003) also studied finding k-disjoint paths for a *given* pair of nodes while minimizing the total node power needed by nodes on these k-disjoint paths.

An excellent survey of some recent theoretical advances and open problems on energy consumption in ad hoc networks can be found in Călinescu *et al.* (2004) and Clementi *et al.* (2002). Calinescu *et al.* (2004) provided a tabular survey of the most recent results about power control for various operations. Their survey is reproduced here in Table 10.1. In the table, NPH means NP-Hard and APXH means APX-hard (there is an $\epsilon > 0$ such that the existence of an approximation algorithm with ratio $1 + \epsilon$ implies that $P = NP$). SCH means the problem is as set-cover-hard, which implies that there is no polynomial-time algorithm with approximation ratio $(1 - \epsilon) \ln n$ for any $\epsilon > 0$, unless $P = NP$. DST means that the problem reduces (approximation-preserving) to a directed steiner tree, and DSTH means that the DST reduces (approximation-preserving) to the problem given by the cell. The best-known approximation ratio for a DST is $O(n^\epsilon)$ for any $\epsilon > 0$, and finding a polylogarithmic approximation ratio remains a major open problem in the field of approximation algorithms.

10.5   Conclusion and Remarks

In this chapter, we study a number of power-assignment-related optimization problems. Notice that power assignment can be viewed as a special case of topology control. Notice that given a power assignment P for all nodes, we can define the induced graph G_P that contains a directed edge uv iff the power needed to support link uv is no more than the power assigned to u. On the other hand, given a structure H, which is a subgraph of the original communication graph, we can induce a power assignment as follows: The power assigned to a node u is the minimum power needed to support the communications to *all* its incident links in H. Obviously, given the structure G_P, the power induced from this structure is the same as the original power assignment P. Thus, to find an optimum power assignment P such that the induced graph G_P has a certain property is the same as

Table 10.1 Upper bounds (UB) and lower bounds (LB) on the complexity of power assignments for various problems.

| | Complexity of the min-total-power-assignment problems | | | | | |
| | Asymmetric power requirements | | Symmetric power requirements | | Euclidean $\alpha \geq 2$ | |
Connectivity Reqs	UB	LB	UB	LB	UB	LB
Strong connectivity	$3 + 2\ln(n-1)$ (Călinescu et al., 2003)	SCH (Călinescu et al., 2003)	2 (Chen and Huang, 1989; Kirousis et al., 1997)	APXH*	2	NPH (Clementi et al., 2001b)
Broadcast	$2 + 2\ln(n-1)$ (Călinescu et al., 2003)	SCH	$2 + 2\ln(n-1)$	SCH (Wan et al., 2001)	12 (Wan et al., 2001)	NPH*
Multicast	DST*	DSTH (Călinescu et al., 2003)	$O(\ln n)$ (Caragiannis and Kaklamanis, 2002)	SCH** (Wan et al., 2001)	18.59 (Wan et al., 2001)	NPH*
Symmetric connectivity	$2 + 2\ln(n-1)$ (Călinescu et al., 2003)	SCH (Călinescu et al., 2003)	$\frac{5}{3} + \varepsilon$ (Althaus et al., 2006)	APXH*	$\frac{5}{3} + \varepsilon$ (Althaus et al., 2006)	NPH*
Biconnectivity		NPH*	4 (Călinescu and Wan, 2003)	NPH*	4	
Symmetric biconnectivity		APXH*	4 (Călinescu and Wan, 2003)	APXH*	4	NPH (Călinescu and Wan, 2003)
k-edge-connectivity		NPH*	$2k$ (Călinescu and Wan, 2003)	NPH*	$2k$	
Symmetric k-edge-connectivity		APXH*	$2k$ (Călinescu and Wan, 2003)	APXH*	$2k$	NPH (Călinescu and Wan, 2003)

Notes: Results marked by ⋆ are the folklore results; references followed by ⋆⋆ indicate that the result is implicitly proved in the respective papers. See text for definitions of the abbreviations used.

finding a structure H that has a certain property while its total power $P(H)$ is minimized. See Călinescu *et al.* (2004) for a summarization of the results for power assignment.

There are still a number of open problems that are not addressed in the literature yet. For example, what is the best approximation ratio achievable for min–total power assignment for strong connectivity and symmetric connectivity when the power needed by links are symmetric? Notice that the MST heuristic will give an approximation ratio of 2 to both problems.

Problems

10.1 Construct a network example of n nodes, the power requirement for links, and a power assignment to nodes such that the induced directed communication graph is strongly connected, whereas the induced undirected communication graph only has one link.

10.2 Prove that when the power requirement is asymmetric; i.e., $\mathbf{w}(uv)$ and $\mathbf{w}(vu)$ may be different, finding a power assignment for nodes such that the resultant network is strongly connected and the total power is minimized is NP-hard.

10.3 Conduct simulations to study all the algorithms discussed in this chapter about min–total power assignment for biconnectivity. Which algorithm has the best performance when the nodes are randomly placed in a region and the cost of a link is related to the link length, say, $\mathbf{w}(uv) = c + \|uv\|^{\alpha}$ for $\alpha \geq 2$ and $c > 0$?

10.4 For symmetric unicast, design an efficient algorithm to find the path with minimum power consumption. What is the time complexity of your implementation? Conduct simulations to study the power-consumption difference of the paths found by optimum symmetric unicast and directed unicast.

10.5 Assume that the power requirement for a link uv is $\|uv\|^{\beta}$ for a constant $\beta \geq 2$. Consider a multicast with source node s and a set of receivers $R \subset V$ in a network $G = (V, E)$. Design an efficient algorithm such that the total power consumption for multicast is within a constant factor of the optimum. What is the theoretical performance guarantee of your method? Conduct simulations to study the actual performance of your method when nodes V are randomly placed in a region.

10.6 Assume that the power requirement for a link uv is arbitrary and given by $\mathbf{w}(uv)$. Consider a multicast with source node s and a set of receivers $R \subset V$ in a network $G = (V, E)$. Design an efficient algorithm such that the total power consumption for multicast is within $\Theta(\log n)$ factor of the optimum. Conduct simulations to study the actual performance of your method when nodes V are randomly placed in a region.

10.7 Given a tree T, we showed that $c_B^S(T) \leq 2c(T)$, where $c(T)$ is the total cost of all links in T. Construct an example of a tree T on n nodes such that $\frac{c_B^S(T)}{c(T)}$ is as large as possible. Can you construct an example such that $\frac{c_{\hat{B}}^{S}(T)}{c(T)} = 2 - o(1)$, where $o(1)$ denotes a sufficiently small number?

10.8   We proved that the broadcast based on a MST structure will consume energy at most a constant factor of the optimum when the power needed to support a link uv is $\|uv\|^{\beta}$ for $\beta \geq 2$. What is the worst-case performance ratio of the MST-based broadcast if the power needed to support a link uv is $c + \|uv\|^{\beta}$ for $\beta \geq 2$ and $c > 0$?

10.9   In most studies, we ignore the idle listening cost and overhearing by nodes that are not intended receivers of a transmission. In other words, when a node u transmits data to a neighboring node v, we ignore the cost of all nodes w that are within the carrier-sensing range of node u's transmission. When node w is not the intended receiver, we can use a TDMA schedule to avoid such idle listening and overhearing. Assume that a node v will have cost $c(u, v)$ of receiving a unit amount of data from a neighboring node u. The cost of node u sending data to node v is $\mathbf{w}(u, v)$. Prove that finding the minimum-cost broadcast tree rooted at a node s is NP-hard. Design an efficient algorithm to find a broadcast tree with minimum total cost. What is the worst-cast performance ratio of your method? Conduct simulations to study the performance of your algorithm when the nodes are randomly placed, the transmitting cost $\mathbf{w}(u, v)$ of a link uv is proportional to $\|uv\|^{\beta}$, and the receiving cost $c(u, v)$ is randomly selected from a range $[c_1, c_2]$.

11 Critical Transmission Ranges for Connectivity

11.1 Introduction

Hundreds of protocols (Bose *et al.*, 2001; Chlamtac and Farago, 1999; Das *et al.*, 2000; Johnson and Maltz, 1996; Ko and Vaidya, 1997; X.-Y. Li *et al.*, 2002a; Maltz *et al.*, 1999; Perkins, 1997a; Ramanathan and Steenstrup, 1996; Royer and Toh, 1999; Stojmenovic and Lin, 2001; Y. Wang and Li, 2002b; Zaruba *et al.*, 2001) that take into account the unique characteristics of wireless ad hoc networks have been developed. Among them, energy efficiency, routing, and MAC layer protocols have attracted the most attention. One of the remaining fundamental and critical issues is to have fault-tolerant network deployment without sacrificing the spectrum-reusing property. In other words, the network should support multiple disjoint paths connecting every pair of nodes. Obviously, we can increase the transmission range of all nodes to increase the fault tolerance of the network. However, increasing the transmission range will cause more signal interference (thus reducing the throughput) and increase the power consumption of every node. Because power is a scarce resource in wireless networks, it is important to save the power consumption without losing the network connectivity. The universal minimum power used by all wireless nodes such that the induced network topology is connected is called the *critical power*. Generally, given any network property, we can define the critical value as follows:

DEFINITION 11.1 *Given a monotone graph property* $\mathbb{P}$ *and a set* V *of* n *static nodes, the critical transmission range (CTR)* $\rho(V, \mathbb{P})$ *is the minimum value* r_n *such that* $G(V, r_n)$ *has the property* $\mathbb{P}$. *Here,* $G(V, r_n)$ *is the graph that has an edge* uv *iff the Euclidean distance* $\|uv\| \leq r_n$.

The critical value for a set of static nodes is closely related to the min–max power-assignment problem studied in Chapter 10. When the network property $\mathbb{P}$ is monotone and polynomial-time testable, we can always find the critical value for a set of nodes in polynomial time. Determining the critical power where the wireless needs are statically distributed was studied by several researchers (Gupta and Kumar, 1998; Ramanathan and Rosales-Hain, 2000; Sanchez *et al.*, 1999). Both Ramanathan and Rosales-Hain (2000) and Sanchez *et al.* (1999) use the power assignment induced by the longest incident edge of the EMST over wireless nodes V.

THEOREM 11.1 *Let V be a set of nodes placed in a Euclidean space. The CTR for connectivity of the network is the length of the longest edge of the Euclidean minimum spanning tree, EMST(V).*

It was proved by Penrose (1997) that given a set of points uniformly and randomly distributed in a unit-area square, the longest edge of the MST asymptotically equals the longest edge of the nearest neighbor graph (NNG). Because the nearest neighbor can be found locally, we can determine the critical power asymptotically by using a localized method instead of constructing the MST if the wireless devices are randomly and uniformly distributed in a unit-area square.

Although determining the critical power for static wireless ad hoc networks is well studied (see Chapter 10 for details), it remains to study the critical power for connectivity for wireless networks in which nodes are randomly deployed or for mobile wireless networks. In Chapter 10, we studied a number of algorithms for assigning power to wireless nodes when the static network is already given (node positions are fixed), such that the resultant network is k-connected for $k \geq 1$. Assume that the power requirement is symmetric and uniform; i.e., the power needed to connect any two nodes depends on only their Euclidean distance. Let V be a set of n wireless nodes distributed in a given region and $r_n(V)$ be the minimum transmission range needed to ensure that the network, in which all nodes have a transmission range $r_n(V)$, is connected. Notice that the minimum of all nodes' power to ensure a k-connected network clearly depends on the network instances. In other words, $r_n(V)$ could be different from $r_n(V')$ for two different sets of n nodes V and V'.

Assume that we want to deploy n sensors in an area by airplane. In this case, we cannot control the exact location of each sensor. Assume that the sensors will be distributed randomly in a given area. Then, one fundamental question that we did not address in Chapter 10 is how to set the transmission range of sensors such that the resulted network is connected. Clearly, setting the transmission range to the length of the field will guarantee connectivity, but this will waste the precious power of sensors and introduce too much interference in communications. When setting the range smaller, we can no longer guarantee that the network is always connected for all possible deployment of sensors. Then, an interesting and important question is whether we can set up a transmission range r_n such that the resulted network is connected almost surely (or with at least a certain probability). If so, what will be the minimum transmission range r_n? In this chapter, we address this fundamental question. Notice that *almost sure* events are those that have zero probability of not occurring though it is still possible that they might occur.

As the wireless nodes move around, it is impossible to have a unanimous critical power to guarantee the connectivity for all instances of the network configuration. Thus, we need to find a critical power, if possible, at which each node has to transmit to guarantee the connectivity of the network almost surely; i.e., with high probability sufficiently close to one. For simplicity, we assume that the wireless devices are distributed in a fixed region, typically a unit square (or disk), according to some distribution function (e.g., uniform distribution or Poisson process). Additionally, we assume that the movement of wireless devices still keeps them at the same distribution (uniform or Poisson process) if nodes are mobile. We discuss the case in which this is not true later. Gupta and Kumar (1998) showed that there is a critical power almost surely when the wireless nodes are randomly and uniformly distributed in a unit-area disk. The result by Penrose (1997) implies the same conclusion. Moreover, Penrose (1997) gave the probability that the network is connected if the transmission radius is set as a positive real number r and the number of nodes n goes to infinity.

Let $G(V, r)$ be the graph defined on V with edges $uv \in E$ if and only if $\|uv\| \leq r$. Here, $\|uv\|$ is the Euclidean distance between nodes u and v. Let $\mathcal{G}_\Omega(\mathcal{X}_n, r_n)$ be the set of graphs $G(V, r_n)$ for n nodes V that are uniformly and independently distributed in a two-dimensional (2D) region Ω, which could be a unit-area disk $\mathcal{D}$ or a unit square $\mathcal{C}$ with center at the origin. Generally, we use $\mathcal{G}_\Omega(\mathcal{F}_n, r_n)$ to denote the set of graphs $G(V, r_n)$, where V is any set of n nodes distributed in region Ω according to a probability distribution $\mathcal{F}$ (with density n if needed).

DEFINITION 11.2 *Given a graph property* $\mathbb{P}$*, the critical range* $\rho_\Omega(\mathcal{F}, \mathbb{P})$ *is the minimum value* r_n *such that* $\mathcal{G}_\Omega(\mathcal{F}_n, r_n)$ *has the property* $\mathbb{P}$ *almost surely; i.e.,*

$$\lim_{n \to \infty} \Pr[G \text{ has property } \mathbb{P} \mid G \in \mathcal{G}_\Omega(\mathcal{F}_n, r_n)] = 1.$$

When it is clear from the context, we drop the subscript Ω in the critical range $\rho_\Omega(\mathcal{F}, \mathbb{P})$; i.e., we denote it as $\rho(\mathcal{F}, \mathbb{P})$ for a probability distribution $\mathcal{F}$.

The problem considered by Gupta and Kumar (1998) is then to determine the critical value for connectivity; i.e., the critical value r_n such that a random graph in $\mathcal{G}_\mathcal{D}(\mathcal{X}_n, r_n)$ is asymptotically connected with a probability of one as n goes to infinity. Let $P_{\Omega,k}(\mathcal{X}_n, r_n)$ be the probability that a graph in $\mathcal{G}_\Omega(\mathcal{X}_n, r_n)$ is k-connected. Then Gupta and Kumar (1998) showed that if $n\pi r_n^2 = \ln n + c(n)$, then $P_{\mathcal{D},1}(n, r_n) \to 1$ iff $c(n) \to +\infty$ as n goes to infinity. The result by Penrose (1997) implies a stronger result: If $n\pi r_n^2 = \ln n + \alpha$, then $P_{\mathcal{D},1}(n, r_n) = e^{-e^{-\alpha}}$ as n goes to infinity.

Fault tolerance is one of the central challenges in designing the wireless ad hoc networks. To make fault tolerance possible, first, the underlying network topology must have multiple disjoint paths to connect any two given wireless devices. Here, the path could be vertex disjoint or edge disjoint. We use the vertex-disjoint multiple paths in this chapter, considering the communication nature of the wireless networks. Thus, formally, the graph $G(V, r_n)$ should be k-node-connected for some positive integer $k > 1$. In this chapter, we are interested in what the condition of r_n is such that the underlying network topology $G(V, r_n)$ is k-connected almost surely when V is uniformly and randomly distributed over a 2D domain Ω. For simplicity, we assume that the geometry domain Ω is a unit square $\mathcal{C}$. Gupta and Kumar (1998) basically studied the connectivity problem for $k = 1$ and Ω as a unit-area disk.

It was shown in X.-Y. Li *et al.* (2004b) that given n points randomly distributed in a unit square $\mathcal{C}$, if the transmission range r_n satisfies $n\pi r_n^2 \geq \ln n + (2k - 1)\ln\ln n - 2\ln k! + \alpha + 2\ln\frac{8k}{2^k\sqrt{\pi}}$, then $G(V, r_n)$ is $(k + 1)$-connected with a probability of at least $e^{-e^{-\alpha}}$ as n goes to infinity. Notice that this result is analogous to the corresponding result for Bernoulli graphs $\mathcal{G}(n, p)$; see Bollobás (2001). A similar result was presented by Penrose (1997, 1999a) for the toroidal model instead of the Euclidean model. He showed that the hitting radius r_n such that the graph $G(V, r_n)$ is $(k+1)$-connected satisfies

$$\lim_{n \to \infty} \Pr\left(n\pi r_n^2 \leq \ln n + k \ln \ln n - \ln k! + \alpha\right) = e^{-e^{-\alpha}}.$$

The toroidal metric is used to eliminate boundary effects.

The theoretical value gives us insight on how to set the transmission radius to achieve k-connectivity with certain probability for a network of n devices, or how many devices are needed to achieve the k-connectivity with certain probability when the transmission range of each device is a fixed value. This result also applies to mobile networks when the moving of wireless nodes always generates randomly (or Poisson process) distributed node positions, although this is hard to achieve based on recent studies on some mobile models. This result has applications in the system design of large-scale wireless networks. For example, for setting up a sensor network monitoring a certain region, how many sensors should we deploy to have a multiple connected network, knowing each sensor can transmit to the farthest range r_0? Notice that most of the results presented in this chapter hold only when the number of wireless devices n goes to infinity, which is difficult to deploy practically. Some simulation studies on the transmission radius achieving k-connectivity with a certain probability for practical settings are reported in this chapter. The relation between the minimum node degree and the connectivity of graph $G(V, r)$ is also studied here.

11.2   Preliminaries

Given an event Y, let $\Pr(Y)$ be the probability of Y. We denote the expected value of a random variable X by $E[X]$; i.e., $E[X] = \sum_x x \Pr(X = x)$ for a discrete variable. As is standard, we write log for a base-2 logarithm and ln for a natural logarithm. We say a function $f(n) \to a$ if $\lim_{n \to \infty} f(n) = a$.

11.2.1   Point Process

A point set process is said to be a *uniform random-point process*, denoted by $\mathcal{X}_n$, in a region Ω if it consists of n independent points, each of which is uniformly and randomly distributed over Ω.

The standard probabilistic model of a *homogeneous Poisson process* is characterized by the property that the number of nodes in a region is a random variable depending only on the area (or volume in higher dimensions) of the region. In other words,

1. The probability that there are exactly k-nodes appearing in any region Ψ of area A is $\frac{(\lambda A)^k}{k!} e^{-\lambda A}$.
2. For any region Ψ, the conditional distribution of nodes in Ψ given that exactly k-nodes in the region is *joint uniform*.

Hereafter, we let $\mathcal{P}_n$ be a homogeneous Poisson process of intensity n on the unit square $\mathcal{C} = [-0.5, 0.5] \times [-0.5, 0.5]$.

11.2.2 Connectivity and Minimum Degree

A graph is called k-vertex-connected (k-connected for simplicity) if, for each pair of vertices, there are k mutually vertex-disjoint paths (except end vertices) connecting them. Equivalently, a graph is k-connected if there is no set of $k - 1$ nodes whose removal will partition the network into at least two components. Thus, a k-connected wireless network can sustain the failure of $k - 1$ nodes. A graph is called k-edge-connected if, for each pair of vertices, there are k mutually edge-disjoint paths connecting them. The *vertex connectivity*, denoted by $\kappa(G)$, of a graph G is the maximum k such that G is k-vertex-connected. The *edge connectivity*, denoted by $\xi(G)$, of a graph G is the maximum k such that G is k-edge-connected. The minimum degree of a graph G is denoted by $\delta(G)$, and the maximum degree of a graph G is denoted by $\Delta(G)$. Clearly, for any graph G,

$$\kappa(G) \leq \xi(G) \leq \delta(G) \leq \Delta(G).$$

We omit the symbol G in the preceding notation if it is clear from the context.

A graph property is called *monotone increasing* if G has such a property then all graphs on the same vertex set containing G as a subgraph have this property. Let $\mathcal{Q}$ be any monotone-increasing property of graphs; for example, the connectivity, the k-edge-connectivity, the k-vertex-connectivity, a minimum node degree of at least k, and so on. The *hitting radius* $\varrho(V, \mathcal{Q})$ is the infimum of all r such that graph $G(V, r)$ has property $\mathcal{Q}$. For example, $\varrho(V, \kappa \geq k)$ is the minimum radius r such that $G(V, r)$ is at least k-vertex-connected; $\varrho(V, \delta \geq k)$ is the minimum radius r at which the graph $G(V, r)$ has a minimum degree of at least k. It is obvious that for any V,

$$\varrho(V, \kappa \geq k) \geq \varrho(V, \delta \geq k).$$

Penrose (1999a) showed that these two hitting radii are asymptotically the same for n points V randomly and uniformly distributed in a unit square and n goes to infinity.

11.3 Critical Range for Connectivity

The CTR for connectivity in dense networks (in which the number of nodes n in a unit area is sufficiently large) can be characterized by use of results from a recent applied probability theory on *geometric random graphs* (GRGs). Notice that for a set of nodes, the CTR for connectivity is always the length of the longest edge of the EMST of this set of nodes. Consequently, studying the CTR for connectivity is equivalent to the study of the longest edge of the EMST of a set V of nodes when V follows a certain distribution such as Poisson distribution or random uniform distribution. The following theorem was proved in Penrose (1997).

THEOREM 11.2 *Assume* n *points are distributed uniformly at random in the 2D unit square and let* M_n *be the random variable denoting the length of the longest edge of EMST built on this set of* n *nodes. Then,*

$$\lim_{n\to\infty} \Pr\left(n\pi M_n^2 - \log n \le \beta\right) = \frac{1}{e^{e^{-\beta}}}$$

for any real number β.

COROLLARY 11.1 *Assume that* n *points* V *are distributed uniformly at random in the 2D unit square. Given the transmission range* r_n, *assume that a function* f(n) *satisfies*

$$r_n = \sqrt{\frac{\log n + f(n)}{n\pi}}.$$

Then, the network $G(V, r_n)$ *is connected almost surely if* $\lim_{n\to\infty} f(n) = +\infty$. *The network* $G(V, r_n)$ *is not connected almost surely if* $\lim_{n\to\infty} f(n) = -\infty$.

Proof. We prove this as follows. From Theorems 11.1 and 11.2, we know that given that $r_n = \sqrt{\frac{\log n + f(n)}{n\pi}}$, the probability that the network is connected is $\frac{1}{e^{e^{-f(n)}}}$. When $\lim_{n\to\infty} f(n) = \infty$, this implies that $\lim_{n\to\infty} e^{-f(n)} = 0$, which implies that $\lim_{n\to\infty} e^{e^{-f(n)}} = 1$. This proves the first part.

When $\lim_{n\to\infty} f(n) = -\infty$, this implies that $\lim_{n\to\infty} e^{-f(n)} = \infty$, which implies that $\lim_{n\to\infty} e^{e^{-f(n)}} = \infty$. This further implies that $\lim_{n\to\infty} \frac{1}{e^{e^{-f(n)}}} = 0$. This proves the second part. ∎

For example, when $f(n) = \log n$, the network is connected with high probability (which is $1/e^{1/n}$ when n is sufficiently large). When $f(n) = \log \log n$, the network is connected with high probability (which is $1/e^{1/\log n}$ when n is sufficiently large). When $f(n) = -\log \log n$, the network is connected only with a probability of about $1/e^{\log n} = 1/n$ when n is sufficiently large.

For some applications, the wireless nodes are not distributed in a two-dimensional space; instead, they are distributed in a line (e.g., vehicular ad hoc networks). In the case of such one-dimensional (1D) networks (in which nodes are placed along a line), we can characterize the CTR by combining Theorem 1 of Holst (1980), Theorem 2 of Penrose (1997), and Theorem 2 of Penrose (1999b); thus, we have the following theorem:

THEOREM 11.3 *Assume that* n *nodes are distributed uniformly at random in a segment of unit length. The CTR for connectivity (and the longest edge of EMST) is*

$$r_n = \frac{\log n + f(n)}{n},$$

where f(n) *is an arbitrary function such that* $\lim_{n\to\infty} f(n) = +\infty$.

It is also not difficult to prove the following theorem:

THEOREM 11.4 *Assume that a set of nodes is distributed in a segment of unit length according to the Poisson distribution with density* n. *The CTR for connectivity (and the longest edge of EMST) is*

$$r_n = \frac{\log n + f(n)}{n},$$

where f(n) *is an arbitrary function such that* $\lim_{n \to \infty} f(n) = +\infty$.

On the other hand, for some networks, the nodes are distributed in three-dimensional (3D) space. In this case, we can derive the CTR for connectivity for a 3D network by combining Theorem 1.4 of Dette and Henze (1989) and Theorem 1.1 of Penrose (1999a).

THEOREM 11.5 *Assume that* n *nodes are distributed uniformly at random in a 3D cube of unit length. The CTR for connectivity (and the longest edge of the EMST) is*

$$r_n = \sqrt[3]{\frac{\log n - \log \log n}{\pi n} + \frac{3}{2} \frac{1.41 + f(n)}{\pi n}},$$

where f(n) *is an arbitrary function such that* $\lim_{n \to \infty} f(n) = +\infty$.

Notice that unlike that of the 2D networks, the formula for the CTR in a 3D network contains an additional asymptotical term, $\log \log n$. It is observed in Dette and Henze (1989) that this term is due to the boundary effect, in which a disk with radius r_n near the boundary could have only a fraction (i.e., one quarter for 2D input and one eighth for 3D input) of it inside the deployment region. This boundary effect turns out to be negligible in the 2D case, whereas it is not negligible for a network in a d-dimensional case when $d \geq 3$.

Notice that when the nodes are distributed according to the Poisson point process, we have similar results for the CTR for connectivity. The proof given for the result in Gupta and Kumar (1998) is somewhat laborious, involving techniques from continuous-percolation theory. To facilitate an understanding of the underlying behavior, Krishna-machari *et al.* (2002) present a simpler 1D model that yields an analogous result. The model works as follows: We have nodes distributed with a Poisson arrival rate λ on R^+. Without loss of generality, let us assume that the first node is located at 0. Then, the question to be answered is this: What is the probability $P_n(R)$ that the first n nodes form a connected network, in which each node may communicate with any other node that is within the communication distance R? Let x_i be the location of the ith node. Then, obviously, we have

$$P_n(R) = \Pr\left[\bigcap_{i=1}^{n-1}(x_{i+1} - x_i \leq R)\right].$$

Because the interarrival distance of a Poisson sequence is i.i.d. exponentially distributed and, hence, memoryless, we know that $P_n(R) = \prod_{i=1}^{n-1}(1 - e^{-\lambda R}) = (1 - e^{-\lambda R})^{n-1}$. For example, when we set $R = \frac{\log n + f(n)}{n}$ for a certain fixed $f(n)$, we have

$$P_n(R) = (1 - e^{-\lambda R})^{n-1} = \left(1 - \frac{e^{-f(n)}}{n}\right)^{\frac{n-1}{e^{-f(n)}}e^{-f(n)}} \simeq e^{-e^{-f(n)}}.$$

Note that the condition $\lambda R_c = \log(n) + f(n)$ is a condition in which each node has $\log(n) + f(n)$ neighbors to its right; hence, $2\log(n) + 2f(n)$ neighbors on average.

11.4   Critical Range for *k*-Connectivity

In this section, we concentrate on the hitting radius for the *k*-connectivity for *n* randomly and uniformly distributed points in a unit-area square $\mathcal{C}$. We build a result based on the result by Penrose (1999a).

For convenience, instead of the random-point process $\mathcal{X}_n$, we consider a homogeneous Poisson point process of rate *n*, denoted by $\mathcal{P}_n$, on a unit-area square $\mathcal{C}$. As in Penrose (1999a), here we let $\mathcal{E}(k, n, r)$ denote the expected number of points of $\mathcal{P}_n$ with degree *k* in a graph of $G(\mathcal{P}_n, r)$. Let $D(\mathbf{x}, r)$ be the disk centered at $\mathbf{x}$ with radius *r*. Given a point $\mathbf{x}$, let $v_r(\mathbf{x})$ be the area of the intersection of $D(\mathbf{x}, r)$ with the unit-area square $\mathcal{C}$. Additionally, let

$$\phi_{n,r,k}(\mathbf{x}) = [n \cdot v_r(\mathbf{x})]^k \frac{e^{-n \cdot v_r(\mathbf{x})}}{k!}.$$

Here, $\phi_{n,r,k}(\mathbf{x})$ is the probability that point $\mathbf{x}$ has degree *k*. It was known from Penrose (1999a) that

$$\mathcal{E}(k, n, r) = n \int_C \phi_{n,r,k}(\mathbf{x}) d\mathbf{x}.$$

Then, Penrose (1999a) (Theorem 1.2) proved the following theorem:

THEOREM 11.6 *Let* α *be any real number. Given any metric* l_p *on* $\mathcal{C}$ *with* $1 < p \leq \infty$ *and any integer* $k \geq 0$*, and* r_n *satisfying the following condition*

$$\lim_{n \to \infty} \mathcal{E}(k, n, r_n) = e^{-\alpha},$$

then we have

$$\lim_{n \to \infty} \Pr(\varrho(\mathcal{P}_n, \delta \geq k + 1) \leq r_n) = e^{-e^{-\alpha}}.$$

Notice that the same theorem is true when the random-point process $\mathcal{P}_n$ is used instead of the homogeneous Poisson point process. The remainder of this section is devoted to estimating the value r_n. Penrose (1999a) agreed that r_n *is not so easy to find because of the dominance of complicated boundary effects.* The estimated radius r_n also makes the graph $G(\mathcal{P}_n, r_n)$ *k*-connected with probability $e^{-e^{-\alpha}}$ when *n* goes to infinity because Penrose (1999a) proved that it is almost sure that $\varrho(\mathcal{X}_n, \kappa \geq k) = \varrho(\mathcal{X}_n, \delta \geq k)$ and $\varrho(\mathcal{P}_n, \kappa \geq k) = \varrho(\mathcal{P}_n, \delta \geq k)$ as *n* goes to infinity.

11.4.1   Lower Bound

We first study the asymptotic lower bound for the hitting radius r_n for the $(k + 1)$-connectivity.

Obviously, $v_r(\mathbf{x}) \leq \pi r^2$ for any point $\mathbf{x}$ inside the unit-area square $\mathcal{C}$. Because $\phi_{n,r,k}(\mathbf{x})$ is a monotone-increasing function of $v_r(\mathbf{x})$, we have

$$\phi_{n,r,k}(\mathbf{x}) = [n v_r(\mathbf{x})]^k \frac{e^{-n \cdot v_r(\mathbf{x})}}{k!} < (n \pi r^2)^k \frac{e^{-n \cdot \pi r^2}}{k!}.$$

We then bound $\mathcal{E}(k, n, r)$ as follows:

$$\mathcal{E}(k, n, r) = n \int_C \phi_{n,r,k}(\mathbf{x}) d\mathbf{x} < n(n\pi r^2)^k \frac{e^{-n\pi r^2}}{k!}.$$

Notice that if we use πr^2 for $v_r(\mathbf{x})$ instead of the actual area $v_r(\mathbf{x})$, the computed radius r is less than the actual required radius. This is because $v_r(\mathbf{x}) < \pi r^2$ for point $\mathbf{x}$ near the boundary of the square. Thus, the probability that there are at least k-neighbors within distance r of point $\mathbf{x}$ is increased when we use πr^2 for $v_r(\mathbf{x})$ for point $\mathbf{x}$ near the boundary. To remedy the approximated area πr^2, the actual value r should be larger than the computed one.

We estimate r when $v_r(\mathbf{x}) = \pi r^2$ is used as the area measurement. Let $y = \pi r^2$. From $\lim_{n\to\infty} \mathcal{E}(k, n, r_n) = e^{-\alpha}$, we have $e^{-\alpha} = \lim_{n\to\infty} n(ny)^k \frac{e^{-n \cdot y}}{k!}$. We relax the condition by ignoring the condition of n going to infinity. In other words, we consider that

$$e^{-\alpha} = n(ny)^k \frac{e^{-ny}}{k!}.$$

It implies that by taking ln on both sides,

$$-\alpha = \ln n + k \ln n + k \ln y - ny - \ln(k!).$$

Thus,

$$-k \ln y + ny = (k + 1) \ln n - \ln(k!) + \alpha.$$

Dividing both sides by k, we have

$$\frac{n}{k} y - \ln y = \frac{k + 1}{k} \ln n - \frac{1}{k} \ln(k!) + \frac{\alpha}{k}.$$

Let $z = \frac{n}{k} y$. Then, $\ln y = \ln z + \ln k - \ln n$. Then

$$z - \ln z = \ln k - \ln n + \frac{k + 1}{k} \ln n - \frac{1}{k} \ln(k!) + \frac{\alpha}{k}$$

$$= \frac{1}{k} \ln n + \ln k - \frac{1}{k} \ln(k!) + \frac{\alpha}{k}.$$

Notice that if $z = \ln z + t$, then $z > t + \ln t$, where $t > 0$. Then, we have

$$z > \frac{1}{k} \ln n + \ln k - \frac{1}{k} \ln(k!) + \frac{\alpha}{k} + \ln \left[\frac{1}{k} \ln n + \ln k - \frac{1}{k} \ln(k!) + \frac{\alpha}{k} \right]$$

$$> \frac{1}{k} \ln n + \ln k - \frac{1}{k} \ln(k!) + \frac{\alpha}{k} + \ln \left(\frac{1}{k} \ln n \right).$$

Consequently, by substituting back $z = \frac{n}{k} \pi r^2$, we have

$$\frac{n}{k} \pi r^2 > \frac{\ln n}{k} + \ln k - \frac{1}{k} \ln(k!) + \frac{\alpha}{k} - \ln k + \ln \ln n,$$

which implies that

$$n\pi r^2 > \ln n + k \ln \ln n - \ln k! + \alpha.$$

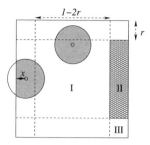

Figure 11.1 The area $v_r(\mathbf{x})$ for a point $\mathbf{x}$. © 2004, ACM.

Notice that the function $(ny)^k \frac{e^{-n \cdot y}}{k!}$ achieves the maximum value when $y = k/n$. It is monotone decreasing for $y > k/n$ and monotone increasing for $y < k/n$. We always assume that k is a fixed constant throughout this chapter. Then, we have the following theorem:

THEOREM 11.7 *Let n wireless nodes V be randomly and uniformly distributed in a unit-area square. If we want the graph $G(V, r_n)$ to be $(k + 1)$-connected with a probability of at least $e^{-e^{-\alpha}}$, the transmission radius r_n satisfies*

$$n \pi r^2 > \ln n + k \ln \ln n - \ln k! + \alpha. \qquad (11.1)$$

Notice that for the toroidal model, Penrose (1999a) gave the same bound for r_n such that the graph is guaranteed to be $(k + 1)$-connected asymptotically. Moreover, the results of Gupta and Kumar (1998) and Penrose (1997) are just special cases in which $k = 0$, if this bound is tight. Notice that in our analysis, we implicitly assume that $k > 0$. Additionally, we could improve the lower bound of our analysis by considering a more tight area estimation for point $\mathbf{x}$ near the boundary of the square, but the analysis would be much more complicated.

11.4.2 Upper Bound

We showed that if we want the network $G(V, r_n)$ to be $(k + 1)$-connected with probability at least $e^{-e^{-\alpha}}$, we have to set the transmission radius r_n to satisfy inequality (11.1) for n points randomly and uniformly distributed in a unit-area square. In this subsection, we continue to study the upper bound of the transmission radius to achieve the same $(k + 1)$-connectivity. The estimated upper bound is different from the lower bound even asymptotically. Again, we derive the upper bound from the equation $n \int_C \phi_{n,r,k}(\mathbf{x}) d\mathbf{x} = e^{-\alpha}$.

We partition the unit square to three regions: region I is $[-0.5 + r, 0.5 - r] \times [-0.5 + r, 0.5 - r]$, region III is the four corners, and the remaining area is region II (see Figure 11.1). We compute the area $v_r(\mathbf{x})$ for point $\mathbf{x}$ located in these three regions separately. Obviously, for any $\mathbf{x}$ in region I, $v_r(\mathbf{x}) = \pi r^2$. For a point $\mathbf{x}$ in region II, assume that its distance to the boundary of C is x; then, the area is

$$v_r(\mathbf{x}) = \pi r^2 - r^2 \cos^{-1}\left(\frac{x}{r}\right) + x\sqrt{r^2 - x^2}.$$

Here, $0 \le x \le r$. Assume that $x = r \cos \theta$, where $0 \le \theta \le \pi/2$. Then, $v_r(\mathbf{x}) = r^2(\pi - \theta + \sin \theta \cos \theta)$. It is easy to show that

$$\frac{\pi r^2}{2}(1 + \cos \theta) \le r^2(\pi - \theta + \sin \theta \cos \theta) \le \frac{\pi r^2}{2} + 2r^2 \cos \theta.$$

By substituting $x = r \cos \theta$, we bound $v_r(\mathbf{x})$ as follows:

$$\frac{\pi r^2}{2} + \frac{\pi r}{2}x \le v_r(\mathbf{x}) \le \frac{\pi r^2}{2} + 2rx.$$

Let $r^\star$ be the solution of $n \int_{\mathcal{C}} \phi_{n,r,k}(\mathbf{x})d\mathbf{x} = e^{-\alpha}$. Let Ω be any subregion of $\mathcal{C}$. Let $w(\mathbf{x})$ be any function such that $w(\mathbf{x}) \le v(\mathbf{x})$ and is monotone increasing of r. Let $\varphi_{n,r,k}(\mathbf{x}) = [nw(\mathbf{x})]^k \frac{e^{-n \cdot w(\mathbf{x})}}{k!}$. Thus, $\varphi_{n,r,k}(\mathbf{x}) \le \phi_{n,r,k}(\mathbf{x})$. Let r' be the solution of $n \int_{\Omega} \varphi_{n,r,k}(\mathbf{x})d\mathbf{x} = e^{-\alpha}$. Then, $r^\star \le r'$. This is because $w(x)$, $v_r(x)$ are monotone-increasing functions of r, and $(ny)^k \frac{e^{-ny}}{k!}$ is a monotone-increasing function when $y \le k/n$. Thus, to bound the transmission radius r from above so that the graph $G(V,r)$ is $(k+1)$-connected, we use the lower bound of $v_r(\mathbf{x})$ and we also compute the integral only for regions I and II. Notice that

$$\int_{\mathcal{C}} (nv_r(\mathbf{x}))^k \frac{e^{-nv_r(\mathbf{x})}}{k!}d\mathbf{x} > \int_{I} (nv_r(\mathbf{x}))^k \frac{e^{-nv_r(\mathbf{x})}}{k!}d\mathbf{x} + \int_{II} (nv_r(\mathbf{x}))^k \frac{e^{-nv_r(\mathbf{x})}}{k!}d\mathbf{x}.$$

Obviously, for region I, we have

$$\int_{I} [nv_r(\mathbf{x})]^k \frac{e^{-nv_r(\mathbf{x})}}{k!}d\mathbf{x} = (n\pi r^2)^k \frac{e^{-n\pi r^2}}{k!}(1 - 2r)^2.$$

The integral over region II is four times the integral over the rectangular region near the boundary, where the length of the rectangle is $1 - 2r$ and the width is r. Assume that the distance of a point $\mathbf{x}$ to the boundary is x. Notice that $v_r(\mathbf{x}) > \frac{\pi r^2}{2} + \frac{\pi r}{2}x$. Let $y = \frac{\pi r^2}{2} + \frac{\pi r}{2}x$. We have

$$\int_{II} [nv_r(\mathbf{x})]^k \frac{e^{-nv_r(\mathbf{x})}}{k!}d\mathbf{x} = 4(1 - 2r) \int_{x=0}^{r} [nv_r(x)]^k \frac{e^{-nv_r(x)}}{k!}dx$$

$$> \frac{8(1 - 2r)}{\pi k! r} \int_{y=\frac{\pi r^2}{2}}^{\pi r^2} (ny)^k e^{-ny}dy = \frac{8(1 - 2r)}{n\pi r} \left(e^{-t/2} \sum_{j=0}^{k} \frac{t^j}{j!2^j} - e^{-t} \sum_{j=0}^{k} \frac{t^j}{j!} \right).$$

Here, $t = n\pi r^2$. The last equation comes from $\int z^k e^{-z}dz = -e^{-z}k! \sum_{j=0}^{k} \frac{z^j}{j!}$. Then, the transmission radius $\varrho(\mathcal{P}_n, \kappa \ge k)$ is bounded from above by the solution of the following equation:

$$e^{-\alpha} = nt^k \frac{e^{-t}}{k!}(1 - 2r)^2 + \frac{8(1 - 2r)}{\pi r} \left[e^{-t/2} \sum_{j=0}^{k} \frac{(t/2)^j}{j!} - e^{-t} \sum_{j=0}^{k} \frac{t^j}{j!} \right]$$

$$< nt^k \frac{e^{-t}}{k!} + \frac{8}{\pi r} k e^{-t/2} \frac{(t/2)^k}{k!}.$$

The inequality comes from $e^{-t/2} \frac{(t/2)^j}{j!} < e^{-t/2} \frac{(t/2)^{j+1}}{(j+1)!}$ for $j < t/2$. Here, we assume that $k < t/2$. Remember that here, $t = n\pi r^2 \ge \ln n$ asymptotically from the lower-bound analysis. By a sequence of estimations, the following theorem was proved in X.-Y. Li *et al.* (2003b, 2004b):

THEOREM 11.8 *Let* n *wireless nodes* V *be randomly and uniformly distributed in a unit-area square. If we set the transmission radius* r_n *to satisfy*

$$n\pi r^2 > \ln n + (2k - 1) \ln \ln n - 2 \ln k! + 2\alpha + 2 \ln \frac{8k}{2^k \sqrt{\pi}},$$

then the graph $G(V, r_n)$ *is* $(k + 1)$*-connected with a probability of at least* $e^{-e^{-\alpha}}$ *when* n *goes to infinity.*

Obviously, if $\alpha \to \infty$, then $e^{-e^{-\alpha}} \to 1$. For example, if we set $\alpha = \ln \ln n$ (i.e., we want the graph $G(V, r_n)$ to be $(k + 1)$-connected with a probability of at least $e^{-1/\ln n} > 1 - \frac{1}{\ln n}$), we have to set the transmission radius r_n to satisfy

$$n\pi r^2 > \ln n + (2k + 1) \ln \ln n - 2 \ln k! + 2 \ln \frac{8k}{2^k \sqrt{\pi}}.$$

If we want the graph $G(V, r_n)$ to be $(k + 1)$-connected with a probability of at least $e^{-1/n} > 1 - \frac{1}{n}$, we have to set the transmission radius r_n to satisfy

$$n\pi r^2 > 3 \ln n + (2k - 1) \ln \ln n - 2 \ln k! + 2 \ln \frac{8k}{2^k \sqrt{\pi}}.$$

Additionally, if $\alpha \to -\infty$, then $e^{-e^{-\alpha}} \to 0$. It implies that the graph $G(V, r_n)$ will be $(k + 1)$-connected with a very low probability if this bound of the hitting radius is tight.

Notice that the preceding analysis of the asymptotic upper bound of the transmission radius can also be used to derive a tighter lower bound on the transmission radius. We use the fact that $\frac{\pi r^2}{2} + \frac{\pi r}{2} x \leq v_r(\mathbf{x})$ to derive the upper bound of the transmission radius. To analyze the lower bound, we have to use the fact that $v_r(\mathbf{x}) \leq \frac{\pi r^2}{2} + 2rx$ to estimate the area $v_r(\mathbf{x})$ for point $\mathbf{x}$ near the boundary. In addition, we have to compute the integral in all three regions. To simplify the analysis, for point $\mathbf{x}$ in region III, we also use $v_r(\mathbf{x}) \leq (\pi r^2/2) + 2rx$ to estimate the area $v_r(\mathbf{x})$. Then, similar to the preceding analysis of the upper bound, the lower bound on t is at least the solution of the following equation:

$$e^{-\alpha} = nt^k \frac{e^{-t}}{k!} (1 - 2r)^2 + \frac{2}{r} \left(e^{-\frac{t}{2}} \sum_{j=0}^{k} \frac{t^j}{2^j j!} - e^{-t} \sum_{j=0}^{k} \frac{t^j}{j!} \right).$$

By tedious computing, we can compute the asymptotic lower bound as

$$t > \ln n + (2k - 1) \ln \ln n - 2 \ln k! + 2\alpha.$$

Remark: Although we have computed the lower and upper bounds for the transmission range r_n such that the graph $G(V, r_n)$ is $(k + 1)$-connected with a probability of at least $e^{-e^{-\alpha}}$, these bounds hold only when n goes to infinity and k is assumed to be a constant. When n is a practical finite number (especially when n is comparable with $k!$), the bounds do not hold anymore. This observation is witnessed by our experimental results.

Recently, Wan and Yi (2004) and Yi (2005) closed the gap between the upper bound and the lower bound for the critical range for k-connectivity in a 2D network. Wan and Yi (2004) gave an exact formula for the probability of connectivity when $n \to \infty$, as follows:

THEOREM 11.9 *Assume that* n *nodes are distributed uniformly at random in a unit square (or a disk of unit area). For* k ≥ 2, *if*

$$r_n = \sqrt{\frac{\log n + (2k - 3)\log\log n + \zeta}{\pi n}},$$

where

$$\zeta = \begin{cases} -2\log\left(\sqrt{e^{(-\beta + \pi^2/16)} - \pi/4}\right) & \text{if } k = 2 \\ 2\log\frac{\pi}{2^k k!} + 2\beta & \text{if } k > 2 \end{cases},$$

then the probabilities of the two events $\rho(\mathcal{X}_n, \delta \geq k)$ *and* $\rho(\mathcal{X}_n, \kappa \geq k)$ *both converge to* $e^{-e^{-\beta}}$ *as* $n \to \infty$.

Notice that here, δ denotes the minimum nodal degree of the graph $G(V, r_n)$ and κ denotes the connectivity of the graph $G(V, r_n)$. When the deployment region is a unit-area square, the preceding theorem still holds but with a different choice of ζ:

$$\zeta = \begin{cases} -2\log\left(\sqrt{e^{(-\beta + \pi/4)} - \sqrt{\pi}/2}\right) & \text{if } k = 2 \\ 2\log\frac{\sqrt{\pi}}{2^{k-1} k!} + 2\beta & \text{if } k > 2 \end{cases}.$$

From the preceding theorem, we can have the following theorem:

THEOREM 11.10 *Assume that* n *nodes are distributed uniformly at random in a unit square (or a disk of unit area). The CTR for* k-*connectivity for any constant* 1 < k < n − 1 *is*

$$r_n = \sqrt{\frac{\log n + (2k - 3)\log\log n + f(n)}{\pi n}},$$

where f(n) *is an arbitrary function such that* $\lim_{n \to \infty} f(n) = +\infty$.

Comparing this formula with the formula for connectivity, we can see that the only difference is the second-order term $(2k - 3)\log\log n$. This means that, asymptotically, k-connectivity with $k > 1$ can be achieved by slightly increasing the transmission range for one-connectivity.

11.5   Connectivity with Bernoulli Nodes

So far, we assume that all wireless nodes will function properly during the course of networking. In certain scenarios, wireless nodes may be faulty with a certain probability. We now study what the CTR is such that the network is k-connected almost surely even the nodes could be faulty with a certain probability $1 - p$. We model this by using Bernoulli nodes: At any time instance, any node in the network is active with a constant probability of $1 \geq p > 0$. We always assume that the node activations are independent events. Again, we denote $G(V, r_n)$ as the communication graph we obtain by setting the transmission range of each node to r_n, and two nodes in V are connected by an edge iff their Euclidean distance is at most r_n, regardless of their status of being active or inactive. We denote $G(V, r_n, p)$ as the subgraph of $G(V, r_n)$ induced by all active nodes in V when a node in V is active with a probability p. Notice that to get $G(V, r_n, p)$, all inactive nodes (and all edges incident to any inactive nodes) will be removed from $G(V, r_n)$. We denote $B(V, r_n, p)$ as the subgraph of $G(V, r_n)$ by removing all links that have at least one inactive end node. Notice that inactive nodes are not even removed in $B(V, r_n, p)$, they are not connected to any active nodes. Structures $G(V, r_n, p)$ and $B(V, r_n, p)$ have different interpretations based on the applications to be used. Recently, there were some studies that focused on the asymptotical behaviors of these two structures, such as the critical range for connectivity and the giant component.

As stated in Wan and Yi (2004) and Yi (2005), these structures can be used to model several network design problems:

1. *Fault-tolerance*: In practice, when we deploy a wireless ad hoc network, we cannot expect all nodes to function perfectly correctly all the time. Because of either internal breakdown or a harsh environment, a node may fail in some situations. The failure nodes will not take part in routing/relaying and thus may affect the connectivity of the network formed by the "good" nodes. Whether a node fails depends on a number of factors. To simplify the analysis, we can assume that a node will fail with some fixed probability p. Thus, if all nodes are modeled with the Bernoulli model with parameter p, the remaining nodes in a network can still form a connected network; i.e., they can tolerate such random failures of other nodes if and only if the graph $G(V, r_n, p)$ is connected.

2. *Randomized construction of virtual backbone*: Backbone has been used in a number of operations for wireless ad hoc networks such as routing, data collection, and node localization. It is often required that the backbone be connected; i.e., the graph induced by all nodes in the backbone is connected and all nodes in the network are connected to at least one node in the backbone. A virtual backbone is a CDS of the network topology. A node is said to be a dominator if it belongs to the virtual backbone and a dominatee otherwise. A number of methods have been proposed for constructing backbones (or CDS); see Chapters 7 and 8 for more details. In addition to these deterministic backbone-construction methods, some backbone-construction methods will randomly select a node to be on the backbone. In a randomized construction of a virtual backbone, each node volunteers to be a dominator independently and uniformly with constant probability p. By modeling all nodes with the Bernoulli model with parameter p, the dominators form a virtual backbone if and only if the graph $B(V, r_n, p)$ is connected.

3. *Randomized broadcast routing*: A broadcast has vast applications in wireless ad hoc networks such as route maintenance, data gathering, and so on. A broadcast sends given information from a given root node to all nodes in the network. Chapter 14 discusses in detail some of the broadcast algorithms and structures used for broadcast. In

randomized broadcast routing, each node will determine whether to relay a broadcast message by probability. If every node volunteers to relay the message independently and uniformly with constant probability p, we can model this as the Bernoulli model with parameter p. In this scenario, the broadcast message can reach all nodes if and only if the graph $B(V, r_n, p)$ is connected.

4. *Randomized wake/sleep management*: Energy consumption is always an important issue for WSNs. A common, widely used approach to conserve energy consumption is to let nodes periodically (or randomly) switch between the states of wake-up and sleep. Although using a deterministic periodical schedule of wakeup/sleep may require a fewer number of nodes to be active for certain services such as coverage or connectivity of the network, it is often difficult to maintain this schedule in a dynamic environment. For the purpose of energy conservation and easy maintenance, a randomized awake/sleep management lets each node wake up independently and uniformly with some probability. When this probability is uniform for all nodes (i.e., with some constant probability p), we again can model this by using the Bernoulli model. To maintain the network connectivity, all awake nodes should form a connected graph and, in some application scenarios, every asleep node is also required to be adjacent to at least one awake node. By modeling all nodes with the Bernoulli model with parameter p, the network connectivity is maintained if and only if the graph $B(V, r_n, p)$ is connected.

The following theorems have been proved by Wan and Yi (2004) and Yi (2005):

THEOREM 11.11 *Assume that* n *nodes are distributed uniformly at random in the disk of the unit area. Let*

$$r_n(\zeta) = \sqrt{\frac{\log n + \zeta}{\pi p n}}$$

for some constant ζ. *Let* ρ_G *(and, respectively,* ρ_B*) be the minimum transmission range such that the graph* $G(V, r_n, p)$ *[respectively,* $B(V, r_n, p)$*] is connected. Then,*

$$\lim_{n \to \infty} \Pr[\rho_G \leq r_n(\zeta)] = e^{-pe^{-\zeta}},$$

$$\lim_{n \to \infty} \Pr[\rho_B \leq r_n(\zeta)] = e^{-e^{-\zeta}}.$$

From this, it is easy to derive the following corollary:

COROLLARY 11.2 *Assume that* n *nodes* V *are distributed uniformly at random in the disk of unit area and assume that the nodes are active with a constant and independent probability* p *with* $0 < p \leq 1$. *The CTRs for connectivity of graph* $G(V, r_n, p)$ *and graph* $B(V, r_n, p)$ *are the same and both equal*

$$r_n = \sqrt{\frac{\log n + f(n)}{p \pi n}},$$

where f(n) *is an arbitrary function such that* $\lim_{n \to \infty} f(n) = +\infty$.

If we compare the expressions of the CTR without and with Bernoulli nodes, the only difference is the additional multiplicative term p at the denominator of the critical range. Notice that this can intuitively be interpreted as follows: Because every node will be active with a probability p, the expected total number of nodes V' that will be active will be $n' = pn$. Thus, the connectivity of the graph $G(V, r_n, p)$ can be treated as the connectivity of graph $G(V', r_{n'})$ instead. Notice that V' can be sort of viewed as a set of pn nodes uniformly at <u>random distributed in the disk.</u> Thus, the critical range for graph $G(V', r_{n'})$ is $r_{n'} = \sqrt{\frac{\log(n') + f(n')}{\pi n'}} = \sqrt{\frac{\log n + \log p + f(np)}{\pi pn}}$, where $f(n')$ is an arbitrary function such that $\lim_{n' \to \infty} f(n') = +\infty$. We can then estimate that the critical range for connectivity of $G(V, r_n, p)$ is $r_n = r_{n'} = \sqrt{\frac{\log n + g(n)}{\pi pn}}$, where $g(n) = \log p + f(np)$ is an arbitrary function such that $\lim_{n \to \infty} g(n) = \lim_{n \to \infty} f(n) = +\infty$.

Notice that Yi (2005) also provided the study of the critical transmission range for connectivity, the number of isolated nodes when each node is up independently with a probability p_1, and each link is up independently with a probability p_2.

THEOREM 11.12 (Yi, 2005) *Suppose that* $\lim_{n \to \infty} p_1 p_2 \ln n = \infty$ *and nodes have the same maximum transmission radius,*

$$r = \frac{\ln n + \epsilon}{np_1 p_2 \pi},$$

for some constant ϵ*. Then, the total number of isolated nodes is asymptotically Poisson with mean* $e^{-\epsilon}$*, and the total number of isolated active nodes is also asymptotically Poisson with mean* $p_1 e^{-\epsilon}$*.*

Notice that to apply the preceding theorem, we need the probability that the links are up to be independent and the probability that the nodes are up should also be independent.

11.6   Practical Performances

We analyzed the theoretical condition for the transmission radius r_n such that the graph $G(V, r_n)$ is k-connected with high probability. To confirm our theoretical analysis, we conduct simulations to see what the practical value of r_n is such that the wireless network $G(V, r_n)$ is k-connected with high probability. Remember that our bounds hold only when n goes to infinity. So nothing is known when n is a small practical integer. Notice that Bettstetter (2002) also recently conducted simulations to study k-connectivity, with a minimum degree of k, and their relations. No explicit expression of r is given in Bettstetter (2002).

System Settings

The geometry domain, in which the wireless nodes are distributed, is a unit square $C = [-0.5, 0.5] \times [-0.5, 0.5]$. As shown by previous results, we know that the random-point process $\mathcal{X}_n$ and the homogeneous Poisson point process $\mathcal{P}_n$ will have the same connectivity behavior asymptotically. For the simplicity of conducting simulations, we

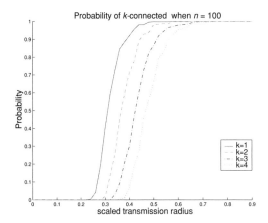

Figure 11.2 Transition phenomena of a graph being k-connected, where $n = 100$. © 2005, ACM.

choose n points that are randomly and uniformly distributed in $\mathcal{C}$. For each randomly generated point set V and a transmission radius r, we construct the graph $G(V, r)$ in a centralized manner. To speed up the construction of $G(V, r)$, we partition the points into grids of size r. Thus, a point p can connect with points from only at most nine grids: one grid containing p and eight adjacent grids.

One of the major steps in conducting the simulations is to compute the connectivity of an induced UDG $G(V, r_n)$. It is easy to test whether a graph is connected by simply checking if a spanning tree contains all n nodes. To test whether the graph $G(V, r_n)$ is k-connected, we use the following observation: It is k-connected if and only if the minimum cut is at least k, which is equivalent to the flow between any pair of nodes being at least k. So, given the graph $G(V, r_n)$, we compute the maximum flow between any pair of nodes by assigning each edge a weight of one. A simpler method is the use of BFS to compute how many disjoint paths connect a node v to a node u. The time complexity of this approach is $O(n^2 m)$, where m is the number of edges in $G(V, r)$, which could be as large as n^2. For unit-capacity flow, there is an $O(\min(m, n^{3/2})m^{1/2})$ time-complexity algorithm (Goldberg and Rao, 1997).

Transmission Phenomena for Connectivity

A graph property of $G(V, r)$ is said to satisfy a transition phenomena if there is a radius r_0 such that the graph $G(V, r)$ almost surely does not have this property when $r < r_0$ and the graph $G(V, r)$ almost surely has this property when $r > r_0$. It was already shown that the property that $G(V, r)$ has the minimum node degree k satisfies transition phenomena; additionally, the fact that the graph $G(V, r)$ is k-connected satisfies transition phenomena. The simulations shown in Figure 11.2 confirm the theoretical results. We found that the transition becomes faster when the number of nodes increases. We test $0.1 \leq r \leq 0.9$ by using an interval of 0.02; i.e., we test a total of 40 different transmission radii. Given a transmission radius r and number of nodes n, we generate 500 sets of random n points in $\mathcal{C}$. We compute the connectivity of each graph $G(V, r)$ and summarize how many are k-connected for $k = 1, 2, 3,$ and 4.

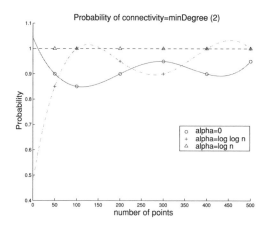

Figure 11.3 The probability that a graph with minimal degree k is k-connected for $k = 2$.
© 2005, ACM.

Connectivity and Minimum Degree

Penrose (1999a) showed that the hitting radius for k-connectivity and the hitting radius for achieving minimum degree k are asymptotically the same for points randomly and uniformly distributed in a unit-area square as n goes to infinity. We conduct extensive simulations on various number of points $n = 50, 100, 200, 300, 400,$ and 500. Given n, k, and α, we select r according to the bound given in Theorem 11.8. Here, the connectivity $k = 2$ and $\alpha \in \{0, \ln \ln n, \ln n\}$. For each case, we generate 500 random-point sets. The simulations illustrated in Figure 11.3 show that the probability that $G(V, r)$ is k-connected when its minimum degree is k is already sufficiently close to one when n is of the order of 50, especially when α is set as $\ln n$. This surprising result implies a fast method to approximate the connectivity of a graph by simply counting the minimum node degrees.

Connectivity for a Small Point Set

Theoretically, we derived an asymptotic bound of the transmission range r_n for n points randomly and uniformly distributed in a unit-area square such that the graph $G(V, r_n)$ is k-connected with certain probability. We have to admit that the result holds only when n is large enough compared with $k!$. We first conduct simulations to measure the gap between the theoretical probability of graph $G(V, r)$ being k-connected and the actual statistical probability of its being k-connected for various radius r. Typically, we set $n\pi r^2 = \ln n + (2k - 1) \ln \ln n - 2 \ln k! + 2\alpha + 2 \ln \frac{8k}{2^k \sqrt{\pi}}$. Then, we test all 54 cases of $n = 50, 100, 200, 300, 400,$ and 500; $k = 1, 2, 3,$ and 4; $\alpha = \ln \ln n$, and $\ln$. The corresponding theoretical k-connectivity probabilities for them are $1/e$, $1 - (1/\ln n)$, and $1 - (1/n)$ when $\alpha = 0$, $\ln \ln n$, and $\ln$, respectively. The probability is computed over 500 different random-point sets. Figure 11.4 illustrates the simulation results.

It is not surprising that the probability found by simulation is much lower than that of the theoretical analysis (denoted by the upper curves). Notice that the theoretical range r is not always monotone increasing of k when n is a small value. This is the reason some curves cross each other in the figures.

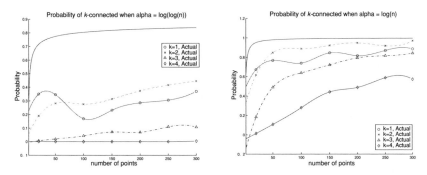

Figure 11.4 Probability that $G(V, r)$ is k-connected if r is set theoretically. © 2005, ACM.

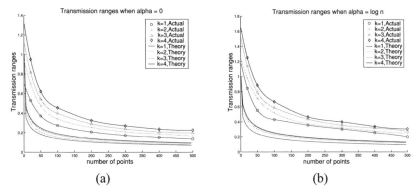

Figure 11.5 (a) Practical range in which $G(V, r)$ is k-connected with probability $1/e$; (b) practical range in which $G(V, r)$ is k-connected with probability $1 - 1/n$. © 2005, ACM.

Practical Transmission Ranges for k-Connectivity

Because the asymptotic bound of the transmission range r_n for n points randomly and uniformly distributed in a unit-area square such that the graph $G(V, r_n)$ is k-connected with certain probability holds only when n is large enough compared with $k!$, we need to study what the actual transmission range is that is required to achieve k-connectivity with certain probability. It is possible to analyze more accurately what the theoretical requirement for r_n is when n is not large enough. However, the analysis is much more complicated because we cannot omit some "constant" terms in any formulas anymore. We leave this tight analysis as possible future work. Alternatively, we conduct simulations to find practical transmission ranges when n is not large enough. See Figure 11.5 for an illustration of our simulation results. It is not surprising that the actual required range is larger than the theoretical bound. However, we found that the actual transmission range takes a similar decreasing pattern as the theoretical result when n goes to infinity.

11.7   Further Reading

The connectivity of random graphs, especially geometric graphs and their variations, has been considered in the random-graph-theory literature (Bollobás, 2001), in the stochastic geometry literature (Dette and Henze, 1990; Penrose, 1997, 1998, 1999a, 1999c), and in the WAN literature (Bettstetter, 2002; Blough *et al.*, 2002; Grossglauser and Tse, 2001; Gupta and Kumar, 1998, 1999; Santi and Blough, 2003; Xue and Kumar, 2004).

Given n nodes V randomly and independently distributed in a unit-area disk $\mathcal{D}$, Gupta and Kumar (1998) showed that $G(V, r_n)$ is connected almost surely if $n\pi r_n^2 \geq \ln n + c(n)$ for any $c(n)$ with $c(n) \to \infty$ as n goes to infinity. Notice that this bound is tight because they also proved that $\mathcal{G}(\mathcal{X}_n, r_n)$ is asymptotically disconnected with positive probability if $n\pi r_n^2 = \ln n + c(n)$ and $\limsup_n c(n) < +\infty$. In other words, the connectedness of the network has transition phenomena when the transition range increases. The wireless network composed of randomly distributed mobile hosts will become connected almost abruptly.

Notice that they actually derived their results for a homogeneous Poisson process of points in $\mathcal{D}$ instead of the independent and uniform point process. They showed that the difference between them is negligible. Additionally, a similar result by Penrose (1997, 1999c) showed that the same result holds if the geometry domain in which the wireless nodes are distributed is a unit-area square $\mathcal{C}$ instead of the unit-area disk $\mathcal{D}$.

Independently, Penrose (1997) showed that the longest edge M_n of the EMST of n points randomly and uniformly distributed in a unit-area square $\mathcal{C}$ satisfies

$$\lim_{n \to \infty} \Pr\left(n\pi M_n^2 - \ln n \leq \alpha\right) = e^{-e^{-\alpha}}$$

for any real number α. Remember that the longest edge of the EMST is always the critical power (Ramanathan and Rosales-Hain, 2000; Sanchez *et al.*, 1999). Thus, the result in Penrose (1997) is actually stronger than that in Gupta and Kumar (1998) because it will give the probability that the network is connected. For example, if we set $\alpha = \ln \ln n$, we have $\Pr(n\pi M_n^2 \leq \ln n + \ln \ln n) = e^{-1/\ln n}$. It implies that the network is connected with a probability of at least $e^{-1/\ln n}$ if the transmission radius of each node r_n satisfies $n\pi r_n^2 = \ln n + \ln \ln n$. Notice that $e^{-1/\ln n} > 1 - \frac{1}{\ln n}$ from $e^{-x} > 1 - x$ for $x > 0$. By setting $\alpha = \ln n$, the probability that the graph $G(V, r_n)$ is connected is at least $e^{-1/n} > 1 - (1/n)$, where $n\pi r_n^2 = 2\ln n$. Notice that the preceding probability is true only when n goes to infinity. When n is a finite number, the probability that the graph is connected is smaller; i.e., we need a transmission radius much larger than r_n to guarantee that the network of n randomly distributed points is connected almost surely. In this chapter, the experimental study of the probability that the graph $G(V, r_n)$ is connected for finite number n was presented. Notice that Bettstetter (2002) also conducted simulations to study k-connectivity, with the minimum degree being k and their relations. However, Bettstetter used the toroidal model instead of the actual Euclidean model.

We now review the results concerning the k-connectivity of a random graph. For general graphs, Bollobás and Thomason [see Theorem 7.5 of Bollobás (2001)] proved that if $c(n) \to \infty$, $c(n) \leq \ln \ln \ln n$, and $p(n) = \frac{\ln n + (k-1)\ln \ln n - c(n)}{n}$, then almost no graph from $\mathcal{B}(n, p(n))$ contains a nontrivial $(k-1)$-separator. Notice that a graph with minimum degree k is k-connected unless it contains a nontrivial $(k-1)$-separator. Thus, this result by Bollobás and Thomason implies that if $p(n) = \frac{\ln n + (k-1) \ln \ln n - c(n)}{n}$, then graphs from $\mathcal{B}(n, p(n))$ almost surely have minimum degree k and thus almost surely are k-connected. Here, a Bernoulli graph $\mathcal{B}(n, p(n))$ defined on n nodes is defined as follows. Consider all edges in a completed graph over n nodes. An edge will be independently selected to $\mathcal{B}(n, p(n))$ with probability $p(n)$.

For a random geometry graph, it was proved by Penrose (1999a) that given any metric l_p with $2 \leq p \leq \infty$ and any positive integer k,

$$\lim_{n \to \infty} \Pr(\varrho(\mathcal{X}_n, \kappa \geq k) = \varrho(\mathcal{X}_n, \delta \geq k)) = 1.$$

The result is analogous to the well-known results in graph theory (Bollobás, 2001) that a graph becomes k-vertex-connected when it achieves the minimum degree k if we add the edges randomly and uniformly from $\binom{n}{2}!$ possibilities. The result by Penrose (1999a) says that a graph of $G(\mathcal{X}_n, r)$ becomes k-connected almost surely at the moment it has minimum degree k by letting r go from 0 to ∞. It implies that if n is large enough, then with high probability, if we start with isolated n random points $\mathcal{P}_n$ in C and add the edges in order of the increasing length to connect the points of $\mathcal{P}_n$, the resulting graph becomes k-vertex-connected at the moment when the minimum degree of the graph becomes k. However, this result does not imply that to guarantee a graph over n points k-connected almost surely, we only have to connect every node to its k-nearest neighbors. Let V be n points randomly and uniformly distributed in a unit square (or disk). Xue and Kumar (2004) proved that to guarantee that a geometry graph over V is connected, the number of nearest neighbors that every node has to connect is asymptotically $\Theta(\ln n)$. Dette and Henze (1990) studied the maximum length of the graph by connecting every node to its k-nearest neighbors asymptotically. In Chapter 12, we study in detail this critical number of neighbors to get a connected network.

Dette and Henze (1990) studied the largest length, denoted here by $r_{n,k}$, of the kth-nearest-neighbor link for n points drawn independently and uniformly from the d-dimensional unit-length cube or the d-dimensional unit-volume sphere. They gave the asymptotic result of this length accordingly as $k < d$, $k = d$, or $k > d$. For the unit-volume cube, they use the norm l_∞ instead of the Euclidean norm l_2. For the unit-volume sphere, their result implies that when $d = 2$ and $k > 2$,

$$\lim_{n \to \infty} \Pr\big[n\pi r_{n,k}^2 \leq \ln n + (2k - 3) \ln \ln n - 2 \ln(k - 1)!$$
$$- 2(k - 2) \ln 2 + \ln \pi + 2\alpha\big] = e^{-e^{-\alpha}}.$$

Notice that Penrose (1999a) had showed that when the domain is a unit-area square, the probability that a random geometry graph $G(V, r_n)$ is k-connected and has a minimum vertex degree k goes to 1 as n goes to infinity. Consequently, we can argue that the transmission radius r_n such that the graph $G(V, r_n)$ is k-connected with high probability satisfies $n\pi r_n^2 \simeq \ln n + (2k - 3) \ln \ln n - 2 \ln(k - 1)! + 2\alpha$. We verified in this chapter that given n random points V over a unit-area square, to guarantee that a geometry graph over V is k-connected, the number of nearest neighbors that every node has to connect is asymptotically $\ln n + (2k - 3) \ln \ln n$.

Similarly, instead of considering $\mathcal{X}_n$, Penrose also considered a homogeneous Poisson point process with intensity n on the unit-area square $\mathcal{C}$. Penrose gave loose upper and lower bounds on the hitting radius $r_n = \varrho(\mathcal{P}_n, \delta \geq k)$ as $\frac{\ln n}{2^{d+1}} \leq nr_n^d \leq d!2\ln n$ for the homogeneous Poisson point process on a d-dimensional unit cube. This result is too loose. More important, the parameter k does not appear in this estimation at all. In this chapter, we derive an exact bound on r_n for 2D n points V randomly and uniformly distributed in $\mathcal{C}$ such that the graph $G(V, r_n)$ is k-connected with high probability.

We also conducted simulations to study the probability that a graph has minimum degree k and has vertex connectivity k simultaneously. Surprisingly, we found that this probability is sufficiently close to one even if n is on the scale of 100. This observation implies a simple method (by just computing the minimum vertex degree) for approximating the connectivity of a random geometry graph.

Penrose (1997, 1999a) also studied the k-connectivity problem for d-dimensional points distributed in a unit-area cube by using the toroidal model instead of the Euclidean model as one way to eliminate the boundary effects. He (1999a) showed that the hitting radius r_n such that the graph $G(V, r_n)$ is $(k + 1)$-connected satisfies $\lim_{n\to\infty} \Pr(n\pi r_n^2 \leq \ln n + k \ln \ln n - \ln k! + \alpha) = e^{-e^{-\alpha}}$.

11.8   Conclusion and Remarks

We considered a large scale of wireless ad hoc networks whose nodes are distributed in a 2D unit-square region. Because fault tolerance is imperative for wireless networks, we showed that to make the graph $G(V, r_n)$ k-connected almost surely, the transmission range r_n should satisfy a certain condition when n goes to infinity. The results presented in this chapter hold also in mobile networks in which the movement of nodes will not destroy the uniform randomness of node locations. Practical transmission ranges were also studied by simulations when n is not a large integer. Recall that a localized method was presented in Chapter 9 to control the network topology given a k-fault-tolerant deployment of wireless nodes such that the resulting topology is still fault-tolerant but with much fewer communication links maintained. We showed that the constructed topology has only a linear number $O(kn)$ of links and is a length spanner.

There are still a number of research questions left for future research. We assumed that the wireless nodes are generated by a random-point process or a Poisson point process. In practical applications, the wireless nodes could have some other estimated distributions such as the inhomogeneous Poisson point process. This is much more complicated than the cases studied from known previous results. Now a challenging question is this: What will be the formula for critical transmission range for k-connectivity if the nodes are distributed in a region following a probability density function $\mathcal{F}$? In this chapter, we studied the critical transmission range for k-connectivity when a node could be active with a fixed uniform probability p. What will be the formula for a critical range for k-connectivity if every link uv with $\|uv\| \leq r_n$ exists only with a predetermined probability p_2? We may first assume that p_2 is uniform for all links. In practice, the probability p_2 could also depend on the Euclidean length $\|uv\|$: Smaller length $\|uv\|$ implies a larger probability p_2 (e.g., $p_2 = 1/e^x$ where $x = \|uv\|$). Another challenging question is what the critical transmission range will be when the link availability probability is *not* independent. The results presented in Yi (2005) assumed that the probability p_e that every link e will be available is independent. In practice, the probability p_e for a link $e = (u, v)$ could depend on nodes u and v (e.g., $p_e = f(u)f(v)$ for some function $f()$ or p_e is independent only between any two links), but not among three links (u, v), (v, w), and (w, u). Then what will be the CTR in these models?

Problems

11.1   Give a rigorous proof that the CTR for connectivity is the maximum edge length in the EMST. Given n nodes in a region whose positions are known, design an efficient algorithm that runs in time $O(n \log n)$ to find the CTR for connectivity.

11.2   Given a set of n nodes in a region, what is the critical transmission range for biconnectivity? Given n nodes in a region whose positions are known, design an efficient polynomial-time algorithm to find the CTR for biconnectivity. What is the running time of your method?

11.3   What is the difference between a random graph and a random geometry graph used to model wireless ad hoc networks and WSNs? Can we simply apply the results from a random graph to a random geometry graph?

11.4   Assume that n nodes will be randomly placed in a square region with side length $a = c\sqrt{\frac{n}{\log n}}$ for some constant $c \leq 1$. Assume that the transmission range of each node is 1. Prove that the network will be connected with high probability. Conduct simulations to study the k-connectivity of the network. In other words, you need to plot the portion of the random networks that are one-connected only, two-connected only, and generally k-connected only.

11.5   Assume that n nodes are randomly placed in a segment of unit length. What is the critical range for two-connectivity?

11.6   Assume that there are n nodes randomly placed in a unit-square region. We built the EMST T connecting these n nodes. Prove that $\sum_{e \in T} \|e\| \leq 2\sqrt{2}\sqrt{n}$.

11.7   Assume that a square region is partitioned into $k \times k$ grid of cells. Further, assume that we randomly place k^2 nodes V in the region. Prove that with high probability, at least half of the cells will contain one node from V.

11.8   Assume that a square region is partitioned into a $k \times k$ grid of cells. Further, assume that we randomly place $ck^2 \log k$ nodes V in the region. Prove that with high probability, each cell will contain one node from V when constant c is chosen properly.

11.9   Assume that n nodes are randomly placed in a square region. There are f channels that can be used. Assume that each node will randomly pick a channel and two nodes can be connected by an edge only if they are within the transmission range r of each other and they use a common channel. What is the CTR r such that the network will be connected?

11.10 Assume that n nodes are randomly placed in a square region. There are f channels that can be used. Assume that each node will randomly pick d channels (with constant $d \in [1, f)$) and two nodes can be connected by an edge only if they are within the transmission range r of each other and they use a common channel. Conduct simulations to study the CTR r such that the network will be connected. Is this result close enough to the formula for the CTR when link availability probabilities are independent?

12 Other Transition Phenomena

12.1 Introduction

In Chapter 11, we essentially studied the critical transmission range (CTR) for k-connectivity in which $k \geq 1$ with or without Bernoulli node mode. Obviously, with the increase of the transmission range r_n, the point graph $G(V, r_n)$ will be connected when r_n reaches a certain value and will remain connected afterwards. This is called the *phase transmission phenomena* (Luczak, 1996) in a random graph generally. Obviously, if a property $\mathbb{P}$ is monotone increasing over a certain parameter (e.g., the transmission range) of the network, this property $\mathbb{P}$ will always have transmission phenomena over this parameter. Besides the CTR, the critical ranges of a number of other graph properties have been studied in the literature. In this chapter, we study the critical range of connectivity over the nodal degree, the critical range of connectivity over the transmission range for sparse networks, the critical range of connectivity over the transmission range for mobile networks, and the critical range of the coverage over the sensing range.

12.2 Critical Node Degree for Connectivity

Given a set V of n nodes distributed in a region, let $\overrightarrow{H_\ell}(V)$ be the directed graph in which each node $v \in V$ is connected to all its ℓ-nearest neighbors. Let $H_\ell(V)$ be the undirected graph in which an edge uv belongs to $H_\ell(V)$ if and only if both directed link $\overrightarrow{uv}$ and directed link $\overrightarrow{vu}$ belong to graph $\overrightarrow{H_\ell}(V)$; i.e., $H_\ell(V)$ is the symmetric version of $\overrightarrow{H_\ell}(V)$. Clearly, the connectivity of the structure $H_\ell(V)$ is a monotone-increasing property over the number ℓ. Thus, there should be a critical value ℓ such that $H_\ell(V)$ is connected when ℓ reaches this critical value. Generally, we define the critical-neighbors number (CNN) as follows:

DEFINITION 12.1 (Critical-neighbor number) *For a monotone-increasing property $\mathbb{P}$, let $\ell(\mathcal{X}_n, \mathbb{P})$ be the critical node degree such that $H_\ell(V)$ has property $\mathbb{P}$ almost surely when V is produced by a point process $\mathcal{X}_n$.*

Xue and Kumar (2004) proved that to guarantee that a geometry graph over V is connected, the number of nearest neighbors that every node has to connect is asymptotically $\Theta(\ln n)$. In other words, we have the following theorem:

THEOREM 12.1 *Assume* n *points are distributed uniformly at random in the 2D unit square and let ℓ_n be the critical number of neighbors such that $H_\ell(V)$ is connected. Then,*

$$\ell_n = \Theta(\ln n).$$

They essentially showed that the structure $H_\ell(V)$ is not connected almost surely if $\ell < c_1 \ln n$ for some small constant $c_1 = 0.074$ and $H_\ell(V)$ is connected almost surely if $\ell > c_2 \ln n$ for some large constant $c_2 = 5.1774 + \epsilon$, where ϵ is an arbitrarily small positive number. Recently, Wan and Yi (2004) showed that we can actually reduce the constant c_2 to about the nature number $e \simeq 2.718$, and the claim is true even for the general k-connectivity requirement. The following theorem was proved in Wan and Yi (2004):

THEOREM 12.2 *For any integer $k \geq 1$ and $\alpha > 1$, the following event will asymptotically almost surely (a.a.s.) occur:*

$$\ell(\mathcal{X}_n, \kappa \geq k) \leq \alpha e \log n.$$

Recall that here, $\kappa \geq k$ represents that the resulting graph has connectivity at least k. To prove this theorem, they essentially showed that for any two constants $1 < \beta < \alpha$, the event $G(\mathcal{X}_n, \sqrt{\frac{\beta \log n}{\pi n}}) \subseteq H_\ell(\mathcal{X}_n)$ a.a.s. occurs.

Blough *et al.* (2003) designed a topology-control method for wireless ad hoc networks based on the k-nearest neighbors. In their protocol, each node will first collect the list of k-nearest neighbors. A node u will keep a link to a neighbor v if and only if v is one of its k-nearest neighbors and u is one of the k-nearest neighbors of node v. To get a connected network topology, the transmission power of each node must be large enough such that each node will have at least $\Theta(\log n)$ neighbors. They also generalize the result to square deployment regions of arbitrary side length. Notice that the results of Xue and Kumar mainly apply to dense networks in which the deployment region has a fixed area and the number of nodes in the region grows to infinity. Blough *et al.* showed that the same result holds for *sparse networks* and for arbitrary node density in general. Thus, it is only the number of nodes in the network, not the area on which the network is deployed, that determines the CNN for connectivity.

To some extent, the theorems by Xue and Kumar (2004) and by Wan and Yi (2004) seem to contradict the results presented in a series of papers that considered the problem of how many neighbors are desirable in a multihop wireless network (Hou and Li, 1986; Kleinrock and Silvester, 1978; Takagi and Kleinrock, 1984). In these papers, the wireless network is modeled as a set of nodes located on the plane according to a Poisson point process and the problem is that of maximizing the one-hop progress of routing in the desired direction under different transmission protocols. Kleinrock and Silvester (1978) showed that the optimal number of neighbors is six for a slotted ALOHA protocol in which all the nodes use the same transmission power. Later, Takagi and Kleinrock revised this magic number to eight (Takagi and Kleinrock, 1984). Hou and Li (1986) considered the same problem by allowing nodes to use different transmission powers. They obtained the magic numbers of six and eight. Different network models and/or different optimization objectives were also later studied by other researchers. However, none of the studies along this line considered the connectivity of the network, whereas the main concern of the study by Xue and Kumar (2004) is the network connectivity.

Notice that the CNN required for connectivity for a random geometry graph is also consistent with the critical number of links for connectivity for a traditional random graph. It was proved in Bollobás (1998) (Theorem 7.3) that if we randomly pick $M = (n/2)[\log n + \beta + o(1)]$ edges from the completed graph of n nodes, the resultant graph G_M is connected with a probability of $e^{-e^{-\beta}}$ when n is sufficiently large. This implies that for a random graph, if the node degree is at least $\log n + \beta$, then the network is connected with a probability of at least $e^{-e^{-\beta}}$.

12.3   Critical Range for Connectivity in Sparse Networks

In Chapter 11, we mainly studied the critical range for k-connectivity when the number of nodes deployed in a fixed region goes to infinity (or is sufficiently large). In practice, it is difficult to have a network with such a sufficiently large number of networking nodes. Thus, the theoretical bounds proved in Chapter 11 give us certain guidance for the CTR for connectivity only when the number of networking nodes is not too large. In this chapter, we study how to characterize the CTR for a *sparse network*, in which the number of nodes deployed in a unit area is not sufficiently large. The basic model we use is adding a further parameter: the side length of the deployment square region. In this new model, ℓ is the independent variable, and the asymptotic value of transmission range r and the number of nodes n that yield a connected network with high probability is then investigated when $\ell \to \infty$. Notice that different from the model of the GRG used in Chapter 11, the node density can either converge to 0 or to some other number, or even diverge as $\ell \to \infty$, depending on the relative values of n and ℓ. Santi and Blough (2003) systematically studied the CTR for this model. The results presented in this section are mainly from Santi and Blough (2003).

THEOREM 12.3 *Assume that* n *nodes* V, *each with transmission range* r, *are placed uniformly at random in a segment of length* ℓ. *Further assume that* rn $= k\ell \log \ell$ *for some constant* k > 0 *and* r $\ll \ell$ *and* n $\gg \ell$.

If k > 2, *or* k $= 2$ *and* r $\gg 1$, *then the resulting communication graph* G(V, r) *is a.a.s. connected.*

If k $\leq 1 - \epsilon$ *and* r $= \Theta(\ell^{\epsilon})$ *for some* $0 < \epsilon < 1$, *then the resulting communication graph* G(V, r) *is a.a.s. disconnected.*

If r *is not of the form* $\Theta(\ell^{\epsilon})$ *but* rn $\ll \ell \log \ell$, *then the resulting communication graph* G(V, r) *is a.a.s. disconnected.*

From Theorem 12.3, we have the following corollary:

COROLLARY 12.1 *For* n *nodes distributed uniformly at random over a segment of length* ℓ, *the CTR for connectivity is*

$$r_n = \zeta \frac{\ell \log \ell}{n},$$

where $1 \leq \zeta \leq 2$ *is a constant.*

It is interesting to compare Corollary 12.1 with Theorem 11.3 (i.e., the analogous theorem for dense networks). First, we observe that the characterization of the CTR for sparse networks is not as exact as that for dense networks because the exact value of ζ is currently unknown for Corollary 12.1. Santi and Blough (2003) conjectured that the value ζ is 1.

THEOREM 12.4 (Santi and Blough, 2003) *Assume that n nodes V, each with transmission range r, are placed uniformly at random in a d-dimensional cube d = 2 or 3 with side length ℓ. Further assume that $r^d n = k\ell^d \log \ell$ for some constant k > 0 and r ≪ ℓ and n ≫ ℓ. If (1) $k > d2^d d^{d/2}$, or (2) $k = d2^d d^{d/2}$ and r ≫ 1, then the resulting communication graph G(V, r) is a.a.s. connected.*

THEOREM 12.5 (Santi and Blough, 2003) *Assume that n nodes V, each with transmission range r, are placed uniformly at random in a d-dimensional cube d = 2 or 3 with side length ℓ. Assume that r ≪ ℓ and n ≫ ℓ. If $r^d n \in O(\ell^d)$, then the resulting communication graph G(V, r) is a.a.s. not disconnected.*

From the upper bound and lower bound for the connectivity just given, Santi and Blough (2003) concluded that the CTR for connectivity might be any function of the following type:

$$\frac{\ell^d f(\ell)}{n},$$

where $f(\ell)$ is a function such that $f(\ell) \in O(\log \ell)$ and $f(\ell) \gg 1$. By means of extensive simulations, Santi and Blough argue that $f(\ell) = \log \ell$ is also a necessary condition for connectivity to a.a.s. occur.

CONJECTURE 12.1 *Assume that n nodes V, each with transmission range r, are placed uniformly at random in a d-dimensional cube d = 2 or 3 with side length ℓ. The critical transmission range for connectivity is*

$$r = \zeta \frac{\ell^d \log \ell}{n},$$

where ζ is a constant with $0 \le \zeta \le 2^d d^{d/2+1}$.

12.4   Critical Range for Connectivity for Mobile Networks

In previous sections, a number of characterizations were presented of the CTRs for connectivity, k-connectivity, and connectivity with Bernoulli nodes. All of these results are based on the model of stationary networks, in which the nodes will *not* move after they are deployed based on a certain distribution probability. In this section, we study the effect of mobility on the CTR. Remember that the definition of the CTR for stationary networks is pretty straightforward: r_n is a CTR for a property $\mathbb{P}$ if $G(V, r)$ will have the property $\mathbb{P}$ asymptotically almost surely for a set V of n nodes uniformly placed at random in a region. Another way to look at a CTR for stationary networks is as follows. Let r_V be the minimum transmission range for a static network $G(V, r)$ such that it has a property $\mathbb{P}$. Then, r_n is a CTR if $\Pr(r_V \le r_n) \to 1$ as n goes to infinity and $\Pr(r_V > r_n) \to 0$ as n goes to infinity.

For mobile networks, the nodes will move around, so $G(V, r)$ will change with time. Let $G_t(V, r)$ be the network topology at time t when n nodes are initially placed as V and nodes will move based a certain mobility model $\mathcal{M}$. This implies that the minimum transmission range such that the network has a certain property also changes over time t. Denote $t_1, t_2, \ldots, t_i, \ldots$, as a sequence of time instants. Let $r_{n,1}, r_{n,2}, \ldots, r_{n,i}, \ldots$, be the sequence of CTR at time t_i. For example, assume that the network property $\mathbb{P}$ is *connectivity*. Then, $r_{n,i}$ is the longest edge of the EMST over V at time t_i. Generally, the sequence $r_{n,i}$ is neither increasing nor decreasing. Several different definitions of a CTR are possible for such a mobile network. For example, one could define the CTR for V as the *largest* $r_{n,i}$ and then define CTR r_n as the asymptotic upper bound for all CTRs of all possible initial deployment V. Clearly, this is very conservative because the maximum of $r_{n,i}$ could be sufficiently large, whereas the majority of $r_{n,i}$ are small. In other words, an occasional extremely large $r_{n,i}$ for one time instant t_i would render the CTR very high. Another way to define the CTR for mobile networks intuitively would be that it will ensure that the network will have this property $\mathbb{P}$ almost at all time instants. A third possible definition for a CTR would be that it will ensure that the network will have this property $\mathbb{P}$ when the node distribution is stabilized based on the mobility model. Here, we implicitly assume that the node distribution will be stabilized under the mobility model, which may not happen for some mobility model, as observed in the literature.

DEFINITION 12.2 *Assume that* n *nodes are initially distributed in a certain geometric region Ω according to a certain probability density function $\mathcal{F}$. After initial deployment, assume that nodes will move according to a certain mobility model $\mathcal{M}$. The asymptotic node spatial distribution generated by mobility mode $\mathcal{M}$ with the initial distribution $\mathcal{F}$ is defined as*

$$\mathcal{F}_\mathcal{M} = \lim_{i \to \infty} \mathcal{F}_i, \tag{12.1}$$

where $\mathcal{F}_i$ is the probability density function modeling the node spatial distribution at time instant t_i. *If the limit on the right-hand side of Eq. (12.1) does not exist, we say that the mobility model $\mathcal{M}$ is* not stable *with initial deployment $\mathcal{F}$. Otherwise, we say that the mobility model $\mathcal{M}$ is* stable *with initial deployment $\mathcal{F}$.*

Notice that in the preceding definition we assume that the node positions at any time instant t_i (including the initial deployment) can be characterized by a certain probability density function that evolves with time. Similar to Santi (2005b), we can define the CTR for mobile networks (when the mobility model stabilizes) as follows:

DEFINITION 12.3 *Assume that* n *nodes are initially distributed in a certain geometric region Ω according to a certain probability density function $\mathcal{F}$. After initial deployment, assume that nodes will move according to a certain mobility model $\mathcal{M}$. The CTR for a graph property $\mathbb{P}$ is defined as the CTR of* n *nodes* V *when* V *is initially deployed in a certain region Ω according to the probability density function $\mathcal{F}_\mathcal{M}$. Here, $\mathcal{F}_\mathcal{M}$ is* *the asymptotic node spatial distribution generated by mobility mode $\mathcal{M}$ with the initial distribution $\mathcal{F}$.*

From this definition of the CTR for mobility model $\mathcal{M}$, we can prove an *ergodic* property of certain mobile networks. Recall that a stochastic process, composed of a sequence of variables $x_1, x_2, x_3, \ldots, x_i, \ldots$ (in the case of CTR for network connectivity, the variable x_i is the length of the longest edge in the EMST at time instant t_i), is *ergodic* if sampling from the sequence of random variables is *statistically equivalent* to repeatedly sampling from a certain, fixed random variable (in the case of CTR for connectivity, the fixed random variable is the length of the longest EMST edge computed when nodes are initially distributed based on density function $\mathcal{F}_\mathcal{M}$ without mobility).

Similar to the Theorem 5.0.3 proved in Santi (2005b), it is not difficult to have the following theorem:

THEOREM 12.6 *Let $\mathcal{M}$ be a stable mobility model. Assume that the mobility model $\mathcal{M}$ is* k-independent; *i.e., there exists a constant integer* k *such that the node positions at time* t_{i+k} *is independent of the node positions at time* t_i *for any* $i \geq 1$. *Then, the mobile network with mobility model $\mathcal{M}$ is* ergodic *with respect to the CTR for any monotone nondecreasing graph property* $\mathbb{P}$ *(e.g., the network connectivity).*

As observed in Santi (2005b), this ergodicity adds a temporal dimension to the definition of the CTR in the presence of mobility. For example, let us assume that we set the transmission range to a value r such that the probability of having a connected network when nodes are placed according to a probability distribution $\mathcal{F}_\mathcal{M}$ is 99%. Further, assume that the mobility model $\mathcal{M}$ is k-independent and stable. By ergodicity, we can state that, on average, the 99% of the longest edges of the EMSTs constructed are at most r. In other words, if we observe the mobile network for a sufficiently long time, we will find that at least 99% of the time, the network $G_i(V, r)$ will be connected.

From the preceding discussion, when the mobility model $\mathcal{M}$ is stable and k-independent for a constant integer k, the problem of characterizing the CTR for a graph property $\mathbb{P}$ with mobility model $\mathcal{M}$ and initial distribution $\mathcal{F}$ can be reduced to the study of the CTR for static networks when nodes are distributed according to a density function $\mathcal{F}_\mathcal{M}$ without mobility. Clearly, when $\mathcal{F}_\mathcal{M}$ is the uniform random distribution or Poisson distribution, the results presented in Chapter 11 can be applied. Surprisingly, some mobility models $\mathcal{M}$ do fall in this category; for example, the Brownian-like mobility model: It was shown by extensive simulations by Blough *et al.* (2003) that the Brownian-like mobility model generates a uniform long-term node spatial distribution.

Thus, we can summarize the steps to study the CTR for a mobile network with mobility model $\mathcal{M}$ and initial nodal distribution $\mathcal{F}$ as follows:

1. Determine whether the mobility model $\mathcal{M}$ is stable with initial spatial distribution $\mathcal{F}$. If so, find the stable density function $\mathcal{F}_\mathcal{M}$.
2. Determine whether the mobility model is k-independent for some integer k.
3. Compute the CTR for a static network with node spatial distribution $\mathcal{F}_\mathcal{M}$. This will also be the CTR for a mobile network with mobility model $\mathcal{M}$ and initial nodal distribution $\mathcal{F}$.

For example, the CTR for connectivity of nodes under a certain spatial distribution $\mathcal{F}_M$ can be computed based on the following theorem proved by Penrose (1999c):

THEOREM 12.7 *Assume that* n *nodes* V *are distributed independently at random in a 2D region, having connected and compact support* Ω *with smooth boundary* $\partial\Omega$*, according to a probability density function* $\mathcal{F}$*. Further assume that* $\mathcal{F}$ *is continuous on the boundary* $\partial\Omega$*. Let* M_n *be the longest edge of the MST built on these* n *nodes* V*. Then, almost surely we have*

$$\lim_{n\to\infty}\frac{\pi n M_n^2}{\log n} = \frac{1}{\min_\Omega \mathcal{F}}.$$

Here, the *support* Ω of a probability distribution function is the set of geometry points in which it has a nonzero probability. A boundary $\partial\Omega$ is *smooth* if and only if the function that represents the boundary $\partial\Omega$ is twice differentiable. Notice that the boundary of a square is *not* smooth, whereas the boundary of a disk is smooth. However, the difficulty of the nonsmooth boundary for a square can be circumvented by use of a corner-rounding technique (first described by Santi): Replace the corner by a quadrant of a disk. This theorem tells us that the critical transmission range for a network is closely dependent on the minimum value of the density distribution $\mathcal{F}$ (and $\mathcal{F}_M$ for mobile networks).

For example, Santi (2005b) presents a detailed study of the random-waypoint (RWP) mobility model. It is known already in the literature (Bettstetter and Krausser, 2001) that the node spatial distribution generated by the RWP model is not uniform (nor is the Poisson distribution); the nodes will somehow concentrate around the center of the deployment region. This phenomenon, called the *border effect*, is because for the RWP model, the waypoint (i.e., the destination of a movement) is selected uniformly at random in a bounded deployment region. Bettstetter *et al.* (2003) first studied the stable density function $\mathcal{F}_{\text{RWP}}$ for a RWP mobility model. They proved the following theorem:

THEOREM 12.8 *The asymptotic spatial density function of a node moving in a unit-area square* $[0, 1] \times [0, 1]$*, according to the RWP mobility model with pause time* t_p *and velocity* v*, is closely approximated by*

$$\mathcal{F}_{\text{RWP}}(x, y) = \begin{cases} P_{\text{pause}} + (1 - P_{\text{pause}})\mathcal{F}_m(x, y), & \text{if } (x, y) \in [0, 1] \times [0, 1], \\ 0, & \text{otherwise} \end{cases}$$

where $P_{\text{pause}} = \frac{t_p}{t_p + \frac{0.521405}{v}}$, $\mathcal{F}_m(x, y) = 0$ *if* $x = y = 0$*, and* $\mathcal{F}_m(x, y) = 6y + \frac{3}{4}(1 - 2x + 2x^2)(\frac{y}{y-1} + \frac{y^2}{x^2-x}) + \frac{3}{2}[(2x - 1)y(1 + y)\log\frac{1-x}{x} + y(1 - 2x + 2x^2 + y)\log\frac{1-y}{y}]$ *if* $\frac{1}{2} \leq x \geq y$*. Notice that the other* $\mathcal{F}_m(x, y)$ *can be deduced from the symmetry* $\mathcal{F}_m(x, y) = \mathcal{F}_m(y, x) = \mathcal{F}_m(1 - x, y) = \mathcal{F}_m(x, 1 - y)$*.*

This theorem implies that the minimum value of $\mathcal{F}_{\text{RWP}}$ is achieved at the corner.

Thus, we have the following theorem (proved in Santi, 2005a):

THEOREM 12.9 *If* n *nodes are originally deployed uniformly at random at a unit square* $[0, 1] \times [0, 1]$ *and move according to the RWP mobility model with pause time* t_p *and velocity* v, *the CTR for connectivity when* $t_p > 0$ *is*

$$r_n^{\text{RWP},t_p} = \frac{t_p + \frac{0.521405}{v}}{t_p} \sqrt{\frac{\log n}{\pi n}}.$$

When $t_p = 0$, the CTR is unknown so far. It was conjectured by Santi (2005a), who used extensive simulations, that

$$r_n^{\text{RWP},0} = \frac{\log n}{4} \sqrt{\frac{\log n}{\pi n}}.$$

Santi also proved that the RWP mobility model is ergodic with respect to the CTR for connectivity.

12.5   Critical Sensing Range for Coverage

In previous sections, we studied mainly the critical *transmission* range such that the network is connected. For WSNs, besides the network connectivity, it is also important to ensure a certain coverage to fulfill its sensing objectives. Assume that each node v_i of the sensor also has a sensing range s_i. The question we study is that, given a set V of n nodes deployed in a certain region, what is the minimum value for a uniform coverage range s such that the sensors provide a complete coverage? The study of the critical coverage range (CCR) problem is motivated by the monitoring applications in WSNs, such as surveillance or habit monitoring. To design such kinds of networks, it is often required that every event happening within a certain region be monitored by one sensor. In some scenarios, to ensure fault tolerance, it is required that every event be monitored by at least a certain number (say, k) of different sensors. When the event could happen anywhere in the region, it is equivalent to requiring that every point of the region be covered by k different sensors.

DEFINITION 12.4 (Critical Coverage Range) *Assume that* n *nodes* V *are deployed into a certain region* Ω. *A point* x *in the region,* Ω *is said to be* k-*covered if there are at least* k *different sensors* $v_{i_1}, v_{i_2}, \ldots, v_{i_k} \in V$ *such that* x *is inside the coverage range* s_{i_j} *of node* v_{i_j}. *A region* $R \subseteq \Omega$ *is said to be* k-*covered if* all *nodes in* R *are* k-*covered. The critical* k-*coverage range problem is to find, given a node deployment, the minimum uniform value of* s_i *such that* R *is* k-*covered. When* k = 1, *we simply call it a CCR problem.*

Notice that the sensors often cannot be deployed by use of some regular patterns because of environment constraints. Furthermore, after the sensors are randomly deployed in a certain region, in most applications, sensors cannot and will not move. Thus, the study of the CCR is often for a randomly deployed set of nodes (e.g., uniformly at random deployment). Observe the strong similarity between the CTR and CCR for a set of wireless nodes. It is widely known in the literature that when the transmission range r_T is at least twice the coverage range r_C and the coverage range ensures the full coverage of the region R, then the network $G(V, r_T)$, where two nodes $u, v \in V$ are connected if their Euclidean distance is at most the transmission range r_T, is connected if V is inside R.

THEOREM 12.10 *Assume that a set of* n *nodes* V *is deployed inside a region* R. *When the transmission range* r_T *is at least twice the coverage range* r_C *and the coverage range ensures the full coverage of the region* R, *then the network* G(V, r_T) *is connected.*

Notice that the reverse of the theorem is not true: There are networks in which $r_T \geq 2r_C$ and the network $G(V, r_T)$ is connected, but V cannot provide the full coverage to all points in R. Philips *et al.* (1989) studied the coverage range for the case of nodes distributed in a square with side length ℓ according to the Poisson point process of a fixed density λ.

THEOREM 12.11 (Philips *et al.*, 1989) *Assume that nodes are distributed in a 2D square* R *with side length* ℓ *according to a 2D Poisson process of density* $\lambda > 0$. *Let* r_C *be the coverage range of each node. If*

$$r_C = \sqrt{\frac{2(1 - \epsilon) \log \ell}{\pi \lambda}}$$

for some constant ϵ *with* $0 < \epsilon < 1$, *then the square region* R *is a.a.s. not covered. If*

$$r_C = \sqrt{\frac{2(1 + \epsilon) \log \ell}{\pi \lambda}}$$

for some constant ϵ *with* $0 < \epsilon$, *then the square region* R *is a.a.s. covered.*

Notice that it is still an open problem whether the square will be covered a.a.s. when $r_C = \sqrt{\frac{2 \log \ell}{\pi \lambda}}$.

OPEN QUESTION 12.1 *Will the square region* R *be a.a.s. covered when* $r_C = \sqrt{\frac{2 \log \ell}{\pi \lambda}}$?

The previous studies concentrated on the one-coverage of the network. In some applications, we may need a k-coverage of all points in the region. Yi (2005) gave some asymptotic bounds for the CCR such that all points in the region are a.a.s. k-covered.

Let $C_{n,r}$ (respectively, $C'_{n,r}$) denote the event that a region (a disk with radius $1/\sqrt{\pi}$ or a square with a unit area) is k-covered by the (open or closed) disks of radius r centered at the points produced by a Poisson point process or a uniform random-point process, respectively. Let $K_{s,n}$ (respectively, $K'_{s,n}$) denote the event that a region (a disk with radius $\sqrt{s}/\sqrt{\pi}$ or a square with side length $\sqrt{s}$) is k-covered by the unit-area (closed or open) disks centered at the points produced by a Poisson point process or uniform random point process in this region with density n. Yi (2005) then studied the probability that these events happen when the coverage radius is set to some specific values and the number of nodes goes to infinity or s goes to infinity.

THEOREM 12.12 (Yi, 2005) *Let*

$$r_n = \sqrt{\frac{\ln n + (2k - 1) \ln \ln n + \xi_n}{\pi n}}.$$

If $\lim_{n\to\infty} \xi_n = \zeta$ for some $\zeta \in R$, then

$$1 - \beta(\zeta) \le \lim_{n\to\infty} \Pr(C_{n,r_n}) \le \frac{1}{1 + \alpha(\zeta)},$$

$$1 - \beta(\zeta) \le \lim_{n\to\infty} \Pr(C'_{n,r_n}) \le \frac{1}{1 + \alpha(\zeta)}.$$

If $\lim_{n\to\infty} \xi_n = +\infty$, then

$$\lim_{n\to\infty} \Pr(C_{n,r_n}) = \lim_{n\to\infty} \Pr(C'_{n,r_n}) = 1.$$

If $\lim_{n\to\infty} \xi_n = -\infty$, then

$$\lim_{n\to\infty} \Pr(C_{n,r_n}) = \lim_{n\to\infty} \Pr(C'_{n,r_n}) = 0.$$

Here, $\alpha(\zeta)$ and $\beta(\zeta)$ are some special functions defined in Yi (2005).

THEOREM 12.13 (Yi, 2005) *Let*

$$\mu(s) = \ln s + 2k \ln \ln s + \xi(s).$$

If $\lim_{n\to\infty} \xi_n = \zeta$ for some $\zeta \in R$, then

$$1 - \beta(\zeta) \le \lim_{n\to\infty} \Pr(K_{s,\mu(s)s}) \le \frac{1}{1 + \alpha(\zeta)},$$

$$1 - \beta(\zeta) \le \lim_{n\to\infty} \Pr(K'_{s,\mu(s)s}) \le \frac{1}{1 + \alpha(\zeta)}.$$

If $\lim_{n\to\infty} \xi_n = +\infty$, then

$$\lim_{n\to\infty} \Pr(K_{s,\mu(s)s}) = \lim_{n\to\infty} \Pr(K'_{n,\mu(s)s}) = 1.$$

If $\lim_{n\to\infty} \xi_n = -\infty$, then

$$\lim_{n\to\infty} \Pr(K_{s,\mu(s)s}) = \lim_{n\to\infty} \Pr(K'_{s,\mu(s)s}) = 0.$$

Probabilistic studies of k-coverage by a random-point process were conducted for $k = 1$ in Hall (1988) and arbitrary integer-valued constant k in Zhang and Hou (2004) but with certain limitations. Both studies assume Poisson point processes on a square and use the toroidal metric, rather than the Euclidean metric, which is more relevant to the applications. The assumption of the toroidal metric technically eliminates the boundary effect under the Euclidean metric. As demonstrated by Yi (2005), the boundary effect is the major technical challenge that requires much delicate and involved analysis.

12.6   Critical Range for Successful Routing

In previous sections, we studied the critical range or the number of neighbors for connectivity or coverage. These properties are still inherent to the geometry graph defined by the nodes. In this section, we study the CTR such that a certain routing algorithm will be successful with high probability. Notice that the connectivity of a

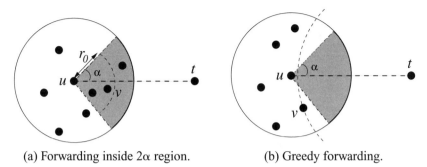

(a) Forwarding inside 2α region.   (b) Greedy forwarding.

Figure 12.1 Illustrations of LEARN: (a) power-efficient forwarding in 2α-sector region; (b) traditional greedy forwarding when the sector region is empty. © 2006, IEEE.

network cannot guarantee that an arbitrary routing scheme will successfully find a path for any pair of nodes. As an example, in this section, we study the routing protocol LEARN (Y. Wang *et al.*, 2006) and the CTR such that it will be a.a.s. successful.

First, a power-efficient localized routing method, called *localized-energy-aware restricted-neighborhood routing* (LEARN) (Y. Wang *et al.*, 2006), is described in detail. We assume that each wireless node v_i can dynamically adjust its transmission power based on the neighboring node with which it wants to communicate. We further assume that the energy needed to support the transmission of a unit amount of data over a link uv is $\mathbf{c}(\|uv\|)$, where $\mathbf{c}(x)$ is a nondecreasing function on x. Let $\mathbf{r}_0$ be the value such that $\frac{\mathbf{r}_0}{\mathbf{c}(\mathbf{r}_0)} = \max_x \frac{x}{\mathbf{c}(x)}$. We call $\frac{\mathbf{r}_0}{\mathbf{c}(\mathbf{r}_0)}$ the maximum *energy mileage* under energy-consumption model $\mathbf{c}(x)$. Here, we assume that the derivative of function $d[\frac{\mathbf{c}(x)}{x}]/dx = \frac{\mathbf{c}'(x)x - \mathbf{c}(x)}{x^2}$ is monotone increasing; thus, $\mathbf{r}_0$ is unique. We assume that the energy mileage $\frac{x}{\mathbf{c}(x)}$ is an increasing function when $x < \mathbf{r}_0$ and is a decreasing function when $x > \mathbf{r}_0$. The definition of energy mileage provides us the insight into designing energy-efficient routing (without considering retransmissions here). Whenever possible, we should use a link that has a larger energy mileage. Additionally, to save energy consumption, the total distance traveled should be as small as possible. Thus, intuitively, the LEARN routing protocol will work as follows. The current intermediate node u with message will first find the "best" neighbor v among all neighbors w such that $\angle wu\mathbf{t} \le \alpha$ for a parameter $\alpha < \pi/3$. Here, we define the "best neighbor" as the node v such that $\frac{\|uv\|}{\mathbf{c}(\|uv\|)}$ is maximum among all such neighbors; the use of the angle α is to bound the total distance of the routing path. Algorithm 34 illustrates our localized energy-aware routing protocol. Recall that $\mathbf{r}_0$ is the best link length that achieves that maximum energy mileage.

To make later analysis easier, we call the routing algorithm LEARN if *no* greedy routing and *no* face routing are used when no node v satisfies $\angle vu\mathbf{t} \le \alpha$. If greedy routing is applied afterward, then the routing protocol is called LEARN-G. Furthermore, if face routing is used at the end to get out of the local minimum, the routing protocol is called LEARN-GF. Figure 12.1 illustrates the localized routing algorithm. If no such neighbor v exists, then we use the traditional greedy routing, as shown in Figure 12.1(b).

The rest of this section is devoted to studying the asymptotic transmission range for the LEARN routing method. In any greedy-routing method, the packet may be dropped

---

Algorithm 34 LEARN: Localized Energy-Aware Restricted Neighborhood Routing

---

Input: A parameter $0 < \alpha < \frac{\pi}{3}$, energy model $\mathbf{c}(x)$, $\mathbf{r}_0$, and two constant parameters $\eta_1 < \eta_2$. (e.g., $\alpha = \frac{\pi}{4}$, $\mathbf{c}(x) = x^2 + c$, $\mathbf{r}_0 = \sqrt{c}$, $\eta_1 = 1/2$ and $\eta_2 = 2$).

1: Assume the uniform transmission range of all nodes is r.

2: **while** node u receives a packet with destination $\mathbf{t}$ **do**

3: **if** $\|t - u\| \leq r$ **then**

4:   Node u forwards the data to $\mathbf{t}$ directly and returns.

5: **end if**

6: **if** $\mathbf{r}_0 < r$ **then**

7:   **if** $\exists v$ with $\eta_1 \mathbf{r}_0 \leq \|uv\| \leq \eta_2 \mathbf{r}_0$ and $\angle vu\mathbf{t} \leq \alpha$ **then**

8:    Node u forwards the packet to such a neighbor v such that $|\|uv\| - \mathbf{r}_0|$ is minimized.

9:   **else if** $\exists v$ with $\|\mathbf{t} - v\| < \|\mathbf{t} - u\|$ and $\angle vu\mathbf{t} \leq \alpha$ **then**

10:    Node u forwards the packet to the node v with the minimum $\|\mathbf{t} - v\|$.

11:   **else if** $\exists v$ with $\|\mathbf{t} - v\| < \|\mathbf{t} - u\|$ **then**

12:    Node u forwards the packet to the node v with the minimum $\|\mathbf{t} - v\|$. In other words, node u applies the traditional greedy routing.

13:   **else**

14:    Node u applies the face-routing method to guarantee the delivery, or simply drops the packet.

15:   **end if**

16: **else**

17:   {Comment: The following is about the case $\mathbf{r}_0 \geq r$}

18:   **if** $\exists v$ with $\|\mathbf{t} - v\| < \|\mathbf{t} - u\|$ and $\angle vu\mathbf{t} \leq \alpha$ **then**

19:    Node u forwards the packet to the node v with the minimum $\|\mathbf{t} - v\|$.

20:   **else if** $\exists v$ with $\|\mathbf{t} - v\| < \|\mathbf{t} - u\|$ **then**

21:    Node u forwards the packet to the node v with the minimum $\|\mathbf{t} - v\|$. In other words, node u applies the traditional greedy routing.

22:   **else**

23:    Node u applies the face-routing method to guarantee the delivery or simply drops the packet.

24:   **end if**

25: **end if**

26: **end while**

---

by some intermediate node u before it reaches the destination $\mathbf{t}$ when node u could not find any of its neighbors that are "better" than itself. Thus, to ensure that the routing is successful for every pair of possible source and destination nodes, each node in the network should have a sufficiently large transmission range such that each intermediate node u will always find a better neighbor. Assume that V is the set of all wireless nodes in the network and each wireless node has a transmission range r. Then, the physical communication network is modeled by a UDG $G(V, r)$, where two nodes u and v are connected in $G(V, r)$ if and only if their Euclidean distance is at most r. A routing method $\mathcal{A}$ is *successful* over a network G if the routing method $\mathcal{A}$ can find a path for any pair of source and destination nodes. Given a routing method $\mathcal{A}$ and a set of wireless nodes V, we define the *critical transmission range*, denoted as $\rho_{\mathcal{A}}(V)$, for successful routing of $\mathcal{A}$ over V as the *minimum* transmission range r such that the routing method $\mathcal{A}$ over the network $G(V, r)$ is successful. The subscript $\mathcal{A}$ is omitted from $\rho_{\mathcal{A}}(V)$ if it is clear from the context.

Previously, several studies (e.g., Bettstetter, 2002; Gupta and Kumar, 1998; X.-Y. Li *et al.*, 2004b; Penrose, 1997) focused on the CTR for certain network properties such as connectivity, k-connectivity, and coverage. Surprisingly, there is not much study for the CTR for certain routing methods, except a recent result (Wan and Yi, 2006) for traditional greedy routing (Bose and Morin, 2001). Obviously, for traditional greedy routing, which selects a neighbor v of u that is closest to the destination node $\mathbf{t}$, and $\|v - \mathbf{t}\| < \|u - \mathbf{t}\|$, the critical transmission range $\rho(V)$ for successful routing is $\max_{u,v} \min_{w \in lune(u,v)} \|w - u\|$ where lune $lune(u, v)$ is the intersection of two disks centered at u and v, respectively, with $\|u - v\|$ as radius. It was proved in Wan and Yi (2006) that $\rho(\mathcal{P}_n) = \sqrt{\frac{\beta_0 \ln n}{\pi n}}$ a.a.s. for $\beta_0 = 1/(\frac{2}{3} - \frac{\sqrt{3}}{2\pi}) \simeq 1.6^2$ and Poisson point process $\mathcal{P}_n$ of density n over a convex compact region Ω with unit area and bounded curvatures.

It is easy to show that, given a set of nodes V already distributed in a region Ω, the CTR $\rho(V)$ for successful routing by restricted greedy routing LEARN is $\max_{u,v} \min_{w:\, \angle wuv \leq \alpha} \|w - u\|$, where α is the parameter used by LEARN. When the destination node is fixed, say node $\mathbf{t}$, the CTR will be $\max_u \min_{w:\, \angle wut \leq \alpha} \|w - \mathbf{t}\|$. Here, we prove a similar result as in Wan and Yi (2006) for our restricted greedy-routing method LEARN. We also assume that the network nodes are given by a Poisson point process $\mathcal{P}_n$ of density n over a convex compact region Ω with unit area and bounded curvatures.

THEOREM 12.14 *The LEARN routing (with parameter $\alpha < \frac{\pi}{3}$) will a.a.s. find a path from the source to the target when the transmission range r_n satisfies $n\pi r_n^2 = \beta \ln n$ for any constant $\beta > \beta_0 = (\pi/\alpha)$.*

Proof. To prove this, it is sufficient to show that when each node has a transmission range r_n satisfying the preceding condition, for every node u and every node v there is always a node $w \in \mathcal{P}_n$ such that $\angle wuv \leq \alpha$ and $\|w - u\| \leq r_n$; i.e., any intermediate node u can find a "better" neighbor w toward the destination node v.

Given a point distribution $\mathcal{P}_n$, let $\mathcal{S}(\mathcal{P}_n, r_n)$ be the minimum number of such neighboring nodes w that can be chosen by any intermediate node u for any possible destination v. As proved in Wan and Yi (2006), it suffices to prove that the cardinality $|\mathcal{S}(\mathcal{P}_n, r_n)| > 0$. We actually will show a much stronger result that $\mathcal{S}(\mathcal{P}_n, r_n) \geq \frac{1}{2}\mathcal{L}(\frac{\beta}{\beta_2}) \ln n$ for any constant $\beta_0 < \beta_2 < \beta$. Here, $\mathcal{L}(x)$ is defined in Wan and Yi (2006) as $\mathcal{L}(x) = x\phi^{-1}(1/x)$ for $x > 0$ and $\phi(x) = 1 + x \ln x - x$ for $x > 0$.

Notice that, given a node u, the area that node u can choose its neighbor to forward data for a given destination t is a sector of disk $\mathcal{D}(u, r_n)$ with angle 2α. Here, $\mathcal{D}(u, r)$ denotes the disk centered at node u with radius r. We use $\mathbf{Y}$ to denote such a sector. Let d denote the diameter of this sector $\mathbf{Y}$. Clearly, $d = r_n$ when $\alpha \leq (\pi/6)$, and $d = 2 \sin \alpha r_n$ when $(\pi/6) \leq \alpha < (\pi/3)$.

Assume that the space is partitioned into grids (quadrates) of side length η, which we call η-tessellation of space. In this section, we consider ϵd-tessellation, where ϵ is a constant to be specified later. A *polyquadrate* is defined as the set of quadrates that intersect with a convex and compact region (e.g., $\mathbf{Y}$). Notice that when the grid partition shifts, we will have different polyquadrates for the fixed region $\mathbf{Y}$. A polyquadrate in an η-tessellation is said to have a *span s* if it can be contained in a square of side length $s\eta$. We are interested only in polyquadrates that have spans of at most $\frac{1}{\epsilon}$ and areas of at least a certain fraction of πr_n^2. Assume that, given $\mathbf{Y}$, there are I_n different such polyquadrates that are completely contained inside, with spans of at most $(1/\epsilon)$ and areas of at least $\delta_1 \pi r_n^2$. Here, δ_1 is a constant to be specified later. For ith such polyquadrates, let X_i denote the number of nodes of V contained inside. Also assume that there are I_n' different such polyquadrates that intersect the boundary of $\mathbf{Y}$ with spans of at most $\frac{1}{\epsilon}$ and areas of at least $\delta_2 \pi r_n^2$. Here, δ_2 is a constant to be specified later. For ith such polyquadrates, let X_i' denote the number of nodes of V contained inside. According to Lemma 4 of Wan and Yi (2006), we have $I_n = O\left(\left(\frac{1}{\epsilon d}\right)^2\right) = O\left(\frac{n}{\ln n}\right), I_n' = O\left(\frac{1}{\epsilon d}\right) = O\left(\sqrt{\frac{n}{\ln n}}\right)$. Furthermore, X_i are Poisson random variables with rates of at least $(\delta_1 \pi r_n^2)n = \beta \delta_1 \ln)n$. Similarly, X_i' are Poisson random variables with rates of at least $(\delta_2 \pi r_n^2)n = \beta \delta_2 \ln n$. Then, according to Lemma 6 of Wan and Yi (2006), we have $\min_{i=1}^{I_n} X_i > \mathcal{L}(\beta') \ln n$ a.a.s. for any $1 < \beta' < \delta_1 \beta$, and $\min_{i=1}^{I_n'} X_i' > \frac{1}{2}\mathcal{L}(2\beta'') \ln n$ a.a.s. for any $1 < \beta'' < \delta_2 \beta$. Thus, we a.a.s. have $\min(\min_{i=1}^{I_n} X_i, \min_{i=1}^{I_n'} X_i') \geq \min[\mathcal{L}(\beta'), \frac{1}{2}\mathcal{L}(2\beta'')] \ln n$ for any $1 < \beta' < \delta_1 \beta$, and $1 < \beta'' < \delta_2 \beta$.

To prove the theorem, it is sufficient to show that

$$\mathcal{S}(\mathcal{P}_n, r_n) \geq \min\left(\min_{i=1}^{I_n} X_i, \min_{i=1}^{I_n'} X_i'\right).$$

In other words, we only need to show that for any $\mathbf{Y}$,

1. It either contains a polyquadrate P that has a span of at most $\frac{1}{\epsilon}$ and an area of at least $\delta_1 \pi r_n^2$ when Y is inside Ω; or
2. It contains a polyquadrate P' that has a span of at most $\frac{1}{\epsilon}$ and an area of at least $\delta_1 \pi r_n^2$ when $\mathbf{Y}$ intersects the boundary of Ω.

First, consider the case in which $\mathbf{Y}$ is contained inside Ω. For the polyquadrate P, we consider the induced polyquadrate (denoted by $P_{-\sqrt{2}\epsilon d}$) formed by all quadrates of P that intersect with region $\mathbf{Y}_{-\sqrt{2}\epsilon d}$. Here, $\mathbf{Y}_{-x}$ denotes the region of $\mathbf{Y}$ that is a distance of at least x from the boundary of $\mathbf{Y}$. The span of polyquadrate $P_{-\sqrt{2}\epsilon d}$ is at most $\left[\frac{d-2\sqrt{2}\epsilon d}{\epsilon d}\right] + 1 < \frac{1}{\epsilon}$. The area of the polyquadrate $P_{-\sqrt{2}\epsilon d}$ is at least the area of $\mathbf{Y}_{-\sqrt{2}\epsilon d}$. Notice that $|\mathbf{Y}_{-\sqrt{2}\epsilon d}| = \alpha(r_n - 2\sqrt{2}\epsilon d)^2$. Thus, it is sufficient to require that $|\mathbf{Y}_{-\sqrt{2}\epsilon d}| = \alpha(r_n - 2\sqrt{2}\epsilon d)^2 \geq \delta_1 \pi r_n^2$.

We now consider the case in which $\mathbf{Y}$ intersects the boundary of Ω. Let $\mathbf{Y}'$ be the part that is fully contained inside Ω. Similarly, we consider the polyquadrate $P'_{-\sqrt{2}\epsilon d}$ induced by $\mathbf{Y}'_{-\sqrt{2}\epsilon d}$; i.e., the quadrates that are contained inside Ω and $\mathbf{Y}$. Clearly, the span of this polyquadrate is also at most $\frac{1}{\epsilon}$. It is not difficult to show that the area of $\mathbf{Y}'$ is at least $\frac{1}{2}$ of the area of $\mathbf{Y}$. Thus, the area $|P'_{-\sqrt{2}\epsilon d}|$ of polyquadrate

$P'_{-\sqrt{2}\epsilon d}$ is at least $|P'_{-\sqrt{2}\epsilon d}| \geq |\mathbf{Y}'_{-\sqrt{2}\epsilon d}| \geq \frac{1}{2}|\mathbf{Y}_{-\sqrt{2}\epsilon d}|$. Thus, it is sufficient to require that $\frac{1}{2}|\mathbf{Y}_{-\sqrt{2}\epsilon d}| = \frac{1}{2}\alpha(r_n - 2\sqrt{2}\epsilon d)^2 \geq \delta_2 \pi r_n^2$.

In summary, we require the following conditions for the parameters δ_1, δ_2, and ϵ:

$$\beta\delta_1 > 1,$$
$$\beta\delta_2 > 1,$$
$$\alpha\left(1 - 2\sqrt{2}\epsilon\frac{d}{r_n}\right)^2 \geq \delta_1\pi,$$
$$\alpha\left(1 - 2\sqrt{2}\epsilon\frac{d}{r_n}\right)^2 \geq 2\delta_2\pi.$$

Clearly, we can choose $\delta_2 = \frac{1}{2}\delta_1$. Notice that we defined $\beta_0 = \frac{\pi}{\alpha}$. Thus, it is equivalent to require that

$$\beta\delta_1 > 1,$$
$$\left(1 - 2\sqrt{2}\epsilon\frac{d}{r_n}\right)^2 \geq \delta_1\beta_0.$$

This clearly has a solution when $\beta > \beta_0$. For example, we can select a constant β_1 such that $\beta_0 < \beta_1 < \beta$ and let $\delta_1 = \frac{1}{\beta_1}$. Then, $\epsilon = \frac{1-\sqrt{\beta_0/\beta_1}}{2\sqrt{2}d_0}$. Here, $d_0 = \frac{d}{r_n}$, which is 1 when $\alpha \leq (\pi/6)$ and is $2\sin\alpha$ when $(\pi/6) \leq \alpha < (\pi/3)$. In this case, we can choose β_2 such that $\beta_0 < \beta_1 < \beta_2 < \beta$ and set $\beta' = (\beta/\beta_2)$ and $\beta'' = \beta'/2$. Then, we have $\mathcal{S}(\mathcal{P}_n, r_n) \geq \frac{1}{2}\mathcal{L}(\beta/\beta_2)\ln n$.

Notice that when $r_n = \sqrt{\frac{\beta\ln n}{\pi n}}$ for some $\beta > \beta_0 = (\pi/\alpha)$, the probability that an intermediate node u cannot find a forwarding node w to a destination $\mathbf{t}$ is $e^{-n\alpha r_n^2}$; i.e., the probability that the sector does not contain any node. Because the path from any source node to any destination node contains at most n hops, the probability that LEARN routing protocol is successful is at least $(1 - e^{-n\alpha r_n^2})^n = (1 - n^{-\frac{\beta}{\beta_0}})^n > 1 - \frac{1}{n^{\beta/\beta_0 - 1}}$, which goes to 1 as $n \to \infty$. ∎

Notice that in the preceding proof, we can find ϵ, δ_1, and δ_2 only if $\beta > \beta_0 = (\pi/\alpha)$. It is thus natural to conjecture that β_0 is the threshold value. It is indeed true based on the following theorem:

THEOREM 12.15 *The LEARN routing (with parameter $\alpha < \frac{\pi}{3}$) will a.a.s. not be able to find a path from the source to the target when the transmission range r_n satisfies $n\pi r_n^2 = \beta\ln n$ for any constant $\beta < \beta_0 = \frac{\pi}{\alpha}$.*

Proof. We basically show that a.a.s., there are two nodes u and v such that we cannot find a node w for forwarding by node u; i.e., there does not exist node w with $\angle wuv \leq \alpha$ and $\|w - u\| \leq r_n$. Recall that $r_n = \sqrt{\frac{\beta\ln n}{\pi n}}$ for $\beta < \beta_0$. Again, we partition the space by using grids, in which each grid has side length ηr_n for a constant $0 < \eta$, to be specified later. Then, it is easy to show that the number of cells, denoted by I_n here, that are fully contained inside the compact and convex region Ω with unit area is $\Theta(\frac{1}{\eta^2 r_n^2}) = \Theta(\frac{n}{\ln n})$. Let $E_{u,v}$ denote the event that no forwarding node w (with $\angle wuv \leq \alpha$ and $\|w - u\| \leq r_n$) exists for node u to reach node v. Then, to prove our theorem, it is equivalent to prove that the probability that at least one of the event $E_{u,v}$ a.a.s. happens; i.e., $1 - \Pr(\text{none of event } E_{u,v} \text{ happens})$. Clearly, the events $E_{u,v}$ are not independent for all pairs u and v. We consider a special subset of events that is

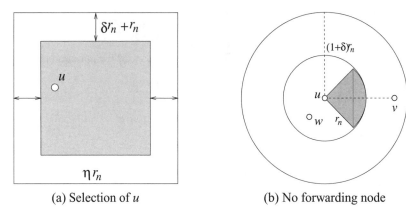

(a) Selection of u (b) No forwarding node

Figure 12.2 Illustrations of the proof of lower bound: (a) a cell and the area where we will select a node u; (b) the event that node u cannot find a forwarding node w to reach a node v. © 2006, IEEE.

independent. Consider any cell produced by the grid partition that is contained inside Ω. See Figure 12.2(a) for an illustration. For each cell, we draw a shaded square with side length $[\eta - 2(1 + \delta)]r_n$, and it has a distance of $(1 + \delta)r_n$ to the boundary of the cell. We consider only the case when node u is located in the shaded square, as in Figure 12.2(a). We also restrict the node v to satisfy that $r_n < \|u - v\| \le (1 + \delta)r_n$; i.e., in the torus area in Figure 12.2(b). Clearly, node v will also be inside this cell, and the shaded sector area [see Figure 12.2(b)] where the possible forwarding node could be located is also inside this cell. Thus, events E_{u_1,v_1} and E_{u_2,v_2} will be independent if u_1 and u_2 are selected as previously from different cells.

Now, for each cell i, we compute the probability that the event E_{u_i,v_i} happens, where u_i is selected from the shaded square of cell i and v_i is selected such that $r_n < \|v_i - u_i\| \le (1 + \delta)r_n$. Recall that for any region with area A, the probability that this region is empty of any nodes (for a Poisson process with rate n) is e^{-nA}. Clearly, the probability that node u_i exists is $1 - e^{-n(\eta-2-2\delta)r_n^2}$ because the shared square has area $(\eta - 2 - 2\delta)r_n^2$; the probability that node v_i exists is $1 - e^{-n(\delta^2+2\delta)r_n^2}$ because the torus has area $(\delta^2 + 2\delta)r_n^2$. Given node u_i and v_i, the probability that event E_{u_i,v_i} happens is $e^{-n\alpha r_n^2} = e^{-\beta/\beta_0 \ln n} = n^{-\beta/\beta_0}$. Consequently, the probability that event $E_{u,v}$ happens for some node pairs u and v is

$$\Pr\left(E_{u_i,v_i}\right) \ge (1 - e^{-n(\eta-2-2\delta)r_n^2})(1 - e^{-n(\delta^2+2\delta)r_n^2})n^{-\beta/\beta_0}$$
$$= (1 - n^{-\beta(\eta-2-2\delta)/\pi})(1 - n^{-\beta(\delta^2+2\delta)/\pi})n^{-\beta/\beta_0}.$$

Thus, the probability that the LEARN routing fails to find a path for some pair of source and destination nodes is

$$\Pr\,(\text{at least one of events } E_{u,v} \text{ happens})$$
$$\ge \Pr\left(\text{at least one of events } E_{u_i,v_i} \text{ happens}\right)$$
$$= 1 - \Pr\left(\text{none of events } E_{u_i,v_i} \text{ happens}\right)$$
$$= 1 - [1 - \Pr(E_{u_i,v_i})]^{I_n}$$
$$= 1 - e^{I_n \ln[1-\Pr(E_{u_i,v_i})]} \ge 1 - e^{-I_n \Pr(E_{u_i,v_i})}.$$

Notice that $I_n\mathrm{Pr}(E_{u_i,v_i}) = \Theta(\frac{n}{\ln n})[1 - n^{-\beta(\eta-2-2\delta)/\pi}][1 - n^{-\beta(\delta^2+2\delta)/\pi}]n^{-\beta/\beta_0} \simeq \frac{n^{1-\beta/\beta_0}}{\ln n}$, which goes to ∞ as $n \to \infty$ when $\beta < \beta_0$, $\eta - 2 - 2\delta > 0$, and $\delta > 0$. This can be easily satisfied; e.g., $\delta = 1$, $\eta = 5$. Thus, $\lim_{n\to\infty} 1 - e^{-I_n\mathrm{Pr}(E_{u_i,v_i})} = 1$. This finishes the proof. $\blacksquare$

Notice that the preceding results assume that the deployment region Ω has a unit area. Generally, we will often have a convex and compact region Ω with area $\mathcal{D}$, and the transmission range r could be fixed (or dynamically changed based on node density). Assume again that the network nodes are produced by a Poisson process with rate n (i.e., the expected number of nodes in a unit area is n, and thus the total number of deployed nodes in the area is $n\mathcal{D}$). Then, by a proper scaling of the distance unit, we have the following theorem:

THEOREM 12.16 *When the transmission range* r_n *and the Poisson process rate* n *satisfy* $n\pi r_n^2 = \beta \ln(\mathcal{D} \cdot n)$ *for any* $\beta > \beta_0$, *our LEARN routing protocol will a.a.s. successfully route the data. When* $n\pi r_n^2 = \beta \ln(\mathcal{D} \cdot n)$ *for any* $\beta < \beta_0$, *our LEARN protocol will* not *be able to a.a.s. route the data.*

So far, we have mainly concentrated on the routing LEARN. Notice that the CTR of our LEARN-G routing protocol will be exactly the same as the traditional greedy-routing (Bose *et al.*, 2001) method because at last we use the greedy routing to find the forwarding node if LEARN fails. There are a number of other localized routing methods developed already and many to be developed in the future. We thus would like to know the general CTR range for successful routing by any localized routing method $\mathcal{A}$. We would like to generalize the preceding theorems. We make the following conjecture:

CONJECTURE 12.2 *For a general localized routing method* $\mathcal{A}$, *the CTR is* $\rho_{\mathcal{A}}(\mathcal{P}_n) = \sqrt{\frac{\beta_{\mathcal{A}} \ln(\mathcal{D}\cdot n)}{n\pi}}$. *Here,* $\beta_{\mathcal{A}}$ *is the ratio of the area of the disk centered at an intermediate node* u *with radius* r_n *over the area of the forwarding region in this disk from where the intermediate node* u *can choose its next neighbor* w; $\mathcal{D}$ *is the area of the convex and compact deployment region; and* n *is the rate of the Poisson point process. We require that the forwarding region of any intermediate node* u *for any target node* t *be* convex and compact *and at least a constant fraction of the forwarding region be contained inside the deployment region* Ω.

Notice that the preceding conjecture applies not only to the routing method, it also applies to the critical range for the connectivity of the network in which $\beta_{\mathcal{A}} = 1$. This is based on the following observation: A network, formed by a set V of n nodes and each node with a transmission range r_n, is connected if and only if the routing method $\mathcal{H}$ that uses the path with the minimum hop number can successfully find a path for every pair of source and destination nodes. For this special routing method $\mathcal{H}$, clearly the area to find the forwarding node w by a node u is the disk $\mathcal{D}(u, r_n)$; i.e., $\beta_{\mathcal{H}} = 1$.

So far, we have assumed that the link is always reliable and the nodes are always awake. This is always an ideal case. To capture the practical aspects of wireless networks, we assume that a wireless link uv is reliable with a *constant* probability $p_1 > 0$ and each node is awake with a *constant* probability $p_2 > 0$. Similarly, if the preceding conjecture is true, we should be able to show that the CTR for a successful routing by a general routing method $\mathcal{A}$ is

$$\rho_{\mathcal{A}}(\mathcal{P}_n) = \sqrt{\frac{\beta_{\mathcal{A}} \ln(\mathcal{D}n)}{n\pi p_1 p_2}}.$$

12.7   Further Reading

It has been a folklore result that there is a CTR for a property $\mathbb{P}$ that is monotone-increasing. In Chapter 11, we studied in detail the transition phenomenon of the network connectivity (and k-connectivity) over the uniform transmission range of all wireless nodes.

One closely related question to the critical transmission radius is the *coverage* problem: Consider disks of radius r that are placed in a 2D unit-area disk $\mathcal{D}$ with centers from a Poisson point process with intensity n; when will these disks cover the unit disk? A result shown by Hall (1988) implies that if $n\pi r^2 = \ln n + \ln \ln n + c(n)$ and $c(n) \to \infty$, then the probability that there is a vacancy area in $\mathcal{D}$ is 0 as n goes to infinity; if $c(n) \to -\infty$, the probability that there is a vacancy in $\mathcal{D}$ is at least $\frac{1}{20}$. This implies that the hitting radius r_n such that $G(V, r_n)$ is covered satisfies $\pi r_n^2 \le 4\frac{\ln n + \ln \ln n + c(n)}{n}$ for $c(n) \to +\infty$. Recently, Yi (2005) provided a more accurate formula for the CCR to achieve k-coverage; i.e., each point in the region is covered by at least k-disks. Surprisingly, Yi's asymptotical bound

$$r_n = \sqrt{\frac{\ln n + (2k - 1)\ln \ln n + \xi_n}{\pi n}}$$

for k-coverage is almost the same as that of the CTR for achieving k-connectivity.

Another closely related problem is this: When will a Bernoulli graph be connected if we increase the probability of the links being chosen? Let $\mathcal{B}[n, p(n)]$ be the set of graphs on n nodes in which each edge of the completed graph K_n is chosen independently with probability $p(n)$. It has been shown that the probability that a graph in $\mathcal{B}[n, p(n)]$ is connected goes to one if $p(n) = \frac{\ln n + c(n)}{n}$ for any $c(n) \to \infty$. Although Yi's asymptotic expressions are the same as those by Gupta and Kumar (1998), we cannot apply these to the wireless model because in wireless networks, the existence of two edges is not independent, and we do not choose edges from the completed graph by using a Bernoulli model. Recently, Yi (2005) and Wan and Yi (2004) provided an exact formula of the CTR for k-connectivity with Bernoulli wireless networks. They also studied the number of isolated nodes in such a network.

Notice that Penrose proved that to get a k-connected network from a random geometry graph, it is asymptotically the same as requiring that each node have a degree of at least k in the random geometry graph. However, this result does not imply that to guarantee that a graph over n points is a.a.s. k-connected, we only have to connect every node to its k-nearest neighbors. Let V be n points randomly and uniformly distributed in a unit square (or disk). Xue and Kumar (2004) proved that to guarantee that a geometry graph over V is connected, the number of nearest neighbors that every node has to connect is asymptotically $\Theta(\ln n)$.

Santi and Blough (2003) provided a detailed study of the CTR for connectivity for sparse networks in which the density will not go to infinity. They also presented some formulas for the CTR when the number of nodes is sufficiently small. When the nodes in a network are mobile, all our previous studies on the CTR cannot be applied directly. In this chapter, we studied several possible definitions of CTRs for mobile networks. Most of the results presented here are from Bettstetter and Krausser (2001), Bettstetter *et al.* (2003), Blough *et al.* (2003), and Santi (2005a, 2005b).

Observe that all these studies so far (e.g., Bettstetter, 2002; Gupta and Kumar, 1998; X.-Y. Li *et al.*, 2003c, 2004b; Penrose, 1997) focused on the CTR for certain network properties such as connectivity, k-connectivity, and coverage. Surprisingly, there is not much study for the CTR for certain routing methods, except a recent result (Wan and Yi, 2006) for traditional greedy routing (Bose *et al.*, 2001). In other words, what is the CTR r_n such that a routing method will be successful with high probability on a random network $G(V, r_n)$? Wan and Yi (2006) were the first to study the CTR for greedy routing on RNGs. Then, Y. Wang *et al.* (2006) studied the CTR for a new routing protocol, LEARN.

12.8   Conclusion and Remarks

In this chapter, we studied the CTR (and CNN and CCR) for a number of graph properties and routing schemes. We first studied the CNN k such that the network is connected with high probability when each node is connected to its k-nearest neighbors. We show that k is asymptotically $\Theta(\log n)$. Notice that this is also consistent with the CTR for network connectivity. When each node has a transmission range CTR $r_n = \sqrt{\frac{\log n + f(n)}{\pi n}}$ for $f(n) \to \infty$, the number of neighbors of each node is $\Theta(\log n)$ with high probability. We also studied the CTR for sparse networks in which the node density will not go to infinity. Another fundamental question studied is that of the CTR for mobile networks. It was shown that as long as the mobility model $\mathcal{M}$ is *stable* with initial deployment $\mathcal{F}$ (the deployment region having connected and compact support Ω with smooth boundary $\partial\Omega$, and $\mathcal{F}$ is continuous on the boundary $\partial\Omega$), then the CTR for connectivity can be reduced to the study of the CTR when all nodes are distributed according to $\mathcal{F}_\mathcal{M}$. When the wireless nodes are required to provide a coverage to a region, the minimum sensing range is called the CCR. It is surprising that the CCR for a network is asymptotically the same as the CTR for connectivity, although it is widely known that for any fixed static network, if the transmission range is at least twice the sensing range, coverage implies connectivity. These previous studies mainly concentrated on the graph properties of $G(v, r_n)$. In this chapter, with LEARN as a running example, we also showed how to study the CTR for a routing protocol. We also generalized this idea to study the CTR of a more general routing scheme as long as it satisfies certain conditions.

Problems

12.1   For random wireless networks, it was proved that each node should have at least $\log n$ neighbors to get a connected network with high probability. Is this bound consistent with the CTR? Recall that in previous chapters on topology control, we developed a number of methods that can get a connected network for which the degree of each node is bounded by a constant. Is this a contradiction to the results presented in this chapter that require each node to have degree $\Theta(\log n)$ to get a connected network with high probability?

12.2   Conduct simulations to study the CNN for random networks of n nodes such that the network is connected with high probability. You should plot this relation between the probability and the CNN when n is fixed. Also plot the other relation between the CNN and the value of n, in which the CNN is defined as the number of neighbors required to get $1 - (1/n)$ of the network samples connected.

12.3   Why could the CTR for mobile networks be much different from the CTR for static random networks?

12.4   Conduct extensive simulations to study the CTR for connectivity in a mobile network of n nodes in which n is sufficiently large. Here, the mobility models include the RWP model, the Brownian-like mobility model, the Markov random walk model, and i.i.d. the mobility model.

12.5   Assume that a set of n nodes is randomly placed in a unit-square region. Each node has a fixed transmission range r_T. From the CTR, decide what the critical node number n (as a function of r_T) is that, with high probability, we can get a connected network.

12.6   Assume that a set of n nodes is randomly placed in a unit-square region. Each node has a fixed coverage range r_C. From the CTR range, decide what the critical node number n (as a function of r_T) is that, with high probability, the region will be covered.

12.7   Assume that a set of n nodes is randomly placed in a unit-square region. Each node has a fixed transmission range r_T and a fixed coverage range r_C. The relation between r_C and r_T could be arbitrary. From the CTR and the CCR, decide what the critical node number n (as a function of r_T and r_C) is that we can, with high probability, get a connected network and the region will be covered.

12.8   Consider a complete graph of n nodes. Among $n^2/2$ edges in the complete graph, we randomly pick M edges and call the result graph $G_M(n)$. Prove that the graph $G_M(n)$ is almost surely connected if $M > n \log n/2$.

Part IV

Wireless Network Routing Protocols

13 Energy-Efficient Unicast Routing

13.1 Introduction

Multihop structures in wireless networks provide enhanced capacity and fault tolerance. This capacity allows the use of wireless nodes as repeaters, and thus not only enhances the range of communication at low power levels but at the same time short-hop communication causes less spatial interference and allows reuse of the bandwidth available on the frequency channels. The ability for nodes to act as intermediate routers builds into the communication system a natural resilience to node and link failures because alternative paths become available for routing of communications. An important requirement of these networks is that they be self-organizing; i.e., data paths or routes are dynamically restructured with changing topology.

One of the critical issues in the implementation of wireless networks is the design of routing structures and routing protocols. Of considerable importance in this context is the design of distributed efficient algorithms that dynamically update the routing structures. Because the geometric location information regarding the nodes is more readily available, routing algorithms that incorporate this information for effective routing form an increasingly important subject of study.

In this chapter and Chapter 14, we study a number of energy-efficient routing protocols for wireless ad hoc networks. Routing protocols can be categorized as *proactive protocols* or *reactive protocols*, depending on when the routing structure is constructed when a routing request is issued from a source node. Routing protocols can also be categorized as *flat routing protocols* and *hierarchical routing protocols*. For flat routing protocols, each node in the network plays the same role, whereas different nodes may play different roles in a hierarchical routing protocol (e.g., the ones based on the backbone or clustering).

Construction of optimal geometric structures, or virtual backbones, like DSs and CDSs, is NP-hard and thus intractable. Therefore, good approximations of the optimal structures are often sought. A number of centralized or distributed algorithms have been proposed in the literature for approximating the CDS, sparse spanners with a variety of constraints. Good structures forming a backbone on which routing is to be performed need to be maintained when nodes are mobile. In mobile networks involving nodes with limited computing power and memory (an example being sensor network nodes), explicit maintenance of routing tables may not be desired. We therefore prefer localized protocols that make routing decisions based on local neighborhood properties.

Naturally, ensuring a certain QoS is of importance. This includes guaranteed delivery within a specified period of time – a daunting task with limited routing information. We study localized protocols keeping in view the issues of QoS, power utilization, and transmissions ranges. We note that the efficient realization of local neighborhoods, say the two-hop neighborhood, becomes of utmost importance in this context.

13.2   Proactive Approaches

In proactive routing protocols (also called table-driven routing protocols), each node attempts to maintain consistent, up-to-date routing information with every other node in the network. These routing protocols require each node to maintain one or more tables to store routing information, and they respond to changes in network topology or another metric such as bandwidth and interference by propagating updates throughout the network in order to maintain a consistent network view.

13.2.1   Destination-Sequenced Distance-Vector Routing

A typical proactive routing protocol is the Destination-Sequenced Distance-Vector (DSDV) routing protocol described in Perkins and Bhagwat (1994b). Notice that the problem of routing is essentially the distributed version of the shortest-path problem, which can be solved efficiently with the classic Bellman–Ford routing mechanism (Cormen *et al.*, 2001). It is thus no surprise that DSDV is based on the Bellman–Ford mechanism. The improvements made to the Bellman–Ford algorithm include freedom from loops in routing tables. Each node will maintain a routing table that stores (1) a preferred neighbor for every possible destination within the network, and (2) the number of hops to each destination node. The main differences among almost all proactive protocols are the manners in which the routing tables are constructed, maintained, and updated. Specifically, in a distance-vector (DV) algorithm, each node i will maintain, for each possible destination node x, a set of distances $\{d_{i,j}^x \mid j$ is a neighbor of node $i\}$. Here, $d_{i,j}^x$ is the distance computed by node i, from node i to node x if node j is chosen as the next hop node. A neighboring node k will be treated as the next hop for destination x if $d_{i,k}^x = \min_j d_{i,j}^x$. The routing from i to a destination node x is done via successive next hops chosen in this matter.

Each entry of the table maintained by the DSDV is also associated with a sequence number assigned by the destination node, which enables the nodes to distinguish the stale route from new routes. The DSDV uses destination sequence numbers to provide loop-free routes at every instant. The routing table is updated periodically by requiring that every node transmit its routing table throughout the network. Several techniques using some special packets (e.g., *full dump* and *incremental*) have been discussed to alleviate the potentially large amount of network traffic generated by this table updating. New routing broadcasts contain the address of the destination node, the best-known hop count of the current node to the destination node, and the sequence number of the information received about the destination. Because a number of different such routing packets may be received by a node u, the node u will always select the route with the most recent sequence number and the route with the smaller metric in the case of two or more routing packets having the same most recent sequence number. Mobile nodes will also delay the broadcast of a routing update by the length of the setting time, which is often the weighted average time that routes to the destination node fluctuate before the best route (with the best metric) is received.

13.2.2 Optimized Link-State Routing

The link-state approach is closer to the centralized version of the shortest-path computation. To utilize the centralized algorithm, each node will maintain a view of the network topology and also the cost for each link in the network. A challenge for link-state routing is to keep consistent views of the network topology by all nodes. Each node will periodically broadcast the link costs of all its own outgoing links to all nodes in the network by using a protocol such as flooding. When a node receives such information, it will update its view of the network topology and apply the shortest-path algorithm to update the routing (i.e., the next-hop node) for each possible destination node. Observe that the network topology information and the cost information viewed by a node at any instant could be *incorrect* for a number of reasons, such as propagation delay. These inconsistencies may lead to loops in the routing paths being found (recall that here, each node stores only the next-hop node on every path it computed, and the actual path will be formed by a chain of such "next-hop" information of a number of related nodes). It has been observed that these loops are shortlived because they will disappear when the correct network information spreads the whole network with time at most proportional to the diameter of the network.

The Optimized Link-State Routing (OLSR) (Clausen *et al.*, 2001) protocol operates as a table-driven and proactive protocol. This protocol is an optimization of the classic link-state algorithm tailored to the requirements of a mobile WLAN. The key concept used in this protocol is that of multipoint relays (MPRs). MPRs are selected nodes that forward broadcast messages during the flooding process. In the OLSR protocol, every node regularly exchanges topology information with other nodes of the network. Some nodes will be selected as a MPR by some of its neighboring nodes. These MPR nodes will announce this topology information periodically in their control messages. In OLSR, link-state information is generated only by nodes elected as MPRs, and only a MPR node announces to the network that it has a path to the nodes that have selected it as a MPR. In route calculation, the MPRs are used to form the route from a given node to any destination in the network. The protocol uses the MPRs to facilitate efficient flooding of control messages in the network, which is similar to the cluster-based (sometimes called backbone or CDS if cluster-heads are also connected via some intermediate nodes) routing protocols. This technique substantially reduces the message overhead compared with a classic flooding mechanism, in which every node retransmits each message when it receives the first copy of the message. As a third optimization, a MPR node may choose to report only links between itself and its MPR selectors. Hence, contrary to the classic link-state algorithm, partial link-state information is distributed in the network. This information is then used for route calculation.

Notice that in OLSR, each node u in the network selects a set of nodes in its symmetric one-hop neighborhood that may retransmit its messages, which is called the MPR set of that node. This set is selected such that it covers (in terms of radio range) all symmetric strictly two-hop nodes of u. The smaller a MPR set, the less control traffic overhead results from the routing protocol. The MPR for a node is also called the *forwarding neighbor* in Călinescu *et al.* (2001). They provide efficient approximation algorithms to find such forwarding neighbors for a node when the network is modeled by a UDG and the positions of all neighboring nodes are known. In other words, they proved that the number of nodes found by their algorithms is within a small constant factor of the optimum. When the network is a general graph, it can be shown that we can have a polynomial-time algorithm with approximation ratio $\Theta(\log n_0)$, where n_0 is the number of nodes within two hops. This can be done by the reduction to the set-cover problem, in which the strictly two-hop neighbors of node u are elements, and each one-hop neighbor of u corresponds to a set whose elements are all strictly two-hop neighbors of u. The neighbors of node u that are *not* in its MPR set receive and process broadcast messages but do *not* retransmit broadcast messages received from node u. Notice that for a general set-cover game, we cannot find a set of sets in polynomial time such that the number of selected sets to cover a universal set of elements is only $o(\log D)$ times of the optimum when certain complexity results hold. Here, D is the size of the largest set. Thus, we conjecture that when the network is modeled by a general arbitrary graph, we cannot approximate the minimum forwarding set for a node with an approximation ratio of better than $\Theta(\log n_0)$.

13.2.3   Metric-Based Routings

Both DSDV routing and OLSR are based on the shortest-path algorithm. When the cost of a link is simply one, then the routing protocol will find a path with the minimum number of hops to connect the source and the destination nodes. Besides the hop count, a number of different measurements have been used to assign a cost for every link in the network, such as the energy required to support the transmission of unit-amount data over this link (Chang and Tassiulas, 2004), the delay of the link, and the vulnerabilities of the link. Several other metrics are also used for determining the quality of the found path, such as the reliability of the link, the expected data volume (or capacity) of this link, and the battery power level of the sending node of the link. Notice that for the latter case, we will find a path such that the minimum metric of all links is maximized among all possible paths. In other words, the quality of the path is determined by the "worst link" on the path; e.g., the expected data volume of a path is determined by the smallest data volume of all links on the path. Observe that we can find such a best path by using a simple binary search on all possible link costs. For example, to find a path with the maximum data volume, we use a binary search, as follows. We sort all possible data volumes of all links in an increasing order, say $C_1, C_2, \ldots, C_{m-1}$, and C_m. We then test whether there is a path connecting the source and the target node by using links whose data volume is at most $C_{m/2}$. If such a path exists, we then repeat the preceding step by searching the best data volume in the range $C_{m/2+1}, C_{m/2+2}, \ldots, C_{m-1}$ and C_m. If such a path does not exist, we then repeat the preceding step by searching the best data volume in the range $C_1, C_2, \ldots, C_{m/2-1}$ and $C_{m/2}$. Several combinations of a variety of different metrics are also used, such as minimum energy routing and the residual battery power routing.

Notice that, so far, we have assumed that we always want to find a path that will optimize one objective function without constraints. In certain scenarios, we may want to find a path that optimizes one objective function while it satisfies some other additional constraints. For example, we may want to find a path with the smallest total energy cost while the delay of the path is at most a certain value. Notice that the total energy cost of a path is the *summation* of the energy cost of all links on the path. This energy-cost metric is called *additive*. Similarly, the delay metric is also additive because the delay of a path is defined as the summation of the delay of all links on the path. Observe that when the data volume (or bandwidth) is used as a metric for a link, the data volume of the path is the *minimum* data volume of all links on the path. In other words, the data volume metric is *not* additive. Some other constraints do not seem to be additive but are equivalent to some additive constraints. For example, assume that each link e is associated with a *reliability* measurement r_e. Then, typically, the reliability of a path P can be defined as the *multiplication* of the reliability of all links on this path; i.e., $\prod_{e \in P} r_e$. Typically, we would like to have a path with the largest reliability (or whose reliability is at least a certain value $0 < r_0 \leq 1$). This constraint can be converted to an equivalent constraint that is additive, as follows. Define a *reliability cost* $c_e = -\log r_e$. Then, finding a path with the maximum reliability is equivalent to finding a path with the minimum reliability cost $\sum_{e \in P} c_e$ because $-\log(\prod_{e \in P} r_e) = -\sum_{e \in P} \log r_e = \sum_{e \in P} c_e$. Unfortunately, it has been shown in the literature (Garey and Johnson, 1979) that it is NP-hard to find a path that optimizes an additive metric (e.g., smallest total energy) and that satisfies one or more constraint defined by an additive metric (e.g., delay). Given any arbitrarily small $\epsilon > 0$, Phillips (1993) and Hassin (1992) provided a centralized polynomial-time algorithm (called PTAS) for such problems with an approximation ratio of $1 + \epsilon$. Their algorithms specially studied the following problem. Given a graph G and a non-negative cost c_e and a non-negative weight w_e for each edge in the graph. We need to find a path P with the smallest cost $\sum_{e \in P} c_e$ while $\sum_{e \in P} w_e \leq W_0$ for a given W_0 that is polynomial of n. The algorithms presented in Phillips (1993) and Hassin (1992) can find a path P (that satisfies the weight constraint), whose cost is at most $1 + \epsilon$ times the optimum cost among all paths satisfying the weight constraint. The running time of their algorithms is a polynomial of the size of the graph G but not a polynomial on $1/\epsilon$.

13.2.4   Multipath Unicast Routing

For wireless networks, the prime path found for routing often may not be able to support the data flow from the source node to the target node because of fluctuating link bandwidths of the links on the path, or even the broken links on the path. Several techniques have been proposed in the literature to improve fault tolerance of the unicast routing. One approach (e.g., Lee and Gerla, 2000) is to use one or several backup paths: When the prime path cannot support the data flow, one of the backup paths will be invoked for routing. To perform such fault-tolerant routing, the underlying wireless networking structure must be fault-tolerant. A number of fault-tolerant structures (e.g., Bahramgiri *et al.*, 2002; Hajiaghayi *et al.*, 2003; N. Li and Hou, 2003; X.-Y. Li *et al.*, 2004b) were discussed in Chapter 9.

Another approach is to use multipath routing. In other words, two or more paths (typically node or link disjoint) will be used for routing data packets from the source node to the target node. When multipath routing is used, we need to find a collection of paths such that some objective function is optimized to improve the overall system performance. Depending on the main concern of the system, a variety of objective functions of multipath routing has been studied in the literature. For example, when energy is a concern, we may need to minimize the total power consumption of the set of multipaths. When throughput is a concern, we may need to find a set of multipaths such that the overall capacity from the source node to the target node is maximized. Notice that some link transmissions cannot happen simultaneously in wireless networks because of wireless interference. For multipath routings, we need to resolve the intrapath interference (for interference caused by links on the same single path) and interpath interference (for interference caused by links on different paths). Or, we may want to find a set of paths such that the overall reliability that they achieve is maximized. Or, we may want to find routing paths such that the overall system lifetime is maximized under certain information-generating conditions (Chang and Tassiulas, 2004).

13.3   Reactive Approaches

Besides the proactive routing protocols that build the routing tables before the routing is performed, there is another set of routing protocols that will build the routing path on demand, which are often called *reactive approaches* in the literature. Typical examples of reactive routing protocols are ad hoc on-demand vector (AODV) routing (Perkins and Royer, 1999) and dynamic source routing (DSR) (Johnson and Maltz, 1996).

The reactive routing protocol is typically composed of two mechanisms, *route discovery* and *route maintenance*, that work together to allow the discovery and maintenance of routes in the ad hoc network. Route discovery is the mechanism by which a source node S wishing to send a packet to a destination node D obtains a route to node D. The route-discovery mechanism is used only when S attempts to send a packet to D and does not already know a route to the destination node D. Route maintenance is the mechanism by which node S is able to detect, while using a route to the destination node D, whether the network topology has changed such that it can no longer use its route to destination node D because one or more links along the route no longer work. When route maintenance indicates that a route is broken, the source node S (or sometimes some intermediate node along the old path) can attempt to use any other route it happens to know to the destination node D, or it can invoke route discovery again to find a new route. Route maintenance is used only when the source node S is actually sending packets to the destination node D. Route discovery and route maintenance each operate entirely on demand.

13.3.1   Ad Hoc On-Demand Vector Routing

The AODV routing protocol was first described in Perkins (1997b) and Perkins and Royer (1999). The AODV protocol builds on the DSDV algorithm previously described and is an improvement on the DSDV. As opposed to maintaining a complete list of routes as in the DSDV algorithm, the AODV protocol will minimize the number of required broadcasts by creating routes on an on-demand basis. In the AODV protocol, nodes that are not on a selected path do not maintain routing information or participate in routing table exchanges (Perkins, 1997b; Perkins and Royer, 1999).

When a source node u wants to send a message to some destination node v and does not have a valid route to that destination yet, the node u initiates a *path-discovery process* to find a route to the destination node v. Every node maintains two separate counters, a *node sequence number* and a *broadcast_id*. The source node initiates path discovery by broadcasting a route request (RREQ) packet to its neighbors. The RREQ contains the following fields:

[source_addr, source_sequence #, broadcast_id, dest_addr, dest_sequence, hop_count]

Along with its own sequence number and the broadcast ID, the source node includes in the RREQ the most recent sequence number it has for the destination. The AODV utilizes destination sequence numbers to ensure that all routes are loop-free and contain the most recent route information. Each node maintains its own sequence number, *source_sequence*, as well as the *broadcast_id*. The broadcast_id is incremented for every RREQ the node initiates and, together with the node's IP address, uniquely identifies an RREQ.

Each neighbor either satisfies the RREQ by sending a route reply (RREP) back to the source node or rebroadcasts the RREQ to its own neighbors after increasing the *hop_count* until either the destination node v is reached or an intermediate node with a "fresh enough" route to the destination node v is found. Notice that a node may receive multiple copies of the same route broadcast packet from various neighbors. When an intermediate node receives a RREQ, if it has already received a RREQ with the same broadcast_id and source address, it drops the redundant RREQ and does not rebroadcast it. If a node cannot satisfy the RREQ, it keeps track of the following information in order to implement the reverse path setup as well as the forward path setup that will accompany the transmission of the eventual RREP:

- destination IP address
- source IP address
- broadcast_id

- expiration time for reverse path route entry
- source node's sequence number *source_sequence*

Intermediate nodes can reply to the RREQ only if they have a route to the destination whose corresponding destination sequence number is greater than or equal to that contained in the RREQ. During the process of forwarding the RREQ, intermediate nodes record in their route tables the address of the neighbor from which the first copy of the broadcast packet is received, thereby establishing a reverse path. As the RREP is routed back along the reverse path, nodes along this path set up forward route entries in their route tables that point to the node from which the RREP came. These forward route entries indicate the active forward route. Associated with each route entry is a route timer that will cause the deletion of the entry if it is not used within the specified lifetime. Because the RREP is forwarded along the path established by the RREQ, the AODV supports the use of only *symmetric links* in which two end nodes can communicate with each other directly.

In addition to the source and destination sequence numbers, other useful information is also stored in the route table entries and is called the *soft state* associated with the entry. Associated with reverse path-routing entries is a timer called the *RREQ expiration timer*. The purpose of this timer is to purge reverse path-routing entries from those nodes that do not lie on the path from the source to the destination. The expiration time depends on the size of the ad hoc network. Another important parameter associated with routing entries is the *route caching time out*, or the time after which the route is considered to be invalid. In each routing table entry, the address of active neighbors through which packets for the given destination are received is also maintained. A neighbor is considered *active* for that destination if it originates or relays at least one packet for that destination within the most recent *active_ time-out period*.

Routes are maintained as follows. The movement of nodes not lying along an active path does not affect the routing to that path's destination node. If the source node moves during an active session, it can reinitiate the route-discovery procedure to establish a new route to the destination. When either the destination or some intermediate node moves, a special RREP is sent to the affected source nodes. Periodic *hello* messages can be used to ensure symmetric links as well as to detect link failures, thus maintaining the local connectivity of a node. However, the use of *hello* messages is not required. Alternatively, and with far less latency, such failures could be detected by use of link-layer acknowledgments (LLACKs). A link failure is also indicated if attempts to forward a packet to the next hop fail. Once the next hop becomes unreachable, the node upstream of the break propagates an unsolicited RREP with a fresh sequence number (i.e., a sequence number that is one greater than the previously known sequence number) and hop count of ∞ to all active upstream neighbors. Those nodes subsequently relay that message to their active neighbors and so on. This process continues until all active source nodes are notified. It terminates because the AODV maintains only loop-free routes and there are only a finite number of nodes in the ad hoc network. On receiving notification of a broken link, source nodes can restart the discovery process if they still require a route to the destination.

13.3.2   Dynamic Source Routing

The DSR protocol is a simple and efficient routing protocol designed specifically for use in multihop wireless ad hoc networks of mobile nodes. DSR allows the network to be completely self-organizing and self-configuring, without the need for any existing network infrastructure or administration. The protocol is composed of two mechanisms, route discovery and route maintenance, that work together to allow nodes to discover and maintain source routes to arbitrary destinations in the ad hoc network. The use of source routing allows packet routing to be trivially loop-free, avoids the need for up-to-date routing information in the intermediate nodes through which packets are forwarded, and allows nodes that are forwarding or overhearing packets to cache the routing information in them for their own future use. All aspects of the protocol operate entirely on-demand, allowing the routing packet overhead of the DSR to scale automatically to only that needed to react to changes in the routes currently in use.

When a mobile node has a packet to send to some destination, it first consults its route cache to determine whether it already has a route to the destination. If it has an unexpired route to the destination, it will use this route to send the packet. Otherwise, it initiates route discovery by broadcasting a RREQ packet. This RREQ contains the address of the destination, along with the source node's address and a unique identification number. Each node receiving this RREQ packet checks whether it knows of a route to the destination. If it does not, it adds its own address to the route record of the packet and then forwards the packet along its outgoing links. To limit the number of RREQs propagated on the outgoing links of a node, a mobile node only forwards the RREQ if the request has not yet been seen by the mobile nodes before (thus, it requires all mobile nodes to store previously seen RREQ packets within a certain time period) and the mobile's address are not in the RREQ packet. Any intermediate node that contains in its route cache an unexpired route to the destination will generate a RREP packet, and thus this intermediate node will *not* forward the RREQ packet to its neighbors. The RREP contains a path from the source to the destination node that is the concatenation of the route from the source node to this intermediate node (known from the RREQ packet received by this intermediate node) and the path from this intermediate node to the destination node that was cached by this intermediate node. When the destination node receives a RREQ packet, it will also generate a RREP packet, which contains a path from the source node to the destination node. Notice that by the time the packet reaches either the destination or such an intermediate node, it contains a route record yielding the sequence of hops taken so far. Furthermore, to return the RREP, the responding node must have a route to the initiator. If it has a route to the initiator in its route cache, it may use that route. Otherwise, if symmetric links are supported, the node may reverse the route in the route record. If symmetric links are not supported, the node may initiate its own route discovery and piggyback the route reply on the new RREQ.

Route maintenance is accomplished through the use of *route-error packets* and acknowledgments. If the packet is retransmitted by some node u to its next-hop node v, the maximum number of times and no receipt confirmation is received from v, this node

u returns a ROUTE ERROR message to the original sender of the packet, identifying the link over which the packet could not be forwarded. When a route-error packet is received, the hop uv that produces such an error is removed from the node's route cache and all routes containing the hop are truncated at that point. In addition to route-error messages, acknowledgments are used to verify the correct operation of the route links. Such acknowledgments include passive acknowledgments, in which a mobile node is able to hear the next hop forwarding the packet along the route. A node forwarding or otherwise overhearing any packet may add the routing information from that packet to its own route cache. In particular, the source route used in a data packet, the accumulated route record in a RREQ, or the route being returned in a RREP may all be cached by any node. Routing information from any of these packets received may be cached, whether the packet was addressed to this node, sent to a broadcast (or multicast) MAC address, or received while the node's network interface is in promiscuous mode.

The ability for nodes to reply to a RREQ based on information in their route caches could result in a possible RREP "storm" in some cases. In particular, if a node broadcasts a RREQ for a target node for which the node's neighbors have a route in their route caches, each neighbor may attempt to send a RREP, thereby wasting bandwidth and possibly increasing the number of network collisions in the area. The following method of addressing this problem was discussed in Johnson and Maltz (1996). If a node can put its network interface into the promiscuous receive mode, it should delay sending its own RREP for a short period, while listening to see if the initiating node begins using a shorter route first. That is, this node should delay sending its own RREP for a random period

$$d = H \times (h - 1 + r),$$

where h is the number of network hops for the route to be returned in this node's RREP, r is a random number between 0 and 1, and H is a small constant delay (at least twice the maximum wireless link propagation delay) to be introduced per hop.

Each RREQ message contains a "hop limit" that may be used to limit the number of intermediate nodes allowed to forward that copy of the RREQ. As the RREQ is forwarded, this limit is decremented, and the REQUEST packet is discarded if the limit reaches zero before finding the target. For example, when the hop limit is set as 0 (the route request packet is called a *nonpropagating RREQ*), we have an inexpensive method of determining whether the target is currently a neighbor of the initiator or whether a neighbor node has a route to the target cached. If no RREP is received after a short time out, we can initiate a RREQ by setting the hop limit larger than 0.

Source routes in use may be automatically shortened if one or more intermediate hops in the route becomes no longer necessary. In some cases, DSR could also potentially benefit from nodes caching "negative" information in their route caches. For example, if a node u caches the fact that a link from a node x to another node y is currently broken (rather than simply removing this hop from its route cache), it can guarantee that no RREP that it receives in response to its new route discovery will be accepted that utilizes this broken link xy.

13.3.3   Opportunistic Routing

Biswas and Morris (2005) proposed an opportunistic routing protocol, ExOR, which is an integrated routing and MAC protocol that increases the throughput of large unicast transfers in multihop wireless networks. Compared with a traditional on-demand routing protocol, ExOR further delays the route construction to the last moment. The AODV and DSR build the routing path on demand: A path from the source node to the destination node is built only when the source node has some packets to be sent to the destination node; whereas, in ExOR, it chooses each hop of a packet's route after the transmission for that hop. This delay choice of next hop has several benefits. First, it avoids the possible links that work fine during the route discovery time period but are broken later. Second, the choice can reflect which intermediate nodes actually received the transmission by utilizing the broadcast nature of wireless signal propagation. Notice that after a node sends out a packet, it is often possible that several neighboring nodes will receive the data packet correctly. This deferred choice gives each transmission multiple opportunities to make progress. As a result, ExOR can use long radio links with high loss rates, which would be avoided by traditional routing. ExOR increases a connection's throughput while using no more network capacity than traditional routing.

However, this benefit does not come free. ExOR's design faces the following challenges. The nodes that receive each packet must agree on their identities and choose one forwarder. In other words, among several neighboring nodes that receive the packet correctly, we need to choose *only* one node to forward the data packet to avoid not only the duplications of the packets but also the increased resource usage and channel competition from potential extra packet transmissions by these neighboring nodes. The agreement protocol must have a low overhead so it will not offset the benefit gained by opportunistic transmitting but must also be robust enough that it rarely forwards a packet zero times or more than once. Finally, to ensure that the actual route formed by these forwarding nodes is efficient, ExOR must choose the forwarder with the lowest remaining cost to the ultimate destination.

ExOR operates on batches of packets in order to reduce the communication cost of agreement. The source node includes in each packet a list of candidate forwarders prioritized by closeness to the destination. Receiving nodes buffer successfully received packets and await the end of the batch. The highest priority forwarder then broadcasts the packets in its buffer, including its copy of the "*batch map*" in each packet. The batch map contains the sender's *best guess* of the highest priority node to have received each packet. The remaining forwarders then transmit in order but only send packets that were *not* acknowledged in the batch maps of higher priority nodes. The forwarders continue to cycle through the priority list until the destination has 90% of the packets. The remaining packets are transferred with traditional routing.

Here, the cost of a node to the destination node is defined as follows. The cost metric is the number of transmissions required to move a packet along *the best traditional route* from the node to the destination, counting both hops and retransmissions. This metric is similar to ETX (Couto *et al.*, 2003). The difference is that ExOR uses only the forward delivery probability to compute the expected number of forward transmissions, whereas ETX uses the forward delivery probability and the acknowledgment delivery probability to compute the expected number of transmissions. ExOR uses knowledge of the complete set of internode loss rates to calculate these ETX values.

Notice that ExOR attempts to schedule the times at which nodes send their fragments so that only one node sends at a time. This scheduling allows higher-priority nodes to send first, which speeds completion and updates lower-priority nodes' batch maps. Scheduling also helps avoid collisions, which is particularly important because ExOR would like to use marginal links on which carrier sense often does not work.

Biswas and Morris (2005) conducted extensive simulations and testbed studies to study the performance of the ExOR routing protocol. Measurements of an implementation on a 38-node IEEE 802.11b testbed show that ExOR increases throughput for most node pairs when compared with traditional routing. For pairs between which traditional routing uses one or two hops, ExOR's robust acknowledgments prevent unnecessary retransmissions, increasing throughput by nearly 35%. For more distant pairs, ExOR takes advantage of the choice of forwarders to provide throughput gains of a factor of two to four.

There are several ways to look at opportunistic routing. Wireless transmissions are not reliable and whether a link is reliable is time-dependent. Assume that an intermediate node u has k neighboring nodes $v_1, v_2, \ldots, v_k$, and each such neighboring node v_i has a perfect link to the destination node D. Let p_i be the reliability of link uv_i for $1 \leq i \leq k$. Then, the traditional routing such as ETX will often select a link that is more reliable. Then, the expected number of transmissions to get the packet so it is reached at the neighboring node v_i is $\frac{1}{p_i}$. Let us see what will be the expected number of transmissions to get the packet so it is reached at *any one* of the neighboring nodes $v_1, v_2, \ldots, v_k$. Let X be the random variable denoting the number of transmissions from u. Then, it is obvious that

$$E(X) = \sum_{t=1}^{\infty} \Pr(X \geq t).$$

The probability $\Pr(X \geq t)$ that it will take at least t transmissions is

$$\prod_{i=1}^{k} q_i^{t-1},$$

where $q_i = 1 - p_i$ is the probability that one transmission from u to v_i will fail. Let $Q = \prod_{i=1}^{k} q_i$. Thus,

$$E(X) = \sum_{t=1}^{\infty} \Pr(X \geq t) = \sum_{t=1}^{\infty} Q^{t-1} = \frac{1}{1 - Q}.$$

After any one of the neighboring nodes gets the packet, it will take only one more additional transmission to reach the destination node. Recall that the expected number of transmissions is $1/1 - q_i$ if we strictly require that the neighboring node v_i must receive the packet and then forward to the destination node. For a simple example when $k = 10$ and $q_i = 0.8$ for $1 \leq i \leq 10$, in the traditional method, the expected number of transmissions is $\frac{1}{1-0.8} = 5$, whereas the expected number of transmissions when the opportunistic transmission is used is $\frac{1}{1-0.8^{10}} \simeq \frac{1}{1-0.1074} \simeq 1.12$.

Opportunistic routing can be viewed as some special geographic routing that is predicated on every node's being aware of its neighbors and their specific locations. In geographic routing, the network layer of a node selects a next-hop forwarder to be the node that is the *best* choice among all its neighbors under a certain criterion toward the destination. For example, some geographic routings will choose the neighboring node that is the closest to the destination node. The ExOR actually chooses the neighboring node that received the packet and has the smallest *expected* cost to the destination node. This information is then sent down to the MAC layer, which waits till it can achieve rendezvous with the selected node. However, in sensor networks, availability of nodes can be disrupted significantly; hence, the MAC layer may suffer a significant delay and energy overhead in retransmitting the packet until it can complete the transmission successfully. Thus, correctly selecting the neighboring node from the set of neighboring nodes that already received the packets will significantly reduce the retransmission overhead. The choice of such a neighboring node with the smallest expected cost to the destination node also will ensure that the cost of these transmissions along the path from the source node to the destination node is not worse.

13.4   Geographic Approaches

The geometric nature of multihop WSNs allows a promising idea: localized geometric routing (or localized routing) protocols. The recent availability of small, inexpensive, low-power GPS receivers and techniques for finding relative coordinates based on signal strengths and the need for the design of power-efficient and scalable networks provided justification for applying position-based routing methods in ad hoc networks. A number of such algorithms were developed recently, in addition to a few basic methods proposed about 10–20 years ago. Here, we study some position-based routing protocols developed for wireless ad hoc networks. We study these protocols in terms of a number of characteristics: loop-free behavior, distributed operation (localized, global, or zonal), path strategy (single path, multipath, or flooding based), metrics used (hop count, power, or cost), memorization (memoryless or memorizing past traffic), guaranteed delivery, scalability, and robustness (strategies to handle the position deviation that is due to the dynamic nature of the network). See Frey and Stojmenovic (2006), Giordano *et al.* (2002), and Mauve *et al.* (2001) for some surveys of position-based routing protocols for WANs.

A routing protocol is *localized* if the decision to which node to forward a packet is based on only the following information:

- *The information in the header of the packet*. This information includes the source and the destination of the packet, but more data could be included, provided that its total length is bounded.
- *The local information gathered by the node from a small neighborhood*. This information includes the set of one-hop neighbors of the node, but a larger neighborhood set could be used provided it can be collected efficiently.

Randomization is also used in designing the protocols. A routing is said to be *memoryless* if the decision to which node to forward a packet is solely based on the destination, current node, and its neighbors within some constant hops. In the literature, localized routing is sometimes called *stateless* (Karp, 2000; Karp and Kung, 2000), *on-line* (Bose and Morin, 1999; Bose *et al.*, 2000), or *distributed* (Stojmenovic and Lin, 2001).

To make the localized geometric routing work, the source node has to learn the current (or approximately current) location of the destination node. The destination node's position will also be included in the data packets forwarded to its neighbors. Other than the destination's position, each intermediate node need know only the relative position of its one-hop neighbors (and maybe its own position for some protocols) in order to decide to which neighboring node to forward packets. Notice that for sensor networks collecting data, the destination node is often fixed; thus, location service is not needed in these applications. However, the help of a *location service* is needed in most application scenarios for mobile wireless networks. Mobile nodes register their locations to the location service. When a source node does not know the position of the destination node, it queries the location service to get that information. In cellular networks, there are dedicated position severs. It will be difficult to implement the centralized approach of location services in WSNs. First, for a centralized approach, each node has to know the position of the node that provides the location services, which is a chicken-and-egg problem. Second, the dynamic nature of the WSNs makes it very unlikely that there is at least one location server available for each node. Recently, algorithms for distributed location services were studied in Basagni *et al.* (1998), Haas and Liang (1999), J. Li *et al.* (2000), and Stojmenovic (1999). Because of space limitations, the location service problem is omitted here. See X.-Y. Li (2003a) for a detailed review.

Because it is not necessary to maintain explicit routes, position-based routing does scale well even if the network is dense and highly dynamic. This is a major advantage in a MANET in which the topology may change frequently. The main prerequisite for position-based routing is that a sender can obtain the current position of the destination node. Therefore, recently proposed location services are also discussed in addition to position-based packet-forwarding strategies.

13.4.1   Simple Heuristics

We first briefly review the simple heuristics proposed in the literature for localized geometric routing protocols. Some of these protocols can guarantee packet delivery and some of these protocols cannot guarantee packet delivery when the nodes are distributed in a certain way such that the local minima occur in routing, in which a node cannot find a neighboring node that is better than itself. Figure 13.1 illustrates some of these localized routing protocols proposed in the literature.

The following routing algorithms on the graphs were proposed recently.

Compass Routing

Let t be the destination node. Current node u finds the next relay node v such that the angle $\angle vut$ is the smallest among all neighbors of u in a given topology. In other

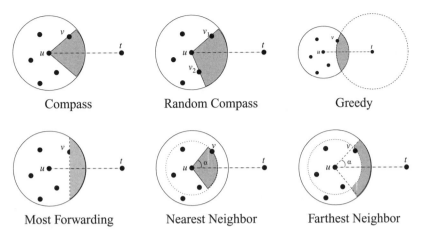

Figure 13.1 Various localized routing methods. Shaded area is empty and v is next node. © 2002, IEEE.

words, the current node u will find the next neighboring node that is *best* in terms of the direction (closest to the direction) to the destination node. See Kranakis *et al.* (1999) for a more detailed description of this method.

Random Compass Routing

Let u be the current node and t be the destination node. Let v_1 be the node on the line above ut such that $\angle v_1 ut$ is the smallest among all such neighbors of u. Similarly, we define v_2 to be nodes below line ut that minimizes the angle $\angle v_2 ut$. Then, node u randomly choose v_1 or v_2 to forward the packet. See Kranakis *et al.* (1999) for further discussion on random compass routing. The reason we randomly choose one of these two nodes for routing is that compass routing cannot guarantee packet delivery in the worst-case scenario. On the other hand, it can be proved that there is a sequence of choices of the random next-hop node such that random compass routing will guarantee the delivery.

Greedy Routing

Let t be the destination node. Current node u finds the next relay node v such that the distance $\|vt\|$ is the smallest among all neighbors of u in a given topology. In other words, the current node u will find the next neighboring node that is *best* in terms of the distance (closest to the destination) to the destination node. See Bose *et al.* (2001) for some discussion of this method.

Most-Forwarding Routing (MFR)

Current node u finds the next relay node v such that $\|v't\|$ is the smallest among all neighbors of u in a given topology, where v' is the projection of v on segment ut. In other words, the current node u will find the next neighboring node that is *best* in terms of the forwarding distance. It is clear to see that this neighboring node v might not necessarily have the shortest distance to the destination node. See Stojmenovic and Lin (2001) for some discussion.

Nearest-Neighbor Routing (NNR)

Given a parameter angle α, node u finds the nearest node v as forwarding node among all neighbors of u in a given topology such that $\angle vut \leq \alpha$. The choice of such a neighboring node is often for reducing the energy consumption of the transmission and also increasing the reliability of the link uv. Notice that for wireless networks, shorter links often imply a higher reliability when the transmission power of the sender is given. In addition, if we assume that the power consumption to support a link uv is a monotone-increasing function of the Euclidean length $\|uv\|$, then shorter links also imply a smaller power consumption for sending data from u to v if we can dynamically adjust the transmitting power of node u. Notice that whether such a protocol will reduce the overall energy consumption from the source to the destination node depends on the exact power model. Recently, Y. Wang *et al.* (2006) proposed a localized routing protocol called LEARN that can theoretically guarantee that the expected power consumption by the route found by LEARN is almost optimum.

Farthest-Neighbor Routing (FNR)

Given a parameter angle α, node u finds the farthest node v as the forwarding node among all neighbors of u in a given topology such that $\angle vut \leq \alpha$. This is similar to MFR. The angle α here typically should be less than $\pi/2$ to ensure that the next-hop node will move toward the destination node.

Greedy-Compass Routing

Current node u first finds the neighbors v_1 and v_2 such that v_1 forms the smallest counterclockwise angle $\angle tuv_1$ and v_2 forms the smallest clockwise angle $\angle tuv_2$ among all neighbors of u with the segment ut. The packet is forwarded to the node of $\{v_1, v_2\}$ with minimum distance to t. See Bose and Morin (1999) and Morin (2001) for more discussion.

Observe that when the current node u finds a neighbor node to forward the packet, it is not necessary that u find such a neighboring node from *all* physical neighboring nodes. The node u often selects the forwarding node from its *logical* neighboring nodes; i.e., the adjacent nodes of u in some network topology (e.g., GG, YG, or Delaunay triangulation).

Notice that it is shown in Bose *et al.* (2001) and Kranakis *et al.* (1999) that compass routing, random compass routing, and greedy routing guarantee to deliver the packets from the source to the destination if Delaunay triangulation (DT) is used as the network topology. They proved this by showing that the distance from the selected forwarding node v to the destination node t is less than the distance from current node u to t. However, the same proof cannot be carried over when the network topology is a YG, GG, RNG, and localized DT. When the underlying network topology is a planar graph, the right-hand rule or face routing is often used to guarantee packet delivery after simple localized routing heuristics fail (Bose *et al.*, 2001; Karp and Kung, 2000; Kuhn *et al.*, 2003a, 2003c; Stojmenovic and Lin, 2001). These are discussed in the next section.

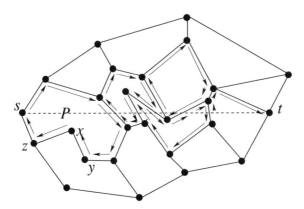

Figure 13.2 An illustration of the face-routing algorithm. © 2005, IEEE.

THEOREM 13.1 (Morin, 2001) *Greedy routing guarantees the delivery of the packets if DT is used as the underlying structure. Compass routing guarantees the delivery of the packets if regular triangulation is used as the underlying structure. There are triangulations (not Delaunay) that defeat these two schemes. Greedy-compass routing works for all triangulations; i.e., it guarantees the delivery of the packets as long as there is triangulation used as the underlying structure. Every oblivious routing method is defeated by some convex subdivisions.*

Here, a triangulation is *regular triangulation* if it is the projection of the lower convex hull of some 3D polytope P into the $X-Y$ plane. DT is a special regular triangulation in which all the vertices of P are on a paraboloid, $z^2 = x^2 + y^2$. Another interesting triangulation is *greedy triangulation*, which is constructed by adding edges in increasing order of their lengths to avoid crossing edges. Morin (2001) also studied localized routing for greedy triangulation. Because the greedy triangulation cannot be constructed locally or very efficiently in a distributed manner, that part is omitted in this book. It is easy to see that there is no memoryless routing method that works in a UDG.

13.4.2 Right-Hand Rule and Face Routing

From the study of some simple greedy-based routing protocols, we know that most of the routing protocols cannot guarantee packet delivery. Then, we need to find some other mechanism to perform routing when some simple routing heuristics fail. The *right-hand rule* is a long-known method for traversing a graph (analogous to following the right-hand wall in a maze), and it has been used in some wireless routing protocols (Bose *et al.*, 2001; Karp, 2000; Karp and Kung, 2000; Stojmenovic and Lin, 2001). The rule states that when arriving at node x from node y, the next edge traversed is the next one sequentially counterclockwise about x from edge xy. In the example shown in Figure 13.2, x will forward the packet to z following the right-hand rule, traversing face P. It is known that the right-hand rule traverses the interior of a closed polygonal region (a face) in clockwise edge order, and it traverses an exterior region in counterclockwise edge order. In general, the right-hand rule is applied in planar graphs (in which no edges intersect each other). Karp (2000) gives a *no-crossing heuristic* to deal with the case in which edges cross.

Applying the right-hand rule in planar graphs, Kranakis *et al.* (1999) proposed a routing protocol called *Face Routing* (in their paper, they call the algorithm *Compass Routing II*). We consider a planar graph G. The nodes and edges of graph G partition the Euclidean plane into contiguous regions called the *faces* of G. The main idea of face routing is to walk along the faces that are intersected by the line segment st between the source s and the destination t. In each face, it uses the right-hand rule to explore the boundaries. On its way around a face, the algorithm keeps track of the points where it crosses the line st. Having completely surrounded a face, the algorithm returns to one of these intersections lying closest to t, where it proceeds by exploring the next face closer to t. Figure 13.2 gives an illustration. See Kranakis *et al.* (1999) and Kuhn *et al.* (2002) for detailed algorithms. They also proved that the face-routing algorithm guarantees to reach the destination t after traversing at most $O(n)$ edges, where n is the number of nodes.

It is also easy to construct an example of a network such that the total number of links traversed by face routing could be as large as $\Theta(n)$ times the smallest number of links needed to connect the source node and the destination node: Nodes s and t are connected by constant hops, and there is a big face with $\Theta(n)$ links crossed by the segment connecting s and t. In addition, although face routing terminates in linear time, it is not satisfactory, as already a very simple flooding algorithm will terminate in $O(n)$ steps. Then, Kuhn *et al.* (2002) proposed a new method, called *Adaptive Face Routing* (AFR), in which restricted search areas are used to avoid exploring the complete boundary of faces. The idea is as follows: The exploration of faces is restricted to an ellipse area. The ellipse size is set to an initial estimate of the optimal path length. If face routing fails to reach the destination (i.e., when it reaches the ellipse or is outside of the ellipse, it has to turn back and start over again), the algorithm will restart with a bounding ellipse of doubled size. They proved that the algorithm will finally find a path to t if s and t are connected. Also, the number of steps of AFR is bounded by $O(c^2(p^*))$, where p^* is an optimal path and $c(p^*)$ is the cost of that path. In their proof, they assumed that the UDG is a civilized graph. Finally, they give a tight lower bound by showing that *any localized* geometric routing algorithm has the worst-case cost $O(c^2(p^*))$. The basic idea of the proof is as follows. Assume that the global optimum path has length $\|p^*\|$. Then, it can be shown that the path p^* is contained inside an ellipse with focus at source node s and destination node t and the circumference is of the order of $\|p^*\|$. The rest of the proof follows from the fact that the method explores the area with an ellipse that is double the size of the previous ellipse.

Recently, Kuhn *et al.* (2003c) extended AFR to a routing algorithm called *Other Adaptive Face Routing* (OAFR). Instead of changing to the next face at the "best" intersection of the face boundary with st, OAFR returns to the boundary point closest to the destination. They proved the cost of OAFR is also bounded by $O(c^2(p^*))$, which is asymptotically optimal. They construct an example to show that every localized and deterministic routing method will have a cost on the order of $\Theta(c^2(p^*))$. See Kuhn *et al.* (2003c) for more details of their example.

13.4.3   Combining Face Routing with Greedy Routing

Greedy routing was used in early routing protocols for wireless networks. However, it is easy to construct a simple example to show that the greedy algorithm will not succeed in reaching the destination but will fall into a local minimum, a node without any "better" neighbors. Then, a natural approach to improving the potential of greedy routing for practical purposes is to combine greedy routing and face routing (or the right-hand rule) to recover the routing after simple greedy routing fails in a local minimum. Many wireless protocols used this approach (Bose *et al.*, 2001; Karp and Kung, 2000; Kuhn *et al.*, 2003a, 2003c; Stojmenovic and Lin, 2001; Yu *et al.*, 2001).

Face routing has been implemented in several variants that differ in the decision when current face traversal has to be interrupted and that face has to be explored next. Frey and Stojmenovic (2006) listed a number of face-routing variants by referring to the well-established names of the entire protocols employing these strategies. It is important to note that the following descriptions explain the face-routing part of these protocols only. The routing path running only the face-routing part of the protocol might be different when running the entire protocol. However, as long as it is clear from the context, we interchangeably use the well-established protocol names for both the entire protocol and its face routing part.

As a general classification, we can distinguish between face-routing strategies that require that the message has to follow a sequence of adjacent faces that are intersected by the straight line *st* connecting the source node *s* with the destination node *t*. These strategies are denoted as *continuative strategies* because they keep the line *st* as a reference during the whole routing process. In contrast, a *volatile strategy* will initialize planar graph routing each time a face change occurs. In other words, the node where a face change has occurred is treated as a planar graph routing start node again. According to this definition, Greedy-Face-Greedy (GFG) and Greedy Perimeter Stateless Routing (GPSR) strategies are continuative while *Greedy Other Adaptive Face Routing (GOAFR+)* are volatile ones.

Bose *et al.* (1999) proposed the GFG routing method that works as follows: As soon as a message encounters an edge that intersects the source destination line *st* at an intersection point *p*, it will change into the face that intersects with the open line segment *pt*. However, only those intersection points are considered that are closer to the destination than the last encountered intersection point where current face traversal was started.

GPSR (Karp, 2000; Karp and Kung, 2000) is one of the famous routing protocols for wireless networks. It uses a RNG or a GG as the planar routing topology and then combines greedy routing and the right-hand rule to forward packets in the network. It works as follows: When a node receives a greed-mode packet, it searches its neighbor table for a neighbor who is closer to destination *t*. If there is one, it will forward it to that neighbor. When no neighbor is closer, the node marks the packet into perimeter mode. GPSR forwards perimeter-mode packets by using a simple planar graph traversal (i.e., right-hand rule). When a packet enters perimeter mode, GPSR records in the packet the location L_p. Then, when receiving a perimeter-mode packet, GPSR will first compare it with a forwarding node's location. GPSR returns a packet to greedy mode if the distance from the forwarding node to t is less than that from L_p to t. For more detail, refer to Karp (2000) and Karp and Kung (2000). GPSR can guarantee delivery of the packets when the underlying network topology is a planar graph.

Recently, Kuhn *et al.* (2003c) proposed a new algorithm to combine greedy routing with their OAFR. They called the new method *Greedy Other Adaptive Face Routing* (GOAFR). The idea is similar to GPSR. When the greedy method falls in a local minimum, GOAFR uses OAFR to recover the routing. As for AFR, they proved that the cost of GOAFR is bounded by $O(c^2(p^*))$, which is asymptotically optimal. In addition, they show that the algorithm is also average-case efficient through extensive simulations. Kuhn *et al.* (2003c) showed simulations of a variety of face-routing algorithms and their combinations with a greedy approach. Notice that unlike GPSR, when face routing is done in GOAFR, it does not return to the greedy method until OAFR completely finishes the exploration of the face. This may affect the efficiency of the routing. Kuhn *et al.* (2003a) used an "early fallback" technique to return to greedy routing as soon as possible. The new algorithm is called GOAFR$^+$. It employs two counters, p and q, to keep track of how many of the nodes visited during the current face-routing phase are located closer (counted by p) and how many are not closer (counted by q) to the destination than the starting point of the current face-routing phase. When a certain fallback condition holds, GOAFR$^+$ directly falls back to greedy mode. This modification makes an obvious improvement for the average-case performance. Their theoretical analysis also proves that GOAFR$^+$ is asymptotically optimal in the worst case.

13.4.4   Routing on Delaunay Triangulation

With respect to localized routing, there are several ways to measure the quality of the protocol. In the analysis of Kuhn *et al.*, they used the number of steps (hops) in a path to measure the quality of their routing methods. Given the scarcity of power resources in WSNs, minimizing the total power used is imperative. A stronger condition is to minimize the total Euclidean distance traversed by the packet. Bose and Morin (1999) and Morin (2001) also studied the performance ratio of previously studied localized routing methods. They proved that none of the previously proposed heuristics guarantees a constant ratio of the traveled distance of a packet compared with the minimum. They gave the first localized routing algorithm such that the traveled distance of a packet from u to v is at most a constant factor of $\|uv\|$ when DT is used as the underlying structure.

Their algorithm is based on the proof of the spanner property of DT (Dobkin *et al.*, 1990). Without loss of generality, let $b_0 = u, b_1, b_2, \ldots, b_{m-1}, b_m = v$ be the vertices corresponding to the sequence of Voronoi regions traversed by walking from u to v along the segment uv. If a Voronoi edge or a Voronoi vertex happens to lie on the segment uv, then choose the Voronoi region lying above uv (see Figure 13.3). Given two nodes u and v, tunnel (u, v) is defined as the collection of triangles that intersects the segment uv. The sequence of nodes $b_i, 0 \leq i \leq m$, defines a path from u to v. In general, Dobkin

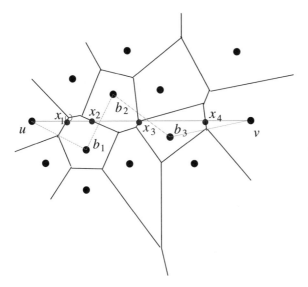

Figure 13.3 A good approximation path using the edges of tunnel (u, v). © 2002, IEEE.

et al. (1990) refer to the path constructed this way between some nodes u and v as the *direct DT path* from u to v.

Assume that line uv is the x axis. The path constructed by Dobkin *et al.* uses the direct DT path as long as it is above the x axis. Assume that the path constructed so far has brought us to some node b_i such that b_i is above uv and b_{i+1} is below uv. Let j be the least integer larger than i such that b_j is above uv. Notice that here, j exists because $b_m = v$ is on uv. Then, the path constructed by Dobkin *et al.* uses either the direct DT path to b_j or takes a *shortcut*. See Dobkin *et al.* (1990) for more details about the condition of when to choose the direct DT path from b_i to b_j, when to choose the shortcut path from b_i to b_j, and how the shortcut path is defined.

Bose and Morin basically use a type of binary search method to find which path is better. Refer to Morin (2001) for more details of finding the path. However, their algorithm needs DT as the underlying structure, which is expensive to construct in wireless ad hoc networks. Bose and Morin (2001) further extend their method to any triangulation that satisfies the diamond property. Here, a triangulation satisfies the diamond property if for every edge uv in the triangulation, either $\triangle uvw_1$ or $\triangle uvw_2$ is empty of other vertices, where w_i satisfies $\angle w_i uv = \angle w_i vu = \frac{\pi}{6}$, for $i = 1, 2$.

X.-Y. Li and Wang (2003) showed that the local DT *PLDel* can be used to approximate the Delaunay triangulation Del almost always when the network is connected and the sensor nodes are randomly deployed. Consequently, the method of Morin (2001) can be used on local DT almost always.

Localized routing protocols support mobility by eliminating the communication-intensive task of updating the routing tables. But, mobility can affect the localized routing protocols in both the performance and the guarantee of delivery. There is no work so far to design protocols with guaranteed delivery when the network topology changes during the routing.

13.4.5   Delivery Guarantee of Localized Routing Protocols

Localized routing protocols can further be classified regarding their *delivery guarantees*. Guaranteed delivery refers to the ability of successfully forwarding a message from source to destination. The definition requires that source and destination be connected by at least one path in the network and that we have an idealized MAC layer in which messages are not lost during any forwarding step. For each localized greedy-routing variant, the message may end up in a node that has to drop the message to prevent a routing loop. Dropping a message might even be necessary, although there exists a path from the source to the destination node. In previous subsections, it was shown how face routing or the right-hand rule can be used to avoid dropping a message; i.e., to get out of the local minimum occurring in a localized routing protocol. Recall that a node faces a *local minimum* if it cannot find any of its neighboring nodes to forward the message that is better than itself in terms of the selection criterion. For applying face routing and right-hand-rule routing, we need a planar structure as the underlying network topology. Notice that this planar structure must be constructed in a localized manner to preserve the *stateless* (or localized routing) property of a localized routing protocol. Typical planar structures used include the GG, RNG, Localized DT, and so on.

Guaranteed delivery of face and several combined greedy–face routing (GFR) schemes in UDGs and quasi-UDGs is a well-established fact. Numerous experimental studies confirm it, and some formal arguments were presented. However, a recent study by Kim *et al.* (2005) claimed that the routing protocols GPSR (Karp and Kung, 2000), GOAFR+ (Kuhn *et al.*, 2003a, 2003c), and the planar graph routing schemes that they use cannot guarantee delivery in arbitrarily undirected planar network graphs. Recently, Frey and Stojmenovic (2006) clarified that this is not the truth in general. They show that specifically in RNGs and GGs recovery from a greedy-routing failure is always possible without a change between any adjacent faces. Guaranteed delivery then follows from guaranteed recovery while traversing the very first face. In arbitrary graphs, however, a proper face-selection mechanism is of importance because recovery from a greedy-routing failure may require visiting a sequence of faces before greedy routing can be restarted again. A prominent approach is to visit a sequence of faces that are intersected by the line connecting the source and the destination node. Whenever encountering an edge that is intersecting with this line, the critical part is to decide whether face traversal has to change to the next adjacent one. Failures may occur from incorporating face-routing procedures that force a change in the traversed face at each intersection. Recently observed routing failures that were produced by the GPSR protocol in arbitrary planar graphs result from incorporating such a face-routing variant. They cannot be constructed by the well-known GFG algorithm that does not force changing the face anytime. Besides methods that visit the faces intersected by the source–destination line, we discuss face-routing variants that simply restart face routing whenever the next face has to be explored. Frey and Stojmenovic gave the first complete and formal proofs that several face routing and combined greedy–face-routing schemes do guarantee delivery in specific graph classes or even any arbitrary planar graphs. They also discuss the reasons that other methods may fail to deliver a message or even end up in a loop.

Table 13.1 Success of face routing applied in its own

	RNG	GG	LDT	Any
GFG	ok	ok	ok	ok
GPSR	loop	loop	loop	loop
Compass routing II	ok	ok	ok	ok
GOAFR+	ok	ok	drop	drop
GOAFR++	ok	ok	ok	ok
GPVFR	?	?	?	loop

Table 13.2 Success of combined greedy–face routing

	RNG	GG	LDT	Any
GFG	ok	ok	ok	ok
GPSR	ok	ok	?	loop
Compass routing II	ok	ok	ok	ok
GOAFR+	ok	ok	drop	drop
GOAFR++	ok	ok	ok	ok
GPVFR	ok	ok	?	loop

Tables 13.1 and 13.2 summarize the delivery guarantee of face routing on a number of known planar structures and the combination of face routing and greedy routing on these planar structures.

In these tables, *ok* denotes guaranteed delivery, *loop* denotes the possibility of a forwarding loop, *drop* denotes the possibility of an incorrect message drop, and *?* denotes that the behavior is not known at the time of writing. In the tables, GPVFR is a localized routing protocol proposed in Leong *et al.* (2005). It works as follows: After starting in a node s, a face change occurs as soon as an intersection of a face with st is found. However, this method does not keep the source–destination line st but rather restarts face exploration at the node that encountered the intersection and treats this node as the new start node of the next source–destination line st.

Notice that the preceding results do not contradict the claim that the face-routing strategies (and combined with some greedy-routing strategies) can guarantee the delivery of the packets. To guarantee delivery, we need to carefully select the next face in the implementation when the currently visited face intersects the line st to avoid a possible loop.

13.4.6 Location-Aided Routing

In previous discussions, we implicitly assume that the position of the destination node is known for routing. In practice, except when the destination node is static, it is often difficult to get the exact *current* position of the destination node when the node decides where to forward the packet. Ko and Vaidya (1998) proposed a location-aided routing (LAR) protocol that will route the packets based on the estimated current position of the destination node from some history information. Consider a node S that needs to find a route to destination node D. Assume that node S knows that node D was at location X at time t_0 and that the current time is t_1. Then, the "expected zone" of node D, from the viewpoint of node S at time t_1, is the region that node S expects to contain node D at time t_1. Node S can determine the expected zone based on the knowledge that node D was at location X at time t_0 and the expected moving speed of node D is v_D. Then, S may assume that the expected zone for node D is the circular region of radius $v_D(t_1 - t_0)$, centered at location X. The expected zone is only an estimate made by node S to determine a region that potentially contains D at time t_1. The actual location of node D could be outside of this disk if the moving speed of node D is larger than v_D.

The current source node S defines (implicitly or explicitly) a *request zone* for the RREQ. A node forwards a RREQ only if it belongs to the request zone (unlike the flooding algorithm). To increase the probability that the RREQ will reach the destination node D, the request zone should include the expected zone. Additionally, the request zone may also include other regions around the request zone. The first scheme of Ko and Vaidya (1998) uses a request zone that is rectangular that has S as one corner and covers the *expected zone* of the destination node D. When an intermediate node receives a RREQ, it discards the request if the node is not within the rectangle specified by the four corners included in the RREQ. To further improve the performance of the routing protocol, we can make the following enhancement:

- When an intermediate node v (inside the request zone of node u for destination node D) receives the RREQ from a neighboring node u, node v replaces the request zone in the packet as the new rectangle defining the request zone for node v.

When this strategy is used, the request zone will typically shrink quickly when the RREQ packet is forwarded toward the destination node. There are two reasons for this: (1) the distance between v and D is smaller and, thus, the diagonal of the rectangle becomes smaller; and (2) the expected zone for destination node D becomes smaller because v is closer to D (thus, v often will have more recent location information of the destination node D).

Their second scheme works as follows: Assume that node S knows the location (x_D, y_D) of the destination node D at some time t_0. Further assume that the time at which route discovery is initiated by node S is t_1, where $t_1 > t_0$. Node S calculates the distance between its own position and location (x_D, y_D), denoted as d_s, and includes this distance with the RREQ message. In addition, the coordinates (x_D, y_D) are also included with the RREQ. When a node v receives the RREQ from another node u, node v calculates its distance from location (x_D, y_D), denoted as d_v. Notice that here, node u may be the source node S or some other intermediate node that forwarded the RREQ by S. Node v forwards the RREQ to its neighbors (with its own distance d_v included in the RREQ packet) only if

$$\alpha \cdot d_u + \beta \geq d_v.$$

Here, α and β are some predetermined constants.

13.4.7   Geocasting

Geocasting (Navas and Imielinski, 1997) is the delivery of packets to nodes within a certain geographic area. For many applications in wireless networks, geocasting is an important and frequent communication service. In sensor networks, geocasting may be required to assign tasks to nodes or to query nodes in a certain area. For example, a user may request all sensors in an area where a fire is spreading to report their temperature. In ad hoc networks, geocasting can be used, for example, to facilitate location-based services by announcing a service in a certain region or sending an emergency warning to a region. The challenging problem in geocasting is distributing the packets to all the nodes within the geocast region with high probability but with low overhead. Observe that the LAR protocol (Ko and Vaidya, 1998) can be viewed as a geocasting routing protocol because the packets will be routed to an estimated region where the destination node could reside. See Jiang and Camp (2002) for a review of some geocasting protocols until 2002 and Yao *et al.* (2004) for performance comparisons of some geocasting protocols until 2004.

Three categories of geocast routing protocols have been developed: flooding-based, routing-based, and cluster-based protocols. Flooding-based protocols use flooding or a variant of flooding to forward geocast packets from the source to the geocast region. Routing-based protocols create routes from the source to the geocast region via control packets. Cluster-based protocols geographically partition a MANET into several disjointed and equal-sized cellular regions and select a cluster-head in each region for executing information exchange.

The location-based multicast (LBM) protocol (Ko and Vaidya, 1999) reduces the forwarding space of geocast packets. A node forwards a geocast packet only if it belongs to the forwarding zone or to the geocast region. Once a packet reaches the geocast region, it is flooded within the geocast region. Two forwarding zone schemes (box and step) are defined for LBM (Ko and Vaidya, 1999). To reduce the overhead further, GeoTORA (Ko and Vaidya, 2000, 2003) uses a unicast routing protocol [temporarily ordered routing algorithm (TORA) (Park and Corson, 1997)] to deliver the packet to the region and then floods within the region. In GeoGRID (Liao *et al.*, 2000), the geographic area of the MANET is partitioned into 2D logical grids. In each grid, one node is elected as the gateway of the grid. Thus, instead of having every node forward data, only gateway nodes forward data.

13.4.8   Location Service

To make localized routing work, the source node has to learn the current (or approximately current) location of the destination node. Notice that for sensor networks collecting data, the destination node is often fixed; thus, location service is not needed in these applications. However, the help of a *location service* is needed in most application scenarios. Mobile nodes register their locations to the location service. When a source node does not know the position of the destination node, it queries the location service to get that information. In cellular networks, there are dedicated position servers. It will be difficult to implement the centralized approach of location services in wireless ad hoc networks. First, for a centralized approach, each node has to know the position of the node that provides the location services, which is a chicken-and-egg problem. Second, the dynamic nature of the wireless ad hoc networks makes it very unlikely that there is at least one location server available for each node. Thus, here we concentrate on distributed location services.

For the wireless ad hoc networks, the location service provided can be classified into four categories: *some-for-all*, *some-for-some*, *all-for-some*, *all-for-all*. Some-for-all service means that some wireless nodes provide location services for all wireless nodes. Other categorizations are defined similarly.

Examples of all-for-all services are the location services provided in the Distance Routing Effect Algorithm for Mobility (DREAM) by Basagni *et al.* (1998). Each node stores a database of the position information for all other nodes in the wireless networks. Each node will regularly flood packets containing its position to all other nodes. Frequency and range of the flooding are used as controls for the cost of updating and the accuracy of the database.

Using the idea of *quorum* developed in the databases and distributed systems, Haas and Liang (1999) and Stojmenovic (1999) developed quorum-based location services for wireless ad hoc networks. Given a set of wireless nodes V, a quorum system is a set of subset $(Q_1, Q_2, \ldots, Q_k)$ of nodes whose union is V. These subsets could be mutually disjoint or often have an equal number of intersections. When one of the nodes requires the information of the other, it suffices to query one node (called the representative node of Q_i) from each quorum Q_i. A virtual backbone is often constructed between the representative nodes using nonposition-based methods such as those in Alzoubi *et al.* (2002c) and Wan *et al.* (2002a). The updating information of a node v is sent to the representative node (or the nearest if there are many) of the quorum containing v. The difficulty of using quorum is that the mobility of the nodes requires the frequent updating of the quorums. The quorum-based location service is often *some-for-some* type.

The other promising location service is based on the quad-tree partition of the 2D space (Amouris *et al.*, 1999). It divides the region containing the wireless network into a hierarchy of squares. The partition of the space in Amouris *et al.* (1999) is uniform. However, we notice that the partition could be nonuniform if the density of the wireless nodes is not uniform for some applications. Each node v will have the position information of all nodes within the same *smallest* square containing v. This position information of v is also propagated to up-layer squares by storing it in the node with the nearest identity to v in each up-layer square containing v. Using the nearest identity over the smallest identity can avoid the overload of some nodes. The query is conducted accordingly. It is easy to show that it takes about $O(\log n)$ time to update the location of v and to query another node's position information.

13.4.9   Node Localization

Recall that all routing protocols are based on the geometry information of nodes or relative positions of the neighbors of the current node holding the data packets. One way to get the location of a node is to use a GPS receiver. However, this approach may be either expensive if the network size is large or inappropriate for certain networking devices such as small sensors. Recently, a number of localization techniques have been developed in the literature to find the geometry positions of nodes. In Chapter 17, we study some recently proposed localization techniques for WSNs.

13.5   Clustering and Hierarchical Routing

Observe that all previously studied proactive and reactive routing protocols essentially assumed that every node in the network plays the same role. The underlying network topology is assumed to be a flat structure. This uniformity has some advantages for designing the routing protocols and easy maintenance of the network. On the other hand, it may lose some optimality in terms of maintenance of the routing structures in mobile networks. For example, the proactive routing protocols will be efficient if the network topology does not change and the link metrics do not fluctuate extensively. It will be a disadvantage to maintaining routing tables in a mobile network in which network topology changes a lot because of fast moving by the nodes. The routing table will often be outdated, and most of the routing information may never be used. On the other hand, the reactive routing protocols will be efficient if the network topology changes frequently. It is a disadvantage to always invoke a RREQ if the network topology rarely changes because most of the RREQ could be a repetition of some historic requests. In the literature, there are a number of routing protocols proposed to address these issues. These protocols can be categorized as *hybrid protocols* and *backbone-based protocols*. In the hybrid protocols, different routing protocols will be used for different situations. In backbone-based routing protocols, the nodes in the network will have different roles. Some nodes will act as forwarders to forward the data packets for other nodes. These nodes typically form a backbone of the wireless network. Thus, all other nodes will be connected to this backbone.

13.5.1   Zone Routing Protocol

Routing protocols for MANETs have to face the challenge of frequently changing topology, low transmission power, and possible asymmetric links. Proactive routing protocols are inefficient for mobile networks for which nodes move frequently and the network topology changes frequently. Thus, proactive routing uses excess bandwidth to maintain routing information. On the other hand, reactive routing protocols are inefficient under these circumstances when the network topology does not change rapidly. Notice that reactive routing involves long RREQ delays. Reactive routing also inefficiently floods the entire network for route determination. The Zone Routing Protocol (ZRP) (Joa-Ng and Lu, 1999; Haas and Pearlman, 1997) combines the advantages of the proactive and reactive approaches by maintaining an up-to-date topological map of a zone centered on each node. Within the zone, routes are immediately available by using essentially the proactive approaches. For destinations outside the zone, ZRP employs a route-discovery procedure, which can benefit from the local routing information of the zones.

One assumption of the ZRP is that the largest part of the traffic is directed to nearby nodes. Therefore, ZRP reduces the proactive scope to a zone centered on each node. In a limited zone, the maintenance of routing information is easier. Further, the amount of routing information that is never used is minimized. Still, nodes farther away can be reached with reactive routing. Despite the use of zones, ZRP has a flat view over the network. In this way, the backbone management overhead related to hierarchical protocols can be avoided. Hierarchical routing protocols depend on the strategic assignment of gateways or landmarks (called cluster-heads sometimes), so that every node can access all levels, especially the top level.

A routing zone is defined for each node separately, and zones of neighboring nodes overlap. The routing zone has a radius ρ expressed in hops. The zone thus includes the nodes whose distance from the node in question is at most ρ hops. The number of nodes in the routing zone can be dynamically adjusted by changing the transmission power of the nodes. Using a smaller transmission power, we reduce the number of nodes within direct reach and thus reduce the overhead of maintaining the routing information for all nodes inside the zone. On the other hand, the number of neighboring nodes should be sufficient to provide adequate reachability and redundancy. Recall that to ensure the network connectivity, in Chapter 12, for n nodes randomly distributed in a 2D region, the number of neighboring nodes a node needs to have should be at least $\Theta(\log n)$. On the other hand, too large a coverage results in many zone members and the update traffic becomes expensive. Further, large transmission coverage adds to the probability of local contention among all neighboring nodes. Notice that if a zone radius of one hop is used, routing is purely reactive and broadcasting degenerates into flood searching. If the radius approaches infinity, routing is reactive. The selection of radius is a trade-off between the routing efficiency of proactive routing and increasing traffic for maintaining the view of the zone.

A node that has a packet to send first checks whether the destination is within its local zone using information provided by an intrazone routing protocol (IARP). In that case, the packet can be routed proactively. Reactive routing is used if the destination is outside the zone. For ZRPs, it also utilizes the fact that the topology of the local zone of each node is known to reduce traffic when global route discovery is needed. Instead of broadcasting packets, ZRP uses a concept called *border casting* that directs a query request to the border of the zone.

13.5.2   Backbone-Based Routing

So far, none of the currently proposed ad hoc routing algorithm seeks to exploit a structure similar to the wired backbone of packet cellular networks. By taking advantage of the backbone of the network, routing algorithms can support a number of additional operations efficiently, such as multicasting, broadcasting, and even fault-tolerant routing for mobile networks. In this subsection, we review some of the backbone-based routing protocols proposed in the literature. Unlike the previously studied protocols, backbone-based routing first imposes a virtual backbone structure on the ad hoc network to support unicast, multicast, and fault-tolerant routing within the ad hoc network. This virtual backbone differs from the wired backbone of cellular networks in two key ways: (1) it may change as nodes move or the status of nodes changes; and (2) it is used primarily not only for routing packets or flows but also for computing and updating routes. The primary routes for packets and flows are still computed by shortest-path computations, typically over the virtual backbone. The backbone, if necessary, can also provide backup routes to handle interim failures.

The backbone of the network typically is simply a CDS of the network (Dai and Wu, 2004; Das and Bharghavan, 1997; Wu, 2002; Wu and Li, 1999). After a backbone is constructed, the routing will be performed as follows. The routing process can be divided into three steps:

1. If the source node is not in the backbone, it forwards the packets to one of its adjacent nodes in the backbone, which is often called the gateway host.
2. This gateway host acts as a new source to route the packets in the backbone only.
3. Eventually, the packets reach a destination node in the backbone that is either the destination host itself or a gateway of the destination host; i.e., a dominator node of the destination node. In the latter case, the destination gateway forwards the packets directly to the destination host.

The gateway host keeps the following information: gateway membership of entire sub-network and the local routing table. The way in which the routing table is constructed and updated on the CDS subnetwork can be different.

Notice that, typically, we construct a backbone as a CDS; i.e., all other nodes in the network are directly connected to some nodes in the backbone. There are some protocols that relax the directed connection to the condition that every node not in the backbone be connected to some node in the backbone within k-hops for some constant $k > 1$. Most of the backbone-based routing protocols differ mainly in the creation of the backbone. Some protocols aim to minimize the size of the backbone (e.g., Y. Wang et al., 2002a), some protocols aim to minimize the energy consumption of the backbone (e.g., Wu et al., 2002), and some protocols aim to improve the stability of the backbone for mobile networks (e.g., Basagni, 1999; Bettstetter and Krausser, 2001; Min et al., 2004) or to minimize some generic cost of the backbone (e.g., Y. Wang et al., 2005b). See Basagni et al. (2006) for some simulation-based comparisons of some backbone-formation methods proposed in the literature. Notice that the backbone-construction method proposed in Y. Wang et al. (2005b) can guarantee that given any source node and any destination node, the shortest path found over the backbone to connect them is at most three times the total cost of the best path connecting them in the original network when all nodes are available for forwarding packets. See Chapters 7 and 8 for the efficient construction of CDS with a theoretical performance guarantee.

For routing by use of a backbone (also called a CDS or DS), we need to have algorithms to maintain the backbone structure for MANETs. For simplicity, we can assume that the processing is fast enough and the movement of nodes can be viewed as a sequence of individual node movement. When a node moves, the following cases are possible:

1. A dominatee node v moves. In this case, we need to update only the dominator node of this dominatee node. If the dominatee is not dominated by any existing dominator node, then this dominatee node also becomes a dominator node itself, and then we need to find connector nodes to connect this new dominator node to the rest of the backbone.
2. A connector node g moves. Recall that a connector node is used to connect two dominator nodes that are within some constant hops away from each other. Different CDS construction algorithms have different methods of choosing connectors (called gateway nodes sometimes). When a connector node moves, we need to find a new connector node to replace the role of this connector node. In addition, when this connector node is not dominated by any existing dominator node, we then need to mark this connector node as a new dominator and find some new connector nodes to connect this new dominator node to the rest of the backbone.
3. A dominator node u moves. Let $N(u)$ be the set of dominatees (and connector nodes) that are neighboring nodes of u. In this case, we need to recompute dominator nodes for $N(u)$ and u; i.e., find nodes that will dominate $N(u)$ and u. Notice that in this computation, we need to actually find dominators only for nodes that are *not* dominated by any other existing dominators. For these new dominators, we need to find new connectors to connect them to the rest of the existing backbone.

13.6   Further Reading

The routing protocols proposed may be categorized as table-driven protocols or demand-driven protocols. A good survey is in Royer and Toh (1999).

Table-driven routing protocols maintain up-to-date routing information between every pair of nodes. The changes to the topology are maintained by propagating updates of the topology throughout the network. Table-driven protocols include the DSDV routing protocol and the *Wireless Routing Protocol (WRP)*. The DSDV protocol is based on the shortest-path algorithm of Bellman–Ford in which the shortest path is determined with respect to the number of hops. The routing tables are updated periodically to keep the routing information correctly. Naturally, only incremental information is required for the update. More details are available from Perkins and Bhagwat (1994b). The wireless protocol, WRP, is also a table-based algorithm that updates link-state information but only with neighbors. These schemes are proactive but suffer a high overhead of communication costs. Another scheme proposed, the ZRP (Joa-Ng and Lu, 1999; Haas and Pearlman, 1997), limits the overhead by explicitly limiting the zone of these updates. Topology updates are not limited to a local neighborhood in this scheme. The topology is subdivided into zones, and different protocols are run to learn about the intrazone and interzone topology. This concept of zoning is a way of introducing a two-level hierarchy into the routing structures.

Other methods of introducing a hierarchical structure into routing have been proposed. *Cluster-Head Gateway Switch Routing (CGSR)* (Chiang, 1997) establishes clusters with cluster-heads that control the nodes within a cluster. Routes are established through the cluster-head. The DSDV methodology is used to establish the routes between cluster-heads. *Fisheye Routing* (Pei *et al.*, 2000a, 2000b) is another methodology that creates zones by sending information more frequently to nearby nodes and less frequently to far away nodes. *Hierarchical State Routing* (Iwata *et al.*, 1999) establishes multilevel clustering.

Source-Initiated On-Demand Routing creates routes only when desired by the source node. At this time, a route-discovery process is initiated within the network. The methodologies that have been proposed include AODV routing (Perkins and Royer, 1999), which discovers routes by using a broadcast mechanism when required. The protocol uses destination sequence to ensure that the routes are loop-free. DSR (Johnson and Maltz, 1996) uses a similar method but maintains a route cache at nodes. This may speed up recovery when route failure occurs. The TORA (Park and Corson, 1997) establishes a directed acryliograph (DAG) rooted at the destination during the route-creation phase and maintains this DAG after every link failure. *Associativity-Based Routing (ABR)* (Toh, 1996) is a different methodology that uses neighborhood associativity or longevity to establish long-lived routes. Similarly, *Signal Stability Routing (SSR)* uses signal strength as a criterion for choosing routes and routes are established only on-demand.

Route discovery can be very expensive in communications costs, reducing the response time of the network. On the other hand, explicit route maintenance can be even more costly in the explicit communication of substantial routing information.

Restricting the size of the network has been found to be extremely important in reducing the amount of routing information. The notion of establishing a subset of nodes that performs the routing has been proposed with *core-extraction distributed ad hoc routing (CEDAR)* (Sinha *et al.*, 1999). The CEDAR protocol dynamically establishes a core of the given network. Route computation is on-demand and is performed by only the core nodes. Other nodes each have a core host that maintains the local topology of nodes of which it is the core host. The core hosts route for each of the non-core nodes. CEDAR establishes the core nodes by using a DS. The use of DSs could be replaced with other graph structures such as a CDS or an IS (Das and Bharghavan, 1997). To reduce the complexity of route discovery, one may also use spanners (Gao *et al.*, 2001b; X.-Y. Li *et al.*, 2001, 2002a): These provide subgraph structures that maintain good approximations to the shortest paths in the original graph.

13.7   Conclusion and Remarks

In this chapter, we briefly reviewed some of the routing protocols proposed in the literature. In the past 10 years, there have been a massive number of routing protocols proposed for wireless ad hoc networks (and WSNs) and also a number of good surveys on routing protocols for wireless ad hoc networks. It is impossible for all interesting routing protocols to be covered here because of space limitations. The routing protocols covered in this chapter are representative of the routing protocols in their categories, such as proactive routing protocols, reactive routing protocols, hybrid routing protocols, and localized (geometric) routing protocols. Although so many protocols were proposed in the literature, there are still a number of challenging problems left for future research: For example, designing an efficient routing protocol that can find a routing path that optimizes one criterion and satisfies a number of other constraints, or designing efficient routing protocols that can theoretically find a *single* routing path with the largest capacity when the intrapath interferences and interpath interferences are considered. We also need to theoretically study the trade-offs between the proactive routing protocols and reactive routing protocols in terms of the node mobility, the node density, the link capacity, and so on. In other words, we need to quantify the conditions when we should apply proactive routing protocols and when we should apply reactive routing protocols.

With geometric position data available via GPS in the current scenario of mobile computing, there is additional geometric information available that can be of use in improving the performance of designing geometric structures and route discovery and maintenance. The natural research problem is the effective use of additional geometric information. In this chapter, we studied geometric graph structures and routing on the structures for mobile ad hoc communications. The problem is how to fully utilize the available geometry information or the geometric structure of the network to design efficient routing protocols. For example, for the maximum independent problem, it is known that we cannot approximate the MIS within a factor of $o(n^{1-\epsilon})$ for a general graph of n vertices, where $\epsilon > 0$ is any given constant. On the other hand, if we know that the input graph is a UDG, then there is a simple method with an approximation ratio of 5 (Hunt *et al.*, 1998). Furthermore, if we know the exact location of every node, there is a PTAS for a MIS when the input graph is a disk graph or some other variation of a disk graph (X.-Y. Li and Wang, 2006). A similar statement holds for the CDS problem. A future problem is to design efficient approximation algorithms for some other important questions by utilizing geometric structure properties.

Although the study of geometric structures is of considerable importance, as previously seen, an additional aspect not covered simply by the maintenance of *good* structures is the use of additional route information to enable quick recovery from faults or topology changes. This will enable the routing to have additional properties of *route recovery* or *fault tolerance*. None of the routing methodologies proposed initially incorporated this aspect into its design. Multiple paths to ensure a degree of fault tolerance were studied (Gafni and Bertsekas, 1981; Lee and Gerla, 2000; Nasipuri and Das, 1999; Park and Corson, 1999; Raju and Garcia-Luna-Aceves, 1999). We studied graph structures such as biconnected subgraphs that can be used to achieve the goals of route recovery. This would not only involve the design of appropriate route information but also the design and maintenance of structures that exhibit fault tolerance. Thus, we need to design efficient algorithms to maintain a good structure for routing for mobile networks.

Problems

13.1   What are the major differences between routing protocols DSDV and OLSR?

13.2   What are the major differences between AODV routing and DSR?

13.3   Consider a network $G = (V, E)$ in which each link e has a reliability $r(e)$. Design an efficient polynomial-time algorithm that runs in time $O(m + n \log n)$ to find a path with the largest reliability from a given source node s to every node in the network. Here, the reliability of the path is defined as the minimum reliability of all links on the path.

13.4   Consider a wireless network $G = (V, E)$. Assume that the delay of each link e in a wireless network is an integer $d(e) \in [1, d_0]$ for a constant d_0. Assume that the energy cost of each link is $c_e \in R^+$. Given a source node s and a target node t, present a polynomial-time algorithm that can find the optimum path connecting s and t with the smallest total cost and its delay is no more than a given bound D. Will your method still work if the delay of each link satisfies $d(e) \in [1, d_0 n]$ for a constant d_0?

13.5   Consider a network $G = (V, E)$ in which each link e has a cost $c(e)$ and a reliability $r(e)$. Assume that the reliability of a path is defined as the minimum reliability of all links on the path. Assume that we want to find a routing path from s to a target node t with minimum reliability r_0. Design a polynomial-time algorithm that can find the minimum total cost path among all paths with a reliability of at least r_0.

13.6   Consider a network $G = (V, E)$ in which each link e has a cost $c(e)$ and a reliability $0 < r(e) \leq 1$. For link-layer reliability, each node u will try up to k times to send data to a node v along link (u, v) if link (u, v) fails. If it cannot get the data to v with k tries, it will discard the data. When transport layer reliability is implemented, the source node will always try to send the data to the target node for up to k_2 tries. In other words, if the data are discarded at some intermediate node, the source node will initiate another round of sending until it has already tried k_2 times. Consider a routing path $\Pi = v_1 v_2 \cdots v_t$ from v_1 to a target node v_t.

Given a routing path $\Pi = v_1 v_2 \cdots v_t$, compute the probability that the target node will receive the data from v_1 if link-layer reliability is implemented. Given a routing path $\Pi = v_1 v_2 \cdots v_t$, compute the expected total cost incurred by all nodes for trying to relay data from v_1 to v_t if link-layer reliability is implemented.

Given a routing path $\Pi = v_1 v_2 \cdots v_t$, compute the probability that the target node will receive the data from v_1 if transport-layer reliability is implemented. Given a routing path $\Pi = v_1 v_2 \cdots v_t$, compute the expected total cost incurred by all nodes for trying to relay data from v_1 to v_t if transport-layer reliability is implemented.

13.7   Continue from the preceding question. Given a source node and a target node, design a polynomial-time algorithm to find a path with the minimum expected cost to connect them.

13.8   Consider a wireless network $G = (V, E)$ and a conflict graph H defined over E (which is the set of all directed communication links) in which two links in E are connected in H if they will cause interferences when they are active simultaneously. Assume that each link in G has a capacity $c(e) > 0$. Prove that given a source node and a target node, it is NP-hard to find the maximum flow that can be supported between them.

13.9   What are the advantages and disadvantages of using a localized routing protocol? What are the typical technical challenges in implementing a localized routing protocol based on geometry?

13.10   What is the local minimum in the localized routing protocol? Besides the planar graph and right-hand rule, what are other possible choices for getting out of the local minimum in the process of localized routing?

13.11   What are the advantages and disadvantages of using a backbone-based routing?

14 Energy-Efficient Broadcast/Multicast Routing

14.1 Introduction

Network-wide broadcasting in MANETs provides important control and route establishment functionality for a number of unicast and multicast protocols. In this chapter, an overview is presented of the recent progress of energy-efficient broadcast and multicast in wireless ad hoc networks.

Notice that, in general, there are four basic techniques for energy-efficient communication (Jones *et al.*, 2001).

1. The first technique is to turn off nonused transceivers to conserve energy. Then, we need to schedule, for every node, when it should sleep, when it should be idle, when it should receive, and when it should transmit such that a networking task is finished in a certain time period while simultaneously saving the energy cost.
2. The second technique is scheduling the competing nodes to avoid wasting energy because of contention. This can reduce the number of retransmissions and increase the nodes' lifetime by turning off the nonused transceivers for a period of time when they are not scheduled to transmit or receive. (This was studied in Chapter 4.)
3. The third technique is to reduce communication overhead, such as to defer transmission when the channel conditions are poor.
4. The fourth technique is to use power control to conserve energy. Each node will dynamically adjust its transmission power based on the downstream neighboring nodes to a level that is sufficient to reach the downstream neighboring node(s). This has the added advantage of reducing interference with other ongoing transmissions.

In this chapter, we mainly focus on the fourth technique. We discuss two energy models that could be used for broadcast.

Nonadjustable Power
If the power consumed at each node is not adjustable, minimizing the total power used by a reliable broadcast tree is equivalent to the minimum-connected-dominating-set (MCDS) problem; i.e., minimizing the number of nodes that relay the message, because all relaying nodes of a reliable broadcast form a connected dominating set (CDS).

Adjustable Power

If the power consumed at each node is adjustable, we assume that the power consumed by a relay node u is $\|uv\|^\beta$, where real number $\beta \in [2, 5]$ depends on the transmission environment and v is the farthest adjacent neighbor of u in the broadcast tree.

For both models, we study several centralized methods that compute broadcast trees such that the broadcast based on them consumes the energy within a constant factor of the optimum if the original communication graph is a UDG. Because centralized methods are expensive to implement, we further study several localized methods that can approximate the minimum-energy broadcast tree for the nonadjustable-power case. For the adjustable-power case, no localized methods can approximate the minimum-energy broadcast tree within a constant factor; thus, we study the several currently best possible heuristics. Several local improvement methods and activity scheduling of nodes (i.e., active, idle, sleep) are also discussed in this chapter.

Broadcasting and Multicasting

Broadcasting is a communication paradigm that allows sending data packets from a source to multiple receivers. In the one-to-all model, transmission by each node can reach *all* nodes that are within radius distance from it, whereas in the one-to-one model, each transmission is directed toward only one neighbor (using, e.g., directional antennas or separate frequencies for each node). Broadcasting in the literature has been studied mainly for the one-to-all model, and we use that model in this chapter. Broadcasting is also frequently referred to as *flooding*.

Broadcasting and multicasting in wireless ad hoc networks are critical mechanisms in various applications, such as information diffusion and wireless networks, and also for maintaining consistent global network information. Broadcasting is often necessary in MANET routing protocols. For example, many unicast routing protocols, such as DSR, AODV, ZRP, and LAR, use broadcasting or a derivation of it to establish routes. Currently, these protocols all rely on a simplistic form of broadcasting called *flooding*, in which each node (or all nodes in a localized area) retransmits each received unique packet exactly one time. The main problems with flooding are that it typically causes unproductive and often harmful bandwidth congestion, as well as inefficient use of node resources. Broadcasting is also more efficient than sending multiple copies of the same packet through unicast. It is highly important to use power-efficient broadcast algorithms for such networks because wireless devices are often powered by batteries only.

Recently, a number of research groups have proposed more efficient broadcasting techniques (Tseng *et al.*, 2002; Williams and Camp, 2002) with various goals such as minimizing the number of retransmissions, minimizing the total power used by all transmitting nodes, minimizing the overall delay of the broadcasting, and so on. Williams and Camp (2002) classified the broadcast protocols into four categories: simple (blind) flooding, probability-based, area-based, and neighbor-knowledge methods. Wu and Lou (2003) classified broadcasting protocols based on neighbor-knowledge information as global, quasi-global, quasi-local, and local. The global-broadcast protocol, centralized or distributed, is based on global state information. In quasi-global broadcasting, a broadcast protocol is based on partial global state information. For example, the approximation algorithm in Alzoubi *et al.* (2002b) is based on building a global spanning tree (a form of partial global state information) that is constructed in a sequence of sequential propagations. In quasi-local broadcasting, a distributed broadcast protocol is based on mainly local state information and occasionally partial global state information. Cluster networks are such examples: Although clusters can be constructed locally most of the time, the chain reaction does occur occasionally. In local broadcasting, a distributed broadcast protocol is based solely on local state information. All protocols that select forward nodes locally (based on a one-hop or two-hop neighbor set) belong to this category. It has been recognized that scalability in wireless networks cannot be achieved by relying on solutions in which each node requires global knowledge about the network. To achieve scalability, the concept of localized algorithms was proposed, because distributed algorithms for which simple local node behavior, based on local knowledge, achieves a desired global objective.

In this chapter, we categorize previously proposed broadcasting protocols into several families: centralized methods, distributed methods, and localized methods. Centralized methods calculate a tree used for broadcasting with various optimization objectives of the tree. In localized methods, each node has to maintain the state of its local neighbors (within some constant hops). After receiving a packet that needs to be relayed, the node decides whether to relay the packet based only on its local-neighborhood information. The majority of the protocols are in this family. In distributed methods, a node may need some information more than a constant hop away to decide whether to relay the message. For example, broadcasting based on a MST constructed in a distributed manner is a distributed method but not a localized method because we cannot construct a MST in a localized manner.

Distributed or Localized Algorithms?

Distributed algorithms and architectures have been terms commonly used for a long time in computer science. Unfortunately, none of the already proposed approaches are applicable to wireless ad hoc networks. To address the needs of distributed computing in wireless ad hoc networks, we have to address how key goals, such as power minimization, low latency, security, and privacy, are affected by the algorithms used. Some common denominators are almost always present, such as the high relative cost of communication to computation in wireless networks.

Because of the limited capability of processing power, storage, and energy supply, many conventional algorithms are too complicated to be implemented in wireless ad hoc networks. Thus, the wireless ad hoc networks require efficient distributed algorithms with low computation complexity and low communication complexity. More important, we expect the distributed algorithms for wireless ad hoc networks to be localized: Each node running the algorithm uses the information of nodes only within a constant number of hops. However, localized algorithms are difficult or impossible to design sometimes. For example, we cannot construct the MST locally.

MAC Specification

Collision avoidance is inherently difficult in MANETs; one often-cited difficulty is overcoming the hidden-node problem, in which a node cannot decide whether some of its neighbors are busy receiving transmissions from an uncommon neighbor. The 802.11 MAC follows a carrier-sense multiple-access/collision-avoidance (CSMA/CA) scheme. For unicast, it utilizes a request-to-send/clear-to-send/data/acknowledgment (RTS/CTS/Data/ACK) procedure to account for the hidden-node problem. However, the RTS/CTS/Data/ACK procedure is too cumbersome to implement for broadcast packets because it would be difficult to coordinate and bandwidth is expensive: A relay node has to perform RTS/CTS individually with all its neighbors that should receive the packets. Thus, the only requirement made for broadcasting nodes is that they assess a clear channel before broadcasting. Unfortunately, clear-channel assessment does not prevent collisions from hidden nodes. Additionally, no resource is provided for collision when two neighbors assess a clear channel and transmit simultaneously. Ramifications of this environment are subtle but significant. Unless specific means are implemented at the network layer, a node has no way of knowing whether a packet was successfully reached by its neighbors. In congested networks, a significant number of collisions occur, leading to many dropped packets. The most effective broadcasting protocols try to limit the probability of collisions by limiting the number of rebroadcasts in the network. Thus, it is often imperative that the underlying structure for broadcasting is degree-bounded and the links are at similar lengths. By use of a power adjustment at each node, the collision of packets and contention for channel will be alleviated. Notice that if the underlying structure for broadcasting is degree-bounded, we can either use the RTS/CTS scheme to avoid the hidden-node problem or we can rebroadcast the dropped packets (such a rebroadcast will be less because the number of intended receiving neighbors is bounded by a small constant).

Reliability

Reliability is the ability of a broadcast protocol to reach all the nodes in the network. It can be considered at the network- or at the medium-access layer. We classify protocols according to their network-layer performance. That is, assuming that the MAC layer is ideal (every message sent by a node reaches all its neighbors), the location-update protocol provides accurate desired information to all nodes about their neighborhood and the network is connected. Broadcast protocols can be *reliable* or *unreliable*. In a reliable protocol, every node in the network is reached, whereas in unreliable broadcast protocols, some nodes may not receive the message at all. It is easy to show that a broadcast protocol is reliable if and only if the set of nodes that forward the data packets forms a CDS of the network.

Message Contents

The broadcast schemes may require different neighborhood information, which is reflected in the contents of messages sent by nodes when they move, react to topological changes, change activity status, or simply send periodically updated messages. For example, the commonly seen *hello* message may contain (all or a subset of) the following information: its own ID, its position, one bit for dominating set status (informing neighbors whether the node itself is in a DS), a list of one-hop neighbors, and its degree. Other content is also possible, such as a list of one-hop neighbors with their positions, or a list of two-hop neighbors, or even global-network information.

The broadcast message sent by the source, or retransmitted, may contain a broadcast message only. In addition, it may contain a variety of information needed for proper functioning of broadcast protocol, such as the same type of information already listed for *hello* messages, some constant bits of the system requirements (e.g., the maximum broadcast delay), or a list of forwarding neighbors of the currently relaying node, informing them whether or not to retransmit the message.

Jitter and Random-Assessment Delay (RAD)

Suppose a source node originates a broadcast packet. Given that radio waves propagate at the speed of light, all neighbors will receive the transmission almost simultaneously. Assuming similar hardware and system loads, the neighbors will process the packet and rebroadcast at the same time. To overcome this problem, broadcast protocols jitter the scheduling of broadcast packets from the network layer to the MAC layer by some uniform random amount of time. This (small) offset allows one neighbor to obtain the channel first, while other neighbors detect that the channel is busy (clear-channel assessment fails) and thus delay their transmissions to avoid collision. Because the node has to back up all received broadcast packets within the random-assessment delay (RAD), the RAD also cannot be too large. On the other hand, if the RAD is small, this node may repeatedly broadcast the same packet, thus causing an infinity loop of rebroadcasting.

Many of the broadcasting protocols require a node to keep track of redundant packets received over a short time interval in order to determine whether to rebroadcast. That time interval, which was termed RAD (Williams and Camp, 2002), is randomly chosen from a uniform distribution between 0 and T_{max} seconds, where T_{max} is the highest possible delay interval. This delay in transmission accomplishes two things. First, it allows nodes sufficient time to receive redundant packets and assess whether to rebroadcast. Second, randomized scheduling prevents transmission collisions.

Performance Measurement

The performance of broadcast protocols can be measured by a variety of metrics. A commonly used metric is the number of message retransmissions with respect to the number of nodes. In the case of broadcasting with adjusted transmission power (and thus an adjusted disk that the message can reach), the total power can be used as performance metrics. The next important metric is reachability, or the ratio of nodes connected to the source that received the broadcast message. Time delay or latency is sometimes used, which is the time needed for the last node to receive the broadcast message initiated at the source. Note that retransmissions at the MAC layer are normally deferred, to avoid message collisions. Some authors consider as an alternative a more restricted indicator;

i.e., whether or not the path from source to any node is always following the shortest path. This measure may be important if used as part of a routing scheme because route paths are created during the broadcast process.

14.2   Centralized Methods

We assume that two energy models could be used for broadcast: One is nonadjustable power and one is adjustable power. In the rest of the section, for these two energy models, several centralized methods are reviewed that can build some broadcast tree whose energy consumption is within a constant factor of the optimum if the original communication graph is modeled by a UDG.

Minimum-energy broadcast/multicast routing in a simple ad hoc networking environment has been addressed by the pioneering work in Clementi *et al.* (2001a, 2001b), Kirousis *et al.* (2000), and Wieselthier *et al.* (2000). To assess the complexities *one at a time*, the nodes in the network are assumed to be randomly distributed in a 2D plane and there is no mobility. Nevertheless, as argued in Wieselthier *et al.* (2000), the impact of mobility can be incorporated into this static model because the transmitting power can be adjusted to accommodate the new locations of the nodes as necessary. In other words, the capability to adjust the transmission power provides considerable "elasticity" to the topological connectivity and, hence, may reduce the need for handoffs and tracking. In addition, as assumed in Wieselthier *et al.* (2000), there are sufficient bandwidth and transceiver resources. Under these assumptions, centralized (as opposed to distributed) algorithms were presented by Das *et al.* (2002), Li and Nikolaidis (2001), Liang (2002), and Wieselthier *et al.* (2000) for minimum-energy broadcast/multicast routing. These centralized algorithms, in this simple networking environment, are expected to serve as the basis for further studies on distributed algorithms in a more practical network environment, with limited bandwidth and transceiver resources, as well as the node mobility.

14.2.1   Centralized Clustering for Nonadjustable Power

We first study the nonadjustable-power-model case in which the transmission power of every node is fixed and cannot be dynamically adjusted based on the set of neighbors to communicate. The set of nodes that rebroadcast messages in a reliable broadcasting scheme define a CDS. Recall that a subset C of V is a CDS if C is a dominating set and C induces a connected subgraph. Consequently, the nodes in C can communicate with each other without using nodes in $V-C$. A broadcasting based on a CDS uses the nodes in the CDS only to relay the message. We first review several methods in the literature to build a CDS.

If every node cannot adjust its transmission power accordingly, then we need find the MCDS to save the total power consumption of the broadcasting protocol. Unfortunately, the problem of finding a CDS of minimal size is NP-complete even for UDGs. Guha and Khuller (1996) studied the approximation of the CDS problem for general graphs. They gave two different approaches; both of them guarantee an approximation ratio of $\Theta[H(\Delta)]$. Because their approaches are for general graphs, they thus do not utilize the geometry structure if applied to the wireless ad hoc networks. One approach is to grow a spanning tree that includes all nodes. The internal nodes of the spanning tree are selected as the final CDS. This approach has an approximation ratio of $2[H(\Delta) + 1]$. The other approach is first approximating the DS and then connecting the DS to a CDS. Guha and Khuller (1996) proved that this approach has an approximation ratio of $\ln \Delta + 3$.

One can also use a Steiner tree algorithm to connect the dominators. This straightforward method gives an approximation ratio of $c[H(\Delta) + 1]$, where c is the approximation ratio for the unweighted Steiner tree problem. Currently, the best ratio is $1 + (\ln 3/2) \simeq 1.55$, which is from Robins and Zelikovsky (2000).

By definition, any algorithm generating a MIS is a clustering method. We first review the methods that approximate the MIS, the minimum dominating set (MDS), and the MCDS. Hunt *et al.* (1998) and Marathe *et al.* (1995) studied the approximation of the MIS and the MDS for UDGs. They gave the first PTASs for the MDS in a UDG. The method is based on the following observations: A MIS is always a DS; given a square Ω with a fixed area, the size of any MIS is bounded by a constant C. Assume that there are n nodes in Ω. Then, we can enumerate all sets with a size of at most C in time $\Theta(n^C)$. Among these enumerated sets, the smallest DS is the MDS. Then, using the shifting strategy, they derived a PTAS for the MDS problem.

Because we have a PTAS for the MDS and the graph Virt$G$ connecting every pair of dominators within at most three hops is connected (Y. Wang and Li, 2002b), we have an approximation algorithm (constructing a MST Virt$G$) for MCDS with an approximation ratio of $3 + \epsilon$. Notice that Berman *et al.* (1998) gave a $4/3$ approximation method to connect a DS and Robins and Zelikovsky (2000) gave a $4/3$ approximation method to connect an IS. Thus, we can easily have an $8/3$ approximation algorithm for the MCDS, which was reported in Alzoubi (2002). Recently, Cheng *et al.* (2003) designed a PTAS for a MCDS in a UDG. However, it is expensive to run their method efficiently in a distributed manner.

THEOREM 14.1 *There is a PTAS for the minimum-energy broadcast in wireless ad hoc networks when each node in the network has a* fixed *uniform transmission power and the energy consumed by a node to receive a message is ignored.*

When the energy consumed by a node to receive a message is not ignored, we can also design an efficient broadcast algorithm with a constant-approximation ratio. Let E_T be the power consumed by a node to send a certain unit amount of data (including the energy used by its own circuit). Let E_R be the power consumed by a node to receive a certain unit amount of data (including the energy used by its own circuit). Notice that, typically, $E_R \leq E_T$ for most wireless devices. For an optimum broadcast scheme over a network with n nodes, let OPT be the set of the nodes forwarding the broadcast data in the optimum broadcast scheme. Here, we assume that the broadcast channel is reliable; i.e., all neighboring nodes of a node u will get the broadcast message $\mathcal{M}$ correctly after node u sends the broadcast message $\mathcal{M}$ only once. Then, the total cost by the optimum broadcast is at least $|\text{OPT}|E_T + nE_R$ because each node needs to receive data for at least once. Recall that the nodes in OPT must form a CDS for the network to ensure that the broadcast can be performed from the source node.

When the network is modeled by a UDG, we apply an efficient CDS approximation scheme from Alzoubi *et al.* (2002a) and Wan *et al.* (2002a). It was proved in Alzoubi *et al.* (2002a) that the CDS constructed by their method has the following properties:

1. The size of the CDS is no more than 10 times the size of the MCDS.
2. For every node v in the network, there is at most a constant, denoted by c, number of neighbors of v selected in the CDS.

Given a CDS, we perform a broadcast as follows. First, the source node sends its data to one of its neighbors that is in the CDS if the source node itself is not in the CDS. When a node receives a broadcast message, it forwards the message to its neighbors only if (1) itself is in the CDS, and (2) it did not forward the same message before. If we perform the broadcast based on this CDS, it is easy to see that the total cost is at most

$$|\text{CDS}|E_T + cnE_R \leq \max(10, c)(|\text{OPT}|E_T + nE_R).$$

Thus, we have the following theorem:

THEOREM 14.2 *There is a polynomial-time (centralized or distributed) algorithm with a constant-approximation ratio for the minimum-energy broadcast in wireless ad hoc networks when each node in the network has a* fixed *uniform transmission power and each node has a fixed energy consumption for receiving a unit amount of data.*

14.2.2   Based on MST and Variations for Adjustable Power

We now study the energy-efficient broadcasting/multicast for the model in which nodes could dynamically adjust their transmission power based on the recipients. The scheme proposed in Marks *et al.* (2002) is built on an alternative search-based paradigm in which the minimum-cost broadcast/multicast tree is constructed by a search process. Two procedures were devised to check the viability of a solution in the search space. Preliminary experimental results show that this method renders better solutions than BIP, though at a higher computational cost. Liang (2002) showed that the minimum-energy broadcast-tree problem is NP-complete and proposed an approximate algorithm to provide a bounded performance guarantee for the problem in a general setting. Essentially, Liang (2002) reduces the minimum-energy broadcast tree problem to an optimization problem on an auxiliary weighted graph and solves the optimization problem so as to give an approximate solution for the original problem. He also proposed another algorithm that yields better performance under a special case. Das *et al.* (2002) proposed an evolutionary approach using genetic algorithms. Das *et al.* (2003) also presented three different integer programming models that can be used to find the solutions to the minimum-energy broadcast/multicast problem. The major drawbacks of optimization-based schemes are, however, that they are centralized and require the availability of global topological information.

Some centralized methods are based on greedy heuristics. Three greedy heuristics were proposed in Wieselthier *et al.* (2000) for the minimum-energy broadcast routing problem: MST, SPT, and BIP.

- The MST heuristic first applies Prim's algorithm to obtain a MST and then orients it as an arborescence rooted at the source node. Notice that here, the weight of a link is the energy cost to send a unit amount of data from the sending node to the source node. It is not the Euclidean distance between the source node and the target node.
- The SPT heuristic applies Dijkstra's algorithm to obtain a SPT rooted at the source node. In other words, we construct the shortest path from the source node to every node in the network, and then union them together to get a tree called the shortest-path tree.
- The BIP heuristic is the node version of Dijkstra's algorithm for the SPT. It maintains, throughout its execution, a single arborescence rooted at the source node. The arborescence starts from the source node, and new nodes are added to the arborescence one at a time on a minimum incremental cost basis until all nodes are included in the arborescence. The cost of adding a new node to the arborescence is the minimum additional power increased by some node in the current arborescence to reach this new node. The implementation of BIP is based on the standard Dijkstra's algorithm, with one fundamental difference on the operation whenever a new node q is added. Whereas Dijkstra's algorithm updates the node weights (representing the current known distances to the source node), BIP updates the cost of each link (representing the incremental power to reach the head node of the directed link). This update is performed by subtracting the cost of the added link pq from the cost of every link qr that starts from q to a node r not in the new arborescence.

They have been evaluated through simulations in Wieselthier *et al.* but little is known about their analytical performances in terms of the approximation ratio. Here, the approximation ratio of a heuristic is the maximum ratio of the energy needed to broadcast a message based on the arborescence generated by this heuristic to the least necessary energy by any arborescence for any set of points.

For a pure illustration purpose, another slight variation of BIP was discussed in detail in Wan *et al.* (2002b). This greedy heuristic is similar to Chvátal's algorithm (Chvátal, 1979) for the set-cover problem and is a variation of BIP. Wan *et al.* (2002b) proposed a new heuristic, the broadcast average incremental power (BAIP). Like BIP, an arborescence, which starts with the source node, is maintained throughout the execution of the algorithm. However, unlike BIP, many new nodes can be added one at a time. Similar to Chvatal's algorithm, the new nodes added are chosen to have the minimal *average* incremental cost, which is defined as the ratio of the minimum additional power increased by some node in the current arborescence to reach these new nodes to the number of these new nodes. In contrast to the $1 + \log m$ approximation ratio of Chvatal's algorithm, where m is the largest set size in the set-cover problem, they showed that the approximation ratio of BAIP is at least $\frac{4n}{\ln n} - o(1)$, where n is the number of receiving nodes.

Wan *et al.* (2002b) showed that the approximation ratios of the MST and BIP are between 6 and 12 and between $(13/3)$ and 12, respectively; on the other hand, the approximation ratios of the SPT and BAIP are at least $(n/2)$ and $(4n/\ln n) - o(1)$, respectively, where n is the number of nodes. We discuss in detail their proof techniques in the next subsection.

The iterative maximum-branch minimization (IMBM) algorithm was another effort (Li and Nikolaidis, 2001) to construct power-efficient broadcast trees. It begins with a basic broadcast tree in which the source directly transmits to all other nodes. Then, it attempts to approximate the minimum-energy broadcast tree by iteratively replacing the maximum branch with less-power, more-hop alternatives.

Both BIP and IMBM operate under the assumption that the transmission power of each node is unconstrained; i.e., every node can reach every other node. Both algorithms are centralized in the sense that they have the following requirements: (a) the source node needs to know the position/distance of every other node; and (b) each node needs to know its downstream, on-tree neighbors so as to propagate broadcast messages. As a result, it may be difficult to extend both algorithms into distributed versions because a significant amount of information is required to be exchanged among nodes.

14.2.3 Theoretical Performance Analysis

Any broadcast routing is viewed as an arborescence (a directed tree) T, rooted at the source node of the broadcasting, that spans all nodes. Let $f_T(\mathbf{p})$ denote the transmission power of the node $\mathbf{p}$ required by T. For any leaf node $\mathbf{p}$ of T, $f_T(\mathbf{p}) = 0$. For any internal node $\mathbf{p}$ of T,

$$f_T(\mathbf{p}) = \max_{\mathbf{pq} \in T} \|\mathbf{pq}\|^{\beta};$$

in other words, the βth power of the longest distance between $\mathbf{p}$ and its children in T. The total energy required by T is $\sum_{\mathbf{p} \in P} f_T(\mathbf{p})$. Thus, the minimum-energy broadcast-routing problem is different from the conventional link-based MST problem. Indeed, although the MST can be solved in polynomial time by algorithms such as Prim's algorithm and Kruskal's algorithm, the minimum-energy broadcast-routing problem cannot be solved in polynomial time unless $P = \mathrm{NP}$ (Clementi *et al.*, 2001a). In its general graph version, minimum-energy broadcast-routing can be shown to be NP-hard (Garey and Johnson, 1979) and, even worse, it cannot be approximated within a factor of $(1 - \epsilon) \log \Delta$ unless $\mathrm{NP} \subseteq D\mathrm{TIME}\left[n^{O(\log \log n)}\right]$, where Δ is the maximal degree and ϵ is any arbitrary small positive constant. However, this hardness of its general graph version does not necessarily imply the same hardness of its geometric version. In fact, as shown later in this chapter, its geometric version can be approximated within a constant factor. Nevertheless, this suggests that the minimum-energy broadcast-routing problem is considerably harder than the MST problem. Recently, Clementi *et al.* (2001a) proved that the minimum-energy broadcast-routing problem is a NP-hard problem and obtained parallel but weaker results compared to those of Wan *et al.* (2002b).

Wan *et al.* (2002b) gave some lower bounds on the approximation ratios of MST and BIP by studying some special instances. Their deriving of the upper bounds relies extensively on the geometric structures of EMSTs. A key result in Wan *et al.* (2002b) is an upper bound on the parameter $\sum_{e \in \mathrm{mst}(P)} \|e\|^2$ for any finite point set P of radius one. Note that the supreme of the total edge lengths of $\mathrm{mst}(P)$, $\sum_{e \in \mathrm{mst}(P)} \|e\|$, over all point sets P of radius one is infinity. However, the parameter $\sum_{e \in \mathrm{mst}(P)} \|e\|^2$ is bounded from above by a constant for any point set P of radius one. They use c to denote the supreme of $\sum_{e \in \mathrm{mst}(P)} \|e\|^2$ over all point sets P of radius one. The constant c is at most 12. The proof of this theorem involves complicated geometric arguments; see Wan *et al.* (2002b) for more details. Note that for any point set P of radius one, the length of each edge in $\mathrm{mst}(P)$ is at most one. Therefore, for any point set P of radius one and any real number $\beta \geq 2$,

$$\sum_{e \in \mathrm{mst}(P)} \|e\|^\beta \leq \sum_{e \in \mathrm{mst}(P)} \|e\|^2 \leq c \leq 12.$$

The next theorem proved in Wan *et al.* (2002b) explores a relation between the minimum energy required by a broadcast and the energy required by the EMST of the corresponding point set.

LEMMA 14.1 (Wan *et al.*, 2002b) *For any point set P in the plane, the total energy required by any broadcast among P is at least $\frac{1}{c} \sum_{e \in mst(P)} \|e\|^\beta$.*

Proof. Let T be an arborescence for a broadcast among P with minimum energy consumption. For any nonleaf node $\mathbf{p}$ in T, let $T_\mathbf{p}$ be an EMST of the point set consisting of $\mathbf{p}$ and all children of $\mathbf{p}$ in T. Suppose that the longest Euclidean distance between $\mathbf{p}$ and its children is r. Then, the transmission power of node $\mathbf{p}$ is r^β, and all children of $\mathbf{p}$ lie in the disk centered at $\mathbf{p}$ with radius r. From the definition of c, we have

$$\sum_{e \in T_\mathbf{p}} \left(\frac{\|e\|}{r} \right)^\beta \leq c,$$

which implies that

$$r^\beta \geq \frac{1}{c} \sum_{e \in T_\mathbf{p}} \|e\|^\beta.$$

Let T^* denote the spanning tree obtained by superposing all $T_\mathbf{p}$'s for nonleaf nodes of T. Then, the total energy required by T is at least $\frac{1}{c} \sum_{e \in T^*} \|e\|^\beta$, which is further no less than $\frac{1}{c} \sum_{e \in \mathrm{mst}(P)} \|e\|^\beta$. This completes the proof. ∎

Consider any point set P in a 2D plane. Let T be an arborescence oriented from some $\mathrm{mst}(P)$. Then, the total energy required by T is at most $\sum_{e \in T_\mathbf{p}} \|e\|^\beta$. From Lemma 14.1, this total energy is at most c times the optimum cost. Thus, the approximation ratio of the link-based MST heuristic is at most c. Together with $c \leq 12$, this observation leads to the following theorem:

THEOREM 14.3 (Wan *et al.*, 2002b) *The approximation ratio of the link-based MST heuristic is at most c and, therefore, is at most* 12.

In addition, they derived an upper bound on the approximation ratio of the BIP heuristic. Once again, the EMST plays an important role.

LEMMA 14.2 (Wan *et al.*, 2002b) *For any broadcasting among a point set P in a 2D plane, the total energy required by the arborescence generated by the BIP algorithm is at most* $\sum_{e \in mst(P)} \|e\|^\beta$.

14.3   Efficient Distributed or Localized Methods

In this section, we study a number of efficient distributed or even localized algorithms that can find an energy-efficient broadcast structure. Again, we separate these algorithms into two different categories: a nonadjustable-power model or an adjustable-power model.

14.3.1   Based on Distributed CDS for Fixed-Power Scenario

A natural structure for broadcasting is the CDS. Many distributed-clustering (or DS) algorithms have been proposed in the literature (Alzoubi *et al.*, 2002c; Amis *et al.*, 2000; Chlamtac and Farago, 1999; Lin and Gerla, 1997a). All algorithms assume that the nodes have distinctive identities (denoted by ID hereafter).

In the rest of this subsection, we interchange the terms cluster-head and dominator. The node that is not a cluster-head is also called a *dominatee*. A node is called a *white* node if its status is yet to be decided by the clustering algorithm. Initially, all nodes are white. The status of a node, after the clustering method finishes, could be a *dominator* with the color *black* or a *dominatee* with the color *gray*. The rest of this subsection is devoted to the distributed methods that approximate the MDS and the MCDS for a UDG.

Clustering Without Geometry Property

For general graphs, Jia *et al.* (2000) described and analyzed some randomized distributed algorithms for the MDS problem that run in polylogarithmic time, independent of the diameter of the network, and that return a DS of a size within a logarithmic factor from the optimum with high probability. Their best algorithm runs in $O(\log n \log \Delta)$ rounds with high probability, and every pair of neighbors exchanges a constant number of messages in each round. The computed DS is within $O(\log \Delta)$ in expectation and within $O(\log n)$ with high probability. Their algorithm works for a weighted DS also.

The method proposed by Das and Bharghavan (1997) and Sivakumar *et al.* (1998) contains three stages: approximating the MDS, constructing a spanning forest of stars, and expanding the spanning forest to a spanning tree. Here, the *stars* are formed by connecting each dominatee node to one of its dominators. The approximation method of a MDS is essentially a distributed variation of the centralized Chvatal's greedy algorithm (Chvátal, 1979) for set cover. Notice that the DS problem is essentially the set-cover problem, which has been well studied. It is then no surprise that the method of Das and Bharghavan (1997) and Sivakumar et al. (1998) guarantees a $H(\Delta)$ for the MDS problem, where H is the harmonic function and Δ is the maximum node degree.

Although the algorithm proposed by Das and Bharghavan (1997) and Sivakumar et al. (1998) finds a DS and then grows it to a CDS, the algorithm proposed by Wu and Li (2001) takes an opposite approach. They first find a CDS and then prune out certain redundant nodes from the CDS. The initial CDS $\mathbb{C}$ contains all nodes that have at least two nonadjacent neighbors. A node u is said to be *locally redundant* if it has either a neighbor in $\mathbb{C}$ with a larger ID that dominates all other neighbors of u or two adjacent neighbors with a larger ID that together dominate all other neighbors of u. Their algorithm then keeps removing all locally redundant nodes from $\mathbb{C}$. They showed that this algorithm works well in practice when the nodes are distributed uniformly and randomly, although no theoretical analysis is given by them for both the worst case and for the average approximation ratio. However, it was shown by Alzoubi et al. (2002c) that the approximation ratio of this algorithm could be as large as $n/2$.

Recently, Dai and Wu proposed several distributed dominant pruning algorithms (Dai and Wu, 2004; Wu and Li, 1999, 2001; Wu et al., 2002). Each node has a priority that can be simply its unique identifier or a combination of remaining battery, degree, or identifier. A node u is "fully covered" by a subset S of its neighboring nodes if and only if the following three conditions hold:

- The subset S is connected.
- Any neighbor of u is a neighbor of at least one node from S.
- All nodes in S have a higher priority than u.

A node belongs to the DS if and only if there is no subset that fully covers it. The advantage of using a CDS as defined in Dai and Wu (2004) and Wu and Li (2001) is that each node can decide whether it is in the DS without any additional communication steps involved, other than those needed to maintain neighborhood information. The neighborhood information needed is either two-hop-neighbor knowledge or one-hop-neighbor knowledge with their position.

Stojmenovic et al. (2002) observed that distributed constructions of a CDS can be obtained following the clustering scheme of Lin and Gerla (1997a). A CDS consists of two types of nodes: cluster-head and border-nodes (also called gateway or connectors elsewhere). The cluster-head nodes are decided as follows. At each step, all white nodes that have the lowest *rank* among all white neighbors are colored black, and the white neighbors are colored gray. The ranks of the white nodes are updated if necessary. The clustering method uses two messages that can be called IamDominator and IamDominatee. A white node claims itself to be a dominator if it has the smallest ID among all its white neighbors, if there are any, and broadcasts IamDominator to its one-hop neighbors. A white node receiving the IamDominator message marks itself as dominatee and broadcasts IamDominatee to its one-hop neighbors. The set of dominators generated by the preceding method is actually a MIS. Here, we assume that each node knows the IDs of all its one-hop neighbors, which can be achieved if each node broadcasts its ID to its neighbors initially. This approach of constructing the MIS is well known. The following rankings of a node are used in various methods: the ID only (Chlamtac and Farago, 1999; Lin and Gerla, 1997a), the ordered pair of degree and ID (Chen *et al.*, 2002), and an ordered pair of degree and location (Stojmenovic and Datta, 2002). Basagni *et al.* (1997) used a general *weight* as a ranking criterion for selecting the node as the cluster-head, in which the weight is a combination of the mentioned criteria and some new ones, such as mobility or remaining energy. After the cluster-head nodes are selected, border-nodes are selected to connect them. A node is a border-node if it is not a cluster-head and there are at least two cluster-heads within its two-hop neighborhood. It was shown by Alzoubi *et al.* (2002c) that the worst-case approximation ratio of this method is also $(n/2)$, although it works well in practice.

Clustering with Geometry Property

Notice that none of the preceding algorithms utilizes the geometry property of the underlying UDG. Recently, several algorithms were proposed with a constant worst-case approximation ratio by taking advantage of the geometry properties of the underlying graph. It is used to connect the cluster-heads constructed as previously described into a CDS with fewer additional nodes. During this second step of backbone formation, some *connectors* (also called *gateways*) are found among all the dominatees to connect the dominators. Then, the connectors and the dominators form a *connected dominating set*. Recently, Wan *et al.* (2002a) and Wu and Lou (2003) proposed a communication-efficient algorithm to find connectors based on the fact that there are only a constant number of dominators within k-hops of any node. The following observation is a basis of several algorithms for a CDS. After clustering, one dominator node can be connected to many dominatees. However, it is well known that a dominatee node can be connected to only at most *five* dominators in the UDG model. Generally, it was shown in Wan *et al.* (2002a), Y. Wang and Li (2002b), and Wu and Lou (2003) that for each node v (dominator or dominatee), the number of dominators inside the disk centered at v with radius k-units is bounded by a constant $\ell_k < (2k + 1)^2$.

Given a dominating set S, let $\text{Virt}G$ be the graph connecting all pairs of dominators u and v if there is a path in the UDG connecting them with at most three hops. Graph $\text{Virt}G$ is connected. It is natural to form a CDS set by finding connectors to connect any pair of dominators u and v if they are connected in $\text{Virt}G$. This strategy was also adopted by Wan *et al.* (2002a) and Wu and Lou (2003). Notice that in the approach by Stojmenovic *et al.* (2002), they set any dominatee node as the connector if there are two dominators within its two-hop neighborhood. This approach is very pessimistic and results in a very large number of connectors in the worst case (Alzoubi *et al.*, 2002c). Instead, Alzoubi *et al.* (2002b) suggested finding only one unique shortest path to connect any two dominators that are at most three hops away.

We briefly review their basic idea of forming a CDS in a distributed manner. Let $\mathsf{P}_{UDG}(u, v)$ be the path connecting two nodes u and v in a UDG with the smallest number of hops. Let's first consider how to connect two dominators within three hops. If the path $P_{UDG}(u, v)$ has two hops, then u finds the dominatee with the smallest ID to connect u and v. If the path $P_{UDG}(u, v)$ has three hops, then u finds the node, say w, with the smallest ID such that w and v are two hops apart. Then, node w selects the node with the smallest ID to connect w and v. Y. Wang and Li (2002b) and Wan et al. (2002a) discussed in detail some approaches to optimizing the communication cost and the memory cost.

The graph constructed by this algorithm is called a CDS graph (or *backbone* of the network). If we also add all edges that connect all dominatees to their dominators, the graph is called an extended CDS, denoted by CDS'. Let opt be the size of the MCDS. It was shown (Marathe et al., 1995) that the size of the computed MIS has a size of at most 4opt + 1. We already showed that the size of the CDS found by the preceding algorithm is at most $\ell_3 k + k$, where k is the size of the MIS found by the clustering algorithm. It implies that the found CDS has a size of at most $4(\ell_3 + 1)\text{opt} + \ell_3 + 1$. Consequently, the computed CDS is at most a $4(\ell_3 + 1)$ factor of the optimum (with an additional constant $\ell_3 + 1$). It was shown in Alzoubi et al. (2002b) and Y. Wang and Li (2002b) that the CDS' graph is a sparse spanner in terms of both hops and length; meanwhile, the CDS has a bounded-node degree.

See Chapters 7 and 8 for a more detailed study of distributed algorithms constructing CDSs.

14.3.2   Localized Low-Weighted Structures for Adjustable-Power Scenario

The centralized algorithms do not consider computational and message overheads incurred in collecting global information. Several of them also assume that the network topology does not change between two runs of information exchange. These assumptions may not hold in practice because the network topology may change from time to time and the computational and energy overheads incurred in collecting global information may not be negligible. This is especially true for large-scale wireless networks in which the topology is changing dynamically because of the changes of position, energy availability, environmental interference, and failures, which implies that centralized algorithms that require global topological information may not be practical.

Flooding is also a good solution for the sake of scalability and simplicity. Several flooding techniques for wireless networks have been proposed, each with respect to certain optimization criterion. However, none of them takes advantage of the feature that the transmission power of a node can be adjusted.

Some distributed heuristics were proposed, as in Cagalj et al. (2002). Most of them are based on the distributed MST method. A possible drawback of these distributed methods is that they may not perform well under frequent topological changes because they rely on information that is multiple hops away to construct the MST. Refer to N. Li et al. (2003) for more details. The RNG, GG, and YG all have $O(n)$ edges and contain the EMST. This implies that we can construct the MST by using $O(n \log n)$ messages.

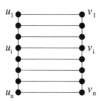

Figure 14.1 An instance in which wireless nodes that every network structure described previously (except MST) have an arbitrarily large total weight. © 2005, IEEE.

Localized minimum-energy broadcast algorithms are based on the use of a locally defined geometric structure, such as a RNG. The RNG consists of all edges uv such that uv is not the longest edge in any triangle uvw. That is, uv belongs to the RNG if there is no node w such that $uw < uv$ and $vw < uv$.

Cartigny *et al.* (2003) proposed a localized algorithm, the related neighborhood-graph-based broadcast-oriented protocol (RBOP) that is built on the notion of a RNG. In RBOP, the broadcast is initiated at the source and propagated, following the rules of neighbor elimination (Stojmenovic *et al.*, 2002), on the topology represented by the RNG. Simulation results show that the energy consumption could be as high as 100% compared with BIP. However, the communication overhead, because of mobility and changes in activity status in BIP, are not considered; therefore, RBOP is superior to BIP in dynamic ad hoc networks. N. Li *et al.* (2003) and Cartigny *et al.* (2005) proposed another localized algorithm, which applies the LMST (localized minimal spanning tree) instead of the RNG as the broadcast topology. In the LMST, proposed in N. Li *et al.* (2003), each node calculates the local MST of itself and its one-hop neighbors. A node uv is in the LMST if and only if u and v select each other in their respective trees. The simulations (Cartigny *et al.*, 2003; N. Li *et al.*, 2003) show that the performance of LMST-based schemes is significantly better than the performance of the RBOP and with about 50% more energy consumption than BIP in static scenarios. Cartigny *et al.* (2005) demonstrated that when $c > 0$ in a power-attenuation model, in which energy consumption for transmitting over an edge uv is $\|uv\|^\beta + c$, there exists an optimal "target" transmission radius so that further energy savings can be obtained if transmission radii are selected near the target radius.

However, as shown in X.-Y. Li (2003b) (also in Figure 14.1), the total weights of a RNG and a LMST could still be as large as $O(n)$ times of the total weight of a MST. Here, in Figure 14.1, $\|u_i v_i\| = 1$ and $\|u_i u_{i+1}\| = \|v_i v_{i+1}\| = \epsilon$ for a very small positive real number ϵ. Given a graph G, let $\omega_b(G) = \sum_{e \in G} \|e\|^b$. Then, $\omega_1(\text{RNG}) = \Theta(n)\omega_1(\text{MST})$ and $\omega_1(\text{LMST}) = \Theta(n)\omega_1(\text{MST})$.

X.-Y. Li (2003b) and X.-Y. Li *et al.* (2004d) described several low-weight planar structures that can be constructed by localized methods with total communication costs $O(n)$. The energy consumption of a broadcast based on those structures are within $O(n^{\beta-1})$ of the optimum; i.e., $\omega_\beta(H) = O(n^{\beta-1})\omega_\beta(\text{MST})$, $\omega_\beta(\text{LMST}_2) = O(n^{\beta-1})\omega_\beta(\text{MST})$, $\omega_\beta(\text{IMRG}) = O(n^{\beta-1})\omega_\beta(\text{MST})$ for any $\beta \geq 1$. This improves the previously known "lightest" structure, the RNG, by a factor of $O(n)$ because, in the worst case, $\omega(\text{RNG}) = \Theta(n)\omega(\text{MST})$ and $\omega_\beta(\text{RNG}) = \Theta(n^\beta)\omega_\beta(\text{MST})$.

We now review these three structures in detail.

Structure Based on RNG′

Although a RNG is a very sparse structure (i.e., the average number of neighbors per node is about 2.5), in some degenerate cases a particular node may have an arbitrarily large degree. This motivated Stojmenovic (2003) to define a modified structure in which each node will have a degree bounded by six. The same structure was independently proposed by X.-Y. Li (2003b) with an additional motivation. Li proved that the modified RNG is the first localized method to construct a structure H with weight $O(\omega(\text{MST}))$ by using total $O(n)$ local-broadcast messages. Note that if each node already knows the positions and IDs of all its neighbors, then no messages are needed to decide which of its edges belong to the (modified) RNG. Notice that, traditionally, the RNG will always select an edge uv even if there is some node on the boundary of lune (u, v). Here, lune (u, v) is the intersection of two disks centered at nodes u and v with radius $\|uv\|$. Thus, the RNG may have an unbounded node degree; e.g., considering $n - 1$ points equally distributed on the circle centered at the nth point v, the degree of v is $n - 1$. Notice that for the sake of lowering the weight of a structure, the structure should contain as few edges as possible without breaking the connectivity. X.-Y. Li (2003b) and Stojmenovic (2003) then naturally extended the traditional definition of the RNG as follows.

We need to make distinct edge lengths. We can achieve this by adding the secondary and, if necessary, the ternary keys for comparing two edges. Each node is assumed to have a unique ID. Then, we consider the record $(\|uv\|)$, $\text{ID}(u)$, $\text{ID}(v)$, where $\text{ID}(u) < \text{ID}(v)$ (otherwise, u and v are exchanged for a given edge). Two edges compare their lengths first to decide which one is longer. If they are the same, they then compare their secondary key, which is their respective lower end-point node's ID. If this is also the same, then the ternary key resolves the comparison (otherwise, we are comparing an edge against itself). This simple method for making a distinct edge length was proposed in N. Li *et al.* (2003). The edge lengths, so defined, are then used in the regular definition of a RNG. It is easy to show that two RNG edges uv and uw going out of the same node must have an angle between them of at least $\pi/6$; otherwise, $vw < uv$ or $vw < uw$, and one of the two edges becomes the longest in the triangle and, consequently, could not be in the RNG. X.-Y. Li denoted the modified RNG structure as RNG′. Obviously, RNG′ is a subgraph of the traditional RNG. It was proved in X.-Y. Li (2003b) and Stojmenovic (2003) that RNG′ still contains a MST as a subgraph. However, RNG′ is still not a low-weight structure.

Notice that it is well known that the communication complexity of constructing a MST of an n-vertex graph G with m edges is $O(m + n \log n)$; the communication complexity of constructing a MST for a UDG is $O(n \log n)$ even under the local-broadcasting communication model in wireless networks. It was shown in X.-Y. Li (2003b) that it is *impossible* to construct a low-weighted structure by using only one-hop-neighbor information.

The localized algorithm given in X.-Y. Li (2003b) that constructs a low-weighted structure with only some two-hop information is as follows.

Obviously, if an edge uv is kept by node u, then it is also kept by node v. The following theorem was proved in X.-Y. Li (2003b):

THEOREM 14.4 (X.-Y. Li, 2003b) *The total edge weight of H constructed by Algorithm 35 is within a constant factor of that of the MST.*

Algorithm 35 Construct a Low-Weight Structure H

1: All nodes together construct the graph RNG′ in a localized manner.
2: Each node u locally broadcasts its incident edges in RNG′ to its one-hop neighbors. Node u listens to the messages from its one-hop neighbors.
3: Assume node u received a message informing existence of edge $xy \in RNG'$ from its neighbor x. For each edge $uv \in RNG'$, if uv is the longest among uv, xy, ux, and vy, node u removes edge uv. Ties are broken by the label of the edges. Here, assume that $uvyx$ is the convex hull of u, v, x, and y.
4: Let H be the final structure formed by all remaining edges in RNG′.

This was proved by showing that the edges in H satisfy the *isolation property*. X.-Y. Li (2003b) also showed that the final structure contains the MST of a UDG as a subgraph.

Clearly, the communication cost of Algorithm 35 is at most $7n$: Initially, each node spends one message to tell its one-hop neighbors its position information; then, each node uv tells its one-hop neighbors all its incident edges $uv \in$ RNG′ (there are at most total $6n$ such messages because RNG′ has at most $3n$ edges). The computational cost of Algorithm 35 could be high because for each link $uv \in$ RNG′, node u has to test whether there is an edge $xy \in RNG'$ and $x \in N_1(u)$ such that uv is the longest among uv, xy, ux, and vy. Then, X.-Y. Li *et al.* (2004c, 2004d) presented some new algorithms that improve the computational complexity of each node while still maintaining low communication costs.

Structure Based on LMST$_k$

The first new method in X.-Y. Li *et al.* (2004d) uses a structure called a *local minimum spanning tree*; let us first review its definition. It was first proposed by N. Li *et al.* (2003). Each node u first collects its one-hop neighbors $N_1(u)$. Node u then computes the minimum spanning tree MST$[N_1(u)]$ of the induced UDG on its one-hop neighbors $N_1(u)$. Node u keeps a directed edge uv if and only if uv is an edge in MST$[N_1(u)]$. They call the union of all directed edges of all nodes the *local minimum spanning tree*, denoted by LMST$_1$. If only symmetric edges are kept, then the graph is called LMST$_1^-$; i.e., it has an edge uv iff both directed edge uv and directed edge vu exist. If ignoring the directions of the edges in LMST$_1$, they call the graph LMST$_1^+$; i.e., it has an edge uv iff either directed edge uv or directed edge vu exists. They prove that the graph is connected and has bounded degree six. X.-Y. Li *et al.* (2004d) also showed that graph LMST$_1^-$ and LMST$_1^+$ are actually planar. Then, they extend the definition to k-hop neighbors; the union of all edges of all minimum spanning tree MST$[N_k(u)]$ is the k *local minimum spanning tree*, denoted by LMST$_k$. For example, the LMST$_2$ can be constructed by the following algorithm.

X.-Y. Li *et al.* (2004d) proved that structures LMST$_2$ (LMST$_2^+$ and LMST$_2^-$) are connected, planar, low-weighted, and have a bounded-node degree at most six. In addition, the MST is a subgraph of LMST$_k$ and LMST$_k \subseteq$ RNG′. Although the constructed structure LMST$_2$ has several nice properties such as having a bounded degree and being planar and low weight, the communication cost of Algorithm 36 could be very

---

Algorithm 36 Construct Low-Weight Structure LMST$_2$ by two-hop Neighbors

---

1: Each node u collects its two-hop neighbor information $N_2(u)$ by using the communication-efficient protocol described in Călinescu (2003).

2: Each node u computes the Euclidean minimum spanning tree MST$[N_2(u)]$ of all nodes $N_2(u)$, including u itself.

3: For each edge $uv \in$ MST$[N_2(u)]$, node u tells node v about this directed edge.

4: Node u keeps an edge uv if $uv \in$ MST$[N_2(u)]$ or $vu \in$ MST$[N_2(v)]$. Let LMST$_2^+$ be the final structure formed by all edges kept. It keeps an edge if either node u or node v wants to keep it. Another option is to keep an edge only if both nodes want to keep it. Let LMST$_2^-$ be the structure formed by such edges.

---

large to save the computational cost of each node. The large communication costs are from collecting the two-hop-neighbor information $N_2(u)$ for each node u. Although the total communication of the protocol described in Călinescu (2003) is $O(n)$, the hidden constant is large.

Combining RNG$'$ and LMST$_k$

We could improve the communication cost of collecting $N_2(u)$ by using a subset of two-hop information without sacrificing any properties. Define

$$N_2^{\text{RNG}'}(u) = \{w \mid vw \in \text{RNG}' \text{ and } v \in N_1(u)\} \cup N_1(u).$$

Our modified algorithm is described as follows:

Algorithm 37 Low-Weight Structure IMRG by Two-Hop Neighbors in RNG$'$

1: Each node u tells its position information to its one-hop neighbors $N_1(u)$ by using a local-broadcast model. All nodes together construct the graph RNG$'$ in a localized manner.

2: Each node u locally broadcasts its incident edges in RNG$'$ to its one-hop neighbors. Node u listens to the messages from its one-hop neighbors.

3: Each node u computes the Euclidean minimum spanning tree MST$[N_2^{\text{RNG}'}(u)]$ of all nodes $N_2^{\text{RNG}'}(u)$, including u itself.

4: For each edge $uv \in$ MST$[N_2^{\text{RNG}'}(u)]$, node u tells node v about this directed edge.

5: Node u keeps an edge uv if $uv \in$ MST$[N_2^{\text{RNG}'}(u)]$ or $vu \in$ MST$[N_2^{\text{RNG}'}(v)]$. Let IMRG$^+$ be the final structure formed by all edges kept. Similarly, the final structure is called IMRG$^-$ when edge $uv \in$ RNG$'$ is kept iff $uv \in MST[N_2^{\text{RNG}'}(u)]$ and $uv \in$ MST$(N_2^{\text{RNG}'}(v))$. Here, IMRG is the abbreviation for *incident MST and RNG graphs*.

Notice that in the algorithm, node u constructs the local minimum spanning tree MST$[N_2^{RNG'}(u)]$ based on the induced UDG of the point sets $N_2^{\text{RNG}'}(u)$. Here, MST(v) is the minimum spanning tree over a set of points V. It is obvious that the communication cost of Algorithm 37 is at most $7n$.

It was shown that structures $IMRG^+$ and $IMRG^-$ have a bounded degree and are still connected, planar, and low-weighted. They are obviously planar and have a bounded degree because both structures are still subgraphs of the modified relative neighborhood graph RNG'. Clearly, the constructed structures are supergraphs of the previous structures (i.e., $LSMT_2+ \subseteq IMRG^+$ and $LSMT_2^- \subseteq IMRG^-$) because Algorithm 37 uses less information than Algorithm 36 in constructing the LMST. It is proved in X.-Y. Li *et al.* (2004d) that Algorithm 37 constructs structures $IMRG^-$ or $IMRG^+$ by using at most $7n$ messages. Structures $IMRG^-$ and $IMRG^+$ are connected, planar, and low-weighted and have a bounded degree. Both $IMRG^-$ and $IMRG^+$ have a node degree of at most six.

Recall that until now, there was no efficient localized algorithm that could achieve all the following desirable features: bounded degree, planar, low weight, and spanner. It is still an open problem.

A Negative Result

X.-Y. Li (2003b) and X.-Y. Li *et al.* (2004c, 2004d) proposed several methods to construct structures in a localized manner such that the total edge lengths of these structures are within a constant factor of the MST. They also showed that the energy consumption of broadcasting based on those structures is within $O(n^{\beta-1})$ of the optimum; i.e., $\omega_\beta(H) = O(n^{\beta-1})\omega_\beta(MST)$, $\omega_\beta(LMST_2) = O(n^{\beta-1})\omega_\beta(MST)$, $\omega_\beta(IMRG) = O(n^{\beta-1})\omega_\beta(MST)$ for any $\beta \geq 1$.

They further showed that it is impossible to design a deterministic localized method that constructs a structure such that the broadcasting based on this structure consumes energy within a factor $o(n^{\beta-1})$ of the optimum. Assume that there is a deterministic localized algorithm to do so: It uses k-hop information of every node u to select the edges incident upon u, and the energy consumption is no more than $O(n^{\beta-1})$ times the optimum. They constructed two set of node configurations such that the k-hop information collected in a special node u is the same for both configurations. In addition, there is an edge uv in both UDGs such that if node u decides to keep edge uv (then, edge uv is kept in both configurations), the energy consumption of one configuration is already more than $O(n^{\beta-1})$ times the optimum; if node u decides to remove edge uv (then, edge uv is removed in both configurations), the structure constructed for another configuration is disconnected. See X.-Y. Li (2004c) for more details. This implies that the low-weighted structures are asymptotically optimum in terms of the worst-case energy consumption for broadcasting among *any* locally constructed topologies when assuming that the energy needed to support a link uv is proportional to $\|uv\|^\beta$.

14.3.3   Combining Clustering and Low Weight

Seddigh *et al.* (2002) specify two more location-based broadcasting algorithms that combine the RNG and internal-node concept (CDS) as follows. PI-broadcast algorithm applies the planar subgraph construction first and then applies the internal-node concept on the subgraph. The result is different from the internal nodes applied on the whole graph. The IP-broadcast algorithm changes the order of concept application compared with the previous algorithm. Internal nodes are first identified in the whole graph, and then the obtained subgraph (containing only internal nodes) is further reduced to a planar one by the RNG construction.

The solution in Seddigh *et al.* (2002) is for a one-to-one communication model, in which a message sent from one node is received by only the targeted neighbor. Li *et al.* (2003a) combined the low-weighted structures and the CDS for energy-efficient broadcasting in traditional one-to-many (omnidirectional antenna) networks. Similarly, they proposed two approaches for combining them, as in Seddigh *et al.* (2002). Notice that the constructed low-weighted structures are a subgraph of RNG; thus, they are still planar graphs. For simplicity, they also call these two combinations *PI-broadcast* and *IP-broadcast*, respectively. They found that the energy consumption of the IP-broadcast schemes in dense networks is significantly less than that of the PI-broadcast schemes. The reason is that in the IP-broadcast schemes, one retransmission by the internal nodes will be received by many noninternal nodes in dense networks; thus, energy consumption is reduced. Several localized improvement heuristics are also applied after the IP-broadcast or PI-broadcast schemes to further improve energy consumption. Ingelrest (see Cartigny *et al.*, 2005) in his master's thesis also combined a LMST with DSs and a target radius idea to derive new minimum-energy broadcast protocols.

14.3.4   Flooding-Based Methods

Selecting Forwarding Neighbors

The simplest broadcasting mechanism is to let every node retransmit the message to all its one-hop neighbors when receiving the first copy of the message, which is called *flooding* in the literature. Despite its simplicity, flooding is very inefficient and can result in high redundancy, contention, and collision. One approach to reducing the redundancy is to let a node forward the message only to a subset of one-hop neighbors who together can cover the two-hop neighbors. In other words, when a node retransmits a message to its neighbors, it explicitly asks a subset of its neighbors to relay the message.

Lim and Kim (2000) proposed a broadcasting scheme that chooses some or all of its one-hop neighbors as rebroadcasting nodes. When a node receives a broadcast packet, it uses a greedy set-cover algorithm to determine which subset of neighbors should rebroadcast the packet, given knowledge of which neighbors have already been covered by the sender's broadcast. The greedy set-cover algorithm recursively chooses one-hop neighbors that cover the most *uncovered* two-hop neighbors and recalculates the cover set until all two-hop neighbors are covered.

Călinescu *et al.* (2001) gave two practical heuristics for this problem (which they called *selecting forwarding neighbors*). The first algorithm runs in time $O(n \log n)$ and returns a subset with a size of at most six times the minimum. The second algorithm has an improved approximation ratio, 3, but with running time $O(n^2)$. Here, n is the number of total two-hop neighbors of a node. When all two-hop neighbors are in the same quadrant with respect to the source node, they give an exact solution in time $O(n^2)$ and a solution with an approximation factor of 2 in time $O(n \log n)$. Their algorithms partition the region surrounding the source node into four quadrants, solve each quadrant by using an algorithm with an approximation factor of α, and then combine these solutions. They proved that the combined solution is at most 3α times the optimum solution. They then gave two different algorithms for finding a disk cover when the two-hop neighbors are restricted to one quadrant with approximation ratio $\alpha = 2$ and 1, respectively.

Their approach assumes that every node u can collect its two-hop neighbors $N_2(u)$ efficiently. Notice that the one-hop neighbors of every node u can be collected efficiently by asking each node to broadcast its information to its one-hop neighbors. Thus, all nodes get their one-hop-neighbor information by using a total of $O(n)$ messages. However, until recently, it was not known how to collect the two-hop-neighbor information with $O(n)$ communications. The simplest broadcasting of one-hop neighbors $N_1(u)$ all neighbors u does let all nodes in $N_1(u)$ collect their corresponding two-hop neighbors. However, the total communication cost of this approach is $O(m)$, where m is the total number of links in the UDG. Recently, Călinescu (2003) proposed an efficient approach to collect $N_2(u)$ by using the CDS (Wan et al., 2002a) as forwarding nodes. Assume that the node position is known. He proved that the approach takes total communications $O(n)$, which is optimum within a constant factor.

Gossip and Probabilistic Schemes
Probabilistic Scheme
The probabilistic scheme from Tseng et al. (2002) is similar to flooding, except that nodes rebroadcast with only a predetermined probability. In dense networks, multiple nodes share similar transmission coverage. Thus, randomly having some nodes that do not rebroadcast saves node and network resources without harming delivery effectiveness. In sparse networks, there is much less shared coverage; thus, nodes will not receive all the broadcast packets with the probabilistic scheme unless the probability parameter is high. When the probability is 100%, this scheme is identical to flooding. Cartigny and Simplot (2003) applied a probability that is a function of the distance to the transmitting neighbor.

Counterbased Scheme
Tseng et al. (2002) show an inverse relationship between the number of times a packet is received at a node and the probability of that node's being able to reach additional area on a rebroadcast. This result is the basis of their counterbased scheme. On reception of a previously unseen packet, the node initiates a counter with a value of one and sets a RAD (which is randomly chosen between 0 and T_max seconds). During the RAD, the counter is incremented by one for each redundant packet received. If the counter is less than a threshold value when the RAD expires, the packet is rebroadcast. Otherwise, it is simply dropped. From Tseng et al. (2002), threshold values above six relate to little additional coverage area being reached.

The overriding, compelling features of the counterbased scheme are its simplicity and its inherent adaptability to local topologies. That is, in a dense area of the network, some nodes will not rebroadcast; in sparse areas of the network, all nodes rebroadcast.

The disadvantage of all counter and probabilistic schemes is that delivery is not guaranteed to all nodes even if ideal MAC is provided. In other words, they are not reliable.

Area-Based Decision

In either probabilistic schemes or the counter-based schemes, a node decides whether to rebroadcast a received packet purely based on its own information. Tseng *et al.* (2002) proposed several other criteria based on the additional coverage area to decide whether the node will rebroadcast the packet. These coverage-area-based methods are similar to the methods of selecting forwarding neighbors, in which a node tries to select a set of one-hop neighbors sufficient to cover all its two-hop neighbors. Area-based methods consider only the coverage area of a transmission; they do not consider whether nodes exist within that area. Two coverage-area-based methods are proposed in Tseng *et al.* (2002): *distance-based scheme* and *location-based scheme*.

In the distance-based scheme, a node compares the distance between itself and each neighbor node that has previously rebroadcast a given packet. On reception of a previously unseen packet, a RAD is initiated and redundant packets are cached. When the RAD expires, all source node locations are examined to see if any node is closer than a threshold distance value. If true, the node does not rebroadcast.

The location-based scheme uses a more precise estimation of expected additional coverage area in the decision to rebroadcast. In this method, each node must have the means to determine its own location (e.g., a GPS). Whenever a node originates or rebroadcasts a packet, it adds its own location to the header of the packet. When a node initially receives a packet, it notes the location of the sender and calculates the additional coverage area obtainable were it to rebroadcast. If the additional area is less than a threshold value, the node will not rebroadcast, and all future receptions of the same packet will be ignored. Otherwise, the node assigns a RAD before delivery. If the node receives a redundant packet during the RAD, it recalculates the additional coverage area and compares that value with the threshold. The area calculation and threshold comparison occur with all redundant broadcasts received until the packet reaches either its scheduled send time or is dropped.

We review also some work related to DSs and broadcasting problems. In Javier *et al.* (2006) a beaconless broadcasting method is proposed. All nodes have the same transmission radius, and nodes are not aware of their neighborhood. That is, no beacons or *hello* messages are sent in order to discover neighbors before the broadcasting process. The source transmits the message to all neighbors. On receiving the packet (together with geographic coordinates of the sender), each node calculates the portion of its perimeter, along the circle of the transmission radius, that is not covered by this and previous transmissions of the same packet. The node then sets or updates its time-out interval, which inversely depends on the size of the uncovered perimeter portion. If the perimeter becomes fully covered, the node cancels retransmissions. Otherwise, it retransmits at the end of the time-out interval. The method is reliable, as opposed to other area-based methods.

Neighbor-Coverage-Based Decision

The method presented in the previous subsection was based on covering an area where nodes could be located. Instead of covering an area, one could simply cover neighboring nodes, assuming that their locations, or existence of their link to a previously transmitting node, are known. The basic method was independently and almost simultaneously (August 2000) proposed in two articles (Peng and Lu, 2000; Stojmenovic and Seddigh, 2002). The methods were called *neighbor elimination* by Stojmenovic and Seddigh (2002), and a similar method, called *scalable broadcast algorithm*, was proposed by Peng and Lu (2000). Two-hop-neighbor information is used to determine whether a node will rebroadcast the packet. Suppose that a node u receives a broadcast data packet from its neighbor node v. Node u knows all the neighbors of node v and, thus, all nodes that are their common neighbors (i.e., already received the data from v). If node u has additional neighbors not reached by node v's broadcast, node u schedules the packet for delivery with a RAD. However, if node u receives a redundant broadcast packet from some other neighbors within RAD, node u will recalculate whether it needs to rebroadcast the packet. This process is continued until either the RAD expires and the packet is then sent or the packet is dropped (when all its neighbors are already covered by the broadcasts of some of its neighbors).

Lipman *et al.* (2002) described the following broadcasting protocol: On receiving a broadcast message(s) from a node h, each node i (that was determined by h as a forwarding node) determines which of its one-hop neighbors also received the same message. For each of its remaining neighbors j (which did not receive a message yet, based on i's knowledge), node i determines whether j is closer to i than any one-hop neighbors of i (that are also forwarding nodes of h) who received the message already. If so, i is responsible for message transmission to j; otherwise, it is not. Node i then determines a transmission range equal to that of the farthest neighbor for which it is responsible.

14.4   Scheduling Active and Sleep Periods

In ad hoc wireless networks, the limitation of power of each host poses a unique challenge for power-aware design. There has been an increasing focus on low cost and reduced node power consumption in ad hoc wireless networks. Even in standard networks such as IEEE 802.11, requirements are included to sacrifice performance in favor of reduced power consumption. To prolong the life span of each node and hence the network, power consumption should be minimized and balanced among nodes. Unfortunately, nodes in the DS in general consume more energy in handling various bypass traffic than do nodes outside the set. Therefore, a static selection of dominating nodes will result in a shorter life span for certain nodes, which in turn result in a shorter life span of the whole network.

Wu *et al.* (2003) studied the dynamic selection of dominating nodes, also called *activity scheduling*. Activity scheduling deals with the way to rotate the role of each node among a set of given operation modes. For example, one set of operation modes is sending, receiving, idling, and sleeping. Different modes have different energy consumption. Activity scheduling judiciously assigns a mode to each node to save overall energy consumption in the networks and/or to prolong the life span of each individual node. Note that saving overall energy consumption does not necessarily prolong the life span of a particular individual node. Specifically, they propose saving overall energy consumption by allowing only dominating nodes (i.e., gateway nodes) to retransmit the broadcast packet. In addition, to maximize the lifetime of all nodes, an activity-scheduling method is used that dynamically selects nodes to form a CDS. Specifically, in the selection process of a gateway node, we give preference to a node with a higher energy level. The effectiveness of the proposed method in prolonging the life span of the network is confirmed through simulation. Source-dependent forwarding sets appear to be more energy-balanced. However, it was experimentally confirmed in Feeney and Nilson (2001) that the difference in energy consumption between an idle node and a transmitting node is not major; the major difference exists between idle and sleep states of nodes. Therefore, the most energy-efficient methods will select a static DS for a given round, turning all remaining nodes to a sleep state. Depending on the energy left, changes in activity status for the next round will be made. The change can therefore be triggered by changes of power status, in addition to node mobility. From this point of view, internal-node-based DSs provide static selection for a given round and more energy efficiency than a forwarding-set-based method that requires all nodes to remain active in all the rounds.

Xu *et al.* (2001) discuss the following sensor sleep node schedule. The trade-off between network lifetime and density for this cell-based schedule was investigated in Blough and Santi (2002). The given 2D space is partitioned into a set of squares (called cells), such that any node within a square can directly communicate with any nodes in an adjacent square. Therefore, one representative node from each cell is sufficient. To prolong the life span of each node, nodes in the cell are selected in an alternate fashion as a representative. The adjacent squares form a 2D grid and the broadcast process becomes trivial. Note that the selected nodes in Xu *et al.* (2001) make a DS, but its size is far from optimal, and also it depends on the selected size of the squares. On the other hand, the DS concept used here has a smaller size and is chosen without using any parameter [size of the square, which has to be carefully selected and propagated with node-relative positioning in the solution (Xu *et al.*, 2001)].

The algorithm proposed by Chen *et al.* (2001) selects some nodes as coordinators. These nodes form a DS. A node becomes coordinator if it discovers that two of its neighbors cannot communicate with each other directly or through one or two existing coordinators. Also, a node should withdraw if every pair of its neighbors can reach each other directly or via some other coordinators (they can also withdraw if each pair of neighbors is connected via possibly noncoordinating nodes, to give other nodes a chance to become coordinators). Because coordinators are not necessarily neighbors, three-hop-neighboring topology knowledge is required. However, the energy and bandwidth required for maintenance of three-hop-neighborhood information was not taken into account in experiments (Chen *et al.*, 2001). On the other hand, if the coordinators are restricted to be neighboring nodes, then the DS definition (Chen *et al.*, 2001) becomes equivalent to one given by Wu and Li (2001). Next, the protocol (Chen *et al.*, 2001) relies heavily on proactive periodic beacons for synchronization, even if there is no pending traffic or node movement. Recent research on energy consumption (Feeney and Nilson, 2001) indicates that the use of such periodic beacons or *hello* messages is an energy-expensive mechanism because of significant start-up costs for sending short messages. Finally, Blough and Santi (2002) observed that the overhead required for coordination with the SPAN tends to *explode* with node density and thus counterbalances the potential savings achieved by the increased density.

14.5   Energy-Efficient Multicast

In certain scenarios, we may need to send the data to a subset of nodes in the network instead of setting the data to all nodes in the network. We thus need to design energy-efficient multicast protocols for connecting the receivers. There are two different scenarios again: One is that every node has its own *fixed* cost of sending one unit amount of data regardless of the receiving nodes; the other scenario is that a node can dynamically adjust its transmission power based on its neighboring nodes in the multicast structure. Here, given a set of receivers $Q \subseteq V$, a multicast structure is a tree (or any other connected graph) that connects all nodes in Q.

14.5.1   Fixed Transmission Power

When every node has a fixed power (not necessarily the same), then the minimum-power multicast problem is the standard node-weighted Steiner tree (NST) problem: Given Q, we need to find a set of Steiner nodes to connect Q while the total weight of the selected Steiner nodes is minimized. When the network is modeled by an arbitrary graph and the node cost could be any arbitrary nonnegative number, the NST can be approximated within $2 \ln k$, where k is the size of Q (Klein and Ravi, 1995).

When the network is modeled as a UDG and the node cost could be an arbitrary fixed value, then the minimum energy of multicast can be approximated within a constant factor (W. Wang and Li, 2004a). The method of constructing a cost-efficient spanning tree for multicast routing works as follows. First, we calculate the pairwise shortest path $\mathsf{LCP}(q_i, q_j, G)$ between any two nodes $q_i, q_j \in Q$ for a network modeled by a node-weighted graph $G = (V, E, c)$, where the node cost vector is c. We then construct a complete edge-weighted graph $K(G, Q, w)$ by using Q as its vertices, where edge $q_i q_j$ corresponds to $\mathsf{LCP}(q_i, q_j, G)$, and its weight $w(q_i q_j)$ is the cost of $\mathsf{LCP}(q_i, q_j, G)$; i.e., $w(q_i q_j) = \|\mathsf{LCP}(q_i, q_j, G)\|$. For later convenience, here the total weight of the least-cost path $\mathsf{LCP}(q_i, q_j, G)$ does not include the cost of two end points q_i and q_j. For convenience of our analysis, we also assume that no two edges in $G = (V, E, c)$ have the same length, and there are no two paths in $G = (V, E, c)$ that have the same length. Dropping this assumption does not change the result of our analysis.

Algorithm 38 Energy-Efficient Multicast Based on Virtual MST (VMST)

1: First, construct the virtual weighted complete graph $K(G, Q)$ on the original network $G = (V, E, c)$.

2: Construct the MST on $K(G, Q)$. The resulting MST is denoted as VMST(G).

3: For each edge $q_i q_j$ selected in VMST(G), we find the corresponding least-cost path LCP(q_i, q_j, G) in G. We mark every internal node v_k on the path LCP(q_i, q_j, G) as a *relay node*.

4: In graph G, build a spanning tree using all nodes marked as *relay node* and all receiver nodes Q, and denote the final spanning tree on G as SVMST(G).

THEOREM 14.5 *SVMST(G) is a five-approximation of the optimal solution in terms of total cost if the wireless network is modeled by a UDG.*

Proof. Assume that the optimal solution is a tree called T_{opt}. Let $V(T_{\text{opt}})$ be the set of nodes used in the tree T_{opt}. Clearly, $\mathbf{c}(T_{\text{opt}}) = \sum_{v_i \in V(T_{\text{opt}})} c_i$. Similarly, for any spanning tree T of $K(G, Q)$, we define $\mathbf{c}(T) = \sum_{e \in T} w(e)$. To prove the theorem, we prove a stronger result: $5\mathbf{c}(T_{\text{opt}}) \geq \mathbf{c}[\text{VMST}(G)]$. Observe that virtual edges selected in VMST(G) could use a relay node multiple times. This implies that $\mathbf{c}[\text{VMST}(G)] \geq \mathbf{c}[\text{SVMST}(G)]$. Remember that a receiver node does not charge for relay transit traffic for other receiver nodes.

First, for all nodes in T_{opt}, when the node weight is disregarded, there is a spanning tree T'_{opt} on $V(T_{\text{opt}})$ with a node degree of at most five because the wireless network is modeled by a UDG. This is due to a well-known fact that there is an EMST with the maximum node degree of at most five for any set of 2D points. Note here that we do not need to explicitly construct such a spanning tree with a maximum degree at most five. Obviously, $\mathbf{c}(T_{\text{opt}}) = \mathbf{c}(T'_{\text{opt}})$. Thus, tree T'_{opt} is also an optimal solution. Furthermore, it has the property that for every node in T'_{opt}, its degree is at most five.

For spanning tree T'_{opt}, we root it at an arbitrary node and duplicate every link in T'_{opt} (the resulting structure is called DT'_{opt}). Clearly, every node in DT'_{opt} has an even degree now. Thus, we can find an Euler circuit, denoted by EC(DT'_{opt}), that visits every vertex of DT'_{opt} and uses every edge of DT'_{opt} exactly once, which is equivalent to saying that every edge in $T'_{\text{opt}}(G)$ is used exactly twice. Consequently, we know that every node v_k in $V(T_{\text{opt}})$ is used exactly $\deg_{T'_{\text{opt}}}(v_k)$ times. Here, $\deg_G(v)$ denotes the degree of a node v in a graph G. Thus, the total weight of the Euler circuit is at most five times the weight $\mathbf{c}(T'_{\text{opt}})$; i.e.,

$$\mathbf{c}[\text{EC}(DT'_{\text{opt}})] \leq 5\mathbf{c}(T'_{\text{opt}}).$$

Notice that here, if a node v_k appears multiple times in EC(DT'_{opt}), its weight is also counted multiple times in $\mathbf{c}[\text{EC}(DT'_{\text{opt}})]$. Each node v_k actually appears exactly $\deg_{T'_{\text{opt}}}(v_k) \leq$ five times.

For Euler circuit EC(DT'_{opt}), we can treat it as a graph defined over the receivers Q: A subpath in EC(DT'_{opt}), connecting two receivers q_i and q_j without any other receiver q_t in between, corresponds to a virtual edge $q_i q_j$ with weight equal to the total cost of all nodes in between, which is clearly at least $\|\mathsf{LCP}(q_i, q_j, G)\|$. Notice that such a graph could have duplicated edges. Thus, $\mathrm{EC}(DT'_{\mathrm{opt}})$ is a graph that spans all receivers Q with total cost of at least $\mathbf{c}[\mathrm{VMST}(G)]$ because $\mathrm{VMST}(G)$ is the MST spanning all receivers and the cost of the edge $q_i q_j$ in $\mathrm{VMST}(G)$ corresponds to the path with the least cost, $\|\mathsf{LCP}(q_i, q_j, G)\|$. In other words,

$$\mathbf{c}[\mathrm{EC}(DT'_{\mathrm{opt}})] \geq \mathbf{c}[\mathrm{VMST}(G)].$$

Consequently, we have

$$\mathbf{c}[\mathrm{VMST}(G)] \leq \mathbf{c}[\mathrm{EC}(DT'_{\mathrm{opt}})] \leq 5\mathbf{c}(T'_{\mathrm{opt}}).$$

This finishes the proof.   ∎

Notice that in the preceding argument, we assume that the cost of the nodes in Q is not counted.

14.5.2   Dynamically Adjustable Transmission Power

Multicast with Asymmetric Communication

We now consider the case in which every node can dynamically adjust its power based on the downstream neighboring nodes in the multicast structure. We assume that the communication will be asymmetric: The cost of the acknowledgment by the receiving node is ignored. The multicast with asymmetric communication is called the min-power asymmetric multicast. The min-power asymmetric multicast then seeks, for any given communication session, an arborescence of minimum total power that is rooted at the source node s and reaches all nodes in Q. As a generalization of min-power asymmetric broadcast routing, min-power asymmetric multicast routing is also NP-hard. Wieselthier (2000) adapted their three broadcasting heuristics to three multicasting heuristics by a technique of pruning, which was called a pruned minimum spanning tree (P-MST), pruned shortest-path tree (P-SPT), and pruned broadcasting incremental power (P-BIP), respectively (Wan *et al.*, 2004a). The idea is as follows: We first obtain a spanning tree rooted at the source of a given multicast session by applying any of the three broadcasting heuristics. We then eliminate from the spanning arborescence all nodes that do not have any descendant in Q.

Notice that for a structural SPT, it has been shown that it has a $\Theta(n)$ approximation ratio for broadcast in the worst-case scenario. Thus, the performance of the P-SPT is also $\Theta(n)$ in the worst case for multicast. For structural MST and BIP, it has been shown in Wan *et al.* (2002b, 2001) that both structures have constant-approximation ratios for broadcast. As a generalization of this, one may expect that the same statement is true for P-MST and P-BIP. Unfortunately, this is not true, as proved by Wan *et al.* (2004a). They show by constructing examples that both structures could have an approximation ratio as large as $\Theta(n)$ in the worst case for multicast. They then further proposed a multicast scheme with a constant-approximation ratio on the total energy consumption. Their protocol for min-power asymmetric multicast routing is based on the Takahashi–Matsuyama–Steiner tree heuristic (Takahashi and Matsuyama, 1980). It can be implemented easily as follows. Throughout the execution, we maintain a tree rooted at the source node:

1. Initially, the multicast tree T contains only the source node.
2. At each iterative step, the multicast tree T is grown by one path from some node in T to some destination node from Q that is not yet in the tree T. The path must have the least total power among all such paths from a node in T to a node in $Q \setminus T$. This path can also be found by collapsing the entire tree T into one artificial node and then applying the single-source shortest-path algorithm.
3. This procedure is repeated until all required nodes are included in T. This heuristic is referred to as Shortest Path First (SPF).

It can be regarded as an adaptation of the MST for min-power asymmetric broadcast routing. Indeed, when the communication session is a broadcast, it acts the same way as the MST. They proved the following theorem for the performance of SPF:

THEOREM 14.6 *For asymmetric multicast communication, the approximation ratio of SPF is between 6 and $2c_0$, which is at most 24. Here, c_0 is defined by the supremum of $\sum_{e \in \text{EMST}(U)} \|e\|^2$ for all finite-sized sets of nodes U located inside a disk with radius 1.*

Multicast with Symmetric Communication

Notice that the preceding results hold for asymmetric communications. Wan *et al.* (2004a) also studied the minimum-energy multicast for symmetric communication. Given a structure H for multicast, the energy needed by a node $v \in H$ for *symmetric communication* is

$$p_{H,S}(v) = \max_{uv \in H} \mathbf{c}(uv).$$

Here, $\mathbf{c}(uv)$ is the energy cost for supporting the communication between links uv, which includes the cost by u and v. The power of H supporting a symmetric multicast is then defined as

$$p_S(H) = \sum_v p_{H,S}(v).$$

Let $\mathcal{A}$ be any polynomial-time approximation algorithm for a Steiner minimum tree in graphs. For any given multicast communication with receivers Q, we apply to obtain a Steiner tree T for $Q \cup \{s\}$, which is used for the symmetric multicast routing. Then, the following theorem was proved in Wan *et al.* (2004a):

THEOREM 14.7 (Wan *et al.*, 2004a) *For any ϱ-approximation algorithm $\mathcal{A}$ for a Steiner minimum tree, the approximation ratio of the algorithm is at most 2ϱ.*

Proof. We denote OPT as the optimum structure for multicast from s to Q and opt as its total cost. Let T be the Steiner tree constructed by algorithm $\mathcal{A}$ for nodes $Q \cup \{s\}$. Let T^* be the optimum Steiner tree for nodes $Q \cup \{s\}$. For a link-weighted structure H, let $\mathbf{c}(H)$ be the total link cost of all links in H. Notice that both T and T^* are link-weighted.

We first prove that opt $> \mathbf{c}(T^*)$. We can do this by orienting tree OPT to a directed tree rooted at the source node s. For each nonroot node v, let v' be its parent node in this rooted tree. Clearly, by definition of $p_{\text{OPT},s}(v)$,

$$p_{\text{OPT},s}(v) \geq \mathbf{c}(vv').$$

Consequently, we have

$$
\begin{aligned}
\text{opt} &= p_S(\text{OPT}) \\
&= p_{\text{OPT},s}(s) + \sum_{v \in \text{OPT}, v \neq s} p_{\text{OPT},s}(v) \\
&\geq p_{\text{OPT},s}(s) + \mathbf{c}(\text{OPT}) \\
&> \mathbf{c}(\text{OPT}) \geq \mathbf{c}(T^*)
\end{aligned}
$$

Next, we prove that $p_S(T) \leq 2\mathbf{c}(T)$ for any tree T:

$$p_S(T) = \sum_{v \in T} \max_{uv \in T} \mathbf{c}(uv) \leq \sum_{v \in T} \sum_{u \mid uv \in T} \mathbf{c}(uv) = 2 \sum_{e \in T} \mathbf{c}(e) = 2\mathbf{c}(T).$$

Thus, we have

$$p_S(T) \leq 2\mathbf{c}(T) \leq 2\varrho\mathbf{c}(T^*) \leq 2\varrho\text{opt}.$$

This finishes the proof of the theorem.   ■

Notice that the correctness of Theorem 14.7 does not depend on the cost model for links. It is valid for any link cost. Currently, the best approximation ratio $\varrho \simeq 1.55$ is due to Robins and Zelikovsky (2000). The more time-efficient method by Takahashi and Matsuyama (1980) has an approximation ratio of 2.

14.6   Further Reading

In the minimum-energy broadcasting problem, each node can adjust its transmission power to minimize total energy consumption but still enable a message originating from a source node to reach all the other nodes in an ad hoc wireless network. The problem is known to be NP-complete. There exists a number of approximate solutions in literature in which each node requires global network information (including distances between any two neighboring nodes in the network) in order to decide its own transmission radius. Three greedy heuristics were proposed in Wieselthier et al. (2000) for the minimum-energy broadcast-routing problem: MST (minimum spanning tree), SPT (shortest-path tree), and BIP (broadcasting incremental power). It was shown that the total energy consumed by MST or BIP methods is no more than 12 times larger than the optimum (Wan et al., 2002b). Cartigny et al. (2003) described a localized protocol in which each node requires only the knowledge of its distance to all neighboring nodes and distances between its neighboring nodes (or, alternatively, geographic position of itself and its neighboring nodes). In addition to using only local information, the protocol is shown experimentally to be competitive even with the best-known globalized BIP solution

(Wieselthier *et al.*, 2000), which is a variation of Dijkstra's shortest-path algorithm. The solution (Cartigny *et al.*, 2003) is based on the use of a RNG that preserves the network connectivity and is defined in a localized manner. The transmission range for each node is equal to the distance to its farthest RNG neighbor, excluding the neighbor from which the message came. Localized energy-efficient broadcasting for wireless networks with directional antennas is described in Cartigny *et al.* (2002) and is also based on a RNG. Messages are sent along only RNG edges, requiring about 50% more energy than the BIP-based (Wieselthier *et al.*, 2000) globalized solution. However, when the communication overhead for maintenance is added, the localized solution becomes superior. Several new structures were proposed to replace RNG to improve energy efficiency, as proposed in Cartigny *et al.* (2005) and N. Li *et al.* (2003). Their simulations show that the performance is comparable to that of BIP. X.-Y. Li (2003b, 2005) and X.-Y. Li *et al.* (2004c, 2004d) recently proposed several methods with further improvements. They described several low-weight planar structures (IMRG and LMST_k) that can be constructed by localized methods with total communication costs $O(n)$, and the simulations showed a significant improvement of energy consumption compared with those of Cartigny *et al.* (2002) and N. Li *et al.* (2003). Although the structures IMRG and LMST_k $(k \geq 2)$ have a total edge length within a constant factor of the MST, the broadcasting based on these locally constructed structures could still consume energy arbitrarily larger than the optimum, for which we assume that the power needed to support a link uv is $\|uv\|^\beta$. It has been proved that broadcasting based on the MST consumes energy within a constant factor of the optimum, but the MST cannot be constructed locally. They also showed that there is no structure that can be constructed locally, and the broadcasting based on it consumes energy within a constant factor of the optimum.

14.7   Conclusion and Remarks

In this chapter, we reviewed several methods for efficient broadcasting for wireless ad hoc networks. There are still many challenging questions left open for further research. So far, all the known theoretically good algorithms either assume that the power needed to support a link uv is proportional to $\|uv\|^\beta$ or is a fixed cost that is independent of the neighboring nodes with which it will communicate. In practice, the energy consumption of a node is neither solely dependent on the distance to its farthest neighbor nor totally independent of its communication neighbor. For example, a more general power-consumption model for a node u would be $c_1 + c_2\|uv\|^\beta$ for some constants c_1 and c_2, where v is its farthest communication neighbor in a broadcast structure. No theoretical result is known about the approximation of the optimum broadcast or multicast structure under this model.

Another important aspect of designing energy-efficient protocols for broadcast and multicast that is often neglected in the literature is the physical constraints of wireless communications. In most of the algorithms, it is assumed that the signal sent by a node will be received by all nodes at one shot it intends to send. However, in practice, it is not the case, which will make designing energy-efficient broadcast and multicast protocols much harder. The first difficulty is that the sender needs to coordinate with the receivers so *all* receivers are ready to receive. It is often difficult, if not impossible, to do so. Then, a natural question is that of how to set the threshold t such that if the number of receivers that are ready is more than t, node u sends the packets; otherwise, node t will not send the packets. Clearly, the setting of the threshold will affect the total actual energy consumption of the broadcast and multicast protocols, in addition to affecting the system performance and the stability of the networks. The second difficulty is that even if some receivers are ready to receive, the receivers cannot always decode the signal correctly because of the signal-strength fluctuation. In other words, the link between two nodes is often probabilistic. We clearly should take this into account when we design energy-efficient broadcast and multicast protocols.

Another important question is how to find efficient broadcast/multicast structures such that the delay from the source node to the last node receiving the message is bounded by a predetermined value while the total energy consumption is minimized. Notice that here, the delay of a broadcast/multicast based on a tree is not simply the height of the tree: Many nodes cannot transmit simultaneously because of interference.

Problems

14.1   Why is broadcast/multicast also important for wireless networks? What are the main challenges in designing efficient broadcast/multicast protocols in wireless networks? What are the differences between the broadcast/multicast protocol and the data-collection and data-gathering protocols for WSNs?

14.2   Conduct simulations to study the performance of broadcast/multicast based on a CDS when all nodes have a uniform transmission power cost and each node has a fixed amount of receiving cost.

14.3   Will Theorem 14.2 still hold if the receiving cost of different nodes could be different? Prove that this is still true as long as the transmission cost is uniform for all nodes.

14.4   Assume that each node has a fixed transmission cost and a fixed receiving cost. However, the transmission cost of different nodes could be different and the receiving cost of different nodes could be different. On the other hand, the transmission range of all nodes is uniform. Design a broadcast scheme such that the total cost of broadcast is within a constant factor of the optimum. What is the best approximation ratio your algorithm can achieve?

14.5   Assume that each node has an adjustable transmission cost and a fixed receiving cost, for which the receiving cost of different nodes could be different. The transmission cost of a node u to another node v is $\|u - v\|^{\alpha}$ for $\alpha \geq 2$. Design a broadcast scheme such that the total cost (including the transmission cost of all transmitting nodes and the receiving cost of all nodes) of broadcast is within a constant factor of the optimum. What is the best approximation ratio your algorithm can achieve?

14.6 Conduct extensive simulations to study the performance of your algorithm when the links could be unreliable.

14.7 From Theorem 14.7 and the more time-efficient method by Takahashi and Matsuyama (1980) for the link-weighted Steiner tree problem, implement a time-efficient and energy-efficient multicast routing algorithm. Conduct simulations to study its practical performances in terms of the approximation ratio and the time complexity.

14.8 Design a message-efficient distributed algorithm for constructing a multicast structure for symmetric multicast such that its total cost is within a constant factor of the optimum. What is the worst-case communication complexity of your distributed algorithm? For multicast, what is the worst-case performance of the best localized algorithm for symmetric multicast?

15   Routing with Selfish Terminals

15.1   Introduction

When designing routing protocols, it is often implicitly assumed that each participant (users or routers) will faithfully follow the prescribed protocols without any deviation—except, perhaps, for a few faulty or malicious ones. For example, in wireless ad hoc networks, it is commonly assumed that each terminal contributes its own resources to forward the data for other terminals to serve the common good and benefits from resources contributed by other terminals to route its packets in return. However, the critical observation that individual users who own these wireless devices are generally *selfish*, aiming to maximize their own benefit instead of contributing to the system, may severely undermine the expected performances of the wireless networks. The limitations of energy supply, memory, and computing resources of these wireless devices raise concerns about the traditional assumption about terminals' conforming to protocols. Sometimes, wireless devices owned by individual users may prefer not to participate in the routing in order to save its energy and resources. Therefore, if all users are selfish, providing incentives is a natural and common way to encourage contribution and thus maintain the robustness and availability of networking systems. The question turns to how to design the proper incentives.

Consider a unicast routing and forwarding protocol in wireless ad hoc networks based on the path with the least total cost, which is called the least-cost path (LCP): Each terminal first declares its cost of forwarding a unit amount of data for other terminals, and the LCP connecting the source and the target terminal is then selected. A very naive incentive is to pay each wireless terminal its declared cost. However, an individual terminal may declare an arbitrarily high cost for forwarding a data packet to other terminals, hoping to increase its payment. Here, we would like to design a payment scheme such that every terminal will report its cost truthfully and forward others' traffic out of its own interest to maximize its profit. This payment scheme is called *strategy proof* (or *truthful* in the literature) because it removes the possible speculation and counterspeculation among terminals when they report their costs. Therefore, always forwarding others' traffic is a *dominant strategy* of each terminal because it maximizes its benefit no matter what other users do.

The most well-known and widely used strategy-proof payment method is the so-called Vickrey–Clarke–Groves (VCG) mechanism family (Clarke, 1971; Groves, 1973; Vickrey, 1961). A VCG mechanism uses an output that maximizes the *social efficiency*;

i.e., the total valuations of participating terminals. Several mechanisms (Anderegg and Eidenbenz, 2003; Feigenbaum *et al.*, 2002; Nisan and Ronen, 1999; Zhong *et al.*, 2004), which essentially all belong to the VCG mechanism family, were proposed in the literature to ensure that each network terminal truthfully reports its cost for unicast. In these mechanisms, the LCP, which maximizes the sum of all terminals' valuations, is used for routing. To support a communication among a group of users, multicast is more efficient than unicast or broadcast because it can transmit packets to destinations using fewer network resources, which in turn increases the social efficiency. A *truthful* multicast routing protocol, which selfish routing terminals will follow, is composed of two components: (1) the tree structure that connects the sources and receivers, and (2) payments to all terminals in the network. Multicast poses a unique challenge in designing strategy-proof mechanisms: It is NP-hard (Guha and Khuller, 1998b, 1999) to find the tree structure with the minimum cost, which in turn maximizes the social efficiency. A range of multicast structures, such as the least-cost path tree (LCPT), the virtual minimum spanning tree (VMST), and the Steiner tree, were proposed to replace the minimum-cost multicast tree (MCMT). In this chapter, we do not redesign the wheel (i.e., new multicast structures); instead, we show how payment schemes can be designed for existing multicast tree structures so that rational selfish terminals will follow the protocols for their own interests.

This chapter focuses on the design of truthful payment schemes for multicast routing in wireless networks with selfish terminals. The main topics to be studied are as follows: First, for each of these widely used multicast structures, we show that a simple application of the VCG mechanism is not strategy-proof: A selfish terminal (router) may have incentives to lie about its cost to increase its profit. This is due to the fundamental difference between unicast and multicast: It is NP-hard to find the MCMT that spans the sources and receivers, whereas the least-cost unicast path can be found in polynomial time. Second, we design strategy-proof payment schemes for these multicast structures. For a given multicast tree structure, we also prove that our payment scheme pays the minimum amount to any terminal among all truthful payment schemes.

15.2   Preliminaries and Network Model

15.2.1   Preliminaries

In designing efficient, centralized, or distributed algorithms and network protocols, the computational agents are typically assumed to be either *correct/obedient*, *faulty*, or *malicious*. In contrast, economists design market mechanisms in which agents are assumed to be *rational*. A standard model for the design and analysis of scenarios in which the participants act according to their own self-interests is as follows. Assume that there are n agents $\{1, \ldots, n\}$, which could be the wireless devices in wireless ad hoc networks or the computers in peer-to-peer networks. For each agent i, it has some *private* information t_i, called *type*; e.g., a router's cost to forward a unit amount data in a network environment. All agents' types define a type vector $t = (t_1, t_2, \ldots, t_n)$. Each agent i chooses a strategy a_i from its strategy space A_i and all agents' strategies define a strategy vector $a = (a_1, a_2, \ldots, a_n) \in A = A_1 \times A_2 \cdots A_n$.

A direct-revelation mechanism is a mechanism such that the only possible strategy available to agent i is to report its type. Notice that an agent may report a type that is different from its actual type t_i in a direct-revelation mechanism. A mechanism $M = (\mathcal{O}, \mathcal{P})$ is composed of two parts: an *output* method $\mathcal{O}(a)$ and a *payment* method $\mathcal{P}(a) = (\mathcal{P}_1(a), \ldots, \mathcal{P}_n(a))$, where $\mathcal{P}_i(a)$ is the monetary number given to the agent i. For each possible output o, agent i's preferences are given by a valuation function v_i that assigns a real monetary number $v_i(t_i, o)$ to output o. Let $u_i(t_i, \mathcal{O}(a))$ denote the *utility* of agent i at the outcome of the game, given its preferences t_i and strategies profile a selected by agents. A common assumption in mechanism-design literature, and one we follow in this chapter, is that agents have quasi-linear utility functions; i.e., they are *quasi-linear* if $u_i(t_i, \mathcal{O}(a)) = v_i(t_i, \mathcal{O}(a)) + \mathcal{P}_i(a)$. An agent i is called *rational* if it always plays the strategy a_i that maximizes its utility. A strategy a_i is called a *dominant strategy* for mechanism $M = (\mathcal{O}, \mathcal{P})$ if it maximizes the utility regardless of what other agents do; i.e.,

$$u_i(t_i, \mathcal{O}(a_i, b_{-i})) \geq u_i(t_i, \mathcal{O}(a_i', b_{-i}))$$

for all $a_i' \neq a_i$ and all strategies b_{-i} of agents other than i. Here, $b_{-i} = (b_1, \ldots, b_{i-1}, b_{i+1}, \ldots, b_n)$ denotes the vector of strategies of every other agent except i, and (a_i, b_{-i}) denotes that agents i play strategy a_i and all other agents j play a strategy b_j. Strategy a is called the *Nash equilibrium* if for each agent i, it maximizes the utility when it uses strategy a_i while other agents use strategy a_{-i}; i.e., fix their strategies. In other words,

$$u_i(t_i, \mathcal{O}(a_i, a_{-i})) \geq u_i(t_i, \mathcal{O}(a_i', a_{-i}))$$

for each i and every possible $a_i' \neq a_i$ for i. Here, $a_{-i} = (a_1, \ldots, a_{i-1}, a_{i+1}, \ldots, a_n)$ denotes the vector of strategies of all other agents except i.

In a multicast routing protocol, the type of terminal k is its private cost c_k of relaying a unit amount of data, and its strategy space A_k in a direct-revelation mechanism is the set of possible costs that terminal k could declare. Its valuation on an output (i.e., a tree spanning the source and all receivers) is 0 if it does not belong to the tree and $-c_k$ if it belongs to the tree. The utility of a terminal k on a multicast tree constructed by a mechanism $M = (\mathcal{O}, \mathcal{P})$ is $\mathcal{P}_k(a) - c_k$.

Hereafter, we consider only direct-revelation mechanisms. Thus, the strategy vector a that is selected by agents becomes the type vector t *declared* by agents. A mechanism has *incentive compatibility* (IC) if reporting that the actual type is a dominant strategy for every agent. Another common property in the literature for mechanism design is *individual rationality* (IR) or *voluntary participation*: The agent's utility of participating in the output of the mechanism is nonnegative. For convenience, let $t|^i b = (t_1, \ldots, t_{i-1}, b, t_{i+1}, \ldots, t_n)$; i.e., each agent $j \neq i$ reports its type t_j except that agent i reports type b. Then, IC implies that for each agent i and any strategy b,

$v_i(t_i, \mathcal{O}(t)) + \mathcal{P}_i(t) \geq v_i(t_i, \mathcal{O}(t|^i b)) + \mathcal{P}_i(t|^i b)$, and IR implies that for each agent i, $v_i(t_i, \mathcal{O}(t)) + \mathcal{P}_i(t) \geq 0$.

Arguably, the most positive result in mechanism design is what is usually called the generalized VCG mechanism by Vickrey (1961), Clarke (1971), and Groves (1973). A direct-revelation mechanism $M = [\mathcal{O}(t), \mathcal{P}(t)]$ belongs to the VCG family if:

1. The output $\mathcal{O}(t)$ maximizes the objective function $g(o, t) = \sum_i v_i(t_i, o)$; and
2. The payment to agent i is $\mathcal{P}_i(t) = \sum_{j \neq i} v_j(t_j, \mathcal{O}(t)) + h_i(t_{-i})$. Here, $h_i()$ is an arbitrary function of t_{-i}.

To make sure that the mechanism satisfies the IR property, $h_i(t_{-i})$ is often set as $-\sum_{j \neq i} v_j(t_j, \mathcal{O}(t_{-i}))$; i.e., the negative of the maximum total valuations achieved if agent i is eliminated. A VCG mechanism is always truthful (Groves, 1973) and, under mild assumptions, VCG mechanisms are the *only* truthful implementations to maximize the total valuations (Green and Laffont, 1977).

Similarly, a weighted version of the VCG mechanism can be implemented as well. A maximization-mechanism-design problem is said to be *weighted utilitarian* if there exists positive real numbers $\beta_1, \ldots, \beta_n$ such that the objective function is $g(o, t) = \sum_i \beta_i v^i(t^i, o)$. A direct-revelation mechanism $m = [o(t), p(t)]$ belongs to the *weighted VCG family* if (1) the output $o(t)$ computed based on the type vector t maximizes the objective function $g(o, t)$; and (2) the payment to agent i is $p^i(t) = \frac{1}{\beta_i} \sum_{j \neq i} \beta_j \cdot v^j[t^j, o(t)] + h^i(t^{-i})$. Here, $h^i()$ is an arbitrary function of t^{-i}. It is proved by Roberts (1979) that a weighted VCG mechanism is truthful.

THEOREM 15.1 (Roberts, 1979) *A weighted VCG mechanism satisfies IC.*

Proof. Assume that $d = (d^1, d^2, \ldots, d^n)$ is the declared cost vector and, for agent i, revealing its truth cost is not its dominant strategy; i.e, $d^i \neq t^i$, so we have

$$v^i[t^i, o(d^i|t^i)] + p^i[t^i, o(d^i|t^i)] < v^i[t^i, o(d)] + p^i(t^i, d).$$

Subtracting $h^i(d^{-i})$ on both sides, we get

$$v^i(t^i, o(d^i|t^i)) + p^i(t^i, o(d^i|t^i)) - h^i(d^{-i}) < v^i(t^i, o(d)) + p^i(t^i, d) - h^i(d^{-i}).$$

Thus,

$$v^i(t^i, o(d^i|t^i)) + \frac{1}{\beta_i} \sum_{j \neq i} \beta_j v^j(t^j, o(d)) < v^i(t^i, o(d)) + \frac{1}{\beta_i} \sum_{j \neq i} \beta_j v^j[t^j, o(d)].$$

Simplifying both sides, we get

$$\sum_{j=1}^{n} \beta_j v^j[t^j, o(d|^i t^i)] < \sum_{j=1}^{n} \beta_j v^j[t^j, o(d)],$$

which is a contradiction. ∎

Although the family of VCG mechanisms is powerful, it has its own limitations. To use the VCG mechanism, we have to compute the *exact* output that maximizes the total valuations of all agents. This makes the mechanism computationally intractable in many cases. Notice that replacing the optimal output with a nonoptimal approximation usually leads to untruthful mechanisms if the payment method is carried out by the VCG mechanism (Nisan and Ronen, 1999). To make the mechanism tractable, the output method $\mathcal{O}()$ and the payment method $\mathcal{P}()$ should be computable in polynomial time. Notice that it is NP-hard to find the tree with the minimum cost for multicast, and finding the minimum-cost tree is equivalent to finding the output that maximizes the total valuations of all agents because we assume that the valuation of an agent k is $-c_k$ if it is in the tree and 0 otherwise. Thus, the VCG mechanism using a minimum-cost tree as its output is not polynomially computable if $P \neq$ NP.

In summary, we want to design strategy-proof multicast protocols for a selfish wireless network with the following properties:

1. **Incentive Compatibility** (IC): An agent will reveal its true cost to maximize its utility no matter what the other agents do.
2. **Individual Rationality** (IR): An agent is guaranteed to have nonnegative utility if it truthfully reports its cost.
3. **Polynomial-Time Computability** (PTC): All computations (the computation of the output and payment) are done in polynomial time.

15.2.2   Communication Model

Power consumption of wireless networks can be divided into two domains: *data communication* and *data processing* (Akyildiz et al., 2002). Compared with that in data processing, a wireless terminal expends more energy in data communication. This involves both data transmission and reception. It has been shown that for short-range communication with low radiation power, transmission and reception energy costs are nearly the same. For example, Feeney and Nilson (2001) studied the power consumption in IEEE 802.11 ad hoc networks. Mixers, frequency synthesizers, voltage-control amplifiers, and so on all consume valuable power in the transceiver circuitry. Shih *et al.* (2001) presented a formulation for average power consumption for a radio that shows that the power (i.e., power consumed by the transmitter/receiver and the output power of the transmitter) is linear to the transmission time. Energy expenditure in data processing is much less compared with data communication, but it is also related to the packet size and data rate. By assuming that the data rate is the same at each terminal, we make a model such that each wireless terminal i in the ad hoc network consumes power g_i to relay a unit-size packet, which may include the energy to receive, process, store, and forward the packet. In practice, for a given unit-size packet, the power consumption of a fixed terminal may vary significantly because of the retransmission caused by collision and bad channel conditions. Thus, more precisely, the g_i could be the expected power consumption to relay a unit-size packet in some circumstances. If the size of the packet is h_i, then the total power consumption is $g_i h$. Notice that because of different devices or transmission power at each terminal, the values of g_i could be various. In this chapter, for the sake of succinctness of our analysis, we normalize the size of the packet to unit size. Terminal i declares a cost c_i that reflects the minimum amount terminal i would ask for to forward the unit-size packet. Here, c_i could include terminal i's actual cost plus the amount it wants to earn. Furthermore, when the link is not reliable, c_i can also be terminal i's expected cost to forward a unit data. When terminal i sends the packet, every terminal that is within a distance of r_i can receive the packet. The distance r_i is called terminal i's transmission range. In the communication network, there is a link between two terminals i and j if and only if the distance between i and j is not greater than $\min(r_i, r_j)$. If every terminal i has the same transmission range, then the communication network is called a UDG. Because of the broadcast property of the wireless terminal, we model the communication network as a node-weighted undirected graph.

Throughout this chapter, we always assume that the network is *biconnected*, which implies that if we remove any terminal, the network is still connected. This assumption is necessary to prevent some terminals from having a monopoly and charging an arbitrary cost in addition to increase network robustness.

It is generally known that it is NP-hard to find the MCMT for a node-weighted graph (Klein and Ravi, 1995). Several multicast structures were proposed in the literature to approximate a MCMT. In practice, two types of multicast structures are used to meet the requirements of different applications: a *source-based multicast tree* and a *share-based multicast tree* (Ballardie *et al.*, 1993; Deering, 1991). For those applications such as on-line movies, they usually have one or only a few senders and lots of receivers. Therefore, we often use a source-based multicast tree in which receivers only receive messages but do not send them. On the other hand, many applications have lots of active senders, such as distributed interactive simulation applications and distributed video gaming (in which most receivers are also senders). In this case, the share-based tree (SBT) is used to increase the scalability.

In this chapter, we study how to design truthful payment schemes for several multicast trees, including source-based trees and SBTs for node-weighted graphs. The following assumptions are adopted in this chapter:

1. Each relay terminal has a privately known cost to relay transit traffic for other terminals, and the cost is *independent* of the number of its children in the multicast tree.
2. The candidate relay terminals (i.e., the terminals in addition to the source and the receivers) will not *collude* with each other to improve their gains.
3. All terminals are rational.
4. The source of the multicast will pay the selected relay terminals.

If we relax any of first three assumptions, we would have to design different mechanisms. If the fourth assumption is not met, we need to design a payment-sharing scheme (W. Wang *et al.*, 2005b) to fairly share the payments to all relay terminals among all receivers. Regarding the collusion, notice that multicast is a special case of unicast. If considering the unicast, W. Wang and Li (2004b, 2005) proved that there is no truthful payment scheme that can prevent *any* two agents from improving their gains by colluding with each other.

15.2.3   Problem Statement

Consider any communication network $G = (V, E, c)$, where $V = \{v_1, \ldots, v_n\}$ is the set of communication terminals, $E = \{e_1, e_2, \ldots, e_m\}$ is the set of links, and c is the cost vector of all terminals. Given a set of sources and receivers $Q = \{q_0, q_1, q_2, \ldots, q_{r-1}\} \subset V$, the multicast problem is to find a tree $T \subset G$ spanning all terminals Q. For simplicity, we assume that $s = q_0$ is the sender of a multicast session if it exists, and then $R = \{q_1, q_2, \ldots, q_{r-1}\}$ is the set of receivers for multicast. Every terminal is required to declare a cost of relaying a unit amount of data. Let $d = \{d_1, d_2, \ldots, d_n\}$ be the declared costs of all terminals; i.e., terminal v_i declares a cost d_i. From the declared-cost profile d, we should construct the multicast tree and decide the payment for the agents. The utility of an agent is its payment received minus its cost if it is selected in the multicast tree. Instead of reinventing the wheel, we still use the previously proposed structures for multicast as the output of our mechanism. Given a multicast tree, we design the strategy-proof payment scheme based on this tree.

Given a structure H, we use $\omega(H)$ to denote the total cost of all terminals in this structure. If we change the cost of any terminal v_i to c_i', we denote the new network as $G' = (V, E, c|^i c_i')$, or simply $c|^i c_i'$. If we remove one terminal v_i from the network, we denote it as $c|^i \infty$. We denote $G \backslash e_i$ as the network without link e_i and denote $G \backslash v_i$ as the network without terminal v_i and all its incident links. For simplicity of notation, we use the cost vector c to denote the network $G = (V, E, c)$ if no confusion is caused.

15.3   Truthful Payment Schemes for Multicast

In this section, we discuss in detail how to conduct truthful multicast routing in wireless networks with selfish terminals. We specifically study the following three structures: the LCPT, the VMST, and the node-weighted Steiner tree (NST). The LCPT, or core-based tree, is the source-based tree commonly used to multicast the packet. Notice that both the VMST and the NST are share-based multicast trees, which implies that the receivers could also be a sender.

15.3.1   Truthful Mechanism for Unicast

To study the design of truthful mechanisms for multicast, we first study truthful mechanisms for unicast.

The unicast routing game has been studied extensively since it was introduced by Nisan and Ronen (1999). They solved the unicast routing game by applying the VCG mechanism in a centralized way, under the assumption that the payment to all relay nodes will be paid by the source, the receiver, or both. They did not address the scenario of how this payment is shared between the source and the receiver and the scenario in which the sender and/or receiver has a valuation on this unicast. In other words, the sender and/or the receiver will not initiate the unicast if the total payment is larger than their budget. Let **d** be the vector of declared costs of all possible relay agents. Let $\mathcal{P}^{VCG}$ denote the

VCG payment for the unicast mechanism and $\mathbf{d}|^k d'_k = (d_1, \ldots, d_{k-1}, d'_k, d_{k+1}, \ldots, d_m)$. The payment to terminal $v_k \in \mathsf{LCP}(s, t, \mathbf{d})$ according to the VCG mechanism is

$$\mathcal{P}_k^{\mathsf{VCG}}(\eta^{=\infty}, \mathbf{d}) = |\mathsf{LCP}(s, t, \mathbf{d}|^k \infty)| - |\mathsf{LCP}(s, t, \mathbf{d}|^k 0)|,$$

where $\eta^{=\infty}$ means that the service requestor (either the sender or the receiver or both) has an infinity valuation. The following fact is a simple property of the VCG mechanism.

FACT 15.1 *For any* $v_i \in \mathsf{LCP}(s, t, \mathbf{d})$, $d_i \leq \mathcal{P}_k^{\mathsf{VCG}}(\eta^{=\infty}, \mathbf{d})$.

A VCG-mechanism-based unicast system is socially efficient and strategy-proof, but its budget-balance factor could be as small as $1/n$: The total payment to relay agents could be n times their total actual costs for a network of n nodes.

Notice that when the LCP is used for routing, the truthful mechanism is uniquely defined under some mild assumptions. In the literature, there are a number of other mechanisms proposed that will not use the LCP for routing. For example, Karlin *et al.* (2005) presented a truthful $\sqrt{\ }$-*mechanism* for the unicast game and proved the following theorem:

THEOREM 15.2 (Karlin *et al.*, 2005) *The frugality of the VCG mechanism is* $\Omega(n)$. *The frugality of the* $\sqrt{\ }$-*mechanism is* $O(\sqrt{n})$, *which is at most* $2\sqrt{2}$ *times the optimal. The frugality of the unicast game is* $\Theta(\sqrt{n})$.

Talwar (2003) proposed measuring the overpayment for a binary-demand game by using the *frugality*, which is defined as the total payment of the mechanism (e.g., VCG) over the total cost of the *second optimal team*, which is the best team that does not intersect with team $\mathbf{T}_{\mathsf{VCG}}$ selected by the mechanism. The frugality notation was then generalized by Karlin *et al.* (2005) for the case in which the second optimal team does not exist or its cost is arbitrarily large even compared with the total VCG payment.

15.3.2   Least-Cost-Path Tree

Constructing a LCPT
First, each terminal v_i reports a cost d_i of forwarding the unit amount size of data, which is collected to the source node. For each receiver $q_i \neq s$, we compute the LCP, denoted by $\mathsf{LCP}(s, q_i, d)$, from the source s to q_i by using the reported cost profile d. The union of all LCPs from the source to receivers is called *least-cost-path tree*, denoted by $\mathsf{LCPT}(d)$. Clearly, we can construct the LCPT in time $O(n \log n + m)$. Next, we discuss how to design a truthful payment scheme while using the LCPT as the output.

VCG Mechanism on LCPT Is Not Strategy-Proof
Intuitively, we would use the VCG payment scheme in conjunction with the LCPT structure as follows: The payment $p_k^{\mathsf{VCG}}(d)$ to each terminal v_k in LCPT is

$$p_k^{\mathsf{VCG}}(d) = \omega(LCPT(d|^k \infty)) - \omega(LCPT(d)) + d_k. \tag{15.1}$$

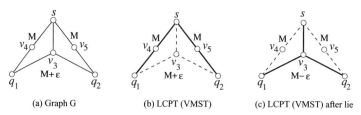

Figure 15.1 The costs of terminals are $c(v_4) = c(v_5) = M$ and $c(v_3) = M + \epsilon$. Here, q_1 and q_2 are the receivers. © 2005, ACM.

We show by an example that the mechanism $M = (\text{LCPT}, \mathcal{P}^{\text{VCG}})$ [whose payment is defined in Eq. (15.1)] is not strategy-proof. In other words, if we simply apply the VCG scheme on the LCPT, a terminal may have incentives to lie about its cost. Figure 15.1 illustrates such an example in which terminal v_3 can lie about its cost to improve its utility. The payment to terminal v_3 is 0 and its utility is also 0 if it reports its cost truthfully. On the other hand, the total payment to terminal v_3 when v_3 reported a cost $d_3 = M - \epsilon$ is $\omega(\text{LCPT}(c|^3\infty)) - \omega(\text{LCPT}(c|^3d_3)) + d_3 = 2M - (M - \epsilon) + M - \epsilon = 2M$ and the utility of terminal v_3 becomes $2M - (M + \epsilon) = M - \epsilon$, which is larger than its original utility 0, when $0 < \epsilon < M$.

Strategy-Proof Mechanism on LCPT

Now, a strategy-proof mechanism is described that does not rely on the VCG mechanism. For each receiver $q_i \neq s$, we compute the LCP from the source s to q_i and compute an intermediate payment $\mathcal{P}_k^i(d)$ to every terminal v_k on $\text{LCP}(s, q_i, d)$ by using the scheme for unicast:

$$\mathcal{P}_k^i(c) = d_k + \omega[\text{LCP}(s, q_i, d|^k\infty)] - \omega[\text{LCP}(s, q_i, d)].$$

Here, $\omega[\text{LCP}(s, q_i, d)]$ denotes the total cost of path $\text{LCP}(s, q_i, d)$. The final payment to a terminal $v_k \in \text{LCPT}$ is

$$\mathcal{P}_k(d) = \max_{q_i \in Q} \mathcal{P}_k^i(d). \tag{15.2}$$

The payment to a terminal not on the LCPT is simply 0.

Before we show that the payment scheme in Eq. (15.2) is truthful, let us illustrate it by a running example of how we pay terminal v_3 in Figure 15.1. If terminal v_3 reports a cost $M + \epsilon$ truthfully, then it gets payment 0 because $v_3 \notin \text{LCPT}$. If terminal v_3 reports a cost $M - \epsilon$, it is now in the LCPT (composed of links sv_3, v_3q_1, and v_3q_2). Its payment then becomes $\max(\mathcal{P}_{v_3}^1, \mathcal{P}_{v_3}^2)$, where $\mathcal{P}_{v_3}^1 = M - \epsilon + \omega(\text{LCP}(s, q_1, d|^{v_3}\infty)) - \omega(\text{LCP}(s, q_1, d)) = M - \epsilon + M - (M - \epsilon) = M$, and $\mathcal{P}_{v_3}^2 = M$ similarly. Then, the profit of terminal v_3 becomes $\max(\mathcal{P}_{v_3}^1, \mathcal{P}_{v_3}^2) - (M + \epsilon) = -\epsilon$, which is less than what it gets by reporting its truthful cost.

THEOREM 15.3 *The payment scheme in Eq. (15.2) is truthful and it is minimum among all truthful payments based on the LCPT.*

Proof. First, we prove that the payment scheme in Eq. (15.2) satisfies IR. When v_k reveals its true cost, if it is not in the LCPT, it has utility 0. Thus, we need to consider only the case in which v_k is in the LCPT. Remember that the LCPT is the union of the LCPs from source to receivers; if v_k is on $\mathsf{LCP}(s, q_i, d)$, then the intermediate payment $\mathcal{P}_k^i(c)$ is based on unicast, which is nonnegative. Thus, the overall payment $\mathcal{P}_k(d) = \max_{q_i \in Q} \mathcal{P}_k^i(d)$ is also nonnegative.

Similarly, because the payment scheme for unicast is truthful, v_k cannot lie about its cost to increase its intermediate payment $\mathcal{P}_k^i(c)$ based on $\mathsf{LCP}(s, q_i, d)$. Thus, it cannot increase $\max_{q_i \in Q} \mathcal{P}_k^i(c)$ by lying about its cost. In other words, the payment scheme is truthful.

We now show that the preceding payment scheme pays the minimum among all strategy-proof mechanisms using the LCPT as the output. Before the optimality of the payment scheme is discussed, some definitions are given. Consider all paths from sender s to receiver q_i. They can be divided into two categories: with terminal v_k or not. The path having the minimum cost among these paths with terminal v_k is denoted as $\mathsf{LCP}_{v_k}(s, q_i, d)$; the path having the minimum cost among these paths without terminal v_k is denoted as $\mathsf{LCP}_{-v_k}(s, q_i, d)$.

Assume that there is another payment scheme $\tilde{p}$ that pays less for a terminal v_k in a network G under a cost profile d. Let $\delta = \mathcal{P}_k(d) - \tilde{p}_k(d)$; then, $\delta > 0$. Without loss of generality, assume that $\mathcal{P}_k(d) = \mathcal{P}_k^i(d)$. Thus, terminal v_k is on $\mathsf{LCP}(s, q_i, d)$, and the definition of $\mathcal{P}_k^i(d)$ implies that $\omega(\mathsf{LCP}_{-v_k}(s, q_i, d)) - \omega(\mathsf{LCP}(s, q_i, d)) = \mathcal{P}_k(d) - d_k$.

Then, consider another cost profile $d' = d|^k(\mathcal{P}_k(d) - \frac{\delta}{2})$, where the true cost of terminal v_k is $\mathcal{P}_k(d) - \frac{\delta}{2}$. Under profile d', because $\omega(\mathsf{LCP}_{-v_k}(s, q_i, d')) = \omega(\mathsf{LCP}_{-v_k}(s, q_i, d))$, we have

$$\omega(\mathsf{LCP}_{v_k}(s, q_i, d'))$$
$$= \omega(\mathsf{LCP}_{v_k}(s, q_i, d|^k 0)) + \mathcal{P}_k(d) - \frac{\delta}{2}$$
$$= \omega(\mathsf{LCP}(s, q_i, d)) + \mathcal{P}_k(d) - \frac{\delta}{2} - d_k$$
$$= \omega(\mathsf{LCP}_{-v_k}(s, q_i, d)) - \frac{\delta}{2}$$
$$< \omega(\mathsf{LCP}_{-v_k}(s, q_i, d)) = \omega(\mathsf{LCP}_{-v_k}(s, q_i, d')).$$

Thus, $v_k \in \mathsf{LCPT}(d')$. From the following Lemma 15.1, we know that the payment to terminal v_k is the same for cost profile d and d'. Thus, the utility of terminal v_k under profile d' by payment scheme $\tilde{p}$ becomes $\tilde{p}_k(d') - d_k' = \tilde{p}_k(d) - d_k' = \tilde{p}_k(d) - (\mathcal{P}_k(d) - \frac{\delta}{2}) = -\frac{\delta}{2} < 0$. In other words, under profile d', when terminal v_k reports its true cost, it gets a negative utility under payment scheme $\tilde{p}$. Thus, $\tilde{p}$ is not strategy-proof. This finishes the proof. ∎

LEMMA 15.1 *If a mechanism based on a structure* T *with a payment method* $\tilde{p}$ *is truthful, then for every terminal* v_k *in the network, if* $v_k \in$ T, *the payment* $\tilde{p}_k(d)$ *to* v_k *is independent of its declared cost* d_k.

Proof. We prove it by contradiction. Suppose that there exists a truthful payment scheme such that $\tilde{p}_k(d)$ depends on d_k. There must exist two valid declared costs x_1 and x_2 such that $x_1 \neq x_2$ and $\tilde{p}_k(d|^k x_1) \neq \tilde{p}_k(d|^k x_2)$. Without loss of generality, we assume that $\tilde{p}_k(d|^k x_1) > \tilde{p}_k(d|^k x_2)$. Now consider the scenario that terminal v_k has an actual cost $c_k = x_2$. Obviously, it can lie about its cost as x_2 to increase its utility, which violates the incentive compatibility property. ∎

Notice that the payment based on $\mathcal{P}_k(c) = \min_{q_i \in Q} \mathcal{P}_k^i(c)$ is not truthful because a terminal may lie about its cost, making it higher so it can discard some low payments from some receivers. In addition, payment $\mathcal{P}_k(c) = \sum_{q_i \in Q} \mathcal{P}_k^i(c)$ is *not* truthful either.

The following lemma is folklore:

LEMMA 15.2 *The total cost of the multicast tree LCPT is at most* r *times the optimum cost, where* r *is the number of receivers.*

Computational Complexity

Assume that there are r receivers; for every terminal q_i, we calculate the payment for every terminal $v_k \in \mathsf{LCP}(s, q_i, c)$ based on $\mathsf{LCP}(s, q_i, c)$ by using the fast payment scheme for unicast (W. Wang and Li, 2004b). This will take $O(n \log n + m)$ time. So, for all terminals, it takes $O(rn \log n + rm)$ to compute the payments. Note that we can construct the LCPT in time $O(n \log n + m)$. A natural question is whether we can reduce the time complexity from $O(rn \log n + rm)$ to $O(n \log n + m)$. We leave it as an open question.

15.3.3   Node-Weighted Steiner Tree

It is well known (Guha and Khuller, 1999; Klein and Ravi, 1995) that it is NP-hard to find the MCMT when given an arbitrary node-weighted graph G. Klein and Ravi (1995) showed that it can be approximated within $O(\ln r)$, where r is the number of receivers.

Constructing the NST

We review the method used in Klein and Ravi (1995) to find a NST. First, some definitions are introduced that are essential for constructing the NST. A *spider* is defined as a tree having at most one terminal of a degree of more than two. Such a terminal (if it exists) is called the center of the spider. Each path from the center to a leaf is called a *leg*. The *cost* of a spider S, denoted as $\omega(S)$, is defined as the sum of the costs of all terminals in spider S. The number of terminals or *legs* of a spider S is denoted by $t(S)$, and the *cost ratio* of a spider S is defined as $\rho(S) = \frac{\omega(S)}{t(S)}$.

THEOREM 15.4 (Klein and Ravi, 1995) *The tree constructed by Algorithm 39 has a cost of at most* $2 \ln r$ *times the optimal.*

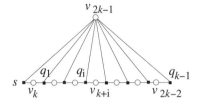

Figure 15.2 Terminals q_i, $1 \leq i < r$, are receivers; the cost of terminal v_{2r-1} is 1. The cost of each terminal v_i, $r \leq i \leq 2r - 2$, is $\frac{2}{2r-i} - \epsilon$, where ϵ is a sufficiently small positive real number. © 2005, ACM.

Algorithm 39 Construct the NST

Input: Graph $G = (V, E, c)$ and a receiver set R.

Output: NST rooted at source s spanning all receivers R.

1: Set the virtual terminal $V_T \leftarrow s$.

2: **repeat**

3:     Find the spider S with minimum cost ratio $\rho(S)$ that connects some receivers and the virtual terminal V_T. {/*For simplicity of the proof, we assume there are no two spiders with the same cost ratio. Dropping the assumption will not change the results.*/}

4:     Contract the spider S by treating all terminals in it as the virtual terminal V_T. The contracted virtual terminal V_T has a weight of zero. We call this one *round*.

5: **until** no receivers left

6: All terminals belonging to the final unique virtual terminal V_T form the NST.

VCG Mechanism on the NST Is Not Strategy-Proof

Again, we may want to pay terminals based on the VCG scheme; i.e., the payment to a terminal $v_k \in \text{NST}(d)$ is

$$\mathcal{P}_k(d) = \omega(\text{NST}(d|^k\infty)) - \omega(\text{NST}(d)) + d_k.$$

It is shown by an example that the payment scheme does *not* satisfy the IR property: It is possible that some terminal has a negative utility. Figure 15.2 illustrates such an example. It is not difficult to show that in the first round, terminal v_r is selected to connect terminals s and q_1 with a cost ratio of $\frac{1}{r} - \frac{\epsilon}{2}$ (and all other spiders have a cost ratio of at least $\frac{1}{r}$). Then, terminals s, v_r, and q_1 form a virtual terminal. At the beginning of round j, we have a virtual terminal, denoted by V_j, formed by terminals v_{r+i-1}, $1 \leq i \leq j - 1$, and receivers q_i, $1 \leq i \leq j$; all other receivers q_i, $j < i < r$, are the remaining terminals. It is easy to show that we will select terminal q_{r+j-1} at round j to connect V_j and q_{j+1} with cost ratio $\frac{1}{r+1-j} - \frac{\epsilon}{2}$. Thus, the total cost of the tree $\text{NST}(G)$ is $\sum_{i=1}^{r-1}(\frac{2}{r+1-i} - \epsilon) = 2H(r) - 2 - (r-1)\epsilon$.

When terminal v_k is not used, it is easy to see that the final tree, $\text{NST}(G \backslash v_r)$, will use only terminal v_{2r-1} to connect all receivers with cost ratio $\frac{1}{r}$ when $\frac{1}{r-1} - \frac{\epsilon}{2} > \frac{1}{r}$. Notice that this condition can be trivially satisfied by letting $\epsilon = \frac{1}{r^2}$. Thus, the utility of terminal

v_r is $\mathcal{P}_1(d) - c(v_r) = \omega(\text{NST}(G\backslash v_r)) - \omega(\text{NST}(G)) = -2H(r) + 3 + (r-1)\epsilon$, which is negative when $r \geq 8$ and $\epsilon = 1/r^2$.

Strategy-Proof Mechanism Based on NST

Notice that the construction of the NST is by rounds. We subsequently show that if terminal v_k is selected as part of the spider with minimum cost ratio under cost profile d in a round i, then v_k is selected before or in round i under cost profile $d' = d|^k d_k'$ for $d_k' < d_k$. If v_k appears before round i, then the conclusion trivially holds. Thus, we can assume that v_k does not appear before round i. Therefore, the graph remains the same for round i after the profile changes. Because the cost of spider $S_i(d)$ under cost profile d becomes $\omega(S_i(d)) - d_k + d_k' < \omega(S_i(d))$ while all other spiders' costs decrease at most $d_k + d_k'$, the spider $S_i^k(d)$ has the minimum cost ratio among all spiders under cost profile d'. Thus, for terminal v_k, there exists a real value $B_k^i(d_{-k})$ such that terminal v_k is selected before or in round i if and only if $d_k < B_k^i(d_{-k})$. If there are r rounds, we have a nondecreasing sequence:

$$B_k^1(d_{-k}) \leq B_k^2(d_{-k}) \leq \cdots \leq B_k^r(d_{-k}) = B_k(d_{-k}).$$

Obviously, terminal v_k is selected in the final multicast tree if and only if $d_k < B_k(d_{-k})$. What follows is the payment scheme based on the NST. For a terminal v_k, if v_k is selected, then it gets payment

$$\mathcal{P}_k(d) = B_k(d_{-k}). \tag{15.3}$$

Otherwise, it gets payment 0.

Regarding this payment, we have the following theorem (whose proof is omitted):

THEOREM 15.5 *The payment scheme in Eq. (15.3) is truthful, and among all truthful payment schemes for a multicast tree based on the NST, the payment is minimal.*

With Theorem 15.5, we need to focus our attention only on how to compute the value $B_k^i(d_{-k})$. Before the algorithm is presented to find $B_k^i(d_{-k})$, we first review in detail how to find the minimum-cost-ratio spider. To find the spider with the minimum cost ratio, we find the spider centered at each terminal v_j with the minimum cost ratio, and over all terminals $v_j \in V$, we choose the minimum among them.

In Algorithm 40, $\omega(L_i(v_j))$ is defined as the sum of the terminals' costs on this branch excluding v_j, and $\Omega_i(L(v_j)) = \sum_{s=1}^{i} \omega(L_s(v_j)) + c_j$. If we remove terminal v_k, the minimum-cost-ratio spider centered at v_j is denoted as $S^{-v_k}(v_j)$ and its cost ratio is denoted as $\rho^{-v_k}(v_j)$. Let $L_1^{-v_k}(v_j), L_2^{-v_k}(v_j), \ldots, L_{r-1}^{-v_k}(v_j)$ be those branches in ascending order before the linear scan.

From now on, we fix d_{-k} and graph G to study the relationship between the minimum cost ratio $\rho(v_j)$ of the spider centered at v_j and d_k. If the minimum-cost-ratio spider with terminal v_k has t legs, then its cost ratio will be a line with a slope of $1/t$. So the ratio-cost function is a piecewise linear function, as shown in Figure 15.3. Observe that the number of the legs of the minimum-cost-ratio spider decreases over d_k. Thus, these line segments have decreasing slopes and there are at most r segments,

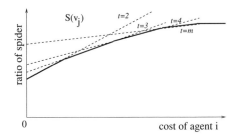

Figure 15.3 A ratio-cost function.

Algorithm 40 Find the Minimum-Cost-Ratio Spider

Input: Graph $G = (V, E, c)$ and a receiver set R.

Output: A minimum-cost-ratio spider S and its cost ratio ρ.

1: **for** every terminal $v_j \in V$ **do**

2:     Compute the LCPT rooted at v_j and spanning all terminals. We call each LCP a *branch*, and its weight is defined as the total cost of the path (excluding the two end terminals).

3:     Sort the branches according to their weights.

4:     **for** every pair of branches **do**

5:        If they have common relay terminals, then remove the branch with the larger weight.

6:     **end for**

7:     Assume that the remaining branches are

$$L(v_j) = \{L_1(v_j), L_2(v_j), \ldots, L_r(v_j)\}$$

    sorted in ascending order according to their weights.

8:     Find the minimum-cost-ratio spider with center v_j by linear scanning: The spider is formed by the first $t \geq 2$ branches such that $\frac{c_j + \sum_{k=1}^{t} L_k}{t} \leq \frac{c_j + \sum_{k=1}^{h} L_k}{h}$ for any $h \neq t$.

9:     Denote the minimum-cost-ratio spider centered at terminal v_j as $S(v_j)$ and its cost ratio as $\rho(v_j)$.

10: **end for**

11: The spider with the minimum cost ratio is $S(v_j)$ such that $\rho(v_j) = \min_{v_i \in V} \rho(v_i)$.

---

where r is the number of receivers. So, given a real value y and fixed d_{-k}, we can find the corresponding cost d_k of v_k in time $O(\log r)$ such that the minimum-cost-ratio spider has a ratio y when v_k has cost d_k. We leave it as an exercise for readers to compute the cut value $B_k^i(d_{-k})$ for each selected node v_k.

Computational Complexity

If we use Algorithm 39 to find NST(d), every round we need time $O(rn \log n + rm)$, where r is the number of receivers. Notice that there are at most r rounds, so the overall time complexity is $O(r^2 \log n + r^2 m)$. For every terminal $v_k \in \mathrm{NST}(d)$, if we apply Algorithm 40 to calculate the payment, it is not difficult to get time complexity $O(rn \log n + rm)$ for each round. Thus, it takes time $O(r^2 n \log n + r^2 m)$ to find the payment for a single terminal $v_k \in \mathrm{NST}(d)$. In the worst case, there could be up to $O(n)$ terminals in $\mathrm{NST}(d)$, so the overall time complexity is $O(r^2 n^2 \log n + r^2 nm)$, which is quite expensive. Finding a more efficient way to reduce the time complexity will be one of the future works.

15.4   Sharing Multicast Costs or Payments Among Receivers

Designing a truthful payment scheme is not the whole story for many practical applications. A natural question to be answered is who will be charged for the payments to the relay agents. A simple solution is that the organization to which the receivers belong pays (W. Wang *et al.*, 2004). However, this solution is not a panacea. In many applications such as video streaming, each individual receiver often has to pay for receiving the data. How to charge the receivers for multicast transmission has been studied extensively in the literature (Adler and Rubenstein, 2002; Cocchi *et al.*, 1993; Feigenbaum *et al.*, 2001a, 2001b, 2003; Herzog *et al.*, 1995). In most of their models, they assumed that (1) every receiver has a valuation for receiving the data and the receiver is selfish; (2) all relay agents are cooperative and will reveal their true costs; and (3) the multicast tree is fixed as the union of the shortest paths from the source to receivers. In a sharp contrast, we take the selfish behavior of the relay agents into account in this chapter. Thus, we model the network differently by assuming that (1) the relay agents are selfish and rational; (2) the receivers always receive the data and pay what they "should" pay in a fair way; and (3) the multicast topology could be any structure, including trees and meshes. We also show the hardness when both the receivers and the relay agents are selfish and rational and each receiver has a privately known valuation.

In this chapter, we focus on the interautonomous system (inter-AS) multicasting instead of intra-AS routing because intra-ASs are usually cooperative instead of non-cooperative. Figure 15.4(a) shows an example of a multicast network topology with end hosts. Figure 15.4(b) shows the corresponding inter-AS multicasting topology.

Here, we model the inter-AS network topology as a graph $G = (V, E, c)$, where $V = \{v_1, \ldots, v_n\}$ is the set of ASs, where $E = \{e_1, e_2, \ldots, e_m\}$ is the set of links between ASs. Usually, in inter-AS routing, each AS actually is an independent economic decision maker that could choose its strategy for financial advantage in routing decisions. We assume that each AS v_i is an individual agent and it has a *fixed* private cost c_i to transmit a unit size of data in multicast. Thus, every AS is called on to declare its cost to the protocol. When the nodes are the selfish agents, we call this network a *node-weighted network*. On the other hand, sometimes we need to treat the selfish agents as links in the network; e.g., the multicast data-gram is sent from one AS to another AS by using an application layer tunneling through other ASs. If links are agents, the network is modeled as a *link-weighted network*. Most of the general techniques in

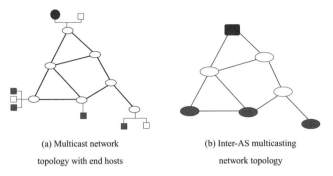

(a) Multicast network

topology with end hosts

(b) Inter-AS multicasting

network topology

Figure 15.4 The solid circle is the source host, the solid squares are the receiving hosts of the multicast group, the solid rectangle is the AS attached with the source host, and the solid ellipses are the ASs attached with the receiving hosts. © 2006, IEEE.

the remaining sections are not specific to one model and thus can be applied to both models.

Given a set of multicast group members, in this chapter, the receivers are the ASs with some attached group members instead of the actual end hosts who are the multicast group members. For the convenience of our analysis, we assume that s is the source AS in one specific multicast and the size of the data is normalized to 1. We also assume that agents in the network will not *collude* to improve their profits together. To prevent monopoly, we assume that the network is biconnected. Given a source node $s = q_0$ and a set of multicast receivers $R = \{q_1, q_2, \ldots, q_r\} \subset V$, we need to design a multicast protocol that:

1. Constructs a topology (e.g., a tree, a mesh, a ring) that spans the source and all receivers.
2. Calculates a payment for each relay AS according to a *payment scheme* that is truthful.
3. Charges each receiver according to a *payment-sharing scheme* that is *fair*. A formal definition of what is fair in given Subsection 15.4.1.

One thing that should be highlighted here is that instead of reinventing the wheel by designing some new multicast structures, we focus on how we can design a truthful payment scheme for certain existing multicast protocols to ensure that they work correctly even in noncooperative networks. From the truthful payment scheme we designed, we further study *how* we charge the receivers in a fair way.

15.4.1 Fair Payment-Sharing Scheme

For a given set of receivers, after we calculate the payment $p_k(d)$ for every relay agent k based on declared costs d, we are ready to study how to share the payments fairly among receivers. Notice that payment-sharing is different from traditional cost-sharing. How to share the multicast cost among the receivers has been studied previously in Cocchi *et al.* (1993), Feigenbaum *et al.* (2001b), Herzog *et al.* (1995), and Moulin and Shenker (2002), with the assumption that the costs of relay agents are public and the multicast topology is a fixed tree. Most of the literature used the *equal link split downstream* (ELSD) pricing scheme to charge receivers: The cost of a link is shared *equally* among all of its downstream receivers. As shown later, if we simply use the ELSD to share the total payment among receivers, it usually is not fair according to some common sense.

Given a set of receivers R, let $\mathcal{P}(R, d) = \sum_k p_k(R, d)$ denote the total payment to all relay agents. For a sharing scheme ξ, let $\xi_i(R, d)$ denote the sharing (also called the charge) of a receiver q_i. Let $\xi(R, d) = \sum_{q_i \in R} \xi_i(R, d)$ be the total payment collected from all receivers. We call a sharing scheme ξ *reasonable* or *fair* if it satisfies the following criteria:

1. **Budget Balance** (BB): The total payment to all agents should be shared by all receivers; i.e., $\mathcal{P}(R, d) = \xi(R, d)$.
2. **Nonnegative Sharing** (NNS): Any receiver q_i's sharing should not be negative; i.e., $\xi_i(R, d) > 0$.
3. **Cross-Monotone** (CM): For any two receiver sets $R_1 \subseteq R_2$ containing q_i: $\xi_i(R_1, d) \leq \xi_i(R_2, d)$. In other words, for a given network, receiver i's sharing does not increase when more receivers require service.
4. **No-Free-Rider** (NFR): The sharing $\xi_i(R, d)$ of a receiver $q_i \in R$ is at least $1/|R|$ of its unicast sharing $\xi_i(q_i, d)$. Thus, the sharing of any receiver will not be too small.

Notice that the definition of "fair" can be changed for different requirements. For example, a common criterion for a multicast sharing scheme is to maximize *network welfare*: select a subset of receivers such that the network welfare is maximized. Here, *network welfare* is defined as the total valuation of all selected receivers minus the cost of the network providing service. Because in our model we do not consider the receiver's valuation, we focus on only the BB instead of maximizing the network welfare. In literature, the Shapley value (Shapley, 1953) is one of the most commonly used sharing schemes to achieve BB and CM. If the total payment $\mathcal{P}(R, d)$ satisfies nondecreasing and submodular property, then the Shapley value minimizes the worst-case network welfare loss among all sharing schemes that achieve BB and CM. Unfortunately, $\mathcal{P}(R, d)$ is not always submodular.

By assuming a universal multicast tree and publicly known link costs, Feigenbaum *et al.* (2001b) proved that the ELSD cost-sharing scheme is fair. Unfortunately, the ELSD scheme is not always fair if it is used to share the payment. Given a link-weighted graph G = (V, E) and a set of terminals Q⊆V, a tree T is called a Link Steiner tree (LST) if it contains all nodes from Q and has the minimum total link weight among all such trees.

LEMMA 15.3 *The ELSD is not a fair payment-sharing scheme for payment $\mathcal{P}$ defined based on a LST.*

Proof. We prove it by presenting a counterexample using the network shown in Figure 15.5(a). When we consider only one receiver in a LST, we have $\mathcal{P}(q_1, c) = 2.6$ and $\mathcal{P}(q_2, c) = 1.4 + 1.5 = 2.9$. See Figure 15.6 for an illustration. For two receivers

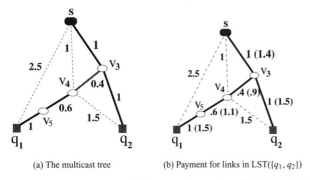

(a) The multicast tree   (b) Payment for links in LST($\{q_1, q_2\}$)

Figure 15.5 Payment calculation based on a LST. © 2006, IEEE.

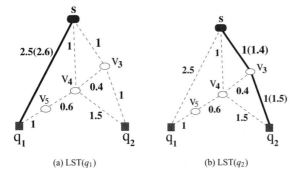

(a) LST(q_1)   (b) LST(q_2)

Figure 15.6 LST(q_1) and LST(q_2) and their corresponding payments. © 2006, IEEE.

q_1, q_2, if we use ELSD to share payments, the sharing by q_1 is $\xi_1(\{q_1, q_2\}, c) = \frac{1.4}{2} + 0.9 + 1.1 + 1.5 = 4.2$, which is larger than its sharing $\xi_1(q_1, c) = 2.6$ when q_1 is the only receiver. Thus, ELSD violates the CM property. It implies that ELSD is not a fair sharing scheme for multicast topology LST. ∎

Furthermore, using the same example, we prove the following lemma:

LEMMA 15.4 *No payment-sharing scheme satisfies both CM and BB for the truthful payment scheme based on a LST.*

Proof. For the sake of contradiction, we assume that a sharing scheme ξ' satisfies both CM and BB. From the property of BB, we have $\xi'_1(q_1, c) = 2.6$, $\xi'_1(q_2, c) = 2.9$, and $\xi'_1(\{q_1, q_2\}, c) + \xi'_2(\{q_1, q_2\}, c) = 6.4$. From CM, we have $\xi'_1(\{q_1, q_2\}, c) \leq \xi'_1(q_1, c) = 2.6$ and $\xi'_2(\{q_1, q_2\}, c) \leq \xi'_2(q_2, c) = 2.9$. Combining these two inequalities, we obtain $6.4 = \xi'_1(\{q_1, q_2\}, c) + \xi'_2(\{q_1, q_2\}, c) \leq 2.9 + 2.6 = 5.5$, which is a contradiction. ∎

Thus, given a certain multicast topology and its corresponding truthful payment scheme, a fair payment-sharing scheme may not exist. It is attractive and important to find the necessary and sufficient condition for the existence of a fair payment-sharing scheme for a given payment scheme.

15.4.2   Truthful Multicast Using Source-Based Tree

This subsection illustrates how to design a truthful multicast protocol with the support of multiprotocol extensions for Border Gateway Protocol-4 (BGP-4) (Bates *et al.*, 2000). We treat every AS i in the network as a node in the graph and assume that it has a fixed cost c_i to relay a unit size of the data-gram for a specific multicast regardless of its downstream links. This could be because the multicast ASs adopt the reverse-path-broadcasting (RPB) scheme or the cost of sending extra copies to other interfaces is negligible. Thus, the network is modeled as a node-weighted graph. All the results presented hereafter also apply to the case in which the network is modeled as a link-weighted graph. We focus on the *source-based tree* in this section and discuss the *shared-based tree* in the next section.

Construct Multicast Tree

Before designing a truthful multicast protocol, we review some technical details of a multiprotocol extension for a BGP (MBGP) including the multicast-tree construction method. MBGP is an extension to the existing border gateway protocol (BGP) (Rekhter and T. Li, 1995). In the BGP, every node v_i stores, for all other nodes v_j, the LCP (the sequence of ASs traversed) from v_i to v_j. Let D be the diameter of the network; i.e., the maximum number of ASs in a LCP. An AS stores $O(nD)$ AS numbers. In the BGP, to perform inter-AS multicast routing, we use the BGP infrastructure that was in place for unicast routing. A multicast routing protocol, such as a multicast independent protocol (MIP) dense mode, uses the multicast BGP database to perform reverse-path forwarding (RPF) lookups for multicast-capable sources.

Thus, given a set of receivers R, the LCP between the source s and each receiver $q_i \in R$ under the reported cost profile d is already in receiver q_i's unicast database. The union of all LCPs between the source and the receivers is called the *least-cost path tree*, denoted by $\mathsf{LCPT}(R, d)$. Every node that is the part of the multicast LCPT has a copy of the tree topology and all data-grams are routed along the tree.

Payment Scheme

It was shown in W. Wang *et al.* (2004) that the direct application of the VCG payment scheme on a LCPT is not truthful. In other words, a node may have incentives to lie about its cost when the VCG payment scheme is used. On the other hand, because the LCPT is formed by the union of the LCPs, by applying Theorem 6 of Kao *et al.* (2005), we can show that the LCPT satisfies the monotone nondecreasing property (MNP). Thus, there exists a truthful payment scheme and the truthful payment can be found accordingly. In Subsection 15.3.2, we studied a design of a truthful mechanism when the LCPT is used for multicast. For each receiver $q_i \in R$, we find the least-cost path $\mathsf{LCP}(s, q_i, d)$ from the source s (say q_0) to q_i and compute an intermediate payment $p_k^{i,0}(d)$ to every node v_k on $\mathsf{LCP}(q_0, q_i, d)$ by using the VCG payment scheme for unicast:

$$p_k^{i,0}(d) = d_k + \omega[\mathsf{LCP}(q_0, q_i, d|^k \infty)] - \omega[\mathsf{LCP}(q_0, q_i, d)].$$

The final payment to a node $v_k \in \mathsf{LCPT}$ is

$$p_k(d) = \max_{q_i \in R} p_k^{i,0}(d). \tag{15.4}$$

The payment to a node is zero if it is not on the LCPT.

Distributed-Payment Algorithm

Remember that the MBGP is only an extension to the BGP that is used for unicast. Usually, the unicast is a dominant activity in the inter-AS routing instead of multicast. Thus, we assume that each AS already implements a truthful payment scheme based on VCG for unicast. Feigenbaum *et al.* (2002) proposed a distributed algorithm to compute the payment $p_k^{i,j}$ for every pair of nodes v_i, v_j and every node v_k on least-cost path $\mathsf{LCP}(v_i, v_j, d)$. Their approach is an extension to the existing BGP routing and converges to a stable state after D_{-k} rounds, where D_{-k} is the maximum possible diameter of graph G after any node k is removed from the network. In their approach, at every node v_i, they store only the length of path $\mathsf{LCP}(v_i, v_j, d)$ for every node v_j, which requires an extra $O(n)$ space. However, in our approach, we require that every node v_i stores all the payments $p_k^{i,j}$ for every possible source node v_j and every node v_k on path $\mathsf{LCP}(v_i, v_j, d)$. Our approach requires an extra space size of $O(\alpha D)$ for every AS, where α is the number of possible source nodes and D is the diameter of the network. Clearly, it avoids the recalculation of every $p_k^{i,j}$ when some nodes' costs are updated. The following algorithm summarizes the distributed payment computing for multicast when $s = q_0$ is the source node:

Algorithm 41 Distributed Payment Computing

1: **for** every receiver q_i **do**
2:     Prepare a control data-gram composed of the payment $p_k^{i,0}$ for every node v_k on path $\mathsf{LCP}(q_0, q_i, d)$.
3:     Sends data-gram containing the payment information to its parent in the tree LCPT.
4: **end for**
5: On receiving a packet containing the payment from its child that originated from receiver q_i, node v_k extracts the payment $p_k^{i,0}$ and sends the data-gram containing all remaining payment information to its parent, if it exists.
6: When a node v_k receives $p_k^{i,0}$ from every downstream receiver q_i, it computes the maximum of them as its final payment.

Now, we discuss the overhead of the distributed multicast payment computation in terms of both communication messages and memory space used in the AS. It is not difficult to observe that every node receives at most r packets of size $O(D)$, where r is the number of the receivers and D is the diameter of the network. For every node v_i, it needs to store for each multicast session S only the final payment p_S, which is negligible. However, sometimes to achieve a high efficiency, node v_k may cache every intermediate payment $p_k^{i,0}$. Even in this case, it only needs an extra $O(r)$ space that is much smaller than the space needed for one session of multicast in a cooperative network. Overall, the

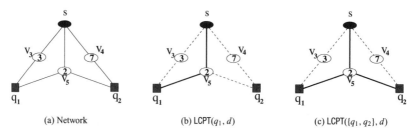

Figure 15.7 ELSD sharing scheme is not fair for payment based on a LCPT. © 2006, IEEE.

overhead needed to calculate the payment is small in terms of both space and network message.

15.4.3   Payment-Sharing Among Receivers

In the literature, the Shapley value (Shapley, 1953) is one of the most commonly used sharing schemes to achieve BB and CM. If the total payment $\mathcal{P}(R, d)$ satisfies a nondecreasing and submodular property, then the Shapley value minimizes the worst-case network welfare loss among all sharing schemes that achieve BB and CM. Here, a payment $\mathcal{P}$ is *submodular* if $\forall R_1 \subseteq Q$ and $R_2 \subseteq Q$, $\mathcal{P}(R_1, d) + \mathcal{P}(R_2, d) \geq \mathcal{P}(R_1 \cup R_2, d) + \mathcal{P}(R_1 \cap R_2, d)$. The *network welfare* is defined as the total valuation of all selected receivers minus the cost of the network providing service. If we apply the Shapley value to multicast payment-sharing, we obtain the following formula for the sharing $\xi_i(R, d)$ of each receiver q_i:

$$\xi_i(R, d) = \sum_{T \subseteq R - q_i} \frac{|T|!(|R| - |T| - 1)!}{|R|!}[\mathcal{P}(T \cup \{q_i\}, d) - \mathcal{P}(T, d)].$$

By assuming a fixed multicast tree and publicly known link costs, Feigenbaum *et al.* (2001b) proved that the ELSD-sharing scheme is the Shapley value. Intuitively, one may want to use ELSD as the payment-sharing scheme. Unfortunately, it will be shown by example that ELSD is not fair when coupled with LCPT. Consider the network shown in Figure 15.7(a). There are two receivers, q_1, q_2. Tree LCPT(q_1, d) is shown in Figure 15.7(b). The total payment to nodes on LCPT(q_1, d) is 3. Consider LCPT$(\{q_1, q_2\}, d)$ illustrated in Figure 15.7(c). The payment to relay only node v_5 is 7. If we apply ELSD to share this payment, the shared payment of receiver q_1 is $(7/2) = 3.5$ when the receiver set is $\{q_1, q_2\}$. Notice that the payment-sharing by q_1 is only 3 when it is the only receiver. Thus, ELSD violates the CM property here. Therefore, some fair sharing scheme other than ELSD should be designed. We can use the Shapley value because of the following lemma:

LEMMA 15.5 *The total payment $\mathcal{P}(R, d)$ for tree LCPT is nondecreasing and submodular with respect to receiver set R.*

Proof. By the definition of LCPT, obviously if $R \subset R' \subseteq Q$, then $\mathsf{LCPT}(d, R) \subseteq \mathsf{LCPT}(d, R')$. Remember that the final payment to a relay agent v_k based on receiver set R is

$$p_k(R, d) = \max_{q_i \in R} p_k^i(d).$$

Observe that $p_k^i(d)$ is not affected by the receiver set R. Thus, for any relay node v_k, if $R \subset R' \subseteq Q$, then $p_k(R, d) \leq p_k(R', d)$. Thus, the total payment to agents on tree $\mathsf{LCPT}(R, d)$ is nondecreasing.

We then prove that the total payment $\mathcal{P}(R, d)$ is a submodular function of set R; i.e., $\forall R_1 \subseteq Q$ and $R_2 \subseteq Q$, $\mathcal{P}(R_1, d) + \mathcal{P}(R_2, d) \geq \mathcal{P}(R_1 \cup R_2, d) + \mathcal{P}(R_1 \cap R_2, d)$. Because $\mathcal{P}(R, d) = \sum_{v_k \in R} p_k(R, d)$, it is sufficient to prove that, $\forall k$,

$$p_k(R_1, d) + p_k(R_2, d) \geq p_k(R_1 \cup R_2, d) + p_k(R_1 \cap R_2, d).$$

We prove this by studying two cases of when the agent v_k is on $\mathsf{LCPT}(R_1 \cap R_2, d)$ or when it is not.

Case 1: Agent v_k is not on $\mathsf{LCPT}(R_1 \cap R_2, d)$. Without loss of generality, assume that v_k is on $\mathsf{LCPT}(R_1 \setminus R_2, d)$. Then, $p_k(R_2, d) = p_k(R_1 \cap R_2, d) = p_k(R_2 \setminus R_1, d) = 0$. Consequently, $p_k(R_1 \cup R_2, d) = \max_{q_i \in R_1 \cup R_2} p_k^i(d) = \max_{q_i \in R_1} p_k^i(d) + \max_{q_i \in R_2 \setminus R_1} p_k^i(d) = \max_{q_i \in R_1} p_k^i(d)$. Therefore, in this case we have

$$p_k(R_1, d) + p_k(R_2, d) = p_k(R_1 \cap R_2, d) + p_k(R_1 \cup x R_2, d).$$

Case 2: Agent v_k is on $\mathsf{LCPT}(R_1 \cap R_2, d)$. Without loss of generality, assume $p_k(R_1, d) \leq p_k(R_2, d)$. Thus,

$$\begin{aligned}
p_k(R_1 \cup R_2, d) &= \max_{q_i \in R_1 \cup R_2} p_k^i(d) \\
&= \max \left\{ \max_{q_i \in R_2} p_k^i(d), \max_{q_i \in R_1 \setminus R_2} p_k^i(d) \right\} \\
&\leq \max \left\{ \max_{q_i \in R_2} p_k^i(d), \max_{q_i \in R_1} p_k^i(d) \right\} \\
&= \max_{q_i \in R_2} p_k^i(d) = p_k(R_2, d).
\end{aligned}$$

On the other hand, we have $p_k(R_2, d) \leq p_k(R_1 \cup R_2, d)$. Thus, $p_k(R_2, d) = p_k(R_1 \cup R_2, d)$. The fact that $R_1 \cap R_2 \subseteq R_1$ implies $p_k(R_1 \cap R_2, d) \leq p_k(R_2, d)$. Therefore, we have

$$p_k(R_1, d) + p_k(R_2, d) \geq p_k(R_1 \cap R_2, d) + p_k(R_1 \cup R_2, d).$$

This finishes the proof. ∎

Consequently, we obtain a sharing scheme satisfying CM and BB by applying the Shapley value. However, for any receiver $q_i \in R$, there are $2^{|R|-1}$ subsets in $R - q_i$. Thus, simply applying the Shapley value directly is computationally intractable when the number of receivers is large. Therefore, another interpretation of the sharing scheme is presented that can be computed efficiently. The basic idea is that a receiver should pay only a proportion of the payment that is due to its existence. Roughly speaking, the payment-sharing scheme works as follows. Notice that a final payment to a node k is the maximum

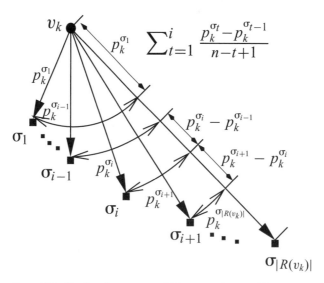

Figure 15.8 Sharing the payment fairly to service providers among receivers.

of payments p_k^i by all receivers. Because different receivers may have different values of payment to agent k, the final payment $\mathcal{P}_k$ should be shared *proportionally* to their values, not *equally* among them (as we do for cost sharing). Figure 15.8 illustrates the payment-sharing scheme that follows. For any node v_k, let $R(v_k)$ be the set of downstream receivers of v_k. Without loss of generality, we assume that $R(v_k) = \{q_{\sigma_1}, q_{\sigma_2}, \ldots, q_{\sigma_{|R(v_k)|}}\}$ such that $0 \leq p_k^{\sigma_1} \leq p_k^{\sigma_2} \leq \cdots \leq p_k^{\sigma_{|R(v_k)|}}$; i.e., $p_k = p_k^{\sigma_{|R(v_k)|}}$. We then divide the payment p_k into $|R(v_k)|$ portions: $p_k^{\sigma_1}, p_k^{\sigma_2} - p_k^{\sigma_1}, \ldots, p_k^{\sigma_i} - p_k^{\sigma_{i-1}}, \ldots, p_k^{\sigma_{|R(v_k)|}} - p_k^{\sigma_{|R(v_k)|-1}}$. Each portion $p_k^{\sigma_i} - p_k^{\sigma_{i-1}}$ is then equally shared among the last $|R(v_k)| - i + 1$ receivers, which have the largest $|R(v_k)| - i + 1$ payments to v_k. Consequently, the portion of the payment to node v_k that should be shared by node q_{σ_i} is

$$\sum_{t=1}^{i} \frac{p_k^{\sigma_t} - p_k^{\sigma_{t-1}}}{n - t + 1}.$$

The total sharing of a receiver q_{σ_i} is then the summation of the total payment-sharing to all relay nodes v_k that are ancestors of this receiver.

We first illustrate how to calculate the payment-sharing by receiver q_1 by using Algorithm 42 for a network represented by Figure 15.7. For node v_5, the two intermediate payments are $p_{v_5}^1 = 3$ and $p_{v_5}^2 = 7$. First, we obtain a rank of these receivers based on the intermediate payments of $\{q_1, q_2\}$. Then, $p_{v_5}^1 = 3$ is equally split between q_1 and q_2, and $p_{v_5}^2 - p_{v_5}^1 = 4$ is charged to q_2 alone. Thus, receiver q_1 is charged $3/2 = 1.5$ and receiver q_2 is charged $1.5 + 4 = 5.5$ in $\mathsf{LCPT}(\{q_1, q_2\}, d)$. Here, q_1's sharing is smaller than sharing 3 when q_1 is the only receiver. This shows that the payment-sharing scheme described in Algorithm 42 is fair for this specific network. Theorem 15.6 shows that the sharing scheme is indeed the Shapley value.

Algorithm 42 Fair Payment-Sharing Scheme for LCPT

1: **for** each node $v_k \in \text{LCPT}(R, d)$ **do**
2: Let $R(v_k)$ be the set of downstream receivers of v_k; i.e., $p_k(d) = \max_{q_i \in R(v_k)} p_k^i(d) = \max_{q_i \in R} p_k^i(d)$.
3: Sort the receivers in $R(v_k)$ according to $p_k^i(d)$ in ascending order. If two or more receivers have the same value, the receiver with the smaller ID ranks first. Let $\sigma = \{\sigma_0, \sigma_1, \ldots, \sigma_{|R(v_k)|}\}$ be the ranking. Here, we add a dummy payment $p_k^{\sigma_0}(d) = 0$ to ranking σ.
4: For a receiver not in $R(v_k)$, its sharing of the payment $p_k(d)$ of node v_k is 0.
5: For a receiver $q_{\sigma_a} \in R(v_k)$, its sharing of the payment $p_k(d)$ to node v_k is

$$f_{\sigma_a}^k(R, d) = \sum_{x=1}^{a} \frac{p_k^{\sigma_x}(d) - p_k^{\sigma_{x-1}}(d)}{|R(v_k)| - x + 1}. \tag{15.5}$$

In other words, for two receivers $q_{\sigma_x}, q_{\sigma_{x+1}}$ who are consecutive in ranking σ, the difference $p_k^{\sigma_{x+1}}(d) - p_k^{\sigma_x}(d)$ is shared by all receivers who rank after $q_{\sigma_{x-1}}$.

6: **end for**
7: The total charge for receiver q_i in LCPT is

$$\xi_i(R, d) = \sum_{v_k \in \text{LCPT}(R, d)} f_i^k(R, d). \tag{15.6}$$

THEOREM 15.6 *The sharing scheme defined by Algorithm 42 is the Shapley value.*

Proof. Remember that the Shapley value for multicast is

$$f_i(R) = \sum_{T \subseteq R \backslash q_i} \frac{|T|!(|R| - |T| - 1)!}{|R|!} [\mathcal{P}(T \cup q_i, d) - \mathcal{P}(T, d)]. \tag{15.7}$$

In other words, the Shapley value of the receiver q_i is $f_i(R)$ given a set of receivers R. Notice that an agent v_k will contribute to $\mathcal{P}(T \cup q_i, d) - \mathcal{P}(T, d)$ if and only if:

1. Agent v_k is an upstream agent of receiver q_i; and
2. $p_k^T(d) < p_k^i(d)$, where $p_k^T(d) = \max_{q_j \in T} p_k^j(d)$.

For fixed T, agent v_k satisfying the preceding two criteria will add a nonnegative value $p_k^i(d) - p_k^T(d)$ to $\mathcal{P}(T \cup q_i, d) - \mathcal{P}(T, d)$. Let $T_{=x}$ be a receiver set with the highest rank in σ that is exactly x. Similarly, we use $T_{<x}$ to denote a receiver set with the highest rank in σ that is less than x. Let $g_k^i(R)$ be payment to agent v_k that is shared by receiver q_i. Assume that q_i is ranked a in the ranking σ when sorting the payment to agent v_k in an increasing order. Then,

$$g_k^i(R) = \sum_{T_{<a} \subseteq R \backslash q_i} \frac{|T_{<a}|!(|R| - T_{<a} - 1)!}{|R|!} p_k^i(d)$$

$$- \sum_{x=0}^{a-1} \sum_{T_{=x} \subseteq R - q_i} \frac{|T_{=x}|!(|R| - |T_{=x}| - 1)!}{|R|!} p_k^{\sigma_x}(d).$$

Let γ be the number of receivers that are not the downstream receivers of v_k. Simplifying the first part of the equation, we get

$$\sum_{T_{<a} \subseteq R - q_i} \frac{|T_{<a}|!(|R| - T_{<a} - 1)!}{|R|!} p_k^i(d)$$

$$= p_k^i(d) \sum_{x=0}^{\gamma+a-1} \frac{x!(|R| - x - 1)!}{|R|!} \binom{a + \gamma - 1}{x}$$

$$= \frac{p_k^i(d)}{|R| - a - \gamma + 1} = \frac{p_k^i(d)}{|R(v_k)| - a + 1}.$$

Simplifying the second part of the equation, we get

$$\sum_{x=0}^{a-1} \sum_{T_{=x} \subseteq R - q_i} \left(\frac{|T_{=x}|!(|R| - |T_{=x}| - 1)!}{|R|!} p_k^{\sigma_x}(d) \right)$$

$$= \sum_{x=0}^{a-1} \left(p_k^{\sigma_x}(d) \sum_{y=0}^{x+\gamma-1} \frac{(y + 1)!(|R| - y - 2)!}{|R|!} \binom{x + \gamma - 1}{y} \right)$$

$$= \sum_{x=1}^{a-1} \frac{p_k^{\sigma_x}(d)}{(|R| - x - \gamma + 1)(|R| - x - \gamma)}$$

$$= \sum_{x=1}^{a-1} \frac{p_k^{\sigma_x}(d)}{(|R(v_k)| - x + 1)(|R(v_k)| - x)}$$

$$= \sum_{x=1}^{a-1} p_k^{\sigma_x}(d) \cdot \left(\frac{1}{(|R(v_k)| - x)} - \frac{1}{(|R(v_k)| - x + 1)} \right)$$

$$= \frac{p_k^{\sigma_{a-1}}(d)}{(|R(v_k)| - a + 1)} - \sum_{x=1}^{a-1} \frac{p_k^{\sigma_x}(d) - p_k^{\sigma_{x-1}}(d)}{(|R(v_k)| - x + 1)}.$$

Combining the preceding two equations, we find that $g_k^i(R)$ is equal to

$$\frac{p_k^i(d)}{|R(v_k)| - a + 1} - \left[\frac{p_k^{\sigma_{a-1}}(d)}{(|R(v_k)| - a + 1)} - \sum_{x=1}^{a-1} \frac{p_k^{\sigma_x}(d) - p_k^{\sigma_{x-1}}(d)}{(|R(v_k)| - x + 1)} \right]$$

$$= \sum_{x=1}^{a} \frac{p_k^{\sigma_x}(d) - p_k^{\sigma_{x-1}}(d)}{(|R(v_k)| - x + 1)}.$$

It shows that the sharing $f_k^i(R)$ computed in Algorithm 42 equals the sharing defined by the Shapley value. ∎

Recall that when the Shapley value is applied to a payment satisfying a submodular and nonincreasing property, the resulting sharing scheme satisfies BB, CM, NNS, and NFR. Thus, we have the following theorem directly:

THEOREM 15.7 *The sharing scheme defined in Algorithm 42 for LCPT satisfies NNS, CM, NFR, and BB.*

15.4.4   Distributed Computing of Payment-Sharing

In practice, we may need to implement a distributed-payment-sharing scheme. In the following, a distributed algorithm is presented that implements the payment-sharing scheme. It requires at most $O(r)$ space for each agent and with $O(rh)$ total messages, where h is the height of the LCPT.

In the distributed algorithm, for any node $v_k \in \mathsf{LCPT}(R, d)$, we not only need its final payment $p_k(d)$, but we also need the intermediate payment $p_k^j(d)$ for every downstream receiver q_j. We assume that this is already available through the distributed-payment-computing scheme (see Algorithm 41). In the distributed-charge scheme, at every node v_k, we use $\vartheta_k[i]$ to store the sum of the charge of v_k's upstream nodes to the receiver q_i. The distributed-payment-sharing scheme is implemented in a topdown fashion from the source to all receivers. It is easy to show that Algorithm 43 indeed correctly computes the payment-sharing of each receiver.

Algorithm 43 Distributed-Payment-Sharing Scheme

1: Initially, source node s sends all its children in LCPT an r-dimensional vector $\vartheta = 0$ for all receivers.
2: Every node v_k in $\mathsf{LCPT}(R, d)$, on receiving a sharing vector $\widetilde{\vartheta}$ from its parent, updates the charge for each of its downstream receivers q_i as $\vartheta_k[i] = \widetilde{\vartheta}[i] + f_k^i[R(v_k)]$. Here, $f_k^i[R(v_k)]$ is calculated according to Algorithm 42.
3: **if** node v_k has at least one downstream receiver, **then**
4:     For every children node v_j, it constructs a charge vector:

$$\vartheta_j = (\vartheta[i_1], \vartheta[i_2], \ldots, \vartheta[i_{|R(v_j)|}]).$$

Here, the charge $\vartheta[i_t]$, $1 \le t \le |R(v_j)|$, is for receiver q_{i_t}, which is a downstream receiver of node v_j. It then sends vector ϑ_j to node v_j.
5: **end if**
6: Every receiver q_i will finally receive a charge $\vartheta[i]$ that is equal to $\xi_i(R, d)$ defined in Eq. (15.6).

15.4.5   Truthful Multicast Using a Share-Based Tree

In the previous subsection, we discussed how to design a truthful multicast protocol by using a MBGP based on a source-based tree LCPT. However, in practice, inter-AS multicast usually uses a shared-based tree (SBT) instead for the following reasons:

1. Multicast routing protocols [e.g., multicast open shortest path first (MOSPF), distance-vector multicast routing protocol (DVMRP), and protocol-independent multicast-sparse mode (PIM-SM)] using a source-based tree are suitable for LAN networks, whereas multicast routing protocols [e.g., PIM-DM (where DM stands for dense mode) and core-based tree (CBT)] using a SBT are more suitable for networks composed of different ASs.

2. The SBT is more scalable than the source-based tree for applications in which every group member could act as a source.

Furthermore, we can show that the size of extra space needed to support the multicast payment calculation could be reduced significantly. Here, we use the PIM-DM as the routing protocol, and the AS should also support the MBGP in order to conduct multicast.

We first review the multicast tree-construction method by the PIM-DM multicast protocol. For a specific multicast group, the PIM-DM protocol specifies a rendezvous point (RP), and the RP maintains a RP tree, which is usually a LCPT that spans all the group members. When any group member wants to send data to the group, it first encapsulates each data packet in a *Register* message and sends it by unicast to the RP for that group. The RP decapsulates the register messages and forwards the enclosed data packet to downstream group members on the shared RP tree. On receiving the data packet from its upstream AS, each intermediate AS further forwards data packets to its downstream ASs. Thus, we can treat the multicast based on a SBT as two separate activities: a unicast from the source to the RP and a multicast with the RP as the virtual source node.

We then discuss how to compute the payment to each relay agent and share these payments among receivers. Let $p_k^R(d)$ denote the payment to a relay node $v_k \in \mathsf{LCPT}(R, d)$ according to our truthful payment scheme [see Eq. (15.2)]. Algorithm 44 presents the truthful payment scheme for multicast based on a SBT.

Algorithm 44 Truthful Payment Scheme for SBT

1: Assume that $s = q_0$ is the RP for a multicast group, and q_i is the source node for a specific multicast session.
2: Let d be the cost vector declared by all relay nodes.
3: Set the receiver set Q as $R \backslash q_i$.
4: Compute the payment $p_k^Q(d)$ for every node v_k on the tree $\mathsf{LCPT}(Q, d)$ rooted at RP s and spanning all receivers Q. Set $p_k^Q(d) = 0$ for other nodes v_k.
5: Calculate the payment $p_k^{i,0}(d)$ for every node v_k on path $\mathsf{LCP}(q_i, q_0, d)$. Set $p_k^{i,0}(d) = 0$ otherwise.
6: **for** each node v_k **do**
7:     $p_k(d) = p_k^Q(d) + p_k^{i,0}(d)$.
8: **end for**

THEOREM 15.8 *The payment scheme defined by Algorithm 44 is truthful.*

The proof of Theorem 15.8 is straightforward and is thus omitted. A distributed-payment-computing protocol similar to Algorithm 41 can be easily designed and is thus omitted here. We now discuss how to share the payments among receivers in Algorithm 45.

THEOREM 15.9 *The payment-sharing scheme defined in Algorithm 45 is fair; i.e., it satisfies NNS, CM, NFR, and BB.*

Algorithm 45 Fair Payment-Sharing Scheme for SBT

1: Set the receiver set $Q = R \backslash q_i$.

2: Share the payment incurred by unicast between q_i and RP equally among all receivers Q. The payment shared by receiver q_k is denoted as $\xi_k^{\text{uni}}(Q, d)$.

3: Share the payment of multicast with source $s = q_0$ and receiver set Q among all receivers according to Algorithm 42. The payment shared by receiver q_k is denoted as $\xi_k^{\text{mul}}(Q, d)$.

4: The final payment shared by the receive q_k is $\xi_k(Q, d) = \xi_k^{\text{uni}}(Q, d) + \xi_k^{\text{mul}}(Q, d)$ when q_i is the source.

The proofs of the correctness of both the preceding method and distributed-payment-sharing computing are similar to the source-based tree case and thus are omitted. Here, we do not consider the source q_i as a receiver, which implies that q_i does not share any payment. If q_i should also be treated as a receiver and share the payment in certain circumstances, we just need to modify the receiver set $Q = R$ instead of $Q = R \backslash q_i$ in Line 1 of Algorithm 45.

15.4.6   Payment-Sharing Among Selfish Receivers

So far, each receiver q_i is assumed to pay its fair share $\xi_i(R, d)$ computed by payment-sharing Algorithm 42. In practice, each individual receiver may have a maximum valuation indicating how much it is willing to pay to receive the information from the source. A receiver will choose to receive the information if and only if the charge is at most its valuation. Furthermore, a receiver could also be *selfish* and *rational*: It will always maximize its profit by manipulating its reported valuation, should it be possible. This makes the multicast design even harder when both the relay agents and the receivers could be selfish. It is well known that a CM *cost-sharing scheme* implies a truthful mechanism for selfish receivers (Moulin and Shenker, 2002). Thus, when each receiver q_i is willing to pay at most ζ_i for the data, we may design a payment-sharing mechanism as follows.

However, we found out that a selected relay agent may have incentives to lie about its relay cost under the payment scheme defined in Algorithm 42. In the following, it is shown that a relay agent could change the payment-sharing of its downstream receivers by either reporting a higher or a lower cost.

Figure 15.9 illustrates such an example of reporting a lower cost. Here, the private valuations of receivers q_1 and q_2 are 12 and 17, respectively. The true costs of links are $c(sv_3) = 5$, $c(sv_4) = 3$, $c(v_3q_1) = 5$, $c(v_4q_2) = 5$, and $c(q_1q_2) = 3$. For the sake of simplicity, we assume that all links (except link v_4q_2) report their costs truthfully in the remaining discussion. Notice that when link v_4q_2 truthfully reports its cost, the multicast tree consists of links sv_3, v_3q_1, sv_4, and v_4q_2, as shown in Figure 15.9(b). In addition, the payments to selected links are $p_{sv_4} = c(sv_3) + c(v_3q_1) + c(q_1q_2) - c(v_4q_2) = 8$, $p_{v_4q_2} = 10$, $p_{sv_3} = 6$, $p_{v_3q_1} = 6$; the payments to all other links are 0. Consider two receivers q_1 and q_2: The payment-sharing by receiver q_1 is $p_{sv_3} + p_{v_3q_1} = 12$, which

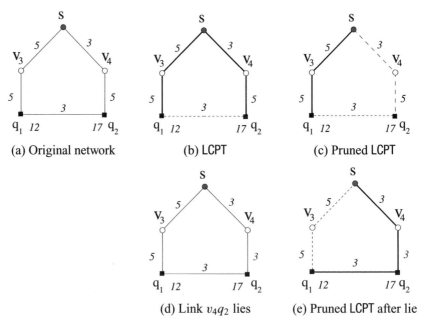

(a) Original network   (b) LCPT   (c) Pruned LCPT

(d) Link v_4q_2 lies   (e) Pruned LCPT after lie

Figure 15.9 A relay agent could lie about its cost (making it lower) to improve its utility using Algorithm 42. © 2006, IEEE.

Algorithm 46 Payment-Sharing for Selfish Receivers R

1: $Q \leftarrow R$.

2: **repeat**

3:    Construct the tree LCPT spanning Q only; i.e., prune out the branches of the original LCPT that do not have receivers in Q.

4:    For each receiver $q_i \in Q$, compute the payment-sharing $\xi_i(Q, d)$ based on the declared costs of all relay agents.

5:    For each receiver $q_i \in Q$, the receiver q_i is removed from Q if $\xi_i(Q, d) > \zeta_i$; i.e., $Q \leftarrow Q - \{q_i\}$ if $\xi_i(Q, d) > \zeta_i$.

6: **until** no receiver is removed in this round

7: All remaining receivers $Q \subseteq R$ will receive the multicast data and pay a sharing $\xi_i(Q, d) \leq \zeta_i$.

---

is not larger than its valuation of 12; the payment-sharing by q_2 is $p_{sv_4} + p_{v_4q_2} = 18$, which is larger than its valuation of 17. Consequently, the receiver q_2 will *not* join the multicast [illustrated in Figure 15.9(c)]. In other words, link v_4q_2 gets payment 0.

Let's see what happens if link v_4q_2 lies about its cost, dropping it down to $3 < c(v_4q_2)$ [illustrated in Figure 15.9(d)]. Figure 15.9(e) shows the multicast tree constructed in this scenario. Notice that when link v_4q_2 reported its cost as 3, the payments to selected links are $p_{sv_4} = 10$, $p_{v_4q_2} = 10$, $p_{q_1q_2} = 4$, and the payments to all other links are 0. It is easy to show that the payment-sharings by receivers q_1 and q_2 are 7 and 16, respectively.

Then, both q_1 and q_2 will join the multicast now. Thus, link v_4q_2 gets a payment of 10 when it lies about its cost, dropping it down to 3.

The preceding example shows that a relay agent could lie about its cost and drop it down to improve its utility. It is not difficult to devise an example such that a relay agent could lie about its cost, moving it up to improve its utility. Notice that it is easy to construct an example such that the payment-sharing and sharing scheme described in Algorithm 46 may return an empty receiver set R although there is a feasible sharing scheme for a nonempty set of receivers. Thus, it is still an open question on how to design a truthful mechanism that also (approximately) maximizes some criteria such as the number of receivers served or the total welfare.

15.5   Existence of Truthful Payment Scheme

Notice that a mechanism-design problem is composed of two parts: The output function $\mathcal{O}$ and a payment function $\mathcal{P}$. For the majority of mechanism-design questions, we often already have some methods $\mathcal{O}$ that will find outputs with certain qualities. A natural question to ask then, is it possible to convert this output method to a truthful mechanism when participating agents are selfish. In other words, given an output method $\mathcal{O}$, is it possible to design a payment scheme $\mathcal{P}$ such that the mechanism $\mathcal{M} = (\mathcal{O}, \mathcal{P})$ is strategy-proof? If such a payment scheme exists, then how is the payment to each agent computed efficiently? This problem has been addressed in several papers (Archer and Tardos, 2001; Archer et al., 2003; Kao et al., 2005; Lehmann et al., 2002).

The first issue is, given an output method, will the VCG payment work? We can generally show that the VCG mechanism not only fails for certain approximation algorithms but also fails for almost all approximation algorithms satisfying certain weak properties. Notice that some algorithms may try to optimize some object function g that is not utilitarian. Thus, $\mu(o, t) = \sum_i v_i(t_i, o)$ is a utilitarian objective function whereas g could be an arbitrary objective function. The following definition is used later.

DEFINITION 15.1 *Let $\mathcal{A}$ be an algorithm that maps the declared type vector into an allowable output. Let T^i be the allowable declared type for agent i and $T = \prod_{i=1}^{n} T^i$ be the space of all possible types. Let $\mathcal{O}^A(T)$ be the range of the output from the space T. We say $\mathcal{A}$ is* local maximal *in its range if, for every pair of t and t' in T such that they only differ in type t_j, $\mu[\mathcal{O}^A(t), t] \geq \mu[\mathcal{O}^A(t'), t]$.*

The following theorem characterizes the VCG mechanism with an approximation algorithm that is truthful:

THEOREM 15.10 *A VCG-based mechanism with an algorithm $\mathcal{A}$ is truthful if and only if $\mathcal{A}$ is* local maximal *in its range.*

Proof. The sufficient condition follows directly from Definition 15.1 about the local maximal. We then prove that if a VCG-based mechanism with an algorithm $\mathcal{A}$ is truthful, then $\mathcal{A}$ is local maximal in its range. Recall that the utility for agent i is $\mu[\mathcal{O}^A(t), t] + h_i(t_{-i})$. From the definition of IC, $\mu[\mathcal{O}^A(t), t] + h_i(t_{-i}) \geq \mu[\mathcal{O}^A(t|^i t_i'), t] + h_i(t_{-i})$.

Thus, $\mu[\mathcal{O}^{\mathcal{A}}(t), t] \geq \mu[\mathcal{O}^{\mathcal{A}}(t|^i t_i'), t]$ for any agent i, which implies the local maximal. ∎

It is not difficult to observe that $\mathcal{A}$ being the local maximal does not imply that it outputs the optimal solution. However, under most circumstances, we can show that the local maximal is closely related to the optimal solution. For example, the following theorem is proved in X.-Y. Li and Wang (2006):

THEOREM 15.11 *If algorithm $\mathcal{A}$ is the local maximal for the set-cover game, then it is either optimal or has an arbitrarily large approximation ratio.*

Because the VCG payment fails for most cases in which the output method is an approximation algorithm, the next natural question is whether we can design a truthful payment scheme coupled with a given output method $\mathcal{O}$. Given the output method $\mathcal{O}$ for a demand game, Kao *et al.* (2005) present a sufficient and necessary condition for the existence of a truthful payment scheme $\mathcal{P}$. Notice that in some applications (e.g., unicast and job scheduling) the valuation v_i of an agent i is of form $-o_i c_i$, where c_i is the cost of the agent when it is selected. Throughout this chapter, we use c_i instead of v_i in the analysis. All results can be easily converted to the case in which v_i is positive, as in an auction.

DEFINITION 15.2 [Monotone nonincreasing property (MNP)] *An output method $\mathcal{O}$ is said to satisfy the MNP if for every agent i and two of its possible costs $c_{i_1} < c_{i_2}$, $\mathcal{O}_i(c|^i c_{i_2}) \leq \mathcal{O}_i(c|^i c_{i_1})$.*

This definition is not restricted to only demand games (Kao *et al.*, 2005). For demand games, this definition implies that if $\mathcal{O}_i(c|^i c_{i_2}) = 1$, then $\mathcal{O}_i(c|^i c_{i_1}) = 1$.

THEOREM 15.12 *For any agent i and any fixed c$_{-i}$ in a demand game with output method $\mathcal{O}$, the following three conditions are equivalent:*

1. *There exists a value $\kappa_i(\mathcal{O}, c_{-i})$ (called a* cut *value) such that $\mathcal{O}_i(c) = 1$ if $c_i < \kappa_i(\mathcal{O}, c_{-i})$ and $\mathcal{O}_i(c) = 0$ if $c_i > \kappa_i(\mathcal{O}, c_{-i})$. When $c_i = \kappa_i(\mathcal{O}, c_{-i})$, $\mathcal{O}_i(c)$ could be either 0 or 1, depending on the tie-breaker of the output method $\mathcal{O}$. Hereafter, we do not consider the tie-breaker scenario in our proofs.*
2. *The output method $\mathcal{O}$ satisfies MNP.*
3. *There exists a truthful payment scheme $\mathcal{P}$ for this demand game.*

Given an output method $\mathcal{O}$ that satisfies the MNP, Kao *et al.* (2005) present several methods to efficiently compute the payment to each individual agent when the output method falls into some categories (e.g., a round-based method, a simple combination of other methods whose payment can be easily computed) and also some recursive compounds of functions.

15.6   Further Reading

15.6.1   Reputation-Based Approaches

Reputation-based methods have been widely proposed to solve the selfishness in networks for several years. The basic idea is that everyone has some credit, and when the user conducts something that benefits the overall network performance, its credit increases and vice versa. Different schemes have methods that have been deployed to monitor the behavior of the network users and to calculate the credit of the users.

How to achieve cooperation among selfish terminals in a network was previously addressed in Blazevic *et al.* (2001), Buttyan and Hubaux (2000, 2003), Jakobsson *et al.* (2003), Marti *et al.* (2000), and Srinivasan *et al.* (2002, 2003). In Marti *et al.* (2000), terminals that agree to relay traffic but do not are termed as *misbehaving*. Their approach is to avoid routing through these misbehaving terminals. In Blazevic *et al.* (2001), Buttyan and Hubaux (2000, 2003), and Jakobsson *et al.* (2003), a secure mechanism to stimulate terminals to cooperate is presented, and the key idea behind these approaches is that terminals providing a service should be remunerated, whereas terminals receiving a service should be charged. Each terminal maintains a counter in a tamper-resistant hardware module, which is decreased by 1 when the terminal originates a packet and increased by 1 when the terminal forwards a packet.

In Marti *et al.* (2000), nodes that agree to relay traffic but do not are termed as *misbehaving*. They used *Watchdog* and *Pathrater* to identify misbehaving users and avoid routing through these nodes. *Watchdog* runs on every node, keeping track of how the other nodes behave; *Pathrater* uses this information to calculate the route with the highest reliability. Notice that this method ignores the reason why a node refused to relay the transit traffics for other nodes. A node will be wrongfully labeled as misbehaving when its battery power cannot support many relay requests and thus refuses to relay. It also does not provide any incentives to encourage nodes to relay the message for other nodes.

Buttyan and Hubaux (2003) focused on the problem of how to stimulate selfish nodes to forward the packets for other nodes. Their approach is based on a so-called *nugget counter* in each node. A node's counter is decreased when sending its own packet and is increased when forwarding other nodes' packets. All counters should always remain positive. To protect the proposed mechanism against misuse, they presented a scheme based on a trusted and tamper-resistant hardware module in each node, which generates cryptographically protected security headers for packets and maintains the nugget counters of the nodes. They also studied the behavior of the proposed mechanism analytically and by means of simulations and showed that it indeed stimulates the nodes for packet-forwarding. In other words, each node has to pay one nugget for sending a packet and is rewarded with one nugget for relaying a packet.

Buttyan and Hubaux (2000) still use a nugget counter to store the nuggets and they also use a fine that decreases the nugget counter to prevent the node from not relaying the packet. They use a packet purse model to discourage the user from sending useless traffic and overloading the network. The basic idea presented in Buttyan and Hubaux (2000) is similar to that of Buttyan and Hubaux (2003) but different in the implementation.

In Srinivasan *et al.* (2002), two acceptance algorithms are proposed. These algorithms are used by the network nodes to decide whether to relay traffic on a per-session basis. Their goal is to balance[1] the energy consumed by a node in relaying traffic for others with energy consumed by other nodes to relay its traffic and to find an optimal trade-off between energy consumption and session-blocking probability. By taking decisions on a per-session basis, the per-packet processing overhead of previous schemes is eliminated. In Srinivasan *et al.* (2002), a distributed and scalable acceptance algorithm called GTFT is proposed. They proved that GTFT results in Nash equilibrium and the system converges to the rational and optimal operating point. It is emphasized here, however, that all the preceding algorithms are based on heuristics and lack a formal framework to analyze the optimal trade-off between lifetime and throughput. More important, they assumed that each path is h hops long and the h relay nodes are chosen with equal probability from the remaining $n - 1$ nodes, which is unrealistic.

Salem *et al.* (2003) presented a novel charging and rewarding scheme for packet forwarding in multihop cellular networks. In their network model, there is a base station to forward the packets. They use symmetric cryptography to cope with the lying. To counter several possible attacks, it precharges some nodes and then refunds them only if a proper acknowledgment is received. Their basic payment scheme is still based on nuggets.

Jakobsson *et al.* (2003) described an architecture for fostering collaboration between selfish nodes of multihop cellular networks. From this architecture, they provided mechanisms based on a per-packet charge to encourage honest behavior and to discourage dishonest behavior. In their approach, all packet originators attach a payment token to each packet, and all intermediaries on the packet's path to the base station verify whether this token corresponds to a special token, called a *winning ticket*. Winning tickets are reported to nearby base stations at regular intervals. The base stations therefore receive both reward claims (which are forwarded to some accounting center) and packets with payment tokens. After verifying the validity of the payment tokens, base stations send the packets to their desired destinations, over the backbone network. The base stations also send the payment tokens to an accounting center. Their method also involves some traditional security methods, including auditing, node-abuse detection, and encryption.

15.6.2   Algorithmic Mechanism Design

Routing has been an important part of the algorithmic mechanism design from the very beginning. Nisan and Ronen (1999) provided a polynomial-time strategy-proof mechanism for optimal unicast route selection in a centralized computational model. In their formulation, the network is modeled as a graph $G = (V, E)$. Each link e in the

---

[1] It is impossible to strictly balance the number of packets a node has relayed for other nodes and the number of packets of this node relayed by other nodes because in a wireless ad hoc network, the majority of the packet transmissions are relayed packets. For example, consider a path of the h hops. $h - 1$ nodes on the path relay the packets for others. If the average path length of all routes is h, then a $1 - 1/h$ fraction of the transmissions is transit traffic.

graph is an agent and has a private type t_e, which represents the cost of sending a unit amount of data along this link. Their mechanism is a VCG mechanism that uses the LCP as its output. Feigenbaum *et al.* (2002) then addressed truthful low-cost routing in a different network model: Each terminal v_k incurs a transit cost c_k for each transit packet it carries. Their mechanism is also a VCG mechanism. They gave a distributed method such that each terminal i can compute a payment $p_{ij}^k > 0$ to terminal k for carrying the transit traffic from terminal i to terminal j if terminal k is on LCP(i, j). Anderegg and Eidenbenz (2003) recently proposed a similar routing protocol for WANs based on the VCG mechanism again. They assumed that each link has a cost and each terminal is a selfish agent. W. Wang and Li (2004b, 2005) proposed a time-optimal method to compute the VCG payment with *all* relay agents in time $O(n \log n + m)$ for unicast in a node-weighted network of n nodes and m links.

Feigenbaum *et al.* (2001b), by assuming a *fixed* multicast structure, designed a strategy-proof mechanism that selects a subset of receivers (each with a privately known willing payment) and then shares the publicly known cost of the multicast tree providing the service among the selected receivers so that BB is achieved. W. Wang *et al.* (2004) studied the payment design for several multicast methods using a number of multicast trees. W. Wang *et al.* (2005b) then presented a general framework for designing truthful mechanisms for multicast and also studied how to share the payments to relay agents fairly among all receivers. Kao *et al.* (2005) proposed several methods to design truthful payment schemes $\mathcal{P}$ when given the output method $\mathcal{O}$ that is monotonic for a binary demand game. Recently, W. Wang *et al.* (2005a) studied the mechanism design for multicast with a differentiated service: The cost of providing a relay service by a relay agent, depending on the quality required by receivers, and each receiver will have its own quality requirement. They showed that previous multicast tree-construction methods do not imply any truthful payment scheme. They then proposed a new DiffServ multicast tree-construction method with the same approximation ratio over the total cost, while at the same time ensuring a truthful payment scheme.

Recently, W. Wang *et al.* (2006a) proposed novel solutions, called optimal unicast routing systems (OURSs), for unicast routing in networks consisting of selfish terminals: To alleviate the inevitable overpayment problem (and, thus, economic inefficiency) of the VCG mechanism, they designed a mechanism that results in Nash equilibria rather than strategy-proofness (using a weakly dominant strategy). In addition, they systematically studied the *unicast routing system* in which both the relay terminals and the service requestor (either the source or the destination nodes or both) could be selfish. To the best of the author's knowledge, this is the *first* paper that presents *socially efficient* unicast routing systems with a proven performance guarantee.

The main contributions of OURS are as follows: (1) For the principal model in which the service requestor is not selfish, they propose a mechanism that provably creates incentives for intermediate terminals to cooperate in forwarding packets for others. Their mechanism substantially reduces the overpayment by using Nash equilibrium solutions as opposed to strategy-proof solutions. They then study a more realistic case in which the service requestor can act selfishly. (2) They first show that if we insist on the requirement of being strategy-proof for the relay terminals, then no system can guarantee that the central authority can retrieve at least $1/n$ of the total payment. (3) They then present a strategy-proof unicast system that collects $1/2n$ of the total payment, which is thus asymptotically optimum. (4) By requiring only Nash equilibrium solutions, they propose a system that creates incentives for the service requestor and intermediate terminals to correctly follow the prescribed protocol. More important, the central authority can retrieve at least half of the total payment.

When VCG mechanisms are applied to complex problems such as multicast, a problem emerges: Even finding the optimal outcomes is computationally intractable. A critical observation made by Nisan and Ronen (2000) and other researchers is that if the optimal output is replaced with a polynomial-time computable one, then the mechanism using the payment based on the VCG mechanism is not guaranteed to be truthful. This phenomenon is almost universal. To address this, Nisan and Ronen (2000) introduced a notion of feasible truthfulness that captures the limitation on agents imposed by their own computational limits. They showed that under reasonable assumptions on the part of the agents, it is possible to turn any VCG-based mechanism into a feasibly truthful one by use of an additional appeal mechanism. In this chapter, we used a totally different approach by using a payment scheme other than the VCG scheme, and we do *not* assume any computational limits on the agents.

15.7   Conclusion and Remarks

In this chapter, we studied how to conduct efficient multicast routing in *selfish* networks by assuming that each terminal will incur a cost when it transmits some data. For each of the widely used structures for multicast, we designed a strategy-proof multicast mechanism such that each agent maximizes its profit when it truthfully reports its cost and when every terminal always forwards others' traffic. The structure based on virtual relay (using MST connecting receivers) is also studied in W. Wang *et al.* (2004). The structures studied in this chapter are the LCPT, the VMST, and the NST. Extensive simulations were conducted to study the practical performances of the proposed protocols.

Notice that in the chapter, only the payment to one session is discussed. When the session is to be repeated, a natural question is: How much we should pay for later sessions? One may argue that we have to pay each agent only its true cost for later sessions. Unfortunately, this will not work for selfish agents. When an agent knows that its payment will be its actual cost for later sessions, it could raise its cost upward. By doing this, it may lose for the first session, but the gains in the later sessions will compensate for the initial loss.

There are many unsolved challenges for designing protocols when participating agents could be selfish. First, we would like to design algorithms that can compute these payments in asymptotically optimum-time complexities. Second, in this chapter, we studied only the tree-based structures for multicast. Practically, mesh-based structures may be more needed for wireless networks to improve the fault tolerance of the multicast. We would like to know whether we can design a strategy-proof multicast mechanism for some mesh-based structures used for multicast. Third, when all of the tree constructions and payment calculations are performed in a centralized way, we would like to study how to design distributed algorithms. Notice that in Feigenbaum *et al.* (2002) and W. Wang and Li (2004b), distributed methods were developed for truthful unicast by use of some cryptography primitives. Another important issue is how to ensure the truthful implementation of the payment computation if the payment computation is done in a distributed way. Several works in the literature have discussed this for the unicast problem (Shneidman and Parkes, 2004, W. Wang and Li, 2005); we would like to extend their ideas and schemes for the multicast case.

This chapter has laid down a building block for further research in designing truthful routing protocols for selfish networks. In all of these protocols, we assume that the source of the multicast will pay the relay terminals to compensate their costs. The source terminal will not charge the receivers for getting the data. As future work, we have to consider the BB of the source if the receivers have to pay the source for getting the data; we also have to consider the fairness of payment-sharing when the receivers will share the total payments with all relay terminals on the multicast structure. Another important task is to study how to implement the protocols proposed in this chapter in a distributed manner.

There are many interesting and important issues that have not been discussed and thus are left for further study. Just a few are listed here.

Collusion

Throughout this chapter, we assume that all agents will not collude together to manipulate the protocol. It is interesting to study what happens when agents can collude and how to find truthful mechanisms that are resistent to collusion. We already know that there is no truthful multicast protocol that can prevent collusion between any pair of relay agents, following a previous result (W. Wang and Li, 2004b) for unicast.

Truthful Distributed Implementation

One thing we should notice is that these agents running the distributed algorithms are indeed noncooperative. How to ensure that they implement the *correct* distributed algorithm we designed also is an important question we have to consider. See W. Wang and Li (2004b) for a previous approach to unicast.

Repeated Games

So far, we assumed that the session is performed once. A natural question is: How should we pay the relay agents and charge the receivers when the multicast game is to be repeated for several sessions? When we know the private cost of each relay agent, should we just pay each relay agent its declared cost starting from the second session? If we do so, clearly the selfish relay agent will increase its declared cost to improve its later benefit, although this may reduce its benefit in the first session.

Nash Design

One thing we should point out is that algorithmic mechanism design is not the only way to deal with selfishness. Many works in the literature use Nash equilibrium, a state on which no agent can improve its utility by unilaterally deviating from its current strategy when other agents keep their strategies. Because Nash equilibrium has a weak requirement, it often can achieve a wider variety of outcomes. We leave it as future work to design multicast protocols using Nash equilibrium instead of truthful algorithmic mechanism design.

Central Authority

One intrinsic assumption of this chapter is that there are some central authorities that pay the relay agents. It is natural to ask this question: What will happen if this kind of central authority does not exist?

Dynamism and Mobility

In this chapter, we fix the network topology and study how to design the multicast routing based on it. In wireless networks, sometimes the user will not only join and leave frequently but also will move according to certain patterns. Thus, we also need to address how to improve the strategy-proof multicast protocol to deal with the dynamism and mobility of the network in the future work.

Problems

15.1 Why could selfish behavior be a problem for wireless ad hoc networks? What are the currently used strategies to prevent selfish behaviors of nodes? What are the advantages of using a game-theoretical approach and disadvantages of using game theory?

15.2 What are the differences between a dominant strategy and Nash equilibrium? What are the differences between pure Nash equilibrium and mixed Nash equilibrium? What is subgame Nash equilibrium?

15.3 What is incentive compatibility and individual rationality?

15.4 Why can we not use the VCG mechanism for an arbitrary question? What are the major constraints for the application of VCG mechanisms?

15.5 In Subsection 15.3.3, we designed a truthful mechanism for multicast by using a NST. Design an efficient polynomial-time algorithm to compute the cut value $B_k^i(d_{-k})$ for each selected node v_k. Assume that you know the cost of each node is an integer in the range $[1, B]$ for some known upper bound B. Can you design an efficient algorithm to compute the payment for each selected agent for multicast by using a NST?

15.6 Prove Theorem 15.5.

15.7 What are the differences between fair sharing of multicast cost and fair sharing of the payment for truthful multicast?

15.8 What is the social efficiency of a mechanism? Can you design a strategy-proof mechanism that is budget-balanced and maximizes the social efficiency?

15.9 Assume that Algorithm 46 is used to select a relay agent and share payment. Devise an example such that a relay agent could lie about its cost and raise it to improve its utility. Design another example to show that Algorithm 46 may return an empty receiver set R although there is a feasible sharing scheme for a nonempty set of receivers.

16   Joint Routing, Channel Assignment, and Link Scheduling

16.1   Introduction

Wireless multihop radio networks such as ad hoc, mesh, or sensor networks are formed of autonomous nodes communicating via radio. Wireless networks have drawn lots of attention in recent years because of their potential applications in various areas. For example, wireless mesh networks (WMNs) are being used as the last mile for extending Internet connectivity for mobile nodes. Many U.S. cities (e.g., Medford, Oregon; Chaska, Minnesota; and Gilbert, Arizona) have already deployed mesh networks. AWA, the Spanish operator of WLANs, will roll out commercial WLANs and mesh networks for voice and data services. Several companies, such as MeshDynamics, have recently announced the availability of multihop, multiradio mesh-network technology. These networks behave almost like wired networks because they have infrequent topology changes, limited node failures, and so forth. For WMNs or WSNs, the aggregate traffic load of each routing node also changes infrequently. A unique characteristic of wireless networks is that the radio sent out by a wireless terminal will be received by all the terminals within its transmission range and also possibly cause signal interference to some terminals that are not intended receivers. In other words, the communication channels are shared by the wireless terminals. Thus, one of the major problems facing wireless networks is the reduction of capacity that is due to interference caused by simultaneous transmissions. Using multiple channels and multiple radios can alleviate but not eliminate the interference. This raises the scalability issue of WMNs.

Gupta and Kumar (1999) studied the asymptotic capacity of multihop wireless networks for two different models. When each wireless node is capable of transmitting at W bits per second using a fixed range, the throughput obtainable by *each* node for a randomly chosen destination is $\Theta(\frac{W}{\sqrt{n \log n}})$ bits per second under a noninterference protocol, where n is the number of nodes. If nodes are optimally assigned and the transmission range is optimally chosen, even under optimal circumstances, the throughput is only $\Theta(\frac{W}{\sqrt{n}})$ bits per second for each node. Similar results also hold for a physical interference model. Recently, several papers (Grossglauser and Tse, 2001; J. Li *et al.*, 2001) further studied the capacity of wireless networks under different models. Kyasanur and Vaidya (2005a) studied the capacity region on random multihop, multiradio, multichannel wireless networks when there are a total c channels available and each node has $m \leq c$ wireless interfaces. In another aspect, several papers (Alicherry *et al.*, 2005; Kodialam and Nandagopal, 2003) recently studied how to satisfy a certain traffic-demand vector from all wireless nodes by joint routing, link scheduling, and channel assignment under certain wireless interference models.

In this chapter, we also study throughput optimization (under certain fairness constraints) via joint routing, link scheduling, and *dynamic* channel assignment. Unlike the previous methods [especially the most related (Alicherry *et al.*, 2005)], we assume that each wireless node has a set of channels that it can operate (because of either the availability of the channels in the neighborhood when a spectrum is used opportunistically or the constraints that are due to the wireless interface cards). Further, we study this problem under a variety of interference models and assume that different nodes may have different transmission radii and interference radii. We prove a constant-approximation ratio of our method for each of these interference models. We also assume that each wireless interface is capable of switching channels between different time slots and is capable of combining multiple consecutive channels into a single channel between different time slots with the new combined bandwidth equivalent to the total bandwidth of the combined channels (Kilpatrick *et al.*, 2006). Furthermore, we studied this problem under two different channel-combining conditions: (1) with channel combining, and (2) without channel combining; and we studied the effect of channel combining on the maximum throughput (under certain fairness constraints).

16.2   System Model and Assumptions

Wireless interference issues have been studied extensively recently because it is widely believed that reducing the interference can increase the overall performance of a wireless network. There are different approaches to reduce the interference, including the scheduling on the MAC layer, route selection on the routing layer, channel assignment if multichannels are available, and power control on the physical layer. In this section, we first introduce our network system model; then, we discuss in detail the interference models that we use and define the problem that we study in this chapter.

16.2.1   Network System Models

In this chapter, we assume that there is a set V of n communication terminals deployed in a plane. The complete communication graph is a *directed* graph $G = (V, E)$, where $V = \{v_1, \ldots, v_n\}$ is the set of terminals and E is the set of possible *directed* communication links. Let $\mathbf{E}^-(u)$ denote the set of directed links that end at node u [i.e., (w, u)], and let $\mathbf{E}^+(u)$ denote the set of directed links that start at node u [i.e., (u, v)].

Every terminal v_i has a transmission range $R_T(i)$ such that the necessary condition for a terminal v_j to receive correctly the signal from v_i is $\|v_i - v_j\| \leq R_T(i)$, where $\|v_i - v_j\|$ is the Euclidean distance between v_i and v_j. Notice that $\|v_i - v_j\| \leq R_T(i)$ is *not* the sufficient condition for $(v_i, v_j) \in E$. Some links do not belong to G because of either the physical barriers or the selection of routing protocols. To the best of my knowledge, only Kumar *et al.* (2005) used a model similar to ours. We always use $\mathbf{L}_{i,j}$ to denote the directed link (v_i, v_j) hereafter. Each terminal v_i also has an interference range $R_I(i)$ such that terminal v_j is interfered by the signal from v_i if $\|v_i - v_j\| \le R_I(i)$ and v_j is not the intended receiver. The interference range $R_I(i)$ is not necessarily the same as the transmission range $R_T(i)$. Typically, $R_T(i) \le R_I(i) \le cR_T(i)$ for some constant c. We call the ratio between them the *interference–transmission ratio* for node v_i, denoted as $\gamma_i = \frac{R_I(i)}{R_T(i)}$. In practice, $2 \le \gamma_i \le 4$. For all wireless nodes, let $\gamma = \max_{v_i \in V} \frac{R_I(i)}{R_T(i)}$.

We assume that the WMN is a multihop, multiradio, and multichannel network. Let $\mathbb{F} = \{\mathbf{f}_1, \mathbf{f}_2, \ldots, \mathbf{f}_K\}$ be the set of K orthogonal channels (typically, frequency channels or CDMA codes) that can be used by all wireless nodes. For example, for IEEE 802.11 networks, $K = 11$. Each wireless terminal u is equipped with $\mathcal{I}(u) \ge 1$ radio interfaces; namely, $\kappa(u, 1), \kappa(u, 2), \ldots, \kappa[u, \mathcal{I}(u)]$. We assume that each radio interface $\mathcal{I}(u)$ is capable of combining the set of consecutive channels $\mathbf{f}_a, \mathbf{f}_{a+1}, \ldots, \mathbf{f}_b$ into one combined channel $\mathbf{f}_{ab}$, where $a \le b \le K$. Channel-combining allows the scheduler to assign multiple channels on a radio interface at a give time slot t to a single user. The scheduler will dynamically change the number of combined channels assigned to a user to achieve fairness and at the same time provide the user with higher throughput. In the case of $a = b$, the scheduler will assign one channel to the user and no channel-combining takes place at the radio interface. Let $\mathcal{F} = \{\mathbf{f}_{a_1 b_1}, \mathbf{f}_{a_2 b_2}, \ldots, \mathbf{f}_{a_m b_m}\}$ be the set of m (not necessarily pairwise orthogonal) channels that can be used by all wireless nodes, where $b_i \ge a_i \, \forall i \in [1, m]$ and $m \le K(K + 1)/2$. $\mathcal{F}$ includes both combined and noncombined channels. For the combined channels, we denote the cocombined-channel interference by $\mathbf{I}_{\text{set}}(\mathbf{f}_{ab}, \mathbf{f}_{pq})$, which is defined as:

$$\mathbf{I}_{\text{set}}(\mathbf{f}_{ab}, \mathbf{f}_{pq}) = \begin{cases} 1, & \text{if } \mathbf{f}_{ab} \text{ and } \mathbf{f}_{pq} \text{ interfere} \\ 0, & \text{otherwise.} \end{cases}$$

For example, when no adjacent channel interference is defined, $\mathbf{f}_{ab}$ and $\mathbf{f}_{pq}$ *do not* interfere iff $a \le b$, $p \le q$, and $(p > b) \vee (a > q)$.

Traditionally, in the literature (e.g., Alicherry *et al.*, 2005; Kodialam and Nandagopal, 2003; Kumar *et al.*, 2005) it is assumed that a wireless interface card can operate on *all* channels from $\mathbb{F}$. However, here we assume a general case in which each wireless interface can operate on only a subset of channels from $\mathbb{F}$ because of the hardware constraints, and the same assumption is valid for $\mathcal{F}$. More specifically, we let $\mathcal{F}(u, i)$ be the set of channels (combined and noncombined) that can be used by the ith wireless interface $\kappa(u, i)$ for node u, where $1 \le i \le \mathcal{I}(u)$. Let $\delta(u, i, \mathbf{f}_{ab}) \in \{0, 1\}$ be the indicator function of whether the ith wireless interface of node u can use channel $\mathbf{f}_{ab}$. Thus, the channels that can be used by a wireless node u are represented by a subset $\mathcal{F}(u) \subset \mathcal{F}$, where $\mathcal{F}(u) = \bigcup_{1 \le i \le \mathcal{I}(u)} \mathcal{F}(u, i)$. Define $\delta(u, \mathbf{f}_{ab}) = \bigvee \delta(u, i, \mathbf{f}_{ab})$, $\forall i \in [1, \mathcal{I}(u)]$ for node u.

For notational convenience, we use $\mathcal{F}(e)$ to denote the set of common channels among $\mathcal{F}(u)$ and $\mathcal{F}(v)$ for any link $e = (u, v)$, and we let $\delta(e, \mathbf{f}_{ab}) \in \{0, 1\}$ be the indicator function of whether a channel $\mathbf{f}_{ab}$ can be used by a link e. Obviously, $\delta(e, \mathbf{f}_{ab}) = \delta(u, \mathbf{f}_{ab})\delta(v, \mathbf{f}_{ab})$ for a link $e = (u, v)$. For each link $e = (u, v)$ operating on a channel $\mathbf{f}_{ab} \in \mathcal{F}(e)$, we denote by $\mathbf{c}(e, \mathbf{f}_{ab})$ the rate for link e. This is the maximum rate at which mesh node u can communicate with mesh node v in one-hop communication by using channel $\mathbf{f}_{ab}$. Clearly, the maximum rate that can be supported by a link e is at

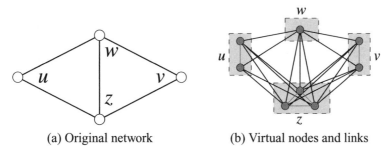

(a) Original network   (b) Virtual nodes and links

Figure 16.1 Virtual nodes and virtual links defined by splitting nodes. The virtual nodes in one shaded region correspond to a node in the original network.

most $\sum_{\mathbf{f}_{ab} \in \mathcal{F}(e)} \mathbf{c}(e, \mathbf{f}_{ab})$. Notice that the links are directed; thus, the capacity could be asymmetric; i.e., $\mathbf{c}[(u, v), \mathbf{f}_{ab}]$ may not be the same as $\mathbf{c}[(v, u), \mathbf{f}_{ab}]$.

Our approach to optimizing throughput is to visualize a multiradio node as a collection of fully connected multiple virtual nodes with infinite bandwidth links between them; i.e., node u with $\mathcal{I}(u)$ radio interfaces can be seen as a group of $\mathcal{I}(u)$ fully connected virtual nodes $\widehat{u}_i, i \in [1, \mathcal{I}(u)]$. Each virtual node $\widehat{u}_i$ has exactly radio interface i to connect to neighboring virtual nodes $\widehat{v}_k$. Figure 16.1 is an illustration of an example in which $\mathcal{I}(u) = 2, \mathcal{I}(w) = 1, \mathcal{I}(v) = 2$, and $\mathcal{I}(z) = 3$. We define two types of virtual links for each virtual node $\widehat{u}_i$:

1. *Directed external virtual links* connect virtual node $\widehat{u}_i$ with other nodes outside its group; e.g., virtual node $\widehat{v}_k$. This type of link has limited capacity and may cause interference with other external virtual links using the same channel.
2. *Internal virtual links* connect virtual node $\widehat{u}_i$ with all virtual nodes $\widehat{u}_q$ in its group, where $q \neq i$ and $q \in [1, \mathcal{I}(u)]$. This link has infinite capacity, does not interfere with any other link, and resembles the internal switching of data from one radio interface to another radio interface by a node.

Now, we define directed link $e(u, v)$ as the superposition of directed virtual links $\widehat{e}_{ij}(u, v)$, where $\widehat{e}_{ij}$ is defined as the external virtual link from $\widehat{u}_i$ to $\widehat{v}_j$ using radio interface i of node u and j of node v, where $1 \leq i \leq \mathcal{I}(u)$ and $1 \leq j \leq \mathcal{I}(v)$. We define $\delta(\widehat{e}_{ij}, \mathbf{f}_{ab}) \in [0, 1]$ as the indicator function of whether virtual link $\widehat{e}_{ij}$ can use $\mathbf{f}_{ab}$ for communication. Let $\mathbb{R}cc_{ij}$ be the set of common channels between radio interface i of node u and radio interface j of node v. Then, $\widehat{e}_{ij}$ can have up to $|\mathbb{R}cc_{ij}|$ choices of channels for communication. For each virtual link $\widehat{e}_{ij} = (u, v)$, we denote by $\mathbf{c}(\widehat{e}_{ij}, \mathbf{f}_{ab})$ the capacity of the virtual link using channel $\mathbf{f}_{ab}$. For simplicity, we also use $\widehat{e}$ to denote such a communication link. We denote the fraction of time that virtual link $\widehat{e}$ will be actively using $\mathbf{f}_{ab}$ by $\alpha(\widehat{e}, \mathbf{f}_{ab})$, the capacity of $\widehat{e}$ using $\mathbf{f}_{ab}$ as $\mathbf{c}(\widehat{e}, \mathbf{f}_{ab})$, the subset of channels that can be used by $\widehat{e}$ as $\mathbb{R}cc(\widehat{e})$, the set of directed external virtual links that end at node u as $\widehat{\mathbf{E}}^-(u)$, and the set of directed external virtual links that start at node u as $\widehat{\mathbf{E}}^+(u)$.

We also assume that among the set V of all wireless nodes, some have gateway functionality and provide the connectivity to the Internet. For simplicity, let $\mathcal{S} = \{\mathbf{s}_1, \mathbf{s}_2, \ldots, \mathbf{s}_g\}$ be the set of g gateway nodes, where $\mathbf{s}_i$ is actually node v_{n+i-g}. All other wireless nodes v_i (for $1 \leq i \leq n - g$) are called *ordinary* wireless nodes. We assume that the gateway nodes will *not* act as a relay node for a pair of ordinary wireless nodes. Each ordinary node u will aggregate the traffic from all its users and then route them to the Internet through some gateway nodes. We use $\ell_O(u)$ to denote the total aggregated outgoing traffic for its users by node u and $\ell_I(u)$ to denote the total aggregated incoming traffic for its users by node u. We mainly concentrate on one of the traffic flows in this chapter, say, incoming traffic. For notation simplicity, we use $\ell(u)$ to denote such a load for node u. Notice that traffic $\ell(u)$ is not requested to be routed through a specific gateway node; neither is it requested to be using a single routing path. Our results can be easily extended to deal with both incoming and outgoing traffic by defining routing flows for both traffic flows separately.

To schedule two links at the same time slot, we must ensure that the schedule will avoid interference. Two different types of interference have been studied in the literature: namely, *primary interference* and *secondary interference*. Primary interference occurs when a node transmits and receives packets at the same time; secondary interference occurs when a node receives two or more separate transmissions. In this chapter, we study throughput maximization under various interference models, such as the protocol-interference model (PrIM) (Gupta and Kumar, 1999), the fixed protocol-interference model (fPrIM) (W. Wang *et al.*, 2006b), the RTS/CTS model, and the transmitter-interference model (TxIM) (Yi *et al.*, 2003).

It is known that joint routing, link scheduling, and channel assignment will improve overall network throughput performance. Traditionally, in wired networks, maximum throughput can be found via the simple maximum-flow solution. This technique cannot be directly applied to wireless networks because wireless interference may cause some flows to be unschedulable. Further complicating the study is that the channel that can be used by a wireless interface of a node could be dynamically or statically assigned.

To study this cross-layer optimization, we model the interference of multichannel, multiradio, multihop wireless networks by a conflict graph. Given a communication graph $G = (V, E)$, we use the *conflict graph* (e.g., Jain *et al.*, 2003) F_G to represent the interference in G. Each vertex (denoted by $\mathbf{L}_{i,j}$) of F_G corresponds to a directed link (v_i, v_j) in the communication graph G. There is an *edge* between vertex $\mathbf{L}_{i,j}$ and vertex $\mathbf{L}_{p,q}$ in F_G if and only if $\mathbf{L}_{i,j}$ conflicts with $\mathbf{L}_{p,q}$ because of interference. Recall that whether two links conflict at a certain time depends on (1) the channels they use for communication, (2) the geometry locations of these two links, and (3) the interference model used underneath (e.g., PrIM or RTS/CTS model).

16.3   Problem Formulation for Cross-Layer Optimization

This section gives a mixed integer-programming (IP) formulation of the necessary and sufficient conditions for when we want to maximize the network throughput or the fairness among the flows. For cross-layer optimization, the flow that can be supported by mesh networks not only needs to satisfy the capacity constraint but also needs to be schedulable by all links without interference. Furthermore, the scheduling of links in multichannel and multiradio mesh networks also needs to satisfy the channel and radio constraints no matter whether dynamic channel assignment or static channel assignment is used.

16.3.1 Maximize Fairness

First, we formulate the routing problem to maximize the fairness of the achieved flow. If virtual link $\widehat{e}$ is assigned channel $\mathbf{f}_{ab}$ for $\alpha(\widehat{e}, \mathbf{f}_{ab})$ fraction of time, then $\alpha(\widehat{e}, \mathbf{f}_{ab})\mathbf{c}(\widehat{e}, \mathbf{f}_{ab})$ is the corresponding achieved flow. Given a routing (and corresponding link and channel scheduling), the achieved fairness λ is defined as the minimum ratio of *achieved flow* over the *demanded load* over all wireless mesh routers. Clearly, the achieved flow at a router u is the difference between the flow going out of node u and the flow coming in to node u; i.e., $\sum_{\widehat{e} \in \widehat{\mathbf{E}}^{+}(u)} f(\widehat{e}) - \sum_{\widehat{e} \in \widehat{\mathbf{E}}^{-}(u)} f(\widehat{e})$. Here, variable $f(\widehat{e})$ is the total traffic scheduled over virtual link $\widehat{e}$ using various channels. We also assume that for each sink $\mathbf{s}_i$, its maximum outgoing capacity is $F_{\text{ISP}}(\mathbf{s}_i)$. We use variable $f_{\text{isp}}(\mathbf{s}_i)$ to denote the actual flow out of this sink $\mathbf{s}_i$. Then, it is easy to get the following mixed IP for maximizing the fairness linear programming:

MAX Fairness LP: $\max \lambda$

$$
\begin{aligned}
\sum_{\widehat{e} \in \widehat{\mathbf{E}}^{+}(u)} f(\widehat{e}) - \sum_{\widehat{e} \in \widehat{\mathbf{E}}^{-}(u)} f(\widehat{e}) &\geq \lambda \ell(u), && \forall u \in V, \\
\sum_{\mathbf{f}_{ab} \in \mathbb{RCC}(\widehat{e})} \alpha(\widehat{e}, \mathbf{f}_{ab})\mathbf{c}(\widehat{e}, \mathbf{f}_{ab}) &= f(\widehat{e}), && \forall \widehat{e}, \\
\alpha(\widehat{e}, \mathbf{f}_{ab}) &\geq 0, && \forall \widehat{e}, \mathbf{f}_{ab}, \\
\sum_{\widehat{e} \in e} \alpha(\widehat{e}, \mathbf{f}_{ab}) &\leq 1, && \forall e, \\
f_{\text{isp}}(\mathbf{s}_i) &\geq 0, && \forall \mathbf{s}_i, \\
f_{\text{isp}}(\mathbf{s}_i) &\leq F_{\text{ISP}}(\mathbf{s}_i), && \forall \mathbf{s}_i,
\end{aligned}
$$

there *exists* an interference-free schedule for $\alpha(\widehat{e}, \mathbf{f}_{ab})$.

16.3.2 Maximize Throughput

In the majority of applications, we not only have to guarantee certain fairness of the achieved flows for all end wireless devices, but we also have to achieve the largest possible throughput under certain fairness constraints. Assume that we have minimum fairness constraints λ_0. Maximum throughput routing is equivalent to solving the following linear programming (**LP-flow-throughput**) for $\alpha(\widehat{e}, \mathbf{f}_{ab})$:

LP-Flow-Throughput: $\max \sum_{i=1}^{g} f_{\text{isp}}(\mathbf{s}_i)$

$$
\begin{aligned}
\sum_{\widehat{e} \in \widehat{\mathbf{E}}^{+}(u)} f(\widehat{e}) - \sum_{\widehat{e} \in \widehat{\mathbf{E}}^{-}(u)} f(\widehat{e}) &= f(u), && \forall u \in V - \mathcal{S}, \\
f(u) &\geq \lambda_0 \ell(u), && \forall u \in V, \\
\sum_{\widehat{e} \in \widehat{\mathbf{E}}^{-}(\mathbf{s}_i)} f(\widehat{e}) + f(\mathbf{s}_i) &= f_{\text{isp}}(\mathbf{s}_i), && \forall \mathbf{s}_i \in \mathcal{S}, \\
\sum_{\mathbf{f}_{ab} \in \mathbb{RCC}(\widehat{e})} \alpha(\widehat{e}, \mathbf{f}_{ab})\mathbf{c}(\widehat{e}, \mathbf{f}_{ab}) &= f(\widehat{e}), && \forall \widehat{e}, \\
\alpha(\widehat{e}, \mathbf{f}_{ab}) &\geq 0, && \forall \widehat{e}, \mathbf{f}_{ab}, \\
\sum_{\widehat{e} \in e} \alpha(\widehat{e}, \mathbf{f}_{ab}) &\leq 1, && \forall e, \\
f_{\text{isp}}(\mathbf{s}_i) &\geq 0, && \forall \mathbf{s}_i, \\
f_{\text{isp}}(\mathbf{s}_i) &\leq F_{\text{ISP}}(\mathbf{s}_i), && \forall \mathbf{s}_i,
\end{aligned}
$$

there *exists* an interference-free schedule for $\alpha(\widehat{e}, \mathbf{f}_{ab})$.

16.3.3   Link Scheduling

Our objective is to give each link $\mathbf{L} \in G$ a transmission schedule $\mathcal{S}(\mathbf{L})$, which is the list of time slots and the corresponding (possibly) combined channels it could send packets to such that the schedule is interference-free and the overall throughput of the network is maximized. Let $X_{\widehat{e},t,\mathbf{f}_{ab}} \in \{0, 1\}$ be the indicator variable that is 1 only when $\widehat{e}$ will transmit at time t using channel $\mathbf{f}_{ab}$. We focus on periodic schedules in this chapter. A schedule is *periodic* with period T if for every virtual link $\widehat{e}$, every channel $\mathbf{f}_{ab}$, and time slot t, $X_{\widehat{e},t,\mathbf{f}_{ab}} = X_{\widehat{e},t+iT,\mathbf{f}_{ab}}$ for any integer $i > 0$. For a virtual link $\widehat{e}$, let $\mathbf{I}(\widehat{e})$ denote the set of links $\widehat{e}'$ that will cause interference if $\widehat{e}$ and $\widehat{e}'$ are scheduled at the same time slot t using the same channel $\mathbf{f}_{ab}$. Notice that a virtual edge $\widehat{e}' \in \mathbf{I}(\widehat{e})$ if $\widehat{e}'$ and $\widehat{e}$ share a common virtual node because any radio can be active only for either transmitting or receiving (but not both) at one specific channel. A schedule $\mathcal{S}$ is *interference-free* if $X_{\widehat{e},t,\mathbf{f}_{ab}} + X_{\widehat{e}',t,\mathbf{f}_{pq}} \leq 1$ for any $\widehat{e}' \in \mathbf{I}(\widehat{e})$, any time slot t, any channel $\mathbf{f}_{ab}$, and any $\mathbf{f}_{pq}$ with $\mathbf{I}_{\text{set}}(\mathbf{f}_{ab}, \mathbf{f}_{pq}) = 1$.

We then mathematically formulate the necessary and sufficient condition for schedulable flow $f(e) = \sum_{i,j} \sum_{\mathbf{f}_{ab} \in \mathbb{F}(e)} \alpha(\widehat{e}_{i,j}, \mathbf{f}_{ab}) c(\widehat{e}_{i,j}, \mathbf{f}_{ab})$. A flow f [equivalently, whether a given vector $\alpha(\widehat{e}_{i,j}, \mathbf{f}_{ab})$ for all virtual edges $\widehat{e}$ and $\mathbf{f}_{ab}$ is schedulable] is schedulable if and only if we can find integer solution $X_{\widehat{e},t,\mathbf{f}_{ab}}$ satisfying the following conditions:

$$
\begin{aligned}
X_{\widehat{e},t,\mathbf{f}_{ab}} + X_{\widehat{e}',t,\mathbf{f}_{pq}} &\leq 1, \quad \forall (\widehat{e}', \mathbf{f}_{pq}) \in \mathbf{I}(\widehat{e}, \mathbf{f}_{ab}), \\
&\qquad\quad \forall \widehat{e}, t, \mathbf{f}_{ab}, \\
\frac{\sum_{1 \leq t \leq T} X_{\widehat{e},t,\mathbf{f}_{ab}}}{T} &= \alpha(\widehat{e}, \mathbf{f}_{ab}), \quad \forall \widehat{e}, \forall \mathbf{f}_{ab}, \\
\sum_{\widehat{e}: u \in \widehat{e}, \mathbf{f}_{ab} \in \mathbb{F}} X_{\widehat{e},t,\mathbf{f}_{ab}} &\leq \mathcal{I}(u), \quad \forall u, \forall t \in [1, T], \\
X_{\widehat{e},t,\mathbf{f}_{ab}} &\leq \delta(\widehat{e}, \mathbf{f}_{ab}), \quad \forall \widehat{e}, \forall \mathbf{f}_{ab}, \\
X_{\widehat{e},t,\mathbf{f}_{ab}} &\in \{0, 1\}, \quad \forall \widehat{e}, \forall t, \forall \mathbf{f}_{ab}.
\end{aligned}
\tag{16.1}
$$

The first condition says that a schedule should be interference-free. The second condition says that the schedule should achieve the required flow $\alpha(\widehat{e}, \mathbf{f}_{ab})$. The third condition specifies that the number of active links (incident upon a node u) using all channels should be at most the number of radios that node u has. Notice that node u can be either the sender or the receiver for at most $\mathcal{I}(u)$ links simultaneously. The fourth condition says that a node can use only the channels that are available and operative by its radios. Observe that the fourth condition is actually implied by the first condition and the fifth condition because the virtual edges using the same radio of a wireless device u will always interfere; i.e., $\widehat{e}_{i,k} \in \mathbf{I}(\widehat{e}_{i,j})$ and $\widehat{e}_{k,j} \in \mathbf{I}(\widehat{e}_{i,j})$ for any k.

Dynamic Channel Assignment

We now formulate the channel assignment that can facilitate link scheduling and also schedulable flows. Dynamic channel assignment (DCA) methods are studied in this chapter, in which we assume that every wireless interface can *dynamically* change the channel (e.g., spectrum or CDMA code) for transmitting signals based on a certain schedule.

Let variable $Y_{\widehat{e},t,\mathbf{f}_{ab}} \in \{0, 1\}$ denote whether virtual link $\widehat{e}$ will be assigned the combined channel $\mathbf{f}_{ab}$ at time t. Let $Y_{u,t,\mathbf{f}_{ab}} \in \{0, 1\}$ denote whether node u uses combined channel $\mathbf{f}_{ab}$ at time t. Notice that in the link-scheduling formulation, we use an indicator variable $X_{\widehat{e},t,\mathbf{f}_{ab}}$ to denote whether virtual link $\widehat{e}$ will be active using a combined channel $\mathbf{f}_{ab}$ at time slot t. This indicator variable can also be directly interpreted as the DCA for nodes: If $X_{\widehat{e},t,\mathbf{f}_{ab}} = 1$ for $\widehat{e} = (u, v)$, then node u will be assigned the channel $\mathbf{f}_{ab}$ to communicate with node v at time slot t, and node v will be assigned this channel $\mathbf{f}_{ab}$ for receiving a signal. In other words, a feasible scheduling $X_{\widehat{e},t,\mathbf{f}_{ab}}$ for all links, all combined channels, and all time slots already equivalently defines a feasible DCA $Y_{\widehat{e},t,\mathbf{f}_{ab}} \leftarrow X_{\widehat{e},t,\mathbf{f}_{ab}}$. Thus, there are not any additional constraints when DCA is used.

Polynomial-Time Schedulable Flows

It is widely known that it is NP-hard to decide whether a feasible scheduling $X_{e,t,\mathbf{f}_{ab}}$ exists when given the flow $f(e)$ [or, equivalently, $\alpha(e, \mathbf{f}_{ab})$] for wireless networks with interference constraints. For some interference models, several papers gave relaxed necessary conditions and relaxed sufficient conditions for schedulable flows that can be decided in polynomial time. For example, without channel combination, for a RTS/CTS model with uniform transmission range $R_T(v_i)$ and uniform interference range $R_I(v_i)$, Alicherry *et al.* (2005) gave a sufficient condition $\alpha(e, \mathbf{f}) + \sum_{e' \in \mathbf{I}(e)} \alpha(e', \mathbf{f}) \leq 1$ and a necessary condition $\alpha(e, \mathbf{f}) + \sum_{e' \in \mathbf{I}(e)} \alpha(e', \mathbf{f}) \leq C(q)$. Here, $C(q)$ is a constant depending on uniform $q = \frac{R_I(u)}{R_T(u)}$.

DEFINITION 16.1 *The set* $\mathbf{I}(\widehat{e}, \mathbf{f}_{ab})$ *denotes the set of all pairs* $(\widehat{e}', \mathbf{f}_{pq})$ *such that if virtual link* $\widehat{e}$ *will use channel* $\mathbf{f}_{ab}$ *for communication (at time slot t) and virtual link* $\widehat{e}'$ *will use channel* $\mathbf{f}_{pq}$ *for communication (at time slot t), then interference will result at some end nodes of these links.*

Notice that for any virtual edge $\widehat{e}' = (\widehat{u}_i, \widehat{v}_j)$ and any $\mathbf{f}_{pq}$, we have $(\widehat{e}', \mathbf{f}_{pq}) \in \mathbf{I}(\widehat{e}, \mathbf{f}_{ab})$ if $\widehat{e}'$ is adjacent to either $\widehat{u}_i$ or $\widehat{v}_j$ because any radio of a node can perform only one operation (either one transmitting or one receiving, but not both). The only other scenario in which $(\widehat{e}', \mathbf{f}_{pq}) \in \mathbf{I}(\widehat{e}, \mathbf{f}_{ab})$ (when $\widehat{e}'$ and $\widehat{e}$ do not share a common virtual node) is $\mathbf{I}_{\text{set}}(\mathbf{f}_{pq}, \mathbf{f}_{ab}) = 1$ and $\widehat{e}' \in \mathbf{I}(\widehat{e})$.

For an interference model $\mathcal{M}$, $\mathbf{I}_{\mathcal{M}}(e) \subseteq \mathbf{I}(e)$ will be defined based on the interference model $\mathcal{M}$ for the purpose of link scheduling. For an example of the PrIM model, $\mathbf{I}_{\mathcal{M}}(\widehat{e})$ is the set of virtual edges in $\mathbf{I}(\widehat{e})$ whose Euclidean length is at least that of $\widehat{e}$. We also define $\mathbf{I}_{\mathcal{M}}(\widehat{e}, \mathbf{f}_{ab})$ as a subset of $\mathbf{I}(\widehat{e}, \mathbf{f}_{ab})$ satisfying a special property depending on the interference model $\mathcal{M}$; i.e.,

$$\mathbf{I}_{\mathcal{M}}(\widehat{e}, \mathbf{f}_{ab}) = \{(\widehat{e}', \mathbf{f}_{pq}) \mid (\widehat{e}', \mathbf{f}_{pq}) \in \mathbf{I}(\widehat{e}, \mathbf{f}_{ab}) \text{ and } \widehat{e}' \in \mathbf{I}_{\mathcal{M}}(\widehat{e})\}.$$

The required property is explained later. For *each* of the interference models discussed in this chapter, a necessary and sufficient condition for schedulable flows is presented later (the proof of Theorem 16.1 is deferred to a later section).

THEOREM 16.1 *Consider the active fraction* $\alpha(\widehat{e}, \mathbf{f}_{ab}) \in [0, 1]$ *of each link using each channel. A sufficient condition that this* α *is schedulable is*

$$\alpha(\widehat{e}, \mathbf{f}_{ab}) + \sum_{(\widehat{e}', \mathbf{f}_{pq}) \in \mathbf{I}_{\mathcal{M}}(\widehat{e}, \mathbf{f}_{ab})} \alpha(\widehat{e}', \mathbf{f}_{pq}) \leq 1, \forall \widehat{e}, \forall \mathbf{f}_{ab}.$$

A necessary condition that this α *is schedulable is*

$$\alpha(\widehat{e}, \mathbf{f}_{ab}) + \sum_{(\widehat{e}', \mathbf{f}_{pq}) \in \mathbf{I}_{\mathcal{M}}(\widehat{e}, \mathbf{f}_{ab})} \alpha(\widehat{e}', \mathbf{f}_{pq}) \leq C_{\mathcal{M}}, \forall \widehat{e}, \forall \mathbf{f}_{ab}.$$

Here, $C_{\mathcal{M}}$ *is a constant depending on the specific interference model and* γ*. Notice that* $\sum_{\widehat{e} \ni \widehat{u}_j, \mathbf{f}_{ab}} \alpha(\widehat{e}, \mathbf{f}_{ab}) \leq 1$ *is always required for any virtual node* $\widehat{u}_j$*.*

Thus, given a constant integer $C \in [1, C_{\mathcal{M}}]$, we replace the condition $X_{\widehat{e}, t, \mathbf{f}_{ab}} + X_{\widehat{e}', t, \mathbf{f}_{pq}} \leq 1, \forall \widehat{e}, \forall \widehat{e}', \forall \mathbf{f}_{ab}, \mathbf{f}_{pq}$ such that $(\widehat{e}', \mathbf{f}_{pq}) \in \mathbf{I}(\widehat{e}, \mathbf{f}_{ab})$ of feasible link scheduling with the following condition:

$$\alpha(\widehat{e}, \mathbf{f}_{ab}) + \sum_{(\widehat{e}', \mathbf{f}_{pq}) \in \mathbf{I}_{\mathcal{M}}(\widehat{e}, \mathbf{f}_{ab})} \alpha(\widehat{e}', \mathbf{f}_{pq}) \leq C, \forall \widehat{e}, \forall \mathbf{f}_{ab}.$$

Notice that when $C = 1$, we will show that the flow f is guaranteed to have a feasible interference-free link scheduling. For any flow f that can be implemented by interference-free link scheduling, we also have $C \leq C_{\mathcal{M}}$.

16.3.4   Integrated Cross-Layer Mixed IP

We now integrate all the conditions that the schedulable flow needs to satisfy into one cross-layer mixed-IP formulation to maximize the fairness among flows or to maximize the total throughout of all flows. The mixed-IP formulation for the joint routing, link–channel scheduling for multiradio, multichannel, multihop wireless networks is as follows:

Mixed-IP Flow Fairness: max λ

$$\sum_{\widehat{e} \in \widehat{\mathbf{E}}^+(u)} f(\widehat{e}) - \sum_{\widehat{e} \in \widehat{\mathbf{E}}^-(u)} f(\widehat{e}) \geq \lambda \ell(u), \qquad \forall u \in V,$$

$$\sum_{\mathbf{f}_{ab} \in \mathbb{R}CC(\widehat{e})} \alpha(\widehat{e}, \mathbf{f}_{ab}) \cdot \mathbf{c}(\widehat{e}, \mathbf{f}_{ab}) = f(\widehat{e}), \qquad \forall \widehat{e},$$

$$f(\widehat{e}) \geq 0, \qquad \forall \widehat{e},$$

$$\sum_{\widehat{e} \in e} f(\widehat{e}) = f(e), \qquad \forall e,$$

$$\alpha(\widehat{e}, \mathbf{f}_{ab}) \geq 0, \qquad \forall \widehat{e},$$

$$\sum_{\widehat{e} \in e} \alpha(\widehat{e}, \mathbf{f}_{ab}) \leq 1, \qquad \forall e, \forall \mathbf{f}_{ab},$$

$$\sum_{\mathbf{f}_{ab}, \widehat{e}: \widehat{u}_i \in \widehat{e}} \alpha(\widehat{e}, \mathbf{f}_{ab}) \leq 1, \qquad \forall \widehat{u}_i,$$

$$\sum_{\widehat{e} \in e} \alpha(\widehat{e}, \mathbf{f}_{ab}) = \alpha(e, \mathbf{f}_{ab}), \qquad \forall e,$$

$$X_{\widehat{e}, t, \mathbf{f}_{ab}} + X_{\widehat{e}', t, \mathbf{f}_{pq}} \leq 1, \qquad \forall (\widehat{e}', \mathbf{f}_{pq}) \in \mathbf{I}(\widehat{e}, \mathbf{f}_{ab}), \\ \forall \widehat{e}, t, \mathbf{f}_{ab},$$

$$\frac{\sum_{1 \leq t \leq T} X_{\widehat{e}, t, \mathbf{f}_{ab}}}{T} = \alpha(\widehat{e}, \mathbf{f}_{ab}), \qquad \forall \widehat{e}, \forall \mathbf{f}_{ab},$$

$$\sum_{\widehat{e}: u \in \widehat{e}; \mathbf{f}_{ab} \in \mathbb{F}} X_{\widehat{e}, t, \mathbf{f}_{ab}} \leq \mathcal{I}(u), \qquad \forall u, \forall t \in [1, T],$$

$$X_{\widehat{e}, t, \mathbf{f}_{ab}} \leq \delta(\widehat{e}, \mathbf{f}_{ab}), \qquad \forall \widehat{e}, \forall \mathbf{f}_{ab},$$

$$X_{\widehat{e}, t, \mathbf{f}_{ab}} \in \{0, 1\}, \qquad \forall \widehat{e}, \forall t, \forall \mathbf{f}_{ab}.$$

Here, for a link $e = (u, v)$, a virtual edge $\widehat{e} \in e$ if $\widehat{e}$ connects a virtual node $\widehat{u}_i$ and a virtual node $\widehat{v}_j$. Observe that in the preceding mixed-IP formulation, we consider both *external* virtual edges and *internal* virtual edges. Remember that $\widehat{\mathbf{E}}^+(u)$ [and $\widehat{\mathbf{E}}^-(u)$]

contains all virtual (external and internal) edges incident at the group of virtual nodes by actual device u. Assigned flow $f(\widehat{e})$ is defined for all virtual edges; $\alpha(\widehat{e}, \mathbf{f}_{ab})$ is defined for only *external* virtual edges. The solution of the internal virtual edge tells us about radio-switching and channel-switching in a node. Although solving the preceding mixed-IP will give us optimum routing and link–channel scheduling, it is generally time expensive to solve this because the original problem is NP-hard. Then, we relax it to linear programming (LP) by getting rid of scheduling variables X. From the previous study, we generally require that, given a constant integer $C \in [1, C_M]$, we need to solve the following LP (**LP-flow-fairness**) for $\alpha(\widehat{e}, \mathbf{f}_{ab})$ such that

$$\textbf{LP-Flow-Fairness:} \qquad \max \lambda$$

$$\sum_{\widehat{e} \in \widehat{\mathbf{E}}^+(u)} f(\widehat{e}) - \sum_{\widehat{e} \in \widehat{\mathbf{E}}^-(u)} f(\widehat{e}) \geq \lambda \ell(u), \qquad \forall u \in V,$$

$$\sum_{\mathbf{f}_{ab} \in \mathbb{R}CC(\widehat{e})} \alpha(\widehat{e}, \mathbf{f}_{ab}) \mathbf{c}(\widehat{e}, \mathbf{f}_{ab}) = f(\widehat{e}), \qquad \forall \widehat{e},$$

$$f(\widehat{e}) \geq 0, \qquad \forall \widehat{e},$$

$$\alpha(\widehat{e}, \mathbf{f}_{ab}) \geq 0, \qquad \forall \widehat{e},$$

$$\sum_{\widehat{e} \in e} \alpha(\widehat{e}, \mathbf{f}_{ab}) \leq 1, \qquad \forall e,$$

$$\sum_{\mathbf{f}_{ab}, \widehat{e}: \widehat{u}_i \in \widehat{e}} \alpha(\widehat{e}, \mathbf{f}_{ab}) \leq 1, \qquad \forall \widehat{u}_i,$$

$$\alpha(\widehat{e}, \mathbf{f}_{ab}) + \sum_{(\widehat{e}', \mathbf{f}_{pq}) \in \mathbf{I}_M(\widehat{e}, \mathbf{f}_{ab})} \alpha(\widehat{e}', \mathbf{f}_{pq}) \leq C, \qquad \forall \widehat{e}, \forall \mathbf{f}_{ab},$$

$$\alpha(\widehat{e}, \mathbf{f}_{ab}) \leq \delta(\widehat{e}, \mathbf{f}_{ab}), \qquad \forall \widehat{e}, \forall \mathbf{f}_{ab}.$$

Recall that here, $\sum_{\widehat{e} \in e} f(\widehat{e}) = f(e)$, $\forall e$, is the total flow assigned to link e. Notice that condition $\sum_{\mathbf{f}_{ab}, \widehat{e}: u \in \widehat{e}} \alpha(\widehat{e}, \mathbf{f}_{ab}) \leq \mathcal{I}(u)$, for any node u is already implied by the condition $\sum_{\mathbf{f}_{ab}, \widehat{e}: \widehat{u}_i \in \widehat{e}} \alpha(\widehat{e}, \mathbf{f}_{ab}) \leq 1, \forall \widehat{u}_i$. Similarly, we can formulate LP for **LP-flow-throughput** such that the solution $\alpha(\widehat{e}, \mathbf{f}_{ab})$ is guaranteed to have a feasible link and channel scheduling.

The rest of the chapter is devoted to designing a polynomial-time method that can find such a link and channel scheduling that satisfies the solution $\alpha(\widehat{e}, \mathbf{f}_{ab})$ from the LP. We also prove that the achieved fairness or throughput is within a constant factor of the optimum.

16.4 Efficient Link, Channel Scheduling

In this section, first an efficient algorithm is presented to find a feasible link scheduling given a flow found by our LP.

16.4.1 Polynomial-Time Scheduling Method

First, centralized scheduling is presented for link transmission. Our method is based on some algorithms presented in W. Wang *et al.* (2006b) on link scheduling for single-channel networks. Assume that T is the number of time slots per scheduling period. Then, we need to schedule $T\alpha(\widehat{e}, \mathbf{f}_{ab})$ time slots for a virtual link $\widehat{e}$ by using channel $\mathbf{f}_{ab}$. Notice that here, we need to schedule the transmission of only *external* virtual edges: what time to transmit and using what combined channel. For simplicity, we assume that the chosen T results in that $T\alpha(\widehat{e}, \mathbf{f}_{ab})$ is an integer for every virtual edge $\widehat{e}$ and $\mathbf{f}_{ab}$. Notice that when we schedule such a pair of (link, channel), we need to ensure that the scheduling

---

Algorithm 47 Centralized Greedy Link Scheduling

Input: A virtual communication graph $G = (V', E)$ of m links, an interference model $\mathcal{M}$, and $\alpha(\widehat{e}, \mathbf{f}_{ab})$ for all external virtual links and for all channels.

Output: Interference-free link scheduling.

1: Sort the *external* virtual links in the virtual communication graph G according to some *special* order based on the interference model $\mathcal{M}$. Let $(\widehat{e}_1, \widehat{e}_2, \ldots, \widehat{e}_m)$ be the sorted list of links.

2: **for** $i = 1$ to m **do**

3:     **for** each possible channel $\mathbf{f}_{ab} \in \mathbb{F}$ **do**

4:         Let $N(\widehat{e}_i, \mathbf{f}_{ab}) = T \cdot \alpha(\widehat{e}_i, \mathbf{f}_{ab})$ be the number of time slots that virtual link $\widehat{e}_i$ will be active using channel $\mathbf{f}_{ab}$.

5:         Assume $\widehat{e}_i = (u, v)$. Set *allocated* $\leftarrow 0$; $t \leftarrow 1$;

6:         **while** *allocated* $< N(\widehat{e}_i, \mathbf{f}_{ab})$ **do**

7:             **if** $X_{e', t, \mathbf{f}_{pq}} = 0$ for every $(\widehat{e}', \mathbf{f}_{pq})$ pair from $\mathbf{I}_{\mathcal{M}}(\widehat{e}_i, \mathbf{f}_{ab})$ **then**

8:                 Set $X_{\widehat{e}_i, t, \mathbf{f}_{ab}} \leftarrow 1$;

9:                 Set *allocated* $\leftarrow$ *allocated* $+ 1$;

10:             **end if**

11:             Set $t \leftarrow t + 1$.

12:         **end while**

13:     **end for**

14: **end for**

---

is interference-free and satisfies the radio and channel-availability constraints of all nodes. Algorithm 47 illustrates our scheduling method. The basic idea of our scheduling is first sorting the external virtual links based on some specific order and then processing the requirement $\alpha(\widehat{e}, \mathbf{f}_{ab})$ for each of the possible channels $\mathbf{f}_{ab}$. Assume that there is a table $\mathbf{Y}(t)$ for each virtual node $\widehat{u}_j$; i.e., the jth radio at node u. The table stores the current assignment for (virtualEdge, channel) pair; i.e., an entry $\mathbf{Y}(t)$ is $(\widehat{e}, \mathbf{f}_{ab})$ means that node u will use NIC j to transmit at time t using channel $\mathbf{f}_{ab}$ for link $\widehat{e}$ if the directed virtual edge $\widehat{e}$ starts from virtual node $\widehat{u}_j$; otherwise, node u will use NIC j to receive at time t using channel $\mathbf{f}_{ab}$ for link $\widehat{e}$ ended at virtual node $\widehat{u}_j$.

For a virtual edge $\widehat{e}_i$, we need to find $N(\widehat{e}_i, \mathbf{f}_{ab}) = T\alpha(\widehat{e}_i, \mathbf{f}_{ab})$ *empty* entries that will not cause interference with other scheduled pairs of (link, channel). If there are consecutive time slots of a radio available, we choose consecutive time slots (to reduce the channel-switching cost). Notice that our algorithm relies on some special sorting of the links, depending on the interference models. We process links in this sorted order. When processing the ith virtual link $\widehat{e}_i$, we process the channels in order and assign link $\widehat{e}_i$ the *earliest* (not necessarily consecutive) $N(\widehat{e}_i, \mathbf{f}_{ab}) = T\alpha(\widehat{e}_i, \mathbf{f}_{ab})$ time slots using channel $\mathbf{f}_{ab}$ that will not cause any interference to already scheduled links and satisfies the radio and channel-availability constraints. It is now explained in detail how the sorting is done for different interference models, and also $\mathbf{I}_{\mathcal{M}}(\widehat{e})$ is defined for a link $\widehat{e}$ with respect to all interference models $\mathcal{M}$ (namely, RTS/CTS model, fPrIM, PrIM, TxIM).

RTS/CTS

One sorting for the RTS/CTS model is to sort all links in decreasing order of their interference radius. For the performance proof, we adopt this sorting. Here, the *interference radius* of a link $e = (v_i, v_j)$ is defined as $r_{i,j} = \max\{R_I(i), R_I(j)\}$. The interference radius of a virtual edge $\hat{e}_{i,j}$ (with actual end nodes u and v) is the same as that of the actual link $e = (u, v)$. The set $I_{\text{RTS/CTS}}(\hat{e})$ is the set of links whose *interference radius* is at least that of $\hat{e}$ and interfering with $\hat{e}$ under the RTS/CTS model. In this case, the constant $C_M \leq 120$ (the proof is similar to that of Lemma 3 of W. Wang *et al.* 2006b).

fPrIM

In the fPrIM, we again consider the (virtual) conflict graph. We choose the vertex, which is the virtual link in the virtual communication graph, with the largest value $d_{i,j}^{\text{in}} - d_{i,j}^{\text{out}}$ in the residue conflict graph; remove the vertex and its incident edges. $d_{i,j}^{\text{in}}$ and $d_{i,j}^{\text{out}}$ are the *in-degree* and *out-degree*, respectively, of vertex $\mathbf{L}_{i,j}$ in the conflict graph for the fPrIM. Repeat this process until there is no vertex in the graph. Then, the links (in the original graph) are sorted by their reverse removal order. The set $I_{\text{fPrIM}}(\hat{e})$ is the set of incoming links of $\hat{e}$ that interfere $\hat{e}$. Here, a virtual link e' is called the incoming link of $\hat{e}$ if the activation of link e' will cause interference at the receiving node of link $\hat{e}$; i.e., the vertex corresponding to e' in the conflict graph F_G^{FP} is the incoming neighbor of the vertex for $\hat{e}$ in the graph F_G^{FP}. In this case, the constant $C_M = \left\lceil \frac{2\pi}{\arcsin \frac{\gamma-1}{2\gamma}} \right\rceil$; e.g., $C_M = 25$ when $\gamma = 2$ (the proof is similar to that of Lemma 6 of W. Wang *et al.*, 2006b).

TxIM

In the TxIM, we sort the links according to the interference radius of the sending node in a nonincreasing order. Notice that links in this chapter are always directed. For a link $e = (v_i, v_j)$, the set $I_{\text{TxIM}}(\hat{e})$ is the set of virtual links $e' = (v_p, v_q)$ such that the interference range $R_I(p)$ of node v_p is at least the interference range $R_I(i)$ of node v_i and $\|v_p - v_i\| \leq R_I(p) + R_I(i)$; i.e., nodes v_p and v_i interfere with each other. In this case, the constant $C_M = 5$ (the proof is similar to that of Claim 2 of Kumar *et al.*, 2005).

PrIM

In the PrIM, we sort links according to their Euclidean length in a nonincreasing order. The set $I_{\text{PrIM}}(\hat{e})$ is the set of links whose *Euclidean distance* is at least that of $\hat{e}$. In this case, the constant $C_M \leq (5 + \frac{4}{\eta})^2$ (the proof is similar to that of Claim 2 of Kumar *et al.*, 2005).

16.4.2   Correctness and Performance Guarantee

We first show that Algorithm 47 indeed finds interference-free scheduling when $\alpha(\hat{e}, \mathbf{f}_{ab})$ is from a feasible solution of linear programmings **LP-flow-throughput** and **LP-flow-fairness**, in which the adjustable constant C is set as 1.

THEOREM 16.2 *Algorithm 47 produces a feasible interference-free link–channel scheduling when* $\alpha(\widehat{e}, \mathbf{f}_{ab})$ *is a feasible solution of LP with* $C = 1$.

Proof. We prove this theorem separately for each of the possible interference models used as follows. From the linear program **LP-flow-fairness** for $\alpha(\widehat{e}, \mathbf{f}_{ab})$, which is applicable to each interference model previously mentioned, we get the solution α. Essentially, we need to show that Algorithm 47 will terminate. Notice that after the algorithm terminates, we know that for every link $\widehat{e}$ and every channel $\mathbf{f}_{ab}$, it is already assigned a fraction $\alpha(\widehat{e}, \mathbf{f}_{ab})$ time slots of all T time slots in a schedule period. Consider a specific link $\widehat{e}$ that is to be processed. From the special sorting used by our algorithm for each interference model, we know that all virtual links $\widehat{e}'$ that have been processed *and* conflict with $\widehat{e}$ (i.e., interfering $\widehat{e}$ or being interfered by link $\widehat{e}$) must be a subset of $\mathbf{I}_{\mathcal{M}}(\widehat{e})$. Notice that it may be not exactly $\mathbf{I}_{\mathcal{M}}(\widehat{e})$ because of possible different tie-breaking of sorting. Recall that in our LP, we had a condition that for every channel $\mathbf{f}_{ab}$, $\alpha(\widehat{e}, \mathbf{f}_{ab}) + \sum_{(\widehat{e}', \mathbf{f}_{pq}) \in \mathbf{I}_{\mathcal{M}}(\widehat{e}, \mathbf{f}_{ab})} \alpha(\widehat{e}', \mathbf{f}_{ab}) \leq 1$. This implies that for each $\mathbf{f}_{ab}$,

$$N(\widehat{e}, \mathbf{f}_{ab}) + \sum_{(\widehat{e}', \mathbf{f}_{pq}) \in \mathbf{I}_{\mathcal{M}}(\widehat{e}, \mathbf{f}_{ab})} N(\widehat{e}', \mathbf{f}_{pq}) \leq T, \ \forall \widehat{e}.$$

Thus, we can always find $N(\widehat{e}, \mathbf{f}_{ab}) = T\alpha(\widehat{e}, \mathbf{f}_{ab})$ time slots among T slots in a period for link $\widehat{e}$ using channel $\mathbf{f}_{ab}$ because all conflict links that already have been processed by Algorithm 47 occupied at most $\sum_{(\widehat{e}', \mathbf{f}_{pq}) \in \mathbf{I}_{\mathcal{M}}(\widehat{e}, \mathbf{f}_{ab})} N(\widehat{e}', \mathbf{f}_{pq}) \leq T - N(\widehat{e}, \mathbf{f}_{ab})$ time slots. Notice that each virtual node $\widehat{u}_i$ has only one radio. Because the total number of time slots needed to be assigned for a virtual node $\widehat{u}_i$ is $\sum_{\widehat{e}: \widehat{u}_i \in \widehat{e}, \mathbf{f}_{ab}} T\alpha(\widehat{e}, \mathbf{f}_{ab}) \leq T$, among T time slots, we can always find time slots for link $\widehat{e}$ using a channel $\mathbf{f}_{ab}$ (after considering all conflicting links scheduled before).

Because $\alpha(\widehat{e}, \mathbf{f}_{ab}) > 0$ for the current virtual link $e = (\widehat{u}_i, \widehat{v}_j)$, we know that $\delta(u, \mathbf{f}_{ab}) = \delta(v, \mathbf{f}_{ab}) = 1$ from $\delta(u, \mathbf{f}_{ab}) \geq \alpha(\widehat{e}, \mathbf{f}_{ab})$ and $\delta(v, \mathbf{f}_{ab}) \geq \alpha(\widehat{e}, \mathbf{f}_{ab})$. In other words, both the end nodes $\widehat{u}_i$ and $\widehat{v}_j$ of the link $\widehat{e}$ can operate on channel $\mathbf{f}_{ab}$. This finishes the proof. ∎

THEOREM 16.3 *Algorithm 47, together with the LP formulation **LP-flow-fairness**, produces a feasible interference-free link–channel scheduling whose achieved fairness is at least* $1/C_{\mathcal{M}}$ *of the optimum, when* $\alpha(\widehat{e}, \mathbf{f}_{ab})$ *is a feasible solution of LP using* $C = 1$.

Proof. Consider an optimum flow assignment defined by $\alpha^*(\widehat{e}, \mathbf{f})$; i.e., the flow supported by a link $\widehat{e}$ is $\sum_{\mathbf{f}_{ab}} \alpha^*(\widehat{e}, \mathbf{f}_{ab}) \mathbf{c}(\widehat{e}, \mathbf{f}_{ab})$. From Theorem 16.1, we know that

$$\alpha^*(\widehat{e}, \mathbf{f}_{ab}) + \sum_{(\widehat{e}', \mathbf{f}_{pq}) \in \mathbf{I}_{\mathcal{M}}(\widehat{e}, \mathbf{f}_{ab})} \alpha^*(\widehat{e}', \mathbf{f}_{pq}) \leq C_{\mathcal{M}}.$$

Define a new flow α' as $\alpha'(\widehat{e}, \mathbf{f}_{ab}) = \frac{\alpha^*(\widehat{e}, \mathbf{f}_{ab})}{C_{\mathcal{M}}}$. Obviously,

$$\alpha'(\widehat{e}, \mathbf{f}_{ab}) + \sum_{(\widehat{e}', \mathbf{f}_{pq}) \in \mathbf{I}_{\mathcal{M}}(\widehat{e}, \mathbf{f}_{ab})} \alpha'(\widehat{e}', \mathbf{f}_{pq}) \leq 1.$$

It is easy to show that the new flow α' satisfies all conditions of our linear programming **LP-flow-fairness**. In other words, α' is a feasible solution for this LP. Consequently, the solution of **LP-flow-fairness** is at least that of α', which is $\frac{1}{C_M}$ of the optimum. This finishes the proof. ∎

It is easy to show that our LP solution for **LP-flow-throughput** also finds a flow assignment whose total flow is at least $\frac{1}{C_M}$ of the optimum when $\lambda_0 = 0$. If we relax the condition $f(u) \geq \lambda_0 \ell(u)$ to $f(u) \geq \frac{\lambda_0}{C_M} \ell(u)$ in the linear programming **LP-flow-throughput**, we can show that our LP will find a flow assignment whose total achieved throughput is at least $\frac{1}{C_M}$ of the optimum and the achieved fairness is at least $\frac{\lambda_0}{C_M}$ (instead of the required λ_0).

The Proof of Theorem 16.1

The rest of the subsection is devoted to the proof of Theorem 16.1, which is based on some theorems proved in W. Wang *et al.* (2006b). The sufficient condition comes from the correctness of Algorithm 47, which gives a valid link–channel schedule. We now show the correctness of that necessary condition. As with Lemmas 3 and 6 of W. Wang *et al.* (2006b), here we can show that we have Lemma 16.1.

LEMMA 16.1 *There exists a constant* C_1 *such that for any time slot* τ, *any valid RTS/CTS interference-free link scheduling* S *must satisfy*

$$X_{\widehat{e},\tau,\mathbf{f}_{ab}} + \sum_{(\widehat{e}',\mathbf{f}_{pq}) \in \mathbf{I}_{RTS/CTS}(\widehat{e},\mathbf{f}_{ab})} X_{\widehat{e}',\tau,\mathbf{f}_{ab}} \leq 2C_1.$$

LEMMA 16.2 *For any time slot* τ, *any valid interference-free link scheduling* S *under the fPrIM must satisfy*

$$X_{\widehat{e},\tau,\mathbf{f}} + \sum_{(\widehat{e}',\mathbf{f}_{pq}) \in \mathbf{I}_{fPrIM}(\widehat{e},\mathbf{f}_{ab})} X_{\widehat{e}',\tau,\mathbf{f}} \leq \left(\frac{2\pi}{\arcsin \frac{\gamma-1}{2\gamma}} \right).$$

Similar to Claim 2 in Kumar *et al.* (2005), here we can show that we have Lemma 16.3.

LEMMA 16.3 *For PrIM or TxIM, there exists a constant* C_M *depending on the interference model such that for any time slot* τ, *any valid interference-free link scheduling* S *satisfies*

$$X_{\widehat{e},\tau,\mathbf{f}_{ab}} + \sum_{(\widehat{e}',\mathbf{f}_{pq}) \in \mathbf{I}_M(\widehat{e},\mathbf{f}_{ab})} X_{\widehat{e}',\tau,\mathbf{f}_{ab}} \leq C_M.$$

Consequently, we can generally show that any interference-free scheduling should satisfy

$$X_{\widehat{e},\tau,\mathbf{f}_{ab}} + \sum_{(\widehat{e}',\mathbf{f}_{pq}) \in \mathbf{I}_M(\widehat{e},\mathbf{f}_{ab})} X_{\widehat{e}',\tau,\mathbf{f}_{ab}} \leq C_M$$

for some constant $C_\mathcal{M}$, depending on the interference model $\mathcal{M}$. Thus, we have

$$\sum_{t=1}^{T} \left(X_{\widehat{e},\tau,\mathbf{f}_{ab}} + \sum_{(\widehat{e}',\mathbf{f}_{pq})\in\mathbf{I}_\mathcal{M}(\widehat{e},\mathbf{f}_{ab})} X_{\widehat{e}',\tau,\mathbf{f}_{ab}} \right) / T \leq C_\mathcal{M}.$$

It is equivalent to saying that for each channel $\mathbf{f}_{ab}$ and virtual edge $\widehat{e}$,

$$\alpha(\widehat{e},\mathbf{f}_{ab}) + \sum_{(\widehat{e}',\mathbf{f}_{pq})\in\mathbf{I}_\mathcal{M}(\widehat{e},\mathbf{f}_{ab})} \alpha(\widehat{e}',\mathbf{f}_{ab}) \leq C_\mathcal{M},$$

which finishes the proof of Theorem 16.1.

16.4.3   Improvement and Implementations

We can also use parametric searching to improve the overall achieved flow. Assume that we replace

$$\alpha(\widehat{e},\mathbf{f}_{ab}) + \sum_{(\widehat{e}',\mathbf{f}_{pq})\in\mathbf{I}_\mathcal{M}(\widehat{e},\mathbf{f}_{ab})} \alpha(\widehat{e}',\mathbf{f}_{pq}) \leq 1$$

with

$$\alpha(\widehat{e},\mathbf{f}_{ab}) + \sum_{(\widehat{e}',\mathbf{f}_{pq})\in\mathbf{I}_\mathcal{M}(\widehat{e},\mathbf{f}_{ab})} \alpha(\widehat{e}',\mathbf{f}_{pq}) \leq C$$

for an integer $C > 1$. Then, the computed flow f may or may not be schedulable. Recall that for any schedulable flow, we should have $\alpha(\widehat{e},\mathbf{f}_{ab}) + \sum_{(\widehat{e}',\mathbf{f}_{pq})\in\mathbf{I}_\mathcal{M}(\widehat{e},\mathbf{f}_{ab})} \alpha(\widehat{e}',\mathbf{f}) \leq C_\mathcal{M}$. In other words, it is sufficient to use integer $C \leq C_\mathcal{M}$ to find all schedulable flows.

We solve the corresponding LP by using an integer C starting from $C_\mathcal{M}$ and finding the α. We then call our scheduling method (Algorithm 47) to find whether or not α is schedulable. If we cannot find a schedule for it, we then decrease C by 1 and then repeat the preceding steps. Notice that the preceding steps can be repeated for at most $C_\mathcal{M}$ rounds. If we can find a valid schedule for an integer $C' \in [1, C_\mathcal{M}]$, the found flow achieves a fairness (or throughput) of at least $\frac{C'}{C_\mathcal{M}}$ times the optimum.

By simulation, we studied the improvement to fairness and maximal throughput in the WMN *with channel combining* when increasing C from 1 to $C_\mathcal{M}$. In this simulation, we varied the maximum number of channels between 2 and 4 and the maximum number of radios between 1 and 2. For the different combinations of number of radios or number of channels, we changed C from 1 to 10 in one-step increments. Figures 16.2(a) and 16.2(b) show the results when the maximum number of channels $= 4$ and maximum number of radios $= 1$ and 2, respectively; each point in the figure is the average of five simulations. The figures show that fairness and throughput are increasing with increasing C, as we expected, and at a certain C_0, the fairness and throughput will reach the maximum demand of all nodes and will not increase with increasing C; in Figures 16.2(a) and 16.2(b), $C_0 = 6$ for a one-radio configuration and $C_0 = 5$ for a two-radio configuration. Also, the figures show that C_0 decreases when the number of radios per node is increased, but an increasing number of channels does not change the value of C_0.

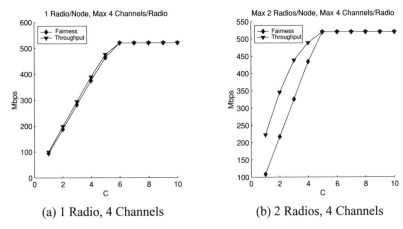

(a) 1 Radio, 4 Channels   (b) 2 Radios, 4 Channels

Figure 16.2 Impact of changing C on fairness and throughput.

16.5   Further Reading

Kodialam and Nandagopal (2004) studied the effect of interference on the achievable-rate region in multihop wireless networks. They treated the interference models as linear constraints and solved the flow problem by using LP. They also considered (2003) the problem of jointly routing the flows and scheduling transmissions to achieve a given rate vector using the protocol model of interference. They developed necessary and sufficient conditions for the achievable rate vector. They formulated the problem as a LP problem and implemented primal–dual algorithms for solving the problem. The scheduling problem is solved as a graph-edge-coloring problem using existing greedy algorithms. They extended their work (Kodialam and Nandagopal, 2005) to multiradio multichannel WMNs. Again, they provided a relaxation of their model to a linear program that gives the necessary conditions for a valid, feasible solution. They used this LP solution to assign channels to the links and also to schedule the time slots in which each link and channel is active by using the greedy approach.

Kumar *et al.* (2005) considered the throughput capacity of wireless networks between given source–destination pairs for various interference models. They developed analytical performance evaluation models and distributed algorithms for routing and scheduling that incorporate fairness, energy, and dilation (path-length) requirements and provide a unified framework for utilizing the network close to its maximum-throughput capacity. Alicherry *et al.* (2005) mathematically formulated the joint channel-assignment and routing problem in multiradio mesh networks and, taking into account the interference constraints, the number of channels in the network and the number of radios available at each mesh router. They established necessary and sufficient conditions under which an interference-free link communication schedule can be obtained and designed as a simple greedy algorithm to compute such a schedule. They used a flow-transformation technique to design an efficient channel-assignment algorithm, and they showed that their algorithm for the joint channel-assignment, routing, and scheduling problem is a constant-factor-approximation algorithm. Notice that the studied network in Alicherry *et al.* (2005) is restricted to a UDG; i.e., the uniform interference range is assumed to be a fixed multiple of the uniform communication range.

Scheduling has been studied extensively (Liu and Lloyd, 2001; Ramanathan, 1999) in the past few years because of its application for assigning time slots in TDMA MAC protocols that eliminate collision and guarantee fairness. Scheduling can be reduced to different coloring problems: *edge-coloring* and *vertex-coloring*.

Edge-coloring, in which every edge corresponds to a valid communication link, is a natural way to capture the link-scheduling problem. An edge-coloring is *valid* if no two incident edges share the same color. Vizing's theorem (Berge, 1973) states that a valid edge-coloring for an *undirected* graph can be obtained by using at most $\Delta + 1$ colors, where Δ is the maximum node degree in the graph. On the other hand, any edge-coloring needs at least Δ colors. Any edge-coloring that uses $\Theta(\Delta)$ colors is close to the optimal. Panconesi and Srinivasan (1992) proposed a randomized distributed-edge-coloring method that uses at most $2\Delta + 1$ colors. To some extent, this captures some transmission restrictions in wireless ad hoc and sensor networks in which no node can receive or send at the same time slot, but it did not address some other interferences such as secondary interference. When one has a valid edge-coloring, it can be easily mapped to a TDMA scheduling. However, it is possible that two communication links sharing the same color still interfere with each other in a wireless network. To remedy this, Gandham *et al.* (2005) proposed using a two-phase scheduling method: In the first phase, a distributed valid edge-coloring is obtained; in the second phase, a valid scheduling that takes into account the secondary interference is obtained. In essence, the model of Gandham *et al.* (2005) is based on the PrIM. The overall scheduling in Gandham *et al.* (2005) provided a performance guarantee only when the conflicting links form a tree. Jain *et al.* (2003) proposed a new concept: a *conflict graph* that captures the interference in wireless networks.

Vertex-coloring is one of the most fundamental NP-hard problems in graph theory and has been thoroughly studied. A vertex-coloring is *valid* iff any two adjacent vertices receive different colors. The minimum number that is needed for a valid vertex-coloring for a graph G is known as the *chromatic number* $\chi(G)$. It is known that for a general graph, the chromatic number cannot be approximated within $n^{1-\varepsilon}$ for any $\varepsilon > 0$, unless $ZPP = NP$ (Feige and Kilian, 1998). For vertex-coloring of a general graph G, it was proved in Szekeres and Wilf (1968) that every graph G can be colored using $\delta(G) + 1$ colors. Then Hochbaum (1983) presented a method to find the value of $\delta(G)$ and color G using $\delta(G) + 1$ colors in $O(|V| + |E|)$ time. Ramanathan (1999) proposed a unified framework for TDMA-, FDMA-, and CDMA-based multihop wireless networks. They also proposed a time-slot assignment to edges; the number of time slots required is at most $O(\theta)$ times the optimum, where θ is the thickness of a graph; i.e., the minimum number of planar graphs into which the network can be decomposed. Krumke *et al.* (2001) proposed efficient approximation algorithms for the distance-2 vertex-coloring problem for various geometric graphs including (r, s)-civilized graphs, planar graphs, and graphs with bounded genus. Kumar *et al.* (2004) studied packet-scheduling under the

RTS/CTS interference model and gave polylogarithmic/constant-factor-approximation algorithms for various families of disk graphs and randomized near-optimal approximation algorithms for general graphs.

Several distributed-scheduling algorithms that use $O(\Delta)$ colors have been proposed in the literature. Recently, Moscibroda and Wattenhofer (2005a) proposed an $O(\Delta)$ distributed-coloring method with time complexity $O(\Delta \log n)$. It is worth pointing out that the coloring in Moscibroda and Wattenhofer (2005a) considered a simple interference model and the time is close to time needed in practice. However, the coloring was based on the assumption that the wireless ad hoc network can be modeled as a UDG; i.e., their method will return a coloring that guarantees that only any adjacent nodes in the UDG will get different colors, but nonadjacent nodes may get the same color. In addition, they assumed that the interference range is the same as the transmission range. This is different from the interference-free scheduling studied in this chapter.

Channel assignment has long been studied in the wireless networking research community. Traditionally, because of the limited number of channels available for a cellular network, channel assignment is used to improve the spatial reuse of channels to eliminate or reduce interference within nearby cells. In graph-theory terminology, this traditional channel assignment is a special case of *list-coloring* problems. List-coloring has been widely studied in both graph theory and network communities (Fotakis *et al.*, 1999; Garg *et al.*, 1996b; Janssen and Narayanan, 1999). A list-coloring of a graph is an assignment of a color to each vertex from a list of choices available at that vertex so that two vertices joined by an edge get different colors.

The channel assignment for ad hoc wireless networks asks for an allocation $\mathcal{A}$ of channels to the interfaces of each node such that the resultant network satisfies a certain property; e.g., being connected. Another reasonable requirement is that the number of assigned channels to a node v_i should be no more than its interfaces $\mathcal{I}(v_i)$; thus, no channel switching is needed at an interface. A number of results (Bahl *et al.*, 2004; Chandra *et al.*, 2004; Draves *et al.*, 2004; Kyasanur and Vaidya, 2004, 2005b; Raniwala and Chiueh, 2005; Raniwala *et al.*, 2004; So and Vaidya, 2004a, 2004b; J. Wang *et al.*, 2006) have been proposed recently for channel assignment and/or routing in multichannel and multiradio wireless networks. Many schemes have been proposed to exploit multiple channels for performance improvement in wireless networks.

One class of work is to enable only one NIC to operate in multiple channels. Multinet (Chandra *et al.*, 2004) exploits the power-saving mode (PSM) in the IEEE 802.11. In SSCH (slotted seeded channel hopping) (Bahl *et al.*, 2004), each node advertises its channel-hopping sequence, and neighbors exchange data packets during the time slot for which they use the same channel. To enable a single interface to access multiple channels, the MMAC (multichannel MAC) protocol uses a scheme similar to that of the PSM (So and Vaidya, 2004b). However, this scheme requires the change in MAC protocols and thus cannot be readily used in current wireless networks. Nasipuri *et al.* (1999) also propose a protocol, which assumes that a node can continuously monitor all available channels. However, in all the preceding schemes that use a single interface, frequent channel-switching introduces extra overhead, and nodes are inherently limited to one reception or transmission at a given time.

Another line of work is to use multiple interfaces available at each wireless node. In the multiradio unification protocol (MUP) (Adya *et al.*, 2003), each node statically assigns a channel to each interface card, and when a node needs to transmit a packet, it checks the channel condition and uses the channel with the best condition at that time. Kyasanur and Vaidya (2005b) classified possible channel-assignment approaches into static and dynamic and proposed a hybrid routing heuristic that combines properties of the two classes. Raniwala *et al.* (2004) considered the joint problem of channel assignment and routing in static mesh networks. They assume that the long-term traffic load between source and destination pairs is known a priori. From this assumption, they present a centralized heuristic for throughput improvement. More recently, they proposed a distributed algorithm (Raniwala and Chiueh, 2005). Draves *et al.* (2004) proposed a new routing metric when nodes have multiple interface cards. They assume that each edge is assigned a fixed channel, but they do not consider the channel-assignment problem. Shin *et al.* (2006) proposed a randomized distributed-channel-assignment method called SAFE that uses one-hop neighborhood information.

16.6   Conclusion

In this chapter, we considered the problem of obtaining good cross-layer multipath routing, interference-aware link–channel scheduling for a multichannel, multiradio, multihop wireless network with and without channel-combining to maximize the fairness (given traffic demands of all nodes) or throughput of the network. We assumed a general model for wireless networks; i.e., nodes could have different transmission ranges and different interference ranges, and a link uv may not exist even if $\|uv\|$ is less than the transmission range of node u. Efficient algorithms were presented that achieve fairness or throughput within a constant factor of the optimum. We also conducted extensive simulations to study the performances of our algorithms and to study the impact of channel-combining to maximize the fairness or throughput of the network. Our results showed that channel-combining indeed increases fairness and throughput of the network and showed that using a large number of radios (channels) to satisfy the network traffic demand is not needed when channel-combining is supported by the network. Our results also showed that when channel-combining is supported, increasing the number of radios will increase the fairness and throughput to a certain point after which increasing the number of radios will start decreasing the fairness and throughput of the network.

There are still a number of challenging questions for future research. The first question is one of how to efficiently collect the information about the interfering links of a given link in a wireless networking environment. This was not an issue in previous studies because they assumed a UDG model and assumed the same interference range for all nodes. The second question is one of how to map efficiently a flow from a mesh router to a given gateway node with a decoupled path from the multipath routes between them. The third question is one of how to study dynamic link–channel scheduling by taking into account the channel-combining cost. Last, but not least, is the question of how to study the link scheduling in a dynamic environment where the traffic load on links could have some small changes.

Problems

16.1   What are the advantages and disadvantages of using a cross-layer design of joint routing, link scheduling, and channel switching?

16.2   Assume that there are K single channels that can be used and each wireless device has at most $\mathcal{I}$ NICs. The wireless network has n wireless devices. How many virtual nodes and virtual links are in the virtual communication graph?

16.3   In the formulation for maximizing fairness or maximizing throughput, we did not exactly specify how multipath routing is performed. Given a solution $\alpha(\widehat{e}, \mathbf{f}_{ab})$ for each possible virtual link $\widehat{e}$ and each possible combined channel $\mathbf{f}_{ab}$, design a polynomial-time algorithm to split the flow into multiple paths connecting the source and the target node. Prove that in the multiple-path routing found by LP, at most m multiple paths are needed to connect the source node and the target node, where m is the number of virtual links.

16.4   In the conditions (defined in Section 16.1) presented for a schedulable flow $\alpha(\widehat{e}, \mathbf{f}_{ab})$, we require that $\sum_{\widehat{e}:u\in\widehat{e}, \mathbf{f}_{ab}\in\mathbb{F}} X_{\widehat{e}, t, \mathbf{f}_{ab}} \leq \mathcal{I}(u)$, $\forall u$, $\forall t \in [1, T]$. Show that this condition can be integrated to the first condition: $X_{\widehat{e}, t, \mathbf{f}_{ab}} + X_{\widehat{e}', t, \mathbf{f}_{pq}} \leq 1$, $\forall (\widehat{e}', \mathbf{f}_{pq}) \in$ $\mathbf{I}(\widehat{e}, \mathbf{f}_{ab})$. In other words, this condition is not needed when $\mathbf{I}(\widehat{e}, \mathbf{f}_{ab})$ is defined appropriately.

16.5   In Theorem 16.1, we present a necessary and sufficient condition for schedulable flows. Can you use other formulas to characterize some necessary conditions and some sufficient conditions such that the flow is schedulable? Consider the conflict graph defined over all virtual links or channels [i.e., $(\widehat{e}, \mathbf{f}_{ab})$] and let $\mathcal{Q} = \{Q_1, Q_2, \ldots, Q_g\}$ be the set of all cliques of the conflict graph. Here, a clique Q_i is a set of $(\widehat{e}, \mathbf{f}_{ab})$ such that any pair of $(\widehat{e}, \mathbf{f}_{ab})$ and $(\widehat{e}', \mathbf{f}_{pq})$ in Q_i will conflict if they are scheduled in the same time slot. Show by example that

$$\sum_{(\widehat{e}, \mathbf{f}_{ab})\in Q_i} \alpha(\widehat{e}, \mathbf{f}_{ab}) \leq 1$$

is a necessary condition only for schedulable flows but not a sufficient condition. In other words, there are flows that satisfy the preceding condition but are not schedulable.

16.6   Implement the joint-routing, link-scheduling, and channel-switching algorithms described in this chapter and compare the throughput achieved with the throughput achieved without using joint optimization. For example, the routing could be done by using AODV routing or some other routing protocols and the link scheduling is done via CSMA/CA MAC.

Part V

Other Issues

17 Localization and Location Tracking

17.1 Introduction

Having location information can be very useful and it has so many applications. It can answer questions such as: Are we almost to the campsite? What lab bench was I standing by when I prepared these tissue samples? How should our search-and-rescue team move to quickly locate all the avalanche victims? Can I automatically display this stock devaluation chart on the large screen I am standing next to? Where is the nearest cardiac defibrillation unit?, and so on. Service providers can also use location information to provide some novel location-aware services. The navigation system based on a GPS is an example. A user can tell the system his destination and the system will guide him there. Phone systems in an enterprise can exploit locations of people to provide follow-me services.

Researchers are working to meet these and similar needs by developing systems and technologies that automatically locate people, equipment, and other tangibles. Indeed, many systems over the years have addressed the problem of automatic location sensing. Because each approach solves a slightly different problem or supports different applications, they vary in many parameters, such as the physical phenomena used for location determination, the form factor of the sensing apparatus, power requirements, infrastructure versus portable elements, and resolution in time and space.

For outdoor environments, the most well-known positioning system is the global positioning system (GPS) (Peng and Mirsa, 1999), which uses 24 satellites set up by the U.S. Department of Defense to enable global 3D positioning services. There are often more than 24 operational satellites as new ones are launched to replace older satellites. GPS provides specially coded satellite signals that can be processed in a GPS receiver, enabling the receiver to compute position, velocity, and time. Four GPS satellite signals are used to compute positions in three dimensions and the time offset in the receiver clock. For civil usage, GPS provides accuracy of around 20–50 m. In addition to the GPS system, positioning can also be done using some wireless networking infrastructures. Taking the Personal Communication System (PCS) cellular networks as an example, the $E911$ emergency service requires determining the location of a phone call via the base stations of the cellular system.

In a GPS, triangulation uses ranges to at least four known satellites to find the coordinates of the receiver and the clock bias of the receiver. For our node-location purposes, we are using a simplified version of the GPS triangulation because we deal only with distances and there is no need for clock synchronization. For the following reasons, GPS is not suitable for WSNs and much work has been dedicated recently to positioning and location-tracking in the area of WSNs:

1. It is not available in an indoor environment because satellite signals cannot penetrate buildings.
2. For more fine-grained applications, higher accuracy is usually necessary in the positioning result. For example, a Standard Positioning Service (SPS) is used by civil users worldwide without charge or restrictions. Its precision has only about 100 m horizontal and 156 m vertical accuracy.
3. Sensor networks have their own battery constraint, which requires special design.

Many applications of sensor networks require knowledge of physical sensor positions. For example, target detection and tracking are usually associated with location information (D. Li *et al.*, 2002). Further, knowledge of sensor location can be used to facilitate network functions such as packet-routing (Bose *et al.*, 2001; Karp and Kung, 2000; J. Li *et al.*, 2000) and collaborative signal processing (Heidemann and Bulusu, 2001). A sensor position can also serve as a unique node identifier, making it unnecessary for each sensor to have a unique ID assigned prior to its deployment. Location information can be used not only to minimize the communication but also to improve the performance of wireless networks and provide new types of services. For example, it can facilitate routing in a wireless ad hoc network to reduce routing overhead. This is known as *geographic routing* (Ko and Vaidya, 2000; Navas and Imielinski, 1997). Through location-aware network protocols, the number of control packets can be reduced. Other types of location-based services include geocast (Ko and Vaidya, 1998), by which a user can request to send a message to a specific area, and temporal geocast, by which a user can request to send a message to a specific area at a specific time. In contrast to traditional multicast, such messages are not targeted at a fixed group of members but rather at members located in a specific physical area.

However, location discovery in WSNs is very challenging. In sensor networks, the capabilities of individual nodes are very limited and nodes are often powered by batteries only. To conserve energy, collaboration between nodes is required and communication between nodes should be minimized. To achieve this goal for nodes in WSNs, we want to determine the location of individual sensor nodes without relying on external infrastructures (e.g., base stations, satellites). Thus, first, the positioning algorithm must be distributed and localized in order to scale well for large sensor networks. Second, the localization protocol must minimize communication and computation overhead for each sensor because nodes have very limited resources (e.g., power, CPU, memory). Third, the positioning functionality should not increase the cost and complexity of the sensor because an application may require thousands of sensors. Fourth, a location-detection scheme should be robust. It should work with accuracy and precision in various environments and should not depend on sensor-to-sensor connectivity in the network.

The localization problem has received considerable attention in the past because in many applications there is a need to know where objects or persons are; hence, various location services have been created. Undoubtedly, the GPS is the most well-known location service in use today. The approach taken by GPS, however, is unsuitable for low-cost sensor networks because GPS is based on an extensive infrastructure (i.e., satellites). Likewise, solutions developed in the area of robotic (Atiya and Hager, 1993; Leonard and Durrant-Whyte, 1991; Tins *et al.*, 2001) and ubiquitous computing (Hightower and Borriella, 2001) are generally not applicable for sensor networks because they require too much processing power and energy. Recently, a number of localization systems have been proposed specifically for sensor networks (Bulusu *et al.*, 2000; Capkun *et al.*, 2001; Niculescu and Nath, 2001). We are interested in truly distributed algorithms that can be employed for large-scale ad hoc sensor networks (100+ nodes). Such algorithms should be:

- self-organizing (i.e., do not depend on global infrastructure)
- robust (i.e., be tolerant to node failures and range errors)
- energy efficient (i.e., require little computation and, especially, communication)

These requirements immediately rule out some of the proposed localization algorithms for sensor networks. In this chapter, we briefly discuss what information is available to the nodes whose locations are unknown and then the methods in which this information can be used to derive the location of the object are discussed. We study some of the newly developed localization methods and the target tracking methods. We also review some experimental location and tracking systems.

17.2 Available Information

To find the position of an object or a device, the basic step is to use reference points (also called anchor points) whose locations are known. The object determines the distance, angle, or both, between itself and the reference point. In 2D space, if an object knows its distance from three reference points whose locations are known, it can calculate its location. In 3D space, four reference points are needed.

On the other hand, if an object knows both its distance and the angle (or the vector in 3D space) between itself and a reference point, then it can easily calculate its location. This has been exploited in the radar systems widely used in military applications. This section describes several such basic approaches. The next subsection discusses how to use multiple reference points jointly to estimate the location of a device.

17.2.1 Time of Arrival (ToA)

Measuring the distance from an object to some point P by using ToA (also known as time of flight) means measuring the time it takes to travel between the object and the point P at a known velocity. The object itself may be moving, such as an airplane traveling at a known velocity for a given time interval, or, as is far more typical, the object is approximately stationary and we are instead observing the difference in transmission and arrival time of an emitted signal. For example, sound waves have a velocity of approximately 344 m/s in 21°C air. Therefore, an ultrasound pulse sent by an object and arriving at point P 14.5 ms later allows us to conclude that the object is 5 m away from point P. Measuring the time of flight of light or a radio is also possible but requires clocks with much higher resolution (by 6 orders of magnitude) than those used for timing ultrasound because a light pulse emitted by the object has a velocity of 299,792,458 m/s and will travel the 5 m to point P in 16.7 ns. Also, depending on the capabilities of the object and the receiver at point P, it may be necessary to measure a round-trip delay corresponding to twice the distance. Ignoring pulses arriving at point P via an indirect (and, hence, longer) path caused by reflections in the environment is a challenge in measuring time of flight because direct and reflected pulses look identical. Active Bats (Ward *et al.*, 1997) and others statistically prune away reflected measurements by aggregating multiple receivers' measurements and observing the environment's reflective properties. Another issue in taking time-of-flight measurements is agreement about the time. When only one measurement is needed, as with round-trip sound or radar reflections, *agreement* is simple because the transmitting object is also the receiver and must simply maintain its own time with sufficient precision to compute the distance. However, in a system like a GPS, the receiver is not synchronized with the satellite transmitters and thus cannot precisely measure the time it took the signal to reach the ground from space. Therefore, GPS satellites are precisely synchronized with each other and transmit their local time in the signal, allowing receivers to compute the difference in time of flight. GPS receivers can compute their 3D position (i.e., latitude, longitude, and elevation) by using four satellites. The satellites are always above the receivers so only three satellites would normally be required to provide distance measurements in order to estimate a 3D position. However, in a GPS, a fourth satellite measurement is required to allow us to solve for the fourth unknown, the error between the receiver clock and the synchronized satellite clocks, a system of four equations (four satellite signals) and four unknowns (X, Y, Z, and transmission time). Refer to Dana (2000) for an excellent summary of GPS theory. To maintain synchronization, each of the 27 GPS satellites contains four cesium/rubidium atomic clocks that are locally averaged to maintain a time accuracy of 1 part in 1013. Furthermore, each satellite gets synchronized daily to the more accurate atomic clocks at the U.S. Naval Observatory by U.S. Air Force GPS ground control. Time-of-flight location-sensing systems include GPS, the Active Bat Location System (Harter *et al.*, 1999), the Cricket Location Support System (Priyantha *et al.*, 2000), Bluesoft (see http://www.bluesoft-inc.com/), and PulsON Time Modulated Ultra Wideband technology (Kelly *et al.*, 2002).

ToA systems typically use signals that move at a slower speed, such as ultrasound, to measure the time of signal arrival. Figure 17.1(a) illustrates this idea. An ultrasound signal is sent from the transmitter to the receiver; in return, the receiver sends a signal back to the transmitter. After this two-way handshake, the transmitter can infer the distance from the round-trip delay of the signals:

$$\frac{[(T_3 - T_0) - (T_2 - T_1)]V}{2}. \tag{17.1}$$

Here, V is the velocity of the ultrasound signals, T_0 is the time the sender first sent the signal, T_1 is the time the receiver received the signal, T_2 is the time the receiver sent

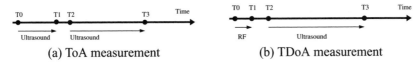

(a) ToA measurement   (b) TDoA measurement

Figure 17.1 ToA and TDoA methods. © 2006, IEEE.

back a signal, and T_3 is the time the sender received the signal from the receiver. The error of such a measurement may come from the processing time of signals [such as computing latency and the unknown delay $(T_2 - T_1)$ at the receiver's side]. Notice that the preceding formula also has several advantages compared with using $(T_3 - T_2)V$ or $(T_1 - T_0)V$:

1. One advantage is that this formula will get rid of the clock difference between the sending node and the receiving node. Notice that if we simply use formula $(T_1 - T_0)V$ to estimate the distance, the time T_1 is measured by the receiving node and time T_0 is measured by the sending node, which may have an arbitrary drift from the "real" time.
2. Another advantage is that using the average of the distance estimation $(T_3 - T_2)V$ or $(T_1 - T_0)V$, we can further reduce the error of the estimation based on large-number theory.

17.2.2   Time Difference of Arrival (TDoA)

Another distance-estimation technique is the time difference of arrival (TDoA). Although similar to the ToA scheme, this method uses two signals that travel at different speeds, such as the RF and ultrasound (US). Figure 17.1(b) shows how TDoA works; transmission in one direction is sufficient. At T_0, the transmitter sends an RF signal, followed by an US signal at time T_2. Assume that at time T_1 and time T_3, the receiver received the RF signal and the US signal, respectively. Let x denote the distance between the transmitter and receiver; then we have

$$\frac{x}{V_{\text{RF}}} = T_1 - T_0, \frac{x}{V_{\text{US}}} = T_3 - T_2,$$

and, thus,

$$x\left(\frac{1}{V_{\text{RF}}} - \frac{1}{V_{\text{US}}}\right) = (T_3 - T_1) - (T_2 - T_0),$$

where V_{RF} and V_{US} are the traveling speeds of RF and US signals, respectively. The receiver can then determine its distance to the transmitter by

$$[(T_3 - T_1) - (T_2 - T_0)]\left(\frac{V_{\text{RF}} \times V_{\text{US}}}{V_{\text{RF}} - V_{\text{US}}}\right). \tag{17.2}$$

For TDoA, in addition to errors caused by processing time, the receiver also must know the precise value of $(T_2 - T_0)$ to determine the distance.

We also discuss another approach to estimate the distance between two nodes. Notice that $\frac{x}{V_{RF}} = T_1 - T_0$ and $\frac{x}{V_{US}} = T_3 - T_2$. If we summarize them, we have $x(\frac{1}{V_{RF}} + \frac{1}{V_{US}}) = (T_3 - T_0) - (T_2 - T_1)$. Thus, we can estimate the distance from the sender to the receiver as

$$x = [(T_3 - T_0) - (T_2 - T_1)] \left(\frac{V_{RF} \times V_{US}}{V_{RF} + V_{US}} \right). \tag{17.3}$$

This formula cannot compensate for the clock drifts of the receiver and the sender.

For location of an object A, TDoA is also used. Assume that the object A will emit signals and that there are several receivers that could receive the signal from A. Observe that the pulse emitted from A will arrive at slightly different times at two spatially separated receiver sites. This TDoA is due to the different distances of each receiver from the sender A. Because the receivers may not know when the sender sent the signal, the receivers will use the difference between the time they got the signal to estimate the location of the sender. In fact, for given locations of the two receivers, a whole series of sender locations would give the same measurement of TDoA. Given two receiver locations and a known TDoA, the locus of possible sender locations is a 3D hyperboloid (i.e., a surface approximately shaped like two cones joined at the points). In simple terms, with two receivers at known locations, the sender can be located on any point of a hyperboloid. Consider now a third receiver at a third location. This would provide a second TDoA measurement and, hence, locate the sender on a second hyperboloid. The intersection of these two hyperboloids defines a curve that the sender must be on, if there is no error in time measurement. If we have a fourth receiver, then we could have three independent hyperboloids that will have a common intersection point. Thus, in three dimensions, having four different receivers v_1, v_2, v_3, and v_4 is enough to find the location of the sender by using the TDoA. In two dimensions, having three different receivers is enough to find the location of the sender by using TDoA.

17.2.3   Angle of Arrival (AoA)

The AoA approach is another commonly used method for positioning (Niculescu and Nath, 2001; Priyantha *et al.*, 2000). Such approaches require an antenna array or an array of US receivers that can determine the angle and orientation of received signals.

Angulation uses angles, instead of distances, to determine the position of an object. In general, 2D angulation requires two angle measurements and one length measurement, such as the distance between the reference points as shown in Figure 17.2.

Angulation implementations sometimes choose to designate a constant reference vector (e.g., magnetic north) as $0°$. In three dimensions, one length measurement, one *azimuth* measurement, and two angle measurements are needed to specify a precise position. Phased antenna arrays are excellent enabling technology for the angulation technique. Multiple antennas with known separation measure the ToA of a signal. Given the differences in arrival times and the geometry of the receiving array, it is then possible to compute the angle from which the emission originated. If there are enough elements

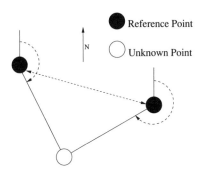

Figure 17.2 This example of 2D angulation illustrates locating an object, denoted by a white disk, using angles relative to a $0°$ reference vector and the distance between two reference points. © 2006, IEEE.

in the array and large-enough separations, the angulation calculation can be performed. The VHF Omnidirectional Ranging (VOR) aircraft navigation system is a different example of the angulation technique. As any pilot knows, VOR stations are ground-based transmitters in known locations that repeatedly broadcast simultaneous signal pulses. The first signal is an omnidirectional reference containing the station's identity. The second signal is swept rapidly through $360°$, like the light from a lighthouse, at a rate such that the signals are in phase at magnetic north and $180°$ out of phase to the south. By measuring the phase shift, aircrafts listening to a VOR station can compute their radial, the compass angle formed by the direct vector to the VOR station and magnetic north, to $1°$. Aircraft location can be computed via angulation by use of two VOR stations. VHF radio signals are limited to line-of-sight reception and the range of the transmitted signals is 40–130 nautical miles.

17.2.4 Signal Strength

Besides using signal traveling time, another distance-estimation technique is to use the property of signal degradation while traveling in a space to determine the mutual distance. Because signals traveling in a space typically reduce in strength with respect to the distance that they travel, the received signal strength (RSS) can be measured at the receiver's side. A mathematical propagation model can be derived to estimate the distance d between a transmitter and a receiver (Rappaport, 1996) as follows:

$$\mathrm{PL}(d) = \mathrm{PL}(d_0) + 10\alpha \log\left(\frac{d}{d_0}\right),$$

where $\mathrm{PL}()$ is the path-loss function with respect to distance measured in decibels, α is a loss exponent that indicates the rate at which loss increases with distance; and d_0 is the reference distance determined from a measurement close to the transmitter. The path-loss exponent α usually ranges from 2 to 4. In other words, it is assumed that the ratio of the received signal strength over the signal strength at distance d_0 is $\propto \left(\frac{d}{d_0}\right)^{\alpha}$. Using path loss may incur significant errors. For example, for IEEE 802.11b, a number of experimental studies show that a trend for the relation between distance and signal strength does exist; however, the curve is unstable in small ranges. The true signal-strength model is complex, and many uncontrollable environmental factors (e.g., shadows and terrain) are present.

To solve the preceding problem, it is necessary to model the error for signal attenuation. One possibility is to include a random variable in the preceding path-loss function as follows:

$$PL(d) = PL(d_0) + 10\alpha \log\left(\frac{d}{d_0}\right) + X_\rho,$$

where X_p is a zero-mean Gaussian random variable with a standard deviation ρ. Because of the existence of such errors, errors will occur as well when a device is positioned based on signal strength. Assuming a similar error model in measuring distances, Slijepcevic *et al.* (2003) further analyzed the location errors in a WSN and proved that the distribution of location error can be approximated by a family of *Weibull distributions*.

The intensity of an emitted signal decreases as the distance from the emission source increases. The decrease relative to the original intensity is the attenuation. Given a function correlating attenuation and distance for a type of emission and the original strength of the emission, it is possible to estimate the distance from an object to some point P by measuring the strength of the emission when it reaches P. For example, a free-space radio signal emitted by an object will be attenuated by a factor proportional to $1/r^2$ when it reaches point P at distance r from the object. In environments with many obstructions such as an indoor office space, measuring distance by use of attenuation is usually less accurate than that measured by time of flight. Signal-propagation issues, such as reflection, refraction, and multipath, cause the attenuation to correlate poorly with distance resulting in inaccurate and imprecise distance estimates. The SpotON (Hightower *et al.*, 2001) ad hoc location system implements attenuation measurement by using low-cost tags. SpotON tags use radio-signal attenuation to estimate intertag distance (Hightower *et al.*, 2000) and exploits the density of tag clusters and correlation of multiple measurements to mitigate some of the signal propagation difficulties.

17.3   Computational Complexity of Sensor Network Localization

The localization problem for sensor networks is to reconstruct the positions of all of the sensors in a network, given the distances between some pairs of sensors, and/or the angle measurements of some pairs of nodes. In the past few years, many algorithms for solving the localization problem were proposed, without knowing the computational complexity of the problem. Aspnes *et al.* (2004) showed that no *polynomial-time* algorithm can solve this problem (given the distances between some nodes) in the worst case, even for sets of distance pairs for which a unique solution exists, unless RP = NP. Recall that in complexity theory, RP (randomized polynomial time) is the complexity class of problems for which a probabilistic Turing machine exists with these properties:

- It always runs in polynomial time in the input size.
- If the correct answer is NO, it always returns NO.
- If the correct answer is YES, then it returns YES with a probability of at least $\frac{1}{2}$ (otherwise, it returns NO).

Notice that P is a subset of RP, which is a subset of NP.

Recently, numerous localization protocols have been proposed in the literature for WSNs using a variety information available from sensors. Recall that, typically, a localization algorithm can use one or some combinations of the following information, depending on the types of sensors:

1. **Link:** Whether two sensors are within the transmission range of each other. Here, we typically assume that the transmission ranges of all sensors are the same.
2. **Anchor nodes:** The geometry information of a subset of nodes, in which this subset of nodes is often static in most scenarios. In some applications, the anchor nodes (also called beacon nodes) could be mobile to provide beacon information to nodes.
3. **Orientation:** The relative orientation of a neighbor v of the current node u; i.e., the line uv has an angle θ from the north.
4. **Angle:** The angle between two neighbors v_1 and v_2 with the current node u as apex; e.g., the angle $\angle v_1 u v_2$ is $\frac{\pi}{5}$.
5. **Distance:** The Euclidean distance from the current node to a neighboring node v_1; e.g., the distance $\|v_1 u\|$ is 20 m.

Notice that the preceding information (e.g., the existence of a link, the distance, the angle, the orientation) is typically not accurate: It could have absolute errors or relative errors. In the literature, some protocols adopt a simple model by assuming that the information is precise without absolute errors or relative errors.

Although the designs of the previous schemes have demonstrated clever engineering ingenuity, and their effectiveness is evaluated through extensive simulations, the focus of these schemes is on algorithmic design, without knowing the fundamental computational complexity of the localization process. In sensor network localization, because only nodes that are within a communication range can measure their relative distances, the graphs formed by connecting each pair of nodes that can measure each other's distance are better modeled as UDGs. Such constraints could have the potential of allowing computationally efficient localization algorithms to be designed.

The localization problem considered here is to reconstruct the positions of a set of sensors given the distances between any pair of sensors that are within some unit disk radius r of each other. Some of the sensors may be beacons, sensors with known positions, but the impossibility results are not affected much by whether beacons are available. To avoid precision issues involving irrational distances, it is assumed that the input to the problem is presented with the distances squared. If we make the further assumption that all sensors have integer coordinates, all distances will be integers as well.

Notice that sometimes the localization problem may have distance information of some pairs of nodes, in addition to the angle information between some pairs of nodes.

In certain cases, the information collected may have some errors, in which the errors could be relative errors (e.g., $\frac{d'}{\|d'-d\|} \leq \delta$ for a given δ, where d' is the measured distance and d is the actual distance) or absolute errors (e.g., $\|d' - d\| \leq \epsilon$ for a given ϵ).

17.3.1   Complexity Results with Distance Information

We first study the computational complexity for localization when we know the distance information between nodes. For the main result, we consider a decision version of the localization problem, which we call UDG reconstruction. This problem essentially asks if a particular graph with given edge lengths can be physically realized as a UDG with a given disk radius in two dimensions. The input is a graph G in which each edge uv of $G = (V, E)$ is labeled with an integer l_{uv}^2, the square of its length, together with an integer r^2 that is the square of the radius of a unit disk. The output is *yes* or *no*, depending on whether there exists a mapping from vertices V to a set of points in R^2 such that the distance between u and v is l_{uv} whenever uv is an edge in G and exceeds r whenever uv is not an edge in G.

Notice that when any pair of nodes within the transmission range forms a link, the wireless network is modeled as a UDG. When a node can measure the distance to some of its neighbors within its transmission range accurately, intuitively we can find the locations of all nodes by solving some equations. For example, Biswas and Ye (2004) show that network localization in UDGs can be formulated as a semidefinite programming problem and thus can be efficiently solved. A condition of their algorithm, however, is that the graphs be densely connected. More specifically, their algorithm requires that $\Omega(n^2)$ pairs of nodes know their relative distances, where n is the number of sensor nodes in the network. Aspnes *et al.* (2004, 2006) and Eren *et al.* (2004) showed that UDG reconstruction is *NP-hard*, based on a reduction from circuit satisfiability. The constructed graph for a circuit with m wires has $O(m^2)$ vertices and $O(m^2)$ edges, and the number of solutions to the resulting localization problem is equal to the number of satisfying assignments for the circuit. In each solution to the localization problem, the points can be placed at integer coordinates, and the entire graph fits in an $O(m)$-by-$O(m)$ rectangle, where the constants hidden by the asymptotic notation are small. The construction also permits a constant fraction of the nodes to be placed at known locations. Formally, we have Theorem 17.1:

THEOREM 17.1 *There is a polynomial-time reduction from circuit satisfiability to UDG reconstruction, in which there is a one-to-one correspondence between satisfying assignments to the circuit and solutions to the resulting localization problem.*

A consequence of this result is Corollary 17.1:

COROLLARY 17.1 *There is no efficient algorithm that solves the localization problem for sparse sensor networks modeled by a UDG in the worst case unless P = NP.*

It might appear that this result depends on the possibility of ambiguous reconstructions, in which the position of some points is not fully determined by the known distances. However, if we allow randomized reconstruction algorithms, a similar result holds even for graphs that have unique reconstructions.

COROLLARY 17.2 *There is no efficient randomized algorithm that solves the localization problem for sparse sensor networks that have unique reconstructions unless RP = NP.*

Finally, because the graph constructed in the proof of Theorem 17.1 uses only points with integer coordinates, even an approximate solution that positions each point to within a distance $\epsilon < 1/2$ of its correct location can be used to find the exact locations of all points by rounding each coordinate to the nearest integer. Because the construction uses a fixed value for the unit disk radius r (i.e., the natural scale factor for the problem), we have Corollary 17.3.

COROLLARY 17.3 *The results of Corollary 17.1 and Corollary 17.2 continue to hold even for algorithms that return an approximate location for each point, provided the approximate location is within $\epsilon \cdot r$ of the correct location, where $0 \leq \epsilon < \frac{1}{2}$ is a fixed constant.*

Notice that it is unknown at present whether these results continue to hold for solutions that have large positional errors but that give edge lengths close to those in the input. The suspicion presented in Aspnes *et al.* (2004) is that edge-length errors accumulate at most polynomially across the graph, but they have not yet carried out the error analysis necessary to prove this. If this suspicion is correct, we would have Conjecture 17.1.

CONJECTURE 17.1 *The results of Corollary 17.1 and Corollary 17.2 continue to hold even for algorithms that return an approximate location for each point, provided the relative error in edge length for each edge is bounded by ϵ / n^c for some fixed constant c.*

Aspnes *et al.* (2006) and Eren *et al.* (2004) provided a theoretical foundation for the problem of network localization in which some nodes know their locations and other nodes determine their locations by measuring the distances to their neighbors. They construct grounded graphs to model network localization and apply *graph rigidity theory* to test the conditions for unique localizability and to construct uniquely localizable networks. They further study the computational complexity of network localization and investigate a subclass of grounded graphs for which localization can be computed efficiently.

Let us begin with a network $G = (V, E)$ in real d-dimensional space (where $d = 2$ or 3) consisting of a set of $m > 0$ nodes labeled 1 through m that represent special *beacon nodes* together with $n - m > 0$ additional nodes labeled $m + 1$ through n that represent ordinary nodes. Recall that the locations of all beacon nodes are already known a priori. Each node is located at a fixed position in $\mathbb{R}^d$ and has associated with it a specific set of *neighboring* nodes. The essential property about neighboring is that the definition of a neighbor is a symmetric relation in the sense that node j is a neighbor of node i if and only if node i is also a neighbor of node j. Thus, the UDG model is a special case of the preceding model. The network localization problem with distance information studied in Aspnes *et al.* (2006) and Eren *et al.* (2004) is to determine the locations p_i of all nodes in d-dimensions, given the graph of the network $G = (V, E)$, the positions of the beacons p_j in d-dimensions for $1 \leq j \leq m$, and the distance $d(i, j)$ between each neighbor pair $e = (i, j) \in E$.

DEFINITION 17.1 *Assume that we are given the network $G = (V, E)$, the positions p_j of all beacon nodes $1 \leq j \leq m$ and distance $d(i, j)$ for each $(i, j) \in E$. The network localization problem formulated is said to be* solvable *if there is* exactly one *set of positions $(p_{m+1}, p_{m+2}, \ldots, p_n)$ (one for each node in the network) consistent with the given data.*

Aspnes *et al.* (2006) and Eren *et al.* (2004) are concerned with the generic solvability of the problem that means, roughly speaking, that the problem should be solvable not only for the given data but also for slightly perturbed but consistent versions of the given data. Let function f be a mapping from

$$(p_1, p_2, \ldots, p_m, p_{m+1}, \ldots, p_n)$$

to

$$(p_1, p_2, \ldots, p_m, \|e_1\|, \|e_2\|, \ldots, \|e_q\|),$$

where $\|e_k\|$ is the Euclidean length of the kth edge in graph $G = (V, E)$ and q is the total number of edges in G. Obviously, given any positions of n nodes, the value of function f is uniquely defined. The localization problem is then to find $(p_1, p_2, \ldots, p_m, p_{m+1}, \ldots, p_n)$, given $(p_1, p_2, \ldots, p_m, \|e_1\|, \|e_2\|, \ldots, \|e_q\|)$.

DEFINITION 17.2 *The localization problem is* solvable *at given information $(p_1, p_2, \ldots, p_n)$ if there is an open neighborhood of $(p_1, p_2, \ldots, p_n)$ on which function f is an injective function. In other words, there is $\epsilon > 0$ such that, given any vector $(p_1, p_2, \ldots, p_m, \|e_1\|, \|e_2\|, \ldots, \|e_q\|)$, there is* at most *one preimage point $(p_1', p_2', \ldots, p_n')$ such that $\|p_i - p_i'\| \leq \epsilon$ for all $1 \leq i \leq n$.*

It is easy to observe that the network localization is related to the graph-rigidity theory. Roughly speaking, a graph (with a given set of vertices each corresponding to a sensor here and an edge corresponding to the link whose distance is known) with given edge distances is *rigid* in d-dimensions if for any two geometric realizations $\mathcal{G}_1$ and $\mathcal{G}_2$ of the graph in d-dimensions there is a distance preserving mapping from $\mathcal{G}_1$ to $\mathcal{G}_2$. It is widely known that a graph is rigid in d-dimensions if and only if, given the positions of $d + 1$ nodes in the graph, the positions of all other nodes are uniquely determined. Thus, we have Theorem 17.2:

THEOREM 17.2 *Suppose that there are at least $d + 1$ beacons in general position for $d = 1, 2,$ and 3. The network-localization problem is solvable if and only if point formation for the graph G is globally rigid:*

A graph G is *redundantly rigid* in d-dimensions if the removal of any single edge results in a graph that is also generically rigid in d-dimensions. The following theorem is crucial in deciding whether the graph G is generically globally rigid:

Table 17.1 Computational complexity for localization in WSNs using various information available.

Graph Model and Available Info	Hardness	Reference
UDG graph only	NP-hard	(Breu and Kirkpatrick, 1998; Kuhn et al., 2004d)
UDG graph with $O(1)$-hop distances	NP-hard	(Aspnes et al., 2004; Eren et al., 2004)
UDG graph with $O(1)$-hop angles	NP-hard	(Bruck et al., 2005)
Noisy $O(1)$-hop distances and angles	NP-hard	(Basu et al., 2006)
Accurate distances and relative angles	NP-hard	(Basu et al., 2006)
$O(1)$-hop angles and distances	in P	(Bruck et al., 2005)
$\Omega(n^2)$ pairs distances	in P	(Biswas and Ye, 2004; So and Ye, 2005)
all pairs angles	in P	(Bruck et al., 2005)

THEOREM 17.3 (Jackson and Jordan, 2005) *A graph G with n $\geq$ 4 vertices is generically globally rigid in two dimensions if and only if it is three-connected and redundantly rigid in two dimensions.*

By a polynomial-time reduction of the set-partition problem (which is NP-hard) to the globally rigid weighted graph realization problem (the network localization problem), Aspnes et al. (2006) an Eren et al. (2004) proved the following theorem:

THEOREM 17.4 *A globally rigid weighted graph realization in the plane is NP-hard.*

Recall that Corollary 17.1 is about the location of a UDG with length information of $o(n^2)$ edges.

17.3.2 Complexity Results with Other Information

We now study the computational complexity of localization when we are given the set of all wireless sensors and the set of *all* links (i, j) where sensor i and j are within the transmission range r of each other. Here, r is not given. This problem is also called *UDG Recognition* (Breu and Kirkpatrick, 1998). Breu and Kirkpatrick (1998) proved that the UDG recognition problem is NP-hard, even if the input graph is planar. Here, a realization of a UDG G is to map each vertex v of the graph to a unit disk centered at a point, say p_v, in d-dimensions (typically, $d = 2$ or 3 for sensor networks) such that there is a link uv in G if and only if $\|p_u - p_v\| \leq r$. Here, r is the uniform radius of all disks.

Besides the connectivity information and link-distance information, angle information has also been used in localization of sensors. Table 17.1 summarizes the currently known computational complexity results when various information is used for localization. In the table, $O(1)$-hop means that the information is only about some pairs of nodes that are at most a constant number of hops away.

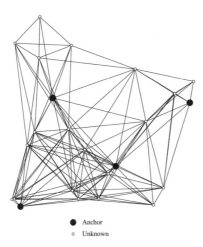

Figure 17.3 Example of a network topology in which the anchor nodes' positions are known. Here, the anchor nodes are marked as black. © 2006, IEEE.

17.4   Progressive Localization Methods

In this section, we first review some localization methods that will find the location of a node v based on some information it collected within a constant number of hops (typically, one hop) of this node v. There are several methods, such as ToA, TDoA, and *signal strength*, with which a wireless node can estimate its distance or its relative location to a reference point. Undoubtedly, the GPS is the most well-known location service in use today. The approach taken by GPS, however, is unsuitable for low-cost, ad hoc sensor networks because a GPS is based on an extensive infrastructure (i.e., satellites).

We assume that only a limited fraction of nodes, which are called *anchor* nodes, have self-location capabilities. Note that in wireless ad hoc sensor networks, there is no fine control over the placement of the sensor nodes when the network is installed (e.g., when nodes are dropped from an airplane). Consequently, we assume that nodes are randomly distributed across the environment. For simplicity and ease of presentation, we limit the environment to two dimensions, but all algorithms are capable of operating in three dimensions. Figure 17.3 shows an example network with 25 nodes; pairs of nodes that can communicate directly are connected by an edge. The connectivity of the nodes in the network (i.e., the average number of neighbors) is an important parameter that has a strong impact on the accuracy of most localization algorithms.

In some application scenarios, nodes may be mobile. Here, however, we focus on static networks, in which nodes do not move for a reasonably short period of time, because this is already a challenging condition for distributed localization. Note that anchor nodes have the same capabilities (e.g., processing, communication, energy consumption) as all other sensor nodes with unknown positions. Ideally, the fraction of anchor nodes should be as low as possible to minimize the installation costs.

Lateration computes the position of an object by measuring its distance from multiple reference positions. Calculating an object's position in two dimensions requires

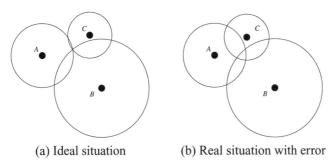

(a) Ideal situation   (b) Real situation with error

Figure 17.4 Trilateration method. © 2006, IEEE.

distance measurements from three noncollinear points, as shown in Figure 17.4(a); in three dimensions, distance measurements from four noncoplanar points are required. Domain-specific knowledge may reduce the number of required distance measurements. For example, if all reference points are above the object, then distance measurements from only three reference points are required. Two different lateration techniques are subsequently described.

17.4.1 Trilateration

Trilateration is a well-known technique in which the positioning system has a number of *beacons* at known locations. These beacons can transmit signals so that other devices can determine their distances to these beacons based on received signals. If a device can hear at least three beacons, its location can be estimated. Figure 17.4(a) shows how trilateration works: A, B, and C are beacons with known locations. From A's signal, one can determine the distance to A and, thus, that the object should be located at the circle centered at A with radius equal to estimated distance. Similarly, from B's and C's signals, it can be determined that the object should be located at some circles centered at B and C, respectively. Thus, the intersection of the three circles is the estimated location of the device. The preceding discussion assumed an ideal situation; however, as mentioned earlier, distance estimation always contains errors that will, in turn, lead to location errors. Figure 17.4(b) illustrates an example in practice. The three circles do not intersect in a common point. In this case, the maximum-likelihood method may be used to estimate the device's location. Let the three beacons A, B, and C be located at (x_A, y_A), (x_B, y_B), and (x_C, y_C), respectively. For any point (x, y) on the plane, a difference function is computed:

$$\sigma_{x,y} = |\sqrt{(x - x_A)^2 + (y - y_A)^2} - r_A| + |\sqrt{(x - x_B)^2 + (y - y_B)^2} - r_B|$$
$$+ |\sqrt{(x - x_C)^2 + (y - y_C)^2} - r_C|,$$

where r_A, r_B, and r_C are the estimated distances to A, B, and C, respectively. The location of the object can then be predicted as the point (x, y) among all points such that $\sigma_{x,y}$ is minimized.

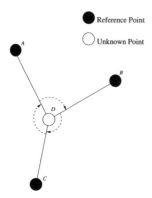

Figure 17.5 Angle measurement from three beacons A, B, and C. © 2006, IEEE.

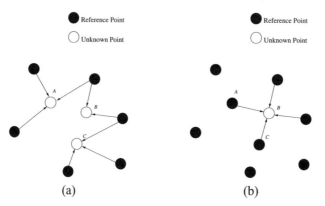

(a)   (b)

Figure 17.6 (a) Atomic multilateration (b) Iterative multilateration. © 2006, IEEE.

In addition to using the ToA approach for positioning, the AoA approach can be used. For example, in Figure 17.5, the unknown node D measures the angles of $\angle ADB$, $\angle BDC$, and $\angle ADC$ by the received signals from beacons A, B, and C. From this information, D's location can be derived (Niculescu and Nath, 2001) when we know the locations of three beacon nodes.

17.4.2 Multilateration

The trilateration method has its limitation in that at least three beacons are needed to determine a device's location. In a sensor network, in which nodes are randomly deployed, this may not be true. Several multilateration methods are proposed to relieve this limitation. The AHLoS (Ad Hoc Localization System) (Savvides *et al.*, 2001) enables nodes to discover their locations by using a set of distributed iterative algorithms. Figure 17.6 shows an example in which, initially, beacon nodes contain only nodes marked as reference points. Device nodes A, B, and C are at unknown locations. In the first iteration, as Figure 17.6(a) shows, the locations of nodes A and C will be determined.

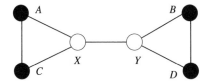

Figure 17.7 An example of a network in which trilateration cannot localize the position of all nodes.

Once the location of a device is estimated, its role is changed to a beacon node so as to help in determining other devices' locations. This is repeated until all hosts' locations are determined (if possible). As Figure 17.6(b) shows, in the second iteration, the location of node B can be determined with the help of nodes A and B, which are now serving as beacons.

If the distance or regular estimation is precise, we can show that the order in which each node determines its location and then serves as beacon node will *not* affect the number of nodes whose positions can be computed. However, when the information is not precise, it does affect this and, further, the precision of the system.

Notice that a simple iterated trilateration is suboptimal in that there are many localizable networks that cannot be localized by this simple iterated trilateration. Only networks called trilateration networks are completely localized by iterated trilateration (Eren *et al.*, 2004). Figure 17.7 illustrates an example of networks that are localizable but not localizable by simple iterated trilateration. Here, nodes A, B, C, D are nodes with known positions. We need to find the positions of nodes X and Y. A link in the figure denotes that the distance between the two end nodes is known by some techniques. Clearly, the positions of X and Y are fixed. However, none of them has distance information to three nodes; thus, their positions cannot be localized using trilateration.

17.4.3　Sweep and Linear-Programming-Based Techniques

Recently, Goldenberg *et al.* (2006) proposed a similar method for localization in sparse networks using the sweep technique. They studied fine-grained localization with the following requirements: (1) compute positions without potential systematic errors; (2) localize uniquely localizable nodes with high probability, even in networks that are not uniformly dense; and (3) for nodes that cannot be uniquely localized but can be localized up to a set of possibilities output the set of possible positions whenever feasible. It is easy to envision that outputting the set of all possible positions of a node can be useful in many applications. They call this generalized localization objective finite localization, as opposed to unique localization, on which most previous localization techniques focus. The idea of the sweeps algorithm is related to simple iterated trilateration. In iterated trilateration, an initial set of three nodes (with some known position) is fixed and used to define a coordinate system. At each stage of the algorithm, there is a set of localized nodes and a set of unlocalized nodes. If an unlocalized node has distance measurements to at least three localized nodes, its position is calculated and it is added to the set of localized nodes. After a node is added to the set of swept nodes, all of its distance measurements to other already swept nodes are considered, potentially eliminating some of its possible positions and some of the possible positions of already swept nodes. Note that there exist examples, such as wheel networks (Eren *et al.*, 2004; Goldenberg *et al.*, 2006), for which there are no chances to eliminate any possibilities until the very last edge is added. In other words, the basic idea of a sweeps algorithm is to process nodes in some order and use information to refine the position possibilities of nodes. To further reduce the growth in possible positions, they propose using a particular sweep ordering. In the so-called shell sweeps, their algorithm performs a breadth-first sweep in which, at each stage, all nodes having a distance measurement to *at least* two already swept nodes are placed earlier in the ordering than all other nodes. Let us use the example in Figure 17.7 again. Node x has distance measurement to two nodes with known positions; it is then added to swept nodes, and its position has two possibilities (and similarly for node y). Then, because node x also knows the distance to node y, it will eliminate one of its position possibilities and result in a correct localization.

Notice that this technique is different from the method used in Basu *et al.* (2006). Basu *et al.* (2006) relaxed the nonconvex constraints of noise data (about distance measurement and angle measurement) to approximating convex constraints and proposed LP for two formulations of the resulting localization problem, which they call the *weak-deployment* and *strong-deployment* problems. These two formulations give upper and lower bounds on the location uncertainty, respectively: No sensor is located outside its weak-deployment region, and each sensor can be anywhere in its strong-deployment region without violating the approximate distance and angle constraints. Although LP-based algorithms are usually solved by centralized methods, they proposed distributed, iterative methods, which are provably convergent to the centralized algorithms solutions.

17.4.4   Distributed Localization Using Noisy Data

Recently, Basu *et al.* (2006) studied the distributed localization of WSNs by using noisy distance and/or angle information. First, if we have the *accurate* distance and angle information of all pairs (or a sufficient number of pairs) of nodes, we can easily compute the location of all nodes (with the knowledge of constant number of anchor nodes) by using semidefinite programming (Biswas and Ye, 2004). Unfortunately, Basu *et al.* (2006) proved that the localization problem becomes essentially NP-hard when we are given *noisy* distance and angle information. Here, distance information is *noisy* if we know only that $d \leq \|u - v\| \leq D$ for some values $d < D$; angle information is *noisy* if we know only that the direction of node v with respect to a node u is $\alpha_1 \leq \angle Nuv \leq \alpha_2$ for some values $\alpha_1 \leq \alpha_2$. Here, N is a node pointing to North. Notice that the angle constraints for node v can be represented by two linear constraints. However, the constraints on the distance cannot be represented by linear conditions (they are actually quadratic). Then, they relax these constraints to linear constraints by using a bounding box to represent the area where node v could be located. See Figure 17.8 for an illustration.

From the relaxation of the feasible region of node v relative to the position of node u, they studied the weak-deployment region and the strong-deployment region of a WSN.

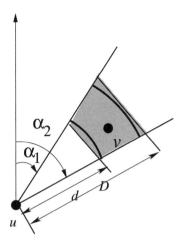

Figure 17.8 The actual feasible region of node v given noisy distance and angle information. The shaded region denotes a relaxed feasible region for node v, which can be represented by linear constraints on the position (x, y) of node v. Here, the angles α_1 and α_2 denote the smallest direction and the largest direction of node v. Distances d and D are the smallest and the largest distances of node v to node u. Notice that this feasible region and relaxed feasible region is only the view of node u.

Here, for the weak-deployment problem, we need to find the (set-theoretic) maximal deployment regions W_i such that for any point $p \in W_i$, there exists a feasible placement of all other nodes $v_j \neq v_i$ that is consistent with placing node v_i at position p. The strong-deployment problem asks us to find deployment regions S_i, of a specified shape, such that *all* constraints implied by the regions of uncertainty for neighboring pairs are satisfied if we locate node v_i at *any* point within the deployment region S_i, for each $i = 1, 2, \ldots, n$. The objective function is to maximize (maximally scale) the size of the smallest deployment region. The motivation behind the preceding definitions is in obtaining upper and lower bounds on the uncertainty of node locations. The weak-deployment region captures all possible locations for a node given the constraints of the given noisy distance and angle information and the positions of some anchor nodes. On the other hand, the strong-deployment region captures sort of the best of what an algorithm can do for localization. For the strong-deployment region, *any combination of points* within these regions simultaneously satisfies *all constraints*, based on the measurement data. One cannot tell where the real node location is within the strong-deployment region. In other words, any location in the strong feasible region for a node could be the real position of the node given the current known information. Notice that the preceding definition is based on the actual feasible region. Basu *et al.* (2006) provide both distributed and centralized algorithms to approximate the weak-deployment region and the strong-deployment region by using the relaxation of the boundary.

Hu and Evans (2004) studied the location for mobile sensor networks for which some mobile nodes will serve as anchor nodes. Although mobility would appear to make localization more difficult, Hu and Evans (2004) introduced the sequential Monte Carlo localization method and argue that it can exploit mobility to improve the accuracy and precision of localization.

17.5   Network-Wide Localization Methods

In this section, we review in detail some of the localization methods proposed in the literature, in which nodes without location information will first collect some information related to the beacon nodes with known location information. From the information collected, they will try to find the location for themselves. The specific method that we study is called an APS. An APS (ad hoc positioning system) (Niculescu and Nath, 2001) is a distributed, hop-by-hop positioning algorithm to provide approximate locations for all nodes in a network for which only a limited fraction of nodes have self-location capability. Also, an APS is appropriate for indoor location-aware applications.

17.5.1   Algorithm

It is not desirable to have the landmarks emit with large power to cover the entire network for several reasons: collisions in local communication, high power usage, coverage problems when moving. Also, it is not acceptable to assume some fixed positions for the landmarks because the applications envisioned by APSs are either in flight deployments over inaccessible areas or possibly involve movement and reconfiguration of the network. In this case, one option is to use hop-by-hop propagation capability of the network to forward distances to landmarks. In general, they aim for the same principle as GPSs, with the difference that the landmarks are contacted in a hop-by-hop fashion, rather than directly, as ephemerides are. Once an arbitrary node has estimates to a number (≥ 3) of landmarks, it can compute its own position in the plane, using a procedure similar to the one used in GPS position calculation described in the previous section. The estimate we start with is the centroid of the landmarks collected by a node. In what follows, we refer to one landmark only because the algorithm behaves identically and independently for all the landmarks in the network. It is clear that the immediate neighbors of the landmark can estimate the distance to the landmark by direct signal-strength measurement. Using some propagation method, the second-hop neighbors then are able to infer their distance to the landmark, and the rest of the network follows in a controlled flood manner, initiated at the landmark. Complexity of signaling is therefore driven by the total number of landmarks and by the average degree of each node.

17.5.2   Distance to Anchors

The APS uses three methods of hop-to-hop distance propagation and examines advantages and drawbacks for each of them. These three methods are **Sum-dist**, **DV-hop**, and **Euclidean**, which are discussed in detail later. Each propagation method is appropriate for a certain class of problems as it influences the amount of signaling, power consumption, and position accuracy achieved.

Nodes that can communicate with anchor nodes directly are able to find their distance to anchor nodes, but this information is not available to all nodes. Nodes share information to collectively determine the distances between individual nodes and the anchors, so that an (initial) position can be calculated. None of the alternatives engages in complicated calculations, so finding the distance to anchors is communication-bounded. Most of distributed localization algorithms share a common communication pattern: Information is flooded into the network, starting at the anchor nodes. A network-wide flood by some anchor A is expensive because each node must forward A's information to its (potentially) unaware neighbors. This implies a scaling problem: Flooding information from all anchors to all nodes will become too expensive for large networks, even with low anchor fractions. Fortunately, a good position can still be derived with knowledge (position and distance) from a limited number of anchors. Therefore, nodes can simply stop forwarding information when enough anchors have been *located*. This simple optimization (with a *flood limit*) has proved to be highly effective in controlling the amount of communication.

Sum-Dist

This method is also known as *DV-distance* (Niculescu and Nath, 2001). The basic idea of this method is that every node will find the *shortest distance* to an anchor node using available measurements. Observe that this distance does not reflect the actual Euclidean distance from this node to the anchor: It is an approximation of the shortest graph distance. The simplest solution for determining the distance to the anchors is simply adding the ranges encountered at each hop during the network flood. Sum-dist starts at the anchors, which send a message including their identity, position, and a path length set to 0. Each receiving node adds the measured range to the path length and forwards (i.e., broadcasts) the message if the flood limit allows it to do so. Another constraint is that when the node has received information about the particular anchor before, it is allowed to forward the message only if the current path length is less than the previous one. The end result is that each node will have stored the position and minimum path length to at least flood-limit anchors.

DV-Hop

A drawback of Sum-dist is that range errors accumulate when distance information is propagated over multiple hops. This cumulative error becomes significant for large networks with few anchors (long paths) and/or poor ranging hardware. A robust alternative is to use topological information by counting the number of hops instead of summing the (erroneous) ranges. This approach was named DV-hop by Niculescu and Nath (2001) and Hop-TERRAIN by Savarese *et al.* (2002). The DV-hop propagation method is the most basic scheme, and it first employs a classical distance-vector exchange so that all nodes in the network get distances, in hops, to the landmarks. DV-hop essentially consists of two flood waves. After the first wave, which is similar to Sum-dist, nodes have obtained the position and minimum hop count to at least flood-limit anchors. The second calibration wave is needed to convert hop counts into distances such that nodes can compute a position. This conversion consists of multiplying the hop count by an average hop distance. Whenever an anchor A_1 infers the position of another anchor A_2 during the first wave, it computes the distance between them and divides that by the number of hops to derive the average hop distance between A_1 and A_2. When

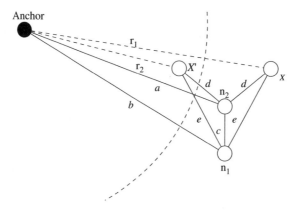

Figure 17.9 Determining distance using the Euclidean method.

calibrating, an anchor takes into account all remote anchors of which it is aware. Nodes forward (broadcast) calibration messages only from the first anchor that calibrates them, which reduces the total number of messages in the network. The main difficulty of using the DV-hop approach is in getting a better estimate of the average hop distance. The second difficulty is that the hop distance could have a very large variance. Similar to Sum-dist, this approach will not directly approximate the Euclidean distance from this node to an anchor.

Euclidean

A drawback of DV-hop is that it fails for highly irregular network topologies, in which the variance in actual hop distances is very large. Niculescu and Nath (2001) proposed another method, called Euclidean, that is based on the local geometry of the nodes around an anchor. Again, anchors initiate a flood, but forwarding the distance is more complicated than in the previous cases. When a node has received messages from two neighbors that know their distance to the anchor and to each other, it can calculate the distance to the anchor. Figure 17.9 shows a node X that has two neighbors n_1 and n_2 with distance estimates (a and b) to an anchor. Together with the known ranges c, d, and e, there are two possible values (r_1 and r_2) for the distance of the node to the anchor. Niculescu describes two methods to decide which, if any, distance to use. The neighbor-vote method can be applied if there exists a third neighbor n_3 that has a distance estimate to the anchor and that is connected to either n_1 or n_2. Replacing n_2 (or n_1) with n_3 will again yield a pair of distance estimates. The correct distance is part of both pairs and is selected by simple voting. Of course, more neighbors can be included to make the selection more accurate.

Node Position

Now, nodes can determine their position by using lateration, min–max (presented by Savvides *et al.*, 2002), or other methods based on the distance estimates to a number of anchors provided by one of the three alternatives (i.e., Sum-dist, DV-hop, or Euclidean). The determination of the node positions does not involve additional communication.

17.6   Target Tracking and Classification

One of the most important areas in which the advantages of sensor networks can be exploited is for tracking mobile targets. Scenarios in which such networks may be deployed can be both military (e.g., tracking enemy vehicles, detecting illegal border crossings) and civilian (e.g., tracking the movement of wild animals in wildlife preserves). Typically, for accuracy, two or more sensors are simultaneously required for tracking a single target, leading to coordination issues. Additionally, given the requirements to minimize the power consumption that is due to communication or other factors, we would like to select the bare essential number of sensors dedicated for the task while all other sensors should preferably be in the hibernation or off state. To simultaneously satisfy the requirements like power-saving and improving overall efficiency, we need large-scale coordination and other management operations. These tasks become even more challenging when one considers the random mobility of the targets and the need to coordinate the assignment of the sensors best suited for tracking the target as a function of time. In this section, managing and coordinating a sensor network for tracking moving targets is discussed.

The power limitations that are due to the small size of the sensors, the large numbers of sensors that need to be deployed and coordinated, and the ability to deploy sensors in an ad hoc manner give rise to a number of challenges in sensor networks. Each of these needs to be addressed by any proposed architecture in order for it to be realistic and practical.

- **Scalable coordination:** A typical deployment scenario for a sensor network comprises a large number of nodes reaching in the thousands to tens of thousands. At such large scales, it is not possible to attend to each node individually because of a number of factors. Sensor nodes may not be physically accessible, nodes may fail, and new nodes may join the network. In such dynamic and unpredictable scenarios, scalable coordination and management functions are necessary, which can ensure a robust operation of the network. In light of target tracking, the coordination function should scale with the size of the network, the number of targets to be tracked, number of active queries, and so forth.
- **Tracking accuracy:** To be effective, the tracking system should be accurate and the likelihood of missing a target should be low. Additionally, the dynamic range of the system should be high while keeping the response latency, sensitivity to external noise, and false alarms low. The overall architecture should also be robust against node failures.
- **Ad hoc deployability:** A powerful paradigm associated with sensor networks is their ability to be deployed in an ad hoc manner. Sensors may be thrown in an area affected by a natural or manmade disaster or air-dropped to cover a geographical region. Thus, sensor nodes should be capable of organizing themselves into a network and achieving the desired objective in the absence of any human intervention or fixed patterns in the deployment.
- **Power constraints:** The available power in each sensor is limited by the battery lifetime because of the difficulty or impossibility of recharging the nodes. As a consequence,

protocols that tend to minimize the energy-consumption or power-aware protocols that adapt to the existing power levels are highly desirable. Additionally, efforts should be made to turn off the nodes themselves if possible in the absence of sensing or coordination operations.

- **Computation and communication costs:** Any protocol being developed for sensor networks should keep in mind the costs associated with computations and communication. With current technology, the cost of computation locally is lower than that of communication in a power-constrained scenario. As a consequence, emphasis should be put on minimizing the communication requirements.

17.6.1   Pattern Matching Using Database

Pattern matching (also known as *fingerprinting*) tries to compare the received signal pattern against the training patterns in the database and to determine the likelihood that the device is currently located in a position. A typical solution has two phases:

- **Off-line phase:** The purpose of this phase is to collect signals from all base stations at each training location; thus, the received signal strengths are recorded in the database. For higher accuracy, one may establish multiple entries in the database for the same training location. From the database, some positioning rules, which form the positioning model, will then be established.
- **Real-time phase:** With a well-trained positioning model, one can estimate a device's location given the signal strengths collected by the device from all possible base stations. The positioning model may determine a number of locations, each associated with a probability.

There are several similar searching methods in the matching process in the literature such as *nearest-neighbor algorithms* (Bahl and Padmanabhan, 2000; Bahl *et al.*, 2000) and *probability-based algorithms* (Roos *et al.*, 2002).

17.6.2   Network-Based Tracking

At the network level, location tracking may be done via the cooperation of sensors. Tseng and colleagues (2003) addressed these issues by using an agent-based paradigm. Once a new object is detected by the network, a mobile agent will be initiated to track the roaming path of the object. Then, the agent invites some nearby slave sensors to cooperatively position the object and inhibit other irrelevant (i.e., farther) sensors from tracking the object. More precisely in their protocol, only three agents will be used for the tracking purpose at any time, and they will move as the object moves. The trilateration method is used for positioning.

Figure 17.10 shows an example. The sensor network is deployed in a regular manner, and it is assumed that each sensor's sensing distance equals the distance between two neighboring sensors. Initially, each sensor is in the idle state, searching for new objects. Once detecting a target, a sensor will transit to the election state, trying to serve as the master agent. The sensor nearest to the target will win. The master agent will then

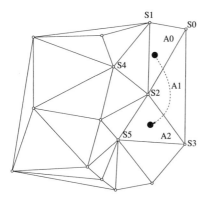

Figure 17.10 Roaming path of an object (dashed curve). © 2006, IEEE.

dispatch two neighboring sensors as the slave agents; master and slave agents will cooperate to position the object. In the figure, the object is first tracked by sensors $\{S_0, S_1, S_2\}$ when resident in A_0, then by $\{S_0, S_2, S_3\}$ when in A_1, by $\{S_2, S_3, S_5\}$ when in A_2, and so forth. The master agent is responsible for collecting all sensing data and performing the trilateration algorithm. It also conducts data fusion by keeping the tracking results while it moves around. At the proper time, the master agent will forward the tracking result to the data center.

17.6.3   Collaborative Signal Processing

Power consumption is a critical consideration in a WSN. The limited amount of energy stored at each node must support multiple functions, including sensor operations, on-board signal processing, and communication with neighboring nodes. Thus, one must consider power-efficient sensing modalities, low sampling rates, low-power-signal processing algorithms, and efficient communication protocols to exchange information among nodes. To facilitate monitoring of a sensor field, including detection, classification, identification, and tracking of targets, global information in both space and time must be collected and analyzed over a specified space–time region. However, individual nodes provide only spatially local information. Furthermore, because of power limitations, temporal processing is feasible over only limited time periods. This necessitates *collaborative signal processing (CSP)* (i.e., collaboration between nodes to process the space–time signal). A CSP algorithm can benefit from the following desirable features:

- **Distributive processing:** Raw signals are sampled and processed at individual nodes but are not directly communicated over the wireless channel. Instead, each node extracts relevant summary statistics from the raw signal, which is typically of a smaller size. The summary statistics are stored locally in individual nodes and may be transmitted to other nodes on request. This requires signal-processing and some simple machine-learning techniques to process raw signals.
- **On-demand processing:** To conserve energy, each node performs only signal-processing tasks that are relevant to the current query. In the absence of a query,

each node retreats into a standby mode to minimize energy consumption. Similarly, a sensor node does not automatically publish extracted information (i.e., it forwards such information only when needed).

- **Information fusion:** To infer global information over a certain space–time region from local observations, CSP must facilitate efficient, hierarchical information fusion and progressively lower bandwidth information must be shared between nodes over progressively large regions. For example, (high-bandwidth) time-series data may be exchanged between neighboring nodes for classification purposes. However, lower-bandwidth CPA (closest point of approach) data may be exchanged between more distant nodes for tracking purposes.

- **Multiresolution processing:** Depending on the nature of the query, some CSP tasks may require higher spatial resolution involving a finer sampling of sensor nodes or higher temporal resolution involving higher sampling rates. For example, reliable detection may be achievable with a relatively coarse space–time resolution, whereas classification typically requires processing at a higher resolution.

17.6.4   Target Tracking Using Space–Time Cells

Each object in a geographical region generates a time-varying space–time signature field that may be sensed in different modalities, such as acoustic, seismic, or thermal. The sensors sample the signature field spatially, and the density of the sensor should be commensurate with the rate of spatial variation in the field. Similarly, each sensor should sample at a rate commensurate with the required bandwidth. Thus, the rate of change of the space–time signature field and the nature of the query determines the required space–time sampling rate. A moving object in a region corresponds to a peak in the spatial signal field that moves with time. Tracking an object corresponds to tracking the location of the spatial peak over time. D. Li *et al.* (2002) developed a space–time cell approach for tracking moving objects.

Using Space–Time Cells

To enable tracking in a sensor network, the entire space–time region is divided into *space–time cells* to facilitate local processing. The size of a space–time cell depends on the velocity of the moving target and the decay exponent of the sensing modality. It should approximately correspond to a region over which the space–time signature field remains nearly constant. In principle, the size of space–time cells may be dynamically adjusted as new space–time regions are created based on predicted locations of targets. Space–time signal averaging may be done over nodes in each cell to improve the signal-to-noise ratio. We note that the assumption of a constant signature field over a space–time cell is at best an approximation in practice because of several factors, including variations in terrain, foliage, temperature gradients, and the nonisotropic nature of source signal. However, such an approximation may be judiciously applied in some scenarios for the purpose of reducing intrasensor communication as well to improve algorithm performance against noise.

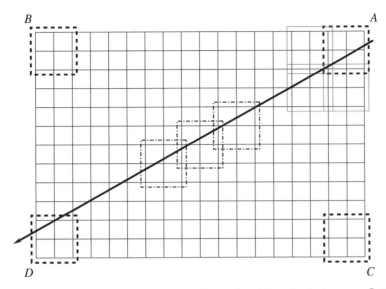

Figure 17.11 Schematic illustrating detection and tracking of a single target. © 2006, IEEE.

Single-Target Tracking

One of the key premises behind the networking algorithms being developed in Wisconsin (D. Li *et al.*, 2002) is that routing of information in a sensor network should be geographic-centric rather than node-centric. In other words, from the viewpoint of information routing, the geographic locations of the nodes, rather than their arbitrary identities, are the critical quantities. Some of the nodes in each cell are designated as *manager nodes* for coordinating signal processing and communication in that cell.

Figure 17.11 illustrates the basic idea of region-based CSP for detection and tracking of a single target. Under the assumption that a potential target may enter the monitored area via one of the four corners, four cells (A, B, C, and D) are created. Nodes in each of the four cells are activated to detect potential targets. Each activated node runs a detection algorithm whose output is sampled at an a priori fixed rate depending on the characteristics of expected targets. Suppose a target enters cell A. Tracking of the target consists of the following five steps:

1. Some and perhaps all of the sensor nodes in cell A detect the target. These nodes are the active nodes and cell A is the active cell. The active nodes also yield CPA time information. The active nodes report their energy detector outputs to the manager nodes at N successive time instants.
2. At each time instant, the manager nodes determine the location of the target from the energy detector outputs of the active nodes. The simplest estimate of target location at an instant is the location of the node with the strongest signal at that instant. However, more sophisticated algorithms for target localization may be used to achieve finer accuracy of localization with possibly higher complexity.
3. The manager nodes use locations of the target at the N successive time instants to predict the location of the target at $M(< N)$ future time instants.

4. The predicted positions of the target are used by the UW-API protocols (D. Li *et al.*, 2002) to create new cells that the target is likely to enter. This is illustrated in Figure 17.11, where the three dotted cells represent the regions that the target is likely to enter after the current active cell (cell *A* in Figure 17.11). A subset of these cells is activated for subsequent detection and tracking of the target.

5. Once the target is detected in one of the new cells, it is designated the new active cell, and the nodes in the original active cell (cell *A* in Figure 17.11) may be put in the standby state to conserve energy.

Steps 1–5 are repeated for the new active cell, and this forms the basis of detecting and tracking a single target. For each detected target, the tracking information, such as the location of the target at certain past times, is usually passed from one active cell to the next one.

Multiple-Targets Tracking

We then discuss the approach by UW-API protocol to track multiple targets. If multiple targets are sufficiently separated in space or time – that is, they occupy distinct space–time cells – essentially the same procedure as described in Subsection 17.6.4 may be used: A different track is initiated and maintained for each target. Sufficient separation in time means that the energy-detector output of a particular sensor exhibits distinguishable peaks corresponding to the CPAs of the two targets. Similarly, sufficient separation in space means that at a given instant, the spatial target signatures exhibit distinguishable peaks corresponding to nodes that are closest to the targets at that instant. The assumption of sufficient separation in space, time, or both may be too restrictive in general. In such cases, classification algorithms are needed that operate on spatiotemporal target signatures to classify them. This necessarily requires a priori statistical knowledge of typical signatures for different target classes.

Target Classification

Here, we focus on single-node (i.e., no collaboration between nodes) classification based on temporal target signatures: A time-series segment is generated for each detected event at a sensor node and processed for classification. Some form of temporal processing, such as a fast Fourier transform (FFT), is performed and the transformed vector is fed to a collection of classifiers corresponding to different target classes. The outputs of the classifiers that detect the target, active classifiers are reported to the manager nodes, as opposed to the energy-detector outputs. Steps 1–5 in this subsection are repeated for all the active classifier outputs to generate and maintain tracks for different classified targets. In some cases, both energy-based CPA information and classifier outputs may be needed.

Classification or, more specifically, statistical classification is a widely studied subject in machine learning. It is a procedure in which individual items are placed into groups based on quantitative information on one or more characteristics inherent in the items (referred to as traits, variables, characters, and so on) and based on a training set of previously labeled items. Formally, the problem can be stated as follows: Given training data $\{(x_1, y_i), \ldots, (x_n, y_n)\}$ produce a classifier $h : \mathcal{X} \to \mathcal{Y}$ that maps an object $x \in \mathcal{X}$ to its classification label $y \in \mathcal{Y}$. For example, in the problem of target tracking studied here, then x_1 is some representation of a target and y is a possible target. Examples of classification algorithms include linear classifiers (e.g., logistic regression and Perceptron), quadratic classifiers, k-nearest neighbor, decision trees, neural networks, Bayesian networks, support vector machines, and hidden Markov models. We later review some of the methods in detail in the context of target classification for WSNs. Given a set of N-dimensional feature vectors $\{x | x \in R^N\}$, we assume that each of them is assigned a class label, $\omega_c \in \Omega = \{\omega_1, \omega_2, \ldots, \omega_m\}$, that belongs to a set of m elements. We denote by $p(\omega_c)$ the prior probability that a feature vector belongs to class ω_c. Similarly, $p(\omega_c | x)$ is the posterior probability for class ω_c given that x is observed.

A *minimum error classifier* maps each vector x to an element in such a way that the probability of misclassification (i.e., the probability that the classifier label is different from the true label) is minimized. To achieve this minimum error rate, the optimal classifier decides x has label ω_i if $p(\omega_i | x) \geq p(\omega_j | x)$ for all $j \neq i$, $\omega_i, \omega_j \in \Omega$. In practice, it may be very difficult to evaluate the posterior probability in a closed form. Instead, one may use an appropriate discriminant function $g_i(x)$ that satisfies $g_i(x) > g_j(x)$ if $p(\omega_i | x) > p(\omega_j | x)$ for $j \neq i$, for all x. Then, the minimum error classification can be achieved: Decide x has label ω_i if $g_i(x) > g_j(x)$ for $j \neq i$. The minimum probability of misclassification is also known as the *Bayes* error, and a minimum error classifier is also known as a Bayes classifier or a maximum a posteriori probability (MAP) classifier. Three classifiers that approximate the optimal Bayes classifier are briefly discussed.

- **k-nearest-neighbor (kNN) classifier:** The kNN classifier uses all the training features as the set of prototypes $\{p_k\}$. During the testing phase, the distance between each test vector and every prototype is calculated, and the k prototype vectors that are closest to the test vector are identified. The class labels of these k-nearest prototype vectors are then combined using majority vote or some other method to decide the class label of the test vector. When $k = 1$, the kNN classifier is called the nearest-neighbor classifier. It is well known that asymptotically (in the number of training vectors), the probability of misclassification of a nearest-neighbor classifier approaches twice the (optimal) Bayes error. Hence, the performance of a nearest-neighbor classifier can be used as a baseline to gauge the performance of other classifiers. However, as the number of prototypes increases, a kNN classifier is not very suitable for actual implementation because it requires too much memory storage and processing power for testing.

- **Maximum-likelihood (ML) classifier:** Using a Gaussian mixture density model, in this classifier, the distribution of training vectors from the same class is modeled as a mixture of Gaussian density functions. That is, the likelihood function is modeled as

$$p(x|\omega_i) \propto G_i(x|\theta_i) = \sum_k |\Lambda_{ik}|^{-N/2} \exp\left[-\frac{1}{2}(x - m_{ik})^T \Lambda_{ik}^{-1}(x - m_{ik})\right], \quad (17.4)$$

where $\theta_i = [m_{i1}, m_{i2}, \ldots, m_{ip}, \Lambda_{i1}, \Lambda_{i2}, \ldots, \Lambda_{ip}]$ are the mean and covariance matrix parameters of the P mixture densities corresponding to class ω_i. These model parameters can be identified by applying an appropriate clustering algorithm, such as the k-means algorithm or the expectation-maximization algorithm, to the training vectors of each class. The discriminant function is computed as $g_i(x) = G_i(x|\theta_i)p(\omega_i)$ in which the prior probability $p(\omega_i)$ is approximated by the relative number of training vectors in class i.

- **Support vector machine (SVM) classifier:** A SVM is essentially a linear classifier operating in a higher-dimensional space. Consider a binary classification problem without loss of generality. Let $\{\varphi(x)\}_{i=1}^M$ be a set of nonlinear transformations mapping the N-dimensional input vector to an M-dimensional feature space ($M > N$). A linear classifier, characterized by the weights $w_1, w_2, \ldots, w_M$, operates in this higher-dimensional feature space $g(x) = \sum_{j=1}^M w_j \varphi_j(x) + b$, where b is the bias parameter of the classifier. The optimal weight vectors for this classifier can be represented in terms of a subset of training vectors, termed the *support vectors* $w_j = \sum_{i=1}^Q \alpha_i \varphi_j(x_i), j = 1, 2, \ldots, M$. Using the preceding representation for the weight vectors, the linear classifier can be expressed as $g(x) = \sum_{i=1}^Q \alpha_i K(x, x_i) + b$, where $K(x, x_i) = \sum_{j=1}^M \varphi_j(x)\varphi_j(x_i)$ is the symmetric kernel representing the SVM. In practice, the SVM discriminant function $g(x)$ is computed using the kernel representation, bypassing the nonlinear transformation into the higher-dimensional space (Alon *et al.*, 2006). The classifier design then corresponds to the choice of the kernel and the support vectors. By appropriately choosing the kernel, a SVM can realize a neural network classifier as well. Similar to neural networks, the training phase can take a long time. However, once the classifier is trained, its application is relatively easy. In general, a different SVM is trained for each class. The output of each SVM can then be regarded as an estimate of the posterior probability for that class and the MAP decision rule can be directly applied.

17.6.5   Target Tracking Based on Cooperative Binary Detection

Unlike other sensor-network-based methods, which depend on determining distances to the target or the AoA of the signal, the cooperative tracking approach requires only that a sensor be able to determine if an object is somewhere within the maximum detection range of the sensor. Cooperative tracking is proposed as a method for tracking moving objects and extrapolating their paths in the short term. By combining data from neighboring sensors, this approach enables tracking with a resolution higher than that of the individual sensors being used. In cooperative tracking, statistical estimation and approximation techniques can be employed to further increase the tracking precision and enable the system to exploit the trade-off between accuracy and time lines of the results. The method reviewed here focuses on acoustic tracking; however, the presented methodology is applicable to any sensing modality where the sensing range is relatively uniform. Here, an acoustic tracking system for WSNs is considered a practical application of the cooperative tracking methodology. Acoustic tracking relies on a network of microphone-equipped sensor nodes to track an object by its characteristic *acoustic signature*.

Cooperative tracking is a solution for tracking objects using sensor networks and may achieve a high degree of precision while meeting the constraints of sensor network systems. The approach uses distributed sensing to identify an object and determine its approximate position and local coordination and processing of sensor data to further refine the position estimate. The salient characteristics of the cooperative tracking approach are that it achieves resolution that is finer than that of the individual sensors being used and that it provides early estimates of the object's position and velocity. Thus, cooperative tracking is useful for short-term extrapolation of the object's path.

System Model

In the real world, objects can move arbitrarily; i.e., possibly changing speed and direction at any time. The representation of such arbitrary paths may be cumbersome and unnecessarily complex for the purpose of tracking the object's path with a reasonable degree of precision. Instead, an approximation of the path can be considered. Cooperative tracking uses a *piecewise linear approximation* to represent the path of the tracked object. Although the object itself may move arbitrarily, its path is considered a sequence of line segments along which the object moves with a constant speed. The degree to which the actual path diverges from its representation depends on several factors, including speed and turning radius of the object itself. For vehicles such as cars driving along highways, the difference is quite small, whereas for a person walking a curved route with tight turns, it may be significant. In either case, accuracy can be improved by increasing the resolution of the sensor network, either through increasing sensor density or by other means.

In cooperative tracking, it is assumed that each node is equipped with a sensor (in the case of acoustic tracking, a microphone) and a radio for communication with nearby nodes. Because these embedded systems are designed to be small and cheap, the sensors they are equipped with are unlikely to be very sophisticated. Traditionally, tracking relies on sensors that are long range and can detect the direction of an object and the distance to it. This is not the case with sensor networks: The microphones used for acoustic tracking are likely to be short-range, nondirectional, and poorly suited for detecting the distance to the sound source. We assume that only binary (on–off) detection can be used. It is possible to generalize this analysis if multilevel detection is feasible. Moreover, without proper calibration, the detection range may be neither uniform nor exact. Figure 17.12 shows the model of a sensor considered in this chapter. Given a sensor with a nominal (i.e., noncalibrated) range R, the object will always be detected if it is distance $R - e$ or less away from the sensor, detected some of the time between $R - e$ and $R + e$, and never detected beyond that range. In some systems, it was found that setting $e = 0.1R$ comes fairly close to the actual behavior of the sensors used in experiments. For simplicity, it is assumed that the object emits sound of a frequency not present in the environment, so there are no false positives. However, the results are fairly robust with respect to intermittent detection (i.e., false negatives) during the period of observations. It is worth noting that the sensor model is generic enough to encompass other sensing modalities beyond acoustic. All that is required is a sensor with a relatively uniform range, as previously defined, which is capable of differentiating the target from the environment. A magnetometer, a device that detects changes in magnetic fields, is one such sensor.

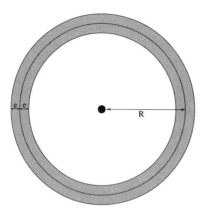

Figure 17.12 Model of a sensor. For nominal sensing range R, the object is always detected when it is $R - e$ away or closer, never detected beyond $R + e$, and has a nonnegative chance of detection between $R - e$ and $R + e$.

Algorithm

The simplest distributed tracking algorithm entails simply recording the times when each sensor detects the object and then performing line-fitting on the resulting set of points. Although simple, this approach is not very precise: It can track only the object with a resolution of the sensor range R. Moreover, if a sensor detects the object more than once as it moves through the sensor's detection range, that information is lost.

The position of a stationary object, or a moving object for that matter, that is determined with this method is not very precise and depends heavily on the number, detection range, and precision of sensors that detect the sound. Instead of looking at a single position measurement, we are interested in the path of a moving object, which is a sequence of positions over a period of time. Combining a large number of somewhat imprecise position estimates distributed over space and time may yield surprisingly accurate results. Cooperative tracking addresses the problem of high-resolution tracking using sensor networks. It improves accuracy by combining information from neighboring sensors. The only requirement for cooperative tracking to be used is that the density of sensor nodes must be high enough for the sensing ranges of several sensors to overlap. When the object of interest enters the region where multiple sensors can detect it, its position can be pinned down with a higher degree of accuracy because the inter-section area is smaller than the detection area of a single node. The outline of a generic cooperative tracking algorithm is as follows:

- Each node records the duration for which the object is in its range.
- Neighboring nodes exchange these times and their locations.
- For each time instant, the object's estimated position is computed as a weighted average of the detecting nodes' locations.
- A line-fitting algorithm is run on the resulting set of points.

Several of these steps require careful consideration. First, the algorithm implicitly assumes that the sensor clocks are synchronized and that the sensors know their locations. Second, we obtain a position reading by a weighted average of the locations of the

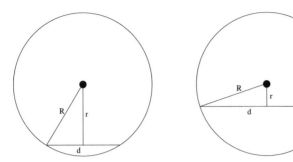

Figure 17.13 If the object's speed is constant, detection time is directly proportional to path segment d and inversely proportional to distance r from the sensor to the object's trajectory.

nodes that detected the sound at a given instant, but the exact weighting scheme is not specified. This is an important issue because selecting an appropriate scheme will improve accuracy, whereas a poor choice might be detrimental to it. The simplest choice is to assign equal weights to all sensors' readings. This effectively puts the estimate of the object's position at the centroid of the polygon with sensors acting as vertices. Intuitively, it should be more accurate than noncooperative tracking. However, it is possible to do even better. Consider Figure 17.13: Sensors that are closer to the path of the target will stay in sensor range for a longer duration. Thus, to increase accuracy, the weight of a sensor's reading should be proportional to some function of the duration for which the target has been in range of that sensor.

Once the individual position estimates are computed, the final step of the line-fitting algorithm can begin. Least-squares regression can be used to find the equation of the line. It is interesting to note that the duration-based weighting scheme for position estimates moves the points closer to the actual path, thereby reducing variance in the least-squares computation. Also important is the fact that the multistep approach enables early estimates of the path to be computed so that continuous refinement is possible as more data points become available. The resulting equation of the line extrapolates the path of the object until it changes course sharply. This information may be used by the system; e.g., for asynchronous wake-up of nodes likely to be in its path.

Data Aggregation

The final step of the algorithm involves performing a line-fitting computation on the set of all the position estimates (or some subset of them). Unlike position estimates, which can be performed in a distributed manner with only local communication, this necessitates collecting sensor readings from many sensor nodes at a centralized location for processing. This process is called *data aggregation*, and it is present in one form or another in virtually all sensor network applications. The main concerns for data aggregation are timeliness and resource usage. Timeliness, with respect to sensor data, is critical to real-time monitoring and control applications in which stale data are useless or even detrimental. Resources, in particular network bandwidth and message buffers, are quite scarce in networked embedded systems. Low bandwidth of small wireless transmitters and the potential for contention with other messages drastically limit the amount of data that can pass through the network.

We assume that some nodes in the sensor network are gateway nodes connected to outside networks such as the Internet. To process the data from the sensor network, it needs to be sent through one of these gateway nodes to the more powerful computers connected to the outside network. To do this efficiently, a tree rooted at each gateway is constructed and spans the entire network. Each sensor node in the tree collects data from its children and sends it up the tree to either the closest or the least-busy gateway. A number of methods have been proposed in the literature to construct a data-aggregation tree that will optimize a variety of objective functions, such as minimizing delay, minimizing the number of messages, or minimizing the total energy cost of data aggregation.

17.6.6   Distributed Prediction Tracking (DPT)

The distributed prediction tracking (DPT) algorithm is specifically aimed at addressing the various challenges outlined in Section 17.6 while accurately tracking moving targets. As the name suggests, this algorithm does not require any central control point, eliminating the possibility of a single point of failure and making it robust against random node failures. The tracking task is carried out distributively by sequentially involving the sensors located along the track of the moving target. DPT assumes a cluster-based architecture for the sensor network, and the choice is motivated by the need to ensure the sensor network's scalability and energy efficiency. Any suitable clustering mechanism from those proposed in the literature may be used; note that DPT does not impose any specific requirements or restrictions on the choice of the clustering algorithm.

Although no assumption is made on the choice of the clustering algorithm, we assume that the cluster-head (CH) has the following information about all sensors belonging to its cluster: (1) sensor identity, (2) location, and (3) energy level. When tracking a moving target and deciding which sensors to use for tracking, the CH's decision-making procedure will be based on this information. The assumptions about the sensors are subsequently enumerated. These assumptions are realistic and targeted at reducing the energy cost and prolonging the whole network's lifetime as well:

1. All sensors have the same characteristics.
2. Sensors are randomly distributed across the whole sensing area with uniform density.
3. Each sensor has two sensing radii: normal beam r and high beam R. The default operation uses the low beam, and the high beam is turned on only when necessary. The following relationship holds between the energy consumed by the low and high beams:

$$\frac{E_{\text{lowbeam}}}{E_{\text{highbeam}}} = \frac{r^2}{R^2}. \tag{17.5}$$

4. A sensor's communication and sensing channels stay in the hibernation mode most of the time, where they consume minimal energy. Each sensor will wake up routinely to receive possible messages from its CH. The sensor will perform sensing according to its CH's requirements.

No specific assumptions are made about the movement pattern of the targets. However, DPT assumes that the targets originate outside the sensing region and then move inside. Also, it is assumed that the movement of each tracked target needs to be forwarded to a central location, which we term the *sink*. In reality, the sink could be either a special node or a terminal associated with a human.

DPT Algorithm

DPT distinguishes between the *border sensors*, sensors located within a given distance of the border, and *nonborder sensors* in terms of their operation. Whereas border sensors are required to keep sensing at all times in order to detect all targets that enter the sensing region, the nonborder sensor's sensing channel hibernates unless it is specifically asked to sense by its cluster-head. Because the target is assumed to move from outside into the sensing area, it will be detected by the border sensors when trespassing the border. As soon as a target is detected, a sequence of tasks in the order of *sense–predict–communicate–sense* are carried out distributively by a series of sensors that are located along the target's track. This forms the essential idea behind the DPT algorithm.

Let $CH_1, CH_2, \ldots, CH_N$ denote the sequence of cluster-heads that become involved with tracking the target as it proceeds from its very first location to the last. The information gathered by each cluster-head is sent all the way back to the sink (either sent intact or after being aggregated) for further processing as well as to the downstream cluster-head CH_{i+1}. The *target identity* is created when the target is first detected. This identity is unique, and all cluster-heads that co-track this target use it to identify the target. To facilitate the smooth tracking of the target, CH_i predicts the future location of the moving target and informs the downstream cluster-head CH_{i+1} ahead of time about this target. The accuracy of the *prediction* is very important if downstream cluster-heads are to be identified accurately and the overall tracking mechanism is to be effective. Many prediction mechanisms are possible; the simplest one is a linear predictor, which uses only the previous two locations to linearly predict the third location. Higher-order predictions can also be adopted that predict the nth location information based on previous $n - 1$ actual locations. A higher-order prediction results in more accurate results, though at the cost of greater energy consumption.

Sensor-Selection Algorithm

After cluster-head CH_i predicts the location of the target, the downstream cluster-head CH_{i+1} toward which the target is headed receives a message from CH_i indicating this predicted location. With information of all sensors belonging to CH_{i+1} available in its database, the search algorithm running at CH_{i+1} is able to locally decide the sensor triplet to monitor the target. The selection rule chooses three sensors (if possible) such that their distances to the predicted location are not only less than the sensor's normal beam r but also are the smallest. After the sensor triplet is chosen, CH_{i+1} sends them a *wake-up message* so that they are ready to monitor the target. If the prediction and selection process succeeds, after sensing, each sensor will send a location message to CH_{i+1}. If CH_{i+1} in unable to find enough sensors eligible for this sensing task with the normal sensing beam, it will try to search for eligible sensors within a distance R, the higher sensing beam, from the predicted location. The selected sensors, whose distance from the predicted location is greater than r and lower than R, will now be contacted and instructed to sense with their high beam, while the rest of the sensors in the triplet use their normal beam. If CH_{i+1} is unable to find enough sensors even with high sensing beams, it asks its neighboring cluster-heads for help.

Failure Recovery

Let's first identify two possible failure scenarios. As described in the previous subsections, each upstream cluster-head sends a message to the expected downstream cluster-head. If the upstream cluster-head does not get any confirmation from the downstream cluster-head after a given period of time, then it assumes that the downstream cluster-head is no longer available and the target has been lost. Another failure scenario occurs when the target changes its direction or speed so abruptly that it moves significantly away from the predicted location and falls out of the detectable region of the sensor triplet selected for the sensing task. In both of these failure scenarios, a straightforward solution is to wake up all sensors within a given area, which is calculated based on the target's previous actual location. The *recapture* radius σ is an important parameter in this process and is decided by the target's moving speed and time elapsed since it was last sensed.

17.7   Experimental Location and Tracking Systems

In this section, several location systems are introduced. Although they may not be specially designed for WSNs, these design concepts and experiences will benefit future implementations of positioning systems in WSNs.

17.7.1   Active Badge and Bat

Efficient location and coordination of staff in any large organization is a difficult and recurring problem. Hospitals, for example, may require up-to-date information about the location of staff and patients, particularly when medical emergencies arise [see Want *et al.* (1992) for detailed information]. A solution to the problem of automatically determining the location of an individual has been to design a tag in the form of an "active badge" that emits a unique code for approximately a tenth of a second every 15 s (a beacon). These periodic signals are picked up by a network of sensors placed around the host building. A master station, also connected to the network, polls the sensors for badge "sightings," processes the data, and then makes it available to clients that may display it in a useful visual form. The badge was designed in a package roughly $55 \times 55 \times 7$ mm and weighs a comfortable 40 g. Pulse-width-modulated infrared (IR) signals are used for signaling between the badge and sensor mainly because IR solid-state emitters and detectors can be made very small and very cheaply (unlike ultrasonic transducers); they can be made to operate within a 6-m range; and the signals are reflected by partitions and therefore are not directional when used inside a small room. Moreover, the signals will not travel through walls, unlike radio signals that can penetrate the partitions found in office buildings. An active signaling unit consumes power; therefore, the signaling rate is an important design issue. First, by emitting a signal only every 15 s, the mean current consumption can be very small with the result that badge-sized batteries will last for about one year. Second, it is a requirement that several people in the same locality be detectable by the system. Because the signals have a duration of only one-tenth of a second, there is approximately a 2/150 chance that two signals will collide when two badges are placed in the same location. For a small number of people, there is good probability that they will all be detected. The active badge also incorporates a light-dependent component that, when dark, turns the badge off to conserve battery life. Reduced lighting also increases the period of the beacon signal to a time greater than 15 s. In ambient lighting conditions in a room, this effect only slightly modifies the period, but it is another factor that ensures that synchronized badges will not stay synchronized very long. If the badge is placed in a drawer during nonoffice hours, on weekends, and during vacation, the effective lifetime of the batteries is increased by a factor of 4. A disadvantage of an infrequent signal from the badge is that the location of a badge is only known, at best, to a 15-s time window. However, because in general a person tends to move relatively slowly in an office building, the information that the active badge system provides is very accurate. An active badge signal is transmitted to a sensor through an optical path. This path may be found indirectly through a surface reflection; for example, from a wall.

The active badge location system was developed at Olivetti Research Laboratory, now AT&T at Cambridge. A successor of the active badge system is the bat system (Addlesee *et al.*, 2001), which consists of a collection of wireless transmitters, a matrix of receiver elements, and a central RF base station. The wireless transmitters, called bats, can be carried by a tagged object and/or attached to equipment. The sensor system measures the time of flight of the ultrasonic pulses emitted from a bat to receivers installed in known and fixed positions, and it uses the time difference to estimate the position of each bat by trilateration. The RF base station coordinates the activity of bats by periodically broadcasting messages to them. The location of the bat can be determined within 3 cm of error in a 3D space at 95% accuracy. This accuracy is quite enough for most location-aware services; however, the deployment cost is high.

17.7.2   Cricket

Cricket is a location-support system for in-building, mobile, location-dependent applications. It allows applications running on mobile and static nodes to learn their physical location by using listeners that hear and analyze information from beacons spread throughout the building. Cricket is the result of several design goals, including user privacy, decentralized administration, network heterogeneity, and low cost. Rather than explicitly tracking user location, cricket helps devices learn where they are and lets them decide to whom to advertise this information; it does not rely on any centralized management or control and there is no explicit coordination between beacons; it provides information to devices regardless of their type of network connectivity; and each cricket device is made from off-the-shelf components and costs less than U.S.$10. See Priyantha *et al.* (2000) for more information about cricket. By not tracking users and services, user privacy concerns are adequately met. It is emphasized that cricket is a *location-support* system rather than a conventional *location-tracking* system that tracks and stores location information for services and users in a centrally maintained database.

Cricket uses a combination of RF and US to provide a location-support service to users and applications. Wall- and ceiling-mounted beacons are spread through the building, publishing location information on a RF signal. With each RF advertisement, the beacon transmits a concurrent ultrasonic pulse. The listeners receive these RF and ultrasonic signals, correlate them to each other, and infer the space in which they are currently. The beacons use a decentralized randomized transmission algorithm to minimize collisions and interference among each other. The listeners implement a decoding algorithm to overcome the effects of US multipath and RF interference (more information about the cricket system can be found at http://cricket.csail.mit.edu/).

17.7.3   RADAR

RADAR (Bahl and Padmanabhan, 2000) is a RF-based system for locating and tracking users inside buildings. RADAR operates by recording and processing signal-strength information at multiple base stations positioned to provide overlapping coverage in the area of interest. It combines empirical measurements with signal propagation modeling to determine user location and thereby enable location-aware services and applications.

RADAR complements the data-networking capabilities of RF WLANs with accurate user location and tracking capabilities, thereby enhancing the value of such networks. RADAR uses signal-strength information gathered at multiple receiver locations to triangulate the user's coordinates. Triangulation is done using both empirically determined and theoretically computed signal-strength information.

Notice that in addition to the preceding review location and tracking system, a number of other systems have been developed; for example, the Ubisense system (see http://www.ubisense.net), NIST Location System, and Place Lab system.

17.8   Conclusion and Remarks

In this chapter, some fundamental techniques in positioning and location tracking have been discussed and several experimental systems were briefly reviewed. We first reviewed which information may be used by a localization method and then studied the computational complexity of localization when certain information is used. Progressive and network-wide localization methods were then briefly studied. We also studied some tracking methods used by WSNs to track some objects. Location information may enable new types of services, especially for WSNs. Accuracy and deployment costs are two factors that may contradict each other, but both are important factors for the success of location-based services. There are still a number of challenging algorithmic questions left for future study. The first question is: What is the critical network density for having a localizable WSN? What is the critical network density such that a given localization method, such as multilateration or sweep, can find the location information of all WSNs within a given error bound? What is the critical number of beacons needed for a given network to be localizable? Notice that WSNs are often deployed randomly in a certain region. Previously, we studied the critical nodal density (or transmission range) such that the resulting WSN formed by a set of randomly deployed sensors is connected with high probability. It will be interesting to study the critical density (or transmission range) such that a random network is localizable (i.e., the positions of all nodes can be determined based on some distance or angle information). The second set of questions is: How will a localization method perform when there are errors on the distance and angle information used? That is, how much will the computed location of every node differ from its actual location?

Problems

17.1 Research the literature. In typical MANETs and WSNs, what is the information available for localization of wireless nodes? What are the advantages and disadvantages of using different information? What is the current accuracy level of the information collected for different wireless networks?

17.2 Given that the majority of node-localization problems are either NP-hard or polynomial solvable, why is node localization still a challenging research problem?

17.3 Consider a network of n nodes that are randomly placed in a unit-square region. Assume that you have some error bound on the distance estimation between any pair of nodes that can communicate directly. Implement a localization algorithm based on the information of distance and error bound on distance. Will the fact that the network nodes are randomly placed (thus, the transmission range is set large enough to get a connected network with high probability) help your algorithm?

17.4 Consider a network of n nodes that are randomly placed in a unit-square region. Assume that you have some error bound ϵ on the distance estimation between any pair of nodes that can communicate directly. Further assume that you will place k^2 anchor nodes uniformly in the network, where anchor nodes will be placed in $k \times k$ grid points of the region. Assume that the position of each anchor node is known and each anchor node can communicate directly with all nodes within distance R. Implement a localization algorithm based on the information of distance and error bound on distance.

17.5 Assume that there is a path $v_1 v_2 \cdots v_{k-1} v_k$ from v_1 to v_k. You already know that the path is the shortest hop path between v_1 and v_k. For each link $v_i v_{i+1}$, you measure its

Euclidean length and get a value x_i. What is the expected Euclidean distance between nodes v_1 and v_k? For simplicity, you can start from $k = 3$ and then try to generalize your analysis.

17.6   Assume that you are given a network $G = (V, E)$ of n nodes located in two dimensions and the exact edge length of k edges from E. What is the smallest k such that it is possible to find the position of all n nodes? What is the minimum k when the nodes are located in three dimensions?

17.7   Assume that you are given a network $G = (V, E)$ of n nodes located in two dimensions and the measured link distance of some links. Here, the distance measurement may have an absolute error up to ϵ units. Further assume that a link (u, v) belonging to E implies that $\|u - v\| \leq r$ for a given constant r. First, write a quadratic programm to test whether the given input data is consistent. In other words, you have to decide whether there is a network deployment that will satisfy all the distance constraints given. If the given data are not consistent, formulate the node-localization problem using quadratic programming with the objective function of minimizing the number of violated constraints. If the given data are consistent, formulate the node-localization problem using quadratic programming with the objective function of minimizing the maximum distance-estimation error.

18 Performance Limitations of Random Wireless Ad Hoc Networks

In previous chapters, especially Chapter 16, we studied what is the maximum throughput achievable by a given wireless network under a certain wireless interference model. In some applications (e.g., WSNs), we often need a rough estimation on the achievable throughput when we randomly deploy n number of wireless nodes in a given region. In this chapter, we study how the capacity of wireless networks scale with the number of nodes in the networks (when given a fixed deployment region) or scale with the size of the deployment region (when given a fixed deployment density) for a various number of operations such as unicast and broadcast. We assume that each wireless node can transmit at W bits/s over a common wireless channel. We will see that it is immaterial to results presented in this chapter if the channel is broken up into several subchannels of capacity $W_1, W_2, \ldots, W_M$ bits/s as long as we have $\sum_{i=1}^{M} W_i = W$. As always, we assume that the packets are sent from node to node in a multihop manner until they reach their final destinations. The packets could be buffered at intermediate nodes while awaiting transmission. In some results, we assume that every intermediate node has an infinite buffer size. For most of the results presented here, the delay of routing is not considered; i.e., the delay in the worst case could be arbitrarily large for some results.

18.1   Introduction

In wireless ad hoc networks, wireless nodes may cooperate in routing each others' packets. Lack of a centralized control of the functionality and possible node mobility give rise to many challenging issues in the network, the medium access and physical layers of a wireless ad hoc network. At the network layer, the main challenging problem is that of routing, which has to deal with time-varying network topology, possible power constraints of wireless nodes, and characteristics of the wireless channel (e.g., unstable, broadcast nature, fading, and so on). The choice of MAC also is restricted by the fact that the network topology is time varying and there is no centralized control. In previous chapters, we studied TDMA, CDMA, and the dynamic assignment of frequency bands to improve network throughput. Notice that for a mobile wireless network, random access appears to be the current favorite protocol because of its simplicity and quick adaption to mobility and dynamic data rate by nodes. For a mobile wireless network, static FDMA is inefficient in dense networks, and CDMA is very difficult to implement because of node mobility and the need for keeping track of spreading codes for nodes in the time-varying neighborhoods. Notice that TDMA has recently been proposed to improve network throughput for some networks or parts of the networks (Ahn *et al.*, 2006), especially for static networks. At the physical layer, an important issue is the power control, which was studied in Chapter 10. A careful selection of the transmission power of nodes will not only improve the nodal life but also will improve the spatial reuse of frequency and, consequently, possibly improve the network throughput.

The main purpose of this chapter is to study the *capacity* of wireless networks when we choose the best protocols for all layers. We assume that a set of n wireless nodes is *randomly* distributed in a fixed region (e.g., a unit square by a proper scaling). Because of spatial separation, several wireless nodes can transmit simultaneously, provided that these transmissions will not cause *destructive* wireless interferences to any of the transmissions. As in the literature, here we mainly consider two types of networks: *arbitrary networks* and *random networks*.

In an arbitrary network, the node locations, destinations of the sources, and traffic demands are all arbitrary. All n nodes are arbitrarily located in the deployment region. Each node has an arbitrarily chosen destination to which it wishes to send traffic at an arbitrary data rate. Each node can choose an arbitrary range or transmission power for *each* transmission. To describe when a transmission is received successfully by its intended recipient, we allow two possible models for a successful one-hop reception: the *protocol model* and the *physical model*.

In a random network, n nodes are randomly located; i.e., independently and uniformly distributed inside a fixed deployment region. Each node has a randomly chosen destination to which it wishes to send data in rate $\lambda(n)$ bits/s. The destination for each node is also independently chosen as the nearest node to a randomly located point; i.e., uniformly and independently distributed. Unlike in the arbitrary network, here we assume that the nodes are homogeneous; i.e., all nodes use the same nominal transmission power.

In this chapter, we mainly study the capacity of some networks. Gupta and Kumar (1999) defined two different capacity measurements: *transport capacity* and *throughput capacity*. We also adopt their definitions in this chapter.

Given any set of successful transmissions taking place over time and space, they say that the network transports 1 *bit-meter* when 1 bit has been transported a distance of 1 m toward the destination. Notice that for unicast, the concept of "transported one distance toward the destination" is well defined. In the case of broadcast or multicast, we need more rigorous definitions. The sum of products of bits and the distance over which they are carried is called the *transport capacity* of a network. Recall that in most results presented here, we assume that the deployment region is scaled to have 1 m^2; thus, when the deployment area has area A m^2, then all the results should be scaled by $\sqrt{A}$.

DEFINITION 18.1 (Transport capacity) *Given an arbitrary network for which the node placement and traffic pattern are fixed, a transport capacity $\beta(n)$ is feasible if there is a choice of transmission power and spatial and temporal schemes for scheduling transmissions such that the sum of products of bits and distance over which they are carried is at least $\beta(n)$.*

Observe that to some extent, the concept of transport capacity is more general than the throughput capacity defined later. Transport capacity actually does not care about to which recipient some packets belong. It somewhat characterizes the total number of packets that can be moved to a 1-m distance by all nodes in the network. For example, if a network has transport capacity $\beta(n)$ and the distance between the source and destination node is about d m, then the source node would obtain a throughput capacity of $\beta(n)/d$ bits/s.

In addition to this transport capacity, Gupta and Kumar (1999) defined the feasible throughput capacity.

DEFINITION 18.2 (Throughput capacity) *A throughput $\lambda(n)$ bits/s for each node is feasible if there is a spatial and temporal scheme for scheduling transmissions such that by operating the network in a multihop fashion and buffering at intermediate nodes when awaiting transmission, every node can send $\lambda(n)$ bits/s on average to its chosen destination node. That is, there is a $T < \infty$ such that in every time interval (with unit seconds) $[(i-1)T, iT]$, every node can send $T\lambda(n)$ bits to its corresponding destination node.*

The throughput capacity or the transport capacity depends on the specific locations of the nodes. For random networks, these locations are random. When we know only that a network is a random network of n wireless nodes, we would like to predict the capacity of any such network instance. Let $\mathcal{N}$ denote all random wireless networks of n wireless nodes. Then, we would like to know what the capacity is of a random network $G \in \mathcal{N}$ of n wireless nodes. Notice that this is similar to the critical range for random networks.

DEFINITION 18.3 (Throughput capacity of random networks) *We say that the throughput capacity of the class of random networks is of the order of $\Theta(f(n))$ bits/s if there are deterministic constants $c > 0$ and $c < c' < +\infty$ such that*

$$\lim_{n \to \infty} \mathbf{Pr}(\lambda(n) = cf(n) \text{ is feasible}) = 1,$$

$$\liminf_{n \to \infty} \mathbf{Pr}(\lambda(n) = c'f(n) \text{ is feasible}) < 1.$$

The results presented in this chapter allow for a perfect scheduling algorithm that will use the known node locations and the known traffic demands between all pairs of nodes to coordinate the wireless transmissions temporally and spatially to avoid packet losses from wireless interference. We always assume that the nodes are static although their positions are randomly selected in a random network. The main results that are presented in this chapter are as follows. The transport capacity of an *arbitrary network* under the protocol model is $\Theta(W\sqrt{n})$ bit-meters/s if the nodes are optimally placed, the traffic pattern is optimally chosen, and the range of each transmission is chosen optimally. For the physical model, the transport capacity $cW\sqrt{n}$ bit-meters/s is feasible, whereas $c'Wn^{\frac{\alpha-1}{\alpha}}$ bit-meters/s is not feasible for some appropriate constants c and c'. For *random networks* with a physical model, a throughput $\lambda(n) = \frac{cW}{\sqrt{n \log n}}$ bits/s is feasible, whereas $\lambda(n) = \frac{c'W}{\sqrt{n}}$ bits/s is not feasible for some appropriate constants c and c', with a probability approaching 1 as $n \to \infty$. Gupta and Kumar (1999) provide some examples of the constants c and c' for both situations. When node-location information and/or traffic demand information is not available and/or the nodes could be mobile, the achievable capacity by any routing/scheduling algorithm could only be even smaller.

In addition to unicast, we study the capacity of a network for broadcast (Keshavarz-Haddad *et al.*, 2006; Zheng, 2006). For a broadcast operation, we assume that there is a given source node s that wants to broadcast data continuously to all nodes in the network. Some nodes in the network will serve as relay nodes. The *broadcast capacity* of an arbitrary network, for a given source node s, is defined as $\lim_{T \to \infty} \frac{D(T)}{T}$, where $D(T)$ is the amount of unique data received by *every* node in the network in a time period T. Here, if a data packet is received by a node multiple times, it is counted as only one data packet. For random networks, we say a random network $\mathcal{G}$ of n nodes has a broadcast capacity of $\Theta(f(n))$ bits/s if there are deterministic constants $c > 0$ and $c < c' < +\infty$ such that the probability that *a randomly selected arbitrary* network G from $\mathcal{G}$ has broadcast capacity $cf(n) = 1$ (or, more precisely, sufficiently close to 1); whereas *a randomly selected arbitrary* network G from $\mathcal{G}$ has broadcast capacity $c'f(n)$ that is less than 1.

18.2   Capacity of Unicast for an Arbitrary Network

In this section, we first review the results by Gupta and Kumar about the capacity of unicast of an arbitrary network. We consider the setting on a planar disk of unit area and the following assumptions as was discussed in Gupta and Kumar (1999):

1. There are n nodes arbitrarily located in the unit disk. The results will carry over to any domain of unit area in two dimensions, which is the closure of its interior.
2. The network transports $\lambda n T$ bits over T second.
3. The average distance between the source and the destination traveled by a bit is $\overline{L}$.
4. Each node can transmit over *any* subset of M channels with capacities W_m bits/s for the mth channel and $\sum_{m=1}^{M} W_m = W$.
5. The time is slotted into lengths of τ seconds and synchronized.
6. The interference model used is the protocol model or the physical model defined as follows:

   For the physical model, suppose that node X_i transmits over the mth subchannel to a node X_j. This transmission is successfully received by node X_j only if

   $$\|X_k - X_j\| \le (1 + \Delta)\|X_i - X_j\|$$

   for every other node X_k that is simultaneously transmitting over the same subchannel.

   For the physical-interference model, let $\mathcal{N}_t$ be the subset of nodes simultaneously transmitting at a time slot t over a certain subchannel. Let P_k be the power used by a node $X_k \in \mathcal{N}_t$ to transmit. Then, the transmission from a node $X_i \in \mathcal{N}_t$ is successfully received by a node X_j only if

   $$\frac{\frac{P_i}{\|X_i - X_j\|^{\alpha}}}{N + \sum_{X_k \in \mathcal{N}_t, k \ne i} \frac{P_k}{\|X_k - X_j\|^{\alpha}}} \ge \beta.$$

Here, N is the ambient-noise power level, α is the path attenuation factor, and β is the minimum SINR required by the receiver node for successful receptions. In other words, we assume that the signal power decays with distance d as $1/d^\alpha$, with $\alpha > 2$. Notice that the preceding requirement in the physical interference model is equivalent to

$$\frac{\frac{P_i}{\|X_i - X_j\|^\alpha}}{N + \sum_{X_k \in \mathcal{N}_t} \frac{P_k}{\|X_k - X_j\|^\alpha}} \geq \frac{\beta}{\beta + 1}.$$

The main results proved in Gupta and Kumar (1999) are as follows:

THEOREM 18.1 (Gupta and Kumar, 1999) *In the PrIM, the transport capacity $\lambda n \overline{L}$ is bounded as follows:*

$$\lambda n \overline{L} \leq \sqrt{\frac{8}{\pi}} \frac{1}{\Delta} W \sqrt{n} \ bit\text{-}meters/s. \tag{18.1}$$

In the physical-interference model, we have

$$\lambda n \overline{L} \leq \sqrt[\alpha]{\frac{2\beta + 2}{\beta}} \frac{1}{\sqrt{\pi}} W n^{\frac{\alpha-1}{\alpha}} \ bit\text{-}meters/s. \tag{18.2}$$

Notice that when the domain of deployment is A m^2 instead of 1 m^2, then all the preceding upper bounds are scaled by $\sqrt{A}$. The preceding theorem only gives the upper bound on the capacity of an arbitrary network. Then, Gupta and Kumar further proposed a constructive lower bound on the transport capacity of an arbitrary network for the PrIM and physical-interference model. For their constructive lower bound, they assume special placement of nodes and assignment of traffic patterns.

THEOREM 18.2 (Gupta and Kumar, 1999) *There is a placement of nodes and an assignment of traffic patterns such that the arbitrary network can achieve $\frac{1}{1+2\Delta} \frac{Wn}{\sqrt{n}+\sqrt{8\pi}} \simeq \frac{1}{1+2\Delta}$, $W\sqrt{n}$ bit-meters/s under the protocol model, and $\frac{1}{(16\beta(2^{\alpha/2}+\frac{6\alpha-2}{\alpha-2}))^{1/\alpha}} \frac{W \cdot n}{\sqrt{n}+\sqrt{8\pi}}$ bit-meters/s under the physical-interference model, where n is a multiple of 4.*

The basic idea of their constructive method is to place nodes evenly (e.g., a grid or cellular) and assign each node a small transmission range that is enough to result in a connected network. The reason for doing so is that the transport capacity of a network is proportional to $1/r$, where r is the transmission range of all nodes. Thus, to maximize the capacity, we need to use the smallest transmission range as long as the network is connected. They also provide further enhancement of the capacity of some specially constructed networks by carefully placing the nodes and assigning the traffic patterns.

Observe that the capacity lower bound achieved by Gupta and Kumar for arbitrary networks requires the careful selection of the node placement and traffic patterns. It is interesting to study the transport capacity of a given network $G = (V, E)$ in which the placement of all nodes is already given and fixed, and the set of sources and target nodes is also given. Then, we need to find an assignment of traffic flows for all pairs of source and target nodes and a scheduling of node transmissions such that the traffic demands are satisfied and the transmissions are interference-free. This has been widely studied recently [see, e.g., Alicherry *et al.* (2005), Kodialam and Nandagopal (2003b), Kumar *et al.* (2005), and W. Wang *et al.* (2006b)], sometimes with slightly changing objectives such as maximizing the fairness instead of maximizing the total transport throughput by all pairs of source nodes and target nodes. See Chapter 16 for more discussion on this topic.

18.3   Capacity of Unicast for Randomly Deployed Networks

In this section, we study the capacity of unicast for random networks. Gupta and Kumar provided a constructive scheme to show that one can spatially and temporally schedule all transmissions in a random network such that when each randomly located source node has a randomly chosen destination node, each source–destination pair can be guaranteed a capacity of $\Theta(\frac{W}{[1+(\Delta)^2]\sqrt{n \log n}})$ bits/s with a probability approaching 1 as $n \to \infty$. The routing scheme they proposed uses the Voronoi tessellation and the routing from the source node to the destination node will be over nearly straight-line paths. Recall that routing based on Voronoi tessellation (and Delaunay triangulation), if properly designed, can guarantee that the routing path has a Euclidean distance of at most a constant factor of the minimum. See Chapter 13 for more details about the routing on Delaunay triangulation and Voronoi diagram.

The method of Gupta and Kumar partitions the deployment region into Voronoi cells by using some artificial points. Essentially, they select a set A of points $A = \{a_1, a_2, \ldots, a_p\}$ such that each Voronoi cell produced based on A:

1. Contains a disk of area $\frac{100 \log n}{n}$ with radius $\rho(n)$, and
2. Is contained in a disk with radius $2\rho(n)$.

Here, we assume that n is large enough such that $\frac{100 \log n}{n}$ is less than 1. Because the random network will be connected, we know from Chapter 11 that the transmission range r will be at least of the order of $\Theta(\frac{\log n}{n})$. Their proof actually used the VC dimension (Vapnik and Chervonenkis, 1971). From this, they proved that with high probability, each produced Voronoi cell will contain at least one wireless node. The routing method used by Gupta and Kumar is similar to the routing method we studied in Chapter 13. Packets originating at a source node X_i will be relayed from the Voronoi cell V_i containing X_i to the adjacent Voronoi cells V_i' in a sequence of hops. In each hop, the packet will be transferred from one Voronoi cell to another Voronoi cell in the order in which they intersect the line formed by the source node X_i and the destination node X_j. For each Voronoi cell V, Gupta and Kumar proved that the mean number of routing paths that pass through V is of the order of $O(\sqrt{n \log n})$. They also prove that with high probability, the *actual* number of routing paths passing through a Voronoi cell is at most $O(\sqrt{n \log n})$. Recall that the traffic handled by a cell (i.e., all wireless nodes contained inside this Voronoi cell) is proportional to the number of routing paths passing through it. From the preceding analysis, Gupta and Kumar proved the following theorem:

THEOREM 18.3 *For random networks on the surface of a sphere in the protocol model, there is a deterministic constant $c > 0$ such that the transport throughput $\lambda(n) = \frac{cW}{(1+\Delta)^2 \sqrt{n \log n}}$ bits/s is feasible with high probability.*

For random networks on the surface of a sphere in the physical-interference model, there are deterministic constants c_1 and c_2 such that

$$\lambda(n) = \frac{c_1}{\{2[c_2\beta(3 + \frac{1}{\alpha-1} + \frac{2}{\alpha-2})]^{1/\alpha} - 1\}^2} \frac{W}{\sqrt{n \log n}} \text{bits/s}$$

is feasible with high probability.

They also provided the upper bound on the throughput capacity of random networks. The key to obtaining the upper bound is to utilize the fact that each transmission will consume a valuable area; i.e., no other nodes inside that area can transmit simultaneously. Thus, under the PrIM, the number of simultaneous transmissions using any specific subchannel is proportional to $1/[\Delta r(n)]^2$, where $r(n)$ is the uniform transmission range of every wireless node in the network. Let $\overline{L}$ be the mean length of the path of packets. Then, the mean number of hops taken by a packet is at least $\overline{L}/r(n)$. Because each source node will generate $\lambda(n)$ bits/s, and there are n source nodes and each bit needs to be relayed on average by at least $\overline{L}/r(n)$ nodes, we thus have that the total number of bits/s served by all networking nodes needs to be at least $\lambda(n)n[\overline{L}/r(n)]$. On the other hand, because the number of simultaneous transmissions is at most $\frac{c}{\Delta^2 r(n)^2}$ for some constant $c > 0$, we have

$$\lambda(n)n[\overline{L}/r(n)] \leq \frac{cW}{\Delta^2 r(n)^2}$$

because the data rates of the channels are W. This further implies that

$$\lambda(n) \leq \frac{cW}{\Delta^2} \frac{1}{\overline{L}nr(n)}.$$

Recall that to ensure that there is no isolated node in a random network of n nodes with high probability, we need $r(n) > c'\sqrt{\frac{\log n}{n}}$ for some constant $c' > 0$. Consequently, the following upper bound on the throughput capacity can be obtained.

THEOREM 18.4 *For random networks deployed on the surface of a sphere, under the PrIM, there is a deterministic constant $c < +\infty$ such that*

$$\lim_{n \to \infty} \mathbf{Pr}\left[\lambda(n) = \frac{cW}{\Delta^2 \overline{L}\sqrt{n \log n}} \text{ is feasible}\right] = 0,$$

where $\overline{L}$ is the average distance the packet needs to travel from all source nodes to all its corresponding destination nodes.

They also provide an upper bound on the throughput capacity of random networks on the sphere surface under the physical-interference model. They basically show that *any* upper bound on the transport capacity for arbitrary networks under the protocol model is also an upper bound on the transport capacity for random networks under the physical-interference model. The conversion is as follows: Assume that under the physical-interference model, node X_i is sending information to node X_j and node X_k is also transmitting simultaneously. Recall that here, we assume that all nodes transmit using the same power P. Then, we have

$$\frac{\frac{P}{\|X_i-X_j\|^\alpha}}{N + \frac{P}{\|X_k-X_j\|^\alpha}} \geq \beta.$$

This implies that $\|X_k - X_j\| \geq \beta^{1/\alpha}\|X_i - X_j\|$. Generally, we require that $\beta > 1$. Then, we have

$$\|X_k - X_j\| \geq (1 + \Delta)\|X_i - X_j\|$$

for $\Delta = \beta^{1/\alpha} - 1 > 0$. In other words, any set of simultaneous transmissions feasible under the physical-interference model is also feasible under the PrIM with some special Δ. Thus, the upper bound on the transport capacity for the latter also holds for the former.

From the results obtained for networks deployed on the surface of a 3D sphere, Gupta and Kumar further provide some constructive lower bounds on the transport capacity for networks deployed in a 2D disk. Although the radius required for network connectivity is still asymptotically the same in this case, different routing methods are needed to avoid creating hotspots in the network where the load of some nodes is still at most $\sqrt{n \log n}$ with high probability. Notice that if we simply draw a line segment from the source node to the destination node, we cannot prove this statement anymore. To circumvent this, they essentially map the disk into a large sphere and then prove that all results will carry over.

18.4   Capacity of Broadcast for an Arbitrary Network

In previous sections, we studied the throughput capacity for unicast of an arbitrary network and random networks of n nodes for which the total channel capacity is W. It is easy to foresee that the capacity bound may change for different operations such as broadcast. In this section and the next section, we study the throughput capacity of an arbitrary network or random networks for *single-source broadcast*. Assume that we have a network $G = (V, E)$ of n wireless nodes. Every node $v_i \in V$ has the same transmission power and thus the same transmission range r. Thus, the network G is modeled by a UDG. Let W bits/s be the data rate that can be supported by the common communication channel. In other words, we assume that nodes will not dynamically change their coding to change the supported data rate based on the distance where the receivers are.

The broadcast capacity of an arbitrary network was studied in Keshavarz-Haddad *et al.* (2006) and Tavli (2006). They essentially show that the broadcast capacity of a given network is $\Theta(W)$ for single-source broadcast and the achievable broadcast capacity per node is only $\Theta(W/n)$ if each of the n nodes will serve as a source node.

THEOREM 18.5 (Keshavarz-Haddad *et al.*, 2006; Tavli, 2006) *Assume that the channel capacity is W bits/s. The broadcast capacity of an arbitrary wireless network is* $\Theta(W)$.

Proof. If there is only one source node s for broadcast, it is easy to show that the upper bound of broadcast is at most W because the number of bits that come out of the source node is at most WT in time T s. Thus, the capacity is bounded from above by W. We show that $\Theta(W)$ is achievable as follows: Given a network G, we first create a connected dominating set (CDS) in which the source node s is part of the CDS. The method we used to create CDS could be any method from Chapter 7. The CDS constructed using those methods has the following nice property: For each node v from the CDS, the number of communication neighbors of v in CDS is bounded from above by a constant; say, c. Assume that the interference range of every node is a constant $(1 + \Delta)$ times of the communication range r. Notice that in practice, Δ typically is around 1. Let H be the interference graph constructed for the CDS: The nodes of H are nodes from the constructed CDS; two nodes u and v from the CDS are connected in the interference graph H only if there is a node $w \in V$ such that $\|u - w\| \le (1 + \Delta)r$ and $\|v - w\| \le r$. Then, it is easy to show that the graph H also has a degree of at most $c(2 + \Delta)^2$. Consequently, we can color H by using at most $1 + c(2 + \Delta)^2$ colors. Let $t(v) \in [1, 1 + c(2 + \Delta)^2]$ be the color assigned to node v from the CDS. In other words, we can schedule the transmissions of all nodes in the CDS without causing any interference, as follows: Node v transmits only at time $t(v) + iT$, where $T = 1 + c(2 + \Delta)^2$. Then, we can perform a broadcast based on the constructed CDS as follows: A node v in the CDS is scheduled to relay the data from its parent node at time $t(v) + iT$. Thus, the achieved data rate using such a broadcast is $(W/T) = \Theta(W)$. Consequently, the achievable broadcast capacity is $\Theta(W)$ (where the lower bound and the upper bound match).

When there are multiple sources in the network, it is not straightforward that the total broadcast capacity of all sources is bounded from above by $\Theta(W)$. Assume that all nodes are deployed in a square [the proof will carry over to the case in which the deployment region is any region Ω such that $\frac{|\Omega \cap D(v,r)|}{|D(v,r)|}$ is bounded from below by some constant for any node $v \in V$, where $D(v, r)$ is a disk centered at v with radius r]. Let n_1 be the maximum number of nodes that can transmit simultaneously. Let A be the area of the deployment region that is covered by disks $\cup_{v \in V} D(v, r)$. Let $A_0 = \pi r^4$. Then, obviously, $n_1(A_0/4) \le A$ because each transmission of a node will occupy at least an area of $A_0/4$ inside the square deployment region. This implies that the total number of bits that can be transmitted in a second all over the network is at most $W\left(\frac{A}{A_0/4}\right)$.

On the other hand, each bit from any source node needs to be relayed by some CDS. Notice that here, a bit from a different source node may use a different CDS. Let us consider the area covered by any fixed instance of CDS; i.e., $\cup_{v \in CDS} D(v, r)$. Notice that here, the CDS denotes a fixed connected dominating set. Notice that every node $u \in V$ is either in a CDS or is adjacent to a node in a CDS. We can cluster nodes into $|CDS|$ clusters: each cluster $\mathcal{C}_i$ contains one node v_i from the CDS and the nodes dominated by this node. It is easy to show that

$$|\cup_{v \in C_i} D(v,r)| \leq \pi(2r)^2 = 4A_0 = 4|D(v,r)|,$$

because all nodes in C_i are in disk $D(v,r)$. Here, $|\mathcal{X}|$ denotes the area of a region $\mathcal{X}$. Thus, the area $\cup_{v \in \text{CDS}} D(v,r)$ covered by all nodes in any CDS is at least $A/4$. In other words, each bit from any source node will be relayed by at least $\frac{A/4}{A_0}$ nodes because each relay node covers an area of at most A_0.

Combining the preceding analyses, we know that the aggregate broadcast capacity that can be supported is at most

$$\frac{W\left(\frac{A}{A_0/4}\right)}{\frac{A/4}{A_0}} = 16W = \Theta(W).$$

Notice that the coefficient 16 can be reduced by a tighter analysis.

On the other hand, we can perform broadcast as before by using a CDS. The only modification is that we need one more time slot in a scheduling period to let some source nodes that are not in the CDS have a chance to upload their data to their dominator in the CDS. Then, the data will be broadcast to the network with the same CDS used for all source nodes.   ■

18.5   Capacity of Broadcast for Randomly Deployed Networks

In this section, we briefly study the broadcast capacity for random networks when Gaussian channel model is used. Zheng (2006) studied the broadcast capacity and the information diffusion rate and derived fundamental bounds in both the extended network model and the dense network model. Essentially, it was shown in Zheng (2006) that the data rate of continuous broadcast is $\Theta((\log n)^{-\frac{\alpha}{2}})$ in extended networks, whereas a direct single-hop broadcast is efficient for dense networks. For random networks, typically there are three ways to increase the number of nodes to infinity:

1. One way is to *fix* the deployment region and then increase the node density to infinity. This is typically called the *dense model*. This model is widely studied; e.g., Gupta and Kumar (1998) studied the critical transmission range using this model. Compared with practical deployment, this model has a drawback in that the minimum power needed for having a connected network will be arbitrarily small when node density n is sufficiently large.

2. Another way is to *fix* the node density to a given constant and increase the deployment region to infinity. This is typically called the *extended model*. This model was also used in several papers to study the critical transmission range or capacity (e.g., Santi and Blough, 2003; Zheng, 2006). Compared with a practical deployment, this model also has a drawback in that the minimum power needed for having a connected network will be sufficiently large when the number of nodes in the network is sufficiently

large. Here, we assume that there is a constant lower bound on the minimum SINR such that the receiver can correctly decode the signal.

3. The third way is to fix the transmission range of all nodes and then increase the node density (asymptotically the same as the node degree when the transmission range is fixed) and the deployment area to increase the number of nodes in the network. We call this model the *fixed-range model*. Assume that n nodes will be deployed. It was proven in Xue and Kumar (2004) that the minimum node degree for connectivity is $\Theta(\log n)$. This implies that the area of the deployment region is at most $\Theta(\frac{n}{\log n})$.

Zheng (2006) mainly studied the broadcast capacity for the extended network model. Zheng constructed a random extended network by placing nodes according to a Poisson point process $\mathcal{P}$ of a unit rate on a 2D plane. Let $B(n)$ denote the box $[0, \sqrt{n}] \times [0, \sqrt{n}]$ and let $\mathcal{P}_n = \mathcal{P} \cap B(n)$ be a Poisson point process of a unit rate on $B(n)$. It was assumed also that nodes are individually power-constrained. Let P_i denote the power used by node i for transmission. Assume that the maximum power for transmission by each node is $P_{\max}$. Assume that the channel follows an ambient Gaussian noise model with power spectral density of $N_0/2$ and the signal attenuation of $d^{-\alpha}$, where d is the distance between the source and the receiver node. When some common information is directly broadcast from a node i to a set of receivers $\mathcal{R}$, capacity-achieving Gaussian channel codes are assumed to support the worst achievable data rate of all receivers; i.e.,

$$r_i = \min_{j \in \mathcal{R}} B \log \left(1 + \frac{P_i d_{i,j}^{-\alpha}}{B N_0 + \sum_{k \in I} P_k d_{k,j}^{-\alpha}} \right),$$

where I is the set of nodes that are simultaneously transmitting with node i using the same channel and B is the bandwidth of the channel. Notice that this model of data rate is *different* from the model of data rate used to study the capacity for unicast in previous sections, where we assume a *fixed* data rate W for the channel. This is also different from the physical model used in previous sections in which we assumed a minimum bound on the SINR, whereas here, we do not have such constraint. In Zheng (2006), it is further assumed that no cooperative relay strategy is used at the physical layer to improve the throughput. Based on the rate definition supported by a node i to its set of downstream children nodes $\mathcal{R}$, every sender node needs to determine the set $\mathcal{R}$ of receivers it needs to reach and chooses a coding scheme and the corresponding transmission power such that the receiver node with the least SINR can also successfully decode the message.

THEOREM 18.6 (Zheng, 2006) *Assume a power-attenuation exponent $\alpha > 2$. In extended networks, with high probability, the broadcast capacity is $\Theta(P_{\max}(\log n)^{-\alpha/2})$, where $P_{\max}$ is the maximum transmission power every node can use.*

To prove the preceding theorem, we show that the broadcast capacity is bounded from above by $\Theta(P_{\max}(\log n)^{-\alpha/2})$ with high probability. Then, we provide a constructive broadcast scheme that can achieve $\Theta(P_{\max}(\log n)^{-\alpha/2})$ broadcast capacity with high probability.

In broadcasting data from a source node to all nodes in the network, messages will be forwarded on a spanning tree (or a collection of spanning trees sometimes). Notice that the broadcast tree can change over time, and a node may use a different transmission power and thus have a different set of downstream children nodes. A node v will *never* be able to receive messages at a rate faster than the capacity of its best incident link. Recall that under the link rate assumption, the maximum data rate that can be received by node v is always from the link uv with the shortest Euclidean length.

Define the nearest neighbor graph NNG(n) as follows: It contains all nodes in the network, and each node v is connected to its nearest neighbor u (with the smallest Euclidean distance). Let M_n be the longest edge of the NNG with nodes produced by a Poisson point process of rate n on a 2D unit area. It was proved in Penrose (1999b) that

$$\mathbf{Pr}\!\left(n\pi M_n^2 - \log n \le \gamma\right) = e^{-e^{-\gamma}}.$$

In other words, with high probability, the longest edge in NNG(n) is at least $\Theta(\sqrt{\frac{\log n}{n}})$ with high probability. When we scale the unit square area to a square of length $\sqrt{n}$, then the Poisson point process of rate n becomes a Poisson point process of rate 1, and the longest edge M_n in the NNG will also be scaled up by a factor $\sqrt{n}$. It is thus natural to conclude that the longest edge of the NNG of nodes produced by the Poisson point process of rate 1 over a square of side length $\sqrt{n}$ is $\Theta(\log n)$ with high probability. This indeed proved to be true in Zheng (2006).

THEOREM 18.7 *Let M_n be the longest edge of the NNG constructed from a Poisson point process of unit rate over a square with side length $\sqrt{n}$. If $\lim_{n\to\infty} f(n) = +\infty$, then*

$$\lim_{n\to\infty} \mathbf{Pr}\!\left[\pi M_n^2 - \log n + f(n) \ge 0\right] = 1,$$
$$\lim_{n\to\infty} \mathbf{Pr}\!\left[\pi M_n^2 - \log n - f(n) \le 0\right] = 1.$$

For example, we can set $f(n) = \log n/2$. Then, it is easy to show that

$$\lim_{n\to\infty} \mathbf{Pr}\!\left(M_n \ge \sqrt{\frac{\log n}{2\pi}}\right) = 1,$$
$$\lim_{n\to\infty} \mathbf{Pr}\!\left(M_n \le \sqrt{\frac{3\log n}{2\pi}}\right) = 1.$$

Assume that uv is the longest edge in the NNG produced; i.e., $\|uv\| = M_n$. To improve the data rate received by a node v, we clearly need to reduce the interference; i.e., only one node is sending and one node v is receiving. The data rate that node v can receive is bounded from above by

$$B\log\left(1 + \frac{P_{\max}/M_n^\alpha}{BN_0}\right),$$

which, with high probability, is at most

$$B\log\left[1 + \frac{P_{\max}/\left(\frac{\log n}{2\pi}\right)^{\alpha/2}}{BN_0}\right] \simeq B\frac{P_{\max}\left(\frac{\log n}{2\pi}\right)^{-\frac{\alpha}{2}}}{B\cdot N_0} = \frac{P_{\max}\left(\frac{\log n}{2\pi}\right)^{-\frac{\alpha}{2}}}{N_0}.$$

This first approximation comes from the fact that $\log(1 + x) \simeq x$ when $x \to 0$ and $(\log n)^{-\alpha/2} \to 0$ when $n \to \infty$ and $\alpha > 2$. This concludes that the maximum broadcast data rate that can be supported by a random extended network is at most $\Theta(P_{\max}(\log n)^{-\alpha/2})$ with high probability.

We then use a constructive method to show that such a broadcast capacity can be achieved with high probability for a random extended network. The basic idea of this constructive method again relies on constructing a good CDS: Each node in the CDS has only a constant number of interfering nodes in the CDS. Thus, we can find a schedule with a constant period, say T, such that each node in the CDS will have at least one time slot to send its data in the schedule period. Another important observation is that the smallest transmission radius needed to have a connected network is also asymptotically $\Theta(\log n)$. This implies that the continuous data rate that can be supported by each node in the CDS is also of the order of $\Theta(P_{\max}(\log n)^{-\alpha/2})$. Combining the fact that each node in the CDS can use at least one time slot to send data every constant T slots, the achieved broadcast capacity is still of the order of $\Theta(P_{\max}(\log n)^{-\alpha/2})$.

Zheng (2006) did not specifically use the CDS structure. It partitions the deployment region into cells and will select one node from each cell as the relay node for broadcast; i.e., these representative nodes will be in the CDS. The side length ℓ of the cell should be carefully selected. There are many ways to select the cell size. One way is to select ℓ such that any node in one cell is a communication neighbor of any node in an adjacent cell. Thus, $\ell = r/\sqrt{5}$, where r is the uniform communication range. It is easy to show that the collection of representative nodes from all cells form a DS and the node degree in the DS is bounded from above by a constant. A CDS can be constructed from this DS, whose size is only a constant time of the size of a DS.

Recall that the broadcast capacity is defined under the physical-interference model. Thus, under a schedule of transmissions, we need to show that the actual data rate supported by a node is indeed $\Theta(P_{\max}(\log n)^{-\alpha/2})$. This is indeed true and proved by the following theorem in Zheng (2006):

THEOREM 18.8 *Assume that* $\alpha > 2$. *For any given integer* k > 0, *under the general physical-interference model, there exists a TDMA scheduling in which one node per square of edge length* ℓ *can transmit to nodes located within a radius* k *in Manhattan distance, with fixed rate* R(k) *given as*

$$R(k) \geq \frac{B}{(2 + 2k)^2} \log\left(1 + \frac{P_{\max}}{B\{N_0[\ell(k + 1)]^\alpha\} + K'P_{\max}}\right),$$

where K' *is a constant independent of k and* ℓ.

Notice that here, K' is some constant depending actually on $\sum_{i=1}^{\infty} \frac{1}{i^{\alpha-1}}$, which converges to a constant $\frac{1}{\alpha-2}$ when $\alpha > 2$. When $\alpha = 2$, K' is not a constant anymore: It is $O(\log n)$ instead. Recall that $\ell = \Theta(r(n)) = \Theta(\sqrt{\log n})$, where $r(n)$ is the minimum transmission range needed for having a connected network with high probability. Theorem 18.8 implies that when k is some constant, then the achievable data rate by nodes in the CDS (representative nodes selected from all cells) is of the order of $\Theta(\frac{P_{\max}}{\ell^\alpha}) = \Theta(\frac{P_{\max}}{(\log n)^{\frac{\alpha}{2}}})$. It is because

$$\lim_{n\to\infty} \frac{B}{(2+2k)^2} \log\left(1 + \frac{P_{max}}{B\{N_0[\ell(k+1)]^\alpha\} + K'P_{max}}\right)$$

$$= \lim_{n\to\infty} \frac{B}{(2+2k)^2} \cdot \frac{P_{max}}{B\{N_0[\ell(k+1)]^\alpha\} + K'P_{max}}$$

$$= \frac{P_{max}}{(2+2k)^2 N_0(k+1)^\alpha \ell^\alpha}$$

When the node deployment follows the fixed-range model, we show that the broadcast capacity is actually $\Theta(1)$ instead of $\Theta(P_{max}(\log n)^{-\alpha/2})$ for $\alpha > 2$. For simplicity, we assume that the transmission range of every node is one unit under the fixed-range model. First, under this fixed-range model, the length of the longest edge in the NNG is $\Theta(1)$ with high probability. Then, as with the extended model, we can show that the broadcast capacity is bounded from above by

$$B \log\left(1 + \frac{P_{max}g}{BN_0}\right),$$

where g is the signal loss ratio at one unit distance; i.e., $P_{max}g$ is the signal power at the receiver's side. Notice that in the preceding bound, all numbers are fixed constants, which implies that the broadcast capacity is bounded from above by a constant. Second, we show that the broadcast capacity is also bounded from below by a constant by constructing a broadcast scheme. The scheme is again to partition the deployment region into cells with side length $\ell = 1$ and select one node from each cell as the broadcast relay node. By applying Theorem 18.8, we know that such a broadcast scheme will achieve a broadcast capacity of at least $\Theta(1)$ because both k and ℓ are constants now in Theorem 18.8. Thus, we have the following theorem:

THEOREM 18.9 *Assume that capacity-achieving Gaussian channel codes are used. The broadcast capacity of random networks with the fixed-range model is $\Theta(B\log(1 + \frac{P_{max}g}{BN_0})) = \Theta(1)$.*

Observe that all the preceding analysis cannot carry over to the dense model. The reason is as follows: In the analysis of the extended model and the fixed-range model, we assume that if a node u sends a signal to a node v with power P, then the power received by node v is $\frac{P}{\|u-v\|^\alpha}$. This will not violate the physical law that the received power is no more than the sending power if $\|u - v\| > 1$. However, for the dense model, this power-attenuation model is no longer valid because the distance is often sufficiently close to $\Theta(\sqrt{\frac{\log n}{n}}) \to 0$ now. In other words, the power-attenuation model $\frac{P}{\|u-v\|^\alpha}$ will result in a nonvalid scenario where the receiving power is larger than the sending power. To remedy this, one possible way is to use one of the following power-attenuation models:

$$\frac{P}{(1 + \|u - v\|)^\alpha} \quad \text{or} \quad \frac{P}{1 + \|u - v\|^\alpha}.$$

18.6   Further Reading

Gupta and Kumar (1999) studied the asymptotic capacity of multihop wireless networks for two different models. When each wireless node is capable of transmitting at W bits/s using a fixed range, the throughput obtainable by *each* node for a randomly chosen destination is $\Theta(\frac{W}{\sqrt{n \log n}})$ bits/s under a noninterference protocol, where n is the number of nodes. If nodes are optimally assigned and the transmission range is optimally chosen, even under optimal circumstances, the throughput is only $\Theta(\frac{W}{\sqrt{n}})$ bits/s for each node. Similar results also hold for the physical-interference model. Notice that the results presented in Gupta and Kumar (1999) did not consider the additional burden in coordinating access to wireless channels, the effect of mobility and link failures, or the effect of the need to route traffic in a distributed way. They also did not address the delay of the route. The delay could be caused by burst traffic or when nodes are mobile and links are not stable. It can also be imagined that using directional antennas or beam-forming will help to improve the spatial concurrency of transmissions and thus the capacity of the networks.

Grossglauser and Tse (2001) showed that mobility actually can help to improve the capacity if we allow an arbitrarily large delay. Their main result shows that the average long-term throughput per source–destination pair can be kept constant even as the number of nodes per unit area increases. Notice that this is in sharp contrast to the fixed network scenario (in which nodes are static after random deployment). The main idea used in Grossglauser and Tse (2001) is to use some intermediate node to serve as ferry node: This node will carry the data from the source node and move around, and it will dump the data to the target node when it is within its communication range. They showed that at any time instance, we can schedule $\Theta(n)$ links to transmit simultaneously without causing interference. To achieve this, they have to require that all nodes can dynamically adjust their transmission power based on receivers. Notice that without dynamic power adjustment, the number of simultaneous transmissions is only at most $\Theta(\frac{n}{\log n})$. However, essentially, the results presented in Grossglauser and Tse (2001) still obey the capacity bound proposed in Gupta and Kumar (1999): The capacity is improved because the average distance $\overline{L}$ a packet has to be transmitted is reduced from $\Theta(1)$ in Gupta and Kumar (1999) to $\Theta(r(n))$ in Grossglauser and Tse (2001). In summary, for random networks under the protocol model, the achievable throughput capacity $\lambda(n)$ and the average travel distance $\overline{L}$ satisfies $\lambda(n)\overline{L} \leq \Theta(\frac{W}{\Delta^2 n r(n)})$. This phenomenon was also observed in J. Li *et al.* (2001). They found that the traffic pattern determines whether the per-node capacity of a wireless network will scale to large networks. They observed that nonlocal traffic patterns in which the average distance grows with the network size result in a rapid decrease of per-node capacity. They also examined the interactions of the IEEE 802.11 MAC and ad hoc forwarding and the effect on the capacity of wireless networks. Although IEEE 802.11 discovers reasonably good schedules, they nonetheless observed capacities markedly less than the optimal even for very simple networks, such as chain and lattice networks, with very regular traffic patterns. This confirms the importance of using a carefully designed transmission schedule to improve the network throughput whenever possible.

Recently, a number of papers studied the trade-offs between capacity and delay for unicast under various mobility models, such as i.i.d. mobility (Lin and Shroff, 2004; Neely and Modiano, 2005; Toumpis and Goldsmith, 2004), RWP mobility (Sharma and Mazumdar, 2004, 2005), Brownian motion (Gamal *et al.*, 2004; Lin *et al.*, 2006; Sharma and Mazumdar, 2004), and Markovian mobility (Neely and Modiano, 2005).

In Gupta and Kumar (1999), the capacities of wireless networks are solved under a number of assumptions, among them point-to-point coding that excludes, for example, the multiaccess and broadcast codes. Gastpar and Vetterli (2002) studied the capacity of wireless networks when network coding can be used to improve the capacity. They essentially considered the same physical model under different traffic patterns (relay traffic patterns). They allow for arbitrary complex network coding. In their model, there is only one source and destination pair and all other nodes will assist this transmission. They show that the capacity of such wireless networks with n nodes under relay traffic patterns behaves like $\log n$ bits per second. This demonstrates the power of network coding: Under the point-to-point coding assumption considered in Gupta and Kumar (1999), the achievable data rate is constant, independent of the number of nodes.

Francheschetti *et al.* (2006) showed that a rate $\Theta(\frac{1}{\sqrt{n}})$ is achievable in networks of randomly located nodes. Hence, there is no gap between the capacity of randomly located nodes and arbitrarily located nodes, at least up to scaling and in the high attenuation regime. The constructive strategy that achieves the rate $\Theta(\frac{1}{\sqrt{n}})$ is based on multihop transmission, pairwise coding and decoding at each hop, and a time-division multiple-access (TDMA) scheme. The proof of the result follows from percolation-theory arguments.

Capacity can also be generalized to the notion of a *capacity region*. For a given statistical description of the network, a set of constraints (e.g., power per node, link capacity), and a list of desired communication pairs, the capacity region is the closure of all rate tuples that can be achieved simultaneously. Here, a rate tuple specifies the rate for each of the desired communications. Kyasanur and Vaidya (2005a) studied the capacity region on random multihop, multiradio, multichannel wireless networks when there are a total c channels available and each node has $m \leq c$ wireless interfaces. In another aspect, several papers (Alicherry *et al.*, 2005; Kadialam and Nandagopal, 2003) recently studied how to satisfy a certain traffic demand vector from all wireless nodes by joint routing, link scheduling, and channel assignment under certain wireless interference models.

Broadcast capacity of an arbitrary network has been studied in Keshavarz-Haddad *et al.* (2006) and Tavli (2006). They essentially show that the broadcast capacity of a given network is $\Theta(W)$ for single-source broadcast, and the achievable broadcast capacity per node is only $\Theta(W/n)$ if each of the n nodes will serve as the source node.

18.7   Conclusion and Remarks

In this chapter, we essentially studied the capacity that can be achieved by some wireless networks. The total end-to-end throughput capacity for unicast is roughly $O(W\sqrt{n})$ for any network of n nodes deployed in a unit area. It is possible to carefully place nodes to achieve $O(W/(\overline{L}\sqrt{n}))$ end-to-end throughput for every source node. Here, $\overline{L}$ is the average distance between the source node and its corresponding destination node. When nodes are randomly placed, then the end-to-end throughput capacity that can be achieved per source node (with a randomly selected target node) is reduced to $\Theta(\frac{W}{\overline{L}\sqrt{n\log n}})$. In other words, the random placement of nodes will further reduce the achievable capacity by $1/\sqrt{\log n}$ factor. Here, we assume that for networks, the transmission range is selected to maximize the achievable throughput (which happens to be the least transmission range needed for connectivity). On the other hand, by using mobile nodes to serve as relay nodes, we can improve the capacity by essentially reducing the $\overline{L}$ from $\Theta(1)$ to the minimum $2r(n)$. Using network coding, we further improve the capacity by another $\Theta(\log n)$ factor. We also reviewed some results about the capacity of wireless networks for broadcast.

An interesting research problem is: What is the throughput capacity for multicast? Assume that for each node v in the network, we randomly select $1 \leq k < n$ nodes as receiver nodes of a multicast from v. What will be the capacity achievable per node? Here, when $k = 1$, we essentially need to solve the capacity for unicast, and when $k = n - 1$, we essentially need to solve the capacity for broadcast. Recently, Li *et al.* (2007) and Shakkottai *et al.* (2007) gave tight asymptotic bounds for capacity of multicast for large-scale ad hoc networks. Li *et al.* generally assumed that there are n_s multicast sessions, each with $k - 1$ receivers from V, and the transmission range r and side-length a of the deployment square satisfying that the resulting random network is connected with high probability. Using the protocol-interference model, they proved the following results: When $n_s \geq \Omega(\log k \cdot \sqrt{\frac{n\log n}{k}})$, the aggregated multicast capacity of n_s multicast sessions is

$$\Lambda_k(n) = \begin{cases} \Theta\left(\frac{a}{r} \cdot \frac{W}{\sqrt{k}}\right) & \text{when } k = O\left(\frac{a^2}{r^2}\right) \\ \Theta(W) & \text{when } k = \Omega\left(\frac{a^2}{r^2}\right) \end{cases} \tag{18.3}$$

It is interesting to study the broadcast capacity when nodes could dynamically adjust its coding to change the data rate based on different link characteristics to different receivers. In other words, we could further improve the capacity by exploiting the user diversity. In practice, the communication link capacity depends on the distance between the source node and the target node. Further, the results summarized in this chapter mainly assumed the protocol-interference model and fixed data rate for each communication link. Fran, cheschetti *et al.* (2006) closed the gap in the capacity of random wireless networks using percolation theory when all nodes transmit at a fixed power P and the link capacity is defined as $B \log(1 + SINR)$, where B is the bandwidth and SINR the is signal-to-interference-noise ratio.

Problems

18.1 In what circumstances can we say it is immaterial to results presented in this chapter if the channel is broken up into several subchannels of capacity $W_1, W_2, \ldots, W_M$ bits/s as long as we have $\sum_{i=1}^{M} W_i = W$?

18.2 What are the differences between the PrIM for an arbitrary network and the PrIM for the random networks used by Gupta and Kumar?

18.3 Prove that $\sum_{i=1}^{\infty} \frac{1}{i^{\theta}}$ converges to a constant when $\theta > 1$.

18.4 Assume that the physical-interference model is used; i.e.,

$$\frac{\frac{P_i}{\|X_i - X_j\|^{\alpha}}}{N + \sum_{X_k \in N_t} \frac{P_k}{\|X_k - X_j\|^{\alpha}}} \geq \frac{\beta}{\beta + 1},$$

where node v_i is the sender and node v_j is the receiver. Here, X_i is the position of node v_i, and P_i is the transmission power of node v_i. Assume that all nodes use the same transmission power P. Prove that there is a constant r_0 depending on P, β, and background noise N such that the distance between the sender and the receiver cannot be more than r_0. Prove that there is a constant R_0 depending on P, β, and background noise N such that the distance between any two simultaneous senders is at least R_0.

18.5 Continue from the previous question. Assume that node v is receiving data from a sender u. Assume that $R_1 = a R_0$ for some constant $a > 1$, where R_0 is a constant computed in the previous question. Let W be the set of simultaneously transmitting nodes with node u. Assume that all transmitting nodes will *not* cause interference in any receiving node. Prove that

$$\sum_{\|w_i - v\| \geq R_1, w_i \in W} \|w_i - v\|^{-\alpha}$$

is bounded by a constant when $\alpha > 2$.

18.6 The binomial distribution is the discrete probability distribution of the number of successes in a sequence of n independent yes/no experiments, each of which yields success with probability p. Assume that a random variable X follows the binomial distribution with parameters n and p; i.e., $X = \sum_{i=1}^{n} X_i$, where all n variables $X_i \in \{0, 1\}$ are independent and $\Pr(X_i = 1) = p$. Assume that $k \leq np$. Prove that

$$\Pr(X \leq k) \leq \exp\left(-2 \frac{(np - k)^2}{n}\right).$$

Assume that $k \leq np$. Prove that

$$\Pr(X \leq k) \leq \exp\left(-\frac{1}{2p} \frac{(np - k)^2}{n}\right).$$

18.7 Assume that for mobile networks of n nodes distributed in a unit-square region, the transmission range of each node is set as $\Theta(\sqrt{\frac{\log n}{n}})$ such that the network is connected with high probability. In other words, nodes will *not* dynamically adjust their transmission power based on the receiver. Prove that in this model, the capacity per node that can be achieved is at most $\Theta(\frac{W}{\log n})$ instead of $\Theta(W)$ for the model in which nodes can dynamically adjust their transmission range.

19 Security of Wireless Ad Hoc Networks

19.1 Introduction

Ensuring the security of both the collected data and the process of data collection is vital for the success of WSNs. Because of the constraints of the particular applications and the resource limitations, the security of WSNs is vastly different from that of conventional wired networks. For the example of military applications, wireless sensor nodes usually are sent to an unattended environment (e.g., the battlefield). In these scenarios, wireless sensor nodes are easier to capture or destroy. Thus, the foremost important thing for WSNs is that they can tolerate the dysfunction of a certain number of nodes; e.g., the network formed by nondestroyed nodes should be always connected. The second possible attack for WSNs is that the enemy could distribute a certain number of faked sensor nodes to disturb or even disrupt the communications of legitimate sensors. It is important for a sensor network to design a security mechanism to protect the sensor nodes from malicious attack or to ensure that the sensor network can "tolerate" the malicious attack to some extent. The second challenge for the design of security mechanisms for WSNs is that sensor nodes are always equipped with limited battery and memory. Thus, the traditional public-key–based schemes, such as the Rivest–Shamir–Adleman (RSA) and Diffie–Hellman (D-H) protocols, are not suitable for WSNs. For example, Mica Mote, produced by UC Berkeley, has $128\,kb$ Flash and $4\,kb$ RAM. Notice that the WSN itself is not stable; e.g., the links could be on or off depending on the transmission environment, battery power of nodes, and the traffic load. This poses another challenge for the design of security mechanisms for WSNs compared with the wired networks.

In this chapter, we mainly focus on some fundamentals of cryptography, some key-predistribution protocols, and some secure routing protocols proposed in the literature. Cryptography provides us with some fundamental tools, such as symmetric-key and asymmetric-key encryption, digital signature, and hash functions, to implement some security protocols. Because symmetric-key encryption is more efficient than asymmetric-key encryption, data confidentiality is often achieved with symmetric-key encryption. To do so, all wireless nodes need to have keys for pairwise communications. These keys can be obtained by using some key-agreement protocols that will let two nodes agree on some keys on-line or by using some key-predistribution protocols that will prestore some keys to every node in the network. Here, the prestored keys to different nodes need not be same. After nodes have keys for pairwise secure communication, we need to design secure routing protocols to transfer data from some source node to the destination node.

A number of secure routing protocols have been proposed in the literature for wireless ad hoc networks, especially WSNs.

19.2   Cryptography Fundamentals

Before a public-key system was developed in the early 1970s, cryptography was concerned solely with message confidentiality; i.e., encryption and decryption. Data encryption converts messages from a comprehensible form into an incomprehensible one. The converted incomprehensible form is often called *ciphertext*. Data decryption converts the ciphertext back again to the original message. The original message is also called *plaintext* in the literature. Data encryption needs to ensure that the ciphertext is unreadable by interceptors or eavesdroppers without secret knowledge (namely, the key needed for decryption). The pair of algorithms that performs the encryption and the reversing decryption is called a *cipher*. The detailed operation of a cipher is controlled both by the algorithm and, in each instance, by a *key*. This is a secret parameter (known only to the end communicants) for the cipher algorithm. Keys are important because ciphers without variable keys are trivially breakable and so are rather less than useful.

Since the early 1970s, the cryptography field has expanded beyond confidentiality concerns to include techniques for authentication of message integrity, or sender/receiver identity, digital signatures, interactive proofs, and secure computation, and so on. Encryption/decryption is used to ensure data confidentiality. Typically, an encryption of a message can be viewed as a function defined as follows:

$$\mathbf{E} : (\mathcal{M} \times \mathcal{K}_E) \longrightarrow \mathcal{C},$$

where $\mathcal{M}$ is the message space (i.e., the set of all possible messages), $\mathcal{K}_E$ is the encryption key space (i.e., the set of all valid keys used for encryption), and $\mathcal{C}$ is the ciphertext space (i.e., the set of all possible valid ciphertexts). To simplify the study, in the literature it is always assumed that the message space, the encryption key space, and the ciphertext space are all subsets of nonnegative integers. Data decryption is then another function that converts a ciphertext, using a decryption key, to a plaintext text; i.e.,

$$\mathbf{D} : (\mathcal{C} \times \mathcal{K}_D) \longrightarrow \mathcal{M},$$

where $\mathcal{K}_D$ is the decryption key space (i.e., the set of all valid keys used for encryption). Data-encryption methods can be divided into two large categories: symmetric-key encryption and asymmetric-key encryption. Symmetric-key cryptography refers to encryption methods in which both the sender and the receiver share the same key (or, less commonly, in which their keys are different but related in an easily polynomial-time computable way). This was the only kind of encryption publicly known until 1976. An asymmetric-key encryption system is also called the public-key-encryption system. For public-key-encryption systems, unlike the symmetric-key-encryption systems, the encryption key and the corresponding decryption key are always different. Furthermore, either the encryption key or the decryption key is made public depending on the application scenarios. The revealed key is often called the *public key*, whereas the privately kept key is called the *private key*. For the purposes of secure communication via encryption, often the encryption key is made public while the decryption key is kept private. For public-key-encryption systems, it is computationally infeasible (sometimes under the assumptions of the hardness of some problems) to compute the public key (or the private key) from the private key (or the public key).

Assume that for an encryption/decryption system, the encryption method is E and the decryption method is D. Further assume that k_e is the key for encrypting a message m. In other words, we have the ciphertext $c = E(m, k_e)$. Given the ciphertext c, assume that k_d is a corresponding decryption key that will be used for decryption. Notice that for some encryption systems, k_d is *not* necessarily unique. Thus, to make sure the decryption will indeed recover the original plaintext, we need

$$m = D(c, k_d) = D[E(m, k_e), k_d]$$

for every valid message m and every pair of encryption key and decryption key (k_e, k_d).

The main classical cipher types are *transposition ciphers*, which rearrange the order of letters in a message, and *substitution ciphers*, which systematically replace letters or groups of letters with other letters or groups of letters. For example, a transposition cipher could encrypt the message *help me* as *ehpl em* in a trivially simple rearrangement scheme. A simple example of a substitution cipher could encrypt a message *fly at once* as *gmz bu podf* by replacing each letter with the one following it in the alphabet. It is easy to observe that the ciphertexts produced by these classical ciphers always reveal some statistical information about the plaintext. This is often used to break the classical ciphers by using *frequency analysis*, discovered by Arab around A.D. 1000. Around A.D. 1467, Leon Battista Alberti invented the polyalphabetic cipher, which uses different ciphers (i.e., substitution alphabets) for various parts of a message (often each successive plaintext letter). Most of the recent symmetric-key ciphers still employ the combination of the substitution and transposition of the input.

In addition to encryption/decryption to ensure secrecy in communications, a number of different approaches have been developed historically. For example, stenography (i.e., hiding even the existence of a message so as to keep it confidential) was also first developed in ancient times. Modern examples of stenography include the use of invisible ink, microdots, and digital watermarks to conceal information.

19.2.1 Conventional Encryption Methods

The modern study of symmetric-key ciphers relates mainly to the study of block ciphers and stream ciphers and to their applications.

Block Ciphers and Stream Ciphers

A block cipher takes as input a block of plaintext and a key and outputs a block of ciphertext of the same size. A block cipher operates on blocks of fixed length, often 64 or 128 bits. Examples of block ciphers include the data-encryption standards (DES) and advanced encryption system (AES). Because messages are almost always longer than a single block, some method of knitting together successive blocks is required. Several knitting methods have been developed, some with better security in one aspect or another than others. They are the modes of operations and must be carefully considered when a block cipher is used in a cryptosystem. Several modes of operation have been invented that allow block ciphers to provide confidentiality for messages of arbitrary length. The earliest modes described in the literature [e.g., electronic codebook (ECB), cipher-block chaining (CBC), output feedback (OFB), and cipher feedback (CFB)] provide only confidentiality and do not ensure message integrity. Other modes have since been designed that ensure both confidentiality and message integrity, such as the combined cipher machine (CCM) mode, EAX mode, and offset codebook (OCB) mode. All these modes (except ECB) require an *initialization vector*, or IV — a sort of "dummy block" to kick off the process for the first real block, and also to provide some randomization for the process. There is no need for the IV to be secret, in most cases, but it is important that it is never reused with the same key.

The simplest encryption mode is the ECB mode. In the ECB mode, each message is divided into blocks of the fixed length equal to the size of the message that can be encrypted by the cipher. For example, for the DES encryption method, the block size is always fixed as 64 bits. Let P_i be the ith block partitioned from the plaintext and let C_i be the ith block in the ciphertext. Assume that K_E is the encryption key and K_D is the decryption key. Then, we have

$$\text{encryption}: C_i = E(P_i, K_E),$$
$$\text{decryption}: P_i = D(C_i, K_D).$$

Clearly, an obvious possible disadvantage of this method is that the identical plaintext blocks will be encrypted into identical ciphertext blocks if the same encryption key is used. In other words, it does not hide the data patterns well. It has been observed that if the ECB mode is used to encrypt a pixel-map version of an image, the encrypted image will reveal some visual information about the image. The ECB mode can also make protocols without integrity protection even more susceptible to replay attacks because each block gets decrypted in exactly the same way. Thus, the ECB mode is often not recommended for use in cryptographic protocols at all.

In the CBC mode, each block of plaintext is xored with the previous ciphertext block *before* being encrypted. Using this approach, each ciphertext block is dependent on all plaintext blocks processed up to that point. In addition, to make each encrypted message unique, an IV must be used in the first block. Then, we have for $i \geq 1$

$$\text{encryption}: C_i = E(P_i \oplus C_{i-1}, K_E), \quad C_0 = \text{IV},$$
$$\text{decryption}: P_i = D(C_i, K_D) \oplus C_{i-1}, \quad C_0 = \text{IV}.$$

CBC has been the most commonly used mode of operation. As with all block ciphers, the message must be padded to a multiple of the cipher-block size. Note that a one-bit change in a plaintext affects *all* following ciphertext blocks. This implies that the main drawback is that encryption is sequential (i.e., it cannot be parallelized). On the other hand, a plaintext block P_i can be recovered from just two adjacent blocks C_i and C_{i-1}

of ciphertext. As a consequence, decryption can be parallelized, and a one-bit change to the ciphertext in block C_i causes complete corruption of the corresponding block of plaintext block P_i and P_{i+1} and inverts the corresponding bit in the following block of plaintext.

The CFB mode, a close relative of CBC, makes a block cipher into a self-synchronizing stream cipher. Operation in the CFB mode is very similar to that of CBC. The encryption and the decryption under the CFB mode work as follows:

$$\text{encryption}: C_i = E(C_{i-1}, K_E) \oplus P_i, \quad C_0 = \text{IV},$$
$$\text{decryption}: P_i = D(C_{i-1}, K_D) \oplus C_i, \quad C_0 = \text{IV}.$$

As in the CBC mode, in the CFB mode, changes in the plaintext propagate to all the ciphertext following and the encryption cannot be parallelized. On the other hand, the decryption can be parallelized and a one-bit change in the ciphertext will corrupt only two blocks in the plaintext. Unlike the CBC mode, a form of pipelining of encryption is possible because the only encryption step that requires the plaintext is the final XOR operation. This property of the CFB mode is useful for applications that require low latency between the arrival of the plaintext and the output of the corresponding ciphertext. An example of such an application is a media stream. Another advantage of the CFB mode over the CBC mode is that the message does not need to be padded to a multiple of the cipher-block size. In some senses, the block cipher under the CFB mode is more like a stream cipher, discussed later.

The OFB mode makes a block cipher into a *synchronous* stream cipher: It generates key-stream blocks, which are then XORed with the plaintext blocks to get the ciphertext. An important property of the OFB mode is that just as with other stream ciphers, flipping a bit in the ciphertext produces a flipped bit in the plaintext at the *same location*. This property allows many error-correcting codes to function normally even when applied before encryption. Because of the symmetry of the XOR operation, encryption and decryption are exactly the same, as follows:

$$\text{encryption}: C_i = P_i \oplus O_i, O_i = E(O_{i-1}, K_E), \quad O_0 = \text{IV},$$
$$\text{decryption}: P_i = C_i \oplus O_i, O_i = E(O_{i-1}, K_D), \quad O_0 = \text{IV},$$

where $K_D = K_E$. Because the plaintext or ciphertext is used for only the final XOR, the block-cipher operations to produce O_i could be performed in advance. This allows the final XOR steps to be performed in parallel for different plaintext blocks and ciphertext blocks once the plaintext or ciphertext is available. This is especially useful when the amount of data to be encrypted (or decrypted) is much larger than the amount of data that can be processed by one processor or for streaming data such as in sensor networks.

Another popular operation mode is the counter mode. Like the OFB mode, the counter mode turns a block cipher into a stream cipher. It generates the next key-stream block by encrypting *successive values* of a "counter." The counter can be any simple function that produces a sequence that is guaranteed not to repeat for a long time. Let N_i be the

ith number produced by the counter function. Then, the encryption and decryption work as follows:

$$\text{encryption}: C_i = P_i \oplus O_i, \, O_i = E(N_i, K_E),$$
$$\text{decryption}: P_i = C_i \oplus O_i, \, O_i = E(N_i, K_D),$$

where $K_D = K_E$. For example, we can use the following counter IV $|i$ (IV concatenated by a number i) as the ith value, which is the simplest and most popular. Here, IV is some initial value and also called *nonce* in the literature. The counter mode has characteristics similar to those of OFB but also allows a *random-access* property during decryption and is believed to be as secure as the block cipher being used. In other words, to get the ith plaintext P_i, the total operations is the only constant. Recall that to get the ith plaintext P_i under the OFB encryption mode, the total operation is $\Theta(i)$.

Stream ciphers, in contrast to the block type, create an arbitrarily long stream of key material, which is combined (typically XORed) with the plaintext bit-by-bit or character-by-character, somewhat like the one-time pad. In a stream cipher, the output stream is created based on an internal state that changes as the cipher operates. That state's change is controlled by the key and, in some stream ciphers, by the plaintext stream as well. Although stream ciphers represent a different approach to symmetric encryption from that of the block ciphers, the distinction is not always clear-cut: In some modes of operation for block ciphers, a block cipher can be used in such a way that it acts effectively as a stream cipher. A stream cipher typically executes at a higher speed than that of a block cipher and has a lower hardware complexity. On the other hand, a stream cipher can be susceptible to serious security weaknesses if used inappropriately. For example, for a stream cipher, the same starting state can never be used twice.

Data-Encryption Standard

The DES is a cipher originally selected as an official FIPS (Federal Information Processing Standard) for the United States in 1976. The DES is now considered to be insecure in many applications. This is mainly due to the small key size: The encryption key for the DES is only 56 bits. With such a small key size, DES keys have been broken in fewer than 24 h. The DES is then enhanced by the 3DES, which will repetitively encrypt the data using DES three times with variant keys. 3DES is believed to be practically secure at the time of writing. The cipher has been superseded by the AES.

The DES encrypts a block of data with 64 bits using a key of 56 bits. It falls into the Feistel network structure. The Feistel structure ensures that decryption and encryption are very similar processes. The only difference is that the subkeys are applied in the reverse order when decrypting. The rest of the algorithm is identical. This greatly simplifies implementation, particularly in hardware, because there is no need for separate encryption and decryption algorithms. In a DES algorithm, there are 16 identical stages of processing, termed *rounds*. There is also an initial and final permutation, termed *IP* and *FP*, which are inverses (IP "undoes" the action of FP and vice versa). To make the same structure work for both encryption and decryption, there is an additional "round" before the final permutation FP that will swap the subblocks produced by the 16th round of the Feistel structure. IP and FP have almost no cryptographic significance. In each round, the block of 64 bits is partitioned into two subblocks, each of 32 bits. For simplicity, assume that the ith block is divided into L_i and R_i, denoting the left subblock and the right subblock, respectively. The operation of a Feistel structure in the DES will encrypt the data as follows:

$$L_{i+1} = F(R_i, K_i),$$
$$R_{i+1} = F(R_i, K_i) \oplus L_i.$$

Here, $F(R_i, K_i)$ could be any function for a Feistel structure, and K_i is a subkey used for the ith round. Notice that for the DES algorithm, some special functions are chosen to get a secure encryption algorithm. To ensure that the same structure can be used for decryption, the function F could be *any* function.

The F function used by DES actually has the following four stages:

1. **Expansion:** The 32-bit half-block R_i is expanded to 48 bits by the expansion permutation, denoted as $E(R_i)$, by duplicating some of the bits. The expansion is always fixed, independent of the key.
2. **Key mixing:** The result $E(R_i)$ is then XORed with the round key K_i. Here, the 48-bit round key K_i is derived from the main key (with 56 bits) using a fixed key schedule algorithm.
3. **Substitution:** The result $E(R_i) \oplus K_i$ is then divided into eight 6-bit pieces, which are used as the input of eight S-boxes. The purpose of each S-box is to map each of its 6-bit input into a 4-bit output according to some complex, nonlinear transformation. The S-box is often implemented as a lookup table and provides the core of the security of the DES. Without the nonlinear transformation of the S-box, the cipher will be a simple linear transformation and XOR, which could be broken easily.
4. **Permutation:** The 32 bits computed by the eight S-boxes are then rearranged (permuted) according to a fixed permutation.

Brute-force attack is the most widely used approach to attack the DES because of its small key size. In 1997, the DESCHALL project, led by Rocke Verser, Matt Curtin, and Justin Dolske, used the idle cycles of thousands of computers across the Internet to break a message encrypted by the DES. The feasibility of quickly cracking the DES was then demonstrated in 1998 when a custom DES cracker was built by the Electronic Frontier Foundation (EFF) at a cost of approximately US\$250,000. The machine brute-forced a key in a little more than a two-day search. There are some attacks known that can break the full 16 rounds of the DES with less complexity than a brute-force search: differential cryptanalysis (DC) and linear cryptanalysis (LC). However, the attacks are theoretical and are infeasible to mount in practice. To break the full 16 rounds, DC requires 2^{47} *chosen plaintexts* and LC needs 2^{43} *known plaintexts*.

The simplest variant of 3DES operates as follows:

$$\text{DES} \{\text{DES}[\text{DES}(M, k_1), k_2], k_3\},$$

where DES denotes the encryption function of the DES standard, M is the message block to be encrypted, and k_1, k_2, and k_3 are keys for one DES encryption. This variant is commonly known as EEE because all three DES operations are encryptions. To simplify interoperability between DES and 3DES, the middle step is usually replaced with decryption (and called the EDE mode):

$$\text{DES}\{\text{DES}^{-1}[\text{DES}(M, k_1), k_2], k_3\},$$

where DES^{-1} denotes the decryption function for the DES. Consequently, a single DES encryption with key k can be represented as 3DES with the EDE mode, where $k_1 = k_2 = k_3 = k$. The choice of decryption for the middle step does not affect the security of the algorithm. The use of three steps in 3DES is essential to prevent meet-in-the-middle attacks that are effective against double DES encryption. In other words, it has been shown that 2DES (using two rounds of the DES) can be broken by brute force in time of the order of 2^{56}, the same as that required for breaking the DES. Because of the meet-in-the-middle attack, the effective security provided by 3DES is only 112 bits [i.e., the brute-force attack requires time in $O(2^{112})$].

Advanced Encryption System (AES)

The AES [whose initial version is known as Rijndael (Daemen *et al.*, 2002)] is a block cipher adopted as an encryption standard by the U.S. government. The AES was announced by U.S. National Institute of Standards and Technology (NIST) as U.S. FIPS PUB 197 (FIPS 197) in November 26, 2001, after a five-year standardization process. It is expected to be used worldwide and analyzed extensively, as was the case with its predecessor, the DES. The AES cipher was initially developed by Joan Daemen and Vincent Rijmen, and the name Rijndael comes from a combination of the names of the inventors. Rijndael is a substitution–permutation network, not a Feistel network. The AES is fast in both software and hardware, is relatively easy to implement, and requires little memory. The AES has a fixed block size of 128 bits and a key size of 128, 192, or 256 bits, whereas the original Rijndael cipher can be specified with key and block sizes in any multiple of 32 bits, with a minimum of 128 bits and a maximum of 256 bits.

Most of AES calculations are done in a special finite field: the Galois field $\text{GF}(2^8)$, which is a characteristic 2 finite field with eight terms. In a finite field with characteristic 2, addition and subtraction are identical and are accomplished by use of the XOR operator. Elements of $\text{GF}(p^n)$ may be represented as polynomials of degree strictly less than n over $\text{GF}(p)$. Operations are then performed modulo a polynomial R, where R is an irreducible polynomial of degree n over $\text{GF}(p)$. Here, a polynomial is *irreducible* if it cannot be written as the production of two polynomials over $\text{GF}(p)$ whose degrees are at least 1. For AES encryption, it employs the following reducing polynomial for multiplication:

$$x^8 + x^4 + x^3 + x + 1.$$

For example, to represent a number 83 (i.e., number {53} in hex format), the following are equivalent representations in a characteristic 2 finite field: $x^6 + x^4 + x + 1$ for a polynomial representation and {01010011} for a binary representation. Addition and subtraction are performed by adding or subtracting two of these polynomials together and reducing the result modulo the characteristic. To avoid confusion, we use $\boxplus$ to denote the summation, $\boxminus$ to denote the subtraction, and $\boxdot$ to denote the product of two polynomials (and also the corresponding numbers represented by these polynomials). Thus,

$$(x^6 + x^4 + x^2 + 1) \boxplus (x^7 + x^6 + x^3 + x) = x^7 + x^4 + x^3 + 1.$$

This corresponds to a special addition in binary format:

$$\{01010011\} \boxplus \{11001010\} = \{10011001\}.$$

Notice that under the regular summation of polynomials, we should have a $2x^6$ term. However, this $2x^6$ term becomes $0x^6$ and is dropped when the characteristic is 2 (i.e., with modulo 2 for all coefficients). For example, for two hex numbers {53} and {CA}, their production is $\{53\} \boxdot \{CA\} = \{01\}$ in the AES encryption system (where the irreducible polynomial is $x^8 + x^4 + x^3 + x + 1$). The reason is as follows: {53} is represented by polynomial $(x^6 + x^4 + x + 1)$ and {CA} is represented by polynomial $(x^7 + x^6 + x^3 + x)$. It is easy to calculate the product of these two polynomials $(x^6 + x^4 + x + 1) \boxdot (x^7 + x^6 + x^3 + x) = x^{13} + x^{12} + x^{11} + x^{10} + x^9 + x^8 + x^6 + x^5 + x^4 + x^3 + x^2 + x$. The polynomial $x^{13} + x^{12} + x^{11} + x^{10} + x^9 + x^8 + x^6 + x^5 + x^4 + x^3 + x^2 + x$ is reduced to 1 under the modulo of polynomial $x^8 + x^4 + x^3 + x + 1$. In this case, {53} and {CA} are called *multiplicative inverses* of one another in the AES system because $\{53\} \boxdot \{CA\} = 1$ under the modulo of polynomial $x^8 + x^4 + x^3 + x + 1$ (or number {11B}).

AES operates on a 44 array of bytes, termed the *state*. For encryption, each round of the AES (except the last round) consists of the following four stages:

1. **AddRoundKey:** Each byte of the state is combined with the round key using bitwise XOR operations; each round key is derived from the cipher key by a key schedule. Thus, each subkey is the same size as the state.
2. **SubBytes:** A nonlinear substitution step in which each byte is replaced with another according to a lookup table. In this step, each byte is updated by an 8-bit S-box. Unlike the S-box used for the DES, the S-box used here is derived from the inverse function over GF(2^8). To avoid attacks based on simple algebraic properties, the S-box is actually constructed by combining the inverse function with an invertible affine transformation. The S-box used by the AES is chosen such that it always avoids any fixed points.
3. **ShiftRows:** A transposition step in which each row of the state is shifted cyclically a certain number of steps. It operates on the rows of the states by cyclically shifting the bytes in each row by a certain offset. For the AES, the ith row is shifted $i - 1$ to the left. After the ShiftRows operation, each column of the output state is composed of the bytes from each column of the input.
4. **MixColumns:** A mixing operation that operates on the columns of the state, combining the four bytes in each column, by using an invertible linear transformation.

Each column is treated as a polynomial over $GF(2^8)$ and is then multiplied a fixed polynomial $c(x) = 3x^3 + x^2 + x + 2$ with modulo $x^4 + 1$. Together with ShiftRows, MixColumns provides diffusion in the cipher.

The final round replaces the MixColumns stage with another instance of AddRoundKey.

19.2.2   Public-Key-Encryption Methods

Symmetric-key cryptosystems typically use the same key (or the decryption key can be efficiently computed from the encryption key and vice versa) for encryption and decryption. A significant disadvantage of symmetric ciphers is the key management necessary to use them securely. Any individual must keep n different keys if it wants to keep secure communications with n different parties. In other words, each distinct pair of communicating parties must share a different key to ensure a secure communication. The number of keys required thus increases as n^2, where n is the number of communicating members. For a large network in which n is sufficiently large, it quickly requires complex key-management schemes to keep them all straight and secret. If the communicating parties do not share a secret key between them for secure communication, they need to find a secret key k such that they can use for secure communication while k is not known by any other party. This presents a chicken-and-egg problem when no secure communication channel exists a priori. This is a considerable practical obstacle for cryptosystem users in the real world and is generally known as the key-agreement problem. This problem remained a challenge until a groundbreaking paper published by Whitfield Diffie and Martin Hellman in 1976. The original result is for two parties to construct a secret key without using any existing secure communication channel. This also triggers the invention of the public-key system. In a public-key system, either the decryption key or the encryption key is made public and the coupled other key remains secret. In most applications, the encryption key is often made public. A public-key system is so constructed that calculation of the private key is computationally infeasible from the public key, even though they are necessarily related. Instead, both keys are generated secretly, as an interrelated pair.

A public-key-encryption system (with the encryption method denoted as E and the decryption method denoted as D) typically works as follows: Any communication party A that uses a public-key system will first announce its public key, say K_A^U. It keeps secret the corresponding private key K_A^R. It must be the case that it is computationally infeasible for any party other than A to compute the private key K_A^R from K_A^U. When another party B wants to send a message M to A, it first encrypts the message to use the public key as $C = E(M, K_A^U)$. B then sends the ciphertext C to A. When A receives the ciphertext C, it decrypts the ciphertext using its own private key K_A^R as $M = D(C, K_A^R)$. Thus, a public-key system must ensure that

$$M = D(C, K_A^R) = D(E(M, K_A^U), K_A^R)$$

for every message M that is valid for the system.

Since the groundbreaking results[1] of Diffie and Hellman, a number of public-key cryptosystems have been proposed. Currently, examples of those most widely used are the RSA cryptosystem (developed in 1977 by Ron Rivest, Adi Shamir, and Len Adleman at MIT), the ElGamal cryptosystem (developed by Taher ElGamal in 1984), and the elliptic curve cryptosystem (suggested independently by Neal Koblitz and Victor S. Miller in 1985). The RSA system is still widely used in electronic commerce protocols and is believed to be secure, given sufficiently long keys and the use of up-to-date implementations. The ElGamal algorithm is used in the free GNU Privacy Guard software, recent versions of PGP, and other cryptosystems. The Digital Signature Algorithm (DSA) is a variant of the ElGamal signature scheme, which should not be confused with the ElGamal algorithm. Elliptic curve cryptography (ECC) is an approach to public-key cryptography based on the algebraic structure of elliptic curves over finite fields. The U.S. National Security Agency has endorsed ECC technology by including it in its Suite B set of recommended algorithms. Although the RSA patent has expired, there are patents in force covering some aspects of ECC.

RSA Cryptosystem

For using the RSA cryptosystem, a user, say Alice, has to generate a large number n, which is the product of two large prime numbers; i.e., $n = pq$. Typically, the prime numbers should be at least 500–1000 bits. Alice will also randomly select a positive integer e as its public key for encryptions. Certain relations between n and e should be enforced, which are discussed in detail later. Alice also selects another positive integer d as her private key for decryption, where d and e should satisfy some relation with regard to n. Alice will announce publicly n and e. Thus, typically (n, e) is called the public key. Alice has to make sure that d and p and q are kept private. Number d is called the private key.

When another user, say Bob, wants to send a message M (considered an integer less than n) to Alice, Bob encrypts the message as

$$c = m^e \mod n,$$

which can be quickly computed using the method of exponentiation by squaring. Bob then transmits c to Alice. When Alice received the ciphertext c from Bob, Alice will decrypt it to recover the message as follows:

$$m = c^d \mod n.$$

From this, we need $(m^e \mod n)^d \mod n = m$ for every possible integer m with $0 \leq m < n$. It is equivalent to require that $m^{ed} = m \mod n$ for $0 \leq m < n$. This will be satisfied (mostly from the Euler theorem) if $ed = 1 \mod \phi(n)$, where $\phi(n) = (p - 1)(q - 1)$ is called the Euler totient function. Consequently, Alice needs to select an integer e such

---

[1] In 1997, it finally became publicly known that an asymmetric-key cryptosystem was invented by James H. Ellis at GCHQ, a British intelligence organization in the early 1970s. Additionally, both the D-H protocol and RSA public-key cryptosystem were also previously developed, by Malcolm J. Williamson and Clifford Cocks, respectively.

that $\gcd(e, \phi(n)) = 1$. Then, Alice computes the private key d from the multiplicative inverse of e based on modulo $\phi(n)$ using the *extended Euclidean algorithm*. In other words, $d = e^{-1} \bmod \phi(n)$, which can be computed in polynomial time of $\log n$.

When used in practice, the RSA system must be combined with some form of a padding scheme so that no values of M result in insecure ciphertexts. The RSA system used without padding may suffer from a number of potential problems: suffering the dictionary attack because RSA encryption is a deterministic encryption without any padding scheme using random numbers; suffering an easy breaking when the encryption key is small and the message is a small value. Standards such as PKCSs (Public Key Cryptography Standards) have been carefully designed to securely pad messages prior to RSA encryption. Because these schemes pad the plaintext m with some number of additional bits, the size of the unpadded message M must be somewhat smaller. RSA padding schemes must be carefully designed so as to prevent sophisticated attacks that may be facilitated by a predictable message structure.

ElGamal Cryptosystem

ElGamal consists of three components: the key generator, the encryption algorithm, and the decryption algorithm. The ElGamal encryption cryptosystem is based on the assumption that the discrete logarithm is computationally infeasible to solve, although at the time of the writing, no publicly available method proves or disproves this assumption. The key generator works as follows: Alice first generates an efficient description of a cyclic group $\mathcal{G}$ of order q with a generator g. Notice that the straightforward application of the group $\mathcal{G}$ as $Z_p = \{1, 2, \ldots, p-1, p\}$ for some large prime number p will not result in a secure cryptosystem. Alice then randomly chooses a random integer x from $\{0, 1, \ldots, q-1\}$. In group $\mathcal{G}$, Alice computes

$$h = g^x$$

and publishes h with the description of $\mathcal{G}$, q, and g as her public key. Alice retains x as her secret key.

The encryption algorithm works as follows: Assume that Bob wants to send a message M to Alice. Bob first converts the message M into an element m of $\mathcal{G}$ and then chooses a random integer y from $\{0, 1, \ldots, q-1\}$. In group $\mathcal{G}$, Bob computes

$$c_1 = g^y, \text{ and } c_2 = mh^y.$$

Bob sends the ciphertext (c_1, c_2) to Alice.

When Alice receives the message, she will decrypt the ciphertext to recover the plaintext as follows. Alice computes, in group $\mathcal{G}$,

$$m = c_2(c_1^{-1})^x,$$

where c_1^{-1} denotes the multiplicative inverse of c_1 in group $\mathcal{G}$.

ElGamal is a simple example of a semantically secure asymmetric-key-encryption algorithm (under reasonable assumptions). It is probabilistic, meaning that a single plaintext can be encrypted to many possible ciphertexts, with the consequence that a general ElGamal encryption produces a ciphertext that is twice the size of that of the plaintext. The security of ElGamal rests in part on the difficulty of solving the discrete logarithm problem in $\mathcal{G}$. If the discrete logarithm in $\mathcal{G}$ can be solved, then the ElGamal cryptosystem will be broken, but solving the discrete logarithm is not a necessity. Breaking ElGamal actually relies on solving the decisional D-H assumption.

The two most popular types of groups used in ElGamal are subgroups of Z_p and groups defined over certain elliptic curves. Following is one popular way of choosing an appropriate subgroup of Z_p that is believed to be secure (although there is no proof at the time of the writing):

1. Choose a random large prime p such that $p - 1 = kq$ for some small integer k and large prime number q. This can be done by choosing q first and then test k starting from 2 and checking if $p = kq + 1$ is prime.
2. Choose a random element $g \in Z_p$ such that $g \neq 1$ and $g^q = 1 \bmod p$; i.e., such that g is of order q.
3. The group $\mathcal{G}$ is the subgroup of Z_p generated by g; i.e., the set of kth residues $\bmod p$. When encrypting a message M, the message must be properly encoded as an element m of $\mathcal{G}$, not just an arbitrary element in Z_p.

19.2.3   Digital-Signature Methods

Digital signature is often used to prove that a message is indeed from an entity. This is especially useful when the message itself need not be encrypted. To implement a digital signature based on a public-key infrastructure, the public key used in the signature scheme is tied to a user by a digital-identity certificate, which is issued by a certificate authority. A typical method of digital signature is to encrypt the hash $h(M)$ of the message M using the sender's private key in a public-key cryptosystem. In such a way, the message M cannot be altered in any way without changing the hash $h(m)$ to match, which, if quality hashing algorithms are properly used, will be quite difficult. When another entity receives the message and its coupled signature, it will decrypt the hash using the sender's public key and check the result against a newly generated hash of the alleged plaintext. The recipient can confirm (with high confidence) that the encryption was done with the sender's private key (and so, presumably, by the user who should have been the only person able to use that key) and that the message has not been altered since it was signed. Notice that here, the recipient can never be guaranteed that the message has not been altered or the message is indeed from the claimed sender.

In addition to using the public-key-encryption/decryption cryptosystem for the purpose of digital-signature (notably the RSA scheme), a number of specially tailored digital-signature schemes have been designed (e.g., the DSA and the ElGamal signature algorithm). When the encryption/decryption algorithm is used for digital signature, it is often required to use separate key pairs for signing and encryption for a number of practical reasons, especially because the digital signature often binds the signer of the documents to the terms therein in many countries. Typically, digital-signature schemes include three algorithms: a key-generation algorithm, a signing algorithm, and a verification algorithm. For more details of these algorithms, refer to books about cryptography.

19.2.4   Key-Agreement and Key-Distribution Protocols

Public-key cryptosystems often enjoy a number of advantages over the symmetric-key cryptosystems. However, public-key-encryption cryptosystems often are much slower than the symmetric-key cryptosystems. Thus, in practice, symmetric-key systems are used for encryption. Then, a challenging question is that of how to let two communicating parties find a key for secure communication when a secure communication channel is not there. This is generally called a key-agreement and key-distribution problem. The D-H key-agreement protocol was the first to address this challenge. D-H key exchange is a cryptographic protocol that allows two parties that have no prior knowledge of each other to *jointly* establish a shared secret key over an *insecure* communications channel. This key can then be used to encrypt subsequent communications by using a symmetric-key cipher. The D-H key-exchange protocol can be implemented in any group. The simplest and original implementation of the protocol uses the multiplicative group of integers modulo p, where p is a large prime number and g is primitive root mod p. Here, an integer g is called the primitive root of mod p if for any $1 \leq i \leq p - 1$, there is an integer $j > 0$ such that $g^j = i \bmod p$. Number g is also sometimes called the generator. Recall that modulo (or mod) means that the integers between 0 and $p - 1$ are used with normal addition, subtraction, multiplication, and exponentiation; after each operation, the result is only the remainder after dividing by p.

The original D-H protocol works as follows: Assume that Alice and Bob agree on a large prime number p and its primitive root g. The numbers p and g need not be kept secret. Alice will randomly select a random positive integer a that is at most $p - 1$ and then compute $S_a = g^a \bmod p$. Alice will keep a secret and send S_a to Bob. Similarly, Bob will randomly select a random positive integer b that is at most $p - 1$ and then compute $S_b = g^b \bmod p$. Bob will keep b secret and send S_b to Alice. Alice then computes the value $(S_b)^a \bmod p$ and uses this as the key. Similarly, Bob then computes the value $(S_a)^b \bmod p$ and uses this as the key. It is easy to show that $(S_a)^b \bmod p = (S_b)^a \bmod p$; i.e., Alice and Bob will have the same key.

It is easy to show that if Alice and Bob did get S_b and S_a correctly, breaking the D-H key-agreement protocol is equivalent to a third party finding $g^{a \cdot b} \bmod p$ given $g^a \bmod p$ and $g^b \bmod p$ for randomly selected a and b. The latter problem is often called the *D-H problem*, which is currently considered to be difficult, although no proof or disproof is reported at the time of writing. An efficient algorithm to solve the discrete logarithm problem would make it easy to compute a or b and solve the D-H problem. The *order* of the group G in which the D-H protocol is used should be prime or have a large prime factor to prevent use of the Pohlig–Hellman algorithm to obtain a or b. The D-H protocol also suffers from the man-in-the-middle attack, in which the attacker intercepts all communications between Alice and Bob and pretends to be Bob when communicating with Alice and pretends to be Alice when communicating with Bob.

The man-in-the-middle attack will not solve the D-H problem, but it will produce a key with Alice and another key with Bob while both Alice and Bob are not aware of this. In other words, the man-in-the-middle establishes two distinct D-H keys, one with Alice and the other with Bob and can then later attack the communication between Alice and Bob by decrypting and re-encrypting messages passed between them. A number of protocols have been proposed to eliminate the man-in-the-middle attack by using some public-key infrastructure, which generally requires the digital signature of g^a and g^b or some variation of it.

Besides the key-agreement protocols, another different approach has also been used in practice to let communicating parties find a common secret key. This approach generally requires that the system will first distribute some information to each communicating party for the purpose of key generation before the secure communication starts. When two communicating parties, say Alice and Bob, want to construct a common key, each of them will perform some operations on its predistributed information and communicate the result with the other. This is generally called key predistribution. The predistributed information could be simply some set of keys or some other information that can be used to generate keys.

An elegant scheme for key predistribution is called the *Blom key-predistribution scheme*. Suppose that there are n users in the network who want to communicate with each other pairwisely and securely. For convenience, we always assume that the keys are chosen from Z_p, where $p > n$ is a large prime number. Let k be an integer with $1 \leq k \leq n - 2$. The value of k is the largest size coalition against which the scheme will remain secure. In the Blom key-predistribution scheme, a trust authority (TA) will transmit $k + 1$ elements from Z_p to *each* user using a secure communication channel. Any pair of users A and B should be able to compute a key $K_{A,B} = K_{B,A}$. The security condition is that *any* set of k users disjoint from A and B will not be able to determine *any* information about the key $K_{A,B}$. The Blom scheme works as follows: First, the TA will randomly select a polynomial $f(x, y)$ with the following form:

$$f(x, y) = \sum_{i=0}^{k} \sum_{j=0}^{k} a_{i,j} x^i y^j \quad \mod p,$$

where $a_{i,j} \in Z_p$ for $0 \leq i, j \leq k$ and $a_{i,j} = a_{j,i}$ for all possible i and j. Assume that every user A in the network is assigned a unique ID a in Z_p. Recall that the total number of users n is at most p. For each user A, the TA computes the polynomial

$$g_a(x) = f(x, a) \quad \mod p$$

and then sends the polynomial $g_a(x)$ to user A over a secure communication channel. When two users A and B want to communicate securely using a symmetric-key encryption, they can compute the common key as

$$\text{key produced by } A : K_{A,B} = g_a(b),$$
$$\text{key produced by } B : K_{B,A} = g_b(a).$$

Notice that $g_a(b) = f(b, a)$ and $g_b(a) = f(a, b)$. Clearly, $g_a(b) = g_b(a)$ because $f(b, a) = f(a, b)$. Notice that A and B do not need to communicate about the polynomial $g_a(x)$ and $g_b(x)$. They need to know only the ID of the other.

Another often used approach to generate keys is as follows: The TA will first randomly select a pool $\mathcal{K}$ of possible keys. For each user i, the TA will randomly select a subset $K_i \subset \mathcal{K}$ and send K_i to user i over a secure communication channel. When users i and j want to have a secure communication, they use a challenge-and-response approach to find if they share a common key k; i.e., $k \in K_i$ and $k \in K_j$. Notice that when the size of K_i is larger than the half of the size of $\mathcal{K}$, then it is guaranteed that any pair of users i and j will share at least one common key. Otherwise, it is possible that they cannot find a common key. Most of the studies in the literature are to select the size of K_i such that a user i can have secure communication with enough number of users with high probability. We have more discussion on this approach in next section. An extension of the preceding approach is the q-composite method: Two nodes will find a secret key for communication only if they share q common keys assigned to them by the TA. The final secret key is computed based on some function of those q (or more) shared common keys.

19.3   Key-Predistribution Protocols

In this section, we study some key-predistribution protocols proposed in the literature, especially for WSNs. For WSNs, because of their limitations on the computational power and their storage, it is often impossible to use any asymmetric-key infrastructure. Thus, key agreement and key predistribution are more important for WSNs than for traditional networks. In the literature, most of the study of key-predistribution protocols is based on random-graph theory. Although random-graph theory captures the randomness of wireless communications, it cannot capture the fact that the wireless communication range is limited: Two wireless nodes can communicate with each other only if they are within a certain distance. Thus, we mainly concentrate on the random-geometry-graph model. The random-geometry graph is more suitable for the sensor network because it treats the connectivity of each link by a probability and it also captures the fact that every node has a fixed geometry location. We show in this section that the q-composite method will not provide us with any extra benefit in improving the security: We show that the best choice for q is one. We also study theoretically what the best setting of parameters is to achieve the highest security level for key-predistribution protocols. We not only prove the best parameters in key-predistributed management theoretically but also show that it is corrected by our simulation results.

Key predistribution for WSNs has drawn considerable research attention recently. Obviously, a number of naive methods could be used for the key-distribution problem.

The simplest one may be the master key approach, in which every sensor node uses the same key in the whole network. This approach clearly is the most memory-efficient but has a low security because compromising one node will compromise the whole network. Notice that the number of sensor nodes is usually large in a WSN. Thus, the probability of one sensor node being captured is not negligible for a massive WSN. We can improve the security by adding tamper-resistant hardware; unfortunately, the sensor node with tamper-resistant hardware is typically very expensive. Thus, the overall cost of using tamper-resistant hardware for all sensor nodes will be huge.

The other approach is to use a pairwise key. Each node stores $n - 1$ keys in the sensor network of n nodes and uses a different key to communicate with different nodes. This approach achieves the highest security because compromising one link will not affect any other links. However, because n is typically large, each sensor node in the network should have large memory to store $n - 1$ keys. This is impractical because a sensor node has a very limited memory. Furthermore, it will be difficult to add a new node to the network after the key deployment because we need to add a new key at every existing sensor node for the secure communication to this new node.

Eschenauer and Gligor (2002) first provided the idea of probabilistic key sharing. There are three steps in their method. The key-predistribution phase is performed off-line. In their scheme, there is a pool of keys that can be chosen by sensor nodes. Each sensor node randomly selects k keys from this key pool, where k is determined by the memory size of the node. The scheme also loads the chosen keys and the node identifier onto a trusted control node. In the shared-key discovery phase, every node discovers the neighbor nodes that share at least one common key with it in the communication range by broadcasting key identifiers. The neighboring nodes sharing a common key can also be found via a challenge-and-response approach: The initiating node s will send some message M using a key e to encrypt it; it then sends the encrypted message to all its neighboring nodes; a neighboring node t can recover the message only if it also has the key e for decryption; if so, node t can respond to the challenge from s by sending back the response message to s using e as the encryption key. The neighboring nodes that share no common key but are within the communication range will be assigned a path key in the path-key-establishment phase. By modeling the underlying physical network as a *random graph*, they studied the probability that the network using secure links is connected, the average hop length between two nodes using secure links, and the effect of compromised nodes. This scheme trades off between security and memory. Based on this scheme, a number of schemes were proposed to improve the key-distribution method.

Du *et al.* (2004) provided a key-distribution method to decrease the useless keys while keeping the same connectivity by using deployment knowledge. They assume that nodes are partitioned into groups. With Gaussian distribution, the nodes within a group and between two nearby groups are more likely to become neighbors. In contrast, the nodes between two groups far away from each other are less likely to become neighbors. So, those nodes that are more likely to be neighbors should have a high probability of sharing a common key and vice visa. In this scheme, the main key pool with size $|S|$ is divided into tn key pools with size $|S_c|$. The horizontal and vertical neighbor key pool shares $a|S_c|$ keys and the diagonal neighbor key pool shares $b|S_c|$ keys.

Chan *et al.* (2003) proposed three schemes for key establishment: (1) In the q-composite key scheme, two nodes that share at least q common keys (with $q > 1$) will establish the security channel between them. This scheme is highly tolerant of attack within a threshold. (2) The multipath-reinforcement scheme uses several different disjoint paths to send key sharing to the target node so as to construct the key. This scheme can strengthen the security of a common key (security channel) but it is difficult to gather these paths. (3) In the random-pairwise key scheme, each node identity is matched up with m other randomly selected distinct node IDs and a pairwise key is generated for each pair of nodes. The key is stored in both nodes' key rings, along with the ID of the other node that also knows the key. This scheme makes it possible for authentication and invocation.

There are several papers to improve the security level by changing keys in the key pool. Du *et al.* (2005) use matrix sharing as the key in the key pool, whereas Liu *et al.* (2005) use polynomial sharing as the key in the key pool. There are also some papers on improving the security level by key-construction strategies (e.g., Pietro *et al.*, 2003, 2004; Zhu *et al.*, 2003). Y. Wu and Li (2006) studied the random key distribution for WSNs using a random-geometry-graph approach. Their key-management scheme is based on a q-composite and the bivariate polynomial methods. Notice that the polynomial share method is highly tolerant of node capture because compromising a polynomial should require compromising at least $t + 1$ nodes (polynomial shares), where t is the degree of the polynomial, whereas the q-composite method also aims to improve the tolerance of node capture. If we combine them, can we get a better result? In this part, they described a random, key-predistribution scheme by combining both. First, the setup server generates a set of bivariate polynomials. Each polynomial will be assigned an identity. For each node u, the server will randomly assign it some polynomial shares by using some scheme. Second, two nodes u and v will check if they have the same polynomial share. If they have at least q same polynomial shares, they will construct a security link. Otherwise, the node u will try to find a path to connect with node v using a secure path when they share less than q common polynomial shares. The path is composed of links xy where nodes x and y share at least q common polynomial shares (i.e., they can build a direct secure communication). They studied in detail the probability that one key space was broken and the probability of the network being connected after one key compromising. They also study how to choose the appropriate key size, appropriate q, and so on, to maximize the security while ensuring a connected network with high probability.

19.4   Secure Routing Protocols

19.4.1   Secure Routing for Sensor Networks

A number of routing protocols have been proposed for WSNs, but few have been designed with security as a goal. It is inherently challenging to design secure routing protocols for sensor networks because sensor nodes have slow processors, limited energy supply, and very little memory and storage. In addition, most sensor networks will most likely be deployed in open, physically insecure, or even hostile environments where node compromise is a nonnegligible possibility. Although the defender has the extreme liabilities of insecure wireless communication, limited node capacities, and possible insider threats, the adversaries can likely use powerful devices with long-range communication to attack the network. Thus, the design of secure sensor-network protocols is an extremely nontrivial task. Notice that power consumption for a sensor node is dominated by communication costs. The power consumed by sending a single bit over the radio is equivalent to executing about 800 instructions. Thus, one possible attack to sensor networks is to let sensors run out of power because of a heavy communication load. Karlof and Wagner (2003) proposed threat models and security goals for secure routing in WSNs. They also introduced two new classes of previously undocumented attacks against sensor networks: sinkhole attacks and HELLO floods. They also studied the relevance of the wormhole attacks and Sybil attacks (previously studied under ad hoc networks) to the sensor networks.

To achieve security in ad hoc networks, a number of protocols are based on public-key encryptions (e.g., Hubaux *et al.*, 2001; Kong *et al.*, 2001). However, the public-key system is well beyond the capacities of current sensor nodes. It typically requires several minutes to generate keys with a Palm Pilot, and even a RSA signature takes tens of seconds. Several secure routing protocols based on symmetric-key encryption have been proposed for WSNs (e.g., Hu *et al.*, 2001, 2002; Basagni *et al.*, 2001). Perrig *et al.* (2001) present two security protocols optimized for use in sensor networks: SNEP and μTESLA. The SNEP protocol provides confidentiality, authentication, and freshness between nodes and the sink. The μTESLA protocol provides an authenticated broadcast. Both are useful building blocks for securing routing protocols in sensor networks.

As discussed in Karlof and Wagner (2003), most network-layer attacks against sensor networks fall into one of the following categories:

1. Spoofed, altered, or replayed routing information
2. Selective forwarding
3. Sinkhole attacks
4. Sybil attacks
5. Wormholes
6. HELLO flood attacks
7. Acknowledgment spoofing

By spoofing, altering, or replaying routing information, adversaries may be able to create routing loops, attract or repel network traffic, extend or shorten source routes, generate false error messages, partition the network, increase end-to-end latency, and so on. These kinds of attacks clearly are not unique to WSNs.

For the majority of routing protocols designed in the literature for WSNs, an assumption is that all intermediate nodes will faithfully forward the packets as required by the routing protocol. In a selective forwarding attack, attacking nodes may only forward some of the packets and simply drop other packets.

In a sinkhole attack, the adversary's goal is to lure nearly all the traffic from a particular area through a compromised node, thus creating a metaphorical sinkhole with the adversary at the center. Sinkhole attacks can enable many other attacks by attracting enough packets. Sinkhole attracts need to utilize the path-selection rule of the routing algorithm: It will make a compromised node more likely to be selected by the routing protocol by its nearby nodes. It can do so by declaring some faked node information or spoofing or replaying an advertisement of having a routing path with an extremely high quality (e.g., a larger residue power, less total path cost, and so on).

In a Sybil attack, a single node will present multiple identities to other nodes in the network. This attack is especially a threat to fault-tolerant schemes that need to use multiple entities. This attack also poses a threat to geographic routing protocols when a single node will be in "more than one place" at a given time instant.

In a wormhole attack, an adversary (or a group of adversaries) will tunnel messages received in one part of the network and replay them in a different part. For example, an adversary A that is close to the base station may completely disrupt the routing by creating a wormhole, as follows: Another adversary B convinces nodes that are actually multiple hops from the base station that they are "very close" to the base station. As a consequence, these neighboring nodes may choose to forward the packets to adversary B. Thus, A and B will attract more traffic that can be used for further attacks.

Many routing protocols require nodes to broadcast HELLO packets to announce themselves to their neighbors, and a node receiving such a packet often assumes that the node sending the HELLO message is within its (normal) radio range. This is typically correct when both nodes have the same or similar transmission power. However, an attacker can utilize this to pose an attack. For example, the attacker can broadcast routing or other information with large enough transmission power and thus convince every node in the network that the adversary was its neighbor. Notice that in practice, this attacker is not within the transmission range of a node. When the attack is carefully articulated, the nodes in the network could be left in a state of confusion.

A number of sensor network routing algorithms rely on implicit or explicit link-layer acknowledgments. A wireless channel is the inherent broadcast medium. Thus, an adversary can spoof link-layer acknowledgments using packets overheard, which are addressed to neighboring nodes. By spoofing the acknowledgment packets, the attacker could convince the sender that a weak link is strong or that a dead or disabled node is alive.

19.4.2   Stochastic Routing Protocols

There are a number of algorithms (e.g., Bohacek *et al.*, 2002a, 2002b; Kodialam and Lakshman, 2003; Lee *et al.*, 2005; Stone, 2000) proposed in the literature to achieve various security methods in wireless networks. Among them, some aim to prevent or reduce the effect of the malicious attack. The stochastic routing is provided for this purpose. In traditional shortest-path routing, the shortest path from the source node to the target node is always chosen in high probability to route the packet. Thus, it is easy to predict the routing path if the routing protocol and the link or node cost is known to all, which results in the network prone to be attacked. The stochastic routing is designed to reduce the probability of successful prediction. Suppose the links chosen for routing will not form a cycle. In stochastic routing, for each link e incident to a node u, there is some probability p_e for the link $e = (u, v)$ to be chosen for routing. Clearly, we need $\sum_{v \in V} p_{(u,v)} = 1$, $\forall u \in V$. Here, (u, v) denotes a link from node u to node v. In this way, the probability that the attacker correctly chooses the routing path will be greatly reduced. The challenge now is to choose the probability p_e appropriately so that we can guarantee certain security performance under *all* possible attacking strategies using given limited attacking resources. Recently, Bohacek *et al.* (2002a, 2002b) studied such problems for the traditional wired networks using a two-person zero-sum game model.

Wireless networks pose some additional challenges and also additional opportunities for designing a saddle-routing policy. The challenge comes from the fact that wireless interference often makes an optimal routing problem NP-hard while the counterpart problem in the wired networks is polynomial-time solvable. A typical example of such problems is to find the largest throughput using a multipath routing between a pair of nodes; see Alicherry *et al.* (2005) and W. Wang *et al.* (2006b) for details. Y. Wu *et al.* (2007) consider a multihop, multichannel wireless networks and assume that the routing policy maker can jointly optimize the multipath routing and the link and channel scheduling. They assume that each node has only one radio because the majority of wireless nodes have only one NIC. For link scheduling, they consider synchronized TDMA because this will achieve more throughput than the CSMA contention-based approach (Alicherry *et al.*, 2005; Kumar *et al.*, 2005; W. Wang *et al.*, 2006b). Notice that TDMA-based link scheduling has some implementation overhead and difficulties such as time synchronization among nodes. Y. Wu *et al.* (2007) adopt the TDMA link-channel scheduling to study what the *best* scheduling is that the system can achieve under the *worst* attacking scenario. To simplify their study, they further assume that the attacker can know the strategy used by the routing policy maker and vice versa. Assume that both the attacker and the policy maker can efficiently compute their own benefits, given the attacking strategies of attacker and the routing (and scheduling) strategies of the policy maker.

Generally, they assume that there are two players: the *routing policy maker* and the *attacker(s)*. The attacker tries to reduce the number of packets received by the target nodes by destroying packets, dropping packets, inserting garbage packets to the network, eavesdropping packets, and so on, whereas the routing policy maker tries to design a routing policy to reduce or to prevent such malicious attacks so that the effective network throughput is improved. Assume that there are two kinds of attacks in the network: the *non-packet-dropping attack* and the *packet-dropping attack*. In a non-packet-dropping attack, the attacker tries to eavesdrop on some nodes or some links, which results in the leakage of the packet information, or the attacker tries to modify some packets, which results in useless packets for target node. The routing policy is thus designed to reduce the number of the dirty packets produced by the attacker. In a packet-dropping attack, the attacker will insert some malicious packet, jam the network, or drop the packets in the network so as to minimize the healthy packets flowing into the target, whereas the routing policy maker will try to minimize the effect of the malicious attack in the network. That is, the routing policy maker wants to maximize the healthy packets flowing into the target node. Given an attacking strategy α (how much each link will be attacked) and a routing flow ℓ (the load assigned to each link) by the routing policy maker, the *effective throughput* $T(\alpha, \ell)$ is defined as the number of healthy packets received at the target node that are initiated by the source node.

Y. Wu *et al.* (2007) mathematically formulate the joint routing, link, and channel scheduling problem by the policy maker under the possible attacks. Assume that if an attacker attacks a link e with effort α_e, it will cost the attacker $\alpha_e C_e H_e$ unit amount of money, where C_e is the capacity of the link and H_e is the cost of attacking a unit amount of data on link e. Assume that the attacker has a bounded budget B. The routing policy maker has to decide the load ℓ_e for each link e on the network. Clearly, $\ell_e \leq C_e$. Another property required by a flow assignment is that the total flow should be schedulable: There is a TDMA link schedule $X_{e,t,f}$ (whether link e is active at time slot t using channel f) such that the achieved flow $\frac{\sum_f \sum_{1 \leq t \leq T} X_{e,t,f} \cdot C_{e,f}}{T}$ is at least ℓ_e. Here, $C_{e,f}$ is the link capacity for link e using channel f and T is the schedule period. Notice that the schedule should be interference-free and satisfy the NIC constraint on each node.

They show that unlike the wired network counterpart (Bohacek *et al.*, 2002a, 2002b), it is NP-hard for the policy maker to find an optimal routing strategy, given the attacking strategy. They then provide a relaxed LP formulation and provide a joint routing, link, and channel scheduling such that the achieved effective throughput is within a constant factor of the optimum, under the *worst* possible attacks by the attacker. Their saddle-routing policy will find a multiple-path routing and corresponding link scheduling such that the total effective throughput achieved under the non-packet-dropping attack or the packet-dropping attack is within a constant factor of the optimum when the policy maker has infinite computation power. They also study the stability of the found strategy pair (routing strategy ℓ, attacking strategy α). They show that given ℓ, the attacker cannot find a better attacking strategy other than α. When α is given, they also prove that the policy maker cannot find a routing strategy that can achieve a significantly larger effective throughput. They found that the solution is also stable with respect to the attacking budget: The routing strategy does not need to change if the budget B increases by only a small amount.

19.5   Further Reading

As discussed before, symmetric-key encryption is more efficient than asymmetric-key encryption. A number of key-predistribution protocols have been proposed in the literature to ensure that some pairs of wireless nodes can communicate with each other using symmetric-key encryption. Since Eschenauer and Gligor (2002) first provided probability-key management, many other key-predistributed managements based on it have been proposed. Du *et al.* (2005) designed a key-management method with additional deployment knowledge. Chan *et al.* (2003) proposed a q-composite method and random-pairwise key scheme. Liu *et al.* (2005) proposed a key-management method by using a key pool with polynomial shares. Pietro *et al.* (2003) constructed the key depending on a set of other nodes, in addition to the key between the two nodes, to improve the security level. They also proposed a method to design a key according to the security level. Source node a chooses a set C, which does not include a or b, and requests the node c in C to send some message based on the key between c and b. Node a XORs all the send-back messages and the direct-path key between a and b. Then, node a sends the result and the set C to b. Node b checks the result. If it is correct, a and b construct the channel with the cooperated key. Otherwise, node b drops the message from node a. Pietro *et al.* (2004) constructed the shared key not only by using the node index but also the key so that leaking the node index will not provide much useful information for the attacker. To achieve a different security level, Zhu *et al.* (2003) proposed a scheme named Leap, which supports four types of keys for each sensor node: (1) an individual key shared with the base station, (2) a pairwise key shared with another sensor node, (3) a cluster key shared with multiple neighboring nodes, and (4) a group key shared by all the nodes in the network.

Usually, the conventional routing protocols are based on the shortest path, such as OSPF (Moy, 1998) and RIP (Malkin, 1998). This makes the path predictable and results in the interception or eavesdropping attack. Multipath can ameliorate this matter while making packet reordering more complicated (Thaler and Hopps, 2000). Some techniques can solve it, such as sophisticated coding techniques (Byers *et al.*, 1999), standard prebuffering techniques (Loguinov and Radha, 2000), and so on. Lee and Misra (2005) proposed a distributed secure multipath solution so that the data are routed by multiple paths. There are also some techniques focusing on routing-level security to detect the DoS attack such as CenterTrack (Stone, 2000) and IP Traceback (Savage *et al.*, 2000). RON (Resilient Overlay Network) (Andersen *et al.*, 2001) is an architecture that improves a current network for allowing the network to recover from outages within several seconds. Kodialam and Lakshman (2003) detected network attack by sampling. The idea is that partial packets in the network are sampled and examined. The authors formulated it as a game-theory problem and resolved it by the duality of the linear program.

In the literature, several approaches have been studied on stochastic routing (e.g., Bohacek *et al.*, 2002a, 2002b; Hespanha and Bohacek, 2001; Reiter and Rubin, 1998). In Reiter and Rubin (1998), users send their request either to the server or to the other users rather than directly to the server. In Hespanha and Bohacek (2001), the authors first proposed SSR (security stochastic routing), which takes multiple paths with some probability instead of single-path routing. Then, Bohacek *et al.* (2002a, 2002b) extended their result in Hespanha and Bohacek (2001) by considering more general attacks. However, all these results are based on wired networks, which do not have the interference constraints when scheduling links for transmission. Notice that as observed in the literature, interference constraints often make many problems intractable, such as network throughput maximization and link scheduling.

19.6   Conclusion and Remarks

In this chapter, we briefly reviewed some fundamentals of cryptography, such as symmetric-key encryption, asymmetric-key encryption, digital signature, and key-agreement and key-predistribution protocols. We then discussed in detail how to implement an efficient key-predistribution protocol for WSNs such that a secure communication can be established from each sensor node to the sink node with high probability. We also reviewed some of the secure routing protocols that are specifically tailored for WSNs. The main challenges in designing security protocols for WSNs are the limited storage of each wireless sensor and the limited communication power by sensor nodes.

Problems

19.1 Prove that $a^{-1} \bmod n$ exists if and only if $gcd(a, n) = 1$.

19.2 Find the smallest positive integer x such that $x4321 = 1 \bmod 2007$.

19.3 Find the smallest integer $x > 0$ such that $13x = 4 \bmod 99$ and $7x = 5 \bmod 101$ are satisfied simultaneously.

19.4 Given an integer $n = p_1 p_2 p_3$, where p_1, p_2, and p_3 are prime numbers larger than 2, prove that there are *exactly* eight integers $x \in [1, n]$ satisfying

$$x^2 = 1 \quad \bmod n.$$

19.5 Let integer p be an odd prime number; p does not divide b. Then, prove the following statements:

(a) b is a quadratic residue of p if and only if $b^{\frac{p-1}{2}} = 1 \bmod p$.

(b) b is a quadratic nonresidue of p if and only if $b^{\frac{p-1}{2}} = -1 \bmod p$.

19.6 This question is about solving the quadratic congruence. Assume that $p > 2$ is a prime number and the positive integer $a = x^2 \bmod p$ for some unknown integer $x \in [1, p - 1]$. Prove the following statements:

(a) There are only two solutions (i.e., two integers in $[1, p - 1]$) for equation $a = x^2 \bmod p$.

(b) If $p = 3 \bmod 4$, then $x_1 = a^{\frac{p+1}{4}} \bmod p$, and $x_2 = p - x_1$ are the only two solutions.

(c) If $p = 5 \bmod 8$ and $a^{\frac{p-1}{4}} = 1 \bmod p$, then $x_1 = a^{\frac{p+3}{8}} \bmod p$, and $x_2 = p - x_1$ are the only two solutions.

(d) If $p = 5 \bmod 8$ and $a^{\frac{p-1}{4}} = -1 \bmod p$, then $x_1 = 2a(4a)^{\frac{p-5}{8}} \bmod p$, and $x_2 = p - x_1$ are the only two solutions.

19.7 An affine cipher encrypts a plaintext $x \in [0, 255]$ as $y = k_1 x + k_2 \bmod 256$. A key (k_1, k_2) with $0 \le k_1, k_2 \le 255$ is valid for an affine cipher if the function $k_1 x + k_2 \bmod 256$ is a one-to-one mapping. How many different valid keys exist for this affine cipher?

19.8   Write a code to compute $a^b \bmod n$, when given integer $a > 0$, $b > 0$, and $n > 0$. Here, the input integers could be up to 1000 bits, so your code should be able to take care of big integers. Use your code to find the last digit of the following number $a^b \bmod n$, when $a = 2^{123} - 1$, $b = 2^{999} - 1$, and $n = 2^{345} + 1$.

19.9   Suppose that for the RSA public key (N, e), an algorithm A exists for cryptanalyzing ciphertexts with probability p exists, with running time T seconds per ciphertext. In other words, for proportion p of the possible ciphertexts, the algorithm succeeds in obtaining the plaintexts, and for all cihpertexts (whether succeeding or failing), the algorithm halts within time T. Describe how you can boost your algorithm to succeed within probability $1 - \epsilon$ for arbitrarily small ϵ. What is the resulting running time?

19.10   For DES encryption, prove that if $Y = E_k(X)$, then $\overline{Y} = E_{\overline{k}}(\overline{X})$. Here, $E_k()$ denotes the encryption function of the DES method and $\overline{X}$ is the complement of X. For example, $\overline{1011101} = 0100010$.

19.11   It was known that 2DES does not provide a much stronger security than DES. Prove in detail why this statement is true (you have to analyze in detail a method attacking 2DES using time that is not much longer than the brute-force attack on DES).

19.12   Assume that Charlie knew that Alice and Bob communicate with Hill cipher $C = KP \bmod 26$, where K is a 3×3 matrix. Assume that Charlie intercepted three cipher texts $C_1 = (1, 2, 3)^T$, $C_2 = (4, 5, 6)^T$, and $C_3 = (7, 8, 2)^T$. Further, Charlie knew that the corresponding plaintexts for these ciphertexts are $P_1 = (3, 5, 7)^T$, $P_2 = (2, 4, 6)^T$, and $P_3 = (1, 8, 9)^T$. Help Charlie find the key K.

19.13   This question is about RSA again. Assume that Bob uses RSA and selects two "large" prime numbers $p = 101$ and $q = 73$. Assume also that Bob uses a public encryption key $e = 1001$. Alice sends Bob a message $M = 2003$. What will be the ciphertext received by Bob? Show the detailed procedure that Bob uses to decrypt the received ciphertext.

19.14   Suppose Bob uses the DSA signature scheme with prime numbers $q = 101$, $p = 7879$, and the primitive root $g = 170$. Bob selects his private key as $x = 75$ and publishes his public key as $y = g^x \bmod p = 4567$.

   (a)   Determine Bob's signature on a message with hashed value $h(M) = 5001$ when using a random number $k = 49$.

   (b)   Assume Alice wants to verify this signature; show the procedure Alice used to verify the signature.

19.15   Prove that having an algorithm to solve the D-H problem in polynomial time is equivalent to having an algorithm breaking the ElGamal Encryption cryptosystem. Notice that the two problems are defined as follows:

   • **Diffie-Hellman Problem:** We are given four positive integers p, α, β_1, and β_2, where p is a prime number, α is a primitive root mod p, $\beta_1 = \alpha^{x_1} \bmod p$, and

$\beta_2 = \alpha^{x_2} \bmod p$. Here, x_1 and x_2 are some integers we do not know. We want to find the value $\alpha^{x_1 x_2} \bmod p$ using the given four positive integers p, α, β_1, and β_2.

- ElGamal Encryption Cryptosystem: We are given five positive integers p, α, β, y_1, and y_2, where p is a prime number, α is a primitive root $\bmod p$, $\beta = \alpha^x \bmod p$, $y_1 = \alpha^k \bmod p$, and $y_2 = m\beta^k \bmod p$. Here, x, k, and m are some positive integers we do not know. We want to find the value m using the given five positive integers p, α, β, y_1, and y_2.

Bibliography

Abbreviations

AdHocNow	Conference on AD-HOC NetwOrks and Wireless, Springer, Berlin
CCCG	The Canadian Conference on Computational Geometry, Springer, Berlin
CCS	ACM Conference on Computer and Communication Security, ACM, New York
COCOON	International Computing and Combinatorics Conference, Springer, Berlin
DIALM	International Workshop on Discrete Algorithms and Methods for Mobile Computing and Communications, ACM, New York
DISC	International Symposium on Distributed Computing, Springer, Berlin
EC	ACM Conference on Electronic Commerce, ACM, New York
ECC	European Conference on Communications, Springer, Berlin
ECCC	Electronic Colloquium on Computational Complexity
ESA	European Symposium on Algorithms, Springer, Berlin
Euro-Par	(not short for anything)
FOCS	IEEE Annual Symposium on Foundations of Computer Science, IEEE, New York
FTDCS	The International Workshop on Future Trends of Distributed Computing Systems, Springer, Berlin
GHC	The Grace Hopper Celebration of Women in Computing
GLOBECOM	IEEE Global Communications Conference, IEEE, New York
HICSS	IEEE Hawaii International Conference on System Science, IEEE, New York
ICC	IEEE International Conference on Communications, IEEE, New York
ICCCN	International Conference on Computer Communications and Networks, IEEE, New York
ICCI	International Conference on Computing and Information, IEEE Computer Society Press, New York
ICNP	IEEE International Conference on Network Protocols, IEEE, New York
ICDCS	IEEE International Conference on Distributed Computing Systems, IEEE, New York
INFOCOM	The IEEE Conference on Computer Communications, IEEE, New York
INMIC	The International Multitopic Conference, Springer-Verlag, London
IPDPS	IEEE International Parallel and Distributed Processing Symposium, IEEE, New York
IPPS/SPDP	International Parallel Processing Symposium/Symposium on Parallel and Distributed Processing, IEEE, New York

IPSN	ACM/IEEE International Conference on Information Processing in Sensor Networks, ACM, New York
ISAAC	International Symposium on Algorithms and Computation, Springer-Verlag, London
ISCC	IEEE Symposium on Computers and Communications, IEEE, New York
I-SPAN	The International Symposium on Parallel Architectures, Algorithms, and Networks, Springer-Verlag, London
LATIN	Latin American Theoretical Informatics Symposium, Springer-Verlag, London
LCN	IEEE Conference on Local Computer Networks, IEEE, New York
LNCS	Lecture Notes on Computer Science
MASS	IEEE International Conference on Mobile Ad Hoc and Sensor Systems, IEEE, New York
MILCOM	Military Communications Conference, IEEE, New York
MMT	IEEE Multiaccess, Mobility, and Teletraffic for Wireless Communications Conference, IEEE, New York
MobiCom	ACM Annual International Conference on Mobile Computing and Networking, ACM, New York,
Mobihoc	The ACM International Symposium on Mobile Ad Hoc Networking and Computing, ACM, New York
MONET	Journal of ACM Mobile Networks and Applications
MSWiM	ACM International Symposium on Modeling, Analysis, and Simulation of Wireless and Mobile Systems, ACM, New York
PODC	ACM Symposium on Principles of Distributed Computing, ACM, New York
POMC	ACM Workshop on Principles of Mobile Computing, ACM, New York
SECON	Annual IEEE Communications Society Conference on Sensor, Mesh, and Ad Hoc Communications and Networks, IEEE, New York
SenSys	ACM Conference on Embedded Networked Sensor Systems, ACM, New York
SIGCOMM	ACM Annual Conference of Special Interest Group on Data Communication, ACM, New York
SIROCCO	Colloquia on Structural Information and Communication Complexity, Springer-Verlag, London
SoCG	ACM Symposium on Computational Geometry, ACM, New York
SODA	Annual ACM-SIAM Symposium on Discrete Algorithms, ACM, New York
SOSP	The ACM Symposium on Operating Systems Principles, ACM, New York
SPDP	IEEE Symposium on Parallel and Distributed Processing, IEEE, New York
SSGRR	International Conference on Advances in Infrastructure for Electronic Business, Science, and Education on the Internet, Scuola Superiore G. Reiss Romoli, Springer-Verlag, London
STACS	Symposium on Theoretical Aspects of Computer Science, Springer-Verlag, London
STOC	Annual ACM Symposium on Theory of Computing, ACM, New York
TCS	Theoretical Computer Science (Elsevier)
VTC	IEEE Vehicular Technology Conference, IEEE, New York
WADS	Workshop on Algorithms and Data Structures, Springer-Verlag, London
WCMC	Wireless Communications and Mobile Computing (Wiley)
WCNC	IEEE Wireless Communications and Networking Conference, IEEE, New York
WMAN	Workshop on Mobile Ad Hoc Networks, Springer-Verlag, London

WMCSA	IEEE Workshop on Mobile Computing Systems and Applications, IEEE Computer Society, New York
WOC	The IASTED International Conference on Wireless and Optical Communications Networks, IASTED/ACTA Press, Calgary
WOCN	The IEEE/IFIP International Conference on Wireless and Optical Communications Networks, IEEE, New York

Organizations

IEE	Institution of Electrical Engineers
IEEE	Institute of Electrical and Electronics Engineers
ACM	Association for Computing Machinery
SIAM	Society for Industrial and Applied Mathematics
IFIP	International Federation for Information Processing
IASTED	International Association of Science and Technology for Development
USENIX	Advanced Computing Systems Association

Adachi, F., Sawahashi, M., and Okawa, K. Tree-structured generation of orthogonal spreading codes with different lengths for the forward link of DS-CDMA mobile radio. *IEE Electron. Lett.* **33** (1997), 27–28.

Addlesee, M., Curwen, R., Hodges, S., Newman, J., Steggles, P., Ward, A., and Hopper, A. Implementing a sentient computing system. *Computer*, **34** (2001), 50–56.

Adler, M. and Rubenstein, D. Pricing multicasting in more practical network models. In *Proceedings of the Thirteenth Annual ACM-SIAM Symposium on Discrete Algorithms* (SODA), Philadelphia, pp. 981–990.

Adya, A., Bahl, P., Padhye, J., Wolman, A., and Zhou, L. A multi-radio unification protocol for IEEE 802.11 wireless networks. Tech. Rep., Microsoft Technical Report, MSR-TR-2003-41, June 2003.

Agrawal, D. P. and Zeng, Q. A. Introduction to Wireless and Mobile Systems, Thomson Learning College, 2002.

Ahn, G.-S., Miluzzo, E., Campbell, A. T., Hong, S. G., and Cuomo, F. Funneling-MAC: A localized, sink-oriented MAC for boosting fidelity in sensor networks. In *Proceedings of the First ACM Conference on Embedded Networked Sensor Systems (SenSys)*, ACM, New York (2006).

Akyildiz, I. F., Su, W., Sankarasubramaniam, Y., and Cayirci, E. Wireless sensor networks: A survey. *Comput. Networks* **38** (2002), 393–422.

Alasti, M. and Farvardin, N. D-PRMA: A dynamic packet reservation multiple access protocol for wireless communications. In *MSWiM '99: Proceedings of the 2nd ACM International Workshop on Modeling, Analysis and Simulation of Wireless and Mobile Systems*, ACM, New York (1999), pp. 41–49.

Alicherry, M., Bhatia, R., and Li, L. E. Joint channel assignment and routing for throughput optimization in multi-radio wireless mesh networks. In *ACM MobiCom '05: Proceedings of the 11th Annual International Conference on Mobile Computing and Networking*, ACM, New York (2005), pp. 58–72.

Alon, N., Moshkovitz, D., and Safra, M. Algorithmic construction of sets for k-restrictions. *ACM Trans. Algorithms* **2**, 2 (2006), 153–177.

Althaus, E., Călinescu, G., Mandoiu, I., Prasad, S., Tchervenski, N., and Zelikovsly, A. Power efficient range assignment in ad-hoc wireless networks. In *Proceedings of the IEEE Wireless Communications and Networking Conference (WCNC03)*, IEEE, New York (2003).

Althaus, E., Călinescu, G., Mandoiu, I. I., Prasad, S., Tchervenski, N., and Zelikovsky, A. Power efficient range assignment for symmetric connectivity in static ad hoc wireless networks. *Wireless Networking* **12** (2006), 287–299.

Alzoubi, K. M. *Virtual Backbone in Wireless Ad Hoc Networks*. Ph.D. dissertation, Illinois Institute of Technology, 2002.

Alzoubi, K., Li, X.-Y., Wang, Y., Wan, P.-J., and Frieder, O. Geometric spanners for wireless ad hoc networks. *IEEE Trans. Parallel Distribut. Process.* **14** (2003), 408–421. Short version in *IEEE ICDCS* 2002a.

Alzoubi, K. M., Wan, P.-J., and Frieder, O. Message-optimal connected dominating sets in mobile ad hoc networks. In *Proceedings of the 3rd ACM International Symposium on Mobile Ad Hoc Networking & Computing (MobiHoc)* (2002b), ACM Press, pp. 157–164.

Alzoubi, K. M., Wan, P.-J., and Frieder, O. New distributed algorithm for connected dominating set in wireless ad hoc networks. In *IEEE Hawaii International Conference on System Science (HICSS)*, IEEE, New York (2002c).

Ambühl, C. An optimal bound for the MST algorithm to compute energy-efficient broadcast trees in wireless networks, ICALP (2005), LNCS, vol. 3580, pp. 1139–1150.

Amico, M. D., Merani, M., and Maffioli, F. Efficient algorithms for assignment of OVSF codes in wideband CDMA. In *Proceedings of the IEEE International Conference on Communications (ICC)*, IEEE, New York (2002).

Amis, A. D. and Prakash, R. Load-balancing clusters in wireless ad hoc networks. In *Proceedings of the 3rd IEEE Symposium on Application-Specific Systems and Software Engineering Technology*, IEEE, New York (2000).

Amis, A. D., Prakash, R., Huynh, D., and Vuong, T. Max-min d-cluster formation in wireless ad hoc networks. In *Proceedings of the Nineteenth Annual Joint Conference of the IEEE Computer and Communications Societies (INFOCOM)*, IEEE, New York (2000), Vol. 1, pp. 32–41.

Amouris, K., Papavassiliou, S., and Li, M. A position-based multi-zone routing protocol for wide-area mobile ad-hoc networks. In *Proceedings of the 49th IEEE Vehicular Technology Conference (VTC)*, IEEE, New York (1999), pp. 1365–1369.

Anderegg, L. and Eidenbenz, S. Ad hoc VCG: A truthful and cost-efficient routing protocol for mobile ad hoc networks with selfish agents. In *Proceedings of the 9th Annual International Conference on Mobile Computing and Networking (MobiCom)*, ACM, New York (2003), pp. 245–259.

Andersen, D. G. and M. F., Kaashoek, H. B., and Morris, R. Resilient overlay networks. In *Proceedings of the 18th ACM Symposium on Operating Systems Principles (SOSP)*, ACM, New York (2001).

Archer, A., Papadimitriou, C., Talwar, K., and Tardos, E. An approximate truthful mechanism for combinatorial auctions with single parameter agents. In *Proceedings of the 14th Annual ACM-SIAM Symposium on Discrete Algorithms (SODA)*. SIAM, Philadelphia (2003), pp. 205–214.

Archer, A. and Tardos, E. Truthful mechanisms for one-parameter agents. In *Proceedings of the 42nd IEEE Symposium on Foundations of Computer Science (EOCS)*, IEEE Computer Society, Washington, DC (2001), pp. 482–491.

Arikan, E. Some complexity results about packet radio networks. *IEEE Trans. Inf. Theory* **30** (1984), 190–198.

Arya, S., Das, G., Mount, D., Salowe, J., and Smid, M. Euclidean spanners: Short, thin, and lanky. In *Proceedings of the 27th ACM Symposium on Theory of Computing (STOC)*, ACM, New York (1995), pp. 489–498.

Arya, S. and Smid, M. Efficient construction of a bounded degree spanner with low weight. In *Proceedings of the 2nd Annual European Sympos. Algorithms (ESA), Volume 855 of Lecture Notes in Computer Science* (1994), pp. 48–59.

Aspnes, J., Eren, T., Goldenberg, D. K., Morse, A. S., Whiteley, W., Yang, Y. R., Anderson, B. D. O., and Belhumeur, P. N. A theory of network localization. *IEEE Trans. Mobile Comput.* **5** (2006).

Aspnes, J., Goldenberg, D., and Yang, Y. R. On the computational complexity of sensor network localization. In *Proceedings of First International Workshop on Algorithmic Aspects of Wireless Sensor Networks*, Springer-Verlag (2004). LNCS 3121.

Atiya, S. and Hager, G. Real-time vision-based robot localization. *IEEE Trans. Rob. Autom.* **9** (1993), 785–800.

Bahl, V., Adya, A., Padhye, J., and Wolman, A. Reconsidering the wireless LAN platform with multiple radios. In *Workshop on Future Directions in Network Architecture*, ACM, New York (2003).

Bahl, P., Balachandran, A., and Padmanabhan, V. Enhancements to the radar user location and tracking system. Tech. Rep., Microsoft Research, 2000. citeseer.ist.psu.edu/bahl00enhancements.html

Bahl, V., Chandra, R., and Dunagan, J. SSCH: Slotted seeded channel hopping for capacity improvement in IEEE 802.11 ad-hoc wireless networks. In *MobiCom*, ACM, New York (2004), pp. 216–230.

Bahl, P. and Padmanabhan, V. N. RADAR: An in-building RF-based user location and tracking system. In *INFOCOM* (2000), pp. 775–784.

Bahramgiri, M., Hajiaghayi, M. T., and Mirrokni, V. S. Fault-tolerant and 3-dimensional distributed topology control algorithms in wireless multi-hop networks. In *Proceedings of the 11th Annual IEEE International Conference on Computer Communications and Networks (ICCCN)* (2002), pp. 392–397.

Baker, D. J. and Ephremides, A. The architectural organization of a mobile radio network via a distributed algorithm. *IEEE Trans. Commun.* **29** (1981), 1694–1701.

Baker, D., Ephremides, A., and Flynn, J. A. The design and simulation of a mobile radio network with distributed control. *IEEE's J. Sel. Areas Commun.* **2**, 1 (1984), 226–237.

Ballardie, T., Francis, P., and Crowcroft, J. Core-based trees (CBT). In *SIGCOMM '93: Conference Proceedings on Communications Architectures, Protocols and Applications* (1993), ACM, New York, pp. 85–95.

Bambos, N. and Kandukuri, S. Power-controlled multiple access (PCMA) in wireless communication networks. In *INFOCOM* (2000).

Bao, L. and Garcia-Luna-Aceves, J. J. Channel access scheduling in ad hoc networks with unidirectional links. In *DIALM '01: Proceedings of the 5th International Workshop on Discrete Algorithms and Methods for Mobile Computing and Communications* (2001), ACM, New York, pp. 9–18.

Bao, L. and Garcia-Luna-Aceves, J. Transmission scheduling in ad hoc networks with directional antennas. In *MobiCom '02* (2002), pp. 48–58.

Bao, L. and Garcia-Luna-Aceves, J. J. Topology management in ad hoc networks. In *MobiHoc* (2003), pp. 129–140.

Bar-Noy, A., Bellare, M., Halldórsson, M., Shachnai, H., and Tamir, T. On chromatic sums and distributed resource allocation. *Inf. Comput.* **140** (1998), 183–202.

Barriere, L., Fraigniaud, P., and Narayanan, L. Robust position-based routing in wireless ad hoc networks with unstable transmission ranges. In *Proceedings of the 5th International Workshop on Discrete Algorithms and Methods for Mobile Computing and Communications*, ACM, New York (2001).

Basagni, S. Distributed clustering for ad hoc networks. In *Proceedings of the IEEE International Symposium on Parallel Architectures, Algorithms, and Networks (I-SPAN)*, IEEE, New York (1999), pp. 310–315.

Basagni, S. Finding a maximal weighted independent set in wireless networks. *Telecommun. Syst. Special Issue on Mobile Computing and Wireless Networks*, ACM, New York **18** (2001), pp. 155–168.

Basagni, S., Chlamtac, I., and Farago, A. A generalized clustering algorithm for peer-to-peer networks. In *Workshop on Algorithmic Aspects of Communication*, Springer-Verlag, Berlin (1997).

Basagni, S., Chlamtac, I., Syrotiuk, V., and Woodward, B. A distance routing effect algorithm for mobility (dream). In *MobiCom '98*. ACM, New York, pp. 76–84.

Basagni, S., Herrin, K., Rosti, E., and Bruschi, D. Secure pebblenets. In *MobiHoc '01* (2001), ACM, New York, pp. 156–163.

Basagni, S., Mastrogiovanni, M., Panconesi, A., and Petrioli, C. Localized protocols for ad hoc clustering and backbone formation: A performance comparison. *IEEE Trans. Parallel Distribut. Syst., Special Issue on Localized Communication and Topology Protocols for Ad Hoc Networks* (S. Olariu, D. Simplot-Ryl, and I. Stojmenovic, editors) (2006), Vol. 17, No. 4, pp. 292–306.

Basagni, S., Mastrogiovanni, M., and Petrioli, C. A performance comparison of protocols for clustering and backbone formation in large-scale ad hoc networks. In *1st IEEE International Conference on Mobile Ad-Hoc and Sensor Systems (MASS)*, IEEE, New York (2004).

Basu, A., Gao, J., Mitchell, J. S. B., and Sabhnani, G. Distributed localization using noisy distance and angle information. In *MobiHoc '06* (2006), pp. 262–273.

Bates, T., Chandra, R., Katz, D., and Rekhter, Y. (2000). See draft at http://tools.ietf.org/id/draft-bates-bgp4-nlri-orig-verif-00.txt

Bates, T., Rekhter, Y., Chandra, R., and Katz, D. Multiprotocol Extensions for BGP-4 RFC 2858, June 2000. http//www.ietf.org/rfc/rfc2858.txt

Behzad, A. and Rubin, I. On the performance of graph-based scheduling algorithms for packet radio networks. In *Proceedings of IEEE Global Telecommunications Conference (GLOBECOM)*, IEEE, New York (2003), Vol. 6, pp. 3432–3436.

Berge, C. *Graphs and Hyper Graphs*. North-Holland, Amsterdam, 1973.

Berman, P., Furer, M., and Zelikovsky, A. Applications of matroid parity problem to approximating Steiner trees. Tech. Rep. 980021, Department of Computer Sciences, UCLA, 1998.

Bettstetter, C. Smooth is better than sharp: A random mobility model for simulation of wireless networks. In *MSWiM* (2001), ACM, New York, pp. 19–27.

Bettstetter, C. On the minimum node degree and connectivity of a wireless multihop network. In *MobiHoc '02* (2002).

Bettstetter, C. and Krausser, R. Scenario-based stability anlysis of the distributed mobility-adaptive clustering (DMAC) algorithm. In *MobiHoc '01* (2001), pp. 232–241.

Bettstetter, C., Resta, G., and Santi, P. The node distribution of the random waypoint mobility model for wireless ad hoc networks. *IEEE Trans. Mobile Comput.* **2** (2003), 257–269.

Bharghavan, V., Demers, A., Shenker, S., and Zhang, L. Macaw: A media access protocol for wireless LANs. In *Proceedings of the Conference on Communications Architectures, Protocols and Applications* (1994), (Sigcomm) ACM, New York, pp. 212–225.

Bianchi, G. Performance analysis of the IEEE 802.11 distributed coordination function. *IEEE J. Sel. Areas Commun.* **18** (2000), 535–547.

Biswas, P. and Ye, Y. Semidefinite programming for ad hoc wireless sensor network localization. In Feng Zhao and Leonidas Guibas, editors, *Proceedings of Third International Symposium on Information Processing in Sensor Networks, Berkeley, CA*, ACM, New York, 2004.

Biswas, S. and Morris, R. ExOR: Opportunistic multi-hop routing for wireless networks. In *SIGCOMM '05* (2005), pp. 133–144.

Blazevic, L., Buttyan, L., Capkun, S., Giordano, S., Hubaux, J. P., and Boudec, J. Y. L. Self-organization in mobile ad-hoc networks: The approach of terminodes. *IEEE Commun. Mag.* **39** (2001).

Blough, D., Leoncini, M., Resta, G., and Santi, P. On the symmetric range assignment problem in wireless ad hoc networks. In *Proceedings of the 2nd IFIP International Conference on Theoretical Computer Science (TCS)*, IFIP, Austria (2002).

Blough, D., Leoncini, M., Resta, G., and Santi, P. The k-neighbor protocol for symmetric topology control in ad hoc networks. In *MobiHoc* (2003).

Blough, D. and Santi, P. Investigating upper bounds on network lifetime extension for cell-based energy conservation techniques in stationary ad hoc networks. In *ACM MobiCom '02* (2002).

Bohacek, S., Hespanha, J. P., and Obraczka, K. Enhancing security via stochastic routing. In *ICCCN* (2002a).

Bohacek, S., Hespanha, J. P., and Obraczka, K. Saddle policies for secure routing in communication networks. In *Decision and Control, 2002, Proceedings of the 41st IEEE Conference*, IEEE, New York (2002b).

Bollobás, B. *Modern Graph Theory*, Springer, New York, 1998.

Bollobás, B. *Random Graphs*, Cambridge University Press, New York, 2001.

Bose, P., Brodnik, A., Carlsson, S., Demaine, E. D., Fleischer, R., Lopez-Ortiz, A., Morin, P., and Munro, J. I. Online routing in convex subdivisions. In *International Symposium on Algorithms and Computation* (2000), Springer-Verlag, London, pp. 47–59.

Bose, P., Devroye, L., Evans, W., and Kirkpatrick, D. On the spanning ratio of Gabriel graphs and beta-skeletons. In *Proceedings of the Latin American Theoretical Infocomatics (LATIN)*, Springer-Verlag, London (2002a).

Bose, P., Gudmundsson, J., and Morin, P. Ordered theta graphs. In *Proceedings of the Canadian Conference on Computational Geometry (CCCG)*, Springer-Verlag, London (2002b).

Bose, P., Gudmundsson, J., and Smid, M. Constructing plane spanners of bounded degree and low weight. In *Proceedings of the European Symposium on Algorithms (ESA)*, Springer-Verlag, London (2002c).

Bose, P. and Morin, P. Online routing in triangulations. In *Proc. of the 10th Annual International Symposium on Algorithms and Computation ISAAC*, Springer-Verlag, London (1999).

Bose, P. and Morin, P. Competitive online routing in geometric graphs. In *Proceedings of the VIII International Colloquium on Structural Information and Communication Complexity (SIROCCO 2001)*, Springer-Verlag, London (2001), pp. 35–44.

Bose, P., Morin, P., Stojmenovic, I., and Urrutia, J. Routing with guaranteed delivery in ad hoc wireless networks. *ACM/Kluwer Wireless Networks* **7** (2001).

Boudec, J.-Y. L. and Vojnovic, M. Perfect simulation and stationarity of a class of mobility models. In *INFOCOM* (2005).

Breu, H. and Kirkpatrick, D. G. Unit disk graph recognition is NP-hard. *Comput. Geometry. Theory and Applications* **9**, 1–2 (1998), 3–24.

Broch, J., Maltz, D., Johnson, D., Hu, Y., and Jetcheva, J. A performance comparison of multi-hop wireless ad hoc network routing protocols. In *MobiCom* (1998).

Bruck, J., Gao, J., and Jiang, A. Localization and routing in sensor networks by local angle information. In *MobiHoc* (2005), pp. 181–192.

Buettner, M., Yee, G., Anderson, E., and Han, R. X-MAC: A short preamble MAC protocol for duty-cycled wireless sensor networks. In *SenSys*, ACM, New York (2006).

Bulusu, N., Heidemann, J., and Estrin, D. GPS-less low-cost outdoor localization for very small devices. *IEEE Personal Communs. Mag.* **7** (2000), 28–34. http://www.isi.edu/johnh/PAPERS/Bulusu00a.html

Burkhart, M., von Rickenbach, P., Wattenhofer, R., and Zollinger, A. Does topology control reduce interference? In *MobiHoc '04* (2004), pp. 9–19.

Burns, J. A. formal model for message passing systems, 1980. Technical Report TR-91, Computer Science Department, Indiana University.

Buttyan, L. and Hubaux, J. P. Enforcing service availability in mobile ad-hoc WANS. In *MobiHoc*, ACM, New York (2000), pp. 87–96.

Buttyan, L. and Hubaux, J. Stimulating cooperation in self-organizing mobile ad hoc networks. *ACM/Kluwer Mobile Networks and Applications* **5** (2003).

Byers, J., Luby, M., and Mitzenmacher, M. Accessing multiple mirror sites in parallel: Using tornado codes to speed up downloads. In *INFOCOM*, IEEE, New York (1999).

Cagalj, M., Hubaux, J.-P., and Enz, C. Minimum-energy broadcast in all-wireless networks: Np-completeness and distribution issues. In *MobiCom*, ACM, New York (2002).

Călinescu, G. Computing 2-hop neighborhoods in ad hoc wireless networks. In *AdHoc-Now: International Conference on AD-HOC Networks and Wireless*, Springer-Verlag, London (2003).

Călinescu, G., Frieder, O., and Wan, P.-J. Power assignment in wireless ad hoc networks, Department of Computer Science, Illinois Institute of Technology (2004).

Călinescu, G., Kapoor, S., Olshevsky, A., and Zelikovsky, A. Network lifetime and power assignment in ad-hoc wireless networks. In *ESA 2003*, Springer-Verlag, London (2003).

Călinescu, G., Măndoiu, I., Wan, P.-J., and Zelikovsky, A. Selecting forwarding neighbors in wireless ad hoc networks. In *DIALM*, ACM, New York (2001).

Călinescu, G. and Wan, P.-J. Range assignment for high connectivity in wireless ad hoc networks. In *AdHoc-Now '03*, Springer-Verlag, London (2003a).

Călinescu, G. and Wan, P.-J. Range assignment for high connectivity in wireless ad hoc networks. *ACM MONET* (2001). See http://www.cs.iit.edu/~calinesc/adhoc10.ps. Preliminary version in *AdHoc-Now* (2003b).

Camp, T., Boleng, J., and Davies, V. A survey of mobility models for ad hoc network research. *Wireless Commun. Mobile Comput. (Special Issue on Mobile Ad Hoc Networking: Research, Trends and Applications)* **2** (2002), 483–502.

Cao, M., Ma, W., Zhang, Q., Wang, X., and Zhu, W. Modelling and performance analysis of the distributed scheduler in IEEE 802.16 (2005), 78–89.

Capkun, S., Hamdi, M., and Hubaux, J.-P. GPS-free positioning in mobile ad-hoc networks. In *HICSS*, IEEE, New York (2001).

Caragiannis, I., Flammini, M., and Moscardelli, L. An exponential improvement on the MST heuristic for minimum energy broadcasting in ad hoc wireless networks, ICALP (2007), LNCS, vol. 4596, pp. 447–458.

Caragiannis, P. K. I. and Kaklamanis, C. New results for energy-efficient broadcasting in wireless networks. In *ISAAC*, Springer-Verlag, London (2002).

Cartigny, J., Ingelrest, F., Simplot-Ryl, D., and Stojmenovic, I. Localized LMST and RNG based minimum energy broadcast protocols in ad hoc networks. *Ad Hoc Networks* **3**, 1 (2005), 1–16.

Cartigny, J. and Simplot, D. Border node retransmission based probabilistic broadcast protocols in ad hoc networks. *Telecommun. Sys.* **22** (2003), 189–204.

Cartigny, J., Simplot, D., and Stojmenovic, I. Localized energy-efficient broadcast for wireless networks with directional antennas. In *Proceedings of the IFIP Mediterranean Ad Hoc Networking Workshop (MED-HOC-NET 2002)*, IFIP, Springer Science-Business Media, Berlin (2002).

Cartigny, J., Simplot, D., and Stojmenovic, I. Localized minimum-energy broadcasting in ad-hoc networks. In *INFOCOM* (2003).

Chan, H., Perrig, A., and Dong, D. Random key-predistribution scheme for sensor networks. In *2003 IEEE Symposium on Security and Privacy (SP'03)*, IEEE, New York (2003).

Chan, T. Polynomial-time approximation schemes for packing and piercing fat objects. *J. Algorithms* **46** (2003).

Chandra, B., Das, G., Narasimhan, G., and Soares, J. New sparseness results on graph spanners. In *Proceedings of the 8th Annual ACM Symposium on Computational Geometry*, ACM, New York (1992), pp. 192–201.

Chandra, R., Bahl, P., and Bahl, P. Multinet: Connecting to multiple IEEE 802.11 networks using a single wireless card. In *INFOCOM*, IEEE, New York (2004).

Chang, J.-H. and Tassiulas, L. Maximum lifetime routing in wireless sensor networks. *IEEE/ACM Trans. Networking* **12** (2004).

Chatterjee, M., Das, S., and Turgut, D. WCA: A weighted clustering algorithm for mobile ad hoc networks. *Journal of Cluster Computing* **5**, 2 (2002), 193–204.

Chen, B., Jamieson, K., Balakrishnan, H., and Morris, R. Span: An energy-efficient coordination algorithm for topology maintenance in ad hoc wireless networks. In *Mobile Computing and Networking* (2001), pp. 85–96.

Chen, G., Nocetti, F., Gonzalez, J., and Stojmenovic, I. Connectivity-based k-hop clustering in wireless networks. In *HICSS '02* (2002), Vol. 7, pp. 2450–2459.

Chen, G. and Stojmenovic, I. Clustering and routing in wireless ad hoc networks. Tech. Rep. TR-99-05, SITE, University of Ottawa, 1999.

Chen, J. C., Yip, L., Elson, J., Wang, H., Maniezzo, D., Hudson, R. E., Yao, K., and Estrin, D. Coherent acoustic array processing and localization on wireless sensor networks. *Proc. IEEE* **91** (2003), 1154–1162.

Chen, L., Low, S., Chiang, M., and Doyle, J. Cross-layer congestion control, routing and scheduling design in ad hoc wireless network. In *INFOCOM*, IEEE, New York (2006).

Chen, L., Low, S., and Doyle, J. Joint congestion control and media access control design for wireless ad hoc networks. In *INFOCOM*, IEEE, New York (2005).

Cheng, S.-W., Jia, X., Hung, F., and Wang, Y. Energy-efficient broadcasting and multicasting in static wireless ad hoc networks. In *AAIM: Algorithm Aspects in Information Management*, Springer-Verlag, London (2005), pp. 16–25.

Chen, W.-I. and Huang, N.-F. The strongly connecting problem on multihop packet radio networks. *IEEE Trans. Commun.* **37**, 3 (1989), 293–295.

Cheng, X., Huang, X., Li, D., and Du, D.-Z. Polynomial-time approximation scheme for minimum connected dominating set in ad hoc wireless networks. *Willey Networks* **42** (2003), 202–208.

Cheriyan, J., Vempala, S., and Vetta, A. Approximation algorithms for minimum-cost *k*-vertex connected subgraphs. In *Proceedings of the 34th Annual ACM Symposium on Theory of Computing (STOC)* (2002), ACM, New York, pp. 306–312.

Chiang, C. Routing in clustered multihop, mobile wireless networks with fading channel. In *Proceedings of IEEE Singapore International Conference of Networks (SICON)*, IEEE, New York (1997), pp. 197–211.

Chlamtac, I. and Farago, A. Making transmission schedules immune to topology changes in multi-hop packet radio networks. *IEEE/ACM Trans. Networking* **2** (1994), 23–29.

Chlamtac, I. and Farago, A. A new approach to design and analysis of peer-to-peer mobile networks. *Wireless Networks* **5** (1999), 149–156.

Chlamtac, I. and Kutten, S. A spatial reuse TDMA/FDMA for mobile multihop radio networks. In *INFOCOM*, IEEE, New York (1985), pp. 389–394.

Chvátal, V. A greedy heuristic for the set-covering problem. *Math. Op. Res.* **4** (1979), 233–235.

Cidon, I. and Mokryn, O. Propagation and leader election in multihop broadcast environment. In *12th International Symposium on Distributed Computing (DISC98)*, Springer-Verlag, London (1998), pp. 104–119.

Clark, B., Colbourn, C., and Johnson, D. Unit disk graphs. *Discrete Math.* **86** (1990), 165–177.

Clarke, E. H. Multipart pricing of public goods. *Public Choice* **11**, 1 (1971), 17–33.

Clausen, T., Jacquet, P., Laouiti, A., Muhlethaler, P., Qayyum, A., and Viennot, L. Optimized link state routing protocol. In *IEEE International Multitopic Conference (INMIC)*, IEEE, New York (2001).

Clementi, A., Crescenzi, P., Penna, P., Rossi, G., and Vocca, P. On the complexity of computing minimum energy consumption broadcast subgraphs. In *18th Annual Symposium on Theoretical Aspects of Computer Science (STACS), LNCS 2010* (2001a), pp. 121–131.

Clementi, A. E. F., Ferreira, A., Penna, P., Perennes, S., and Silvestri, R. The minimum range assignment problem on linear radio networks. In *ESA* (2000b), pp. 143–154.

Clementi, A. E., Huiban, G., Penna, P., Rossi, G., and Verhoeven, Y. C. Some recent theoretical advances and open questions on energy consumption in ad-hoc wireless networks. In *3rd Workshop on Approximation and Randomization Algorithms in Communication Networks*, Carleton Scientific, Waterloo (2002).

Clementi, A., Penna, P., and Silvestri, R. The power range assignment problem in radio networks on the plane. In *XVII Symposium on Theoretical Aspects of Computer Science (STACS '00), LNCS 1770* (2000a), 651–660.

Clementi, A., Penna, P., and Silvestri, R. On the power-assignment problem in radio networks. *Electronic Colloquium on Computational Complexity* (2000b). Preliminary results in APPROX'99 and STACS'2000.

Clementi, A., Penna, P., and Silvestri, R. On the power-assignment problem in radio networks. *Electronic Colliquium on Computation Complexity (ECCC)*, 054 (2000c).

Cocchi, R., Shenker, S., Estrin, D., and Zhang, L. Pricing in computer networks: Motivation, formulation, and example. *IEEE/ACM Trans. Network* **1** (1993), 614–627.

Colvin, A. CSMA with collision avoidance. *Comput. Commun.* **6** (1983), 227–235.

Cormen, T. J., Leiserson, C. E., Rivest, R. L., and Stein, C. *Introduction to Algorithms*. MIT Press, McGraw-Hill (2001).

Couto, D. D., Aguayo, D., Bicket, J., and Morris, R. A high-throughput path metric for multi-hop wireless routing. In *MobiCom* (2003).

Czumaj, A. and Zhao, H. Fault-tolerant geometric spanners. In *Proceedings of the 19th Conference on Computational Geometry (SoCG)*, ACM, New York (2003), pp. 1–10.

Daemen, J., Borg, S., and Rijmen, V. The design of Rijndael: AES—the advanced encryption standard, 2002. ISBN 3-540-42580-2.

Dai, F. and Wu, J. An extended localized algorithm for connected dominating set formation in ad hoc wireless networks. *IEEE Trans. Parallel Distribut. Syst.* **15**, 10 (2004), 908–920.

Dana, P. H. Global positioning system overview. http://www.colorado.edu/geography/gcraft/notes/gps/gps.html (2000).

Das, A. K., Alazemi, H. M. K., Vijayakumar, R., and Naguib, A. Optimization models for fixed channel assignment in wireless mesh networks with multiple radios. In *Second Annual IEEE Communications Society Conference on Sensor and Ad Hoc Communications and Networks (SECON)* (2005).

Das, B. and Bharghavan, V. Routing in ad-hoc networks using minimum connected dominating sets. In *ICC'97* (1997a), Vol. 1, pp. 376–380.

Das, A. K., Marks, R. J., El-Sharkawi, M., Arabshahi, P., and Gray, A. Minimum power broadcast trees for wireless networks: An ant colony system approach. In *Proc. IEEE International Symposium on Circuits and Systems*, IEEE, New York (2002).

Das, A. K., Marks, R. J., El-Sharkawi, M., Arabshahi, P., and Gray, A. Minimum power broadcast trees for wireless networks: Integer programming formulations. In *INFOCOM* (2003).

Das, G., Narasimhan, G., and Salowe, J. A new way to weigh malnourished Euclidean graphs. In *ACM Symposium of Discrete Algorithms* (SODA) (1995), pp. 215–222.

Das, S., Perkins, C., and Royer, E. Performance comparison of two on-demand routing protocols for ad hoc networks. In *INFOCOM* (2000).

Das, B., Sivakumar, R., and Bharghavan, V. Routing in ad hoc networks using a spine. In *ICCCN97* (1997b).

Datta, S., Stojmenovic, I., and Wu, J. Internal node and shortcut based routing with guaranteed delivery in wireless networks. *Cluster Comput.* **5** (2002), 169–178.

Deb, S., Médard, M., and Choute, C. Algebraic gossip: A network coding approach to optimal multiple rumor mongering. *IEEE/ ACM Trans. Network*, **14**, SI (2006), 2486–2507.

Deering, S. E. *Multicast Routing in a Datagram Internetwork*. Ph.D. dissertation, Stanford University, 1991.

Dette, H. and Henze, N. The limit distribution of the largest nearest-neighbor link in the unit d-cube. *J. Appl. Probab.* **26** (1989), 67–80.

Dette, H. and Henze, N. Some peculiar boundary phenomena for extremes of rth nearest-neighbor links. *Stat. Probab. Lett.* **10** (1990), 381–390.

Diffie, W. and Hellman, M. E. New directions in cryptography. *IEEE Transactions on Information Theory* **22** (1976), 644–654.

Dobkin, D., Friedman, S., and Supowit, K. Delaunay graphs are almost as good as complete graphs. *Discrete Computational Geometry*, **5**, 4 (1990), pp. 399–407.

Dousse, O., Thiran, P., and Hasler, M. Connectivity in ad hoc and hybrid networks, IEEE INFOCom (2002), 1078–1088.

Draves, R., Padhye, J., and Zill, B. Routing in multi-radio, multi-hop wireless mesh networks. In *MobiCom* (2004).

Du, W., Deng, J., Han, Y. D., Chen, S., and Varshney, P. K. A key-management scheme for wireless sensor networks using deployment knowledge. In *INFOCOM* (2004).

Du, W., Deng, J., Han, Y. S., Varshney, P. K., Katz, J., and Khalili, A. A pairwise key-predistribution scheme for wireless sensor networks. *ACM Trans. Inf. Syst. Security* **8** (2005), 228–258.

Ephremides, A. and Mowafi, O. A. Analysis of a hybrid access scheme for buffered users–probabilistic time division. *IEEE Trans. Software Eng.* **SE-8** (1982), 52–61.

Ephremedis, A. and Truong, T. Scheduling broadcasts in multihop radio networks. *IEEE Trans. Commun.* **38** (1990), 456–460.

Eren, T., Goldenberg, D., Whitley, W., Yang, Y., Morse, S., Anderson, B., and Belhumeur, P. Rigidity, computation, and randomization of network localization. In *INFOCOM* (2004).

Erlebach, T., Jansen, K., and Seidel, E. Polynomial-time approximation schemes for geometric graphs. In *SODA* (2001).

Eschenauer, L. and Gligor, V. A key-management scheme for distributed sensor networks. In *ACM Conference on Computer and Communications Security (CCS)*, ACM, New York (2002), pp. 41–47.

Even, S., Goldreich, O., Moran, S., and Tong, P. On the NP-completeness of certain network testing problems. *Networks* **14** (1984).

Faloutsos, M. and Molle, M. Creating optimal distributed algorithms for minimum spanning trees. Tech. Rep. CSRI-327, 1995.

Fantacci, R. and Nannicini, S. Multiple access protocol for integration of variable bit rate multimedia traffic in UMTS/IMT-2000 based on wideband CDMA. *IEEE J. Sel. Areas Commun.* **18** (2000), 1441–1454.

Feeney, L. and Nilson, M. Investigating the energy consumption of a wireless network interface in an ad hoc networking environment. In *INFOCOM* (2001), pp. 1548–1557.

Feige, U. A threshold of LNN for approximating set cover. *J. ACM* **45** (1998), 634–652.

Feige, U. and Kilian, J. Zero knowledge and the chromatic number. *J. Comput. Syst. Sci.* **57** (1998), 187–199.

Feigenbaum, J., Krishnamurthy, A., Sami, R., and Shenker, S. Approximation and collusion in multicast cost-sharing (abstract). In *ACM Conference*, On Electronic Commerce (EC) (2001a).

Feigenbaum, J., Krishnamurthy, A., Sami, R., and Shenker, S. Hardness results for multicast cost sharing. *Theor. Comput. Sci.* **304** (2003), 215–236.

Feigenbaum, J., Papadimitriou, C., Sami, R., and Shenker, S. A BGP-based mechanism for lowest-cost routing. In *Proceedings of the 2002 ACM Symposium on Principles of Distributed Computing* (PODC) (2002), pp. 173–182.

Feigenbaum, J., Papadimitriou, C. H., and Shenker, S. Sharing the cost of multicast transmissions. *J. Comput. Syst. Sci.* **63** (2001b), 21–41.

Flammini, M., Klasing, R., Navarra, A., and Perennes, S. Improved approximation results for the minimum energy broadcasting problem. In *ACM DIALM-POMC*: The Joint Workshop on Foundations of Mobile Computing, ACM, New York (2004).

Fotakis, D., Pantziou, G., Pentaris, G., and Spirakis, P. Frequency assignment in mobile and radio networks. In *Networks in Distributed Computing, DIMACS Series in Discrete Mathematics and Theoretical Computer Science*, American Mathematical Society (1999).

Francheschetti, M., Dousse, O., Tse, D., and Thiran, P. Closing the gap in the capacity of random wireless networks. *IEEE Transactions on Information Theory* **53**(3) (2006), 1009–1018.

Frey, H. and Stojmenovic, I. On delivery guarantees of face and combined greedy-face routing in ad hoc and sensor networks. In *MobiCom '06* (2006), pp. 390–401.

Fullmer, C. Collision-avoidance techniques for packet radio networks (1998). Ph.D. Dissertation, UC Santa Cruz.

Fullmer, C. L. and Garcia-Luna-Aceves, J. J. Floor acquisition multiple access (FAMA) for packet-radio networks. In *SigComm*, ACM, New York (1995), pp. 262–273.

Gabriel, K. and Sokal, R. A new statistical approach to geographic variation analysis. *Syst. Zool.* **18** (1969), 259–278.

Gafni, E. and Bertsekas, D. Distributed algorithms for generating loop-free routes in networks with frequently changing topology. *IEEE Trans. Commun.* **C-29** (1981), 11–18.

Gamal, A. E., Mammen, J., Prabhakar, B., and Shah, D. Throughput delay trade-off in wireless networks. In *INFOCOM* (2004).

Gandham, S., Dawande, M., and Prakash, R. Link scheduling in sensor networks: Distributed edge-coloring revisited. In *INFOCOM* (2005).

Gao, J., Guibas, L. J., Hershberger, J., Zhang, L., and Zhu, A. Discrete mobile centers. In *SoCG* (2001a), pp. 188–196.

Gao, J., Guibas, L. J., Hershberger, J., Zhang, L., and Zhu, A. Geometric spanner for routing in mobile networks. In *MobiHoc* (2001b).

Garcia-Luna-Aceves, J. J. and Fullmer, C. L. Floor acquisition multiple access (FAMA) in single-channel wireless networks. *Mobile Network Appl.* **4** (1999), 157–174.

Garcia-Luna-Aceves, J. J. and Tzamaloukas, A. Reversing the collision-avoidance handshake in wireless networks. In *Proceedings of the 5th Annual ACM/IEEE International Conference on Mobile Computing and Networking* (1999), pp. 120–131.

Garcias, R. and Garcia-Luna-Aceves, J. J. Floor acquisition multiple access with collision resolution. In *MobiCom* (1996), pp. 187–197.

Garey, M. R. and Johnson, D. S. *Computers and Intractability*. Freeman, New York (1979).

Garg, N., Papatriantafilou, M., and Tsigas, P. Distributed list-coloring: How to dynamically allocate frequencies to mobile base stations. In *Symposium on Parallel and Distributed Processing (SPDP)*, IEEE Computer Society, Washington, DC (1996a), pp. 18–25.

Garg, N., Papatriantafilou, M., and Tsigas, P. Distributed list-coloring: How to dynamically allocate frequencies to mobile base stations. *Max-Planck Institut für Informatik-Report-MPI I*, 10 (1996b).

Gastpar, M. and Vetterli, M. On the capacity of wireless networks: The relay case. In *INFOCOM* (2002).

Gerla, M. and Tsai, J. T.-C. Multicluster, mobile, multimedia radio network. *Wireless Networks* **1** (1995), 255–265.

Giordano, S., Stojmenovic, I., and Blazevic, L. Position-based routing algorithms for ad hoc networks: A taxonomy, 2002. Survey: http://www.site.uottawa.ca/ivan/routing-survey.pdf

Goldberg, A., Plotkin, S., and Shannon, G. Parallel symmetry-breaking in sparse graphs. In *STOC '87* (1987), pp. 315–324.

Goldberg, A. and Rao, S. Flows in undirected unit capacity networks. Tech. Rep. 97-103, NEC Research Institute (1997).

Goldenberg, D. K., Bihler, P., Yang, Y. R., Cao, M., Fang, J., Morse, A. S., and Anderson, B. D. O. Localization in sparse networks using sweeps. In *MobiCom '06* (2006) pp. 110–121.

Green, J. and Laffont, J. J. Characterization of satisfactory mechanisms for the revelation of preferences for public goods. *Econometrica*, **45**, 2 (1977), 427–438.

Gronkvist, J. and Hansson, A. Comparison between graph-based and interference-based STDMA scheduling. In *MobiHoc* (2001), pp. 255–258.

Grossglauser, M. and Tse, D. Mobility increases the capacity of ad-hoc wireless networks. In *INFOCOM* (2001), vol. 3, pp. 1360–1369.

Groves, T. Incentives in teams. *Econometrica* **41**, 4 (1973), 617–631.

Grünewald, M., Lukovszki, T., Schindelhauer, C., and Volbert, K. Distributed maintenance of resource efficient wireless network topologies. In *Proceedings of the 8th European Conference on Parallel Computing (Euro-Par '02)*, Springer-Verlag, London (2002).

Guha, S. and Khuller, S. Approximation algorithms for connected dominating sets. In *ESA '96* (1996), pp. 179–193.

Guha, S. and Khuller, S. Approximation algorithms for connected dominating sets. *Algorithmica* **20** (1998a), 374–387.

Guha, S. and Khuller, S. Improved methods for approximating node-weighted Steiner trees and connected dominating sets. In *Foundations of Software Technology and Theoretical Computer Science*, Springer-Verlag, London (1998b), pp. 54–65.

Guha, S. and Khuller, S. Improved methods for approximating node-weighted Steiner trees and connected dominating sets. *Inf. Comput.* **150** (1999), 57–74.

Gupta, P. and Kumar, P. R. Critical power for asymptotic connectivity in wireless networks. *Stochastic Analysis, Control, Optimization and Applications: A Volume in Honor of W. H. Fleming*, W. M. McEneaney, G. Yin, and Q. Zhang (Eds.), Springer-Verlag, London (1998).

Gupta, P. and Kumar, P. Capacity of wireless networks. *IEEE Trans. Inf. Theory* **46** (2000) 388–404. (Tech. Rep., University of Illinois, Urbana-Champaign, 1999).

Haas, Z., Halperson, J., and Li, L. Gossip-based as hoc routing. In *INFOCOM* (2002).

Haas, Z. and Liang, B. Ad-hoc mobility management with uniform quorum system. *IEEE/ACM Trans. Network.* **7** (1999), 228–240.

Haas, Z. J. and Perlman, M. R. The zone routing protocol (ZRP) for ad hoc networks. In *Internet draft—Mobile Ad hoc NETworking (MANET), Working Group of the Internet Engineering Task Force (IETF)* (November 1997).

Haas, Z. J. and Perlman, M. R. The zone routing protocol (ZRP) for ad hoc networks. In *Internet draft—Mobile Ad hoc NETworking (MANET), Working Group of the Internet Engineering Task Force (IETF)* (June 1999).

Hajiaghayi, M., Immorlica, N., and Mirrokni, V. S. Power optimization in fault-tolerant topology control algorithms for wireless multi-hop networks. In *MobiCom* (2003), pp. 300–312.

Hale, W. Frequency assignment: Theory and applications. *Proc. IEEE* **68** (1980).

Hall, P. *Introduction to the Theory of Coverage Processes*. Wiley, New York, 1988.

Harter, A., Hopper, A., Steggles, P., Ward, A., and Webster, P. The anatomy of a context-aware application. In *MobiCom* (1999), pp. 59–68.

Hassin, R. Approximation schemes for the restricted shortest-path problem. *Math. Oper. Res.* **17** (1992).

Hayashi, T., Nakano, K., and Olariu, S. Randomized initialization protocols for packet radio networks. In *IPPS'99/SPDP '99: Proceedings of the 13th International Symposium on Parallel Processing and the 10th Symposium on Parallel and Distributed Processing*, IEEE Computer Society, Washington, DC (1999), p. 544.

Heidemann, J. and Bulusu, N. Using geospatial information in sensor networks, 2001. *Proceedings of the Workshop on Intersections Between Geospatial Information and Information Technology*, National Research Council, Arlington, VA.

Heinzelman, W. R., Chandrakasan, A., and Balakrishnan, H. Energy-efficient communication protocol for wireless microsensor networks. In *HICSS* (2000), p. 8020.

Heinzelman, W. R., Kulik, J., and Balakrishnan, H. Adaptive protocols for information dissemination in wireless sensor networks. In *MobiCom* (1999), pp. 174–185. http://citeseer.nj.nec.com/heinzelman99adaptive.html

Herzog, S., Shenker, S., and Estrin, D. Sharing the cost of multicast trees: An Axiomatic Analysis. In *Proceedings of the Conference on Applications, Technologies, Architectures, and Protocols for Computer Communication* (1995), ACM, New York, pp. 315–327.

Hespanha, J. P. and Bohacek, S. Preliminary results in routing games. In *Proceedings of the 2001 American Control Conference*, IEEE, New York (2001), pp. 1904–1909.

Hightower, J. and Borriella, G. Location systems for ubiquitous computing. *IEEE Trans. on Comput.* **34** (2001), 57–66.

Hightower, J., Vakili, C., Borriello, G., and Want, R. Design and calibration of the Spoton ad-hoc location sensing system, 2001. UW-CSE 01–08, University of Washington, Tech-Report.

Hightower, J., Want, R., and Borriello, G. Spoton: An indoor 3d location sensing technology based on RF signal strength. Tech. Rep. UW-CSE 00-02-02, University of Washington, Department of Computer Science and Engineering (2000).

Hochbaum, D. Efficient bounds for the stable set, vertex cover, and set packing problems. *Discrete Appl. Math.* **6** (1983), 243–254.

Hochbaum, D. S. and Maass, W. Approximation schemes for covering and packing problems in image processing and VLSI. *J. ACM* **32** (1985), 130–136.

Holst, L. On multiple covering of a circle with random arcs. *J. Appl. Probab.* **16** (1980), 284–290.

Hong, X., Gerla, M., Pei, G., and Chiang, C. A group mobility model for ad hoc wireless networks. In *MSWiM '99* (1999), pp. 53–60.

Hou, T. and Li, V. Transmission range control in multihop packet radio networks. *IEEE Trans. Commun.* **34** (1986), 38–44.

Hsiao, P., Hwang, A., Kung, H. T., and Vlah, D. Load-balancing routing for wireless access networks. In *INFOCOM* (2001).

Hu, L. Topology control for multihop packet radio networks. *IEEE Trans. Commun.* **41**, 10 (1993).

Hu, L. and Evans, D. Localization for mobile sensor networks. In *MobiCom '04* (2004), pp. 45–57.

Hu, Y.-C., Johnson, D. B., and Perrig, A. SEAD: Secure efficient distance vector routing for mobile wireless ad hoc networks. In *Proceedings of the 4th IEEE Workshop on Mobile Computing Systems and Applications (WMCSA)*, IEEE, New York (2002), pp. 3–13.

Hu, Y.-C., Perrig, A., and Johnson, D. B. Ariadne: A secure on-demand routing protocol for ad hoc networks. Tech. Rep. TR01-383, Department of Computer Science, Rice University, December 2001.

Hubaux, J., Buttyan, L., and Capkun, S. The quest for security in mobile ad hoc networks. In *MobiHoc '01* (2001).

Huiban, G. and Verhoeven, Y. C. A self-stabilized distributed algorithm for the range assignment in ad-hoc wireless networks. *Soumis Parallel Process. Lett.* (2004).

Hunt, H. B. H., Marathe, M. V., Radhakrishnan, V., Ravi, S. S., Rosenkrantz, D. J., and Stearns, R. NC-approximation schemes for NP- and PSPACE-hard problems for geometric graphs. *J. Algorithms* **26** (1998), 238–274.

Ingelrest, F. and Simplot-Ryl, D. Localized broadcast incremental power protocol for wireless ad hoc networks. In *Proceedings of the 10th IEEE Symposium on Computers and Communications, ISCC 2005* (2005), pp. 28–33.

Ingelrest, F., Simplot-Ryl, D., and Stojmenovic, I. Optimal transmission radius for energy-efficient broadcasting protocol in ad hoc and sensor networks. *IEEE Trans. Parallel Distribut. Syst.* **17** (2006).

Intanagonwiwat, C., Govindan, R., Estrin, D., Heidemann, J., and Silva, F. Directed diffusion for wireless sensor networking. *IEEE Trans. Network.* **11** (2003), 2–16.

Iwata, A., Chiang, C.-C., Pei, G., Gerla, M., and Chen, T.-W. Scalable routing strategies for ad hoc wireless networks. *IEEE J. Sel. Areas Commun.* **17** (1999), 1369–1379.

Jackson, B. and Jordan, T. Connected rigidity matroids and unique realizations of graphs. *J. Comb. Theory Ser. B* **94** (2005).

Jain, K., Padhye, J., Padmanabhan, V. N., and Qiu, L. Impact of interference on multi-hop wireless network performance. In *MobiCom '03* (2003), pp. 66–80.

Jain, N., Das, S., and Nasipuri, A. Multichannel CSMA with signal power-based channel selection for multihop wireless networks. In *Proceedings of the IEEE Vehicular Technology Conference (VTC)* (2000), pp. 6–20.

Jain, N., Das, S., and Nasipuri, A. A multichannel CSMA MAC protocol with receiver-based channel selection for multihop wireless networks. In *ICCCN* (2001).

Jakobsson, M., Hubaux, J.-P., and Buttyan, L. A micro-payment scheme encouraging collaboration in multi-hop cellular networks. In *Proceedings of Financial Cryptography*, Springer (2003).

Janssen, J. and Narayanan, L. Approximation algorithms for channel assignment with constraints. In *ISAAC*, Springer (1999), pp. 327–336.

Jardosh, A., Belding-Royer, E. M., Almeroth, K. C., and Suri, S. Towards realistic mobility models for mobile ad hoc networks. In *MobiCom* (2003), pp. 217–229.

Jaromczyk, J. and Toussaint, G. Relative neighborhood graphs and their relatives. *Proc. IEEE* **80** (1992), 1502–1517.

Javier, F., Martinez, O., Nayak, A., Stojmenovic, I., Carle, J., and Simplot-Ryl, D. Area-based beaconless reliable broadcasting in sensor networks. *IJSNet* **1**(1/2) (2006), 20–33.

Jetcheva, J. G., Hu, Y.-C., Maltz, D. A., and Johnson, D. B. A simple protocol for multicast and broadcast in mobile ad hoc networks. ID draft-ietf-manet-simple-mbacst-01.txt (2001).

Jia, L., Rajaraman, R., and Scheideler, C. On local algorithms for topology control and routing in ad hoc networks. In *Proceedings of the 15th Annual ACM Symposium on Parallel Algorithms and Architectures*, ACM, New York (2003).

Jia, L., Rajaraman, R., and Suel, T. An efficient distributed algorithm for constructing small dominating sets. In *PODC* (2000).

Jiang, X. and Camp, T. A review of geocasting protocols for a mobile ad hoc network. In *Proceedings of the Grace Hopper Celebration of Woman in Computing (GHC)* (2002).

Joa-Ng, M. and Lu, I.-T. A peer-to-peer zone-based two-level link state routing for mobile ad hoc networks. *IEEE J. Sel. Areas Commun.* **17** (1999), 1415–1425.

Johnson, D. B. and Maltz, D. A. Dynamic source routing in ad hoc wireless networks. In *Mobile Computing*, Imielinski and Korth, Eds. Kluwer Academic, New York (1996), vol. 2, 353.

Jones, C. E., Sivalingam, K. M., Agrawal, P., and Chen, J. C. A survey of energy-efficient network protocols for wireless networks. *Wireless Networks* **4** (2001), 343–358.

Kachirski, O. and Guha, R. Intrusion detection using mobile agents in wireless ad hoc networks. In *Proceedings of the IEEE Workshop on Knowledge Media Networking*, IEEE, New York (2002).

Kao, M.-Y., Li, X.-Y., and Wang, W. Towards truthful mechanisms for binary demand games: A general framework. In *Proceedings of the ACM Conference on Electronic Commerce*, ACM, New York (2005).

Kapoor, S. and Li, X.-Y. Geometric proximity graphs. In *Workshop on Algorithms and Date Structures (WADS)*, Springer (2003).

Karavelas, M. I. and Guibas, L. J. Static and kinetic geometric spanners with applications. In *ACM SODA*, ACM, New York (2001), pp. 168–176.

Karlin, A., Kempe, D., and Tamir, T. Beyond VCG: Frugality of truthful mechanisms. In *IEEE Annual Symposium on Foundations of Computer Science (FOCS)*, IEEE, New York (2005).

Karlof, C. and Wagner, D. Secure routing in wireless sensor networks: Attacks and counter-measures. In *First IEEE International Workshop on Sensor Network Protocols and Applications*, IEEE, New York (2003), pp. 113–127.

Karn, P. MACA — A new channel access method for packet radio. In *ARRL/CRRL Amateur Radio 9th Computer Networking Conference*, Springer (1990), pp. 134–140.

Karp, B. *Geographic Routing for Wireless Networks*. Ph.D. dissertation, Harvard University (2000).

Karp, B. and Kung, H. GPSR: Greedy perimeter stateless routing for wireless networks. In *MobiCom* (2000).

Keil, J. and Gutwin, C. The Delaunay triangulation closely approximates the complete Euclidean graph. In *Proc. 1st Workshop Algorithms Data Structure (LNCS 382)*, Springer-Verlag, London (1989).

Keil, J. M. and Gutwin, C. A. Classes of graphs which approximate the complete Euclidean graph. *Discrete Comput. Geom.* **7** (1992).

Kelly, D., Reinhardt, S., Stanley, R., and Einhorn, M. PulsON second generation timing chip: enabling UWB through precise timing.In *Proceedings of the IEEE Conference on Ultra Wideband Systems and Technologies*, IEEE, New York (2002), pp. 117–121.

Keshavarz-Haddad, A., Ribeiro, V., and Riedi, R. Broadcast capacity in multihop wireless networks. In *MobiCom '06* (2006), pp. 239–250.

Khanna, S. and Kumaran, K. On wireless spectrum estimation and generalized graph coloring. In *INFOCOM* (1998).

Khuller, S. and Vishkin, U. Biconnectivity approximations and graph carvings. *J. ACM* **41** (1994), 214–235.

Kilpatrick, J. A., Cyr, R. J., Org, E. L., and Dawe, G. New SDR architecture enables ubiquitous data connectivity. *RF Design Magazine, http://rfdesign.com/mag/601RFDF3.pdf* (2006), 34–38.

Kim, Y.-J., Govindan, R., Karp, B., and Shenker, S. On the pitfalls of geographic face routing. In *Third ACM/SIGMOBILE International Workshop on Foundation of Mobile Computing, DIAL-M-POMC*, ACM, New York (2005).

Kirousis, L., Kranakis, E., Krizanc, D., and Pelc, A. Power consumption in packet radio networks. In *STACS '97* (1997).

Kirousis, L. M., Kranakis, E., Krizanc, D., and Pelc, A. Power consumption in packet radio networks. *Theor. Comput. Sci.* **243** (2000), 289–305.

Klein, P. and Ravi, R. A nearly best-possible approximation algorithm for node-weighted Steiner trees. *J. Algorithms* **19** (1995), 104–115.

Kleinrock, L., and Silvester, J. Optimum transmission radii for packet radio networks or why six is a magic number. In *Proceedings of the IEEE National Telecommunications Conference*, IEEE, New York (1978), pp. 431–435.

Kleinrock, L. and Tobagi, F. A. Packet switching in radio channels: Part I—carrier sense multiple-access modes and their throughput-delay characteristics. *IEEE Trans. Commun.* **23** (1975), 1400–1416.

Ko, Y.-B. and Vaidya, N. H. Using location information to improve routing in ad hoc networks. Tech. Rep., Department of Computer Science, Texas A&M University (1997).

Ko, Y.-B. and Vaidya, N. H. Location-aided routing (LAR) in mobile ad hoc networks. In *MobiCom* (1998), pp. 66–75.

Ko, Y. and Vaidya, N. Geocasting in mobile ad hoc networks: Location-based multicast algorithms. In *WMCS* (1999), pp. 101–110.

Ko, Y. and Vaidya, N. H. Geotora: A protocol for geocasting in mobile ad hoc networks. In *Proceedings of the IEEE International Conference on Network Protocols (ICNP)*, IEEE, New York (2000), pp. 240–250.

Ko, Y. and Vaidya, N. Any casting-based protocol for geocast service in mobile ad hoc networks. *Comput. Networks J.* **4**, 6 (2003), pp. 743–776.

Kodialam, M. and Lakshman, T. V. Detecting network intrusions via sampling: A game theoretic approach. In *INFOCOM* (2003).

Kodialam, M. and Nandagopal, T. Characterizing achievable rates in multi-hop wireless networks: The joint routing and scheduling problem. In *MobiCom '03* (2003), pp. 42–54.

Kodialam, M. and Nandagopal, T. The effect of interference on the capacity of multi-hop wireless networks. In *Proceedings of the IEEE Symposium on Information Theory*, IEEE, New York (2004).

Kodialam, M. and Nandagopal, T. Characterizing the capacity region in multi-radio multi-channel wireless mesh networks. In *MobiCom '05* (2005), pp. 73–87.

Kong, J., Zerfos, P., Luo, H., Lu, S., and Zhang, L. Providing robust and ubiquitous security support for mobile ad-hoc networks. In *Protocols (ICNP)*, IEEE, New York (2001), pp. 251–260.

Kozat, U. C., Kondylis, G., Ryu, B., and Marina, M. Virtual dynamic backbone for mobile ad hoc networks. In *ICC* (2001).

Kranakis, E., Singh, H., and Urrutia, J. Compass routing on geometric networks. In *CCCG* (1999), pp. 51–54.

Krishnamachari, B., Wicker, S., Béjar, R., and Pearlman, M. Critical density thresholds in distributed wireless networks. In *Communications, Information and Network Security*. Kluwer, (2002).

Krumke, S. O., Liu, R., Lloyd, E. L., Marathe, M. V., Ramanathan, R., and Ravi, S. S. Topology control problems under symmetric and asymmetric power thresholds. In *AdHoc-Now* (2003), pp. 187–198.

Krumke, S., Marathe, M., and Ravi, S. Models and approximation algorithms for channel assignment in radio networks. *Wireless Networks* **7** (2001), 567–574.

Kubicka, E. and Schwenk, A. J. An introduction to chromatic sums. In *Proceedings of the ACM Computer Science Conference*, ACM, New York (1989), pp. 39–45.

Kuhn, F., Moscibroda, T., and Wattenhofer, R. How the hidden-terminal problem affects clustering in ad hoc and sensor networks. Tech. Rep. TR 442, ETH Zurich, Department of Computer Science (2004a).

Kuhn, F., Moscibroda, T., and Wattenhofer, R. Initializing newly deployed ad hoc and sensor networks. In *MobiCom '04* (2004b), pp. 260–274.

Kuhn, F., Moscibroda, T., and Wattenhofer, R. Radio network clustering from scratch. In *ESA* (2004c).

Kuhn, F., Moscibroda, T., and Wattenhofer, R. Unit disk graph approximation. In *Proceedings of the 2004 ACM Joint Workshop on Foundations of Mobile Computing*, ACM, New York (2004d).

Kuhn, F., Moscibroda, T., and Wattenhofer, R. Fault-tolerant clustering in ad hoc and sensor networks. In IEEE *ICDCS* (2006).

Kuhn, F. and Wattenhofer, R. Constant-time distributed dominating set approximation. In *Proceedings of the 22nd Annual Symposium on Principles of Distributed Computing*, ACM, New York (2003), pp. 25–32.

Kuhn, F. and Wattenhofer, R. On the complexity of distributed graph coloring. In *PODC* (2006).

Kuhn, F., Wattenhofer, R., Zhang, Y., and Zollinger, A. Geometric ad-hoc routing: Of theory and practice. In *PODC* (2003a).

Kuhn, F., Wattenhofer, R., and Zollinger, A. Asymptotically optimal geometric mobile ad-hoc routing. In *DIALM* (2002), pp. 24–33.

Kuhn, F., Wattenhofer, R., and Zollinger, A. Ad-hoc networks beyond unit disk graphs. In *DIALM-POMC* (2003b).

Kuhn, F., Wattenhofer, R., and Zollinger, A. Worst-case optimal and average-case efficient geometric ad-hoc routing. In *(MobiHoc* (2003c).

Kumar, V. S. A., Marathe, M. V., Parthasarathy, S., and Srinivasan, A. End-to-end packet-scheduling in wireless ad-hoc networks. In *SODA '04* (2004), pp. 1021–1030.

Kumar, V. S. A., Marathe, M. V., Parthasarathy, S., and Srinivasan, A. Algorithmic aspects of capacity in wireless networks. *SIGMETRICS Perform. Eval. Rev.* **33** (2005), 133–144.

Kyasanur, P. and Vaidya, N. H. Routing in multi-channel multi-interface ad hoc wireless networks. Tech. Rep., Computer Science Department, UIUC Technical Report (2004).

Kyasanur, P. and Vaidya, N. H. Capacity of multi-channel wireless networks: Impact of number of channels and interfaces. In *MobiCom '05* (2005a), pp. 43–57.

Kyasanur, P. and Vaidya, N. H. Routing and interface assignment in multi-channel multi-interface wireless networks. In IEEE WCNC (2005b).

Langendoen, K. and Reijers, N. Distributed localization in wireless sensor networks: A quantitative comparison. *Comput. Networks* **42** (2003). (Special issue on Wireless Sensor Networks).

Lee, P., Misra, V., and Rubenstein, D. Distributed algorithms for secure multipath routing. In *INFOCOM* (2005), pp. 1952–1963.

Lee, S. J. and Gerla, M. AODV-BR: Backup routing in ad hoc networks. In *WCNC* (2000), pp. 1311–1316.

Lehmann, D., Ocallaghan, L. I., and Shoham, Y. Truth revelation in approximately efficient combinatorial auctions. *J. ACM* **49** (2002), 577–602.

Leiner, B. M., Nielson, D. L., and Tobagi, F. Issues in packet radio network design. *Proc. of IEEE* **75** (1987), 6–20.

Leonard, J. and Durrant-Whyte, H. Mobile robot localization by tracking geometric beacons. *IEEE Trans. Robot. Automat.* **7** (1991), 376–382.

Leong, B., Mitra, S., and Liskov, B. Path vector face routing: Geographic routing with local face information. In *ICNP* (2005).

Levcopoulos, C., Narasimhan, G., and Smid, M. Efficient algorithms for constructing fault-tolerant geometric spanners. In *Proceedings of the 30th Annual ACM Symposium on Theory of Computing*, ACM, New York (1998).

Levcopoulos, C., Narasimhan, G., and Smid, M. Improved algorithms for constructing fault-tolerant geometric spanners. *Algorithmica* (2000).

Li, D., Wong, K., Hu, Y., and Sayeed, A. Detection, classification, and tracking of targets. *IEEE Signal Process. Mag.* **19** (2002).

Li, F., and Nikolaidis, I. On minimum-energy broadcasting in all-wireless networks. In *Proceedings of the IEEE 26th Annual IEEE Conference on Local Computer Networks (LCN'01)*, IEEE, New York (2001).

Li, J., Blake, C., Couto, D. S. J. D., Lee, H. I., and Morris, R. Capacity of ad hoc wireless networks. In *MobiCom* (2001).

Li, J., Jannotti, J., De Couto, D., Karger, D., and Morris, R. A scalable location service for geographic ad-hoc routing. In *MobiCom* (2000), pp. 120–130.

Li, L., Halpern, J. Y., Bahl, P., Wang, Y.-M., and Wattenhofer, R. Analysis of a cone-based distributed topology control algorithm for wireless multi-hop networks. In *PODC* (2001).

Li, N. and Hou, J. C. BLMST: A scalable, power efficient broadcast algorithm for wireless sensor networks (2003). QSHINE: First International Conference on Quality of Service in Heterogenous Wired/Wireless Networks (2004), pp. 44–51.

Li, N. and Hou, J. C. FLSS: A fault-tolerant topology control algorithm for wireless networks. In *MobiCom* (2004).

Li, N., Hou, J. C., and Sha, L. Design and analysis of a MST-based topology control algorithm. In *INFOCOM* (2003).

Li, X.-Y. Algorithmic, geometric, and graph issues in wireless networks. In *Wiley Wireless Communications and Mobile Computing (WCMC)*, Wiley, New York (2002).

Li, X.-Y. Topology control in wireless ad hoc networks. In *Ad Hoc Networking*, IEEE, New York (2003a).

Li, X.-Y. Approximate MST for UDG locally. In *COCOON: Annual International Computing and Combinatorics Conference*, Springer, Berlin (2003b).

Li, X.-Y. Localized construction of low-weighted structure and its applications in wireless ad hoc networks. *ACM/Kluwer Wireless Networks (WINET)* **11**, 6 (2005).

Li, X.-Y., Călinescu, G., and Wan, P.-J. Distributed construction of planar spanner and routing for ad hoc wireless networks. In *INFOCOM* (2002a), Vol. 3.

Li, X.-Y., Călinescu, G., Wan, P. J., and Wang, Y. Localized Delaunay triangulation with application in wireless ad hoc networks. *IEEE Trans. Parallel Distribut. Process.* **14** (2003), 1035–1047. Short version appeared at *IEEE INFOCOM* (2002b).

Li, X.-Y., Moaveni-Nejad, K., Song, W.-Z., and Wang, W. Interference-aware topology control for wireless sensor networks. In *SECON '05* (2005a).

Li, X.-Y., Moaveni-Nejad, K., and Wang, Y. Low-weighted structures and internal node-based broadcasting schemes for wireless ad hoc networks. Manuscript (2003a).

Li, X.-Y., Song, W.-Z., and Wang, W. A unified energy-efficient topology for unicast and broadcast. In *MobiCom* (2005b).

Li, X.-Y., Song, W.-Z., and Wang, Y. Efficient topology control for wireless ad hoc networks with non-uniform transmission ranges. *ACM Baltzer Wireless Network (WINET)* **11**, 3 (2005c).

Li, X.-Y., Song, W.-Z., and Wang, Y. Localized topology control for heterogeneous wireless ad hoc networks. *ACM Trans. Sensor Networks* (2005d).

Li, X.-Y., Stojmenovic, I., and Wang, Y. Partial Delaunay triangulation and degree limited localized Bluetooth multihop scatternet formation. *IEEE Trans. Parallel Distribut. Syst.* **15** (2004a), 350–361. The short version appeared at AdHoc-Now (2002).

Li, X.-Y., Tang, S.-J., and Frieder, O. Multicast capacity of large-scale wireless ad hoc network. In *Proceedings of ACM Annual International Conference on Mobile Computing and Networking (MobiCom)* (2007).

Li, X.-Y., Wan, P.-J., and Wang, Y. Power efficient and sparse spanner for wireless ad hoc networks. In *ICCCN* (2001), pp. 564–567.

Li, X.-Y., Wan, P.-J., Wang, Y., and Frieder, O. Sparse power efficient topology for wireless networks. In *HICSS* (2002c).

Li, X.-Y., Wan, P.-J., Wang, Y., Yi, C.-W., and Frieder, O. Robust deployment and fault-tolerant topology control for wireless ad hoc networks. *Wiley J. Wireless Commun. Mobile Comput.* **4** (2004b), 109–125. The short version appeared in *MobiHoc* (2003).

Li, X.-Y. and Wang, W. Approximation algorithms and algorithm mechanism design. In *Approximation Algorithms and Metaheuristics*. Chapman & Hall/CRC (2007).

Li, X.-Y. and Wang, Y. Simple heuristics and PTASS for intersection graphs in wireless ad hoc networks. In *ACM DIALM* (2002).

Li, X.-Y. and Wang, Y. Localized routing for wireless ad hoc networks. In *IEEE ICC* (2003).

Li, X.-Y. and Wang, Y. Efficient construction of low-weight bounded degree planar spanner. *Int. J. Comput. Geometry Appl.* **14** (2004), 69–84.

Li, X.-Y. and Wang, Y. Simple approximation algorithms and PTASS for various problems in wireless ad hoc networks. *J. Parallel Distribut. Comput.* **66** (2006).

Li, X.-Y., Wang, Y., and Song, W.-Z. Applications of k-local MST for topology control and broadcasting in wireless ad hoc networks. *IEEE Trans. Parallel Distribut. Syst.* **15**, 12 (2004c), pp. 1057–1069.

Li, X.-Y., Wang, Y., Song, W.-Z., Wan, P.-J., and Frieder, O. Localized low-weight graph and its applications in wireless ad hoc networks. In *INFOCOM* (2004d).

Li, X.-Y., Wang, Y., Wang, P.-J., and Yi, C.-W. Fault-tolerant deployment and topology control for wireless ad hoc networks. In *MobiHoc* (2003b).

Liang, B. and Haas, Z. Predictive distance-based mobility management for PCS networks. In *INFOCOM* (1999), pp. 1377–1384.

Liang, B. and Haas, Z. J. Virtual backbone generation and maintenance in ad hoc network mobility management. In *INFOCOM* (2000), vol. 3, pp. 1293–1302.

Liang, W. Constructing minimum-energy broadcast trees in wireless ad hoc networks. In *MobiHoc* (2002), pp. 112–122.

Liao, W.-H., Tseng, Y.-C., Lo, K.-L., and Sheu, J.-P. Geogrid: A geocasting protocol for mobile ad hoc networks based on grid. *J. Internet Technol.* **1** (2000), 23–32.

Lim, H. and Kim, C. Multicast tree construction and flooding in wireless ad hoc networks. In *MSWiM* (2000).

Lin, C. R. and Gerla, M. Adaptive clustering for mobile wireless networks. *IEEE J. Sel. Areas Commun.* **15** (1997a), 1265–1275.

Lin, C. R., and Gerla, M. MACA/PR: An asynchronous multimedia multihop wireless network. In *INFOCOM* (1997b).

Lin, J.-C., Yang, S.-N., and Chern, M.-S. An efficient distributed algorithm for minimal connected dominating set problem. In *Proceedings of the Tenth Annual International Phoenix Conference on Computers and Communications 1991*, IEEE, New York (1991), pp. 204–210.

Lin, X., Sharma, G., Mazumdar, R. R., and Shroff, N. B. Degenerate delay-capacity tradeoffs in ad hoc networks with Brownian mobility. *IEEE Trans. Inf. Technol.* **52** (2006).

Lin, X. and Shroff, N. B. The fundamental capacity-delay tradeoff in large mobile ad hoc networks. In *Proceedings of the 3rd Annual Mediterranean Ad Hoc Networks Workshop*, Springer, Berlin (2004).

Lipman, J., Boustead, P., and Judge, J. Efficient and scalable information dissemination in mobile ad hoc networks. In *AdHoc-Now* (2002), pp. 119–134.

Liu, D., Ning, P., and Li, R. Establishing pairwise keys in distributed sensor networks. *ACM Trans. Inf. Syst. Secur.* **8** (2005), 41–77.

Liu, H. and Gupta, R. Selective backbone construction for topology control. In *MASS* (2004).

Liu, R. and Lloyd, E. L. A distributed protocol for adaptive link scheduling in ad-hoc networks. In *Proceedings of the IASTED International Conference on Wireless and Optical Communications (WOC2001)*, ACTA Press, Calgary (2001).

Liu, Z., Joy, T., and Thompson, R. A dynamic trust model for mobile ad hoc networks. In *Proceedings of the 10th IEEE International Workshop on Future Trends in Distributed Computing Systems (FTDCS 2004)*, IEEE, New York (2004).

Lloyd, E. L., Liu, R., Marathe, M. V., Ramanathan, R., and Ravi, S. S. Algorithmic aspects of topology control problems for ad hoc networks. In *MobiHoc* (2002), pp. 123–134.

Lloyd, E. L. and Ramanathan, S. On the complexity of distance-2 coloring. In *ICCI '92: Proceedings of the Fourth International Conference on Computing and Information*, IEEE Computer Society, Washington, DC (1992), pp. 71–74.

Loguinov, D. and Radha, H. End-to-end Internet video traffic dynamics: Statistical study and analysis. In *INFOCOM* (2002).

Lu, S., Bharghavan, V., and Srikant, R. Fair scheduling in wireless packet networks. *IEEE/ACM Trans. Network.* **7** (1999), 473–489.

Luczak, T. The phase transition in a random graph. *Combinatorics, Paul Erdos is Eighty 2*, Janos Bolyai Mathematical Society (1996), 399–422.

Lukovszki, T. New results of fault-tolerant geometric spanners. In *WADS* (1999a), pp. 193–204.

Lukovszki, T. *New Results on Geometric Spanners and Their Applications*. Ph.D. thesis, University of Paderborn (1999b).

Lund, C. and Yannakakis, M. On the hardness of approximating minimization problems. *J. ACM* **41** (1994), 960–981.

Malesinska, E. List coloring and optimization criteria for a channel assignment problem. Tech. Rep. 458/1995, Technische Universitat, Berlin.

Malkin, G. RIP version 2. *RFC 2453* (1998).

Maltz, D., Broch, J., Jetcheva, J., and Johnson, D. The effects of on-demand behavior in routing protocols for multi-hop wireless ad hoc networks. *IEEE J. Sel. Areas Commun.* (1999).

Marathe, M. V., Breu, H., Hunt, H. B., Ravi, S. S., and Rosenkrantz, D. J. Simple heuristics for unit disk graphs. *Networks* **25** (1995), 59–68.

Marco, G. D. and Pelc, A. Fast distributed graph coloring with O(&DGR;) colors. In *SODA* (2001), pp. 630–635.

Marks, R. J., Das, A., El-Sharkawi, M., Arabshahi, P., and Gray, A. Minimum power broadcast trees for wireless networks: Optimizing using the viability lemma. In *Proceedings of the IEEE International Symposium on Circuits and Systems* (2002), pp. 245–248.

Marti, S., Giuli, T. J., Lai, K., and Baker, M. Mitigating routing misbehavior in mobile ad hoc networks. In *MobiCom* (2000).

Matula, D. W. and Beck, L. L. Smallest-last ordering and clustering and graph coloring algorithms. *J. Assoc. Comput. Mach.* **30** (1983), 417–427.

Mauve, M., Widmer, J., and Hartenstein, H. A survey on position-based routing in mobile ad hoc networks. *IEEE Network* **15**, 6 (2001), pp. 30–39.

McDiarmid, C. and Reed, B. Channel assignment and weighted coloring, *Networks* (Aug. 2000), vol. 36, No. 2, pp. 114–117.

Medepalli, K., and Tobagi, F. A. Throughput analysis of IEEE 802.11 wireless LANS using an average cycle time approach. In *IEEE GLOBECOM* (2005).

Min, M., Wang, F., Du, D.-Z., and Pardalos, P. M. A reliable virtual backbone scheme in mobile ad-hoc networks. In *MASS* (2004).

Minn, T. and Siu, K.-Y. Dynamic assignment of orthogonal variable spreading factor codes in w-CDMA. *IEEE J. Sel. Areas Commun.* **18** (2000), 1429–1440.

Mishra, A., Banerjee, S., and Arbaugh, W. Weighted coloring based channel assignment for WLANS. *SIGMOBILE Mobile Comput. Commun. Rev.* **9**, 3 (2005), 19–31.

Moaveninejad, K., Song, W.-Z., and Li, X.-Y. Robust position-based routing for wireless ad hoc networks. *Elsevier Ad Hoc Networks* **3** (2005), 546–560.

Monks, J., Bharghavan, V., and Hwu, W.-M. A power-controlled multiple access protocol for wireless packet networks. In *INFOCOM* (2001).

Morin, P. *Online Routing in Geometric Graphs*. Ph.D. dissertation. Carleton University School of Computer Science, 2001.

Moscibroda, T., and Wattenhofer, R. Coloring unstructured radio networks. In *Proceedings of the 17th Annual ACM Symposium on Parallelism in Algorithms and Architectures*, ACM, New York (2005a), ACM Press, pp. 39–48.

Moscibroda, T. and Wattenhofer, R. Minimizing interference in ad hoc and sensor networks. In *DIALM-POMC* (2005b).

Moulin, H. and Shenker, S. Strategy-proof sharing of submodular costs: Budget balance versus efficiency. In *Economic Theory* (2002). Available in preprint form at http://www.aciri.org/shenker/cost.ps

Moy, J. OSPF version 2. *RFC 2328* (April 1998).

Naor, M. and Stockmeyer, L. What can be computed locally? In *ACM Conference on Theory of Computing*, ACM, New York (1993), pp. 184–193.

Narayanan, L. Channel assignment and graph multicoloring. In *Handbook of Wireless Networks and Mobile Computing*, Wiley (2002), pp. 71–94.

Nasipuri, A. and Das, S. On-demand multipath routing for mobile ad hoc networks (October 1999) ICCCN, IEEE, New York, pp. 64–76.

Nasipuri, A., Zhuang, J., and Das, S. R. A multichannel CSMA MAC protocol for multihop wireless networks. In *WCNC* (1999).

Navas, J. C. and Imielinski, T. Geocast: Geographic addressing and routing. In *MobiCom* (1997), pp. 66–76.

Neely, M. and Modiano, E. Capacity and delay tradeoffs for ad-hoc mobile networks. *IEEE Trans. Inf. Theory* **51** (2005), 1917C193.

Nelson, R. and Kleinrock, L. Spatial-TDMA: A collision-free multihop channel access protocol. *IEEE Transactions on Communications* **33**, 9 (1985), 934–944.

Niculescu, D. and Nath, B. Ad hoc positioning system (APS). In *GLOBECOM* (2001).

Nieberg, T., Hurink, J., and Kern, W. A robust PTAS for maximum weight independent sets in unit disk graphs. In *Proceedings of the 30th International Workshop on Graph-Theoretic Concepts in Computer Science (WG04), Lecture Notes in Computer Science 3353*, Springer, Berlin (2004), pp. 214–221.

Nisan, N. and Ronen, A. Algorithmic mechanism design. In *STOC* (1999), pp. 129–140.

Nisan, N. and Ronen, A. Computationally feasible VCG mechanisms. In *ACM Conference on Electronic Commerce*, ACM, New York (2000), pp. 242–252.

Niyato, D. and Hossain, E. Queue-aware uplink bandwidth allocation and rate control for polling service in IEEE 802.16 broadband wireless networks. *IEEE Trans. Mobile Comput.* **5** (2006), 668–679.

Nocetti, F. G., Gonzalez, J. S., and Stojmenovic, I. Connectivity based k-hop clustering in wireless networks. *Telecommun. Syst.* **22** (2003), 205–220.

Orecchia, L., Panconesi, A., Petrioli, C., and Vitaletti, A. Localized techniques for broadcasting in wireless sensor networks. In *DIALM-POMC* (2004).

Ovalle-Martinez, F., Nayak, A., Stojmenovic, I., Carle, J., and Simplot-Ryl, D. Area-based beaconless reliable broadcasting in ad hoc and sensor networks. *Int. J. Sensor Networks* **2** (2006), 147–159.

Panconesi, A. and Rizzi, R. Some simple distributed algorithms for sparse networks. *Distribut. Comput.* **14** (2001), 97–100.

Panconesi, A. and Srinivasan, A. Improved distributed algorithms for coloring and network decomposition problems. In *STOC* (1992), pp. 581–592.

Park, V. D. and Corson, M. S. A highly adaptive distributed routing algorithm for mobile wireless networks. In *INFCOM* (1997).

Park, V. and Corson, M. Temporally-ordered routing algorithms, Internet Draft, IETE (October 1999).

Pei, G., Gerla, M., and Chen, T.-W. Fisheye state routing: A routing scheme for ad hoc wireless networks. In *ICC* (2000a).

Pei, G., Gerla, M., and Chen, T.-W. Fisheye state routing in mobile ad hoc networks. In *Proceedings of Workshop on Wireless Networks and Mobile Computing ICDCS*, IEEE, New York (2000b).

Peleg, D. *Distributed computing: A locality-sensitive approach.* SIAM, Philadelphia (2000).

Peng, W. and Lu, X. On the reduction of broadcast redundancy in mobile ad hoc networks. In *MobiHoc* (2000).

Peng, W. and Mirsa, P. Special issue on global positioning system. In *Proceedings of the IEEE* (January 1999), vol. 87, pp. 3–15.

Penrose, M. The longest edge of the random minimal spanning tree. *Ann. Appl. Probab.* **7** (1997), 340–361.

Penrose, M. Extremes for the minimal spanning tree on normally distributed points. *Adv. Appl. Probab.* **30** (1998), 628–639.

Penrose, M. On k-connectivity for a geometric random graph. *Random Struct. Algorithms* **15** (1999a), 145–164.

Penrose, M. A strong law for the largest nearest-neighbor link between random points. *J. London Math. Soc.* **60** (1999b), 951–960.

Penrose, M. A strong law for the longest edge of the minimal spanning tree. *Ann. Probab.* **27** (1999c), 246–260.

Perkins, C. Ad hoc on-demand distance vector routing. In *IEEE Military Communications Conference (MILCOM)*, IEEE, New York (1997a).

Perkins, C. Ad hoc on demand distance vector (AODV) routing. In *Internet draft, draft-ietf-manet-aodv-00.txt* (November 1997b).

Perkins, C. and Bhagwat, P. Highly dynamic destination-sequenced distance-vector routing. In *SIGCOMM* (1994a).

Perkins, C. and Bhagwat, P. Highly dynamic destination-sequenced distance-vector routing (DSDV) for mobile computers. *Comput. Commun. Revi.* **4** (1994b), pp. 234–244.

Perkins, C. and Royer, E. M.. Ad hoc on demand distance vector routing. In *WMCSA* (1999), pp. 90–100.

Perrig, A., Szewczyk, R., Wen, V., Culler, D., and Tygar, J. Spins: Security protocols for sensor networks. In *MobiCom* (2001).

Philips, T. K., Panwar, S. S., and Tantawi, A. N. Connectivity properties of a packet radio network model. *IEEE Trans. Inf. Theor.* **35**, 5 (1989), pp. 1044–1047.

Phillips, C. A. The network inhibition problem. In STOC, ACM, New York (1993), pp. 776–785.

Pietro, R. D., Mancini, L. V., and Mei, A. Random key-assignment for secure wireless sensor networks. In *1st ACM Workshop Security of Ad Hoc and Sensor Networks Fairfax*, ACM, New York (2003), 62–70.

Pietro, R. D., Mancini, L. V., and Mei, A. Efficient and resilient key discovery based on pseudo-random key predeployment. In *Proceedings of the 18th IEEE International Parallel and Distributed Processing Symposium (IPDPS'04)*, IEEE, New York (2004).

Polastre, J., Hill, J., and Culler, D. Versatile low power media access for wireless sensor networks. In *SenSys* (2004).

Prakash, R. Unidirectional links prove costly in wireless ad hoc networks. In *DIALM '99* (1999), pp. 15–22.

Priyantha, N. B., Chakraborty, A., and Balakrishnan, H. The cricket location-support system. In *Mobile Computing and Networking* (2000), pp. 32–43.

Raghunathan, V., Schurgers, C., Park, S., and Srivastava, M. Energy-aware wireless microsensor networks. *IEEE Sig. Process. Mag.* **19** (2002), 40–50.

Rajaraman, R. Topology control and routing in ad hoc networks: A survey. *SIGACT News* **33** (2002), 60–73.

Raju, J. and Garcia-Luna-Aceves, J. A new approach to on-demand loop-free multipath routing. In *Proceedings of the International Conference on Computer Communications and Networks (ICCCN)*, IEEE, New York (1999), pp. 522–527.

Ramachandran, K. N., Belding-Royer, E. M., Almeroth, K. C., and Buddhikot, M. M. Interference-aware channel assignment in multi-radio wireless mesh networks. In *INFOCOM* (2006).

Ramanathan, R. and Rosales-Hain, R. Topology control of multihop wireless networks using transmit power adjustment. In *INFOCOM* (2000), pp. 404–413.

Ramanathan, S. A unified framework and algorithm for channel assignment in wireless networks. *Wireless Network* **5** (1999), 81–94.

Ramanathan, S. and Lloyd, E. L. Scheduling algorithms for multi-hop radio networks. In *SIG-COMM* (1992), pp. 211–222.

Ramanathan, S. and Lloyd, E. Scheduling algorithms for multi-hop radio networks. *IEEE/ACM Trans. Network.* **1** (1993), 166–172.

Ramanathan, S. and Steenstrup, M. A survey of routing techniques for mobile communication networks. *ACM/Baltzer Mobile Networks and Applications* (1996), 89–104.

Ramaswami, R. and Parhi, K. K. Distributed scheduling of broadcasts in a radio network. In *(INFOCOM)* (1989), pp. 497–504.

Raniwala, A. and Chiueh, T. Architecture and algorithms for an IEEE 802.11-based multi-channel wireless mesh network. In *INFOCOM* (2005).

Raniwala, A., Gopalan, K., and Chiueh, T. Centralized channel assignment and routing algorithms for multi-channel wireless mesh networks. *ACM Mobile Comput. Commun. Rev.* **8** (2004).

Rappaport, T. *Wireless Communications: Principles and Practice.* Prentice-Hall, Englewood Cliffs, NJ (1996).

Reiter, M. K. and Rubin, A. D. Crowds: Anonymity for web transactions. *ACM Trans. Inf. Syst. Secur.* **1** (1998), pp. 66.

Rekhter, Y. and Li, T. *A Border Gateway Protocol 4 (BGP-4)*, RFC 1771 (March 1995) http//www.ietf.org/rfc/rfc1771.txt

Resta, G. and Santi, P. An analysis of the node spatial distribution of the random waypoint model for ad hoc networks. In *POMC* (2002), pp. 44–50.

RFC 1771. *Cisco System, A Border Gateway Protocol 4 (BGP-4)*.

Rhee, I., Warrier, A., Aia, M., and Min, J. ZMAC: A hybrid MAC for wireless sensor networks. In *SenSys* (2005).

Roberts, K. The characterization of implementable choice rules. In *Aggregation and Revelation of Preferences* (J. J. Laffont, ed.) North-Holland, Amsterdam (1979), pp. 321–349. Papers presented at the 1st European Summer Workshop of the Econometric Society.

Robins, G. and Zelikovsky, A. Improved Steiner tree approximation in graphs. In *SODA* (2000), pp. 770–779.

Rodoplu, V. and Meng, T. H. Minimum energy mobile wireless networks. In *ICC'98* **3** (1998).

Roos, T., Myllymaki, P., Tirri, H., Misikangas, P., and Sievanen, J. A probabilistic approach to WLAN user location estimation. *Int. J. Wireless Inf. Networks* **9**, 3 (2002), pp. 155–164.

Royer, E., and Toh, C.-K. A review of current routing protocols for ad hoc mobile wireless networks. *IEEE Pers. Commun.* (1999).

Rührup, S., Schindelhauer, C., Volbert, K., and Grünewald, M. Performance of distributed algorithms for topology control in wireless networks. In *Proceedings of the 17th International Parallel and Distributed Processing Symposium (IPDPS 2003)*, IEEE, New York (2003).

Salem, N. B., Buttyan, L., Hubaux, J.-P., and Kakobsson, M. A charging and rewarding scheme for packet forwarding in multi-hop cellular networks. In *MobiHoc* (2003).

Salonidis, T. and Tassiulas, L. Distributed dynamic scheduling for end-to-end rate guarantees in wireless ad hoc networks. In *MobiHoc '05* (2005), pp. 145–156.

Sanchez, M., Manzoni, P., and Haas, Z. Determination of critical transmission range in ad-hoc networks. In *Multiaccess, Mobility and Teletraffic for Wireless Communications (MMT '99)* Springer, Berlin (1999).

Santi, P. The critical transmitting range for connectivity in mobile ad hoc networks. *IEEE Trans. Mobile Comput.* **4** (2005a), 310–317.

Santi, P. *Topology Control in Wireless Ad Hoc and Sensor Networks*. Wiley, New York (2005b).

Santi, P. and Blough, D. M. An evaluation of connectivity in mobile wireless ad hoc networks. In *Proc. IEEE DSN* (2002), pp. 89–98.

Santi, P. and Blough, D. M. The critical transmitting range for connectivity in sparse wireless ad hoc networks. *IEEE Trans. Mobile Comput.* **2** (2003), 25–39.

Sarkar, R., Zhu, X., and Gao, J. Double rulings for information brokerage in sensor networks. In *MobiCom* (2006), pp. 286–297.

Savage, S., Wetherall, D., Karlin, A., and Anderson, T. Practical network support for IP traceback. In *SIGCOMM* (2000).

Savarese, C., Rabay, J., and Langendoen, K. Robust positioning algorithms for distributed ad-hoc wireless sensor networks, Usenix Technical Annual Conference, 2002.

Savvides, A., Han, C.-C., and Srivastava, M. Dynamic fine-grained localization in ad-hoc networks of sensors. In *MobiCom* (2001), pp. 166–179.

Savvides, A., Park, H., and Srivastava, M. The bits and flops of the n-hop multilateration primitive for node-localization problems. In *Proceedings of the 1st ACM International Workshop on Wireless Sensor Networks and Applicaiton*, ACM, New York (2002), pp. 112–121.

Schindelhauer, C., Volbert, K., and Ziegler, M. Spanners, weak spanners, and power spanners. In *ISAAC* (2004).

Schindelhauer, C., Volbert, K., and Ziegler, M. Geometric spanners with applications in wireless networks. *Computational Geometry: Theory and Applications* **36**, 3 (2005), pp. 197–214.

Seddigh, M., Gonzalez, J. S., and Stojmenovic, I. RNG and internal node-based broadcasting algorithms for wireless one-to-one networks. *ACM Mobile Comput. Commun. Rev.* **5** (2002), 37–44.

Sen, A. and Huson, M. L. A new model for scheduling packet radio networks. In *INFOCOM* (1996), pp. 1116–1124.

Sen, A. and Huson, M. L. A new model for scheduling packet radio networks. *ACM/Baltzer Journal Wireless Networks* **3** (1997), 71–82.

Sen, A. and Malesinska, E. Approximation algorithms for radio network scheduling. In *Proceedings of 35th Allerton Conference on Communication, Control, and Computing*, University of Illinois, Champaign (1997), pp. 573–582.

Shah, R. C. and Rabaey, J. M. Energy-aware routing for low-energy ad hoc sensor networks. In *WCNC* (2002).

Shakkottai, S., Liu, X., and Srikant, R. The multicast capacity of ad hoc networks. In *Proceedings of the ACM International Symposium on Mobile Ad Hoc Networking and Computing (MobiHoc)* (2007).

Shapley, L. S. A value for *n*-person games. In *Contributions to the Theory of Games*. Princeton University Press, Princeton, NJ (1953), pp. 31–40.

Sharma, G. and Mazumdar, R. On achievable delay/capacity trade-offs in mobile ad hoc networks. In *Proc. Workshop Model. Optimization Ad Hoc Mobile Networks*, Springer, Berlin (2004).

Sharma, G. and Mazumdar, R. Delay and capacity trade-off in wireless ad hoc networks with random way-point mobility. Tech. Rep., School of ECE, Purdue University, W. Lafayette, IN (2005).

Shih, E., Cho, S.-H., Ickes, N., Min, R., Sinha, A., Wang, A., and Chandrakasan, A. Physical layer driven protocol and algorithm design for energy-efficient wireless sensor networks. In *MobiCom* (2001), pp. 272–287.

Shin, M., Lee, S., and Kim, Y.-A. Distributed channel assignment for multi-radio wireless networks. IEEE MASS, 2006.

Shneidman, J. and Parkes, D. Specification faithfulness in networks with rational nodes. In *PODC* (2004).

Sinha, P., Sivakumar, R., and Bharghavan, V. CEDAR: Core extraction distributed ad hoc routing, IEEE JSAC **17**, 8 (1999), pp. 1454–1465.

Sivakumar, R., Das, B., and Bharghavan, V. The clade vertebrata: Spines and routing in ad hoc networks. In *ISCC98* (1998).

Slijepcevic, S., Megerian, S., and Potkonjak, M. Characterization of location error in wireless sensor networks: Analysis and applications. In *Int. Workshop Inf. Process. Sensor Networks (IPSN)* (2003), pp. 625–641.

Smaragdakis, G., Matta, I., and Bestavros, A. SEP: A stable election protocol for clustered heterogeneous wireless sensor networks. In *Second International Workshop on Sensor and Actor Network Protocols and Applications (SANPA 2004)* Springer, Berlin (2004).

So, A. M.-C. and Ye, Y. Theory of semidefinite programming for sensor network localization. In *SODA* (2005), pp. 405–414.

So, J. and Vaidya, N. H. Routing and channel assignment in multi-channel multi-hop wireless networks with single-nic devices. CS Dept. UIUC Tech. Rep. (2004a).

So, J. and Vaidya, N. Multi-channel MAC for ad hoc networks: Handling multi-channel hidden terminals using a single transceiver. In *MobiHoc* (2004b).

So, J. and Vaidya, N. H. A routing protocol for utilizing multiple channels in multi-hop wireless networks with a single transceiver. Tech. Rep., University of Illinois at Urbana-Champaign, (2004c).

Song, W.-Z., Wang, Y., Li, X.-Y., and Frieder, O. Localized algorithms for energy-efficient topology in wireless ad hoc networks. In *MobiHoc* (2004).

Song, W.-Z., Wang, Y., Li, X.-Y., and Frieder, O. Localized algorithms for energy-efficient topology in wireless ad hoc networks. In *MONET* (2005), 911–923.

Srinivas, A. and Modiano, E. Minimum energy disjoint path routing in wireless ad-hoc networks. In *MobiCom* (2003), pp. 122–133.

Srinivasan, V., Nuggehalli, P., Chiasserini, C. F., and Rao, R. R. Energy efficiency of ad hoc wireless networks with selfish users. In *European Wireless Conference 2002 (EW2002)*, Springer, Berlin (2002).

Srinivasan, V., Nuggehalli, P., Chiasserini, C. F., and Rao, R. R. Cooperation in wireless ad hoc wireless networks. In *INFOCOM* (2003).

Stevens, D. and Ammar, M. Evaluation of slot-allocation strategies for TDMA protocols in packet radio networks. In *MILCOM* (1990), pp. 835–839.

Stojmenovic, I. A routing strategy and quorum-based location update scheme for ad hoc wireless networks. Tech. Rep. TR-99-09, Computer Science, SITE, University of Ottawa, 1999.

Stojmenovic, I. Degree-limited RNG. In *Workshop on Wireless Networks* (2003).

Stojmenovic, I. and Datta, S. Power and cost aware localized routing with guaranteed delivery in wireless networks. In *ISCC* (2002).

Stojmenovic, I. and Lin, X. Loop-free hybrid single-path/flooding routing algorithms with guaranteed delivery for wireless networks. *IEEE Trans. Parallel Distribut. Syst.* **12** (2001).

Stojmenovic, I. and Seddigh, M. Broadcasting algorithms in wireless networks. In *Proceedings of the International Conference on Advances in Infrastructure for Electronic Business, Science, and Education on the Internet, SSGRR*, Springer, Berlin (2000).

Stojmenovic, I., Seddigh, M., and Zunic, J. Dominating sets and neighbor-elimination-based broadcasting algorithms in wireless networks. *IEEE Trans. Parallel Distribut. Syst.* **13** (2002), 14–25.

Stone, R. Centertrack: An IP overlay network for tracking DOS floods. *9th USENIX Security Symposium* USENIX, Berkeley (2000).

Szekeres, G. and Wilf, H. An inequality for the chromatic number of a graph. *J. Comb. Theor.* **4** (1968), 1–3.

Takagi, H. and Kleinrock, L. Optimal transmission ranges for randomly distributed packet radio networks. *IEEE Trans. Commun.* **32** (1984), 246–257.

Takahashi, H. and Matsuyama, A. An approximate solution for Steiner problem in graphs. *Math. Jpn.* (1980), 573–577.

Talwar, K. The price of truth: Frugality in truthful mechanisms. In *STACS* (2003), pp. 608–619.

Tang, Z., and Garcia-Luna-Aceves, J. J. Hop-reservation multiple access (HRMA) for ad-hoc networks. In *INFOCOM* (1999).

Tavli, B. Broadcast capacity of wireless networks. *IEEE Commun. Lett.* **10** (2006).

Thaler, D. and Hopps, C. Multipath issues in unicast and multicast next-hop selection. *RFC 2991* (November 2000).

Thurwachter, Charles N. *Wireless Networking*, Prentice Hall, 2002.

Tins, R., Navarro-Serment, L., and Paredis, C. Fault-tolerant localization of teams of distributed robots. In *Proceedings of IEEE/RSJ International Conference on Intelligent Robots and Systems*, IEEE, New York **2** (2001), pp. 1061–1066.

Tobagi, F. Modeling and performance analysis of multi-hop packet radio networks. In *Proc. IEEE* **75** (1987), pp. 135–155.

Toh, C.-K. A novel distributed routing protocol to support ad-hoc mobile networks. In *Proceedings of the 1996 IEEE 15th Annual International Phoenix Conference on Computing and Communication* (1996), pp. 480–486.

Toumpis, S. and Goldsmith, A. Large wireless networks under fading, mobility, and delay constraints. In *INFOCOM* (2004).

Toussaint, G. T. The relative neighborhood graph of a finite planar set. *Pattern Recog.* **12** (1980), 261–268.

Tseng, Y.-C., Kuo, S.-P., Lee, H.-W., and Huang., C.-F. Location tracking in a wireless sensor network by mobile agents and its data fusion strategies. In *IPSN* (2003).

Tseng, Y.-C., Ni, S.-Y., Chen, Y.-S., and Sheu, J.-P. The broadcast storm problem in a mobile ad hoc network. *Wireless Networks* **8** (2002), 153–167. Short version in MobiCom 99.

Turgut, D., Das, S., Elmasri, R., and Turgut, B. Optimizing clustering algorithm in mobile ad hoc networks using genetic algorithmic approach. In *GLOBECOM 2002* (2002).

Tzamaloukas, A. and Garcia-Luna-Aceves, J. A receiver-initiated collision-avoidance protocol for multi-channel networks. In *INFOCOM* (2001).

Vaidya, N. H., Bahl, P., and Gupta, S. Distributed fair scheduling in a wireless LAN. In *MobiCom* (2000), pp. 167–178.

van Dam, T. and Langendoen, K. An adaptive energy-efficient MAC protocol for wireless sensor networks. In *SenSys* (2003).

Vapnik, V. and Chervonenkis, A. On the uniform convergence of relative frequencies of events to their probabilities. *Theor. Probab. Appl.* **16** (1971), 264–280.

Vazirani, V. *Approximation Algorithms*. Springer, New York (2001).

Vickrey, W. Counterspeculation, auctions, and competitive sealed tenders. *J. Finance* **16**, 1 (1961), pp. 8–37.

Wan, P.-J., Alzoubi, K. M., and Frieder, O. Distributed construction of connected dominating set in wireless ad hoc networks. In *INFOCOM* (2002a).

Wan, P.-J., Călinescu, G., Li, X.-Y., and Frieder, O. Minimum-energy broadcast routing in static ad hoc wireless networks. In *INFOCOM* (2000a).

Wan, P.-J., Călinescu, G., Li, X.-Y., and Frieder, O. Minimum-energy broadcast routing in static ad hoc wireless networks. *ACM Wireless Networks* **8** (2002b), 607–617. A preliminary version appeared in *IEEE INFOCOM* (2000b).

Wan, P.-J., Călinescu, G., and Yi, C.-W. Minimum-power multicast routing in static ad hoc wireless networks. *IEEE/ACM Trans. Network.* **12** (2004c), 507–514.

Wan, P.-J., Li, X.-Y., and Song, W.-Z. Theoretically good distributed OVSF-CDMA code assignment in wireless ad hoc networks. In *COCOON* (2005).

Wan, P.-J. and Yi, C.-W. Asymptotic critical transmission radius and critical neighbor number for *k*-connectivity in wireless ad hoc networks. In *MobiHoc* (2004), pp. 1–8.

Wan, P.-J. and Yi, C.-W. Asymptotic critical transmission radius for greedy forward routing in wireless ad hoc networks. In *MobiHoc* (2006).

Wan, P.-J., Yi, C.-W., Jia, X., and Kim, D. Approximation algorithms for conflict-free channel assignment in wireless ad hoc networks. *Wiley J. Wireless Commun. Mobile Comput.* **6**, 2 (2006), pp. 201–211.

Wang, J., Fang, Y., and Wu, D. A power-saving multi-radio multi-channel MAC protocol for wireless local area networks. In *INFOCOM'06* (2006).

Wang, W., Eidenbez, S., Wang, Y., and Li, X.-Y. OURS–optimal unicast routing systems–in noncooperative wireless networks. In *MobiCom* (2006a).

Wang, W. and Li, X.-Y. Low-cost multicast in selfish and rational wireless ad hoc networks. In *MASS* (2004a).

Wang, W. and Li, X.-Y. Truthful low-cost unicast in selfish wireless networks. In *4th International Workshop on Algorithms for Wireless, Mobile, Ad Hoc and Sensor Networks (WMAN) of IPDPS* (2004b).

Wang, W. and Li, X.-Y. Low-cost routing in selfish and rational wireless ad hoc networks. *IEEE Trans. Mobile Comput.* (2005).

Wang, W., Li, X.-Y., Moaveninejad, K., Wang, Y., and Song, W.-Z. The spanning ratios of *beta*-skeleton. In *CCCG* (2003).

Wang, W., Li, X.-Y., and Sun, Z. Design differentiated service multicast with selfish agents. *IEEE J. Sel. Areas Commun.* **24**, 5 (2005a).

Wang, W., Li, X.-Y., Sun, Z., and Wang, Y. Design multicast protocols for noncooperative networks. In *INFOCOM* (2005b).

Wang, W., Li, X.-Y., and Wang, Y. Truthful multicast in selfish wireless networks. In *MobiCom* (2004).

Wang, W., Wang, Y., Li, X.-Y., Song, W.-Z., and Frieder, O. Efficient interference aware TDMA link scheduling for static wireless mesh networks. In *MobiCom* (2006b).

Wang, Y. *Efficient Localized Topology Control for Wireless Ad Hoc Networks.* Ph.D. dissertation, Department of Computer Science, Illinois Institute of Technology, 2004.

Wang, Y. and Li, X.-Y. Distributed spanner with bounded degree for wireless ad hoc networks. In *IPDPS* (2002a).

Wang, Y. and Li, X.-Y. Geometric spanners for wireless ad hoc networks. In *ICDCS* (2002b).

Wang, Y. and Li, X.-Y. Localized construction of bounded degree planar spanner for wireless networks. In *DIALM-POMC* (2003).

Wang, Y., Li, X.-Y., and Frieder, O. Distributed spanner with bounded degree for wireless networks. *Int. J. Foundations Comput. Sci.* **14**, 2 (2003), pp. 183–200.

Wang, Y., Song, W.-Z., Wang, W., Li, X.-Y., and Dahlberg, T. A. LEARN: Localized energy aware restricted neighborhood routing for ad hoc networks. In *SECON* (2006).

Wang, Y., Wang, W., and Li, X.-Y. Distributed low-cost weighted backbone formation for wireless ad hoc networks. In *MobiHoc* (2005a).

Wang, Y., Wang, W., and Li, X.-Y. Efficient distributed low-cost weighted backbone formation for wireless ad hoc networks. *IEEE Trans. Parallel Distribut. Syst.* **17** (2005b), pp. 681–693.

Want, R., Hopper, A., Falcão, V., and Gibbons, J. The active badge location system. Tech. Rep. 92.1, Olivetti Research Ltd. (ORL), 24a Trumpington Street, Cambridge CB2 1QA (1992).

Ward, A., Jones, A., and Hopper, A. A new location technique for the active office. *IEEE Personal Communi.* **4**(5) (1997), 42–47.

Wattenhofer, R., Li, L., Bahl, P., and Wang, Y.-M. Distributed topology control for wireless multihop ad-hoc networks. In *INFOCOM* (2001).

Wattenhofer, R. and Zollinger, A. XTC: A practical topology control algorithm for ad-hoc networks. In *WMAN* (2004).

Wieselthier, J., Nguyen, G., and Ephremides, A. On the construction of energy-efficient broadcast and multicast trees in wireless networks. In *INFOCOM* (2000), pp. 586–594.

Williams, B. and Camp, T. Comparison of broadcasting techniques for mobile ad hoc networks. In *MobiHoc* (2002), pp. 194–205.

Wongthavarawat, K., and Ganz, A. Packet scheduling for QoS suport in IEEE 802.16 broadhand wireless access systems. *Int. J. Commun. Syst.* **16** (2003), 81–96.

Woo, A. and Culler, D. A transmission control scheme for media access in sensor networks. In *MobiCom* (2001), pp. 221–235.

Wu, J. Extended dominating-set-based routing in ad hoc wireless networks with unidirectional links. *IEEE Trans. Parallel Distribut. Syst.* **22** (2002), 327–340.

Wu, J. and Dai, F. Broadcasting in ad hoc networks based on self-pruning. In *INFOCOM* (2003).

Wu, J., Dai, F., Gao, M., and Stojmenovic, I. On calculating power-aware connected dominating sets for efficient routing in ad hoc wireless networks. *IEEE/KICS, J. Commun. Networks* **4** (2002), 59–70.

Wu, J. and Li, H. On calculating connected dominating set for efficient routing in ad hoc wireless networks. In *Proc. of the Third International Workshop on Discrete Algorithms and Methods for Mobile Computing and Communications* (1999), pp. 7–14.

Wu, J. and Li, H. Domination and its applications in ad hoc wireless networks with unidirectional links. In *Proc. of the International Conference on Parallel Processing 2000* (2000), pp. 189–197.

Wu, J. and Li, H. A dominating-set-based routing scheme in ad hoc wireless networks. *Telecommun. Syst. J.* **3** (2001), 63–84 (special issue on wireless networks).

Wu, J. and Lou, W. Forward-node-set-based broadcast in clustered mobile ad hoc networks. *Wireless Commun. Mobile Comput.* **3** (2003), 155–173.

Wu, J., Wu, B., and Stojmenovic, I. Power-aware broadcasting and activity scheduling in ad hoc wireless networks using connected dominating sets. *Wireless Commun. Mobile Comput.* **3** (2003), 425–438.

Wu, S.-L., Lin, C.-Y., Tseng, Y.-C., and Sheu, J.-P. A new multi-channel MAC protocol with on-demand channel assignment for multi-hop mobile ad hoc networks. In *International Symposium on Parallel Architectures, Algorithms and Networks*, IEEE, New York (2000).

Wu, Y. and Li, X.-Y. Random distribution of keys for wireless sensor networks. Manuscript (2006).

Wu, Y., Li, X.-Y., and Wang, W. *Stochastic security in wireless mesh networks via saddle routing policy*, International Conference on Wireless Algorithms, System and Applications (WASA), IEEE, New York (2007).

Xie, H., Tabbane, S., and Goodman, D. J. Dynamic location area management and performance analysis. In *VTC* (1993).

Xu, Y., Heidemann, J., and Estrin, D. Geography-informed energy conservation for ad hoc routing. In *MobiCom* (2001), pp. 70–84.

Xue, F. and Kumar, P. R. The number of neighbors needed for connectivity of wireless networks. *Wireless Networks* **10** (2004), 169–181.

Yao, A. C.-C. On constructing minimum spanning trees in *k*-dimensional spaces and related problems. *SIAM J. Comput.* **11** (1982), 721–736.

Yao, P., Krohne, E., and Camp, T. Performance comparison of geocast routing protocols for a MANET, 2004. In *ICCCN* (2004).

Ye, F., Luo, H., Cheng, J., Lu, S., and Zhang, L. A two-tier data dissemination model for large-scale wireless sensor networks. In *MobiCom* (2002), pp. 148–159.

Ye, W., Heidemann, J., and Estrin, D. Medium access control with coordinated adaptive sleeping for wireless sensor networks. *IEEE/ACM Trans. Network* **12** (2004), 493–506.

Yi, C.-W. *Probabilistic Aspects of Wireless Ad Hoc Networks*. Ph.D. dissertation, Illinois Institute of Technology (2005).

Yi, S., Pei, Y., and Kalyanaraman, S. On the capacity improvement of ad hoc wireless networks using directional antennas. In *MobiHoc* (2003), pp. 108–116.

Yoon, J., Liu, M., and Noble, B. Random waypoint considered harmful. In *INFOCOM* (2003).

Yu, Y., Govindan, R., and Estrin, D. Geographical and energy-aware routing: A recursive data dissemination protocol for wireless sensor networks. Tech. Rep., UCLA Computer Science Department, Technical Report UCLA/CSD-TR-01-0023, 2001.

Zaruba, G., Basagni, S., and Chlamtac, I. Bluetrees—Scatternet formation to enable Bluetooth-based ad hoc networks. In *ICC* (2001).

Zhai, H. and Fang, Y. Physical carrier sensing and spatial reuse in multirate and multihop wireless ad hoc networks. In *INFOCOM* (2006).

Zhang, H. and Hou, J. On deriving the upper bound of α-lifetime for large sensor networks. In *MobiHoc* (2004), pp. 121–132.

Zhao, Q., Tong, L., and Swami, A. Decentralized cognitive MAC for dynamic spectrum access. In *Proceedings of IEEE Symposium on New Frontiers in Dynamic Spectrum Access Networks*, IEEE, New York (2005).

Zheng, R. Information dissemination in power-constrained wireless networks. In *INFOCOM* (2006).

Zheng, R., He, G., Gupta, I., and Sha, L. Time indexing in sensor networks. In *MASS* (2004).

Zhong, S., Li, L., Liu, Y., and Yang, Y. R. On designing incentive-compatible routing and forwarding protocols in wireless ad-hoc networks—an integrated approach using game theoretical and cryptographic techniques. Tech. Rep. YALEU/DCS/TR-1286, Yale University, Computer Science Department (2004).

Zhou, D. and Lai, T. Analysis and implementation of scalable clock-synchronization protocols in IEEE 802.11 ad hoc networks. In *MASS* (2004).

Zhou, D. and Lai, T.-H. A compatible and scalable clock-synchronization protocol in IEEE 802.11 ad hoc networks. In *ICPP* (2005a).

Zhou, D. and Lai, T.-H. A scalable and adaptive clock-synchronization protocol for IEEE 802.11-based multihop ad hoc networks. In *MASS* (2005b).

Zhu, S., Setia, S., and Jajodia, S. Leap: Efficient security mechanisms for large-scale distributed sensor networks. In *CCS* (2003), pp. 62–72.

Zou, Y. and Chakrabarty, K. Sensor deployment and target localization in distributed sensor networks. *ACM Trans. Embedded Comput. Syst.* (2003).

Index